6th International Brazing and Soldering Conference 2015

Long Beach, California, USA
19-22 April 2015

ISBN: 978-1-5108-8729-9

Printed from e-media with permission by:

Curran Associates, Inc.
57 Morehouse Lane
Red Hook, NY 12571

Some format issues inherent in the e-media version may also appear in this print version.

Printed by Curran Associates, Inc. (2019)

For permission requests, please contact American Welding Society
at the address below.

American Welding Society
8669 NW 36 Street, #130
Miami, Florida 33166-6672
USA

Phone: 800-443-9353
305-443-9353

www.aws.org

Additional copies of this publication are available from:

Curran Associates, Inc.
57 Morehouse Lane
Red Hook, NY 12571 USA
Phone: 845-758-0400
Fax: 845-758-2633
Email: curran@proceedings.com
Web: www.proceedings.com

TABLE OF CONTENTS

SESSION 1: SOLDER JOINING TECHNOLOGIES

SESSION 2: MATERIALS AND PROCESS DESIGN 1

SESSION 3: SPECIAL/ADVANCED PROCESS 1

SESSION 4: CORROSION, JOINT RELIABILITY

SESSION 5: LEAD FREE SOLDERS

SESSION 6: NANO-MATERIALS

SESSION 7: MATERIALS & PROCESS DESIGN 2

SESSION 8: CERAMIC, COMPOSITE, GLASS JOINING 1

SESSION 9: FILLER METAL PROPERTIES 1

SESSION 10: FURNACE BRAZING AND FIXTURE

SESSION 11: SPECIAL/ADVANCED PROCESS 2

SESSION 12: MODELING AND SIMULATION FOR BRAZING

SESSION 13: FILLER METAL PROPERTIES 2

SESSION 14: CERAMIC, COMPOSITE, GLASS JOINING 2

SESSION 15: ALUMINUM AND ALUMINUM MATRIX COMPOSITE

SESSION 16: BRAZED JOINT PROPERTIES

Reliability Analysis of Pin-in-Hole Solder Joints by Computational Modeling

P.T. Vianco* and M.K. Neilsen
Sandia National Laboratories
Albuquerque, NM USA
**505-844-3429; ptvianc@sandia.gov*

Abstract

A reliability assessment was made of the thermal mechanical fatigue (TMF) resistance of defective, through-hole solder joints on an encapsulated printed wiring assembly (PWA). The specific defect was incomplete hole fill. The Sandia computational solder joint model COMPSiR© was used to calculate TMF lifetimes. The evaluated configurations included a baseline condition of fillets on both sides and 100% hole fill as well as the anomalous geometries having a single fillet: (a) 100% hole fill; (b) 75% hole fill; (c) 50% hole fill; and (d) the 50% hole fill, but with the encapsulant absent from under the bottom printed wiring board (PWB). The environmental conditions included the service environment, which was comprised of two transportation-and-handling segments plus the actual use segment. Additional analyses predicted TMF lifetimes for an Acceptance test, as well as the combination of Acceptance test and Environmental test (E-test). The latter combination was applied to a percentage of the population prior to being placed into service. A "laboratory" case was also investigated that was comprised of Acceptance test plus a Destructive test (D-test). Fielded hardware exposed to the harshest pre-service test conditions (Acceptance test plus E-test) retained sufficient TMF life, regardless of the failure criteria, for all joint configurations. Only the case of absent encapsulant caused a cautionary risk to the reliability of anomalous interconnections that entered service. The model confirmed that the D-test was not sufficiently severe to bring the solder interconnections into a wear-out failure mode. This study also exemplified the development of a predictive model that provided suitable fidelity, yet at the same time, was tailored to optimize computational efficiency.

Introduction

Printed Wiring Assembly

A printed wiring assembly (PWA) was used to perform the control function for a high-reliability application. Figure. 1 shows a schematic diagram of a portion of the PWA. The construction of the PWA was non-typical because it was constructed as a "sandwich" of two, through hole, printed wiring boards (PWBs). The components – resistors, capacitors, etc. – were located between the two PWBs. Some components had the leads soldered to one PWA; others had their leads soldered to each of the two PWBs. Though not shown, the internal volume as well as the external regions around the PWBs, were encapsulated with a glass micro-balloon-filled epoxy encapsulant to enhance its resistance to mechanical shock.

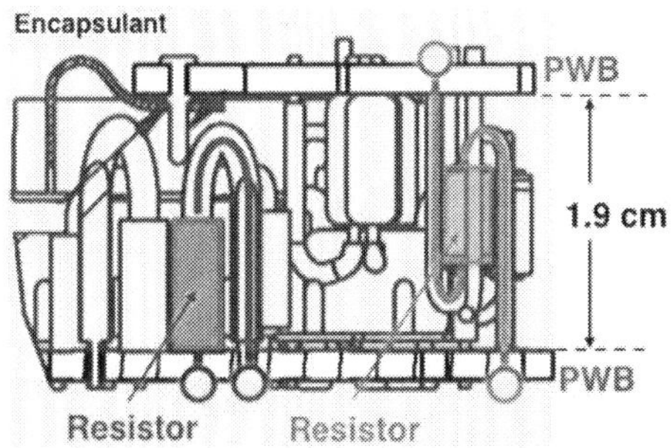

Figure 1 Schematic diagram shows the PWA. The components are attached to one, or both, of the PWBs. The interior as well as the volume above and below the PWBs, are filled with an encapsulant.

During the assembly of the PWA, it was observed that several solder joints exhibited workmanship defects. One such defect was incomplete hole-fill by the solder. An example of this phenomenon is shown in Fig. 2 and identified by the red circle. Of the defects that were recorded, poor solder joint fill generated the greatest concern because it posed an unknown risk to the long-term reliability of the interconnections and PWA function, as-a-whole. The two PWBs had already been soldered to the components so it was not possible to determine the precise depth of the hole fill or that the solder joints complied with the minimum 75% hole fill standard. Therefore, given this complex configuration of two sandwiched PWBs and, moreover, having an encapsulant between them, it became necessary to understand any changes to thermal mechanical fatigue (TMF) performance that resulted from the poor filling of the solder joints. There was lesser concern with respect to the shock and vibration reliability because the encapsulant surrounded the components.

There were two approaches that could have been used to fill the gap in reliability information. The first approach would be to develop an extensive empirical study with which to build a reliability data set. This avenue was impractical due to the absence of available hardware and schedule limitations. The second approach, which was selected for this study, was to use the Sandia solder fatigue computational models, COMPSiR© and TurboSiP© to predict the TMF reliability of the interconnections [1, 2]. The predictive modeling approach was selected for this project, utilizing the COMPSiR© model.

Figure 2 Photograph shows the defect of an incompletely filled solder joint (red circle).

The computational modeling analysis was performed in two parts. The first part examined the stand-alone solder joints, that is, those without the component body and in the absence of an encapsulant. This analysis determined whether the TMF of the solder joint, by itself, was the controlling factor, which would significantly reduce the complexity of the finite element model. In the event that the solder joint structure, alone, did not control TMF, then it would be necessary to develop the finite element mesh at a higher assembly level that included the component body, dual PWB structure, and encapsulant. An important facet of this study was the optimization steps used to develop a model that provided the necessary fidelity, but one that also minimized computational time. In addition, this study took into account the multiple temperature cycling conditions to which the assembly was exposed in test, handling and transportation, as well as the use condition.

Model of the Stand Alone Solder Joints

Finite Element Model

The stand alone solder joint models are shown in Fig. 3. Three conditions were analyzed, which represented different degrees of hole fill between the pin and circuit board. All three cases had a single, full fillet on only one side of the joint. The baseline case in Fig. 3a had complete hole fill. The two other cases, Figs. 3b and 3c, had the hole fill reduced to 75% and 50%, respectively.

The metric of TMF deformation that is computed by the model is the equivalent plastic strain (EQPS). The EQPS is a dimensionless, scalar parameter that is based upon the Von Mises strain that accumulates with each fatigue cycle.

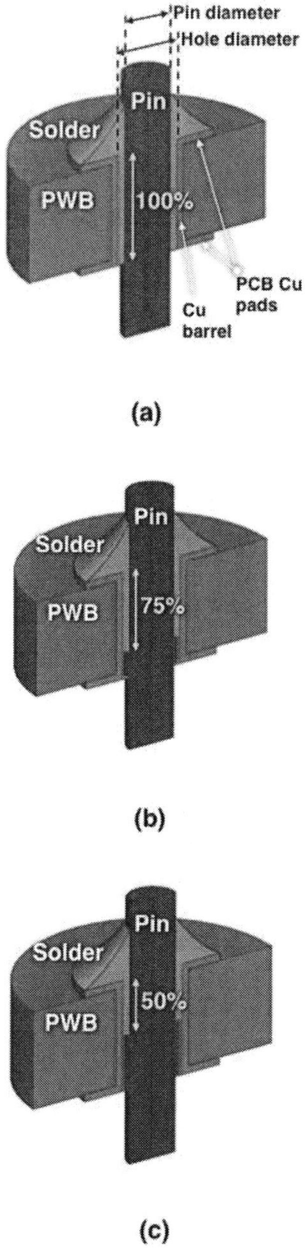

Figure 3 Solid models show the stand alone solder joint configurations based on a single fillet, but varying degrees of hole fill: (a) 100% (baseline); (b) 75%; and (c) 50%.

A preliminary analysis indicated that the finite element model could take advantage of cylindrical symmetry of the problem. The findings showed that the full 360° model could be reduced to a 10° "pie slice" and still retain excellent fidelity to the predictions. The resulting finite element models are shown in Fig. 4.

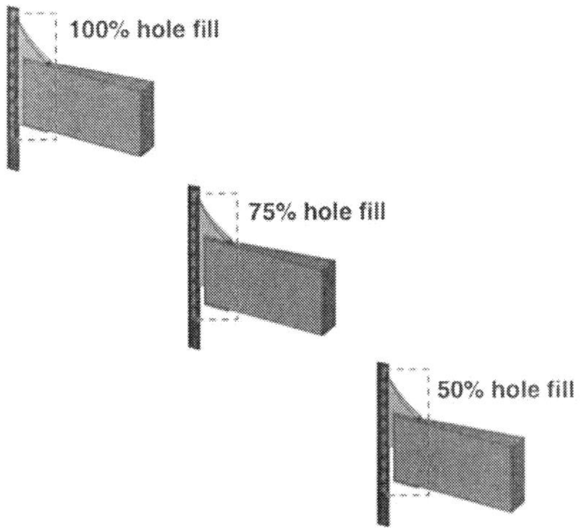

Figure 4 Solid models show the 10° "pie slice," finite element models for the *100%; 75%;* and *50%* hole fill conditions.

The materials set and geometries are listed below:

- Printed wiring board (PWB): 1.6 mm thick; polyimide-glass laminate;
- PWB Cu pads: 0.05 mm in Cu pad;
- PWB Hole: 0.91 mm diam.; 0.025 mm Cu barrel
- Pins: Cu; 0.64 mm diam.
- Solder: 63Sn-37Pb (wt.%)

Temperature Cycle Conditions

The temperature cycling environment was established from a separate study, but was suitable for the present analysis. The profile parameters were: (a) minimum temperature, 5.2°C; (b) maximum temperature, 24.1°C; and a sinusoidal period of one year [4]. The temperature cycle is shown in Fig. 5.

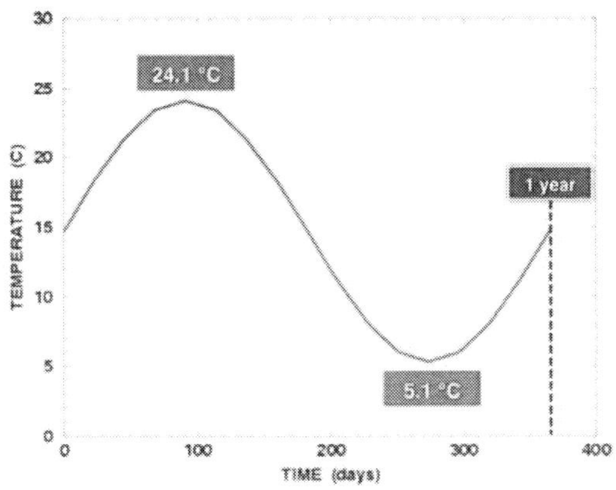

Figure 5 Graph shows the time-temperature profile used to assess the TMF of the stand alone solder joints.

Model Results

The three cases are depicted in Fig. 6. Both the 10° slice, finite element model and the one-cycle, equivalent plastic strain (EQPS) contour maps are provided for each of the three hole fill configurations (one fillet). The single cycle, EQPS contours adequately represents the strain distribution that develops in the solder joint leading up to initiation of a TMF crack. This single-cycle, EQPS contour format is used throughout this report to describe the spatial distribution of TMF strain in the interconnections.

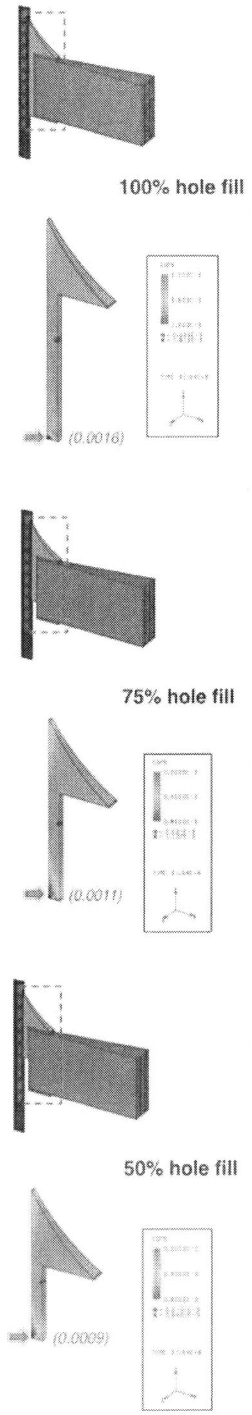

Figure 6 The 10° slice, finite element models are shown with their respective, single cycle, EQPS contours for each of the hole fill configurations (single fillet). The magenta arrow indicates the location of maximum EQPS in the joint, the value of which is shown by the numeral.

The magenta arrow indicates the location of the maximum EQPS value where initiation of a TMF crack would take place. The maximum EQPS value is provided adjacent to the arrow. The first case of the 100% hole shows a maximum EQPS at the bottom of the gap. The asymmetry of the joint has forced the maximum TMF to reside at this location. When both fillets are present, the location of maximum EQPS is located at the top of the fillet for the same temperature cycle. The value of the maximum EQPS is 0.0016, which is more than twice that of baseline, double fillet case (0.0007).

In the case of the 75% hole fill, the location of maximum EQPS is also found at the bottom limit of the gap. But, the maximum value is 0.0011, which is *less than* that for a 100% hole fill (0.016). There is a lesser amount of shear displacement in the 75% hole fill joint due to the shorter length of the solder joint. The same trend is recorded for the 50% hole fill. The maximum EQPS, again where crack initiation is most likely to occur, has the same location; however, the magnitude is reduced to nearly half of the case of 100% hole fill. The results presented in Fig. 6, although appearing to be counter-intuitive at first, clearly show that reducing the hole fill increases the TMF lifetime of the *stand alone* solder joints in the presence of a single fillet.

The computational model predictions were made against two failure criteria: (a) cycles to TMF crack initiation and (b) cycles to 100% cracking. Between all of the cases presented in Fig. 4, the number of cycles required to initiate a crack were in the range of 500,000 – 2,400,000. The number of cycles required to reach 100% cracking was approximately 5,000,000. These predictions imply that the three solder joint configurations have nearly “infinite TMF lifetimes”.

In summary, it was established that the stand-alone solder joints were unlikely to control TMF behavior of the interconnections in the PWA, regardless of workmanship quality. It was concluded that it was necessary to construct the more complex model that includes the component body, both PWBs, and encapsulant structures (Fig. 1). Nevertheless, an important trend that was documented in that initial model, was that in the presence of a single fillet, a reduced hole fill would generate a lower, maximum TMF strain in the joint.

Assembly Model of a Resistor Component

Finite Element Model

The second analysis examined the assembly that was shown in Fig.1. The volume between the two PWBs was filled with the encapsulant. A review of the assembly drawing indicated that, in general, there were two solder joint configurations. There were those cases represented by the blue-colored resistor that had both solder joints (yellow circles) located on the same PWB. Then, there were resistors like the green-colored case that had one lead soldered to one PWB and the second lead, solder to the second PWB. The blue resistor was selected for the analysis because the increased constraint was likely to generate a greater degree of TMF in its solder joints than would occur in the interconnections of the green resistor, which were more capable of flexing under the cyclic loads.

The development of the finite element model for the blue resistor began with the three-dimensional rendition shown in Fig. 7a. The model shows the resistor body, the solder joints, and the (bottom) PWB. The two solder joints are labeled “1” and “2”. The view is along a single symmetry plane defined by the diameters of the resistor body and two joints.

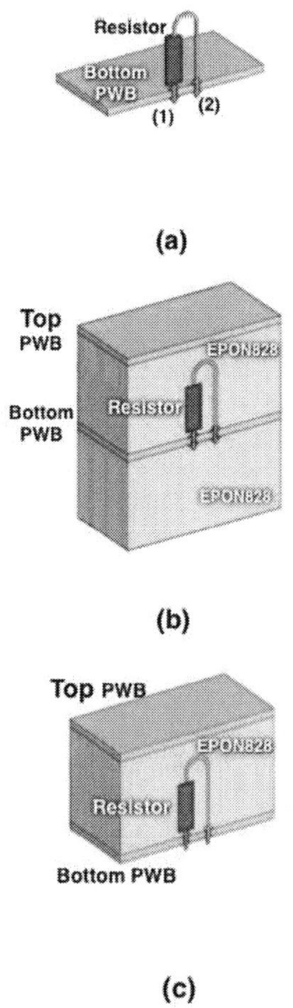

Figure 7 (a) Solid model shows the single symmetry plane of the resistor, its solder joints, and the PWB. The solder joints are labeled “1” and “2.” (b) Solid model of the resistor includes both PWBs and the encapsulant located between them and underneath the bottom PWB. (c) The same model as presented in (b) except that the encapsulant was absent from underneath the bottom PWB.

The addition of the second (top) PWB and the encapsulant results in the configuration that is shown in Fig. 7b. Note that encapsulant is located between the PWBs as well as underneath the bottom PWB that carries the resistor solder joints. The encapsulant was left out from on top of the overhead PWB to simplify the model. It was concluded that the encapsulant there would have minimal effect on the TMF of the solder joints.

However, the presence or absence of encapsulant *under the bottom PWB* was presumed to have an effect on TMF of the interconnections. Therefore, the computational model was exercised to predict the solder joint TMF for the scenario where by the encapsulant were missing from underneath the bottom PWB. The latter configuration is shown in Fig. 7c. This case is explored because it is possible for an encapsulant to contain voids after the filling process. The voids generate an asymmetry to the loading of the PWB that alter the magnitude of TMF in the solder joints.

There are two solder joints for the resistor. It was suspected that the solder joint "1" under the component would experience a greater degree of TMF than its counterpart, "2". This presumption was based upon the interaction between the resistor body and the encapsulant. An analysis was performed with the solder fatigue model to determine whether this stipulation was correct. If valid, it allowed the analysis to consider only one solder joint, which would greatly reduce the number of elements in the finite element model, thereby decreasing the computational time needed to obtain the predictions.

The model was constructed to investigate the relative degrees of TMF that occurred in either joints "1" or "2". There were two configurations that represented the limiting cases of TMF: (a) full fillets, both sides and fully encapsulated (low TMF) and (b) full fillet, one side; 50% hole fill; and encapsulant absent from underneath the bottom PWB (high TMF). The accelerated aging, thermal cycling environment, which was chosen for the evaluation, had the following parameters: (a) temperatures limits of -40°C and 60°C; (b) hold times of 30 min.; and (c) a total cycle time of 93 min.

The solid models that represent the two limiting cases are shown in Figs. 8a and 8b, respectively. Below the solid models are the EQPS contours that were predicted by the computational model, based upon a single thermal cycle. Also, dashed arrows indicate the 75% and 50% hole fill levels for reference. The locations of maximum EQPS are pointed out by the solid arrows, which are accompanied by the corresponding values in parentheses. The solder joint "1", which is directly under the resistor body, experienced nearly an order of magnitude greater TMF strain than did the location of highest EQPS on solder joint "2". The contour colors similarly confirmed that solder joint "1" underwent substantially more TMF deformation, in general, than did joint "2". The same trend is observed in Fig. 8b. Here, the difference of EQPS magnitudes was considerably less, being a little more than a *factor* of two, yet still sufficiently large to allow the model to focus solely on the joint "1".

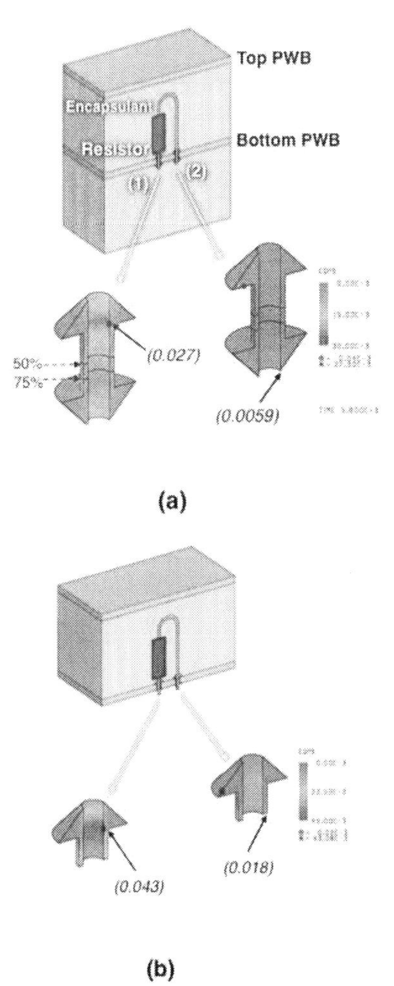

Figure 8 (a) Finite element model shows the two resistor solder joints "1" and "2". The solder joints have full fillets on both top and bottom. Below the finite element model are shown the predicted EQPS strain contours following one thermal cycle (-40°C and 60°C; 30 min. hold times; and 93 min. cycle time). The solid arrow indicates the location of highest EQPS; the EQPS value is provided in parenthesis. The 75% and 50% hole fill extents are shown by the dashed arrows. (b) The finite element model is shown at top as well as EQPS contours is below it (with the maximum, single cycle strain) are shown for the case of 50% hole fill and the absence of encapsulant from underneath the bottom PWB for the same thermal cycle.

In summary, the results presented in Fig. 8 confirmed that solder joint "1" directly under the resistor body experienced the greater amount of TMF deformation. Because that interconnection controlled the reliability of the resistor, the computational model was developed solely for joint "1".

The evaluation of a single solder joint allowed for a further reduction to the size of the finite element model. A single solder joint has a cylindrical symmetry that allows for the full 3-D, 360° configuration to be reduced to a "pie slice" geometry. A 10° slice was selected for the finite element model. This approximation was analyzed along side the full

360° to confirm that there was not a significant loss of fidelity to the model predictions.

Shown in Fig. 9a is the 3-D model of the solder joint having full fillets on both top and bottom sides as well as complete encapsulation. This configuration generates the lower bounds of TMF strain. A modified temperature cycle was used for this analysis (-40°C/60°C; 30 min. hold times; and 93 minute total cycle time). The location of maximum EQPS is indicated by the arrow, together with the numerical value of 0.0273 that resulted from a single thermal cycle. The 10° model is shown to the right in Fig. 9b. The EQPS contour (single cycle) is shown below the finite element model and the maximum value of 0.0282 is in parentheses. The location of maximum EQPS and the latter's value are nearly identical to the corresponding properties of the full 3-D model.

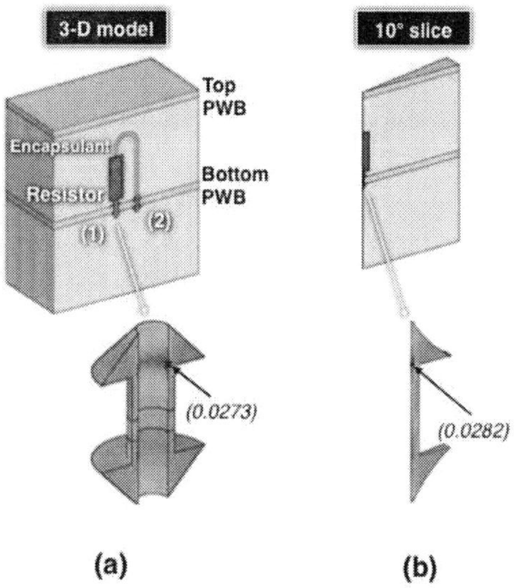

Figure 9 (a) The 3-D finite element model is shown together with the predicted EQPS contours below it for the solder joint "1" of the resistor following one accelerated aging thermal cycle (-40°C/60°C; 30 min. hold times; and 93 minute total cycle time). The solder joint configuration has full fillets on both top and bottom. The solid arrow indicates the location of highest EQPS; that value is provided in parentheses. (b) The finite element model; single EQPS contours; and maximum value of EQPS are shown for the same solder joint, but using the 10° slice model.

This analysis was also performed for the case having a single fillet; 50% hole fill; and encapsulant absent from under the bottom PWB. This circumstance represents the upper limit of TMF strain. The corresponding finite element models; single cycle EQPS contours; and maximum EQPS strain are shown in Fig. 10. As was the case above, both the location and value of maximum EQPS were similar between the two model representations.

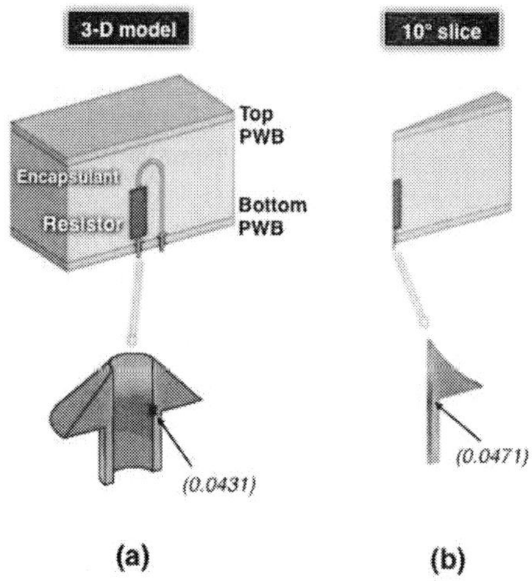

Figure 10 (a) The 3-D finite element model is shown together with the predicted EQPS contours below it for the solder joint "1" of the resistor following one accelerated aging thermal cycle (-40°C/60°C; 30 min. hold times; and 93 minute total cycle time). The solder joint configuration has a full fillet on only one side; 50% hole fill; and the absence of encapsulant under the bottom PWB. The solid arrow indicates the location of highest EQPS; that value is provided in parentheses. (b) The finite element model; single EQPS contours; and maximum value of EQPS are shown for the same solder joint, but using the 10° slice model.

In summary, an analysis determined that, given the cylindrical symmetry of the solder joint, the finite element model of the latter could be reduced to a 10° "pie slice" without a loss of fidelity to the TMF predictions. The result was a significant reduction in the model size and complexity, which reduced the computation time.

The finite element construct is shown in Fig. 11 that was used in the computational model to perform the TMF predictions of the *baseline case*. That latter configuration has full fillets on *both* sides of the joint. The starting, 3-D solid model is shown in Fig. 11a; note the cylindrical configuration. There is one symmetry plane, which passes through the diameters of the solder joints and resistor body. Figure 11b provides a magnified view of the solder joint region. The 10° slice, finite element mesh is presented in Fig. 11c. The mesh size is finer next to the solder joint in order to optimize the prediction fidelity.

There were four cases that were analyzed by the computational model to address the incomplete hole fill problem. The corresponding finite element meshes are shown in Fig. 12 based upon 10° slice geometry. The particular geometries are: (a) 100%, (b) 75%, and (c) 50% hole fill. Figure 12d shows the case of encapsulant being absent from under the PWB and 50% hole fill. Note in Figs. 12b – 12c that the missing solder was replaced by encapsulant (magenta arrows).

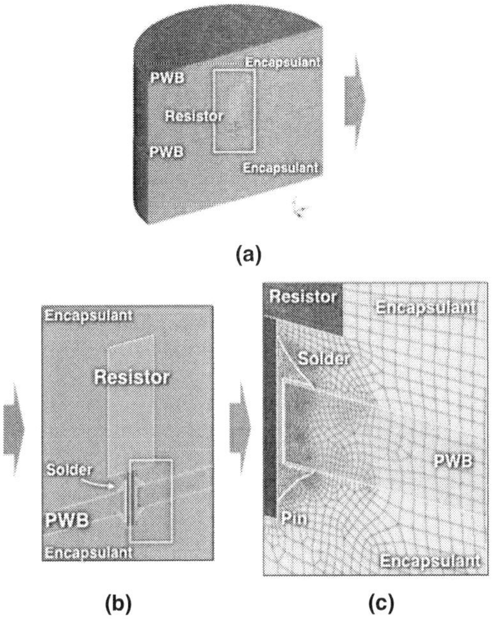

Figure 11 (a) Three-dimensional, solid model shows the cylindrical configuration of the blue resistor and the single solder joint underneath the body. This case represents the baseline configuration that had fillets on both sides and 100% hole fill in the gap. The model is cut in half by the (diameter) symmetry plane. (b) Enlarged image of the resistor and solder joint portions of the model. (c) The finite element mesh shows the 10° slice, finite element model of the solder joint.

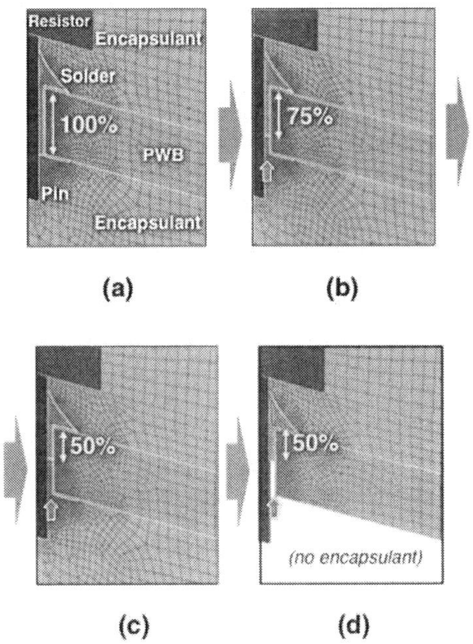

Figure 12 The 10° slice, finite element models are shown for the three cases examined by the computational model for one fillet and percent hole fill: (a) 100%, (b) 75%, and (c) 50% hole fill. (d) The fourth variant has encapsulant absent from under the PWB and 50% hole fill. The magenta arrow signifies that missing solder was replaced with encapsulant.

The finite element meshes in Figs. 12b – 12d were compared to Fig. 2. In the latter case, the absent solder was at the top of the joint, next to the component. On the other hand, the model has the opposite configuration with absent solder on the bottom side of the interconnection. The two configurations produced nearly the same TMF predictions for the following reasons. First of all, the two geometries were symmetric with respect to cyclic loads. Secondly, the TMF contribution that was made by the encapsulant, was derived largely through the loads that it applied to the component body. Those loads were primarily axial (up-and-down) so that they affected the fillets, equally, for either configuration. Bending loads, which would be sensitive to whether the joint was on the top or bottom, were confirmed to present a second-order effect in this study.

Temperature Cycle Conditions

There were four generalized temperature cycling conditions that were responsible for TMF of the solder joints: service condition, Acceptance testing, Environmental testing (E-test), and Destructive testing (D-test). The discussion begins with the service condition, which itself is comprised of three segments. There are two (2) transportation and handling segments and the one use segment. The details of the three segments are listed below, which include minimum and maximum temperature limits; hold times at temperature limits; time of ramp between limits; and number of cycles:

- **Transportation and handling segment #1**: -29°C/49°C; 10 hour holds; 2 hours/ramp; 15 cycles;
- **Transportation and handling segment #2**: -18°C/49°C; 10 hour holds; 2 hours/ramp; 15 cycles;
- **Use segment**: 16°C/33°C; 1 cycle/year; *sinusoidal* between limits; 60 cycles (years).

The following methodology was use to address the combination of several different environments. The explanation begins by considering the transportation-and-handling segment #1 (-29°C/49°C; 24 hour period). The solder joint geometry in Fig. 11c was used to illustrate the analysis. Crack initiation is the failure criterion, which is designated with the subscript, i. The computational model calculates the number of cycles required to reach crack initiation under the transportation-and-handling segment #1, $N_{i,\#1}$. The value of $N_{i,\#1}$ is computed to be 3600 cycles. But, the solder joints experience only 15 cycles under that condition. Therefore, the fraction of TMF life used up by the 15 cycles is 15/3600 = 0.0042. The same computation is made for transportation and handling segment #2 as well as for the use segment.

The total percent of TMF life that is consumed by the service environment is simply the sum of the percent lifetimes used up by the three segments. This methodology is referred to as Miner's Rule. The Miner's rule approach assumes that each of the fatigue steps is independent of one-another. Mathematically, that summation is expressed by equation (1):

$$\text{Total \% Life Used (service)} = \left[(15 / N_{i;\#1}) + (15 / N_{i;\#2}) + [60 / N_{i;\,use}]\right] \times 100 \quad (1)$$

Equation (1) was applied to each of the five solder joint configurations represented in Fig. 12.

A similar equation was developed for the failure criterion of a 100% crack, that is, a crack that propagates through 100% of the joint, which leads to an electrical failure.

At this point, a brief background explanation is presented of the failure criterion. There are three metrics, and associated with each of them an analysis approach, that were used to describe TMF failure. Those three approaches are listed below as "Failure criterion / Analysis approach:"

- EQPS / Coffin-Manson
- Crack Initiation / Crack Simulation Model
- 100% Crack (Open) / Crack Simulation Model

The EQPS/Coffin-Manson method establishes "failure" based upon the first element in the mesh that accumulates sufficient strain so that it meets the Coffin-Manson equation for solder fatigue failure. This approach would approximate TMF crack initiation when used conservatively.

Improved computational techniques have led to the development of numerical models that can predict crack propagation. This tool is the Crack Simulation Model (CSM). This model computes the number of cycles needed to initiate the fatigue crack based upon a fatigue damage criterion within the Unified Creep-Plasticity (UCP) constitutive equation that, in turn, underlies the predictive model. Furthermore, the CSM computes the number of cycles required to propagate a crack through the joint. The failure criterion in this study is the number of cycles to cause 100% cracking of the joint, which would result in an electrical failure. It is important to recognize that the cycles to crack initiation and cycles to 100% cracking do not necessarily scale with one-another. The path to crack initiation is entirely by TMF *deformation* in the mesh cell. However, the path to 100% cracking is a combination of TMF deformation plus crack propagation based upon the loss of load-bearing capacity at the mesh cell.

It was noted earlier that, besides the service condition, there were three tests to which the solder joint was exposed when part of the next-assembly. The details of the three test histories are described below:

- **Acceptance testing**: -32°C/60°C; 4 hour holds; 12°C/min maximum; 2 cycles;
- **Environmental testing ("E-test")**: -32°C/60°C; 4 hour holds; 12°C/min maximum; 5 cycles;
- **Destructive testing ("D-test")**: -32°C/60°C; 4 hour holds; 12°C/min maximum; and 26 cycles.

The test conditions were the same except that they differed by the number of cycles.

The three tests served the following functions. The Acceptance test was meant to screen for those defects responsible for infant mortality failures in order to remove those units from the product line. The E-test had, as its goal, to "stress" the parts to a slightly greater degree than the Acceptance test, in order to bring out the defects responsible for latent failures. Lastly, the D-test confirmed the long-term reliability of the unit by generating wear-out failures. Note that the D-test was developed for the assembly *as-a-whole,* not specifically for the solder interconnections. The computational model was used to determine the amount of TMF life that is removed from the solder joints by the D-test regimen. Since the D-test is expected to remove a signification portion of the assembly life, the tested units are not allowed back into the product stream.

The test regimens were combined with the service conditions (two transportation segments plus the use segment) to form three possible lifetime sequences for the partially-filled solder joint:

- Acceptance test + *Service*,
- Acceptance test + E-test + *Service*, and
- Acceptance test + D-test.

Of the two sequences whereby the solder joints would enter service, only the second one was assessed by the computational model because it represented the more severe circumstance. Therefore, the model predictions presented in this report originated from the latter two bulleted sequences.

The total TMF lifetimes that were consumed in each case, whether by the crack initiation criterion or the 100% crack criterion, were calculated using the Miner's rule approach as was illustrated earlier for the service condition. To reiterate that process, the CSM predicts the number of cycles to failure for each of the individual Acceptance test, E-test, and D-test conditions under one of the failure criterion – crack initiation or 100% crack. Secondly, the fractions of TMF life are calculated, which are consumed within each segment. Lastly, those individual fractions are added up to give the total fraction of TMF life consumed by the solder joint. That value is multiplied by 100 to give the percentage. This process is summarized, below, for the first history, Acceptance test + *Service*, using the crack initiation criterion (i):

$$\text{Total \% Life Used (service)} = \left\{(2 / N_{i;\,Accept.}) + (5 / N_{i;\,E\text{-}test}) + \left[(15 / N_{i;\,\#1}) + (15 / N_{i;\,\#2}) + (60 / N_{i;\,use]})\right]\right\} \times 100 \qquad (2)$$

A similar equation would be in place for the 100% cracking criterion.

Results – Model Predictions

Service Environment

The extent of TMF was assessed for the five variants of the resistor interconnections that are shown in Figs. 11c and 12a – 12d when considering only the service environment. Recall that the service condition is comprised of the two transportation and handling segments together with the use segment. It is noted here that an ancillary calculation was performed to determine the effect of the encapsulant on solder joint TMF versus its complete absence from the assembly. The presence of the encapsulant caused a three- to four-fold increase in TMF strain. The additional strain was due explicitly to the interaction between the encapsulant and the resistor body.

The results are presented in Table 1. The first row is the baseline condition. The model predicts that 0.66% of the solder joint's TMF life is used up by the time that the interconnection reaches the end of service, based upon the crack initiation criterion. When the failure criterion is 100% cracking, then 0.18% of the fatigue life is consumed at the end of service. The percentage decreases between the two criteria because the number of cycles is the same within each of the three segments comprising the service environment (i.e., the "numerator"); however, more cycles are required to reach 100% crack failure ("denominator") beyond crack initiation.

Table 1 Percent of Life Used Through the Service Environment Based Upon the Two Failure Criteria of Crack Initiation and 100% Cracking.

Through-Hole Configuration	Percent of Life Used to Crack Initiation (%)	Percent of Life Used to Complete Crack (%)
Full fillet, both sides	*0.66*	*0.18*
100% hole-fill	**0.65**	**0.20**
75% hole-fill	**0.73**	**0.28**
50% hole-fill	**0.94**	**0.33**
50% hole-fill; *No bottom encapsulant*	**3.9**	**1.7**

Clearly, these percentages, as well as those of the other configurations presented below them in Table 1, are very small, indicating that these solder joint configurations have ample margin to resist TMF failure under the service environment. This point is valid, even when a 2X or 3X safety factor is added to the computations as is often required by high-reliability applications.

The assessment turned to the other configurations that are shown in Table 1. The case of 100% hole-fill (which has the single fillet on one side) had very similar TMF lifetime as did the baseline with two fillets. Thus, the loss of the bottom side fillet was relatively inconsequential to the TMF lifetime. When the hole fill was reduced to 75% and 50%, some trends began to develop, albeit, the magnitudes of TMF were still very small. For example, the fractional change of TMF life calculated between the 100% and 50% hole fill conditions differed between the crack initiation and 100% cracking criteria. This behavior implies that the hole fill affected differently, the EQPS that leads to crack initiation versus the combination of EQPS and crack propagation that are needed for a 100% open interconnection. Physically, there was an increase of EQPS per cycle when proceeding from 100% hole fill to 50% hole fill. The latter case of 50% hole fill had the same change of EQPS and crack initiation site. However, there was the shortened crack path to a complete failure. These trends confirm the earlier premise that the TMF *deformation* (EQPS per cycle), which leads to crack initiation does *not* scale the same as the TMF *deformation-plus-cracking* process that leads to a fully cracked interconnection.

The absence of encapsulant from underneath the bottom PWB (Fig. 12d) caused nearly an order-of-magnitude increase of TMF damage to the joints. Although the 3.9% and 1.7% loss of fatigue life for the two respective failure criteria appear to be small, they could become significant when multiplied by the aforementioned safety factors. More importantly, however, these results reaffirm the need to consider the impact of an incomplete encapsulation of the PWA on the TMF of the individual, through-hole interconnections.

The computational model also provided the TMF strain distribution within the solder joint geometry. Shown in Fig. 13 are the single-cycle, EQPS contours for the baseline solder joint (full fillets, both sides). The diagrams (a) – (c) resulted from the transportation and handling segments as well as the use segment as designated below them. The diagram in Fig. 13d shows the case in which the use segment was allowed to cycle so as to cause complete cracking of the joint (white elements along side the red arrows). The magenta arrows in Figs. 13a – 13d indicate the locations of maximum TMF strain. The EQPS per cycle value is noted by the number in parentheses.

Several findings were compiled from the images presented in Fig. 13. First of all, the location of greatest TMF strain is near the top of the joint, not at the mid-plane location. This behavior is caused by the interaction between the resistor body and the encapsulant, which results in this asymmetry to strain distribution within the joint structure. A second observation is that the cycle in Fig. 13a results in a greater TMF strain (0.021) than does that in Fig. 13b (0.017). The source of this difference is a greater ΔT for the transportation and handling #1 segment when compared to that of the #2 segment. Third, a very small strain (0.0027) is observed in Fig. 15c due to the reduced ΔT of the use segment, in spite of the longer cycle duration (1 year). In summary, transportation and handling procedures can generate a *greater* degree of TMF damage than does the actual use segment and, as such, must be included in a computational analysis when considering the overall lifetime of a solder interconnections.

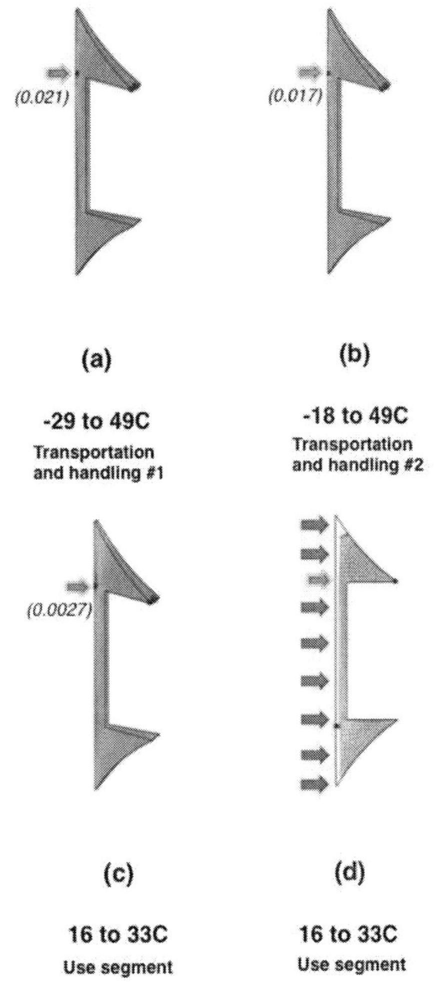

Figure 13 (a – c) Computational model generated the TMF strain contours for the baseline configuration (full fillet, both sides) resulting from a single cycle of the temperature profiles noted below each image. The magenta arrow indicates the location of greatest strain; the latter's magnitude in noted in parentheses. (d) The CSM prediction shows the case of the 100% crack (electrical open) resulting from the TMF that began as shown in (c), the use segment. The white elements, which are indicated by the red arrows, represent the cracked material.

The case represented in Fig. 13d is that of Fig. 13c, but with the CSM prediction of a 100% cracked joint. The crack propagation began at the magenta arrow location and continued in *two directions*. The crack grew towards the top of the fillet as well as to the bottom terminus. The crack path was entirely along the pin/solder interface.

The analysis of the single-cycle, EQPS behavior continues with the focus shifted to the other solder joint variants. The contour maps are shown in Fig. 14. The diagrams were derived from computations based on the use segment temperature cycle (16°C/33°C; 1 cycle/year), only. The trends are representative of the other segments within the service environment. The location of highest strain was unchanged between all four cases (magenta arrows). The EQPS values increased approximately 10% per hole fill condition from the 100% (a) to 50% (b). The location and strain magnitude in Fig. 14a, which describes the 100% hole fill, one fillet case, were the same as the case of both fillets in Fig. 13c. This observation corroborates the earlier comparison made between these two geometries in reference to Table 1, that the absence of the bottom fillet had a limited effect on TMF behavior.

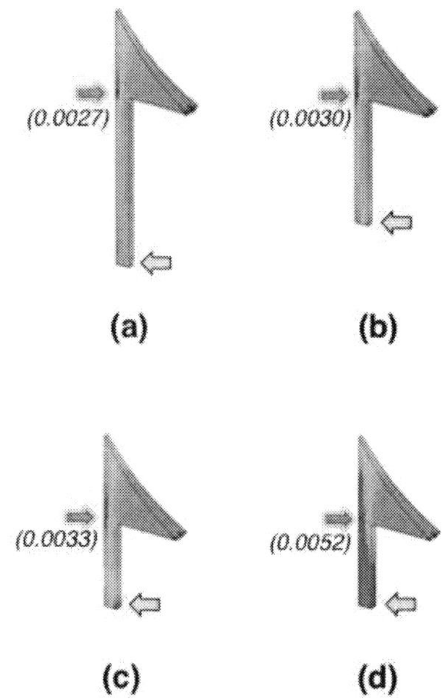

Figure 14 Computational model generated the single cycle, EQPS contours of (a) 100%, (b) 75%, and (c) 50% hole fill conditions (full fillet, one side). The strain was computed for the use segment temperature cycle (16°C/33°C; 1 cycle/year). (d) The EQPS contour is shown for the case of 50% hole fill, but in the absence of encapsulant under the PWB. The magenta arrows indicate the location of greatest EQPS; the latter's magnitude for the single cycle in noted in parentheses. The cyan arrows point out the bottom terminus of the hole fill.

Figure 14d shows the EQPS contours when the encapsulant was absent from underneath the PWB. The resulting asymmetry caused a significant increase in the TMF strain in the overall joint structure. In addition, there was a substantial TMF strain formed at the bottom terminus of the joint (cyan arrow). This localized strain raises the likelihood that crack propagation could initiate there as well as at the locations marked by the magenta arrows.

It is important to recognize that the EQPS that formed at the bottom of each joint (cyan arrows) was also capable of causing significant fatigue damage to the adjacent Cu barrel, albeit, that strain is not shown in these images. The consequence is the potential for a loss of electrical continuity in the interconnection by means of fatigue cracks in that Cu barrel. Empirical studies have confirmed that a Cu barrel fatigue failure can occur well before a TMF crack propagates through 100% of the solder material [3].

The finite element models are shown in Fig. 15 that correspond to the 100% crack path in each of the four configuration. Figure 15a shows that the 100% crack had propagated in two directions from the location of maximum EQPS (magenta arrow), following the solder/pin interface to the top of the fillet and to the bottom terminus (cyan arrow). A small, localized strain at the latter location, which is indicated by the red contours, did not affect the crack path.

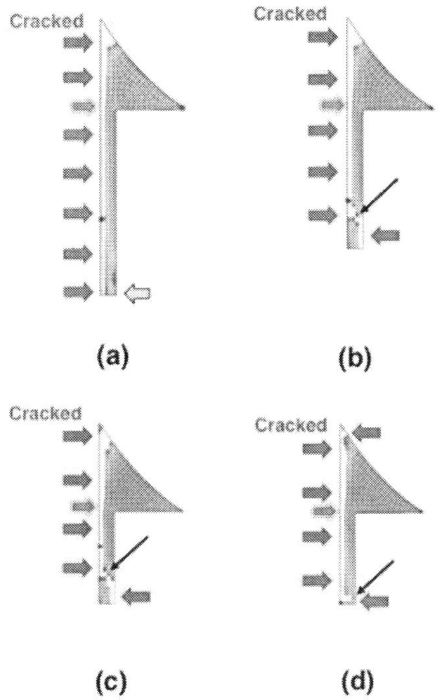

Figure 15 Computational model generated for 100% cracking of the (a) 100%, (b) 75%, and (c) 50% hole fill conditions (full fillet, one side). The computations were based on the use segment (16°C/33°C; 1 cycle/year). The magenta arrow is the site of maximum EQPS. The cyan arrow in (a) signifies the bottom terminus of the joint. (d) The 100% crack morphology is shown for the case of 50% hole fill and the absence of encapsulant under the PWB. The magenta arrows indicate the location of greatest EQPS. The black arrow in (b) – (d) indicates the location where the primary crack and the second crack, which originated from the bottom, joined together.

The 100% crack path is shown in Figs. 15b and 15c for the cases of 75% and 50% hole fill. While the predominant crack path remained at the solder/pin interface, the aforementioned EQPS at the bottom terminus increased in magnitude to the extent that it generated a second crack at the solder/Cu barrel interface. That crack then joined up with the primary crack at the locations denoted by the black arrow. The number of cycles to 100% cracking reflects the combined contributions of the two cracks. Although not performed here, the model analysis can discern the individual contributions made by multiple crack paths to a complete electrical open.

Lastly, the 100% crack diagram in Fig. 15d shows nearly the same crack morphology as that presented in Fig. 15c. The secondary crack had a smaller path when the encapsulant was removed from under the PWB. Therefore, the absence of the encapsulant had more of an effect on propagation of the primary crack along the solder/pin interface than it did on the second crack that originated from the bottom terminus.

In summary, the data in Table 1, regardless of the failure criteria, indicated that the magnitudes of TMF experienced by the variants of the through-hole solder joint were relatively small. Only the case of the absent encapsulant from under the PWB and coincident with the 50% hole fill, would accelerate the TMF to an extent that, with safety factors, could potentially cause a significant loss of fatigue life. Secondly, although the enhanced EQPS strain that occurred at the bottom terminus grew in magnitude with decreasing hole fill, it still did not control the TMF failure of the solder joint, per se. *However, those cyclic strains are imparted on the barrel of the PWB through hole and, potentially cause an electrical open failure due to TMF of the Cu, prior to a solder failure.*

Lastly, a significant amount of discussion was given to the EQPS and crack behaviors observed in Figs. 13, 14, and 15. The goal was to describe the *physical response* by the solder joint to the different hole fill configurations as well as the absence of encapsulant. The analysis indicated that the location of maximum EQPS as well as the crack paths, differed very little between the two transportation and handling segments as well as between the latter and the use segment. These different environments affect the *magnitude of the EQPS* introduced into the solder joints rather than changes to the *spatial distribution* of the TMF strain and similarly, the manner of crack propagation.

Acceptance Test + E-Test + Service Environment

The model predictions were examined for the case of the harshest condition that would be experienced by *fielded hardware*. That condition pertained to the small percentage of parts that experienced the E-test prior to being placed into service. That sequence is shown below:

- Acceptance test + E-test + *Service*

Table 2 lists the percentages of TMF life used up to reach crack initiation and 100% cracking of the five solder joint configurations. The percentages of TMF life used up to crack initiation remained very small at < 2% for all cases except the geometry with encapsulant absent from under the PWB, which was higher, but still small, at 6.7%. The values representing crack initiation increased by 97%, 85%, 92%, and 81% in the order of baseline and hole fills of 100%, 75%, and 50%, respectively, when compared to the data in Table 1 (service condition, only). The data show that the Acceptance test-plus-E-test regimen has a measureable effect on *magnitude* of solder joint TMF life. However, relative rankings of TMF performance did not change between Tables 1 and 2.

The trends were analyzed for the complete (100%) cracking predictions listed in Table 2. In the order of the baseline, 100%, 75%, and 50% hole fill, the percent changes between Tables 1 and 2 were 83%, 70%, 68%, and 61%, respectively. These relative changes differ with respect to those of crack initiation calculated between the same two tables. The important finding confirms that there is not necessarily a direct correlation – e.g., a proportional relationship – between TMF cycles leading to crack initiation and those that include propagation of TMF cracks.

Table 2 Percent of Life Used Through the Combination of Acceptance Test, E-test, and Service Environment Based Upon the Two Failure Criteria of Crack Initiation and 100% Cracking.

Through-Hole Configuration	Percent of Life Used to Crack Initiation (%)	Percent of Life Used to Complete Crack (%)
Full fillet, both sides	*1.3*	*0.33*
100% hole-fill	**1.2**	**0.34**
75% hole-fill	**1.4**	**0.47**
50% hole-fill	**1.7**	**0.53**
50% hole-fill; *No bottom encapsulant*	**6.7**	**2.7**

The case is examined of the 50% hole fill, but with encapsulant absent from under the PWB. Under the crack initiation criterion, there is a 6.7% loss of TMF lifetime under the combined conditions of Table 2, which represents a 72% increase when compared to the same geometry in Table 1. Furthermore, a 3X safety factor results in a fatigue life loss of 20%, which is considered to be "cautionary" in terms of long-term reliability. When the failure criterion is 100% cracking, there is a 2.7% loss of TMF lifetime, which is up by 59% over the prediction in Table 1. The lost TMF lifetime is 8.1% when a 3X safety factor is applied to the prediction. This analysis reaffirms the impact to the TMF lifetime when there is incomplete application of the encapsulant.

In summary, the data in Table 2 indicate that, although the absolute values remained small, the combination of Acceptance test-plus-E-test removed greater percentages of TMF lifetime versus the Service condition, alone. The absence of encapsulant (and 50% hole fill) caused a larger loss of TMF lifetime, 6.7%. The latter value raises some concern when a 3X safety factor is applied against it.

Acceptance Test + D-Test

The analysis continued by examining the case of solder joints that would be exposed to the Acceptance test followed by the D-test. Recall that the objective of the D-test is to cause wear-out failure modes in the entire assembly. The test conditions were not developed, specifically, for the solder joints. Therefore, the model computations were used to determine the extent of TMF lifetime lost by the D-test and whether the latter was sufficiently severe to cause wear-out of the solder interconnections.

The model predictions are shown in Table 3. The percentages of TMF life used up to reach crack initiation under the solder joint configurations were 2.5 %, 2.2%, 2.8%, and 2.9% for the baseline (two fillets) as well as single-fillet variants: 100%, 75%, and 50% hole fill, respectively. These values do not differ significantly between one-another, which indicates a relative insensitivity of TMF to the geometry. As anticipated, the lost TMF lifetime increased when encapsulant was absent from underneath the PWB, in this case, to 11%. The relative ranking down Table 3 was the same as that of the other two tables. The D-test did not alter the spatial distribution of EQPS between joint configurations, which was established by a review of the contour maps. The regimen simply changed the magnitude of the final strain by the greater number of cycles.

Table 3 Percent of Life Used Through the Combination of Acceptance Test and D-test, Based Upon the Two Failure Criteria of Crack Initiation and 100% Cracking.

Through-Hole Configuration	Percent of Life Used to Crack Initiation (%)	Percent of Life Used to Complete Crack (%)
Full fillet, both sides	*2.5*	*0.61*
100% hole-fill	**2.2**	**0.55**
75% hole-fill	**2.8**	**0.75**
50% hole-fill	**2.9**	**0.78**
50% hole-fill; *No bottom encapsulant*	**11**	**3.7**

The 100% cracking criteria predictions were also listed in Table 3. The used-up fatigue life increased over the previous scenarios, yet remained very small at less than 1% in all cases except the configuration with absent underfill, which had reached 3.7%.

The predictions in Table 3 indicate that the D-test did not cause the solder interconnections to come close to their wear-out lifetimes by either failure criterion. This outcome is acceptable in terms of an evaluation of the entire assembly, as long as solder joint TMF was not the life-limiting property vis-à-vis the other structures – e.g., components, PWB, connectors, etc. On the other hand, if solder joint fatigue was the life-limiting factor in the assembly design, then the D-test would require revision to further accelerate TMF, such as more cycles, reduced minimum temperature, or an increase in the maximum temperature.

Conclusions

1. An application arose that required that predictions be made of the thermal mechanical fatigue (TMF) of through-hole solder joints that are partially filled with solder using the COMPSiR© and TurboSiP© computational models.
2. The model predictions were made, based upon the sequence of environments experienced by both fielded hardware as well as non-fielded assemblies exposed to more severe temperature cycles.

3. There were two failure criteria: (a) cycles to reach TMF crack initiation and (b) cycles to cause a 100% crack of the solder that is necessary for an electrical failure.
4. The TMF lifetime used up in the service environment (two transportation segments and the use segment) was less than 1% for the more conservative, crack initiation criterion. An absence of encapsulant from under the PWB caused a 3.9% loss of fatigue life, which is still acceptable with a 3X safety margin.
5. Although the EQPS strains increased at the bottom terminus with decreasing hole fill, they did not control the TMF failure of the solder joint, per se. *However, those cyclic strains are imparted on the adjacent barrel that could potentially cause an electrical open due to fatigue of the Cu.*
6. The harshest environment that fielded hardware would experience – Acceptance test + E-test + service – used a very limited amount of TMF lifetime from the interconnections: < 2% at crack initiation. The exception was the case when encapsulant was absent under the PWB and 50% hole fill where 6.7% of TMF lifetime was used to reach crack initiation, which when combined with a 3X safety factor, would sound a cautionary note in terms of long-term reliability.
7. There is not necessarily a direct correlation – e.g., a proportional relationship – between TMF cycles leading to crack initiation and those leading up to 100% cracking.
8. The D-test was not sufficiently severe to cause solder joint TMF to reach its wear-out failure regime.
9. Three high-level findings were obtained from this analysis. First of all, the TMF *deformation* (EQPS per cycle), which leads to crack initiation, does *not* scale the same as the TMF *deformation-plus-cracking* process that leads to a fully cracked interconnection. Second, incomplete encapsulation can accelerate, significantly, the TMF of through-hole interconnections. Third, transportation and handling procedures can generate a *greater* degree of TMF damage than does actual use segment.

Acknowledgments

The authors wish to thank Brian Wroblewski for his careful review of the manuscript. Sandia is a multiprogram laboratory operated by Sandia Corporation, a Lockheed Martin Company, for the United States Department of Energy's National Nuclear Security Administration under contract DE-AC04-94AL85000.

References

[1] Neilsen, M. and Vianco, P., "UCPD Model for Pb-free Solder," *J. Elect. Pack.* (2014) DOI:10.1115/1.4026851.

[2] Neilsen, M. and Vianco, P., "Simulating Solder Fatigue Crack Initiation and Growth in a Surface Mount Package," *IPC Conference on Soldering and Reliability.*, Costa Mesa, CA., Nov. 2013, CD-ROM.

[3] Susan, D., Kilgo, A, Neilsen, M., and Vianco, P., "Accelerated Aging and Thermal Mechanical Fatigue Modeling of Cu-Plated Through-Holes with Partial Solder Filling," *Int. J. of Materials and Structural Integrity*, Vol. 2 (2008), pp. 138 – 163.

Active Solder Joining... A Technology and Application Update

Ronald W. Smith
S-Bond Technologies, LLC, Lansdale, PA USA
(T) 215-631-7114 info@s-bond.com

Abstract

Active solders have been growing in application, filling an important gap in the joining of dissimilar materials, including ceramics, metals and composites. Active solders incorporate active elements such as Ti, Hf, or Zr along with rare earth element such as Ce or Lu to enable solders to wet and adhere to oxides or other compounds that form or are part of a materials surface. Such active element additions to Sn-Ag, In-Sn, Bi-Sn and Zn-Al based solders have been shown to enable active solders to adhere to many metals, ceramics, and glass. Active solder technology was first introduced commercially in mid-1990's. This paper will review and provide updates to the technology, present on their operating bonding mechanisms on metals, ceramics and glasses while reviewing process technologies that have been developed to enable these active solders to bond a wide variety of materials and material combinations. A range of applications will be presented to demonstrate the utility of active soldering technology and provide demonstration of how active solders can be used in a wide variety of components from aluminum to silicon carbide. The paper presents several applications and the cases for their adopting active solder technology.

Introduction

Active Solder Technology

With their ability to adhere to and wet most metals, many ceramics, carbides and even graphite, active solder filler metals are finding application in many areas, including thermal management components, solar cells, batteries, sensors, optical and electronic components. Most active solders are lead-free modified formulations and are similar to conventional solders with the exception of titanium and rare earth elemental additions. The addition of these elements enable the solder filler metals to wet and bond a greater range of metals, ceramics, glasses, and carbon products than standard solders. This is accomplished without flux, nickel plating, or pre-stripping surface oxide layer.

Many applications for these new materials are in high thermal conductivity situations, where solder materials are needed to improve the rate of heat dissipation.

Mechanical Activation

Active solder fillers do not require flux to adhere to adhere to metals. In conventional soldering, such as in aluminum soldering, the fluxes used are aggressive chemical based cleaning agents that facilitate soldering, by removing oxidation layers from the metals that are being joined. However in active soldering, these filler metals will adhere to the base metal oxides but not without first disrupting the actives solders own rare earth modified oxides that form on the active filler metal as it melts.

Active solder metal fillers have to be disrupted to enable the active elements to come in contact with the underlying base layers. When the molten active solder layers are sufficiently disrupted via mechanical rubbing, brushing or even ultrasonics, then they will wet and flow over metal, ceramic and carbide materials... the first physical step of solder joining — wetting. Figure 1 indicate the disruption of the transparent and thin (~ 20-30 Å) stable oxide films that form on molten active solder.

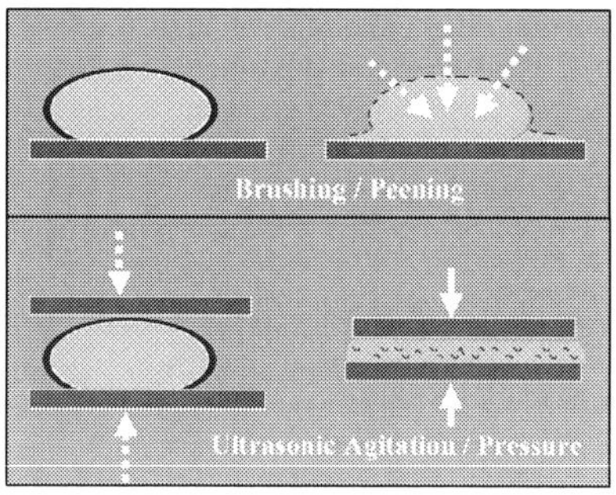

Figure 1. Illustration of the process of surface wetting by active solders.

S-Bond materials are placed onto surfaces by various means of mechanical agitation, also termed "activation", as shown in the figures below. Figure 2 illustrates how the S-Bond alloy can be transferred without flux to aluminum using a rubbing process, while the S-Bond alloy is molten on the part surfaces. Transfer to other geometries such as tubes, flanges, edges can easily be adapted using various tools and techniques. Brushing is another way to transfer S-Bond to joining surfaces. It is particularly effective on irregular surfaces and foam surfaces.

Figure 2. Mechanical activation of active solder filler metal on aluminum surfaces by rubbing/scraping.

Ultrasonic Soldering

Ultrasonic application effectively agitates active alloys onto component surfaces. Ultrasonic soldering is very well suited as a mechanical activation process for fluxless soldering with active solders. Such solders rely on reactive elements such as titanium (Ti) and rare earth elements such as Cerium (Ce) to enable direct wetting of metals, ceramics and glass without the need for chemical fluxes or plating. The cavitation from an ultrasonic solder iron tips disrupt the stable oxide films that form on the molten active solders. Figure 3 illustrates how an ultrasonically activated heated tip enables the active solder fillers to wet and adhere directly to aluminum. The same process enables active solders to wet and bond to works on all metals and ceramics.

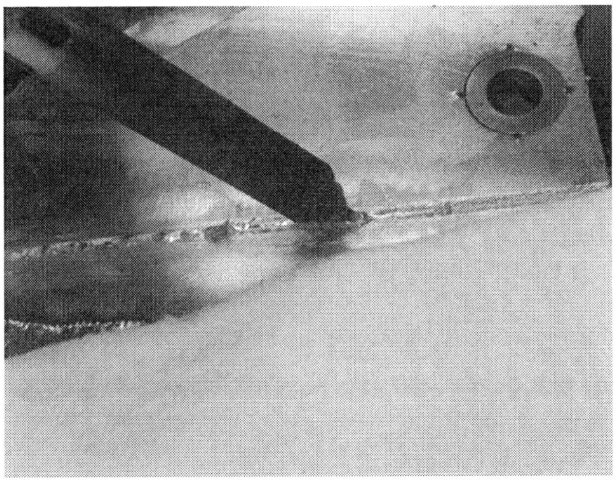

Figure 3. Picture of ultrasonic soldering with active solder alloy Sn-3.8Ag-2.8Ti-0.1Ce-0.1Ga.

The horn tip does not normally touch the part surface and relies on cavitation driven by the 20 - 30 kHz acoustic waves to mechanically agitate and break up the oxides that form on the molten active solder layers. Ultrasonic soldering tips can be narrow and suited for soldering narrow lines and/or point soldering as shown in Figure 3, or larger heated ultrasonic horns (Figure 4) can be used to activate a much larger area to deposit an active solder layer as the first step in soldering two surfaces together.

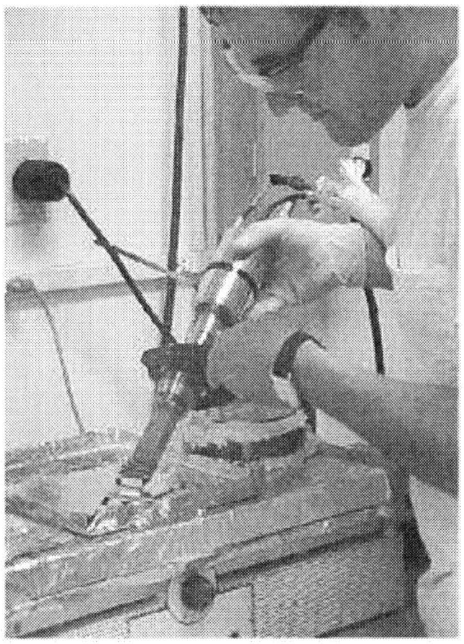

Figure 4.Ultrasonic soldering to wet larger surface areas. (EWI Joining Technology Center)

Ultrasonic soldering of active solder is an excellent high volume production process. Figure 5 illustrates a wire fed robot operated soldering iron that can be used in spot, line or large area ultrasonic activation of active solders. Ultrasonic soldering and active solder permit the precise placement of solder filler metal.

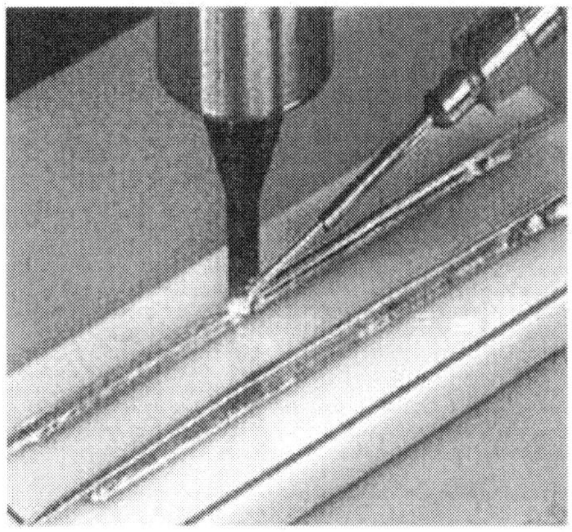

Figure 5. Wire fed robotic ultrasonic soldering head. (photo courtesy of Japan Unix)

Applications

There are many applications that are adopting active soldering as the preferred joining method. The ability to join all metals, most ceramics and composites of the same, without flux and without plating offer many advantages. Some of the more important applications are reviewed below.

Aluminum Bonding

Aluminum bonding/ soldering is done with active solders by melting the filler metals, mechanically spreading them on the surface via brushing, rubbing, or via ultrasonically activated spreaders to wet, adhere and provide a base for bonding. In a subsequent steps, faying surfaces with molten pre-tinned active solder layers are pressed or slid together to activate a strong solder bond. Advantages of active soldering of include

- Lower temperature bonding (from 120 – 250°C) lowers thermal expansion mismatch issues.
- Permits the joining of aluminum to copper and other metals and ceramics.
- No additional metal plating
- Flux free joining; no entrapment and no post cleaning
- Offers repair and re-manufacture of aluminum parts since solders remelt at temperatures much below aluminum melting temperatures.

Applications where active solders show their advantages include a wide range of thermal management components (finned heat exchangers, cold plates, and heat spreaders). Figures 1 – 4 illustrates just some of the many successful examples. Thermal management is a class of applications that include heat exchangers, cold plates, heat spreaders, heat sinks and heat pipes.

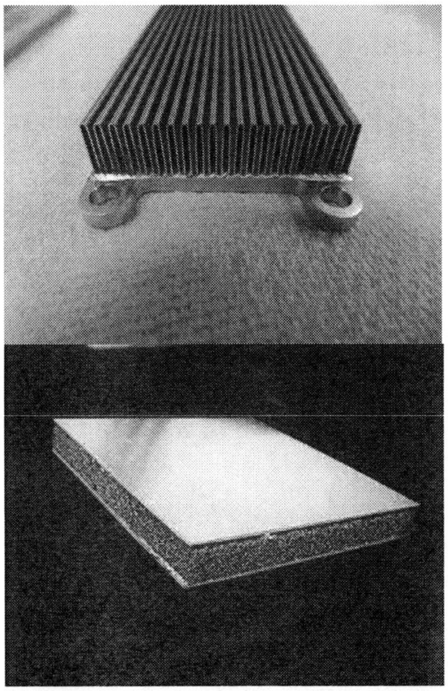

Figure 6. Active solder bonded aluminum for heat exchangers and aluminum clad graphite cored heat spreader.

Figure 7. Copper heat pipes active soldered (Bi-42Sn-2.2Ag-1.8Ti-0.05Ce-0.05Ga) into aluminum housing.

Ceramic Metal Bonding

Ceramic to metal bonding used in many components, including sensors, electronic packaging and in power electronics. From sensor windows where quartz and/or sapphire (single crystal aluminum oxide which is transparent) to aluminum oxide or aluminum nitride are used as insulating bases where high voltages need to be isolated. Two challenges limit the use of direct soldering of ceramic to metals… 1) ceramics are not able to be directly wetted and 2) ceramics and metals have largely different coefficients of thermal expansion (CTE).

Active solders can join ceramics, (oxides, nitrides, carbides, carbon/graphite, silicides, and others) to metals to produce reliable, hermetic joints with all metals…including steel, stainless steels, titanium, nickel alloys, copper and aluminum alloys, provided thermal expansion mismatch at joining temperatures is managed. The advantages of active soldering ceramic to metal include:

- Lowers joining temperatures compared to conventional active brazing.
- Enables larger and more complex assemblies to be fabricated without ceramic cracking or metal distortion.
- Does not require plating of ceramic or metal in the assemblies.
- Eliminates flux residue
- Produces a reworkable ceramic – metal joint that can be taken apart and reliably joined again

Two different processes can be used in active solder ceramic to metal bonding. One method is the "mechanically activated" joining at near the active solder filler melting temperature, (e.g. for S-Bond 220, that is 250ºC). This is done by spreading, rubbing, or brushing the molten alloys onto heated surfaces and assembling "hot" in a way that the active solder alloy surfaces are agitated sufficiently to break the thin oxide skins that form while molten. Such joints on ceramics and many metals are adhesive, but have no chemical bond. An example of the bond is shown in Figure 7. Active solder alloys do bond, but their joint strengths are nominally below 21 MPa

(3,000 psi) in shear. The figure to the right illustrates the adhesive nature of the bond.

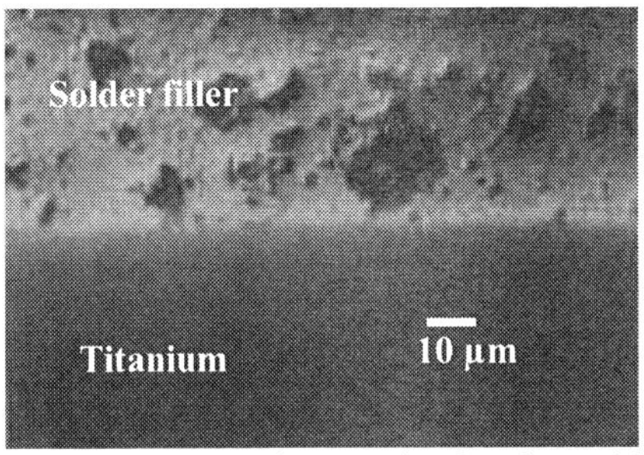

Figure 8. Photomicrograph of active solder (Sn-3.5Ag-2.8Ti-0.2Ce-0.1Ga) bonded to titanium as an adhesive bond.

The other active solder joining process is thermally activated soldering at temperatures over 800C° in vacuum, which develops a chemical bond to the surface, through reactions of the active elements in active solder alloy. At the elevated temperatures, the active elements (Ti, Ce, etc.) in active solders react with the ceramic to develop a chemical bond, as shown in the Figure 9.

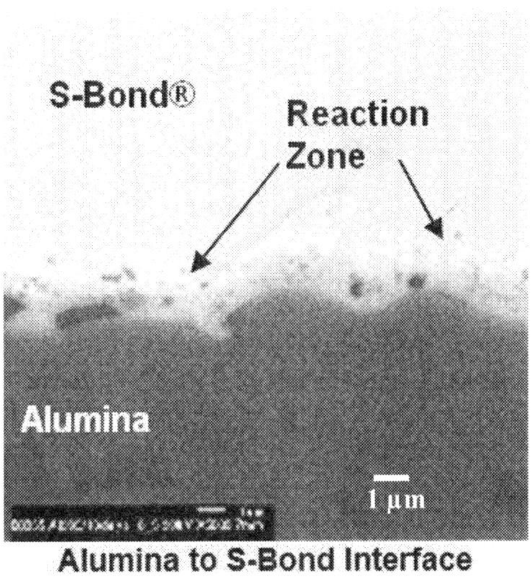

Figure 9. Photomicrograph of active solder (S-Bond®) reactively bonded to aluminum oxide.

This chemical bond with the active solder layer does, in subsequent soldering steps, provide a much higher level of joint strength between the active solder filler metal and the ceramic.

Graphite/Carbon Bonding

Graphite, carbon and carbides have been bonded with active solders using an initial active metallization treatment followed by active soldering to most metals, ceramics or glass. Active soldering in this manner eliminates the need for chemical fluxes which can be absorbed and contaminate the graphite components.

Figure 10 illustrates the structure of the S-Bond soldered interfaces on solid and on foamed graphite. The reactions with active solders' reactive elements, generate a reacted "solderable" surface to which metals, glass, ceramics and other crystalline materials can be bonded.

Figure 10 a). Graphite foam cored phase change material heat sink with aluminum interfaces and tubing.

Figure 10 b). Metallographic cross Graphite foam

Figure 10 c). Photograph of graphite/wire bonded electrode set for water conductivity sensor.

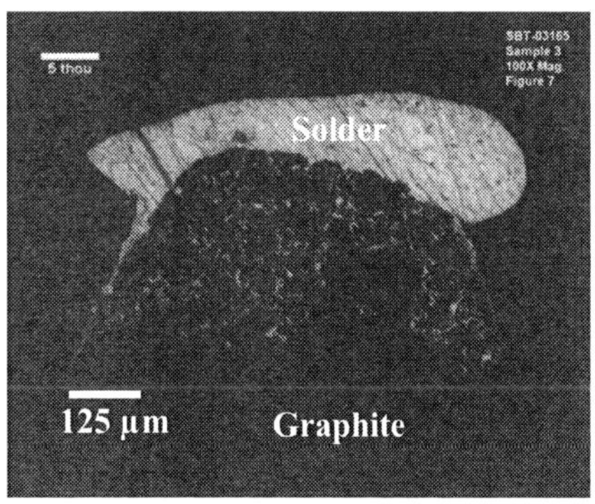

Figure 10 d). Photomicrograph of active solder "metallizing" the top of the graphite electrode pin, illustrating bonding.

Brazing (> 450°C) is another graphite bonding method that use molten active braze filler metals with Ti, Hf or Zr additions.

These braze processes are normally done at temperatures above 800˚C and in vacuum so not to oxidize or decompose. Active soldering offers the advantages of enabling bonding to aluminum (Figure 10a) or bonding on electrical leads (Figure 10c). Lower temperature soldering permits metallic joining to graphite/carbide components used in electrical connections, thermal management and wear components.

Glass Bonding

Glass-metal seals are used in many sensor applications where either optical windows isolate an enclosure or where an insulating feed through is needed for electronic or telecommunication packages. Active solders have the ability to bond to glass… Ti, Ce, Lu and Ga additions to Sn-Ag, Bi-Sn and In-Sn solders enable the filler metals to bind to glass and create a strong joint. The titanium and rare earth elements during the soldering process have been measured to interdiffuse into the glass on the order of 0.001 µm, enough to create a strong, hermetic joint between the glass and the active solder filler metal. Figure 11 illustrates active solder adhering to glass showing how phases in the active solder nucleate on the glass surface, creating a Van-der-Vaals type bond.

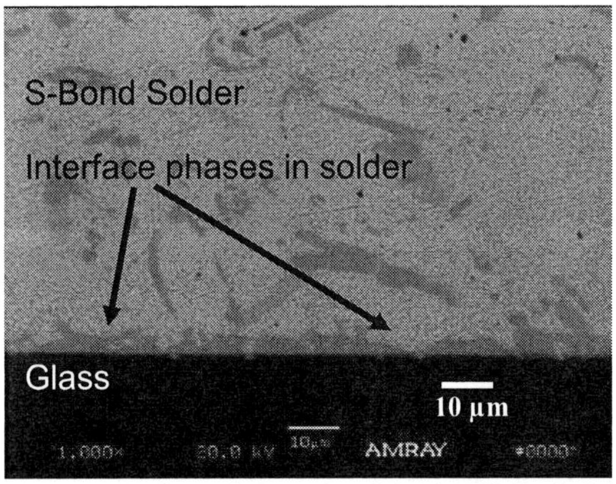

Figure 11. Photomicrograph of Sn-Ag-Ti-Ce-Ga-Mg active solder bonding with glass.

Active solders have shown to be very effective of directly adhering to glass based oxides, as illustrated in Figure 11. Such joining capability demonstrates how active solders can be effectively used to bond glass based windows for sensors and other applications.

Soldering Composites

Metal and ceramic matrix composite materials (MMC and CMC, respectively) have been bonded using active solders such as Sn-Ag-Ti-Ce and Bi-Sn-Ti-Ce. Al-SiC are the most common metal matrix materials that have been bonded at soldering temperatures from 140 – 400˚C. Such Al-SiC materials are used in electronic and optical packaging where the low CTE of the Al-SiC provides good thermal conductivity. In metal matrix composite (MMC) bonding, active solders metallurgically bond to the aluminum metal matrix while the solder wets and adheres to the SiC particles. Figure 12 illustrates the bonding and wetting of an active Zn-Al-Ag-Ce active solder bonded to Al – 60 v/o SiC particle composite. The active solder interacts and dissolves some of the aluminum matrix and the SiC particles enter the solder and are wetted integrated into the solder joint. Figure 13 illustrates and electronic packaging application for active solder joining of Al-SiC to Al-SiC, making a seal for a water cooled electronic heat sink.

Figure 12. Photomicrograph of Zn-Al-Ag-Ce solder bonded to AlSiC - 60 volume % SiC.

Figure 13 shows a water cooled electronic package sealed with active solders.

Sensors

Sensors incorporate a lot of different elements, materials and soldering opportunities. Some applications are seals on windows or feed throughs, others are simpler electrical connections and others solder bonding of active elements such as piezoelectric crystals or silicon based strain gages. Sensor windows can be optical and are usually sapphire, quartz or glass. Other windows can be for x-ray and/or other radiation so windows that seal enclosures can be tungsten, titanium, or other crystals such as CaF_2 or $LiBO_2$. Active solders are finding use in creating hermetic seals, as seen in Figure 14 sealing the internal sensor crystals from the environment.

Figure 14. Sapphire window (~ 25 mm D) active solder metallized. a) after vacuum furnace treatment and b) after reheating to 250C and removing excess metallization pastes

Sapphire is an excellent window for optical sensors, it IR transmittance, strength and low expansion makes it a natural choice for sensor sealing enclosures. Active solder alloys make the bonding of sapphire to metals, first chemically bonding the active solder's Ti to the ceramic in the presence of the tin (Sn) solder matrix.

Figure 15 illustrates the active solder alloys directly on the sealing edges of a sapphire lens. It is these layers later in a second operation join sapphire to metal.

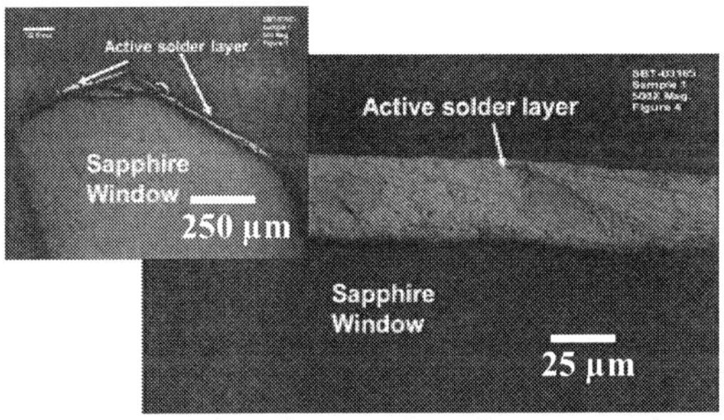

Figure 15. Microphotographs of cross section of sapphire window with reacted active solder "pretinned" layers.

Sapphire is an excellent window for sensors as seen Figure 16 which illustrates sapphire windows active solder sealed in titanium tubes to create hermetic sealed enclosures for scintillation crystals used in gamma ray detection.

Figure 16. Sapphire windows active solder sealed titanium tubes gamma ray detector enclosures.

Figure 10c shows a water conductivity sensor electrode with copper wire leads active soldered directly to graphite electrodes. For active sensors, actives solders are showing their utility in bonding piezo-crystals (lead zirconate, PZT) to metal and / or glass.

Figure 17 shows a PZT crystal active soldered to a metal frame.

Figure 17. Photo of a PZT crystal that is active solder bonded to a metal frame for use in detector.

Solar

Photovoltaic, poly-silicon solar panels function with soldered electrical connections to the back of each solar cell, connecting them in parallel to create a solar panel at the rated voltage and power. The backside of the poly-silicon solar cell has a thermally diffused aluminum collector layer in which the electrons, displaced by the photons in the doped solar cell silicon semiconductor. Copper tabbing wires need to be solder attached to this aluminized layer. Active solders with their ability to bond to aluminum, copper and even silicon, the active soldering of copper tabbing leads to the back of solar cells to make full solar panels is being adopted as standard practice. Figure 18 illustrates the ultrasonic active soldering process. In this process, active solders either applied to the back of the aluminized cell, on the tabbing wire and/or on the ultrasonic soldering iron tip is "scrubbed" to be mechanically activated to produce a flux free solder joint.

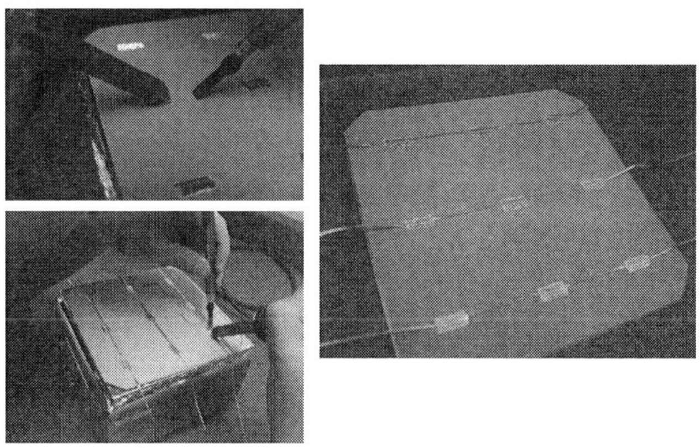

Figure 18. Photographs illustrating the ultrasonic active solder bonding of tabbing interconnect leads on back of solar cells, building a solar panel.

Thermal Management

Heat exchangers, cold plates, heat pipes and vapor chambers heat sinks and special phase change heat sinks are all examples where thermally conductive metal bonds are advantageous. Active solder bonding with its ability to bond aluminum, copper, graphite, even diamond and also bond to AlN, and SiC make it a suitable process and solder filler metal to fabricate a range of thermal management components. Cold plates that use aluminum and/or copper tubing, as seen in Figure … show how active solders can be used in bonding dissimilar metal combinations. Passive heat sinks, bonding copper heat spreaders into aluminum electronic enclosures or active heat sinks that utilize soldered in heat pipes or solder bonded graphite foam infiltrated with phase change materials (e.g. paraffin) are illustrated in Figures 7 and 10, respectively. In these applications lower temperature (140˚C) Bi-Sn-Ti active solders have been developed to permit bonding of thermally sensitive heat pipes and to permit aluminum backed high brightness LED's to vapor chamber heat spreaders, as shown in Figure 19.

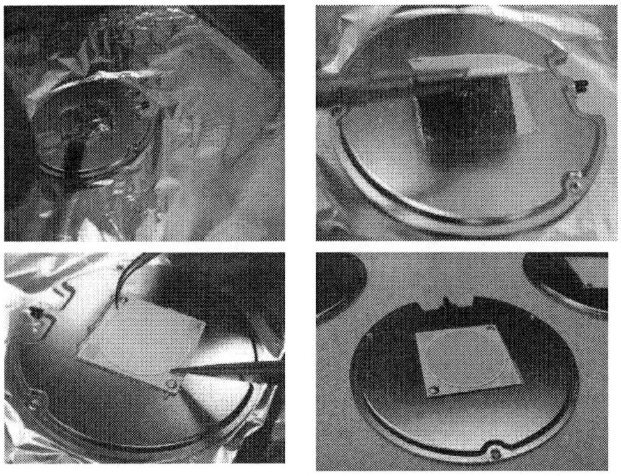

Figure 19. Photographs of the lower temperature (140˚C) active soldering of an aluminum backed LED package to a Ni-plated copper vapor chamber heat sink.

Illustrating another thermal management component, Figure 10a shows a graphite foam cored heat sink where the graphitic foam surfaces was made solderable using an active Sn-Ag-Ti solder treatment at over 850˚C in vacuum after which the treated graphite foam blocks were active soldered into aluminum enclosures. These solder bonded graphitic foams were used in conjunction with phase change materials (paraffin) which were infiltrated into the bonded foams and the soldered interfaces thermally connected the graphite foam cell walls to the aluminum enclosure such that small reservoirs of paraffin were able to quickly absorb and release high amounts of heat as the paraffin melted and solidified using its high heats of fusion to temporarily store high heat loads from heat sources such as electronics mounted onto the aluminum enclosure.

Conclusions

In this paper, active solders have been shown to be able to join a wide range of materials used in sensors, thermal management, electronic packaging, solar and cells. Active solders are increasingly being applied in many dissimilar materials combinations of metals, ceramics and composites all without flux or plating. Such actives solders benefit from the use of mechanical agitation to activate active solders, thus their processing differs significantly from conventional soldering, requiring mechanical agitation instead of chemical fluxes to disrupt the oxides that form on the molten filler metals. Mechanical activation methods that effectively disrupt the active solder oxides include rubbing, brushing, sliding and ultrasonic soldering. Active solders in their melting ranges will not be compatible with reflow soldering, hence special processed and tooling have been developed for each application to assure mechanical agitation can be employed to activate the active solder filler metals for bonding,

In particular active solders were shown to be very compatible with ultrasonic soldering which utilizes 20 – 60 kHz excitation of heated soldering iron tips to disrupt the very stable rare earth oxide films that form on actives solders as they melt and spread on metals, ceramic and composites.

Active solders' ability to wet and adhere to a wide range of materials, commensurate with their lower bonding temperatures make them well suited for bonding dissimilar materials whose differing thermal expansion coefficients many times limit brazing.

Interfacial Reaction of the Ultrasonic-assisted Soldering Copper and Sn-3.0Ag-0.5Cu Solder Alloys

Tan Ai Ting, Farazila Yusof, and Mohd Hamdi Abd Shukor
Department of Mechanical Engineering & Centre of Advanced Manufacturing & Material Processing (AMMP), Faculty of Engineering, University of Malaya, 50603 Kuala Lumpur, Malaysia
Email: aiting2602@gmail.com Tel: +60163467525

Abstract

In this paper, ultrasonic-assisted soldering method was used in the joining of pure copper (Cu) using Sn-3.0Ag-0.5Cu (SAC305) solder alloy. The effect of vibration time on the formation of intermetallic compounds (IMCs) in the solder joints was investigated. The samples were heated to 250°C and hold for 5 minutes, followed by air-cooled in normal atmosphere. The top substrate was exposed to high-amplitude-low-frequency ultrasonic vibration for vibration time of 1s to 4s during the peak temperature holding time. The soldered samples were cross-sectioned and the joint microstructures were examined by SEM and EDS. Both Cu_6Sn_5 and Cu_3Sn IMC layers were observed at the joint interface. However, the formed IMC layers were found to be thinner in ultrasonic soldered samples. The thinnest IMC layers were found in the sample soldered with vibration time of 3s, where the Cu_6Sn_5 IMC layer decreased by 55.97% and the Cu_3Sn IMC layer decreased by 43.45%. At the same, these joints obtained the highest shear strength which is 38.14 MPa. The present work revealed that the growth of IMC was suppressed due to the acoustic cavitation effect of ultrasonic vibration. However, prolonged vibration time can cause excessive dissolution of Cu elements from the substrates which in turn, promoting the growth of both IMC layers and reducing the joint strength.

Introduction

As the world is demanding miniaturized and multifunctional inbuilt electronic devices, the size of the solder joints have reduced dramatically and higher power, stress and heat are imposed on these tiny joints [1]. This causes the excessive growth of brittle intermetallic compounds (IMCs) in the solder joints during service, leading to weakening and cracking along these brittle regions [2], therefore, prompting the need to improve the reliability of current conventional solder joints.

One of the recent approaches used to improve the joint reliability is adding nanoparticles into the conventional Sn-based lead-free solder such as Sn-Ag-Cu solder alloy. These nanoparticles served as reinforcing phases which strengthen the solder matrix by uniform dispersion and refined the IMCs formed at the joint interface. The reinforcing nanoparticles can be metallic particles (such as Ni, Zn, Al) [3, 4], ceramic particles (such as TiO_2, Al_2O_3, ZrO_2) [5-7] and carbon nanotubes [8]. These particles can refine and suppress the formation and growth of IMCs in the solder joint but the efficiency may be reduced significantly by the agglomeration of nanoparticles [5, 9].

Instead of changing the solder alloy composition, different approach such as introducing ultrasonic waves (USW) into solder joint during solidification stage has been proposed in order to homogenize the IMCs formed at the joint interface and refine their grains [1, 2]. USW would induce acoustic cavitation to the solidifying molten solder alloy which can generate effects such as acoustic streaming, shock waves and micro-jets. These could increase the mass and thermal transportation to accelerate the IMCs formation and fasten the nucleation of such IMC phases throughout the joint. Joints with more refined microstructure and higher mechanical properties such as hardness and strength could be produced. Unfortunately, the research conducted on joining of Sn-based lead-free joint with the aid of USW has been relatively scarce and the process parameters other than USW power intensity have not studied in details as well. Therefore, in the present work, using high amplitude and high power USW, the influence of ultrasonic vibration time on the microstructure of Cu/SAC305/Cu solder joint was investigated.

Experiment

Materials

The base metal substrates were Cu plates (99% purity) with the dimension of 20mm x 20mm x 3mm (bottom plate) and 15mm x 15mm x 5mm (top plate). The solder alloy used was Sn-3.0Ag-0.5Cu (SAC305) solder paste. The schematic diagram of the joint configuration is illustrated in Fig. 1.

Experimental Procedures

Prior to soldering, the substrates were ground with 30μm, 15μm, 10μm and 5μm silicon carbide papers and then polished by 1μm diamond suspension and fine polished with 0.05μm colloidal silica suspension. The substrates were ultrasonically cleaned for 15 min in an acetone solution. 0.2g solder paste was sandwiched in between the substrates and the joint clearance was controlled by the load of ultrasonic horn (approximately 4kg).

The assemblies were heated by hot plate (attached to a programmable controller) at the ambience atmosphere. The assemblies were preheated to 170°C and further heated up to 250°C. The assemblies were held for 5 mins at the peak temperature. During the holding time, the ultrasonic horn was loaded on the top Cu plate and vibrated for 0, 1, 2, 3 and 4s. The assemblies were then cooled down to room temperature in ambience air after the holding time. The ultrasonic vibration was 20kHz with amplitude of 30μm.

The joint specimens were cross-sectioned by a slow diamond cutter and were mounted in epoxy. The surface of the cross-sectioned specimens were prepared by using the same metallography method used to prepare the Cu substrates. The microstructure of the specimens were examined by scanning electron microscope (SEM, EVO-MA10, Zeiss) equipped for energy-dispersive x-ray spectroscopy (EDS). The intermetallic compounds (IMCs) present at the joint interface were determined and Image J software was used to measure the thickness of IMCs. The shear strength of the joint specimens was tested by Instron Universal Testing machine (Model 3369). As shown in Fig. 1(b), the top Cu plate was clamped and the shear force was exerted on the bottom Cu plate at the shear speed of 0.5mm/min.

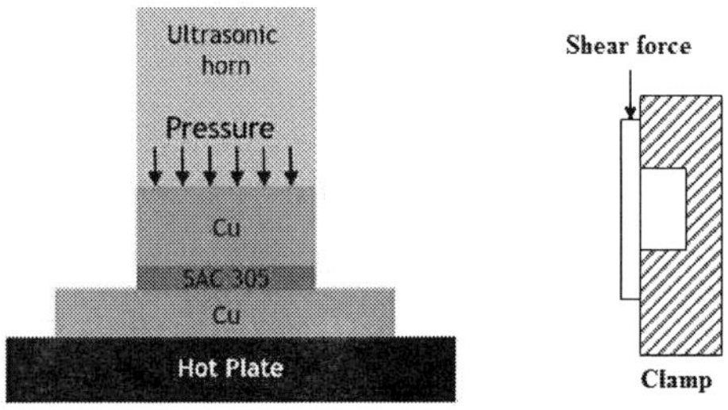

(a) (b)

Figure 1. Schemetic of (a) Ultrasonic soldering setup; (b) Shear test arrangement

Results

Microstructure Observation

Figure 2 shows the microstructures of the Cu/SAC305 joint interface in which treated by various ultrasonic vibration time. There are two layers of IMC formed at the joint interface and they have been identified as Cu_6Sn_5 and Cu_3Sn IMC phases by the EDS result (not shown). The morphology of Cu_6Sn_5 IMC grains changed from irregular scalloped shape to more rounded and uniform scalloped shape when the solder joint was treated with increasing vibration time. More channels between the Cu_6Sn_5 IMC were observed in ultrasonic-treated joint. Furthermore, the size and thickness of the grain reduced when the vibration time increased from 0 to 3s but slightly increased when vibration time is above 3s. For Cu_3Sn IMC, its morphology remained planar even treated with USW. Coarse Cu_6Sn_5 IMC particles were observed to be dispersed in the solder matrix in the non-ultrasonic-treated joint. These particles became finer after treated up to 2s of USW. When the joint is exposed to USW above 2s, these IMC particles elongated and became rod-like at 4s of USW.

The thickness of both IMCs is plotted as a function of vibration time in Fig. 3 and 4. Both Cu_6Sn_5 and Cu_3Sn IMC layer thickness reduced significantly when treated with USW. The maximum IMC reduction for Cu_6Sn_5 and Cu_3Sn IMC are 55.97% and 43.45% respectively and both occurred at vibration time of 3s. This indicates that USW has significant effect on the growth of both IMCs.

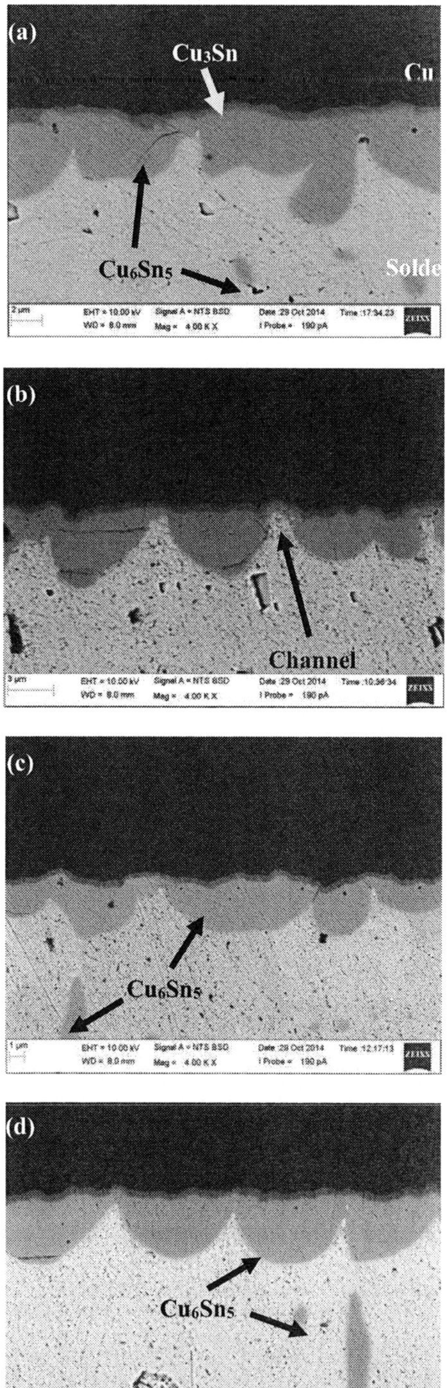

Figure 2. SEM cross-sectional image at Cu/SAC305 interface treated by vibration time of (a) 0s, (b) 1s, (c) 3s and (d) 4s

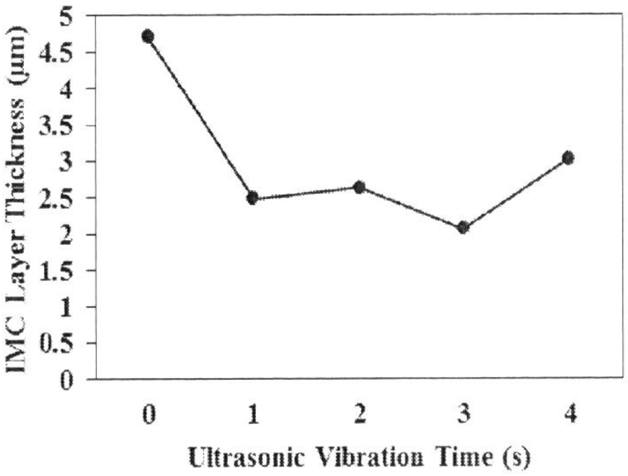

Figure 3. Cu_6Sn_5 IMC thickness in the joint interface at different vibration time

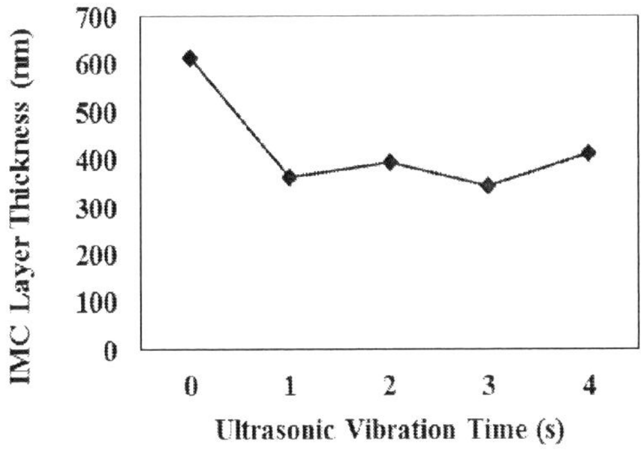

Figure 4. Cu_3Sn IMC thickness in the joint interface at different vibration time

Shear Strength

The average shear strength of the joint treated at varied ultrasonic vibration time are plotted in Fig. 5. The results showed improvement in solder joint after treated with USW. The average shear strength of the joint without USW treatment is 27.07 MPa. The strength continued to increase to 38.14 MPa when exposed to USW up to 3s, which represents an increase of 40.89% from the non-ultrasonic-treated samples. When the samples were further exposed to 4s of USW, the shear strength decreased to 32.66 MPa, which decreased by 14.37% from the maximum shear strength obtained at 3s.

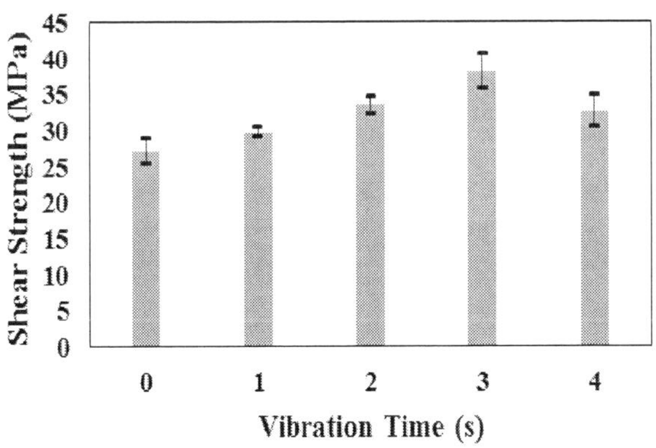

Figure 5. Joint shear strength at various ultrasonic vibration time

Discussion

The reduction of IMCs thickness in the ultrasonic-treated joint is attributed by the acoustic cavitation and streaming effect induced by USW in molten solder alloy. Microbubbles were generated in the molten solder due to the pressure fluctuations that occurred in the USW. When the microbubbles which preferably formed at the solid Cu/liquid solder interface collapsed, high local turbulence (causes acoustic streaming) and release of heat energy (causes erosion) were generated [1, 10]. Eventually, the Cu_6Sn_5 particles and Cu atoms that precipitated at the Cu_6Sn_5/solder interface were flushed away and then be transferred into molten solder matrix due to the acoustic streaming effect. This is evident by the change in Cu_6Sn_5 IMC morphology and thickness before and after ultrasonic treatment (shown in Fig. 2). As the formation of Cu_6Sn_5 IMC at the interface was reduced by the acoustic streaming, the formation of Cu_3Sn IMC was reduced correspondingly based on the reaction stated as follows.

$$Cu_6Sn_5 + 9Cu \rightarrow 5Cu_3Sn \quad \text{(Equation 1)}$$

Therefore, the IMC suppression effect becomes more significant with increasing vibration time and it is correspond to the results demonstrated in Fig. 3. The diffusion mechanism due to streaming effect is illustrated in Fig. 6.

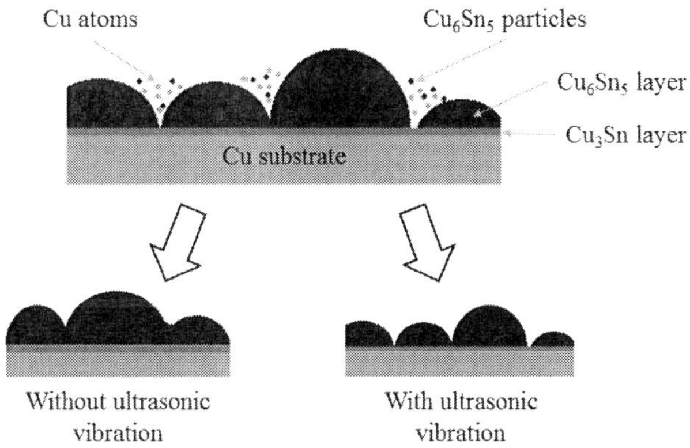

Figure 6. Streaming effect of USW on the IMC formation at the Cu/SAC305 interface

As mentioned earlier, collapsing of microbubbles at the joint interface resulted in releasing excessive heat energy into the molten solder. This would accelerate the nucleation and formation of IMC particles (such as Cu_6Sn_5 and Ag_3Sn particles) in the solder matrix [1, 2]. Furthermore, with the aid of acoustic streaming, the temperature distribution and concentration gradients could be homogenized, thus forming IMC particles with finer microstructure. Meanwhile, they can be uniformly dispersion in the solder matrix due to the streaming effect. The dispersion of these IMC particles would strengthen the solder matrix and resulted in higher shear strength of the joint. This is in good agreement with the shear test results in which the joint shear strength increased gradually when exposed to USW for 0 to 3s.

When the molten solder was exposed to USW up to 4s, excessive dissolution of Cu atoms from the substrates into the molten solder may be induced and causes the additional growth of IMCs at the joint interface as well as the IMC particles dispersed in solder matrix. It is possible that the elongated rod-like Cu_6Sn_5 particles are the broken fragments of Cu_6Sn_5 IMC whiskers formed at the joint interface due to acoustic streaming. These elongated brittle IMC fragments may be the crack initiation site when the joint is subjected to shear force. This may be explained the sudden drop of average shear strength after vibration time of 3s.

Conclusions

In summary, the effect of ultrasonic vibration time would influence the IMC formation and shear strength of the Cu/SAC305/Cu solder joints was demonstrated. When the molten solder was exposed to USW, acoustic cavitation and streaming causes the morphology of Cu_6Sn_5 IMC grains changed from irregular scalloped to uniform and more rounded scalloped-shape. The thickness of both Cu_6Sn_5 and Cu_3Sn IMC reduced gradually with increasing vibration time up to 3s. Meanwhile, the Cu_6Sn_5 IMC particles that dispersed in the solder matrix were found to be finer in shape. Due to the dispersion strengthening effect, the average shear strength of the joint gradually increased with increasing ultrasonic vibration time. However, when the vibration time was further increased to 4s, the IMC layers at the joint interface grow thicker and elongated rod-like Cu_6Sn_5 IMC phases were dispersed in the solder matrix. These dispersed IMC phases may have been the crack initiation site and thus, the shear strength of the joint decreased after 3s of vibration time.

At the vibration time of 3s, the joint achieved the thinnest IMC layer thickness and highest average shear strength as compared to non-ultrasonic-treated solder joint. The thickness of Cu_6Sn_5 IMC layer decreased by 55.97% and the Cu_3Sn IMC layer decreased by 43.45%. The joint strength obtained is 38.14 MPa which increased by 40.89%. The results presented suggest that fine tuning on the ultrasonic vibration time could bring significant improvement on the microstructure and mechanical properties of the solder joint.

Acknowledgments

This work was supported by Ministry of Higher Education, Malaysia with the high impact research (HIR) grant number of HIR-MOHE-16001-00-D000001 and Postgraduate Research Fund (PPP) of University Malaya with Grant No. PG069-2013B.

References

[1] Ji, H., et al., "Ultrafine-Grain and Isotropic Cu/SAC305/Cu Solder Interconnects Fabricated by High-Intensity Ultrasound-Assisted Solidification," Journal of Electronic Materials, 2014. 43(7): p. 2467-2478.

[2] Chinnam, R.K., et al., "Evolution of the microstructure of Sn–Ag–Cu solder joints exposed to ultrasonic waves during solidification," Acta Materialia, 2011. 59(4): p. 1474-1481.

[3] Gain, A.K. and Y.C. Chan, "The influence of a small amount of Al and Ni nano-particles on the microstructure, kinetics and hardness of Sn–Ag–Cu solder on OSP-Cu pads," Intermetallics, 2012. 29: p. 48-55.

[4] Chan, Y.H., M.M. Arafat, and A. Haseeb, "Effects of reflow on the interfacial characteristics between Zn nanoparticles containing Sn-3.8Ag-0.7Cu solder and copper substrate," Soldering & Surface Mount Technology, 2013. 25(2): p. 91-98.

[5] Tang, Y., G.Y. Li, and Y.C. Pan, "Influence of TiO2 nanoparticles on IMC growth in Sn–3.0Ag–0.5Cu–xTiO2 solder joints in reflow process," Journal of Alloys and Compounds, 2013. 554: p. 195-203.

[6] Tsao, L.C., et al., "Effects of nano-Al2O3 particles on microstructure and mechanical properties of Sn3.5Ag0.5Cu composite solder ball grid array joints on Sn/Cu pads," Materials & Design, 2013. 50: p. 774-781.

[7] Gain, A.K., Y.C. Chan, and W.K.C. Yung, "Effect of additions of ZrO2 nano-particles on the microstructure and shear strength of Sn–Ag–Cu solder on Au/Ni metallized Cu pads," Microelectronics Reliability, 2011. 51(12): p. 2306-2313.

[8] Yang, Z., W. Zhou, and P. Wu, "Effects of Ni-coated carbon nanotubes addition on the microstructure and mechanical properties of Sn–Ag–Cu solder alloys," Materials Science and Engineering: A, 2014. 590(0): p. 295-300.

[9] Ko, Y.-K., et al., "Fabrication and interfacial reaction of carbon nanotube-embedded Sn–3.5Ag solder balls for ball grid array," Journal of Alloys and Compounds, 2014. 583(0): p. 155-161.

[10] Leong, T., M. Ashokkumar, and K. Sandra, "The fundamentals of power ultrasound - A review," Acoustics Australia, 2011. **39**(2): p. 54-63

Solder Development and Fabrication Techniques for Coolant Tube Bonding in Lengthy High Current Conductors

S. Z. Jurczynski, H. Schneider
Princeton Plasma Physics Laboratory, Princeton, NJ, USA

Abstract

The National Spherical Torus Experiment (NSTX) is being enhanced to significantly expand plasma conditions that are essential for next step Spherical Torus (ST) development. Improvements to this leading, high performance ST include doubling the magnetic toroidal field (TF) and plasma current with a tenfold increase to the plasma discharge duration. This requires a Center Stack Upgrade (CSU) with double the current capability of the TF inner conductors. Bonding water cooling tubes in these new TF Conductors required developing nonionic solder flux and controlled, repeatable heating.

Keywords—nonionic flux, conductive heating, reducing flame, cooling tube, SnAg solder, eutectic, copper conductor, Ettingshausen-Nernst effect, heating ramp rate, Rockwell B Hardness, thermocouple probes, thermal conduction, thermal expansion, heater control, data acquisition.

Introduction

The National Spherical Torus Experiment (NSTX), a leading high performance Spherical Torus (ST) research experiment, is being enhanced with upgrades aimed at achieving long-pulse high-performance operation. (see Fig. 1)

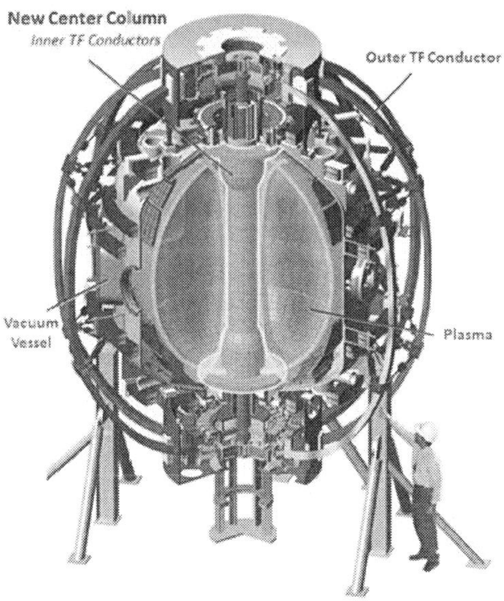

Figure 1 National Spherical Torus Experiment Upgrade

A new, improved inner toroidal field (TF) bundle will utilize 36 identical wedge shaped silver bearing oxygen-free high thermal conductivity (OFHC) copper conductors. High strength copper chromium zirconium (Cu-Cr-Zr) flags are friction stir welded at each conductor end to facilitate a bolted flex connection to the outer TF legs. This single layer design simplifies the cooling paths, increases coolant path diameter and reduces voltage potential between inner quadrant conductors which have the thinnest insulation. Conductor bonding is improved with high shear bond strength vacuum pressure impregnated (VPI) three-component epoxy over glass tape. (see Fig. 2)

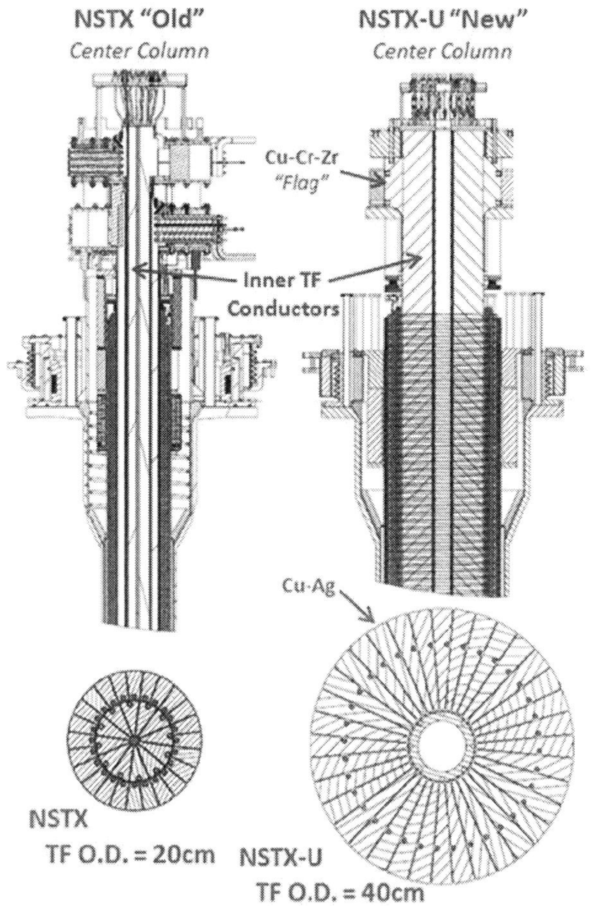

Figure 2 NSTX Center Column Cross Sections

Water cooling for each inner TF conductor is provided by an electrolytic tough pitch copper tube soldered in a channel on one side of the wedge surface. Solder paste with nonionic flux was developed to eliminate possible insulation degradation which could lead to potential carbon tracking between TF conductors. [1] (see Fig 3, 4)

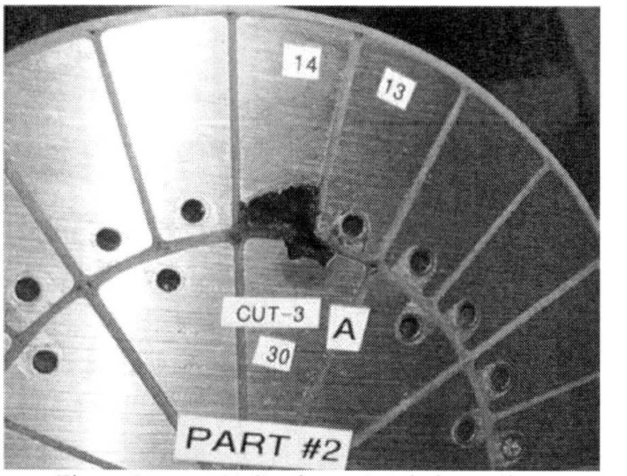

Figure 3 Inner TF Conductor Insulation Failure

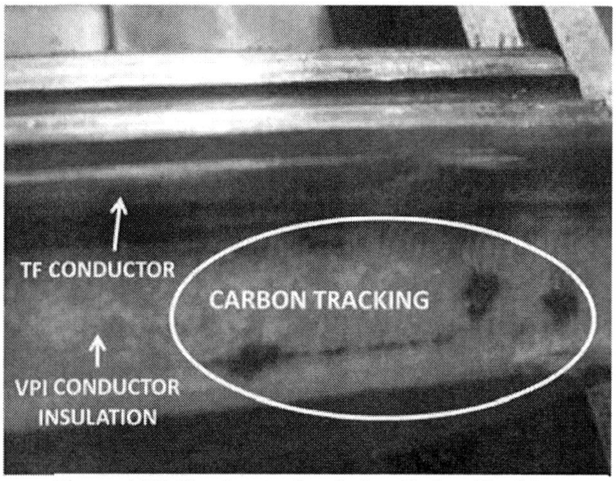

Figure 4 TF Conducotor Insulation Carbon Tracking Degridation

A repeatable and uniform heating source was developed to insure solder paste flux effectiveness for a homogeneous and fully wetted joint and to limit peak temperatures to maintain hardness of the TF conductors, flags and cooling tubes. Fabrication safety, budget and schedule constraints required a system to melt the solder without an oven, burners, or induction heating. To overcome these limitations a stainless steel "hot plate" with heating current supplied by a power supply modified to operate from a closed loop capable controller based system was conceptualized, prototyped and implemented.

Temperature profile repeatability required uniform heater contact and reliable temperature measurement. However, machining of the long, narrow, wedge shaped TF conductors induced material stresses creating flatness and bowing variability. Extremely tight cooling tube clearances, thermal expansion and bowing motion during heating required simple, novel fixturing and temperature measurement methods. Temperature, voltage and current data acquisition was utilized for real time control, temperature profile development, off line data analysis and archiving. This paper presents solutions to these issues and includes a cost effective bonding technique solution for future work.

Solder Paste Development

Flux Formulation

Post soldering operations to insulate and bond the TF conductors include bakeout and curing temperatures to 170°C. With a 221°C melting point, a 96% tin and 4% silver (96Sn/4Ag) eutectic solder in a 325 mesh powder form was selected. Rosin based flux is typically used for tin based solders when soldering copper and electrical components. Prototype soldering with samples of various mass and lengths were performed on hot plates, the development heater and the production stainless steel heater. It was quickly evident that a rosin based flux was not suitable as the reducing acetylene flame carbonized the flux making cleaning and decontamination very difficult. High temperature solder masking, as a damming material, was tried for containing the solder to the tube and channel area. The mask was also carbonized by the reducing flame and contaminated the molten solder. The mask was consequently not used for production soldering. (Fig. 5)

Figure 5 Solder Paste & Flux Development Issues

The solder paste with flux utilized for production was developed through prototyping of soldered samples and formulated as 10% Glycerol Monostearate (GMS - an emulsifier), Tergitol™ (a nonionic surfactant) and 3% Succinic acid (a weak organic acid). This flux allows for efficient water cleansing. Performance of the solder paste was verified by sectioning and tensile testing the prototype solder joints. The tensile strength approached the book value of 110MPa (16KSI) very closely and no sign of solder voids were evident. (see Fig. 6)

Figure 6 Tensile Strength Testing of Production Solder Paste

With the production heater and power supply, the maximum heating rate is 16°C per minute - less than the nominal 25°C per minute for good flux performance. Additional flux was applied before solder paste on the first production TF conductor. Additional flux proved ineffective as the slow heating rate dried out the flux losing its ability to effectively disassociate surface oxides of copper, tin and silver.

Reducing Flame Use

In order to produce a homogeneous solder and fully wetted joint area, a reducing flame was required to disassociate the remaining oxides. (see Fig. 7) Spent residual flux carbonized by the reducing flame is then removed by skimming scale off the molten solder surface leaving only exposed solder (Fig. 8).

Figure 7 Reducing Flame Disassociating Copper, Tin & Silver Oxides

Figure 8 Scale Skimming of Molten Solder

The first production conductor required tube replacement due to areas where the tube was exposed during solder clean up. This grinding operation also uncovered numerous gas pockets which were formed during solder solidification. (see Fig. 9) Decomposing flux releases carbon monoxide, carbon dioxide and water. Three issues were noted; a need for additional hold down clamps, insufficient joint clearance and an abundance of liquid flux trapped below the solder surface (see Fig. 10). Nothing could be done with joint clearance other than a thorough abrading of the tube pocket. A closer look at the boiling temperatures of the flux components; Succinic acid 235°C, Tergitol™ 250°C, Glycerol Monostearate which disassociates into glycerol that boils at 290°C allowed for an increase in soldering temperature. (see Fig. 11)

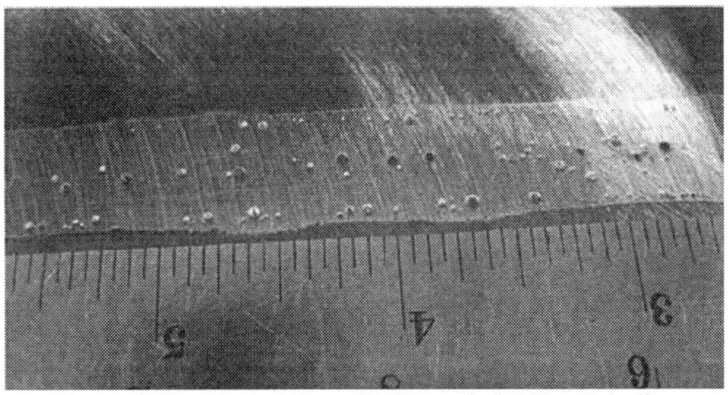

Figure 9 Gas Pocket Porosity Revealed After Grinding

Figure 10 Tube Bottom with Residual Flux

Figure 11 Replication of Residual Flux Conditions

The original soldering temperature profile maintained a minimum temperature of 270°C. This was increased to 300°C to ensure the boiling point of the glycerol was reached. In addition, tube clamps were kept loose and the tube was manually agitated to help expel all flux components. After reduction flaming and scale skimming the tube clamps are tightened. Skimming leaves a clean solder surface and flux is not needed when filling in tube clamp dimples after solidification. The reducing acetylene flame proved adequate for this operation. (see Fig. 12)

Figure 12 Clamp Dimple Filled without Flux Using Reducing Flame

TF conductors were randomly tested before and after soldering operations for Rockwell B Hardness. This became necessary due to the required increase in soldering temperature. The annealing temperature of silver bearing OFHC begins at approximately 375°C. The temperature range across the TF conductor during soldering operations was typically between 300°C and 350°C. Hardness before and after soldering maintained 43 HRB – 44 HRB indicating no softening of the TF conductors occurred.

Heating Development

Testing TF Conductor self-heating, via high current, for soldering operations was requested and an existing power supply provided. This method quickly proved impractical and a platen heating element was conceptualized and prototyped with a 10cm (4 inches) x 2cm (3/4 inches) x 183cm (72 inches) type 304 stainless steel bar. (see Fig. 13)

Figure 13 Solder Paste and Heating Development Station

Temperature gradient Reduction

During initial platen heater testing, a 150°C temperature gradient was measured across the heater. Because there is only one heater and one power source, temperature adjustments were only practical by tailoring the heater's resistance along its length. Increasing resistance where temperature was coolest by drilling holes in the heater at its center reduced the gradient to 40°C. (Fig. 14)

Figure 14 Heater Temperature Profile Tailoring

Schedule pressures and budget limitations necessitated using stainless bar stock already on site. While the length and width were dictated by TF conductor size, variations in heater thickness could have been utilized to adjust overall resistance to provide utilization of the supply's maximum power. (Fig. 15)

Figure 15 Production Soldering Station Assembly

To maintain a homogeneous resistance, two sections of bar stock were welded void free to assemble the 19mm (3/4") x 152mm (6") x 7.3m (24') production heater with 0.0019 ohms resistance. Temperature varied by 70°C across the production heater with an acceptable 50°C across actual TF conductors during soldering and resistance tailoring was not required. (see Fig. 16)

Figure 16 Completed Production Soldering Station

Heating Profile Performance

Obtaining a heating ramp rate of approximately 25°C per minute is desirable for maximum flux performance. Sample bars with similar TF conductor mass and surface area were not available for prototyping. Solder testing was necessary with copper bars having less mass and much less surface area. Only 6°C per minute on the production heater with the initial 6VDC and 4000 ampere power supply was obtained. Even though the samples had significantly less heater contact area than actual TF conductors, it was apparent a power supply with higher voltage output would be required. A 12VDC and 10,000 ampere power supply was then installed and configured for control and instrumentation. At the heater's resistance of 0.0019 ohms, the maximum power available is 12VDC and 5000 amperes which produced ramp rates of 10°C per minute. With the increased surface contact area, the TF conductors during production soldering consistently reached a heating rate of 15°C to 16°C per minute. A thicker heater with lower resistance could have increased current output providing more heating power. Maximizing TF conductor heater contact and flux effectiveness would be required to obtain acceptable soldering.

The production soldering station was characterized and qualified under no load heater conditions. The 24 foot production heater obtained a temperature of 700°C in 30 minutes. No anomalous issues were evident. Heater temperature readings were comparable to visual verification of heater color. In addition, the welded area of the heater was verified to provide homogenous temperature. (see Fig. 17)

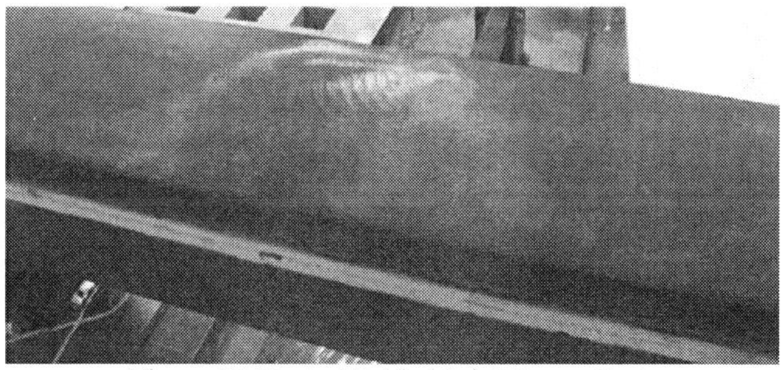

Figure 17 Heater Welded Joint Power Test

Figure18 TF Flag with Heater Tape

Supplemental Heating

The friction stir welded flags are two slightly different sizes. Both are 23cm (9 inches) long and the largest extends 15cm (6 inches). Since they hang over the edge of the TF conductor heating element, both were insulated to minimize them acting like heat sinks. This was insufficient as there was an almost 40°C temperature difference between tube thermocouples near the larger flag. This was corrected by wrapping the flags with 520 watt heater tapes using open loop control. Flags are preheated to approximately 55°C, solder paste is applied to the middle of the conductor first, and then to each end before the platen heater is used. (see Fig. 18)

Production Setup

Due to space and budget limitations, existing welding tables were utilized for heater support, flux fume exhaust system and solder grinding operations. Stanchions cut at a 10° pitch were welded to thin steel "C" channel for a heater support base. Three layers of 6.25mm (1/4 inch) kaolin millboard insulation between the support base and heater provide electrical isolation and thermal insulation. This also provides a "slip plane" for approximately 3.2cm (1.25 inches) of thermal expansion of the heater and TF conductor at soldering temperatures.

Figure 19 Production Station with Fume Exhust Hoods

To prevent current from flowing in the TF conductor, electrical isolation is required from the heater. Muscovite mica sheeting 0.4mm (0.016 inch) thick was utilized for its cost effectiveness, temperature rating and abrasion resistance while being thin enough to provide acceptable thermal conductivity. Flux contamination required mica replacement about every other soldering operation.

Moderate flux fumes are exhausted by a dedicated venting system. Movable fume hoods were fabricated to provide access for TF conductor placement, preparation, soldering and removal (see Fig. 19)

Heater Leveling

End-to-end and localized heater leveling is critical for proper solder fill. Due to the length of the soldering components any variation from level will produce an area where solder may pool depleting adjacent areas of solder exposing the cooling tube. Normally this would not be a major concern. However, the non-ionic flux in the production solder paste has very mild fluxing action. Exposed tubing would then become oxidized, hindering solder wetting during the touch-up phase of the process. While precision optical leveling devices were unavailable, a simple water level enabled excellent overall pitch determination and adjustment of the heater. For localized adjustments an inexpensive laser level and target block were utilized to determine "high" and "low" heater areas. These areas were then adjusted by using bottle jacks and shims under the stanchions. Once a TF conductor was ready for soldering, its leveling was further verified by observing the subtle velocity changes of a steel ball bearing rolled along the coolant tube channel.

Weight of the five cables caused a slight sag to the end of the heater because the supporting structure is slightly shorter, and a very slight inboard bowing occurred. This was remediated with a flexible support structure to accommodate fine height adjustment and allow for thermal expansion of the heater.

TF Conductor & Tube Fixturing

Unless the cooling tube is held in place it will float when the solder liquefies. A tool was fabricated to snug the TF conductor around the tube using small "pinches". This technique was rejected for fear of possible copper fatigue caused by the pinches in the TF conductor. (see Fig. 20)

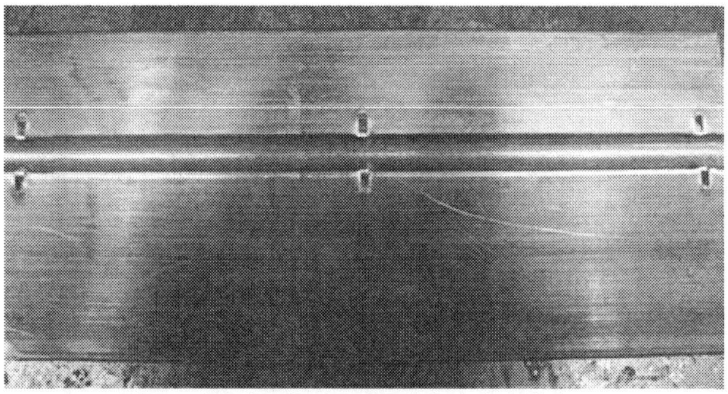

Figure 20 Coolant Tube Retaining Crimps

Measurement of the solder and TF conductor temperatures are also needed so hold-down toggle-clamps were modified with K-type thermocouple probes as the clamping point (see Fig. 3). The thermocouple probes were brazed into drilled out threaded rod mimicking the clamp's original adjustable neoprene tip spindle. (see Fig. 21)

Figure 21 Crimps Replaced with Adjusable Clamps (Right)

The clamps were welded to brackets on the heater support beam every 61cm (2 feet) across the length of the TF conductor. (see Fig. 22) During initial checkout of the production heater, large temperature errors were induced only in the tube clamp thermocouples while the heater thermocouple readings were nominal. Errors of 50°C or more were only present with heater power on. While the thermocouple junction is electrically isolated from its probe sheathing, the probes make contact with the clamp, heater support, stanchions, and base tables. The threaded rods were then insulated from the clamps with machined G10 fiberglass washers to break what was thought to be eddy current paths electromagnetically inducing common mode voltages exceeding the thermocouple signal conditioner's limits. The insulating washers eliminated the temperature errors induced while the heater supply is providing current. However, it is now believed that the large errors are primarily due to thermocouple conductor thermal gradient while in a perpendicular magnetic field which is called the Ettingshausen-Nernst effect [2].

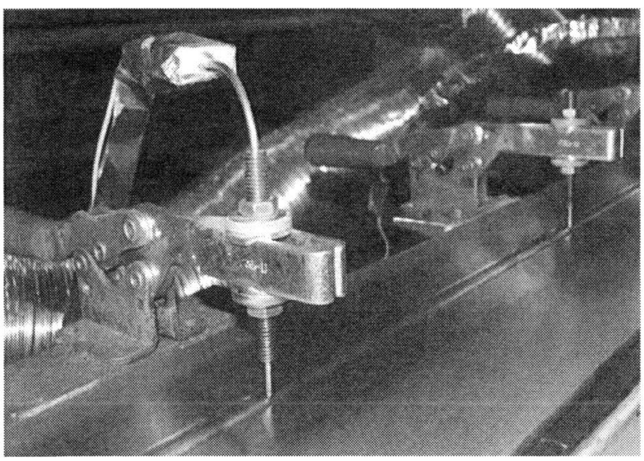

Figure 22 Integrated Clamp with Thermocouple

To prevent liquid solder from running out of the channel ends, holes were punched into high temperature silicon stoppers which were slid on the tube ends. High temperature Room Temperature Vulcanizing (RTV) silicone is applied between the stopper and TF conductor to insure the solder seal and to prevent tube movement during solder solidification. Clearance between the top of the TF conductors and top of the tubes varied between 0.2mm (0.008 inches) and 0.5mm (0.020 inches). This leaves no margin of error when removing excess solder for a flush conductor surface so small weights that were placed on the ends of the tubes as extra insurance of good seating.

It was unknown until the tubes were nicked during solder grinding operations that the weights caused tubes to bow up slightly inboard of the TF conductor end which acted as a fulcrum. Additional clamps with thermocouple like probes were then added between existing clamps. Weights were no longer used on the tube ends and it was hoped the stoppers attached to the TF conductors with RTV would prevent movement during solder solidification. Rated to an intermittent 340°C, RTV softened slightly and was not able to keep the tube firmly seated in its channel. This, again, was not known until tubes were nicked once again during solder grinding.

Machining of the TF conductors created flatness and bowing variations which the vendor alleviated by mechanical manipulation. A roller clamping method was utilized to provide good heater contact and enable thermal expansion of the TF conductors during heating. (see Fig. 23)

Figure 23 Roller Clamp Maximizes Heat Transfer

The first shipment of TF conductors only needed their ends clamped to the heater. As TF conductors with more undulations were received, the number of clamps increased to five – three in the middle of the conductor and one on each end. Because the RTV would soften and allow the tube to move slightly while the solder solidified another tube clamp was needed at each end. With conductor roller clamps already at the ends, no room was available for another separate tube clamp. An integral TF conductor roller clamp and tube clamp assembly was developed and utilized to maintain good thermal conduction and tube seating. (see Fig. 24)

Figure 24 Additional End Clamp with Rollers

Nicked tubes required removal and replacement. A portable vacuum was modified with a stainless steel flexible hose, replaceable tube tip and a steel solder catch can. The TF conductor was heated just enough to melt the solder which was then vacuumed off. After the tube was lifted out of the channel any remaining solder was vacuumed out. Very little solder made its way to the catch can as it solidified on contact with the tube tip. When almost clogged tube tips were removed, solder slugs punched out and the tube tip replaced.

Heater Control, Data Acquisition & Analysis

To repeatedly measure, control, troubleshoot, analyze and archive TF conductor temperatures, a portable control and signal conditioning system was modified for soldering operations (see Fig. 25). Twenty isolated thermocouple signal conditioning modules provide outputs to a loop process controller. (see Fig. 26) A spreadsheet for off-line data analysis was adequate and was used throughout soldering operations.

The prototyping and production power supplies were modified for voltage control and utilized an analog conversion of the controller's digital heat control output. Open loop temperature control was required for several factors. Closed loop temperature control was precluded by one heating element, acceptable temperature gradients across TF conductors, reduction flame flux manipulation, and thermal heating lag variations due to differences in TF conductor heater contact.

Process controller to personal computer communication provided the operator with real-time power control and plots of heater and TF conductor temperature as well as heater power, voltage and current.

Figure 25 Heater Control & Data Acquisition System

Soldering Process

Vendor shipments contained four or five TF Conductors in wooden crates. As a production time saving method, TF conductors were soldered then ground in these groups of four or five. Due to the tight working space, removable plastic sheeting was hung to contain the grinding dust and debris. Soldered TF conductors were crane lifted back into the shipping container for storage to await grinding operations. Once a complete shipment was soldered this task area was cleaned, maintenance performed and prepared for soldering the next shipment. The grinding area was then setup and utilized to finish all the soldered TF conductors.

Soldering a TF conductor is approximately six hours from preparation through cool down to allow handling. While the grinding operations are being performed, data analysis of the soldering operations is completed in preparation for the next shipment of TF conductors.

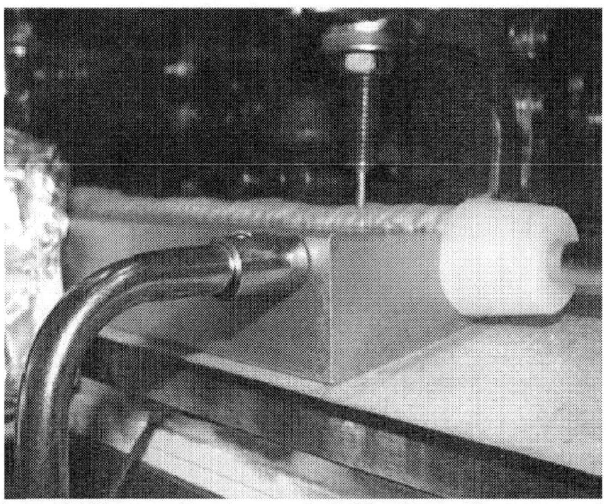

Figure 26 Heater, Conductor, & Solder Paste Thermocouples

Major soldering steps involved include:

Fresh mica, if needed, cut and placed on the heater
TF conductor crane lifted out of shipping box onto the heater
Coolant tubes are inspected and one without damage selected
Tube and channel are cleaned with abrasive pads
Final cleaning of tube and channel with ethanol
Install Flag heaters and wrap with insulation
Roller clamp TF conductor at both ends
Determine high spots of TF conductor with feeler gauge
Place roller clamps accordingly to maximize heater contact
Adjust tube clamps at top of tube with moderate pressure
Pre heat flags to approximately 55°C
Apply solder paste in middle of TF conductor then each end (see Fig. 27)
Verify heater and TF conductor are electrically isolated
Turn heater on, step up to full power, monitor temperatures (see Fig. 28)
At 220°C use reducing flame, add solid solder where needed
Skim scale off top of molten solder
Agitate tubes to expel trapped flux (see Fig. 29)
At 280°C power to 50% continue temperature monitoring
Assess heating rate to anticipate power to reach 300°C
If temperature will reach 300°C minimum - turn power off
Fully clamp tubes and continue temperature monitoring
Allow TF conductor to cool to 160°C
Fill dimples and low spots with solder using reducing flame
Allow TF conductor to cool to 50°C before handling (Fig. 30)
Crane soldered TF conductor to grinding area
Grind, scrape, buff until soldered surface is flush with copper
Crane TF conductor into shipping box (see Fig. 31)

Figure 27 Apply Solder Paste with Pneumatic Applicator

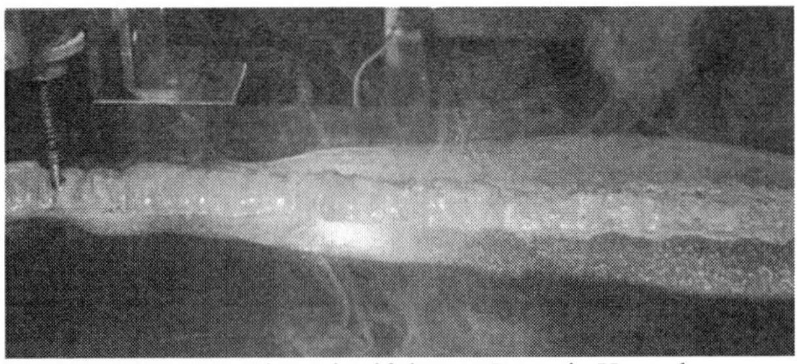

Figure28. Flux Liquifying as Paste is Heated

Figure 29 Solder Skimmed & Tubes Agitated

Figure 30 TF Conductor Ready for Grinding Operation

Figure 31 Completed Conductor Loaded Into Transport Crate

Grinding Operation

The original profile of the TF conductor surface must be maintained for proper clearances during insulating and bonding operations. With proper fill the solder exhibits a convex surface after cooling. This solder must be removed carefully as the top of the coolant tube is only a maximum of 0.5mm (0.020 inches) below the TF conductor surface. If a coolant tube is nicked during grinding operations it must be removed and replaced. A three step technique was developed to minimize the risk of nicking a tube while maximizing the flatness of the TF conductor surface. With a hand grinder the majority of the solder is removed utilizing a coarse wheel. A fine grinding wheel was found to build up too quickly with solder. A coarse wheel used with wax lubricant to further reduce buildup required changing much less frequently. For better control in final solder removal a hand scrapper with a carbide blade is used next. Scraping the surface flatness is monitored with a straight edge. Final surface finishing is accomplished with an orbital sander using a fine grit. This entire grinding process takes just under four hours.

Conclusion

Soldering cooling tubes in large copper conductors presented many different challenges; solder paste development, a single conduction heating element, one power source and fixturing. Even after developing a well-defined soldering process, unforeseen challenges were frequent as each TF conductor behaved slightly different.

Acknowledgements

Chemicals & Metals Technology Inc. provided the solder paste and paste development support. The space loaned by the Princeton University Physics Department and their hospitality is very much appreciated. Special thanks to all the PPPL Tech shops for fabricating and/or installing custom devices and systems.

This work was supported by the USE DOE Contract No. DE-AC02-09CH11466

References

[1] Dudek, L. E., et al, "Identifying the Cause of the NSTX TF Coil Bundle Failure" Report PPPL-4896, Princeton Plasma Physics Laboratory, Princeton, NJ

[2] T.G. Kollie, R.L. Anderson, J.L. Horton, M.J. Roberts, "Large Thermocouple Thermometry Errors Caused by Magnetic Fields" *Review of Scientific Instruments,* Vol. 48, No. 5 (1977), pp. 501-511.

The Formation of Interfacial Reaction between Sn-Ag-Cu Solder and Cu Substrate by Microwave Hybrid Heating

Maisarah Lutfi, Farazila Yusof, Ramesh Singh, Nashrah Hani Jamadon, Mohd Hamdi Abd Shukor
Centre of Advanced Manufacturing and Material Processing, Department of Mechanical Engineering, Faculty of Engineering, University of Malaya, 50603 Kuala Lumpur, Malaysia

Tadashi Ariga
Department of Materials Science, School of Engineering, Tokai University, 1117 Kitakaname, Hiratsuka-Shi, Kanagawa 259-129, Japan

Abstract

In this study, the effect of microwave heating in the reflow soldering of Sn-3.0Ag-0.5Cu (SAC) solder alloys and copper substrate were investigated. The soldering was carried out using a domestic microwave oven at 2.45 GHz frequency and 1 kW. The interface of SAC solder/Cu layers were observed by FESEM analysis facilitated with EDX. Crystal structure of the intermetallic phases was examined using an X-ray diffractometer (XRD). The preliminary result has shown that the formation of Cu_6Sn_5 and Cu_3Sn IMCs at the Sn-3.0Ag-0.5Cu solder/Cu interface after 150s and 180s microwave heating time.

Introduction

The solder plays an important role in microelectronic interconnection for the electrons and thermal energy transfer [1]. Since environmental regulations forbid the use of Pb-containing solder alloy due to toxicity of Pb and health risk to human, many Pb-free solder alloys have been studied as replacements for Sn–Pb solder [2]. Among of them, Sn–Ag–Cu based alloys are leading lead-free candidate solders for electronic manufacturing application [3]. It has the lowest melting temperature compared to Sn-Ag and Sn-Cu based alloys, superior mechanical properties and good solderability [4].

During the soldering process, the metallurgical reaction between Sn–Ag–Cu solder and Cu substrate result in an intermetallic compound (IMC) layer at Sn–Ag–Cu/Cu interface. This IMCs layer is one of the critical parameter to the reliability of the solder joints [5]. It is quite brittle and the thermal mismatch between the solder and the substrate will weaken the solder joint strength and eventually cause the fatigue failure. Adding alloying element in the solder alloys may improve the performance of the interfacial IMCs layer [6]. Another way to minimize the excessive growth of the IMC layer is through faster heating and shortening holding time at the peak temperature in the reflow process [7]. The faster and uniform heating is possible by using microwave energy.

Recent literature has shown very limited research in the soldering process of metallic materials by using microwave energy. Successful soldering of the Sn-3.5Ag and Sn-3.8Ag-0.7Cu on Cu surface using Variable Frequency Microwave (VFM) was reported by Moon et al. [7]. The authors showed that the Cu_6Sn5 and Ag_3Sn IMCs formed at the solder-pad interface by microwave heating at frequency of 6.425 GHz. The IMCs formed had a larger and coarser structure compared to conventional reflow process.

However, no details on characterisation of Sn-3.0Ag-0.5Cu solder/Cu interface through domestic microwave oven were reported. The present paper reports on the results of an on-going project on solder reflow process and characterisation of Sn-3.0Ag-0.5Cu using microwave energy in a multimode applicator. The weight of the susceptor and microwave heating times were varied, while the microwave power and amount of solder paste were fixed. The morphologies of IMCs formation were evaluated and discussed.

Experimental Details

Sn-3.5Ag-0.5Cu (wt. %) solder paste with 12-50 μm particle size and copper (purity ~ 99.99%) plates having dimension 20 mm x 10 mm x 2.5 mm were used for these experiment. The 0.05 g solder paste was placed on the Cu surface at 5 mm in diameter. Then, the copper plate and solder paste were put in the center of the alumina boat which contained the graphite powders as a susceptor. In order to avoid the reflection of microwave from the Cu substrate, the ceramic box refractory fiber was prepared to seal the bulk metallic sample.

Principles of microwave hybrid heating were effectively used to join metallic pieces. Graphite powders were used as a susceptor medium to provide initial heating before starts coupling with microwaves. The graphite powders were placed as near as possible to the joint area so as to induce selective heating in the sample. The solders alloy gets heated through conventional modes of heat transfer from the heated graphite. Subsequently, the solder alloys start to couple with microwaves leading to further rise in temperature and melted.

Figure 1 illustrates a schematic view of the experimental setup of the microwave hybrid heating used for joining of metallic materials. A domestic microwave oven (1 kW, 2.45 GHz, Panasonic NN-CD997S) was used to heat the samples at different heating times with different weight of susceptor. Microwave processing parameters and other details are summarized in Table 1.

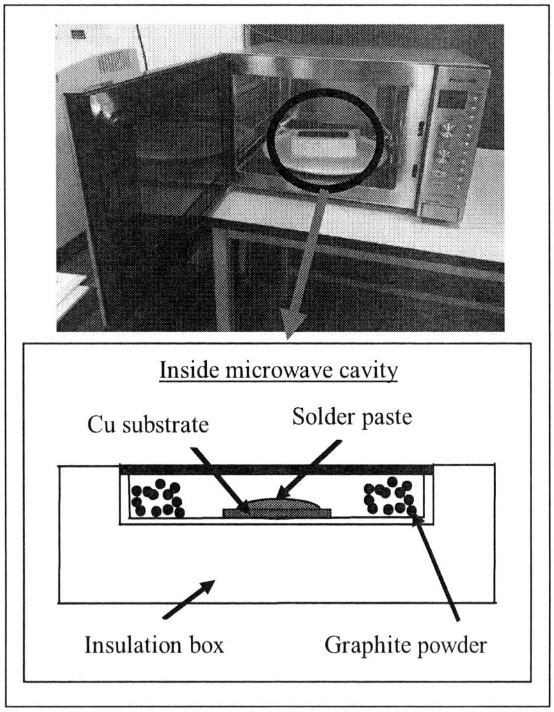

Figure 1 The experimental setup of microwave hybrid heating.

After soldering the specimen was cut by using a high speed diamond cutter. The specimen then was mounted in epoxy, grinded by an appropriate grit of SiC and polished with diamond suspension. The cross sectional of soldered specimens and elemental reaction at the interface were analyzed using field emission scanning electron microscopy (FESEM) coupled with energy dispersive X-ray (EDX). Crystal structure of the intermetallic phases was examined using an X-ray diffractometer (XRD).

Table 1 Microwave specification and parameters used in microwave heating process.

Parameter	
Applicator	Multimode
Microwave frequency	2.45 GHz
Exposure power	1 kW
Heating time	120 s
	150 s
	180 s
Susceptor materials	Graphite powder
Weight of susceptor	2g
	4g
	6g

Results and Discussion

During soldering, Cu will start to dissolve into the SAC solder. As a result, the SAC solder becomes supersaturated with Cu and a layer of intermetallic compound (IMC) was formed. From XRD analysis, two types of intermetallic phase found at the interface of the soldered joint were monoclinic ή-Cu_6Sn_5 and orthorhombic ε-Cu_3Sn. According to the phase

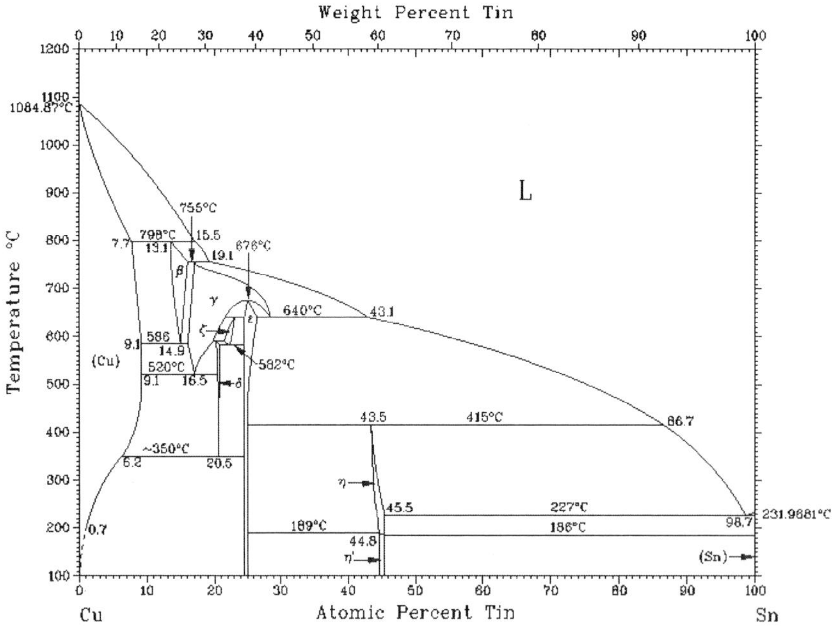

Figure 2 Phase diagram between Cu and Sn [8].

diagram of Cu and Sn as shown in Fig. 2, hexagonal η-Cu_6Sn_5 is the intermetallic compound found at a higher temperature while ή-Cu_6Sn_5 is the intermetallic phase at a lower temperature. However, η-Cu_6Sn_5 phase does not found in this study. This indicated that η-Cu_6Sn_5 fully transformed to ή-Cu_6Sn_5 during the cooling period of the soldered joint [9].

Figure 3 and 4 illustrate the microstructural evolutions of Sn-3.5Ag-0.5Cu solder/Cu interface at 150s and 180s of heating time for different weight of susceptor respectively. It can be seen that two layers of IMCs were formed; top region is Cu_6Sn_5 and the layer beneath is Cu_3Sn. Figure 3a and 3c reveals the unique morphology of Cu_6Sn_5 after soldering at 150s in microwave hybrid heating. It is found that the structure is different with normal scallop-like morphology as shown in Fig. 3b which is scatter-like and whiskers-like morphology respectively.

In this study, the mechanism of scatter-like and whisker-like morphology of Cu_6Sn_5 after soldered in microwave is still unknown. It was happened possibly because of the increase in chemical activity of Cu element with the SAC solder or it might be related with the microwave heating mechanism which is fast heating and cooling process.

The weight of susceptor also has influence on the formation of Cu_6Sn_5 morphology. It is speculated that as the amount of susceptor increased, the heat absorbed by SAC solder inside the insulation box also increased. Thus, overheat molten SAC solder caused an abnormal growth of Cu_6Sn_5 IMC formed at the interface as shown in Fig. 3c. However, this phenomenon only found in soldered sample at 150s microwave heating time. At 180s microwave heating time, the Cu_6Sn_5 morphology was formed as heating in conventional methods; scallop-like morphology (Fig. 4a and 4b) except for figure 4c which had angular-trapezoid morphology. The scallop-like morphology that is formed at 3.5Ag-0.5Cu solder/Cu interface is similar to the scallop-like shape as reported by Niwat Mookam et al [9].

From a thermodynamics viewpoint, Cu_6Sn_5 crystalline are formed because a large driving force for the chemical reaction between Cu and Sn atoms exists at the metastable composition [10]. The ripening process contributes to the formation of the scallop-like structure of Cu_6Sn_5, which results in the coarsening of the scallops as its number is decreasing [11]. Meanwhile Cu_3Sn layer between Cu_6Sn_5 and Cu substrate was formed by solid solution diffusion and its thickness usually is smaller than Cu_6Sn_5. [1]. It was very thin and had more uniform thickness and a layer-like morphology. Previous study [12,13] declared that the formation of Cu_3Sn was inhibited because of the prevention of Sn diffusion through the Cu_6Sn_5 layer, which forms at soldering process.

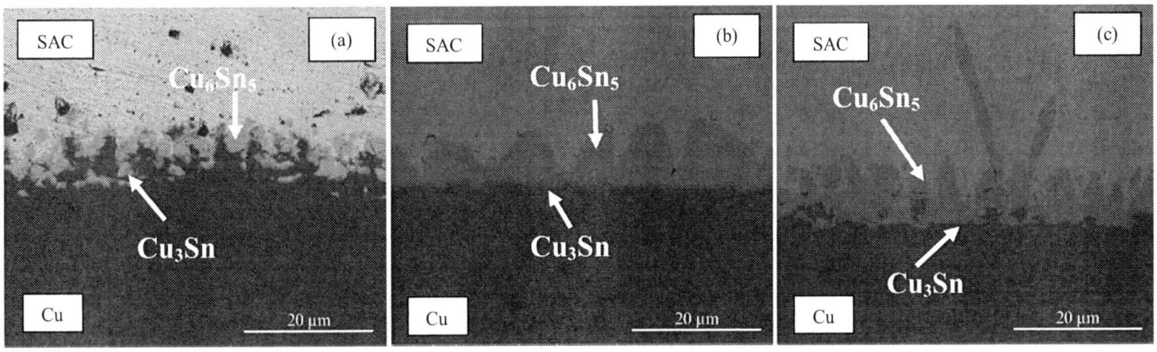

Figure 3 Microstructure of Sn-3.0Ag-0.5Cu solder/Cu interface at 150s heating time for different weight of susceptor; (a) 2g, (b) 4g and (c) 6g

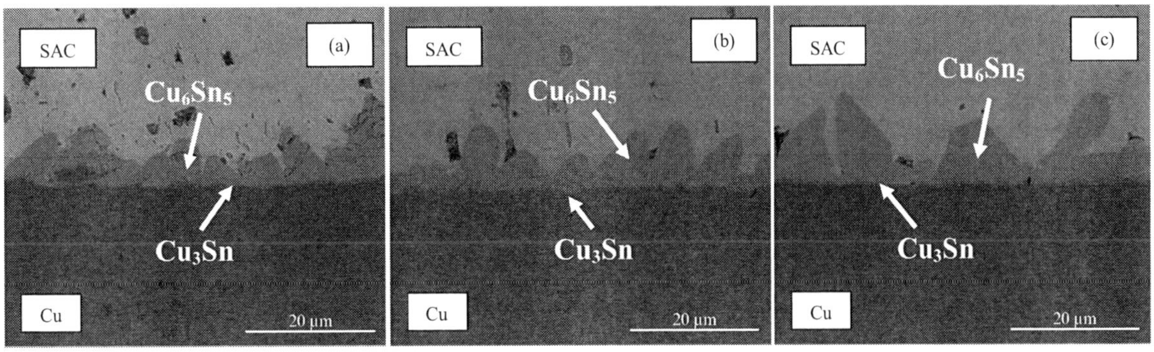

Figure 4 Microstructure of Sn-3.0Ag-0.5Cu solder/Cu interface at 180s heating time for different weight of susceptor; (a) 2g, (b) 4g and (c) 6g

Conclusions

The microstructure of the intermetallic compounds formed at the Sn-3.0Ag-0.5Cu solder/Cu interface through microwave hybrid heating technique has been investigated and the results are summarized as follow:

1. Formation of monoclinic ή-Cu_6Sn_5 and orthorhombic ε-Cu_3Sn.IMCs are observed at the Sn-3.0Ag-0.5Cu solder/Cu interface after heating in different weight of susceptor, 2g, 4g and 6g.

2. The Cu_6Sn_5 IMC scatter-like and whiskers-like morphology were formed at 150s microwave heating time in 2g and 6g of graphite powder respectively.

3. The weight of susceptor has influence on the formation of Cu_6Sn_5 morphology, as the amount of susceptor increased, an abnormal growth of Cu_6Sn_5 IMC formed at the interface.

Acknowledgments

The authors would like to acknowledge the University of Malaya for providing the necessary facilities and resources for this research. This research is supported by UM Fundamental Research Grant Scheme (FP025-2013A) and Postgraduate Research Fund (PG069-2013B) from the Ministry of Higher Education Malaysia and also special thanks to Madam Hartini Baharom for technical and laboratory assisted.

References

[1] Tsao, L. C. *et al*, "Study of Interfacial Reaction between Sn3.5Ag0.5Cu Composite Alloys and Cu Substrate," *Microelectronic Engineering*, Vol. 88 (2011), pp. 2964-2969.

[2] Hirose, A. *et al*, "Joint Strength and Interfacial Microstructure between Sn-Ag-Cu and Sn-Zn-Bi Solders and Cu Substrate," *Science and Technology of Advanced Materials*, Vol. 5 (2004), pp. 267-276.

[3] Kim, K. S. *et al*, "Effects of Intermetallic Compounds on Properties of Sn–Ag–Cu Lead-Free Soldered Joints," *Journal of Alloys and Compounds*, Vol. 352, No. 1-2 (2004), pp. 226-236

[4] Wu, C. *et al*, "Properties of Lead-Free Solder Alloys with Rare Earth Element Additions," *Materials Science and Engineering: R: Reports*, Vol. 44, No. 1 (2004), pp. 1-44.

[5] Abtew, M. *et al*, "Lead-Free Solders in Microelectronics," *Materials Science and Engineering: R: Reports*, Vol. 27, No. 5-6 (2000), pp. 95-141.

[6] Zeng, G. *et al*, "A Review on the Interfacial Intermetallic Compounds between Sn–Ag–Cu Based Solders and Substrates," *Journal of Materials Science: Materials in Electronics*, Vol. 21, No. 5 (2010), pp. 421-440.

[7] Moon, K. *et al*, "Lead-Free Interconnect Technique by using Variable Frequency Microwave," *Journal of Electronic Materials*, Vol. 34, No. 7 (2005), pp. 1081-1088.

[8] Massalski, T. B., et al, Binary Alloy Phase Diagram, 2nd Edition, ASM International (1990).

[9] Niwat Mookam, *et al*, "Effect of Soldering Condition on Formation of Intermetallic Phases Developed between Sn-0.3Ag-0.7Cu Low-Silver Lead-Free Solder and Cu Substrate," *Journal of Alloys and Compounds*, Vol. 509 (2011), pp. 6276 - 6279.

[10] Laurila, T., *et al*, "Interfacial Reactions between Lead-Free Solders and Common Base Materials," *Materials Science and Engineering R*, Vol. 49 (2005), pp. 1- 60.

[11] Lie Mei Lee, *et al*, "Interfacial Reaction of Sn-Ag-Cu Lead-Free Solder Alloy on Cu: A Review," *Advanced in Materials Science and Engineering,* (2013), pp. 1 – 11.

[12] Ahat, S., *et al*, "Effects of Static Thermal Aging and Thermal Cycling of Sn95.5Ag3.8- Cu0.7 Solder Joints", *Journal of Materials Research*, Vol. 6 (2001), pp. 2914 – 2921.

[13] Lee, H. T., *et al*, "'Influence of Intermetallic Compounds on the Adhesive Strength of Solder Joints", *Materials Science and Engineering A*, Vol. 333, No. 1-2 (2002), pp. 24-34.

High Strength Materials for High Temperature Processing: A University-Industry Design Project Collaboration

Mike Powers, Sabyasachi Sen, David Anderson, Angelo Carino, Terry Chen, Robert Ireland, Samuel Kass, Raymond Lin, Roger Lin, Melissa Lum, Peter Saeterun, Sidthi Sanjawala, Matthew Schmitt, Cameron Tracy, Simon Tsui, Corey Wolin, Hanry Yang
University of California at Davis
Davis, California

Abstract

A team of engineering students at the University of California Davis collaborated on a design project to identify, investigate and develop reliable high strength meal alloys for use in high temperature processing. The students considered the required material properties, as dictated by the component assembly design, and how the properties are affected by materials selection and a particular joining process. The team conducted tensile testing and microstructural assessment at UC Davis in order to evaluate the candidate materials. The team also conducted heat treatment experiments at Keysight Technologies (formerly Agilent Technologies) facilities in Santa Rosa, California. Through applied research and testing the student team identified several eclectic metal alloys that display ultimate tensile strengths greater than 900 MPa after exposure to a temperature of 1100°C, promising candidates for high temperature processing.

Introduction

University accreditation criteria and escalating employer expectations are driving new requirements for engineering students to obtain "real world" design experience prior to graduation. To enhance this experience for Materials Science and Engineering majors at the University of California, Davis, the Department of Chemical Engineering and Materials Science offers a senior level capstone design course called Materials Design Project (EMS 188). The central feature of the course is a collaborative design project that is taught jointly by a department professor and an engineer from industry. During the winter and spring quarters of 2011, students in the class worked together as a development team to identify, justify and test reliable high strength metal alloys for use in high temperature processing, such as brazing, welding and diffusion bonding (a "real" project). To define and test their proposed materials selections, the student project team considered the desired material properties (e.g. modulus, yield strength, tensile strength, hardness, surface roughness), as dictated by the component assembly design and processing, and how the critical materials properties are impacted by specific manufacturing processes. In order to justify the project, the team was expected to address economic, intellectual property, regulatory and environmental/safety issues associated with the proposed materials selection and its integration into device applications. The team was responsible for designing and conducting a comprehensive experimental plan to evaluate the proposed material selections and resulting material properties.

The course is structured such that the students are required to make two oral presentations to "management," one during the winter quarter and the other during spring quarter. The first oral presentation is a description of the process development approach that the team intends to follow, essentially a justification proposal where the team solicits management approval for the concept and scope of the project. The second oral presentation is a final project summary with an in-depth description of the process design factors and issues, the analytical and experimental procedures employed, and the conclusions resulting from the investigation. In addition to the oral presentations, the team is required to write a final report which includes an assessment of the suitability of the materials selection for the intended applications and an indication of the potential risks associated with production implementation of the material, along with recommended mitigation and contingency plans.

What follows is essentially an abridged version of the final report that was submitted by the students at the end of the spring 2011 term.

Background

Rapid advances in technological areas such as energy, aerospace, electronics, photonics and chemical analysis are driving a requirement for metal alloys that maintain reasonably high strength when exposed to elevated temperatures [1]. Some of these applications involve the use of the materials at high temperature (i.e. on the order of two thirds of the material's absolute melting temperature), while other applications require materials to be processed at elevated temperatures and subsequently used at room temperature. The latter case is of particular interest in this study, particularly for applications that require high temperature soldering, brazing, welding or diffusion bonding as one or more of the processing steps in the fabrication of a device or assembly.

The primary objective of this project was to research, identify, investigate and test reliable high strength metal alloys that exhibit an ultimate tensile strength (UTS) of 900 MPa or greater after exposure to a processing temperature of 1100°C. This requirement is of particular interest for application to diffusion bonding of high strength materials.

Consideration of the limitations imposed on material performance by a particular diffusion bonding process is crucial in the determination of proper materials selection criteria for the intended application. While the primary requirement is a post-fabrication tensile strength in excess of 900 MPa, simply selecting a metal from a materials database based on this constraint will likely yield unfavorable results, as doing so ignores the substantial effects of high temperature heat-treatment on the mechanical behavior of the material. Processes such as grain growth are largely unavoidable during high temperature excursions and can significantly diminish the material's tensile strength. Additionally, other strengthening mechanisms which do not rely on Hall-Petch effects may be unsuitable due to their detrimental effects on the bonding process. For example, inhomogeneities at or near a material interface, such as precipitates formed during aging or martensite laths formed via quenching, can weaken the diffusion bonded joint through the formation of internal stresses [2]. Further constraints on the viability of strengthening mechanisms for a diffusion bonding application arise from unique geometries and tight tolerances of the component features, which preclude certain aggressive treatments such as quenching.

Based on these unique design limitations, a design paradigm was formulated in which favorable materials behaviors, rather than merely as-received material properties, were identified and used to screen candidate materials. Primary consideration was given to strengthening mechanisms that might be used on a given candidate to either reduce strength losses from grain growth or recover such losses via post-bonding treatment. Greater weight was placed on those mechanisms that are not expected to interfere with the ability of a material to form strong joints during diffusion bonding.

Three fundamental strengthening mechanisms are germane to this work: grain refinement, martensitic transformation, and precipitation hardening. Emphasis was given to those mechanisms which are not expected to negatively affect the diffusion bonding process. In this way, the two most likely failure modes of diffusion bonded components, namely delamination at bond interfaces or bulk deformation, were addressed in the selection of a particular material for testing.

Grain Refinement

The contribution to a metal's tensile strength form Hall-Petch effects is inversely proportional to the square root of its grain size [3]. Therefore, the recrystallization and grain growth that typically occurs during the diffusion bonding process results in substantial strength losses. Grain refinement refers to treatment of a metal in such a way as to minimize grain growth and retain the maximum possible Hall-Petch relationship. To this end, dispersions of intermetallics, carbides, or nitrides are commonly used in advanced stainless steels [4]. These dispersoids serve to pin grain boundaries that migrate during heat treatment, reducing their mobility and effectively minimize net grain growth in the material.

While such particle dispersions may reduce the tensile strength losses associated with the diffusion bonding process, they can also negatively affect the bonding process itself. To perform as pinning agents, such dispersoids must be present in the material during diffusion bonding, possibly leading to the impediment of the joining process by these inhomogeneities.

Martensitic Transformation

Martensite is a high-hardness phase that forms in steels via a diffusionless transformation. The presence of this phase in a steel, usually in the form of plate-like laths in an austenitic or ferritic matrix, results in an increase in strength and hardness [5]. Additionally, the martensitic transformation is a nucleation process with significant interfacial effects, indicating a propensity to significantly alter a material's microstructure. This behavior is attractive since such microstructural evolution may partially offset the effects of grain growth on the material's tensile strength. However, the presence of martensite laths can also inhibit the diffusion bonding process. This means that, ideally, any martensite present in the final component must be formed after diffusion bonding.

The most common method of inducing a martensitic transformation in stainless steels is through a phase transition brought about by austenitization followed by quenching. During the diffusion bonding process the material is heated to over 1000°C, which is above the austenitization temperature of typical stainless steels. However, quenching of the resulting austenite is not a viable treatment, as the large thermal stresses that would result are incompatible with the tight dimensional tolerances of the components being joined. For this reason, an alternate mechanism of martensite formation, one which does not require quenching of the material, is desirable.

In a certain class of meatastable stainless steels, nucleation and growth of martensite can occur through an isothermal process. This transformation is a rate controlled process that depends on both temperature and time, similar to diffusional transformations [6]. Isothermal martensitic transformations generally exhibit C-curve kinetics, with the fastest transformation occurring at temperatures below 0°C. This transformation mechanism offers two main benefits for inclusion in this study. First, the low temperatures at which the transformation occurs allows for hardening of the material without any detrimental effects on the diffusion bonding process resulting from martensite-induced microstructural discontinuities (i.e. the sudden reorientation of C and Fe atoms from the FCC solid solution of γ-Fe to BCT martensite). In other words, the transformation would not be induced until after the interface has been bonded. Additionally, the resulting martensite will form a lath structure, replacing part of the coarse post-diffusion bonded microstructure with plate-like martensite grains. In this way, some degree of grain refinement can be achieved, both increasing the tensile strength due to Hall-Petch effects and introducing grain boundaries at which precipitates will preferentially form.

Precipitation hardening

The presence of various brittle intermetallic compounds (IMC), carbides, or nitrides in a metal can contribute to significant strengthening effects. These particles act as barriers to the motion of dislocations along slip planes by pinning the dislocations and inhibiting the slip deformation mechanism. However, because their presence at or near an interface during diffusion bonding can result in a weakened joint, an age hardenable material is preferable to one in which dispersoids might be present during diffusion bonding. For example, a mechanically alloyed oxide dispersion strengthened (ODS) steel, in which nano-scale oxide particles are milled into the metal and remain thermodynamically stable during austenitization, will likely form a much weaker diffusion bond than an age-hardened steel in which any dispersoids present are taken into solid solution during austenitization. The use of precipitation hardened steel in which the precipitation reaction can be elicited in a post-diffusion bonding aging treatment will allow for both good bonding characteristics and a high component tensile strength.

Experimental Procedure

Based on the materials selection design paradigm, seven materials were identified for evaluation and testing. The materials were benchmarked against the type 304 austenitic stainless steel commonly used in diffusion bonding applications. Although the titanium materials were not expected to meet the ultimate tensile strength target, they are of interest due to their biocompatibility. Table 1 lists the materials investigated in this study.

Table 1 – Material, type, UNS designation and manufacturer

Material	Type	Designation	Manufacturer
304 stainless	austenitic	UNS S30403	Datum Alloys
Custom 465	martensitic, age-hardenable	UNS S46500	Carpenter Technology
Inconel 718	Ni-based, age-hardenable	UNS N07718	Special Metals
Chromflex	tempered martensitic	UNS S42010	Sandvik
Nanoflex	martensitic, age-hardenable	UNS S46910	Sandvik
PH 15-7 Mo	semi-austenitic, age-hardenable	UNS S15700	AK Steel
Ti Grade 2	commercially pure alpha Ti	UNS R50400	Kobe Steel
Ti-6Al-4V	alpha-beta Ti alloy	UNS R56401	Timet

Heat Treatment

In order to simulate the diffusion bonding process, some of the stainless steel tensile test specimens were heated for two hours at 1100°C in a custom vacuum furnace at < 5 x 10^{-6} torr. Ti and Ti alloy specimens were heated in vacuum for two hours at 850°C. Test specimens were stacked and separated by thin ceramic sheets and weights were placed on top of the stack to inhibit curling of the thin specimens during heat treatment. Some of the Nanoflex® specimens were subsequently aged at 475°C for three hours in a nitrogen atmosphere.

Tensile Testing

The various materials were procured in 0.2mm thick flat sheets. Tensile test specimens were fabricated using wire electric discharge machining to ASTM Standard E8M-09 Test Methods for Tension Testing of Metallic Materials, as shown in Fig. 1.

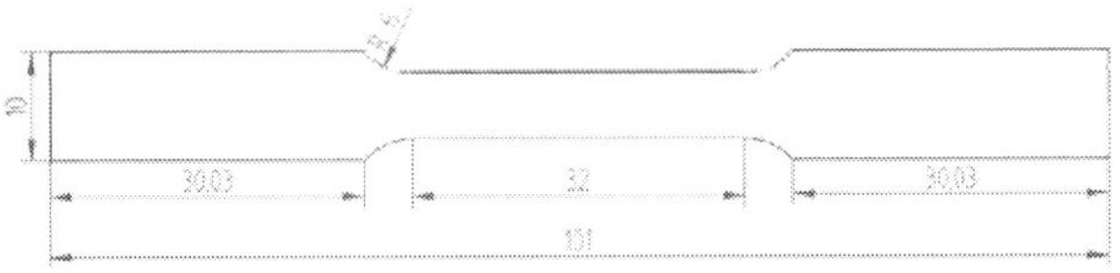

Figure 1 – Tensile test specimen schematic (mm)

Tensile testing was performed on both heat treated (HT) and as-received (AR) samples using an Instron 4202 Universal Testing System equipped with a 50kN load cell at a crosshead speed of 2.54mm/minute and no pre-load.

Microscopy

Microstructural characterization was conducted using both optical and scanning electron microcopy. Specimens were prepared by first cutting them to size using a Buehler® IsoMet® 4000 precision saw with a series 15 HC diamond saw blade. Blade speeds ranged from 750-4000 rpm, depending on the material. Cut samples were then mounted and cured in Buehler EpoMet® fine grain molding compound using a Buehler SimpliMet® 1000 automatic mounting press. Samples were then progressively ground using 320 then 600 grit SiC paper on a Buehler HandiMet® 2 roll grinder. Final polishing was accomplished with a Buehler EcoMet® 3000 variable speed polisher equipped with an AutoMet® 2000 power head using 1 micron diamond paste then 0.05 micron silica paste. Steel alloys were etched using either Aqua Regia (75 mL HCl, 25 mL HNO_3, 50 mL H_2O) or Ralph's Reagent (50 mL H_2O, 50 mL Ethanol, 50 mL Methanol, 50 mL HCl, 1g $CuCl_2$, 3.5g $FeCl_3$, 2.5 mL HNO_3). Ti-based materials were etched using Kroll's reagent (2 mL HF, 6 mL HNO_3, 92 mL H_2O). Optical microscopy was performed using a Nikon Labophot 2A microscope equipped with a high resolution digital capture system. Scanning Electron Microscopy (SEM) employed an FEI XL30 FEG equipped with an energy dispersive x-ray spectrometer.

Results and Discussion

The average ultimate tensile strength for both as-received and heat treated (i.e. simulated diffusion bonded) materials is shown in Table 2. The data is the mean of at least three trials for each material and condition.

Table 2 – Mean UTS for material test specimens

Material	As Received UTS (MPa)	Heat Treated UTS (MPa)	Aged UTS (MPa)
304 stainless	1126	628	-
Custom 465	1000	963	-
Inconel 718	678	656	-
Chromflex	1543	238	-
Nanoflex	1335	1004	1336
PH 15-7 Mo	750	682	-
Ti Grade 2	283	337	-
Ti-6Al-4V	907	760	-

It is interesting that for some of these materials, the UTS displays a significant drop as a result of exposure to the high temperature associated with the diffusion bonding process, while the UTS of others dropped only modestly.

For the case of commercially pure grade 2 Ti, the average UTS increased by approximately 18% as a result of the simulated diffusion bonding thermal excursion. This increase in tensile strength is attributed to α-casing, where the high oxygen affinity of Ti combined with the relatively high solid solubility of oxygen in Ti results in a continuous layer of oxygen stabilized α phase at the surface of the component, which serves to strengthen the α phase [7]. Even under the hard vacuum of the simulated diffusion bonding thermal treatment, there is sufficient oxygen present in the furnace chamber to enable the α-case surface reaction. Figure 2 shows the stress strain curves for as-received and heat treated commercially pure grade 2 Ti.

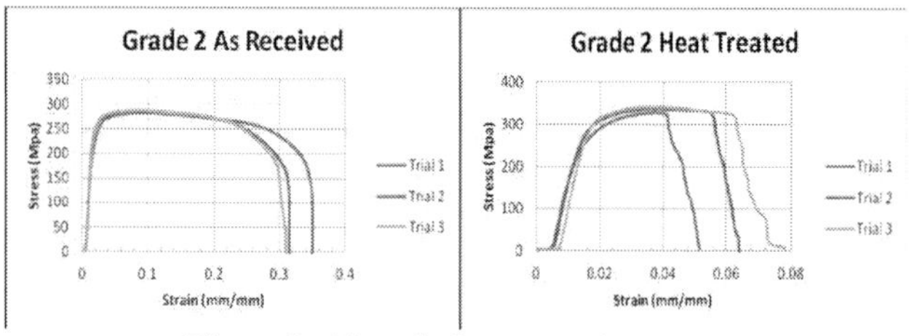

Figure 2 – Ti grade 2 stress-strain curves

It is clear from the area under the stress-strain curves that the toughness of the as-received material is considerably higher than that of the heat treated material. So, although the α-case reaction serves to increase the UTS of α Ti and make it stronger, it also decreases the ductility of the α-case surface layer and lowers the overall toughness.

Although the average tensile strength of the Ti-6Al-4V (grade 23) dropped about 16% after heat treatment, the post "diffusion bonded" UTS is favorable when compared to the benchmark 304 stainless steel material, which displayed an average UTS after heat treatment of 628 MPa. As such, the Ti-6Al-4V material may be attractive for applications that require biocompatibility or where a component must be inert with respect to organic test samples.

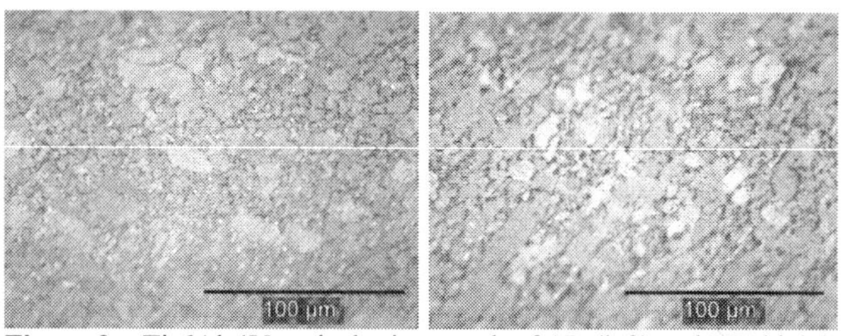

Figure 3 – Ti-6Al-4V optical micrograph of AR (left) and HT (right)

Figure 3 shows polarized bright field optical micrographs at 40x magnification comparing the grain size of as-received and heat treated Ti-6Al-4V alloy. As a result of the thermal treatment, the grain size coarsened from approximately 4.5 to 5.6 μm and correlates with the observed decrease in strength.

The as-received tensile strengths of Inconel® 718 and PH 15-7 Mo were in the same range and both materials displayed a modest drop in UTS as a result of thermal treatment, with the Inconel decreasing by about 3% and the PH 15-7 Mo decreasing by about 9%. Both of these are precipitation hardened materials, where the optimum precipitation hardening thermal sequence should return a UTS on the order of 1500 MPa. The fact that the UTS realized by both materials following the simulated diffusion bonding heat treatment was less than half of the optimum value can be attributed to overaging. Since the diffusion bonding sequence precludes the prescribed ideal thermal treatments and since it is not practical to solutionize, quench and then age the components after diffusion bonding, these materials were not considered as viable candidates for the intended application.

The Chromflex® material displayed the highest as-received tensile strength of any material in this study (over 1.5 GPa). However, it also showed the largest decrease in UTS as a result of exposure to high temperature (~ 85% drop). This can be attributed to the recrystallization and grain growth that occurred during heat treatment. Figure 4 shows an SEM micrograph of the Chromflex material after heat treatment where grains on the order of 100 μm are evidenced.

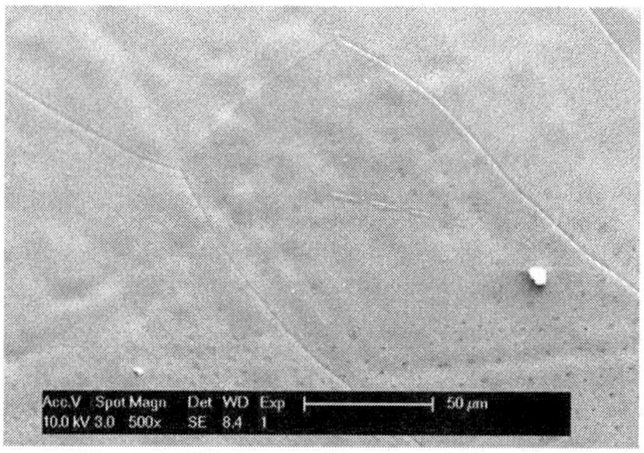

Figure 4 – SEM image of heat treated Chromflex

This sample was not etched prior to imaging, so the pronounced grain boundaries are interesting. This was also the case for the 304 stainless steel sample (see Fig. 5).

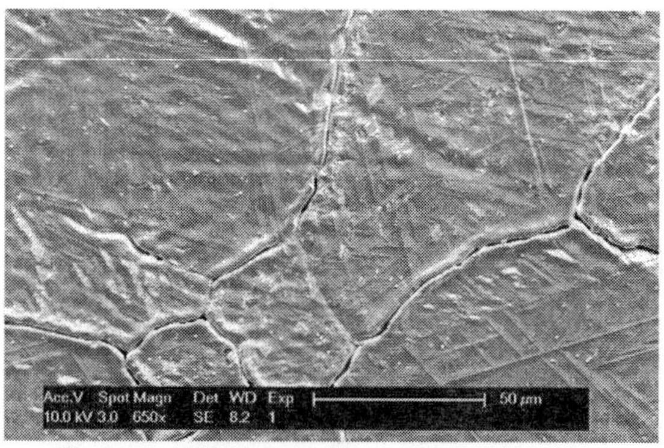

Figure 5 – SEM image of heat treated 304 stainless

Both of these materials display a phenomenon known variously as grain boundary or thermal grooving [8,9], where the intersection of grain boundaries with the free surface of a polycrystalline material forms grooves along the grain boundary intersections in response to high temperature. A schematic profile of an ideal thermal grove is shown in Fig. 6.

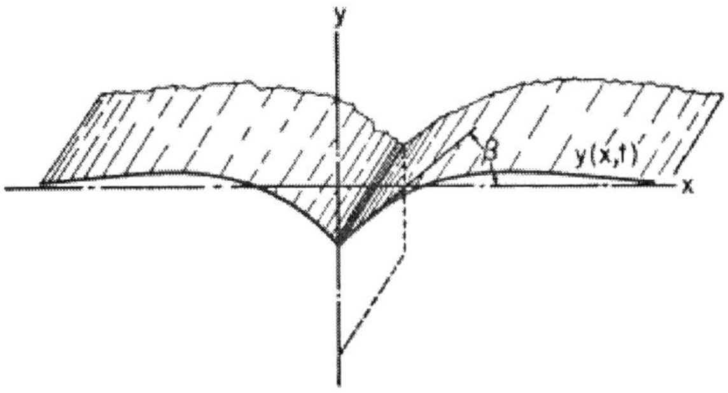

Figure 6 – Profile of an ideal thermal groove (after W. Mullins)

The driving force for thermal grooving is the reduction of the combined surface and grain boundary energies. The equilibrium angle β is governed by Eq. (1),

$$2\gamma_s \sin\beta = \gamma_b \qquad (1)$$

where γ_s and γ_b are, respectively, the surface and boundary free energies. The rapid surface diffusion of atoms in the near region of the intersection produces sharp ridges on either side of the boundary. The ridges tend to flatten by localized evaporation and surface diffusion, which in turn disrupts the equilibrium angle and causes the groove to deepen. Since this effect is associated with free surfaces, it is not considered an issue for the diffusion bonding process where the surfaces are in intimate contact.

The most promising results were achieved with the Nanoflex and Custom 465® materials. Both alloys demonstrated tensile strengths around 1 GPa after simulated diffusion bonding. The UTS of the heat treated Custom 465 material decreased about 3.7% to an average of 963 MPa, while the UTS of the Nanoflex material decreased about 11.5% to an average of 1004 MPa. Since both alloys are maraging type steels, they are similar in composition and mechanical behavior.

Maraging steels are strengthened by generating high percentages of martensite along with precipitation of intermetallics at martensite lath boundaries during an aging process. They offer high strength and toughness, and are generally used in applications in which both formability and strength are desired attributes. The principal alloying element in maraging steels is Ni, with secondary alloying elements such as Ti, Mo, Cu, Al and Co. For the case of the Nanoflex material, the Mo and Cu enable the precipitation of nanoscale dispersoids that can substantially increase the strength of the material after aging [10]. The nickel content allows these materials to form martensite during a subsequent isothermal cooling process and precludes the extensive formation of ferrite when the material is cooled from the autenitizing temperature. Around room-temperature, the austenite becomes thermodynamically unstable and undergoes an isothermal martensitic transformation where the martensite phase nucleates near grain boundaries [11]. The growth of the martensite phase is nucleation-controlled, showing little dependence on the growth of existing laths.

The rate of the martensitic transformation and the achievable maximum martensite fraction are dependent on both the temperature to which the metal is heated prior to cooling and the rate at which the metal is cooled following austenitization. Both factors must be considered in assessing the suitability of maraging steels for the intended application. However, favorable martensite formation can be achieved following thermal excursion to diffusion bonding temperatures. A large volume fraction of martensite can be manifested in slow cooled alloys that are then held at room temperature for twenty-four hours [12]. The high strength of maraging steels, along with the fact that the strength can be enhanced by precipitation reactions after the complete austenitization that occurs during high temperature processing, makes them very attractive candidates for the intended application.

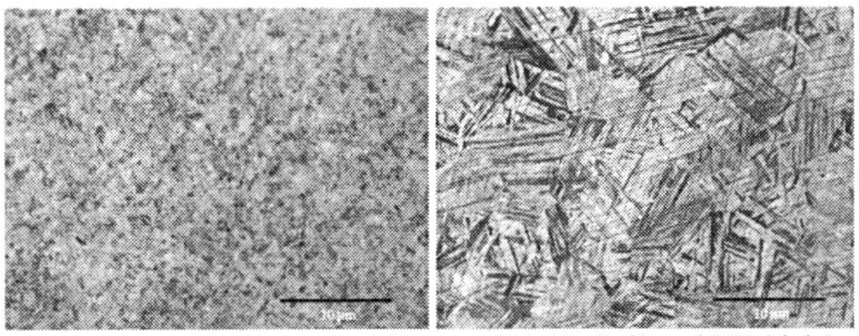

Figure 7 – Optical images of Custom 465 AR (left) and HT (right)

Figure 7 shows optical images of Custom 465 at relatively low magnification. The as-received material is fine grained, with a modest amount of martensite, whereas the heat treated material has much larger grains and a significant degree of martensitic transformation following simulated diffusion bonding.

Similarly, the Nanoflex alloy shown in Fig. 8 compares the fine grain size of as-received material with the coarse grained morphology of the heat treated material. Laths of martensite are evident in the right hand image, which is the heat treated sample.

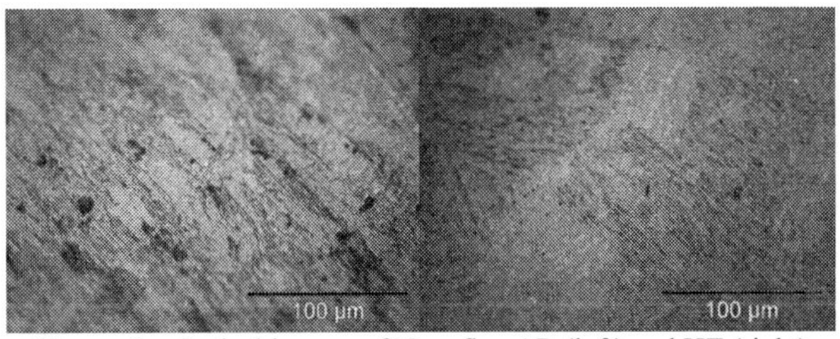

Figure 8 – Optical images of Nanoflex AR (left) and HT (right)

Following an isothermal hold at room temperature for twenty-four hours, some of the Nanoflex samples were then aged at 475°C for three hours, in order to elicit a precipitation reaction. The heat treated, un-aged and heat treated then aged stress-strain curves for Nanoflex are shown in Fig. 9.

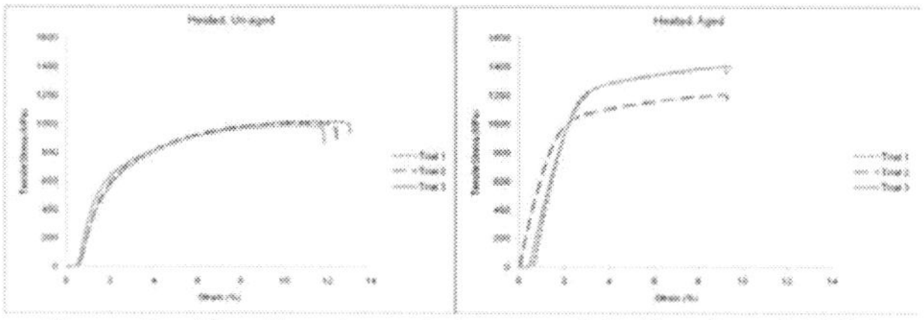

Figure 9 – Nanoflex stress-strain, unaged (left) and aged (right)

The average tensile strength of Nanoflex after aging slightly exceeded that of the as-received samples and increased 33% (over 330 MPa) compared to the samples that were subjected to the diffusion bonding simulation, but not the aging treatment. This increase in tensile strength as a function of aging can be understood by reference to Fig. 10.

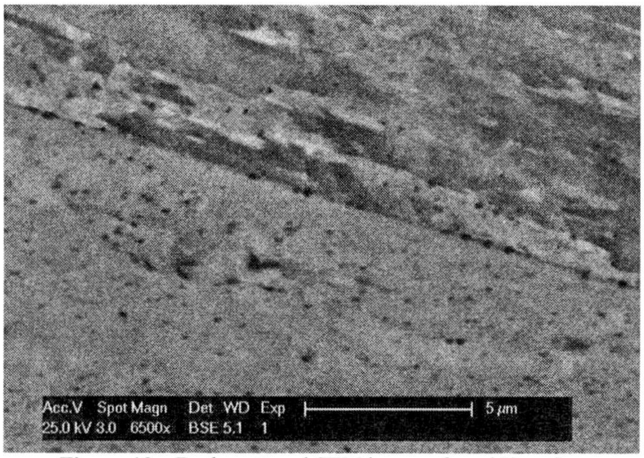

Figure 10 – Back scattered SEM image of aged Nanoflex

Figure 10 shows the boundary between martensite (upper half) and austenite (lower half) with a preponderance of small black dispersoids that are likely nano-scale precipitates. A significant amount of research has been dedicated to the study of precipitation hardening in Nanoflex using transmission electron microscopy, atom probe tomography, and atom probe field ion microscopy [13]. Based on this work, it is suggested that the precipitate dispersion present in the aged samples consists primarily of Cu-rich and Ni-rich IMCs located near martensite lath boundaries. The Cu-rich IMCs form within a few minutes and the Ni-rich IMCs from one to two hours after exposure to 475°C, along with some Mo-rich IMCs. Coarsening of the particles is slow, so it is unlikely that they exceed 10 nm in diameter, on average.

A property of particular interest in Nanoflex, and possibly the Custom 465, is the continuous hardening that occurs over aging times in excess of 1000 hours. The metal alloy does not overage because its high alloy character causes sequential formation of a wide variety of precipitates, leading to the atomic redistribution of previously formed intermetallics, rather than their coarsening. In this way, precipitate evolution occurs mainly due to compositional changes and re-nucleation, with only minimal growth of the precipitates [14]. This aging behavior is interesting because due in part to the large number of different phases precipitated during heat treatment, the net precipitate character of the material is strongly dependent on aging time. For example, Cu-rich intermetallics that form first tend to decrease in density as aging proceeds while, simultaneously, the Cu content of the Ni-rich precipitates tends to increase with time. This behavior may be exploited in future work to achieve higher tensile strengths. As long as the increase in strength that accompanies a longer aging process is determined to be worth the additional process cost, the precipitation hardenable maraging steels may be able to achieve hardening well beyond what was observed in this work.

Conclusions

Based on the results of this study, the martensitic, age-hardenable alloys (Custom 465 and Nanoflex) are considered viable and attractive high strength materials for applications that require processing steps at elevated temperatures. Both of these maraging steels boasted ultimate tensile strengths in excess of 900 MPa after thermal treatment at 1100°C. The Ti-6Al-4V is also considered to be a viable high strength material where the additional constraint of biocompatibility is an issue. Future work will be focused on functional testing of metal alloy devices fabricated from these candidate materials, along with optimization of the aging treatment to maximize ultimate tensile strength.

Acknowledgements

The authors would like to gratefully acknowledge Fred Hayes (ICEM Facility Manager), Professor Kazuo Yamazaki (Mechanical Engineering), Peter Fonda, Troy Topping, Jonathan Kurniawan, Jon Nugyen, Tainbo Liang and Enoch Leung at UC Davis and Floyd Bishop, Tom Kaminska, Jon James, Leyla Hashimi, Matt Lewis, Martin Cisneros and Reid "Turtle" Brennen at Agilent Technologies, for their generous support of this project.

References

[1] Eberhart, M. E., "Why Things Break," Scientific American, October (1999), pp. 66-73.

[2] Powers, M., Sen, S., Nguyentat, T., Knio, O., and Weihs, T., "Bonding Processes," CRC Materials Processing Handbook, J. Groza, M. Powers, E. Lavernia and J.F. Shackelford, Eds., Talyor and Francis Group, (New York, 2007), pp. 21-13 – 21-19.

[3] Meyers, M., and Chawla, K., Mechanical Behavior of Materials, 2nd Ed., Cambridge University Press, (Cambridge, 2007), pp. 348-350.

[4] Olson, G. B., "Advances in Theory: Martensite by Design", Materials Science and Engineering A, 438-440 (2006), pp. 48-54.

[5] Shewmon, P. G., Transformations in Metals, J. Williams Book Company, (Jenks, OK, 1983), pp. 323-354.

[6] Holmquist, M., Nilsson, J. O., and Stigenberg, A. H., "Isothermal Formation of Martensite in a 12Cr-9Ni-4Mo Maraging Stainless Steel," Scripta Metallurgica et Materialia, 33 (1995), pp. 1367-1373.
[7] Luterjering, G. and Williams, J. C., Titanium: Engineering Materials and Processes, Springer-Verlag, (Berlin, 2003), pp. 48-49.
[8] Min, D., Morphological Evolution of Grain-boundary Grooving, VDM Publishing, (Saarbrücken, Germany, 2009), pp. 1-5.
[9] Mullins, W. W., "Theory of Thermal Grooving," Journal of Applied Physics, 28 (1957), pp. 333-339.
[10] Hattestrand, M., Nilsson, J. O., Stiller, K., Liu, P. and Anderson, M., "Precipitation Hardening in a 12%Cr-9%Ni-4%Mo-2%Cu Stainless Steel," Acta Materialia, 52 (2004), pp. 1023-1037.
[11] Datta, K., Delhez, R., Bronveld, P. M., Beyer, J., Geijselaers, H. J. M. and Post, J. "A Low Temperature Study to Examine the Role of ε-Martensite During Strain Induced Transformations in Metastable Austenitic Stainless Steels," Acta Meterialia, 57 (2007), pp. 455-460.
[12] Martin, D. S., Aarts, K. W. P., Rivera Diaz del Castillo, P. E. J., van Dijk, N. H., Bruck, E. and van der Zwaag, S., "Isothermal Martensitic Transformation in a 12Cr-9Ni-4Mo-2Cu Stainless Steel in Applied Magnetic Fields," Journal of Magnetism and Magnetic Materials, 320 (2008), pp. 1722-1728.
[13] Stiller, K., Danoix, R. and Hattestrand, M., "Mo Precipitation in a 12Cr-9Ni-4Mo-2Cu Maraging Steel," Materials Science and Engineering A, 250 (1998), pp. 22-26.
[14] Stiller, K., Hattestrand, M. and Danoix, F., "Precipitation of 9Ni-12Cr-2Cu Maraging Steels, Acta Materialia, 46 (1992), pp. 1027-1034.

Surface Reactions of Single Crystal Diamonds with Chromium Containing Nickel-based Braze Materials

Jussi Laurila and Petri Vuoristo
Tampere University of Technology, Department of Materials Science
Korkeakoulunkatu 6, 33720 Tampere, Finland
jussi.laurila@tut.fi

Olli Tuovinen and Mikko Uusitalo
Valmet Technologies Inc. Tampere, Finland

Abstract

In this study, single crystal diamonds were brazed on to stainless-steel plate using a nickel-based braze material, BNi-7. This braze material contains phosphorous and silicon in order to lower its melting point and to improve its fluxing behavior. The braze material also contains chromium for improved corrosion resistance. The brazing was performed in a vacuum furnace at a temperature range of 1000°C - 1050°C. In brazing, both silicon, and more particularly chromium, react with the surface of the diamond crystal, which affects the microstructure of the braze layer and forms desirable thin carbide layers on the diamond surfaces for improved wetting and bonding of the diamond to the braze material. The brazed diamonds were removed from the braze layer by electrolytic polishing. The specimens were then studied with scanning electron microscopy (SEM). Cross-sectional specimens were also prepared from the brazed coating structures, and the microstructures were characterized with SEM and energy dispersive X-ray microanalysis (EDX). Specimens for transmission electron microscope (TEM) studies were also prepared. Carbon from the diamond single crystal particles was detected to have dissolved into the braze layer, which explains why the shapes of the individual diamond particles become rounded under the high-temperature brazing process. It was also found that carbon was removed layer by layer from the diamond surfaces. No evidence of the weakening of the structure inside the diamond could be found from the cross-sectional fracture surfaces in the vicinity of the original diamond surfaces. Two different carbide layers, Cr_3C_2 and Cr_7C_3, were found at the interfaces between the diamond and the braze material.

Introduction

The unique properties of diamond, such as its extremely high hardness and wear resistance, make the material interesting in tool manufacturing. During many manufacturing processes, diamond crystals often adhere to the base materials used in brazing. Braze materials can be based on silver, copper, titanium, aluminum, palladium and nickel [1]. In this study, single crystal diamonds are brazed on to a stainless steel base material using nickel-based braze material BNi-7. This braze material is recognized as being less expensive than silver-based braze materials, in addition to which it also has good corrosion resistance in applications where simultaneous corrosion loads are present. [1]

Nickel-based braze materials, such as Ni–Cr–P and Ni–Cr–B–Si, are widely used brazing alloys. There is a general acceptance among researchers that the reason for the high bonding strength between diamond and an Ni–Cr brazing alloy is the reaction and formation of carbides at the diamond/braze material interface during the brazing process. The formation of carbides such as Cr_3C_2 and Cr_7C_3 has been widely reported [2, 3, 4]. It has also been reported that there is no graphite formation when chromium is present in the braze material, but that graphite can be formed when nickel braze materials are used. It is also noted in the same source [5] that Cr_3C_2 and Cr_7C_3 type carbides are formed on the diamond surfaces. In this study, these surface layers and the structure of the diamond surface itself are analyzed, and the internal defects in the diamond structure near the surface are characterized in detail.

Experimental

A BNi-7 type nickel-based braze material was used to join single crystal diamonds on to a molybdenum-alloyed stainless steel surface. The composition of the braze material is presented in Table 1. The chromium content of the braze material is high, in the order of 13.0-15.0 weight -%. The phosphorous content is near to the eutectic point of the solution, as can be seen Figure 1 a), being about 10 weight %. This is to ensure a low melting point, and to improve the fluxing and wetting behavior of the braze material. The brazing temperature used in these studies was 1000-1050°C. The brazing was done in a high vacuum. Figure 1a) presents the binary phase diagram of the nickel-phosphorous system.

According to this phase diagram, the melting point of the binary alloy braze material 11 weight % of P is 870°C. Figure 1b) presents the corresponding phase diagram for a chromium-phosphorous system. From these diagrams it is clear that chromium in the braze material increases the brazing temperature. The phase diagrams also show that the solidified braze forms two different phases: a face-centered cubic nickel-based phase with about 0.17 w-% of P and an X_3P phase where X can be nickel and chromium. The crystal structure of the Ni_3P phase is tetragonal, space group I$\bar{4}$; the dimensions for the unit cell being a=8.954Å and c=4.386Å. The Cr_3P phase has the same crystal structure I$\bar{4}$ with unit cell dimensions of a=9.185Å and c=4.560Å. [6, 7, 8]

The interaction of the braze material with the diamond was studied with a Philips XL30 scanning electron microscope (SEM) equipped with an EDAX EDX-analyzer and with a Zeiss Ultra Plus high resolution field-emission scanning electron microscope. The specimens were characterized by secondary and backscattered electrons. The specimens for these observations were prepared by cutting and mechanically polishing the cross-sectional surfaces with diamond paste. Specimens were also prepared by electropolishing the diamonds off from the braze surface. These diamond specimens were also crushed mechanically in order to obtain fracture surfaces for both surface morphology studies and surface cross-section studies with TEM. After crushing, the SEM specimens were glued onto the specimen support stand.

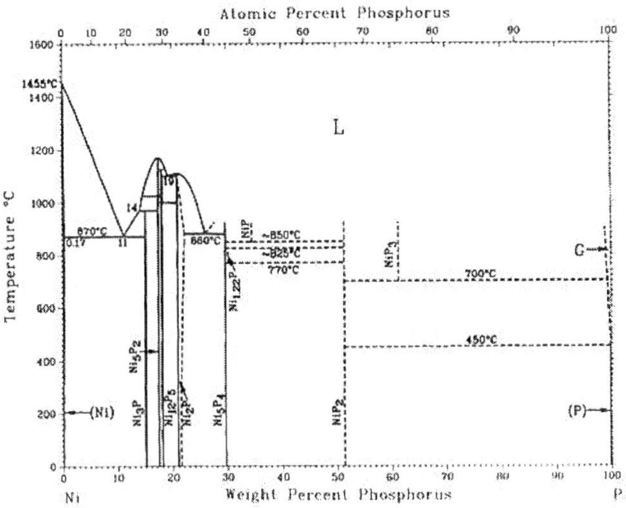

a)

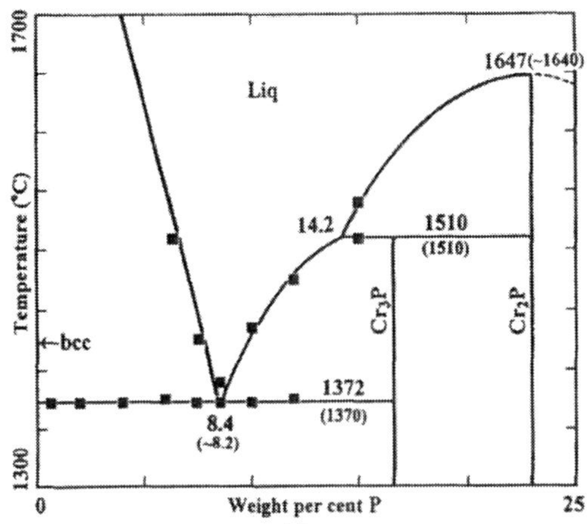

b)

Figure 1: Binary phase diagrams for a) Ni-P and b) Cr-P systems. [6, 7]

The cross-sections of the brazed diamond surfaces were studied with a JEOL 2010 transmission electron microscope (TEM) to study the structure of the diamond crystal at the braze joint. The specimens for the TEM studies were prepared by crushing the diamonds and setting these on a 3 mm diameter copper grid covered with thin carbon film with holes in it.

Table 1: Composition of the nickel based BNi-7 type brazing material [6].

AWS class.	Composition, weight -%						
	Ni	Cr	B	Si	Fe	C	P
BNi-7	Rem.	13.0–15.0	0.02	0.1	0.2	0.06	9.7–10.5
	S	Al	Ti	Mn	Zr	Co	Se
	0.02	0.05	0.05	0.04	0.05	0.1	0.005

Results and discussion

Figure 2 shows one of the rounded diamond single crystals as illustrated by SEM after electropolishing the diamond surface visible under the braze and carbide layers. The surface of the diamond has some roughness and the original sharp corners of the diamond single crystal have been rounded due to carbon dissolving into the braze matrix during the brazing process.

Figure 2: A rounded diamond single crystal after brazing and electropolishing the braze material from the surface.

The structure of the carbide layers formed at the interface between the diamond and the braze material was studied from the polished cross-sectional specimens. Backscattered electrons give higher contrast to the elements which have different mass fractions from the atomic elements. The backscattering coefficient η is largely independent of the primary electron energy, E, in the range of 10-100 keV, so the level of the acceleration voltage used in these studies has no effect on the results. In these studies, the accelerating voltage used was 20 kV.

The measured backscattering coefficient of multi-component targets can be fitted to the formula:

$$\eta = \Sigma_i c_i \eta_i, \qquad [9],$$

where η_i are the backscattering coefficients of the pure elements and the c_i values are their mass fractions. The η coefficients for three types of chromium carbides, $Cr_{23}C_7$, Cr_7C_3 and Cr_3C_2, are calculated and presented in Table 2 together with the coefficients of the pure elements chromium and carbon. It has been reported previously [3, 4] that the two last-mentioned carbides are present at the diamond-braze material interface when chromium is present in the braze material.

Table 2: The backscattering coefficients η for three types of chromium carbides, $Cr_{23}C_7$, Cr_7C_3, Cr_3C_2 and pure elements chromium and carbon

Element	η	Formula	η_i
C:	5.5	$Cr_{23}C_7$:	26.523
Cr:	28.2	**Cr_7C_3:**	**25.974**
		Cr_3C_2:	**24.998**

According to Table 2, the differences in the calculated values of the backscattering coefficient η for the chromium carbides is about 1.0 for the Cr_7C_3 and Cr_3C_2 type carbides; this difference can be seen clearly in Figure 3 as different shades of grey near the diamond surface. Near the surface there is a thin layer of small grey carbides, (see the arrow in Figure 3) surrounded by the larger light grey carbides. Figure 4a) shows an EDX-analysis curve measured from the lighter carbide layer and Figure 4b) a similar EDX analysis curve measured from the darker carbide layer near the diamond surface. Note that the Kα-peak of carbon measured from the dark carbide layer is higher than the peak measured from the lighter carbide, indicating the higher carbon content near the diamond crystal. Note also the presence of molybdenum in the darker carbide layer. The molybdenum has most probably dissolved from the stainless-steel base material.

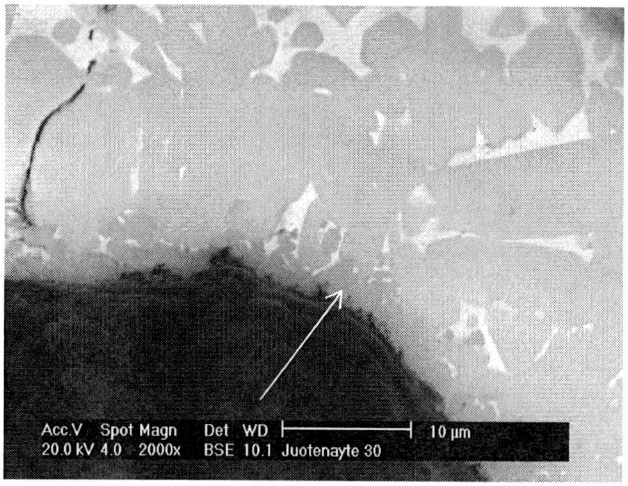

Figure 3: Layer on the diamond: angular carbides and nickel-based matrix between the carbide grains. The difference in content can be easily seen as the difference in the shade of grey. The carbide near the diamond surface, which contains more carbon, is darker than the larger carbides further away from the diamond.

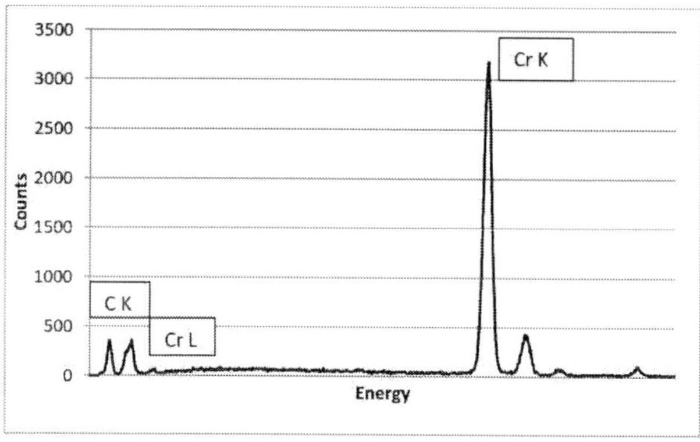

a)

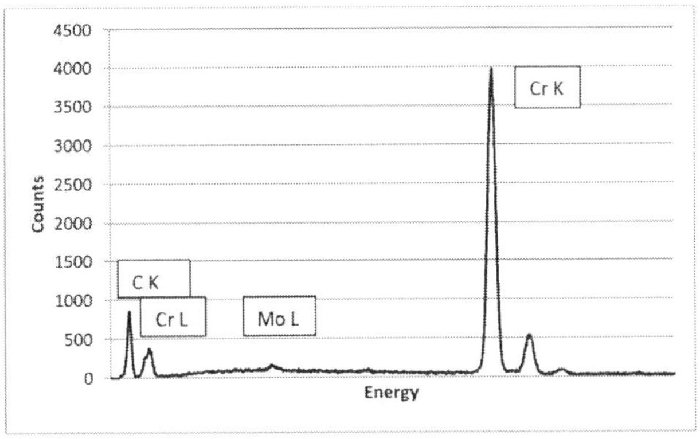

b)

Figure 4: EDX analysis curves measured from the carbide layers presented in Figure 3. The upper curve, a) is an EDX analysis curve measured from the lighter carbide layer and the lower curve, b), is an EDX analysis curve measured from the darker carbide layer nearer the diamond surface.

Figure 5 shows a cross-sectional artificial fracture surface of the diamond single crystal at two different magnifications. The original surface of the diamond after brazing and electropolishing the diamonds off from the braze surface can be seen in the upper parts of the two images and the artificial fracture surfaces in the lower parts. As can be observed, the fracture surfaces are free from defects and therefore there is no evidence of the crystal's internal degradation during the brazing process at the fracture surface near the original surface of the diamond crystal.

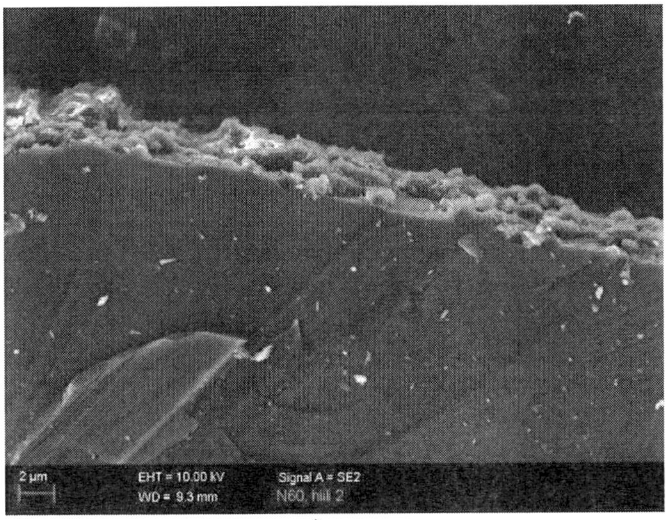

a)

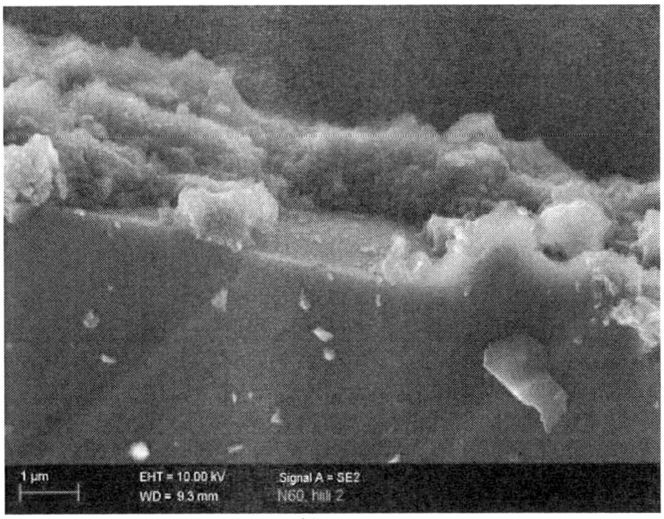

b)

Figure 5: Cross-sectional fracture surfaces of the diamond single crystal at two different magnifications. At both levels of magnification, the original surface of the diamond is shown in the upper part of the image and the fracture surface in the lower part.

Figure 6a) presents a secondary electron image and Figure 6b) a backscattered electron image taken from the same location on the brazed diamond surface. The braze material has been removed mechanically from the diamond surface and the diamond surface is exposed from the braze layer. Some of the braze material still remains on the diamond surface, and this appears as the light parts in Figure 6b). Parts of the diamond crystal layers of type {111} have been dissolved into the braze metal during the brazing procedure. Figure 6c) presents an EDX-analysis curve measured from the area on the diamond surface marked with a circle in Figure 6a). This detailed area was found to be pure carbon. The images and results are a clear example of the degradation mechanism of the diamond during the brazing procedure with nickel-based braze materials. Diamond was found to be removed layer by layer, starting from the surface.

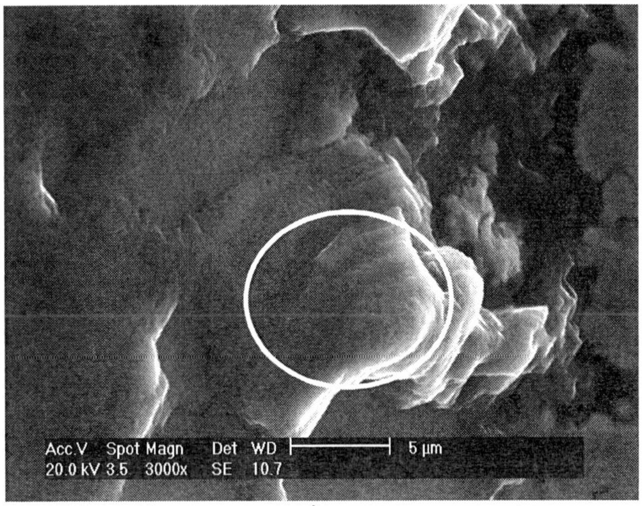

a)

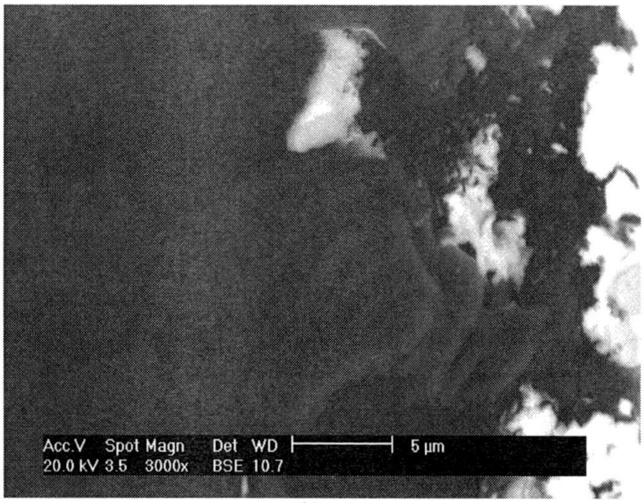

b)

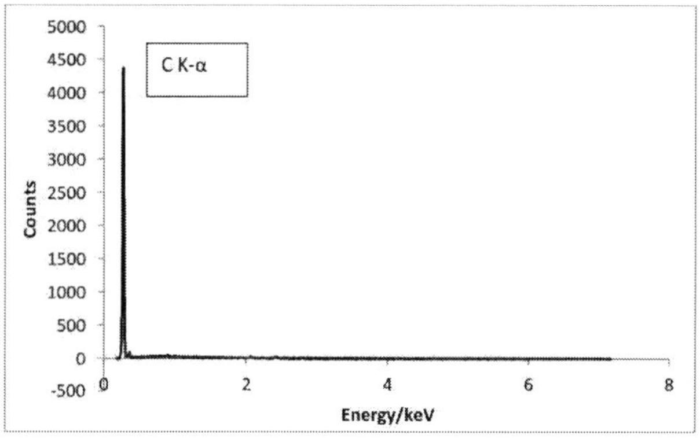

c)

Figure 6: Details of the electropolished diamond surface, a) a secondary electron image and b) a backscattered electron image from the brazed diamond surface. Parts of the diamond crystal layers of type {111} have been destroyed and the carbon has dissolved into the braze metal during the brazing procedure. Figure c) presents the EDX curve measured from the detail seen on the diamond surface.

A TEM micrograph of the internal structure of the diamond below its surface is presented in Figure 7. The specimen was prepared by crushing the diamond after it had been electropolished off from the brazed surface. The bright field image has been taken in two beam conditions with a g-vector of (111); the direction of the g vector can be seen from the selected area diffraction image (SAED). Some structure from the layers on the original diamond surface is still visible and is marked by the circle in Figure 7b. Some fine internal structure can also be seen inside the diamond: twins and other crystal defects. The TEM observation shows that {111} type planes have been partly dissolved into the melting braze material.

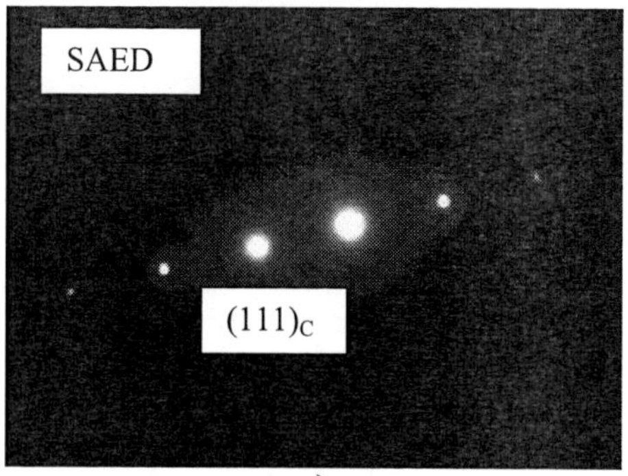

a)

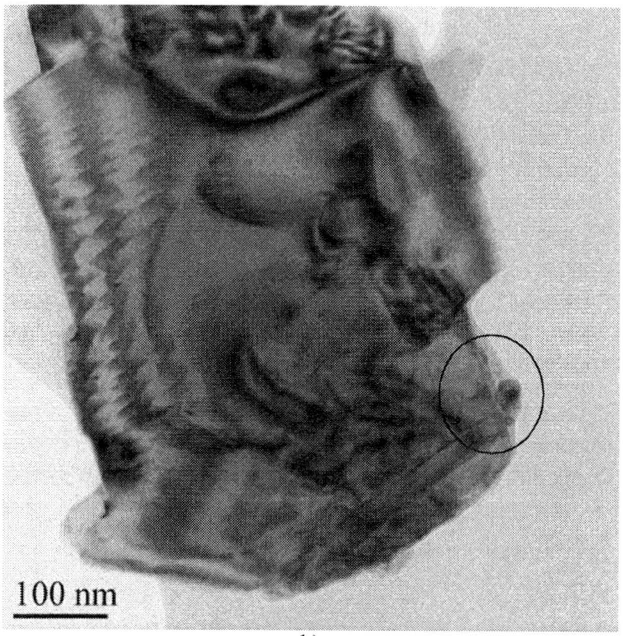

b)

Figure 7: TEM micrograph of the internal structure of the diamond under the surface: a bright field image with two beam condition of g-vector (111) can be seen in the diffraction image a). Some structure from the layers on the original diamond surface is still visible (inside the circle in 7b). There is also some fine internal structure inside the diamond: twins for example.

Conclusions

Two types of chromium carbides were found to be formed during the brazing process and these could be identified from the cross-sectional specimens with SEM using backscattered electron images. The thin carbide layer with higher carbon content near the diamond surface contained molybdenum dissolved from the base material. Molybdenum was not detected in the thicker carbide layer farther from the diamond surface.

It was found that carbon was dissolved layer by layer from the {111} type diamond planes. No evidence of degradation of the diamond's internal structure was detected with TEM, or by close study of the fracture surfaces of the diamond crystals.

This is a positive result with regard to the mechanical strength of the diamond single crystal.

Acknowledgements

The work was carried out under the HYBRIDS program in FIMECC Ltd. The researchers gratefully acknowledge the financial support of Tekes and the participating companies.

References

[1] Advances in brazing, Science, technology and applications, P. Sekulic, Woodhead Publishing, Cambridge, UK, 2013

[2] Interfacial microstructure and performance of brazed diamond grits with Ni–Cr–P alloy, C.Y. Wang, Y.M. Zhou, F.L. Zhang, Z.C. Xu, Journal of Alloys and Compounds 476 (2009) 884–888

[3] H. Trenker, Seidemann, Ind. Diamond Rev. 592 (2002) 49–51.

[4] A.K. Chattopadhyay, L. Chollet, H.E. Hintermann, Surf. Coat. Technol. 45 (1990) 293–298.

[5] Carbon reactivity of binder metals in diamond–metal composites – characterization by scanning electron microscopy and X-ray diffraction W. Tillmann, M. Ferreira , A. Steffen, K. Rüster, J. Möller , S. Bieder, M. Paulus, M. Tolan, Diamond & Related Materials 38 (2013) 118–123

[6] ASM Handbooks, vol. 3, Phase diagrams, American Society for Metals, Metals Park, Ohio, USA, 2014

[7] Thermodynamic description of Cr-P and Fe-Cr-P systems at low phosphorus contents, J. Miettinen, Calphad, Vol. 23, No. 1 , pp. 141-154, 1999

[8] Constitution of binary alloys, F. Shunk, McGraw-Hill, New York, N.Y, USA, 1969

[9] Scanning Electron Microscopy, Ludvig Reimer, Springer Verlag, Berlin, Germany, 1985

[10] Backscattering coefficients for low energy electrons, A. M. D. Assa'd and M. M. El Gomati, Scanning Microscopy Vol. 12, No. 1, 1998 (Pages 185-192)

In Situ Observation of Brazing Processes with SEM

E. Hofmann, J. Zschetzsche, U. Füssel
Technische Universität Dresden, Dresden, Germany
evelyn.hofmann@tu-dresden.de

M. Türpe
MAHLE Behr GmbH & Co. KG, Stuttgart, Germany

Introduction

Wetting is one key parameter in joining processes. It is very sensitive to surface properties. Therefore the examination of different surface properties is of high interest. For adhesive bonding contact angle measurement is a widely used method to evaluate the quality of a surface. In brazing and welding processes, the additionally applied process heat may lead to oxidation of the surface, thus resulting in wetting difficulties. The examination on a macroscopic scale can be carried out easily, whereas flowing and wetting on the surface itself can hardly be seen. For microscopic observation an SEM was used and therefore a comparison with furnace brazing is obvious.

Methods for examining wetting properties

In order to evaluate wetting on different surfaces, standardized tests mostly refer to dip soldering [1]. Standardized tests for brazing are not mentioned. One easy test is propagation measurement of the molten filler metal. This method can be used for wire applications.
Propagation measurements use very clean copper and brass surfaces to measure the efficiency of fluxes [2]. The filler metal melts by dip soldering.
For filler metal applications like inserted or cladded filler metals no common tests are available. Additionally to this aspect none of the methods characterizes the wetting and spreading behaviour in situ. Both contact angle / height of filler metal and filler metal propagation are measured after the experiment [3].
The test parameters concern temperature and time, but not e.g. atmospheres and are only standardized for soldering.

Experimental procedure

[4] and [5] show the importance of in situ experiments to understand the wetting behaviour. Both methods use a heating stage and are carried out under process atmospheres (vacuum or shielding gas).
The experiments described in this paper were carried out in an alterated SEM where the specimen mount was replaced by a heating stage (Figure 1). The heating stage is cooled with water.

Figure 1: Experimental set-up in SEM

The sample size is limited to a diameter of about 5 mm. After cutting and cleaning the specimen, it is placed in a ceramic crucible. The triangular heat protection (Figure 1) prevents heat radiation into the vacuum chamber, but also prevents EDX analysis during the experiment. The heating is set and controlled by an external control unit. The thermal element measures the heating below the ceramic crucible. The maximum temperature is 1000 °C.
First wetting experiments have been carried out using filler metal foil placed flat on different substrates (Table 1).

Table 1: Materials used in the experiments

Substrate	Foil
EN AW-3103	EN AW-4004
1.4301	EN AW-4045
1.4301	Cuprobraze

Both substrate materials form an oxide layer that is difficult to wet, but is known to be brazed under vacuum. Substantial problems are due to the thickness of the substrate material and to ensuring the connection between foil and substrate. Because of the position of the thermal couple (TC), the temperature is not equate with the actual sample temperature and therefore labeled with (TC). The samples were heated up to 1000 °C.
In order to achieve a homogeneous temperature in the sample, distinct temperatures were held for about 10 minutes. Then SE-pictures were taken.

Results

Set-up 1: EN AW-3103/EN AW-4004
At room temperature the surfaces of both filler metal foil and substrate material show surface roughness and cutting edge from sample preparation (Figure 2). Up to 680 °C (TC) no wetting or flowing of the filler metal occurs. Only structural changes on the filler metal surface can be observed.

Figure 2: SE-picture at 20 °C (TC)

Starting at 580°°C (TC) (Figure 3), first lighter coloured spots at the substrate surface develop.

Figure 3: SE-picture at 580 °C (TC)

On pictures taken at 700 °C (TC) (Figure 4) partial wetting and partially bigger dispersion are visible.
Further heating shows further wetting propagation. The observed area shows a ca. 100 µm wide wetting band. The surface of the band and filler metal foil are similar. After cooling down the cutting edge of the foil is still clearly visible (Figure 5). Both changes in surfaces structure as well as volume of the filler metal are visible.

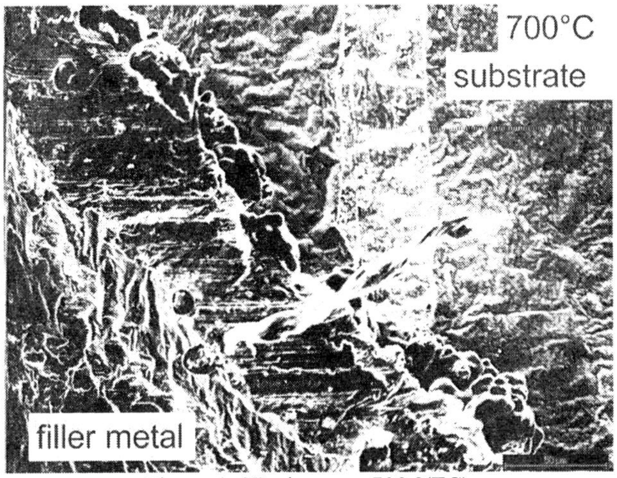

Figure 4: SE-picture at 700 °(TC)

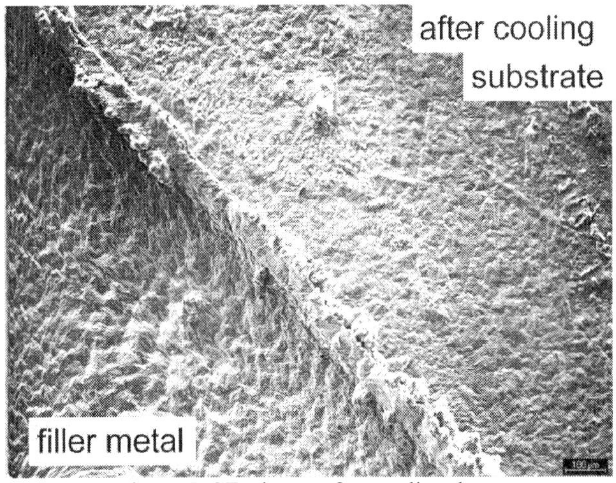

Figure 5: SE-picture after cooling down

Set-up 2: 1.4301/EN AW-4045
Preparation and set-up were carried out the same as before (Figure 6).

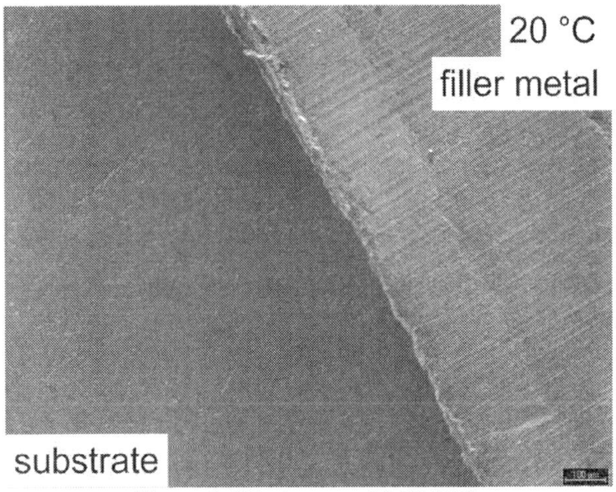

Figure 6: SE-picture at 20 °C (TC)

The foil started fluting during heating up at 830 °C (TC) and contracts (Figure 7). With ongoing heating this effect intensifies.

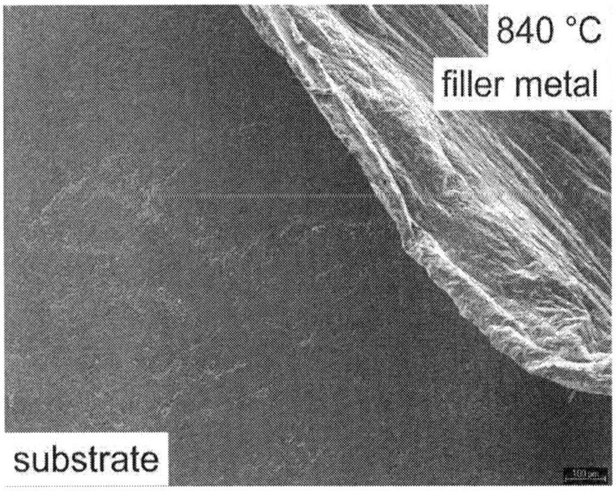

Figure 7: SE-picture at 840 °C (TC)

There is no obvious wetting on the substrate surface. Examination after the experiment shows that the filler metal being directly in contact with the substrate developed small melt spheres on the surface (Figure 8).

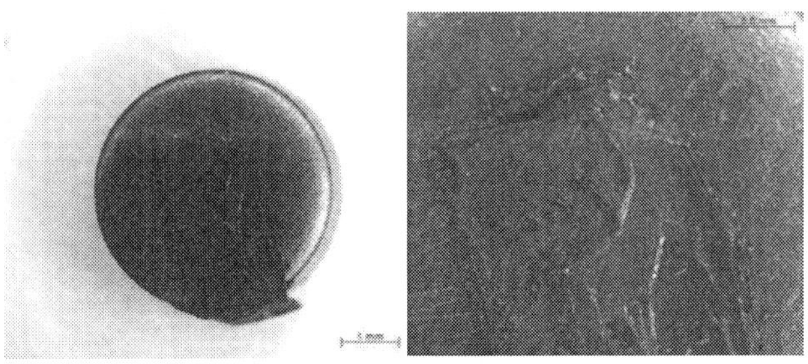

Figure 8: Sample after experiment

Set-up 3: 1.4301/Cuprobraze

For this set-up a small piece of Cuprobraze foil was put directly on the steel surface in order to achieve more contact (Figure 8). During heat up the foil only expands at places without contact, whereas contact areas show melting.

Up to 800 °C (TC) there was no visible change. During the next temperature steps the foil leveled out along the substrate surface.

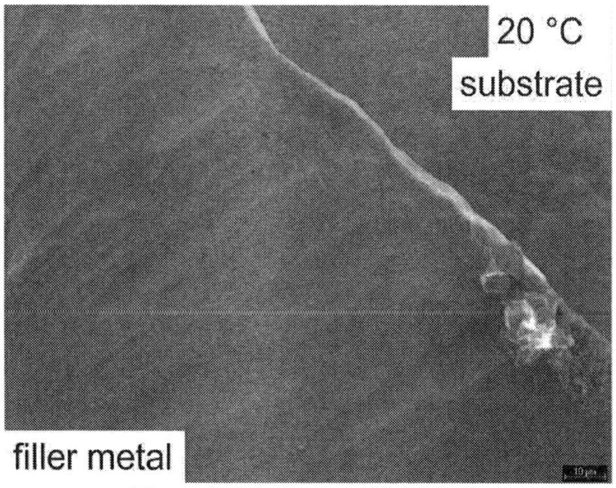

Figure 8: SE-picture at 20 °C (TC)

Wetting of the surface can be seen at 880 °C (TC). The filler metal flow is even up to 920 °C (TC).

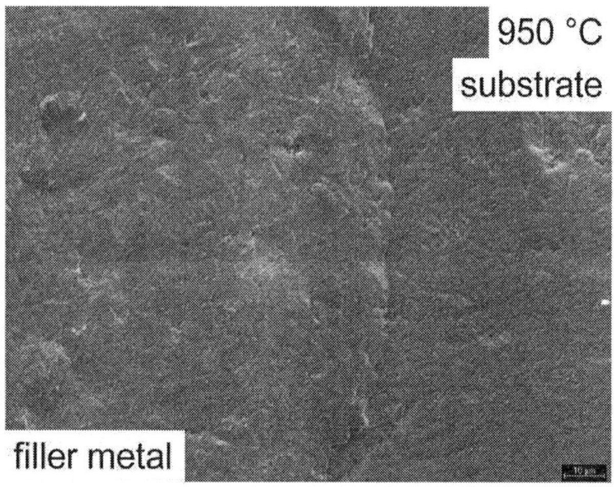

Figure 9: SE-picture at 950 °C

Higher temperatures show no changes in the filler flow.

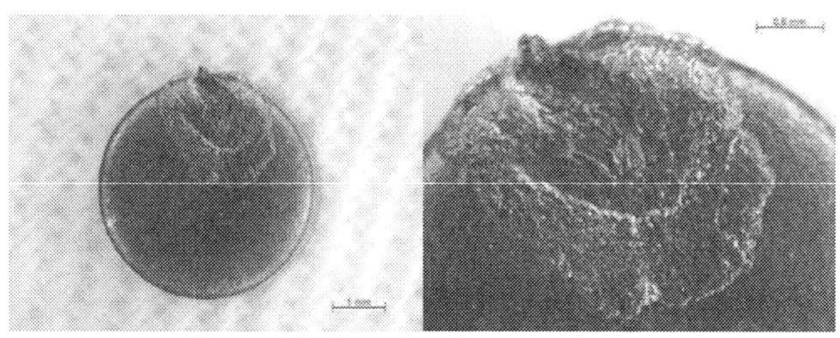

Figure 10: Sample after experiment

In comparison to set-up 2, the wetting of Cuprobraze shows better propagation (Figure 10), but also a smaller wetting area.

Discussion

Understanding wetting mechanisms can be assisted by using SEM techniques. It is especially interesting to see the propagation during heating. Filler metal foils are challenging: in vacuum conditions contact between substrate and foil has to be ensured to ensure thermal conduction. Because of the high magnifications, changes in the substrate and filler metal microstructure can be observed.

Difficulties still exist regarding temperature measurement. Due to the small diameter of the ceramic crucible it is difficult to insert another thermal element.

As observed in the first two experiments, materials whose oxide layers are stable are difficult to wet. Due to the different chemical composition of the third filler metal, very good wetting was possible, which could also be assisted with real brazing samples.

The experiments were the first to be carried out in the altered test arrangement. They show that by using in situ tests transformations on the surfaces can be seen. They can be grain growth and wetting and flowing of the filler metal. With a different temperature measurement these tests will allow to get a connection between process and surface transformation.

Further experiments will be carried with other filler metal applications.

Acknowledgments

We acknowledge Kurt-Schwabe-Institut for providing us the heating stage.

References

[1] DIN 32506-1:1981-07, Lötbarkeitsprüfung für das Weichlöten-Teil 1 Benetzungsprüfungen

[2] DIN EN ISO 9455-10:2013-01, Flussmittel zum Weichlöten-Prüfverfahren-Teil 10: Bestimmung der Wirksamkeit des Flussmittels

[3] Shapiro, A.E.: Determining solder alloy and base metal compatibility. Welding Journal, New York, 2007, 86. Jg., Nr. 9, S. 33

[4] McGurran, B.; Nicholas, M.G. A Study of Aluminum Brazing Filler Metals Using Hot Stage Scanning Electron Microscopy. Welding Journal, 1984, 63. Jg. Nr. 10, S. 295-299

[5] Zähr, J., et al. Anwendbarkeit von Benetzungsversuchen für verschieden Metalle. Metall, 2008, 62. Jg. Nr. 10, S. 611

An Investigation on Thermal Integrity of Brazed Heat Transfer Devices

Hui Zhao[1], **Stefan Elbel**[1, 2], **Pega Hrnjak**[1,2]
[1]Creative Thermal Solutions, Inc., Urbana, IL 61802, USA
[2] University of Illinois at Urbana-Champaign, Urbana, IL 61801, USA
Email: hui.zhao@creativethermalsolutions.com

Abstract

Brazing and soldering technologies are commonly used in fabrication of heat transfer devices made by aluminum and other metals. A well designed air side fin structure can significantly improve heat transfer performance by utilizing the concept of flow boundary layer disruption. Good bonding between base plate and extended fins ensures both mechanical and thermal integrities of the device. However, it can be a challenging task to develop a reliable and cost effective joining technology for novel heat transfer surface designs. For example, open cell metal foams are considered as potential alternative to traditional straight fins for air side heat transfer enhancement, but brazing of foam ligaments with base plate requires careful consideration of filler metal selection as well as brazing conditions. In this paper, heat sink samples are brazed with both aluminum foams and straight fins using a controlled atmosphere furnace. Various forms of filler metals such as paste and foil are used for brazing tests. The brazed joint integrity is examined by standard metallurgical analysis. Thermal performance of brazed heat sink samples are tested in a wind tunnel facility. The influences of brazing process on heat sink thermal performance are discussed in detail.

Introduction

Heat exchangers are used in applications where thermal energy needs to be transferred from one medium to another. One of the most widely used materials for heat exchangers is aluminum and its alloys. In applications where air is used as heating or cooling fluid, it is often necessary to enhance the heat transfer by adding extended fin surfaces to the primary heat exchange surface to increase the air side heat transfer areas. Over the years, many efforts have been devoted to improving heat transfer in heat exchangers by optimizing fin structures. Fin designs are carefully guided by the concept of air flow disruptions based on boundary layer theory in fluid dynamics. Figure 1 illustrates two examples of fin structures that are used for increasing heat transfer in convective flow of air or other fluids. Figure 1(a) is a conventional offset fin design that utilizes a periodical offset of the fin's leading edge to interrupt the fluid flow in favor of higher heat transfer without a significant increase of air side pressure drop [1]. Figure 1(b) shows the novel open cell metal foam which is also used in heat transfer devices owing to its high surface area to volume ratio [2].

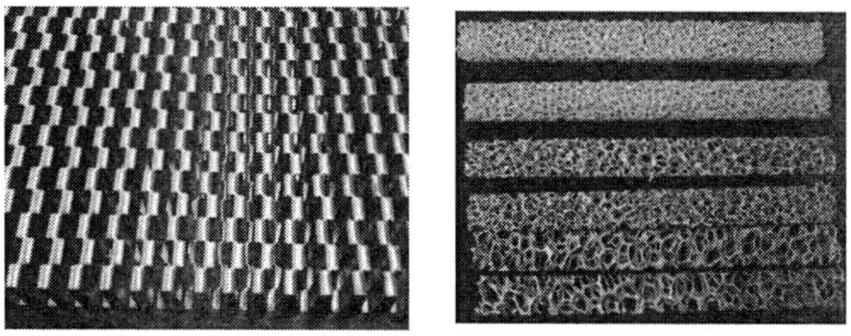

(a) (b)

Figure 1. Examples of fin structures for aluminum heat exchangers: (a) offset fin; (b) open cell metal foam

A good thermal contact between fin and substrate surface is essential to fully utilize the potential of these well-organized extended surfaces. While modern technologies provide many options, e.g., casting, extrusion, and 3-D printing, for manufacturing the aluminum heat exchangers with extended fin structures, brazing remains to be one of the most economic processes in bonding various heat exchanger components including fins to the primary surface. In brazing, the connection between fin and substrate surface is provided by metallurgical bond with filler metals. A successful joint formation ensures both mechanical and thermal integrities of the brazed heat exchangers. Assuming both brazing condition and material selection are well controlled which ensures good wetting and capillary flow of liquid filler metal to fill the gap between fin and substrate, the question is, what is the proper amount of filler metal to be supplied. In this case, the amount of filler metal is directly related to the joint size after brazing. Intuitively, a larger joint provides more connecting areas between fin and substrate surface, therefore a better thermal performance may be expected. However, increasing the amount of filler metal in production is not only costly, but can also lead to serious problems such as base metal erosion due to the interaction between liquid filler metal and solid substrate during brazing [3]. Excessively large aluminum fin/tube brazed joints may not be necessary because aluminum is such a good thermal conductor that provides sufficiently low thermal resistance between fin and tube. A theoretical and numerical study has shown that for serpentine fin/flat tube joints in an aluminum heat exchanger, the change of joint size in a certain range does not have a significant impact on the heat transfer performance [4]. The American Welding Society (AWS) specification for aluminum brazing (AWS C3.7/C3.7M, 2011) requires that a class A (highest quality) joint has less than 15% unbonded areas compared to the total joint area [5], which indicates a certain range of tolerance for the brazed joint integrity in practice. However, experimental evaluation on thermal integrity of brazed joints is rarely found. In this paper, a preliminary experimental study on thermal performance of brazed fin/plate structures is provided. Both aluminum offset fin and metal foam fins are brazed to flat aluminum plates using controlled atmosphere brazing (CAB) process [6]. For each type of fins, different amounts of filler metal are applied to each sample which leads to a substantial difference in brazed joint sizes between two samples. The brazed samples are experimentally tested in a thermal testing facility to evaluate the heat transfer performance. Such an experimental testing method can be readily applied to any heat transfer device that is made of other materials such as copper, steel and titanium.

Brazing materials and experiments

AA3003 aluminum alloy is one of the most commonly used aluminum alloys in manufacturing heat exchangers. In this study, commercially available AA3003 plate is used as the substrate surface material. The offset fins are also made of the AA3003 alloy. Duocel® open cell metal foams (ERG Materials and Aerospace Corporation, CA) are made of AA6101 alloy with a porosity of 10 PPI (pores per inch). Filler metal used in brazing tests is AA4047 aluminum alloy which has a close to eutectic composition (Al-12wt%Si). Melting point of the AA4047 alloy is approximately 577 ^{o}C. During the assembling of the brazing samples, filler metal is sandwiched between fin and base plate in the form of foil or paste depending on the fin structure. The surface area of the AA3003 plate to be bonded with the fin structure is approximately 130mm x 100mm. The brazing sample setup is illustrated in Figure 2.

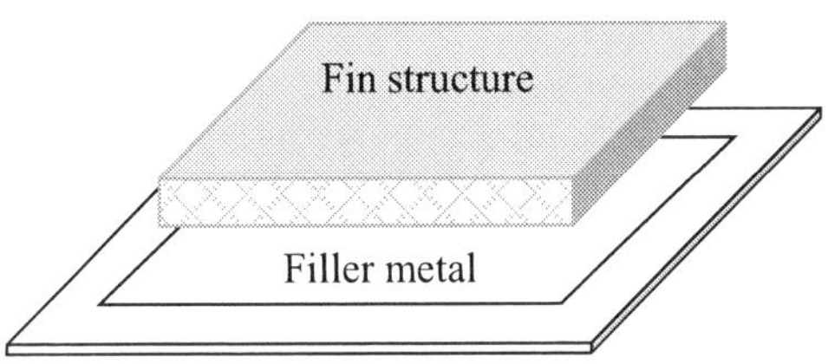

Figure 2. Sample setup for fin/plate brazing assembly

Commercially available thin foils of AA4047 are used for brazing offset fins with the base plates. A thicker foil (0.005″) is applied in Sample 1 and a thinner foil (0.003″) is used for Sample 2. For the task of brazing aluminum foam with the base plate, filler metal paste (sample provided by Solvay Fluor GmbH, Germany) is applied to ensure a good contact between filler metal and the foam surface to be bonded. The filler metal paste is a mixture of AA4047 powder, Nocolok® (a registered trademark of Solvay Fluor GmbH, Hannover Germany) flux and organic binder and carrier. A study has shown that due to the random distribution of the foam ligaments and sometimes non-perfect contact between individual ligaments with the flat substrate surface, application of filler metal in form of flowable paste is a good choice to ensure maximum contact between foam and substrate surface [7]. Another advantages of using paste in this case is that AA6xxx series alloy is used as the foam material which contains Mg as main alloying composition. The addition of Mg in aluminum may impose challenges for brazing (poisoning effect on flux) [6]. A relatively large quantity of flux in the paste mixture can assist the CAB brazing process of Mg-bearing aluminum alloy. Sample 3 was supplied with a relatively large amount of filler metal while a smaller amount of paste was applied in Sample 4. Table 1 lists material combinations for the four samples that are brazed and to be tested for thermal performance. The filler amount represents the weight of applied filler metal divided by the base plate surface area.

Table 1 Material combinations for brazing tests

Sample #	Fin structure	Filler form	Filler amount
Sample 1	Offset fin	Foil (0.005″)	0.35 kg/m^2
Sample 2	Offset fin	Foil (0.003″)	0.21 kg/m^2
Sample 3	Metal foam	paste	0.24 kg/m^2
Sample 4	Metal foam	paste	0.12 kg/m^2

(a)

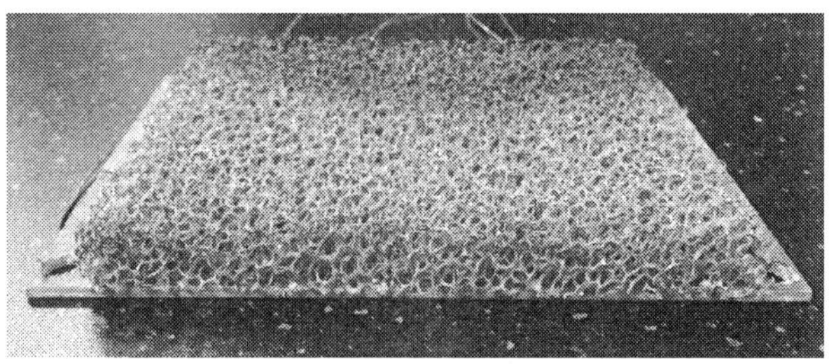

(b)

Figure 3. Brazed plate fin assembly: (a) offset fin (b) open cell aluminum foam

All samples were brazed under a typical CAB brazing conditions in a batch type lab furnace. High purity nitrogen gas flow is supplied from a liquid nitrogen storage tank. Nocolok® flux is applied to the samples to assist wetting of the filler metal on the substrate surfaces. For brazing tests the peak temperature is approximately 605 ^{o}C. Figures 3(a) and 3(b) illustrate a brazed offset fin sample and a brazed metal foam fin sample, respectively. From visual examination, all four samples show satisfactory brazing results, i.e. obvious joint formation between fin surface/ligament to the flat plate can be observed.

The cross-section images of the brazed joints from offset fin Sample 1 and Sample 2 are provided in Figures. 4(a) and 4(b), respectively. In both cases, the filler metal completely fills the gap between the bottom fin surface and the flat plate. As a result of the difference in filler metal amounts supplied, Sample 1 features a much larger joint meniscus at the wedge T location (formed by the vertical fin and the horizontal plate) than Sample 2.

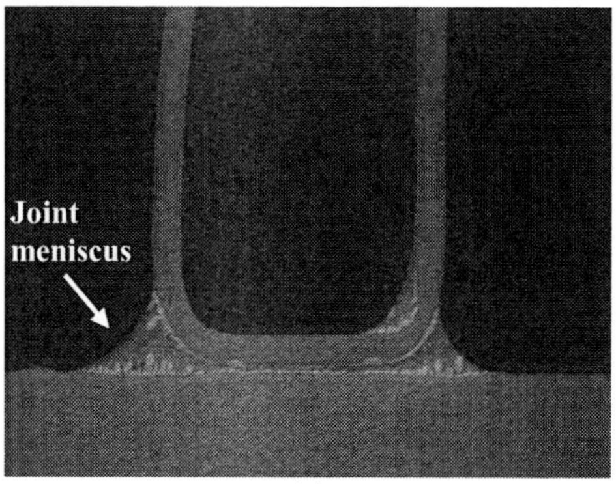

(a)

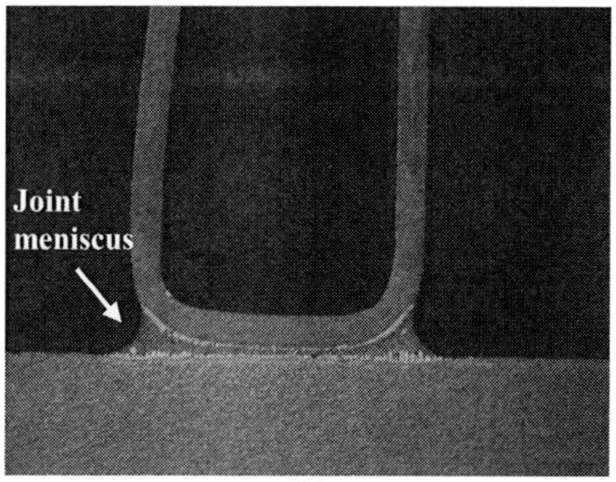

(b)

Figure 4. Cross-section images of brazed offset fin samples: (a) brazed Sample 1; (b) brazed Sample 2

Figure 5 provides close-up views of the brazed joints between foam ligaments and the flat substrate surface. It is apparent that the brazed joints of Sample 3, shown in Figure 5(a), are much larger than those of Sample 4 (Figure 5(b)) due to the difference in the amount of paste filler metal applied.

It is important to learn how such differences in fin/tube joint size can impact heat exchanger performance such as mechanical integrity and heat transfer capacity. This paper focuses on evaluating the thermal performances of the brazed samples as a function of the amount of filler metal supplied. An experimental facility for the purpose of testing thermal integrity of the brazed fin/plate assembly is designed and built in-house. A detailed description of the thermal testing facility and a discussion of results are presented in the next section.

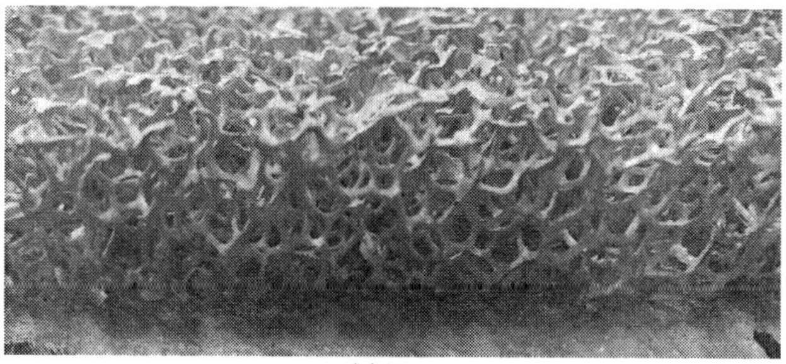

(a)

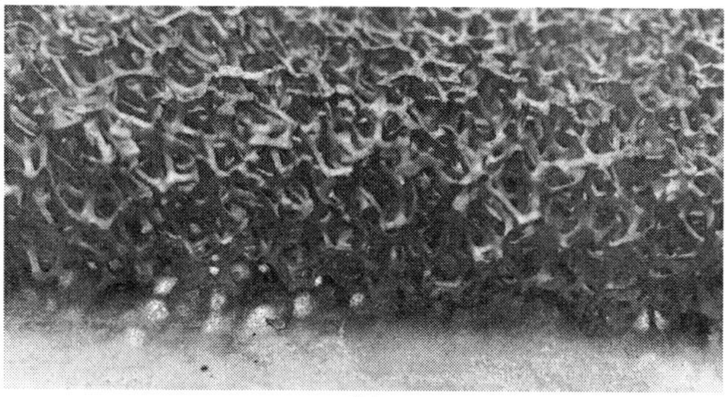

(b)

Figure 5. Brazed joint between metal foam ligaments and the base plate: (a) Sample 3; (b) Sample 4

Thermal performance of brazed samples

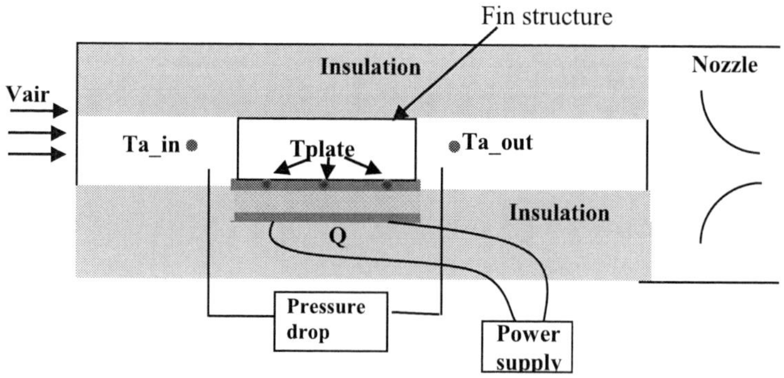

Figure 6. Thermal performance testing facility

All four brazed samples are tested for thermal performance in a wind tunnel testing facility to evaluate their heat transfer capabilities. Figure 6 illustrates a schematic of the thermal testing facility setup. A fan and a nozzle are used to generate convective flow of air through the fin structure. The brazed fin/plate assembly is attached to an aluminum block which serves as heat spreader. Thermal grease was applied between the fin plate and the aluminum block to ensure good thermal contact. A flat heater is attached to the bottom of the aluminum block. All experiments were performed at a constant heating power of 100W. Air flow rate is varied by adjusting the fan speed. Measured experimental parameters include: air inlet and outlet temperatures, air pressure drop across the fin structure, air flow rate and average temperature of the fin base plate.

To evaluate thermal performance of the brazed samples, overall thermal conductance (UA) of each sample is calculated from the heating capacity Q and the logarithmic mean temperature difference LMTD:

$$UA = \frac{Q}{LMTD}$$

The LMTD is calculated from measured temperature values as follows:

$$LMTD = \frac{\Delta T_{max} - \Delta T_{min}}{\ln \frac{\Delta T_{max}}{\Delta T_{min}}}$$

$$\Delta T_{max} = T_{plate} - T_{a\_in}$$

$$\Delta T_{min} = T_{plate} - T_{a\_out}$$

where T_{a_in}, T_{a_out}, and T_{plate} are temperatures of air flow inlet, air flow outlet and the bottom of the brazed sample plate, respectively.

The comparison of thermal performance is first made between Sample 1 and Sample 2, because material selection and geometry parameters for these two samples are all the same except for the physical dimensions of the brazed joint. The experimental results are represented by dimensionless overall thermal conductance vs. air pressure drop, as illustrated in Figure7. $(UA)_{max}$ is the maximum value of the overall thermal conductance among all the testing points for both Samples 1 and 2. In general, air side pressure drop and heat transfer increase as the air flow rate increases. A high heat transfer vs. pressure drop ratio indicates better thermal performance of the heat transfer device. From Figure 7, the difference of the UA value between Samples 1 and 2 at the same pressure drop is less than 2%. Such results indicate a negligible influence of the brazed joint size on thermal performance of the offset fin samples.

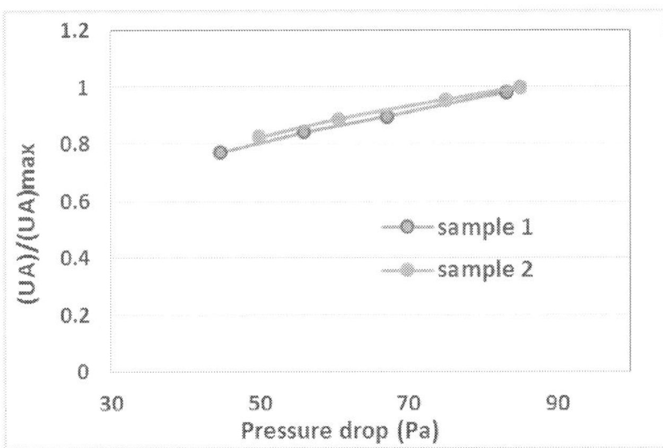

Figure 7. Thermal performance of brazed offset fin samples

A similar comparison of thermal performance is performed between Sample 3 and Sample 4. The dimensionless UA vs. air pressure drop relations for both samples are presented in Figure 8. The $(UA)_{max}$ value in this plot corresponds the maximum overall thermal conductance among all testing points for Samples 3 and 4. From Figure 8 it is found that Sample 3 shows slightly higher thermal conductance than Sample 4 at the same pressure drop (the difference is less than 5%). Such a small difference may be related to the fact that during foam brazing each individual foam ligament needs to be in direct contact with paste filler metal for bonding with the substrate. In other words, there is no continuous capillary gap to distribute the liquid filler metal as in the case of traditional serpentine fins. When a smaller amount of the filler metal paste is used for Sample 4, the percentage of unbonded foam ligaments may be higher, which may lead to a reduced thermal performance. However, slight difference in foam ligament distribution from sample to sample may also contribute to the difference shown in Figure 8.

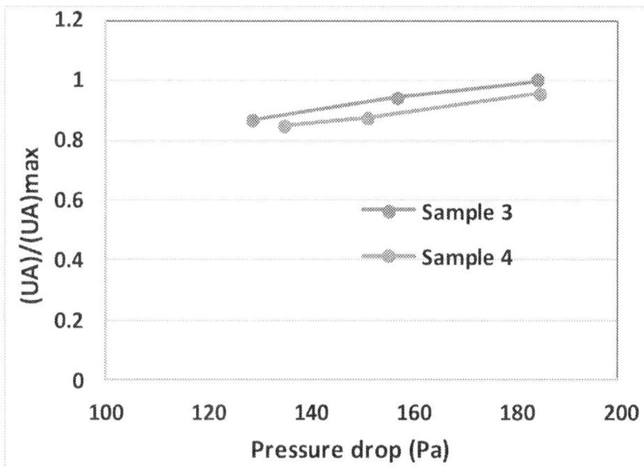

Figure 8. Thermal performance of brazed foam/plate samples

From thermal test results for both offset fin and open cell metal foam fin structures, it is found that the application of excessive amounts of filler metal does not lead to a noticeable enhancement of heat transfer performance. However, the current study is limited to a small range of tested samples. In-depth investigations are needed to study the minimum joint size (filler metal amount) required to maintain the thermal performance of a brazed heat exchanger. In addition, it is important to address questions such as how the material properties (e.g. thermal conductivity) and joint porosity will influence the thermal performance of a brazed heat exchanger. These continuing studies will be presented in future publications.

Conclusions

In this paper, brazing tests on two types of aluminum fin structures with various amounts of filler metal application are performed using controlled atmosphere brazing technology. It is demonstrated that under same brazing conditions, a larger amount of filler metal application leads to larger joint size between fin and substrate surfaces. A preliminary study of the thermal performance of these brazed fin/plate samples shows that thermal performances of these samples are not significantly affected by the size of the joints, provided that the gaps between fin and substrates are completely filled by filler metal after brazing. An optimization on the amount of filler metal application vs. thermal performance of the fin/tube assembly will be further explored using the same evaluation method introduced in this paper.

References

[1] Webb, R.L., Principles of enhanced heat transfer, John Wiley & Sons (New York, 1994).

[2] Ejlali, A. A. et al, "Application of high porosity metal foams as air cooled heat exchangers to high heat load removal systems," International communications in heat and mass transfer, Vol. 36, No. 7 (2009), pp. 674-679.

[3] SOLVAY document, "filler metal management", http://www.aluminium-brazing.com/sponsor/nocolok/Files/PDFs/31352.pdf (accessed 2014/12/05)

[4] Zhao, H., et al, 2009, "Analysis of the Fin-Tube Joints in a Compact Heat Exchanger", Heat transfer Engineering, Vol. 30, No. 12 (2009), pp. 931-940.

[5] AWS C3.7M/C3.7 , Specification for aluminum brazing, American Welding Society (Maimi, Florida, 2011).

[6] Swidersky, H. W. et al, "Fundamental of Controlled Atmosphere Aluminum Brazing", 1st International Congress Aluminum Brazing Technologies for HVAC&R Aluminum-Verlag, Düsseldorf/Germany, (2009)

[7] Sekulic D. P. et al, "Aluminum foam compact heat exchanger: brazing technology development vs. thermal performance", Proceedings Cellmet 2008 (Ed.: G. Stephani and B. Kieback), (2008), pp.5-10.

Joining of Ornament-Cut Diamond to Silver Plate Using Nano-Particle Paste of Silver by Condensed Infrared Ray Heating

Moe Mochizuki, Takahisa Yamazaki, Naoto Tanaka, Toshi-Taka Ikeshoji, Akio Suzumura
Tokyo Institute of Technology, 2-12-1, O-okayama, Meguro-ku, Tokyo, 152-8552, Japan
E-mail: mochizuki.m.af@m.titech.ac.jp, Tel: +81(0)3-5734-3236

Abstract

It is conventional for jewelry to have mounted diamonds on a silver base using claws. We attempted to assemble diamond-base structures using silver nano-ink, which is a paste mixture of nano- and micro-particles of silver dispersed in 1-decanol ($C_{10}H_{22}O$). It is difficult to fix the diamonds at first due to the poor wettability of the diamonds on the molten brazing filler. Therefore, to join the diamond and silver base plate, we applied vanadium hydride to the diamond surface. Using a gold pipe mirror, a condensed infrared ray (IR) could irradiate the pavilion because ornamentally cut diamonds reflect irradiation applied from top of the diamond and because IR cannot penetrate the silver paste. A joining test was conducted under vacuum to analyze the outgas during the cleaning process of the diamond surface using a mass spectrometer. The joined specimen was observed by an optical microscope to verify the reflection from the silver interface. In the case of irradiation to an artificial rectangular solid diamond, the diamond was fixed to a Ag-Cu eutectic alloy plate. The diamond was considered to be held by the formation of vanadium carbide (VC) and the sintered silver particles on the VC. In the case of irradiating natural ornamentally cut diamonds, the diamond was not fixed completely to the pure silver plate. However, the IR irradiation cleaned the diamond surface. At 500 °C, the silver atoms diffuse from the nano-particles, and the vanadium atoms ionize. Moreover, we successfully reconstructed and cleaned the surface of the diamond. Specifically, a joinable surface was formed at 350 °C using the condensed IR heating, which was caused by the molecular vibration on the diamond surface. Moreover, vanadium carbide was formed on the diamond surface.

Introduction

Ornamentally cut diamonds used in jewelry require a large quantity of brightness. If enough light is introduced to the reflection surfaces of the diamond (pavilion), the brightness of the diamond increases. Ornamentally cut diamonds for jewelry are held with claws. However, the claws firmly gripping the diamond prevent reflected light from reflecting brilliantly at a glance. This problem led us to investigate a method of fixing a diamond without being gripped by claws.

Diamond brightness comes from its high reflective index. If a diamond could be brightened by any other method, additional value would be added to the diamond. As one such possibility, we considered that silver mirror reflectance might effectively improve the diamond. If the silver made contact with all the diamond pavilion surfaces, shining of the diamond was observed. It was suspected that using a silver particle ink paste would enable diamonds to shine better. Silver ink paste consists of silver nano- and micro-particles. When a mixture of alcohol and the dispersed silver particles is volatilized, these nano- and micro-particles begin to be sintered and take on the metallic luster of silver. The micro-particles become recrystallized with the nano-particles filling the gaps. Subsequently, the silver particles become bulk silver that is adjacent to the diamond surface. Moreover, the silver ink paste, instead of claws, can be used to set a diamond. Liquid metal does not wet the diamond, and it causes difficulty in positioning the diamond precisely on a base using a joining process. By using silver ink paste, a diamond can be arrayed and positioned without dewetting.

Although silver rarely reacts to diamonds, silver may not adhere to the diamond surface densely when using transient metal, which forms a carbide with the diamond that enables the formation of a dense adhesion between diamond and silver, that is, active metal brazing. Active brazing filler that contains titanium as the active metal may erode the diamond surface because of its good reactivity [1] [2]. Therefore, titanium is not suitable for brazing ornamental diamonds. A brazing filler containing vanadium as the active metal instead of titanium is sufficient to join a diamond without erosion [3] [4]. Silver atoms can be deposited on vanadium carbide because they have matching properties [5] [6].

Experiment

Specimens

We produced two different setting types: specimens A and B. Figure 1 shows the schematic diagram of pre-fixed specimen A. A silver-copper eutectic alloy plate that was 0.1 mm thick was cut into 15 x 70 mm rectangles, and the appropriate amount of the silver nano-ink, which is listed in Table 1, was applied to the plate. Vanadium hydride powder was screenprinted on the plate, and an artificial single crystal diamond (111) was mounted on the ink. Subsequently, the diamond was applied with a light load so that the diamond surface contacted the paste perfectly, and the specimen was baked at 150 °C for 30 minutes in an ambient air atmosphere to degas the 1-decanol solvent, which is the binder used in the silver ink paste.

Figure 2 shows how joint specimen B was made for this study. The substrate was a pure silver base with a 2.8 mm diameter hole in the center and was polished using diamond grids. A natural ornamentally cut diamond weighing 0.118 ct with spacers was set on top of a heater plate without heating, as shown in Figure 2 (a). Vanadium powder was adhered to the pavilion surface as shown in Figure 2 (b). A silver plate with a hole was put on the spacers as shown in Figure 2 (c). Silver ink paste was poured into the hole in the silver plate. The plate and the ink were baked at 150 °C for 30 minutes in an ambient air atmosphere to degas the 1-decanol, which is the binder of the ink paste and this process adhered the plate to the diamond.

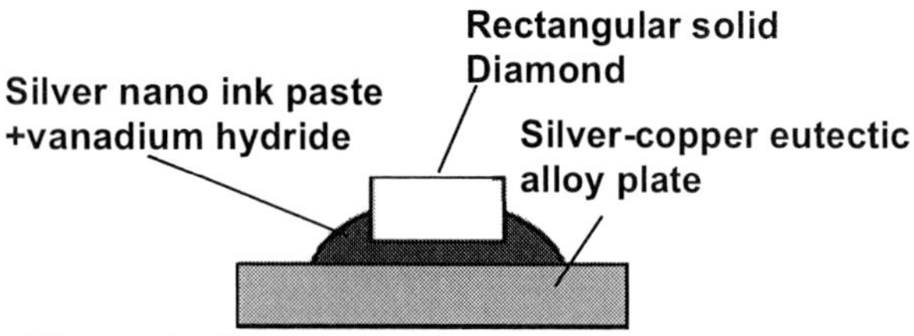

Figure 1: Schematic diagram of specimen A fixed the rectangular solid diamond to the alloy plate by ink.

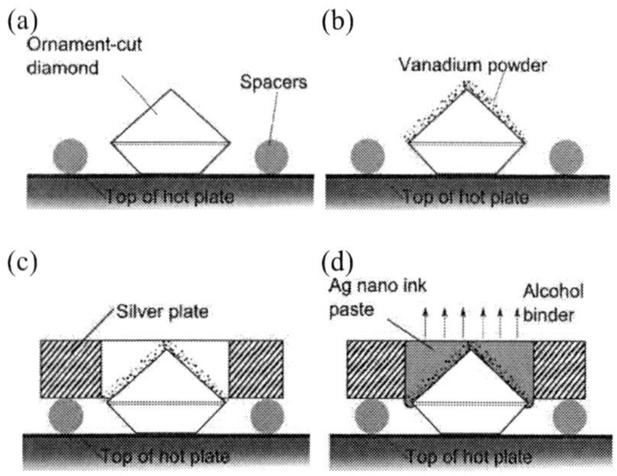

Figure 2: Schematic diagram of procedures (a) (b) (c), and (d) to produce pre-fixed specimen B

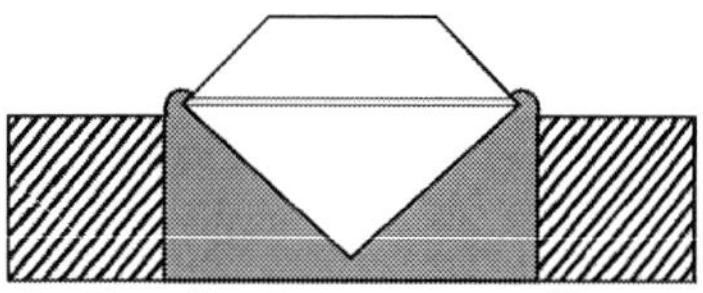

Figure 3: Pre-fixed ornament diamond on the plate by silver ink before IR heating.

Table 1: Specimen properties

	Substrate	Vanadium	Nano-ink	Diamond
A	Ag-Cu Eutectic	VHx 400 mesh	Micro-Ag: 1μm Nano-Ag: -100nm Binder: 1-decanol Nano-particle coat: amine-based Ag: 85 mass%	Artificial single crystal rectangular solid 1.4 x 1.4 mm
B	Pure silver	Same above	Same above	Natural ornament cut 0.118ct

Experimental procedure

Figure 4 shows a schematic diagram of the IR heating device used in the surface modification process. Specimens were heated by condensed IR irradiation through the specimen diamond. IR irradiation from the lamp was reflected by a gold elliptical mirror and irradiated the specimen through the quartz viewport.

In the case of heating specimen A, the focus point of the IR irradiation was at the ink paste. In the case of heating specimen B, the focus point was the top surface of the diamond. The specimen was set on a sample stage in a vacuum chamber, and the chamber was evacuated until it achieved 10^{-4} Pa; the specimen was the irradiated using the condensed IR. A thermocouple was fixed on the quartz stage to make contact with the backside of the eutectic plate in the case of specimen A or the ink paste in the case of specimen B to measure the elevated temperature without being directly irradiated by the IR.

When specimen B was heated, two different heating methods were used. One method involved only putting the sample on a sample stage (B1), and the other method involved irradiating the sample with a quartz pipe mirror (B2). We thought that the IR irradiation density on the pavilion surfaces of the ornamental diamond would not have been enough because IR reflection and dispersion occurs at the diamond. Figure 5 shows the gold pipe mirror settings.

Figure 6 shows the thermal cycles used on specimens A and B. The temperature of specimen A achieved approximately 500 °C in 60 s, and the temperature was then maintained. The temperature of specimen B1 was elevated to approximately 480 °C. In the case of B2, the temperature was elevated to approximately 380 °C over 150 s but was not maintained. The heating rate was faster for specimen A than for specimen B because IR is reflected well on the silver plate. The temperature of specimen B2 was less than that of sample B1 because the pipe mirror limited the IR irradiation that was delivered to the sample stage, so the specimen could not be heated from the sample stage.

The joining process was conducted to analyze the outgas by a mass spectrometer. Strength tests were performed using a tensile strength tester to evaluate the shear strength of the joints of specimen A. Figure 7 shows the strength test apparatus for the shear strength test. The load was applied to the diamond specimen through contact with the strength test apparatus.

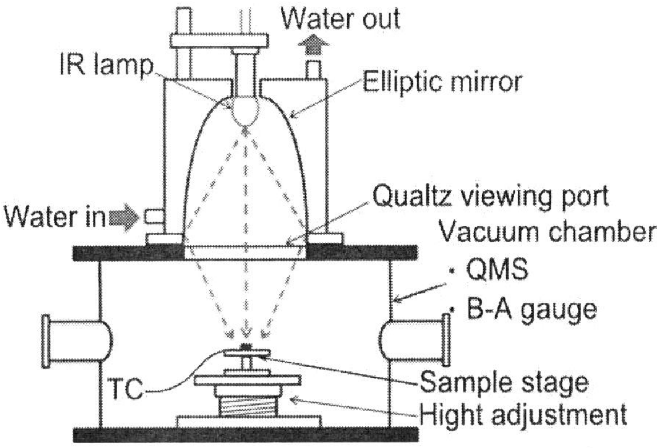

Figure 4: Schematic diagram of IR heating device.

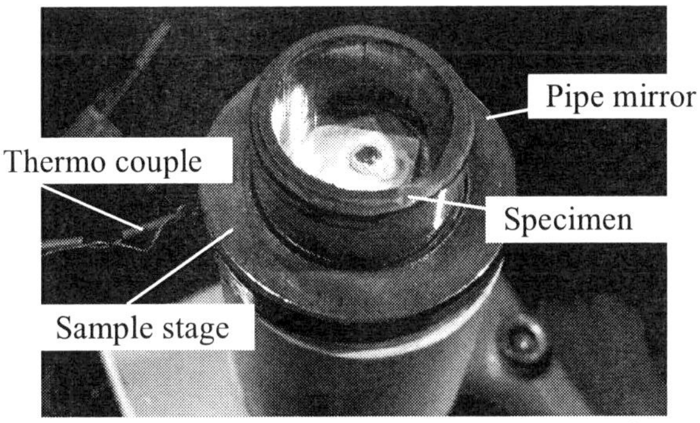

Figure 5: Setting gold pipe mirror for heating ornament cut diamond specimen B2.

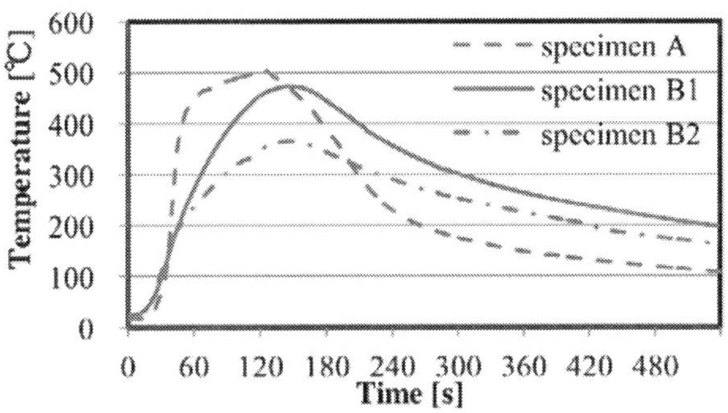

Figure 6: Thermal history of samples by irradiation heating once.

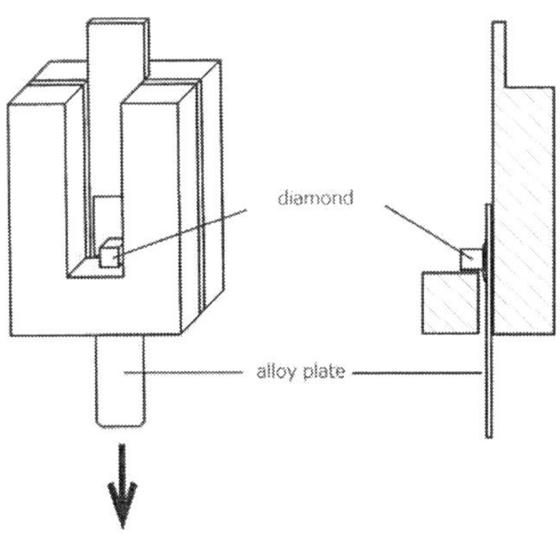

Figure 7: Setting of the joint specimen A for strength test to a tensile strength test device.

Results and Discussions

Figure 8 shows the appearance of specimen A after IR heating. The artificial diamond is placed at the center and is surrounded by a metallic luster of silver ink. Nano- and micro-particles of silver continued the recrystallization. The greatest shear strength of specimen A was 21.07 MPa. The diamond surface of specimen A was slightly modified.

Figure 9 shows the fractured diamond surfaces of specimen A after the strength test. The specimen was fractured at the interface between the diamond and the silver paste. Figure 10 shows the adherence of the diamond specimen. Metallic luster was observed at the adherence. It is possible that silver particles might be bonded to the diamond due to the formation of vanadium carbide.

Figure 11 shows the mass spectra of specimen A prior to IR irradiation and after IR irradiation. The mass spectra were analyzed using a Quadrupole Mass Spectrometer (QMS). The mass spectra of 1-decanol, which was the nano-particle binder and protective agent, was observed.

For the adherence of the diamond of specimen A, XRD analysis of the diamond was conducted. Figure 12 shows the X-ray diffraction pattern. We had hypothesized that V_2C and V_4C_3 would form, but the patterns did not appear. The silver pattern was observed and is shown in Figure 12, and the vanadium peak can be observed. The peak of the diamond (111) did not appear because the angles of the peak were not included in the data. The peaks of Ag (200), Ag (220) and Ag (311) were observed in the diffraction pattern; thus, the adherence of silver particles was considered to be polycrystalline.

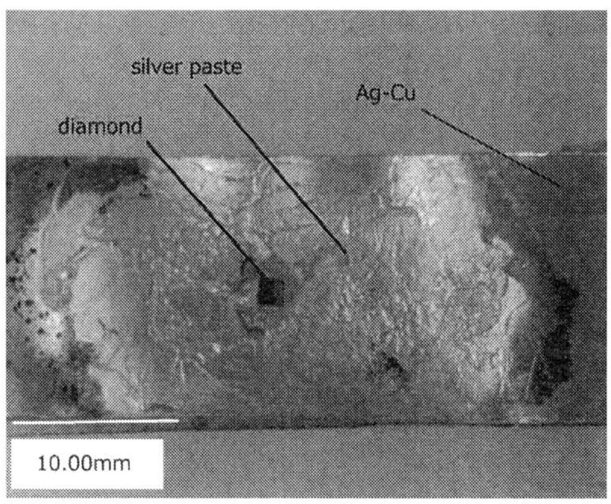

Figure 8: Appearance of joints specimen A after infrared heating.

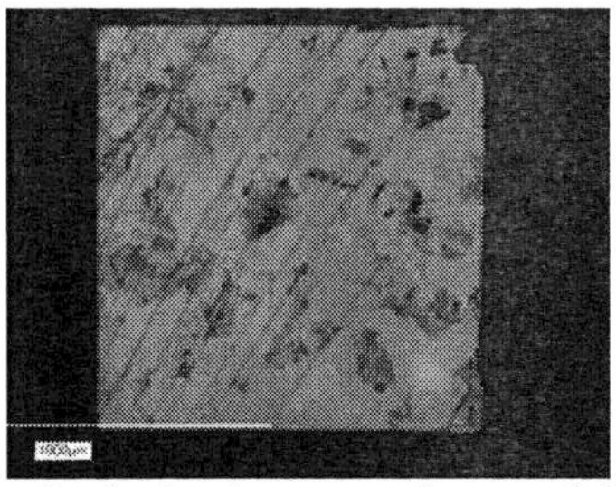

Figure 9: Optical microscope image of fractured diamond surface of specimen A after strength test.

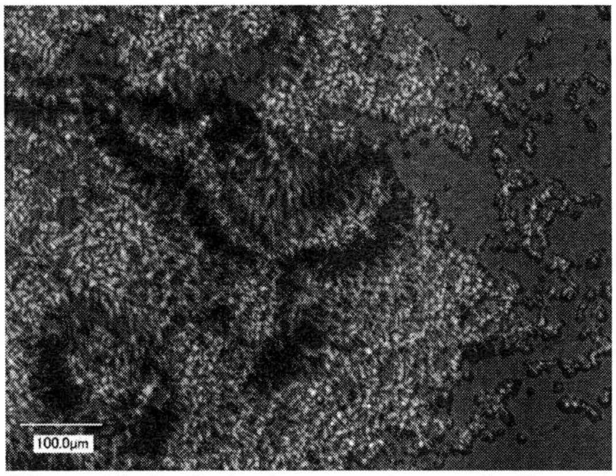

Figure 10: Optical microscope image of adherent on diamond specimen A heating thirty times.

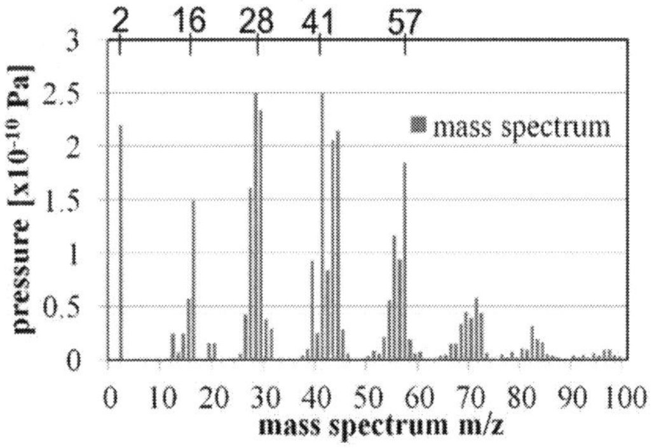

Figure 11: Mass spectrum of difference in before and after IR irradiated specimen A.

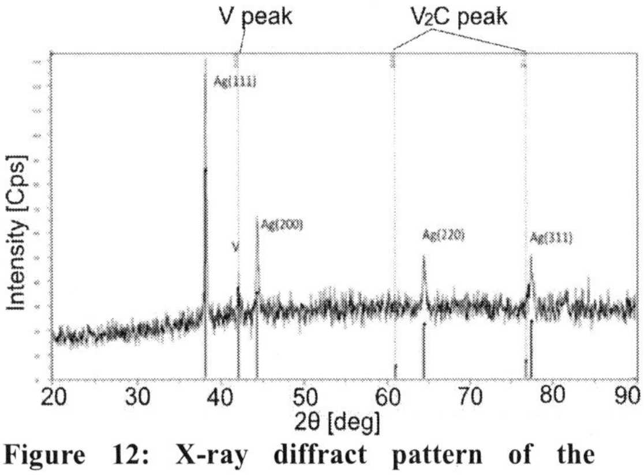

Figure 12: X-ray diffract pattern of the adherent on the diamond for specimen A

Figure 13 shows the appearance of specimen B1. The left portion of the figure is a specimen before irradiation, and the right portion of the figure is a specimen after 10 repeated doses of irradiation. Specimen B1 was not sintered. The diamond dropped slightly when a small shock was applied when it was put on the desk. The silver paste, which adhered to the diamond side, peeled off, as shown in the right portion of the figure.

Figure 14 shows the appearance of specimen B2. The left portion of the figure is a specimen before irradiation, and the right portion of the figure is a specimen after 10 repeated doses of irradiation. Specimen B2 was not sintered enough as a result of this heating cycle. However, the silver paste that adhered to the diamond did not peel off from the diamond. We assume this result meant that the diamond reacted with the vanadium or the paste.

Figure 15 shows the diamond pavilion after irradiation and the loss of adherence from paste. The adherences were not lost after ultrasonic washing with ethanol. The adherences would not be metal vanadium because vanadium powders in Figure 16 are larger than the adherences. Therefore, the adherences may be the silver particle.

Figure 17 shows the mass spectra of the differences in specimen B2 before irradiation and after irradiation. Mass spectra peaks at 14 and at 28 were observed to have high intensities. The peak at 14 is considered to be some type of hydrocarbon from the nano-particle protecting agent. The peak at 28 is considered to be CO, which consisted of carbon from the diamond and oxygen from the diamond termination. Therefore, the diamond surface was assumed to be cleaned by the IR irradiation.

The absorption peak of the nano-Ag paste was at approximately 400 nm. There was no absorption at the near infrared wavelength of 900 nm used in this experiment. For that reason, the nano-Ag paste was not able to be sintered to specimen B. The paste of specimen A was assumed to sintered by the heat transfer from the heated Ag-Cu eutectic alloy plate. When heating specimen B above 500 °C with irradiation from the condensed IR, the paste might have been able to sinter and recrystallize.

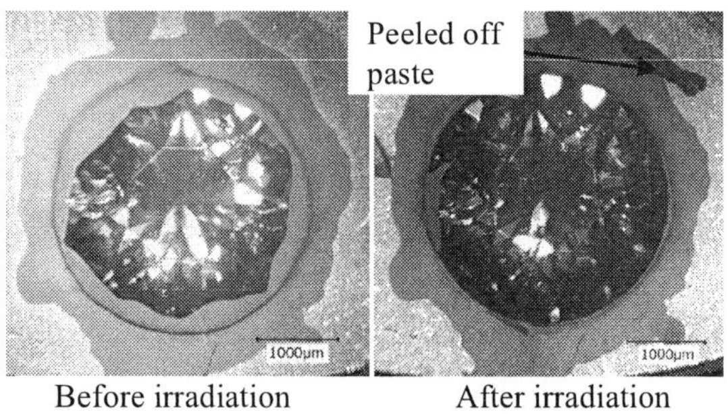

Figure 13: Appearance of specimen B1.

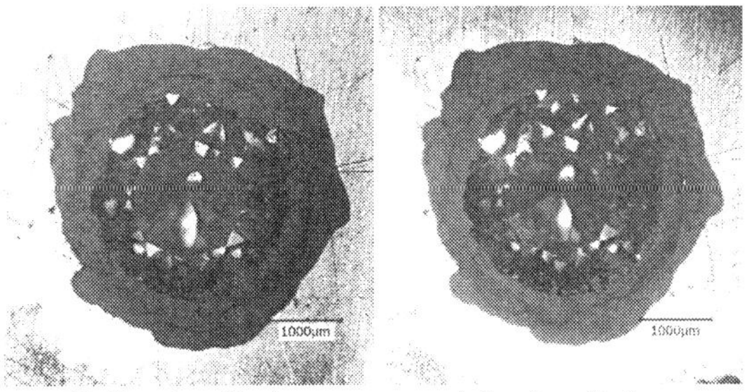

Before irradiation After irradiation

Figure 14: Appearance of specimen B2.

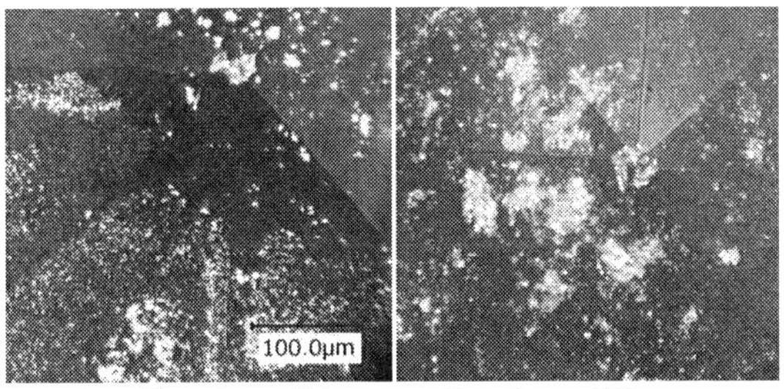

Specimen B1 Specimen B2

Figure 15: Optical microscope image of adherent on diamond specimen B1 and B2.

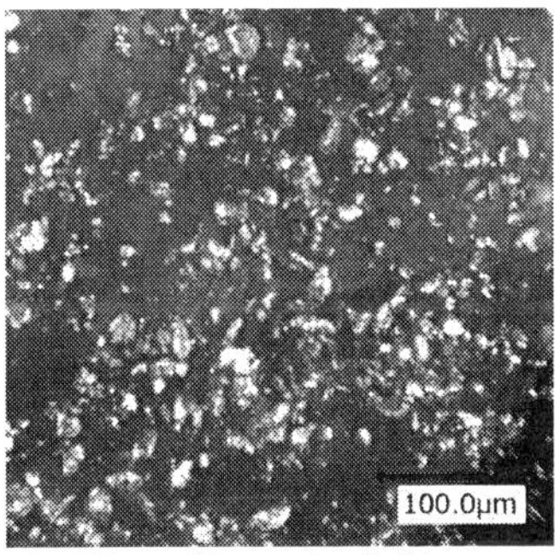

Figure 16: Vanadium hydride powders on diamond pavilion in process of Figure 2 (b).

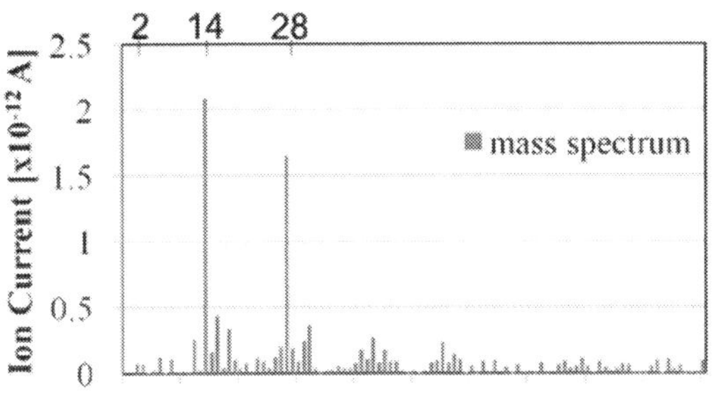

Figure 17: Differential mass spectrum of outgas of before and after irradiation from the specimen B2.

Figure 18 shows schematic diagrams of the particle sintering process. The nano-silver particles are coated in an amine-based protective agent. Silver particles were disperse at room temperature in decyl alcohol (a). Then, the alcohol was evaporated when the mixture was heated to 150 °C (b). The protective agent of the nano-silver particles became gas and dispersed into the atmosphere at 250 °C. The nano-silver particles were adhered to micro-particles, and the nano-particles were sintered when they were IR irradiated at 250 °C (c). When the silver ink was heated above 500 °C, whole particles were recrystallized.

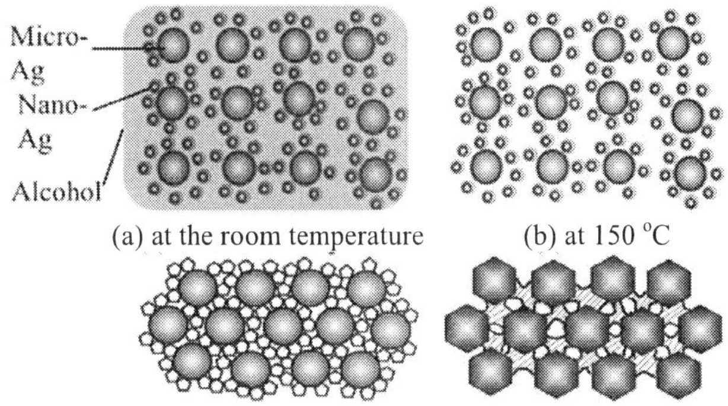

Figure 18: Sintering and recrystalizing process of particles in nano-ink.

It is suspected that vanadium carbide formed at the interface between the diamond and the sintered silver. There are hydrogen-terminated and oxygen-terminated molecules on the diamond surface. When heating the diamond by IR, H and O were removed. The C of the diamond surface was removed at temperatures above 300 °C as CO_2 and CO [6]. The vanadium contact cleaned the diamond surface, and the portion of the surface that was removed becomes vanadium carbide by reacting with vanadium. The silver atoms are deposited on the diamond's surface because properties of the silver and vanadium carbide match [8].

Conclusions

We attempted to assemble and join an ornamentally cut diamond as jewelry to a metal substrate using a silver ink paste that can adhere to the diamond. By using condensed IR irradiation applied above the diamond to the silver ink that was in contact with the diamond pavilion, the reflective surfaces of the diamond could be cleaned. Additionally, vanadium carbide, formed from a mixture of the metal and the diamond, was formed on the diamond surface due to the IR irradiation.

In the case of the artificial rectangular solid diamond, the temperature of the sample reached 500 °C due to the IR irradiation, and then the recrystallization of the silver particles occurred. The maximum shear strength was 21 MPa. In the case of natural ornamentally cut diamond, recrystallization did not occur. The specimen temperature reached 400 °C, although this temperature was not high enough to cause the silver particles to recrystallize. However, from the result of the mass spectra outgas from the specimen, the diamond surface was cleaned by the IR irradiation. Adherence that was considered to be silver particles was observed on the diamond surface at the interface with the silver ink. Additional heating may promote recrystallization of the silver ink.

Acknowledgments

The authors thank the Center for Advanced Materials Analysis of the Tokyo Institute of Technology for their support.

References

[1] Yu. V. Naidich, V. P. Umanskii, I. A. Lavrinenko, Strength of the Diamond-Metal Interface and Soldering of Diamonds, translated by E. Fujita, pp. 15-19, (1992) in Japanese

[2] James C. Sung and Michael Sung, "The brazing of diamond," Int. Journal of Refractory Metals & Hard Materials, Vol. 27, (2009), pp. 382-393.

[3] T. Ishiguro, T. Yamazaki, A. Suzumura, T-T. Ikeshoji, "Relation between fraction of metalizing surface over diamond surface and coherent interface of ionic product" *Proc. 81th Japan Welding Society*, Nagano, Japan, September, 2007, pp. 90-91.

[4] T. Yamazaki, S. Shimizu, A. Suzumura, T-T. Ikeshoji, S. Yamada, "Wettability improvement of diamond in soldering by Sn-In-Al-Si-V" *Proc. 81th Japan Welding Society*, Fukui, Japan, August, 2005, pp. 228-229.

[5] A. Suzumura, T. Yamazaki, K. Takahashi, T. Onzawa, "Solidification Phenomena and Bonding Strength at the Interface of Diamond and Active-Metal-Brazing-Filler" Quarterly Journal of the Japan Welding Society, Vol. 12, (1994), pp. 509-514.

[6] T. Yamazaki, A. Suzumura, "Role of the reaction product in the solidification of Ag-Cu-Ti filler for brazing diamond", Journal of materials science, Vol. 33, pp. 1379-1384, (1998)

[7] Y. Sato, "A Perspective on Diamond Surface Studies" Journal of the Surface Science Society of Japan, Vol. 19, (1998), pp. 407-411.

[8] T. Hokazono, K. Maekawa, K. Yamasaki, "Laser sintering characteristics of Ag nanoparticles on an insulation substrate" *Proc. 15th The Japan Society of Mechanical Engineers*, Tokyo, Japan, March, 2009, pp. 477-478.

Wide Gap Braze Repairs for Gas Turbine Components

C. Hawk and S. Liu
Colorado School of Mines Center for Welding, Joining and Coatings Research, Golden, CO, United States

S.C. Kottilingam
GE Power and Water, Greenville, South Carolina, USA

Abstract

Superalloy gas turbine components are prone to develop cracks while in service. The cost of replacing these parts is expensive, which leads to many cracks being repaired on site. Welding is an unattractive repair method due to the poor weldability of superalloys and the large heat input near the joint that alters the underlying microstructure. Brazing is a viable option because the microstructure surrounding the crack is not affected by the braze whereas welding would alter the surrounding microstructure. Also, the braze composition can be customized to generate the desired microstructures and properties. This work investigates the effect of varying braze composition and brazing temperature on joint strength and ductility. To achieve a microstructure with the desirable mechanical properties, a superalloy powder was combined with a commercial braze powder at varying weight percents. All samples containing BNi-2, which contains boron and silicon as melting point depressants, form a eutectic constituent in the braze joint. This eutectic constituent is harder than the surrounding matrix and is the primary cause of joint failure. Samples containing BNi-5, which contains only silicon as the melting point depressant, form a dispersed eutectic constituent along the grain boundaries. More voids were observed at the base metal/braze alloy interface with braze alloys containing BNi-5. Stress concentration surrounding the voids often causes a crack to propagate and coalesce with other voids, leading to joint failure.

Introduction

Gas turbine components, such as blades and vanes, may develop cracks while in service. Cracks undermine the structural integrity of the part, which inherently reduces the operating efficiency and endangers other components within the turbine [1]. Components are usually made with a nickel or cobalt superalloy, containing alloying elements such as Mo, Ta, and W [1]. Replacement costs are often too expensive, so damaged components are usually repaired [2]. Welding is not a desirable repair method because the weld and heat affected zone are susceptible to cracking [3]. Brazing is a practical repair solution because the braze alloy and associated processing steps do not damage the base metal [3, 4, 5]. Also, customizing the braze composition is relatively easy in order to achieve the desired mechanical properties.

This study examines the effects of braze alloy composition on the quality of the resulting braze joint. The braze alloy is prepared by mixing a commercial braze powder, either BNi-2 or BNi-5, with superalloy powder, MAR-M 247. The commercial braze alloys have a significantly lower melting temperature than the superalloy powder. When the braze alloy reaches the brazing temperature, the liquid commercial braze will suspend the solid superalloy powder. When the braze solidifies, the superalloy is dispersed in the commercial braze matrix. This method is referred to as activated diffusion healing (ADH) [1, 7]. A schematic of ADH is illustrated in Fig.1.

The superalloy powder serves three purposes. (i) Superalloys contribute alloying elements which strengthen the joint. (ii) The superalloy serves as an anchor onto which the commercial braze liquid adheres. (iii) The superalloy particles act as a boron and silicon sink, as seen in Fig.1. The melting point depressants from the commercial braze powder will diffuse into the superalloy particles allowing the commercial braze liquid to isothermally solidify [1, 9].

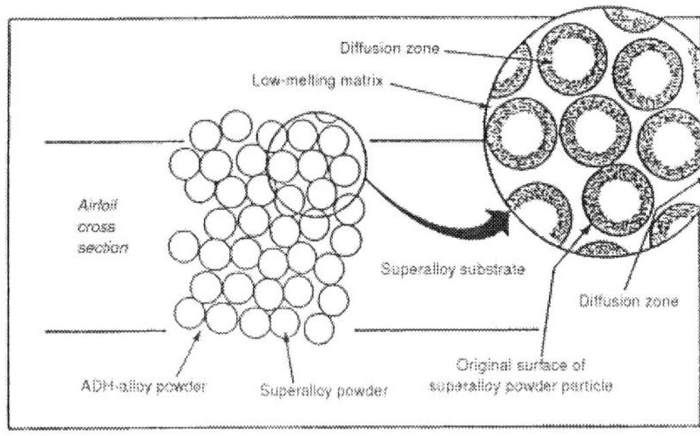

Figure 1: In the activated diffusion healing (ADH) repair process, a superalloy powder is mixed with a filler metal. The joint shows the superalloy with the filler metal before brazing. The inset shows the two powders after brazing. Source: ref 4.

Wide gap brazing is difficult because the braze can no longer rely solely on capillary action [8]. Improper filling of the joint will leave a deficit of braze alloy in the center of the joint. For example, if too much braze alloy is added, the excess braze will spread to the base metal outside of the joint, leaving the joint under filled in the center. The superalloy powder provides extra surface for the braze powder to wet causing complete filling of the braze joint.

The braze retains its structural integrity at high temperature due to transient liquid phase (TLP) bonding. TLP bonding is dependent on melting point depressants (MPD) such as boron or silicon diffusing out of the commercial braze. When MPDs diffuse from the braze alloy into the base metal and superalloy particles, the melting temperature of the braze alloy increases, allowing for the joint to solidify isothermally [9, 10, 11]. Isothermal solidification of the joint is critical; if a different solidification mechanism takes place, then there is a shift in the compositional gradient causing a eutectic constituent to form [12]. The high hardness and brittleness of the eutectic constituent causes the joint to fail under a lower applied stress compared to those where isothermal solidification was the dominant solidification mechanism [5, 13]. The eutectic formation can be attributed as the cause of most joint failures.

Experimental Procedures

Three powders were used to create the various braze alloy compositions: superalloy MAR-M 247, commercial BNi-2, and commercial BNi-5. Six compositions were created by mixing 40, 50, and 60 weight percent of the commercial braze powders with a balance of MAR-M 247. Powders were mixed using a LabRam Ultrasonic Acoustic mixer. Table 1 shows the compositions of the individual powders.

Table 1: Compositions of the nickel based powders and René 108 alloy given in weight percent.

Element	MAR-M 247	BNi-2	BNi-5	René 108
B	-	3.0	-	-
Al	5.0	-	-	5.5
Si	-	4.2	10.0	-
Ti	1.0	-	-	0.7
Cr	8.0	7.0	19.0	8.4
Fe	-	3.0	-	-
Co	10.0	-	-	9.5
Ni	Balance	Balance	Balance	Balance
Mo	-	-	-	0.5
Hf	1.0	-	-	1.5
Ta	3.0	-	-	3.0
W	10.0	-	-	9.5

Superalloy René 108 was used as the base metal and the composition is also shown in Table 1 [14]. Small plates were machined using wire electric discharge machining (EDM) with dimensions of 76 x 25 x 2 mm with a 1.5 x 1.5 mm groove machined down the length of the plate. Plate surfaces were ultrasonically cleaned in acetone, sandblasted with NicroBlast grit, and then cleaned in acetone. The braze alloy was mixed with Vitta Binder BrazeGel to form a paste which was applied to the groove. The plates were brazed under vacuum (10^{-6}-10^{-7} torr) following the brazing cycle illustrated in Fig. 2. The plates were preheated to remove excess moisture and to vaporize the binder. Two different brazing temperatures were used, 1200°C (2192°F) and 1232°C (2250°F). After the brazing hold, samples were cooled to the diffusion hold where the joint is allowed to homogenize. Finally, the joint undergoes an aging step.

Metallography was performed on transverse cross-sections of the specimens following ASTM Standard E3-11 [15]. Samples were polished with 0.05 μm alumina and Vickers microhardness indentations were measured following ASTM Standard E384-11[16]. Samples were etched using glyceregia before obtaining light-optical micrographs. Following ASTM Standard E568-11, volume fractions of the microstructural features were calculated for each composition using Image J software [17].

At least five bend coupons with dimensions 25 x 2 x 1.5 mm were machined out of the braze plates. Metallography was performed on the face of each sample and macrographs were collected before and after bend testing. Four-point bend tests were performed on the coupons with a crosshead speed of 5 x 10^{-4} in·min^{-1} under room temperature conditions. The ultimate bend strength and angular deflection were calculated according to ASTM Standard D6272-10 [18].

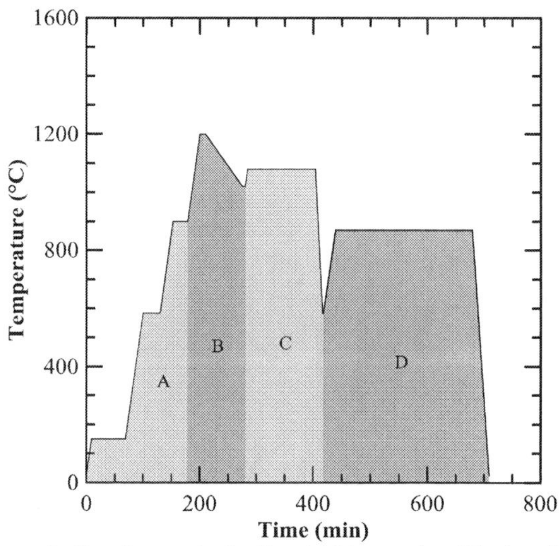

Figure 2: Brazing cycle for René 108 samples. The brazing cycle is made up of 4 sections. A.) Preheat, removal of water and the binder. B.) Brazing. C.) Diffusion hold. D.) Aging hold.

Results

Microstructure

Braze alloys containing BNi-2 demonstrated varying microstructures and changes in the relative volume fraction of constituent phases. The distribution of these phases is important and will be addressed later. The microstructural features that are present in the joint include: a γ matrix, Ni-rich eutectic constituents, Cr-rich phases, large W-rich intermetallic phases, small W-rich intermetallic phases, and small Ta-rich intermetallic phases. Figure 3 illustrates the joint microstructure of 60 wt.% BNi-2 brazed at 1200°C (2192°F). The γ matrix showed dendritic solidification as shown in Fig. 3. Boron and silicon diffused from the liquid BNi-2 into the MAR-M 247 particles, allowing for the BNi-2 to isothermally solidify.

Small W-rich intermetallic phases reside in the center of the matrix grains. The excess liquid that did not isothermally solidify cooled quickly, leaving the Ni-rich eutectic constituent, suspected to be a nickel boride [5, 19]. Figure 4 shows the matrix with the small W-rich intermetallic phases, larger W-rich intermetallic phases, and small Ta-rich phases that form near the eutectic constituent.

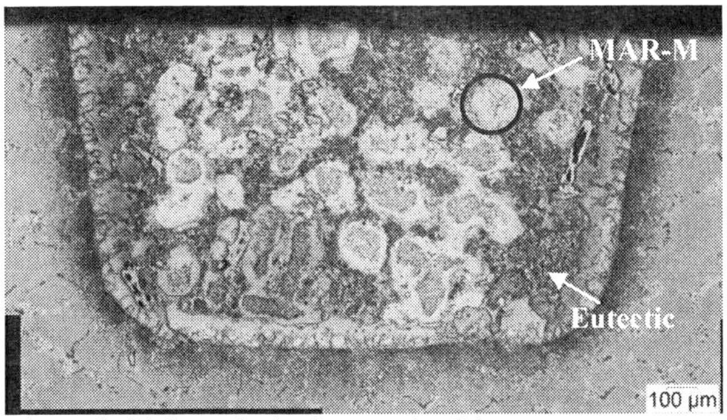

Figure 3: Light optical micrograph of the transverse cross-section of the braze joint containing 60 wt.% BNi-2 as-brazed at 1200°C (2192°F).

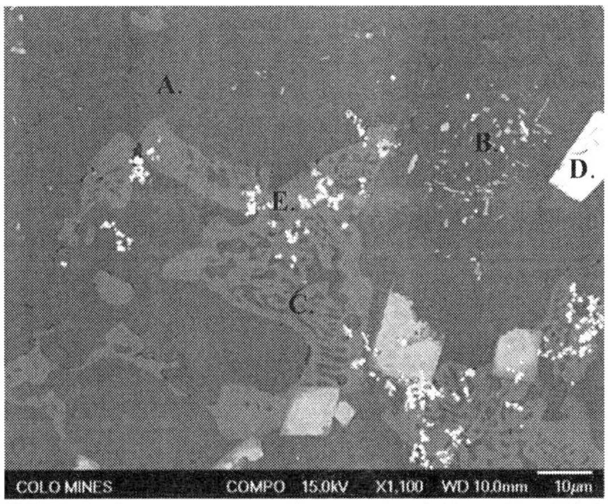

Figure 4: Backscatter SEM micrograph of 40 wt.% BNi-2 as-brazed at 1200°C (2192°F). A.) γ Matrix. B.) Small W-rich phase. C.) Eutectic Constituent. D.) Large W-rich phase, E.) Ta-rich phase.

Braze alloys containing BNi-5 had a finer joint microstructure than joints containing BNi-2. Therefore, braze alloys containing BNi-5 are expected to exhibit higher bend strengths than joints containing BNi-2. Though the microstructure is finer, there are more voids present as seen in Fig. 5. These voids were found at the base metal/braze alloy interface and within the joint body. The voids at the interface are mostly due to non-wetting at the base metal interface. In the body of the joint, the voids are mostly due to solidification shrinkage as indicated by the circle in Fig. 5. Voids are stress risers, so joint quality is expected to degrade with the appearance of voids.

Figure 6 illustrates the microstructural features that develop in BNi-5 joints. The microstructural features include a γ-Ni matrix, W-rich intermetallic phases, and a phase both rich in Ni and Si, hereby referred to as the Si-rich phase. The Si-rich phase forms near the γ matrix grain boundaries. Since silicon has a high solubility in nickel, it is possible that most of the silicon diffused into the γ matrix during isothermal solidification, and the remaining liquid BNi-5 solidified into a dispersed nickel silicide eutectic constituent. This is a possible reason why the eutectic constituent in BNi-2 differs from the eutectic constituent in BNi-5.

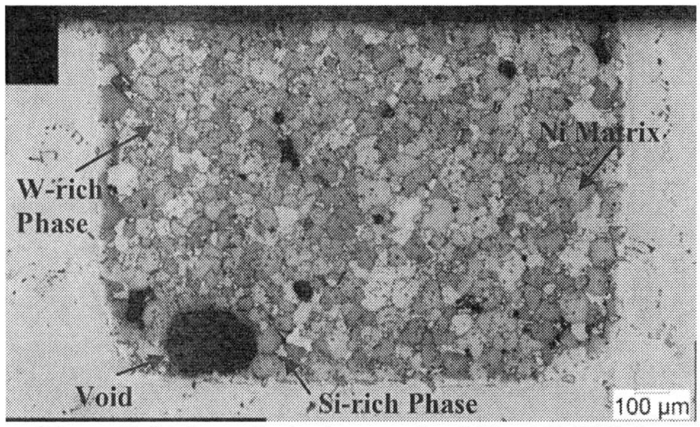

Figure 5: Light optical micrograph of 60 wt.% BNi-5 as-brazed at 1200°C (2192°F).

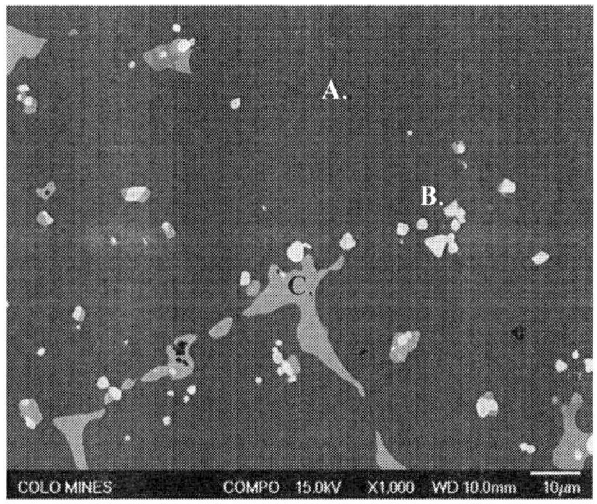

Figure 6: Backscatter SEM micrograph of 40 wt.% BNi-5 as-brazed at 1200°C (2192°F). A.) Matrix phase. B.) W-rich phase. C.) Si-rich phase.

Phase Volume Fraction

The strength of the joint is dependent on the volume fraction of the phases present and the distribution of those phases within the joint. In the case of BNi-2, the eutectic constituent is the primary cause of joint failure. The volume fraction of the eutectic phase and the distribution of it throughout the joint determines the bend strength. Samples containing 50 wt.% BNi-2 were found to have the highest volume fraction of the deleterious eutectic constituent. As such, these samples are expected to have the lowest bend strength shown in Fig.7.

Joints containing 40 wt.% BNi-2 had the least amount of the eutectic constituent, suggesting that these joints will perform the best during bend testing. Figure 7 would suggest that joints brazed at 1200°C (2192°F) should have the best bend performance, but the graph does not take the distribution of these phases into account. The distribution of the eutectic constituents heavily impacts joint quality.

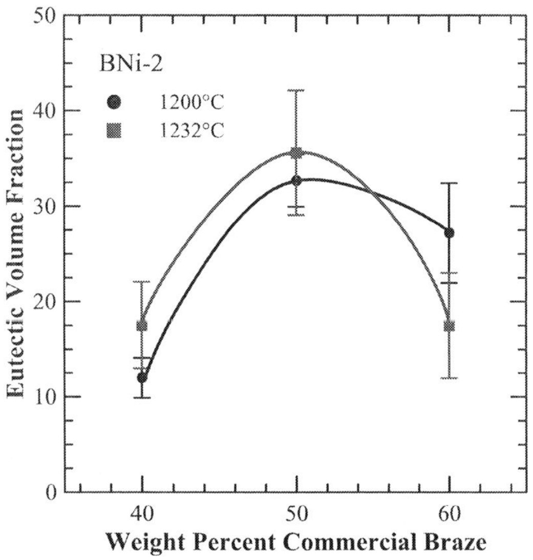

Figure 7: Volume fraction of the eutectic constituents found in samples containing BNi-2 as the commercial braze powder.

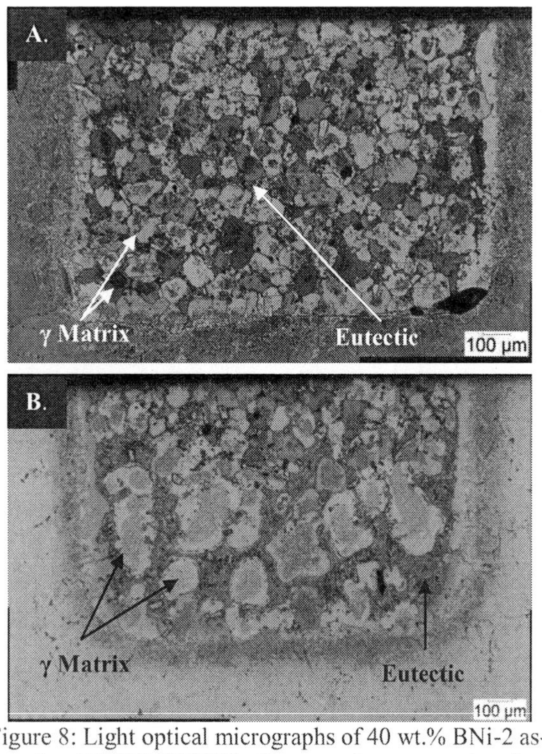

Figure 8: Light optical micrographs of 40 wt.% BNi-2 as-brazed at 1232°C (2250°F). Cross sections were taken from different sections of the same braze plate. A.) Exhibits an even distribution of the phases. B.) Exhibits uneven distribution of the phases with large grains at the bottom and small grains at the top.

Volume fractions of phases varies throughout the joint in samples containing BNi-2. Figure 8 shows two cross sectional areas cut from different regions of the same braze plate. Figure 8A shows an even distribution of the γ matrix, with similar grain sizes and a small amount of the eutectic phase distributed evenly throughout the joint. This distribution of the phases leads to a higher bend strength. Figure 8B was taken from the same braze plate, but from a different section. The microstructure developed an uneven distribution of the γ matrix and eutectic phases in the joint. The bottom of the joint is dominated by larger γ matrix grains surrounded by the eutectic constituent. The top of the joint has small γ matrix grains with small amounts of the eutectic constituent in the grain boundaries. The uneven distribution of these phases will adversely affect the structural integrity of the braze.

Variations in the phase volume fraction seen in the BNi-2 samples are unlikely to be caused by differences in cooling. Braze plates containing BNi-2 and BNi-5 were brazed simultaneously in the vacuum furnace, but only BNi-5 exhibited a consistent distribution of phases. This suggests that the varying distribution of phases in BNi-2 is caused by composition, not cooling. Additionally, sample plates are only 2 mm thick, which is not thick enough to maintain the drastic thermal gradient required to cause the observed variations in microstructure.

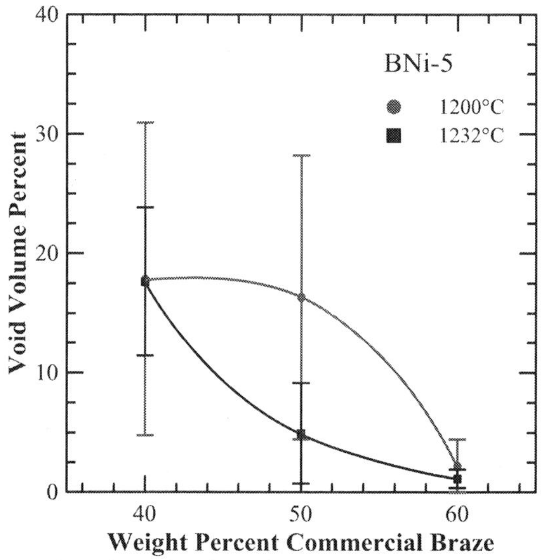

Figure 9: Volume fraction of the voids present in samples containing BNi-5 as the commercial braze powder.

Failure in BNi-5 braze alloys is highly dependent upon the voids present. The amount of voids present in the joint is shown in Fig. 9. For 40 and 50 wt% BNi-5, the voids are mostly due to lack of fusion to the base metal/braze alloy interface. For that reason, 40 and 50 wt.% BNi-5 brazed at both temperatures will not be discussed in the bend testing results. When the brazing temperature is increased to 1232°C (2250°F), wetting of the interface is increased resulting in fewer voids. Therefore, samples containing 60 wt.% BNi-5 brazed at 1232°C (2250°F) are expected to have higher strengths than the other braze alloys.

Microhardness

Vickers microhardness indentations were taken on each sample. The average hardness values for BNi-2 are shown in Table 2. The eutectic constituent and the large W-rich intermetallic phases exhibit significantly higher hardness values when compared to the matrix. The discrepancy in hardness between these phases and the matrix means that uniform deformation cannot take place and a crack may be expected to propagate along the eutectic/matrix interface. In fact, strain was observed to build up along the eutectic constituent using digital image correlation (DIC) before fracturing.

Table 2: Average Vickers microhardness values for the base metal and the phases found in braze alloys containing BNi-2.

Phase	Average Hardness (HV)
René 108	402 ± 27
Interface	416 ± 13
Matrix	420 ± 47
Eutectic	597 ± 89
Intermetallic	757 ± 43

Microhardness values for BNi-5 samples were consistent throughout the joint except for areas surrounding voids. Figure 10 shows the microhardness values of 40 wt.% BNi-5. The average hardness of the braze alloy was 400 Vickers hardness; similar to the microhardness of the René 108. Sudden decreases in hardness are strongly correlated with void formation.

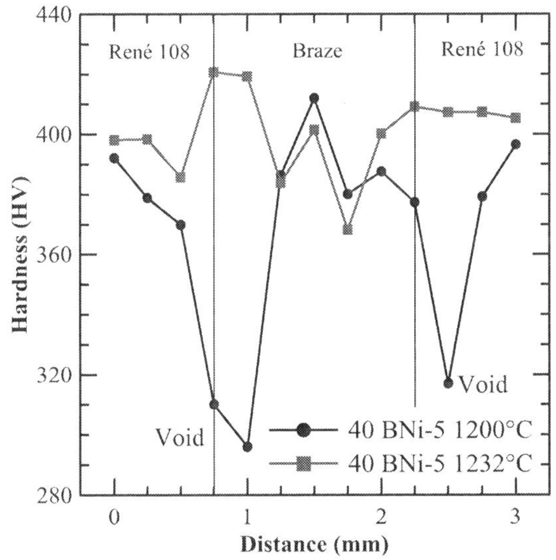

Figure 10: Average Vickers microhardness across the 40 wt.% BNi-5 braze joint. The major decrease in hardness is due to voids.

Bend Testing

Four-point bend tests were performed on bend coupons as seen in Fig. 11. Micrographs of the joints were taken before and after bend testing to conclude how the microstructure affected joint properties. All the samples that contained BNi-2 failed in the eutectic constituent. Cracks appear to propagate through thin films of the eutectic constituent near the base metal/braze alloy interface. Figure 12 shows the microstructure of a joint before and after bend testing. Figure 12A shows that the eutectic constituent was thicker on the left side of the joint than on the right. Since the eutectic constituent is significantly harder than the surrounding matrix, cracks appear to propagate through the thin eutectic constituent on the right as shown in Fig. 12B. Cracks avoid propagation through grains or intermetallic phases, as seen in Figure 12B.

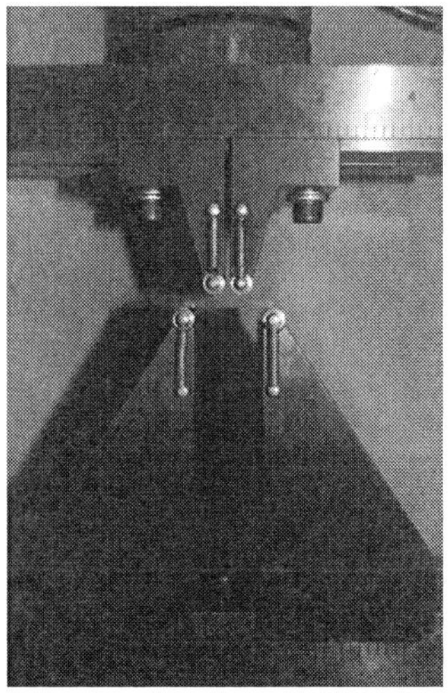

Figure 11: Four-point bend test grips.

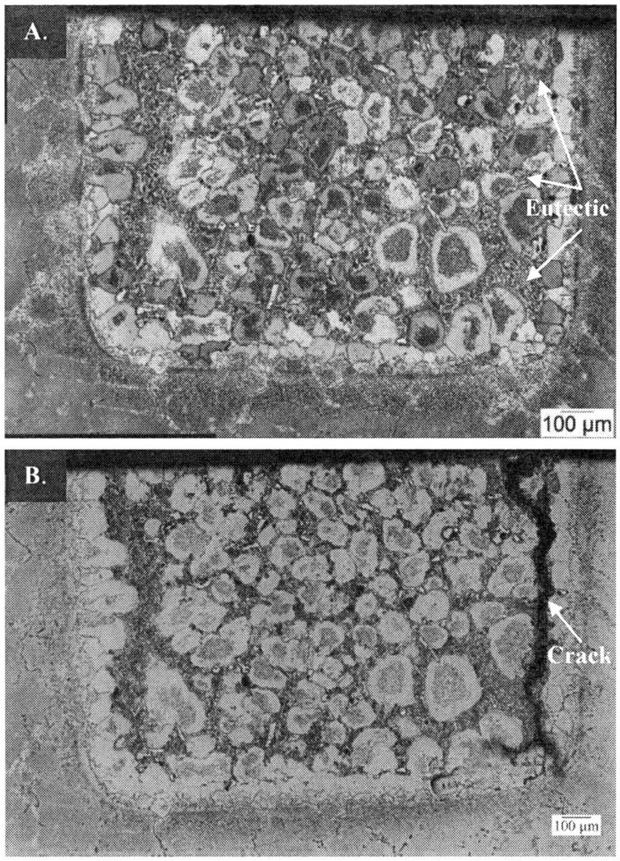

Figure 12: Light optical macrographs of 50 wt% BNi-2 as brazed at 1200°C (2192°F). A.) Before bend testing. B.) After bend testing.

The strengths of the braze can be related to the eutectic volume fractions and the distribution of those phases throughout the joint. Table 3 shows the average strength and angle of deflection for each braze alloy. Joint strength increases with brazing temperature, corresponding to the distribution of the eutectic phases. As expected, 40 wt.% BNi-2 brazed at 1232°C (2250°F) performed the best with the highest bend strength and angle of deflection because it had the least amount of the eutectic constituent and had an even distribution of the eutectic constituent and γ matrix. As expected, 50 wt.% BNi-2 performed the poorest for both braze temperatures due to the large volume fraction of the eutectic constituent.

Table 3: Average strength and average angle of deflection of samples containing BNi-2.

1200°C (2192°F)	40 wt.%	50 wt.%	60 wt.%
Strength (MPa)	587 ± 29	450 ± 59	610 ± 23
Angle (°)	3 ± 0.4	2 ± 0.2	4 ± 0.1
1232°C (2250°F)	**40 wt.%**	**50 wt.%**	**60 wt.%**
Strength (MPa)	658 ± 101	532 ± 36	606 ± 29
Angle (°)	4 ± 0.7	3 ± 0.4	4 ± 0.2

Joints containing BNi-5 all failed at the base metal/braze alloy interface. Instead of the braze alloy deforming, the base metal deformed. Slip bands were formed in the René 108 near the crack propagation. Cracks propagated through the interface to join with other voids. Figure 13 shows the micrograph of a joint with a large void in the bottom left hand corner. The crack then propagated down the interface to join with that large void along with other small voids along the way.

Table 4: Average strength and average angle of deflection of samples containing BNi-5.

1200°C	40 wt.%	50 wt.%	60 wt.%
Strength (MPa)	170 ± 5	244 ± 40	713 ± 86
Angle (°)	12 ± 0.4	4 ± 2	5 ± 1
1232°C (2250°F)	**40 wt.%**	**50 wt.%**	**60 wt.%**
Strength (MPa)	435 ± 97	705 ± 182	745 ± 99
Angle (°)	2 ± 0.6	6 ± 3	5 ± 1

As shown in Table 4, the strength of the braze joints increases with temperature. Strength also increases as weight percent of BNi-5 increases, as expected from the volume fraction analysis, because there are fewer voids. BNi-5 joints exhibited higher strengths and bend angles than BNi-2 joints, most likely due to the smaller grain sizes and even distribution of the phases.

The strength of the joints increased with increasing angles of deflection as show in Fig.14. Normally, strength and ductility are inversely related. Since these joints are brittle and can only sustain small amounts of deformation, the strain-ductility relationship is proportional. Note that the discrepancy of data points is due to large voids or lack of fusion and are considered outliers.

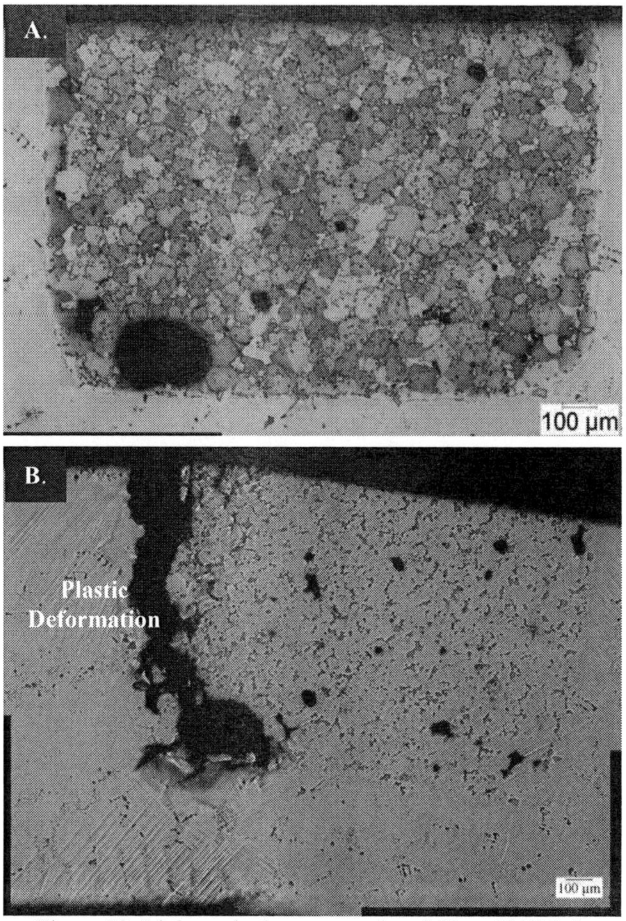

Figure 13: Light optical macrographs of 60 wt.% BNi-5 brazed at 1200°C. A.) Before bend testing. B.) After bend testing.

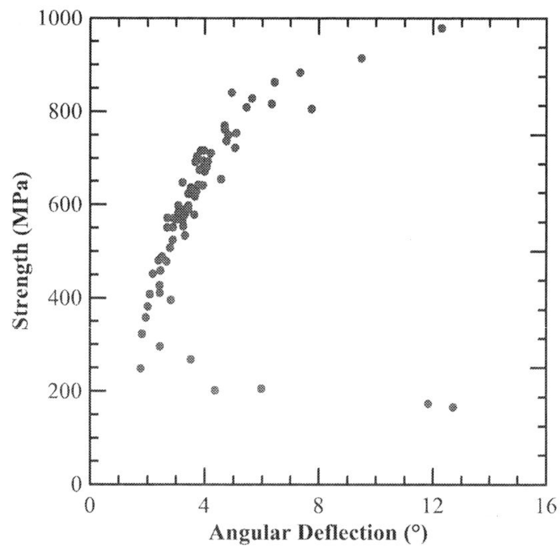

Figure 14: Ultimate bend strength and angular deflection at fracture of each bend coupon.

Conclusions

Joints containing BNi-2 are highly dependent on the volume fraction and distribution of the eutectic constituent. Brazes exhibit higher bend strengths with a low volume fraction of the eutectic constituent, distributed evenly throughout the joint. Braze alloys containing 40 wt.% BNi-2 perform better than the other braze alloys because they contain the least amount of the eutectic constituent. Joints brazed at 1232°C (2250°F) performed better than those brazed at 1200°C (2192°F) because the eutectic constituent was evenly distributed throughout the joint, resulting in 40 wt.% BNi-2 brazed at 1232°C (2250°F) as having the best combined strength and ductility.

Joints containing BNi-5 are highly dependent on the volume fraction of voids. Cracks propagated along the base metal/ braze alloy interface, before coalescing with larger voids. Increasing the temperature to 1232°C (2250°F) strengthens the joint because the amount of voids decreases. Similarly, increasing the weight percent of BNi-5 increases the strength because the amount of voids also decreases. Resulting in 60 wt.% BNi-5 brazed at 1232°C (2250°F) as having the best strength and ductility.

Optimization of processing parameters and braze alloy composition can strongly reduce the eutectic constituents and voids. Work is ongoing to control the distribution of the eutectic constituent in order to increase the bend strength of the joint.

Acknowledgments

The authors would like to thank GE Power and Water for supporting this research. The authors would also like to thank NSF - Center for Integrative Materials Joining Science for Energy Applications for the opportunity to perform this work.

References

[1] X. Huang and W. Miglietti, "Wide Gap Braze Repair of Gas Turbine Blades and Vanes—A Review," *J. Eng. Gas Turbines Power*, vol. 134, no. 1, p. 010801, 2012.

[2] W. Miglietti, "Wide Gap Diffusion Braze Repairs of Nozzel Segments Cast from FSX-414 Co-based Superalloy," in *International Brazing and Soldering Conference*, 2006, pp. 377–384.

[3] M. a. Arafin, M. Medraj, D. P. Turner, and P. Bocher, "Transient liquid phase bonding of Inconel 718 and Inconel 625 with BNi-2: Modeling and experimental investigations," *Mater. Sci. Eng. A*, vol. 447, no. 1–2, pp. 125–133, Feb. 2007.

[4] J. H. G. Mattheij, "Role of brazing in repair of superalloy components advantages and limitations," vol. 1, no. August, pp. 608–612, 1985.

[5] a. Ghoneim and O. a. Ojo, "On the influence of boron-addition on TLP bonding time in a Ni3Al-based intermetallic," *Intermetallics*, vol. 18, no. 4, pp. 582–586, Apr. 2010.

[6] K. Tokoro, N. P. Wikstrom, O. a. Ojo, and M. C. Chaturvedi, "Variation in diffusion-induced solidification rate of liquated Ni–Cr–B insert during TLP bonding of Waspaloy superalloy," *Mater. Sci. Eng. A*, vol. 477, no. 1 2, pp. 311 318, Mar. 2008.

[7] M. Schwartz, *Brazing*, 2nd Editio. Materials Park: ASM International, 2003.

[8] D. M. Jacobson and G. Humpston, *Principles of Brazing*. Materials Park: ASM International, 2005.

[9] W. F. Gale and D. a. Butts, "Transient liquid phase bonding," *Sci. Technol. Weld. Join.*, vol. 9, no. 4, pp. 283–300, Aug. 2004.

[10] G. O. Cook and C. D. Sorensen, "Overview of transient liquid phase and partial transient liquid phase bonding," *J. Mater. Sci.*, vol. 46, no. 16, pp. 5305–5323, May 2011.

[11] X. Wu, R. S. Chandel, and H. Li, "Evaluation of transient liquid phase bonding between nickel-based superalloys," vol. 6, pp. 1539–1546, 2001.

[12] M. Pouranvari, a. Ekrami, and a. H. Kokabi, "TLP bonding of cast IN718 nickel based superalloy: Process–microstructure–strength characteristics," *Mater. Sci. Eng. A*, vol. 568, pp. 76–82, Apr. 2013.

[13] M. Pouranvari, a. Ekrami, and a. H. Kokabi, "Effect of bonding temperature on microstructure development during TLP bonding of a nickel base superalloy," *J. Alloys Compd.*, vol. 469, no. 1–2, pp. 270–275, Feb. 2009.

[14] G. K. Bouse, J. C. Schaeffer, M. F. Henry, and G. E. Company, "Optimizing SC René N4 Alloy for DS AFT-Stage Bucket Applications In Industrial Gas Turbines," pp. 99–108, 2008.

[15] ASTM Standard E3-11, "Standard Guide for Preparation of Metallographic Specimens 1," International, West Conschohocken, PA, pp. 1–12, 2014.

[16] ASTM Standard E384-11, "Standard Test Method for Knoop and Vickers Hardness of Materials 1," ASTM International, West Conschohocken, PA, pp. 1–43, 2014.

[17] ASTM Standard E562-11, "Standard Test Method for Determining Volume Fraction by Systematic Manual Point Count," ASTM International, West Conschohocken, PA, pp. 1–7, 2014.

[18] ASTM Standard D6272-10, "Standard Test Method for Flexural Properties of Unreinforced and Reinforced Plastics and Electrical Insulating Materials by Four-Point Bending 1," ASTM International, West Conschohocken, PA, pp. 1–9, 2014.

[19] J. a. Ajao, "Phase transitions in some nickel-rich nickel–boron–titanium hard alloys," *J. Alloys Compd.*, vol. 493, no. 1–2, pp. 314–321, Mar. 2010.

Load-Capable Design of Arc Brazing Joints for Different Applications in Automotive Engineering

Prof. Dr.- Ing. Uwe Reisgen, Dipl.-Ing. Matthias Angerhausen, Dipl.-Ing. Christoph Geffers, Dipl.-Ing. Alexandros Pipinikas
Welding and Joining Institute of the RWTH Aachen, Germany

Abstract

For a reduction of the vehicle weight, joints made of high-strength steel materials and steel-aluminum dissimilar metals are, to an increasing degree, applied in modern car industry. Digitally controlled short arc processes are particularly suitable for the low-energy brazing of these materials. The short arc process is an arc welding process whereby the droplet transfer takes place during the short circuit. For the joining of high-strength steel materials, often copper-based materials are used as brazing filler wire. It is possible to use zinc-based brazing filler metals for the low-energy joining of steel-aluminum dissimilar metals.
The missing knowledge concerning the evaluation of the resilience of brazed solder joints under static and dynamic strain makes a broad application of brazed sheet structures made of copper and zinc-based soldering materials a challenging task. There is a lack of comprehension regarding the specific input parameters and the interaction between the joining parameters, the seam geometry, the filler material, the process stability, the reproducibility and the component's behaviour.
In the course of this presentation, the influence of the joining parameters on the seam geometry and the resulting mechanical-technological properties is determined by implementing brazing joints. Moreover, the test results about the influence of the geometrical design of the seam on the dynamic and static joint strength are presented.

Introduction

In modern vehicle construction, conventional steels are replaced to an increasing degree by higher-strength steels and lightweight construction materials, such as aluminum. A study of Ducker Worldwide, made in cooperation with European Aluminium Association (EAA), predicts that the aluminum content in a vehicle car body will rise to 180 kg by 2020. The aluminum content used in the year 2006 has, by comparison, been 121 kg [1].
According to forecasts, the content of higher-strength steels in vehicle car body production will be 20% by 2015. In the year 2006, just 7% of the car body materials were manufactured from steels with a strength of more than 600 MPa [2].

The pursued concept of multi-material design serves, on the one hand, the purpose of a clear reduction of the vehicle weight, figure 1. The application of different materials, on the other hand, serves the purpose of adapting the material properties to the strived-for component functions [3]. The requirements, as for example made to the strength of the joint, the corrosion resistance or the economic viability must, however, remain unchanged.

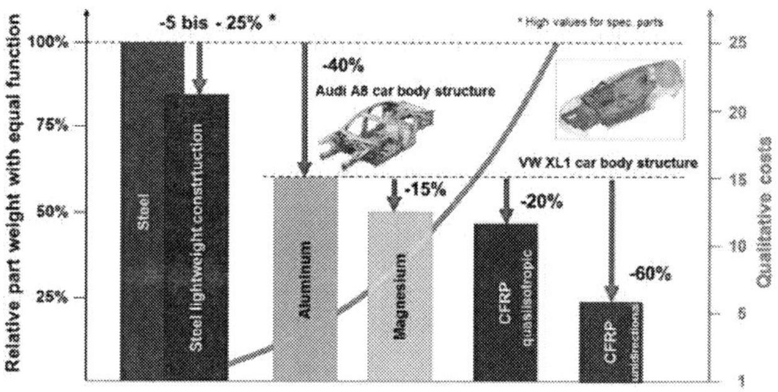

Figure 1: Relative part weight and qualitative costs of different materials [4]

Low-energy, digitally controlled short-arc processes have proven to be suitable methods for the joining of different materials with a view to modern multi-material design. Here, the joining of steel and aluminum presents a major challenge due to the different physical and metallurgical properties.
The selection of adequate wire filler materials, moreover, offers decisive potentials for achieving reproducible joining results with favourable mechanical-technological properties. Zinc-based wire alloys are highly suitable for arc joining of similar high-strength steels and also for the production of dissimilar material joints consisting of zinc-coated steels and aluminum. Copper-based wire filler materials are also suitable candidates for the joining of high-strength steels.
The potential of zinc- and copper-based brazing filler metals has, however, not been used to its full extent yet. This is mainly because the interactions between the individual joining parameters and the wire filler material, and also their influence on the achieved joining result and its mechanical-technological properties, have not yet been clarified to a sufficient degree.

Problem and Approach to Solution

When conventional GMA welding processes are used for the joining of high-strength steel materials, the comparatively high process energy during brazing exerts a negative influence on the material microstructure. This is accompanied by the deterioration of the mechanical-technological properties of the component. Another problem is the preservation of the surface coating of the sheets. If it is damaged by too high a process energy, reliable corrosion protection in the seam area is no longer ensured [5]. If the heat input is too high the result is the inevitable thermal distortion of the component which often necessitates the application of complex and expensive clamping devices for production [6, 7].

The combination of steel and aluminum also presents a major challenge due to the metallurgical incompatibility and the strongly differing physical properties of the two materials [8-10]. The main difficulty lies in the restriction of the formation of the metallurgical unavoidable intermetallic phases to a minimum, since these have a negative effect on the mechanical-technological properties of the joint. For this reason, the melting of the steel sheet and thus the accompanying uncontrolled growth of the intermetallic phases must be avoided during the joining process. Increased heat input into the joining zone supports, thus, the formation of intermetallic phases [11]. In the literature, an intermetallic phase thickness of 10 µm has often been specified as the critical limit [12]. In this regard, figure 2 depicts a micro-section with a crack along the inadmissibly large intermetallic phase of a steel-aluminum dissimilar material joint which has been produced using a short-arc process and aluminum filler wire.

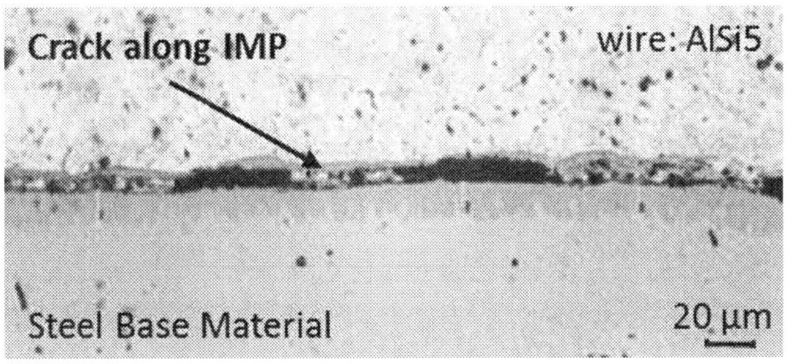

Figure 2: Micro-section of a steel-aluminum dissimilar material joint using an aluminum filler material with a crack along the intermetallic phase

Due to the afore-mentioned challenges, the integration of a heat intensive arc process into vehicle mass production is highly complex, particularly when aluminum filler wire materials for the production of steel-aluminum dissimilar material joints are used. Especially the often necessary use of fluxes for the improvement of the wetting behavior of the molten material is opposed to the economic efficiency of this technology. The thermal distortion which occurs in heat intensive arc processes necessitates, moreover, often complex and thus expensive clamping devices. The parameter window for a stable and reproducible joining process is strongly limited, compared to other joining processes which are used in modern automotive production.

Another challenge to production technology is the necessary determination of the intermetallic phase thickness values since those are decisive for the mechanical-technological properties of the joint. The analysis of the intermetallic phase formation is, so far, only possible via destructive testing and it is, therefore, economically unfeasible.

The application of low-energy, digitally controlled short arc brazing processes allows to meet these challenges. The energy input during joining of high-strength steel materials, which is lower compared to conventional GMA welding, exerts hardly any influence on the material structure and the base metals strengths are largely retained. The corrosion protection in the form of surface coating remains also further active. Compared with conventional GMA welding it is also possible to clearly reduce distortion. During the production of steel-aluminum dissimilar material joints, the low heat input allows to reduce the formation of intermetallic phases to a minimum.

The used brazing filler materials make a considerable contribution towards mastering the challenges. The application of copper-based brazing filler metals is well suited, particularly for the joining of high-strength steels [13]. Despite the different copper-based alloys and high-strength steels, the arc brazing processes are characterized by considerable advantages, particularly in the case of large differences in the thickness of plates, in low distortion joining and with regard to corrosion resistance. The metallurgical composition of the brazed seam has also considerable influence on the joint properties of the component.

The use of zinc-based brazing filler materials allows the joining of steels and also of steel-aluminum dissimilar material joints. The processing of these brazing filler metals with a low-energy arc brazing process enables the strong reduction of the intermetallic phase formation and of the crack formation which is often a result thereof.

Arc Joining of St/St- and St/Al-Joints by Means of Zinc-Based Brazing Filler Metals

The primary objective of automobile manufacturing is the reliable and economic joining of car-related steel materials and steel aluminum dissimilar material joints under production-related requirements or restrictions. For a determination of the user expectations about the arc brazing process, an analysis of demand has been carried out by the ISF of the RWTH Aachen University. This analysis was targeted at representatives from vehicle production and also at material manufacturers. As a conclusion of the analysis of demand, the largest intersection of requirements and restrictions has been created. Above all, requirements such as the maintenance of certain production tolerances or the reduction of component distortion have been specified within the scope of the analysis. As far as the production of steel-aluminum dissimilar material joints is concerned, the main goal was to identify a process which reduces the negative influence of brittle intermetallic phases on the mechanical-technological properties of the joining result.

To achieve the targets of the analysis of demand, a strategy has been pursued to keep the energy input into the joining zone as low as possible and thus to reduce distortion and the formation of brittle intermetallic phases. Low-energy short arc processes and low-melting zinc-based brazing filler metals which had been specially developed for this joining task have been applied.

Materials and Experimental Details

For the investigations DX56D+Z140, which is a hot-dip galvanised deep-drawing steel and is frequently used in automotive production, has been used. As the lightweight construction material, the aluminum alloy EN AW-6016T4 which is typical for outer paneling applications has been used. Steel-steel joints and also steel-aluminum joints have been produced in overlap joint since this joint type is and will further be the most important joint type in automobile production as far as design and construction are concerned, table 1.

Table 1: Investigated base metal combinations

Material 1	Material 2
DX56+Z140 (0,8 mm)	DX56+Z140 (0,8 mm)
EN AW6016-T4	DX56+Z140 (0,8 mm)

Different zinc-based brazing wires have been used in particular the widely used alloy ZnA14 as well as the newly developed alloy ZnAl5Cu3,5 which will achieve a strength of up to 550 MPa when magnesium is added, table 2. The inert gas argon has been used as shielding gas.

Table 2: Applied zinc-based brazing filler metals

Wire Material	R_m
ZnAl4	129 MPa
ZnAl5Cu3,5	257 MPa
ZnAl5Cu3,5 +Mg	550 MPa

For the tests, the CMT-equipment "TransPuls Synergic 4000 CMT" by Fronius has been used. This device allows similar to other controlled short arc processes the reduction of the input energy-per-unit length down to a minimum. Moreover, the process allows a controlled drop transfer and a specific heat control.

In comparison with the conventional short arc process, the short-circuit current is reduced to a minimum at the time of short-circuit. The surface tension of the molten metal and the reversal of the wire feed direction cause the transfer of the droplet into the molten pool.

Conduct of the Investigations and Results

Initially the only available experience was with the processing of aluminum based filler materials. Therefore it has been problematic at the beginning of the investigations to process the zinc-based alloys reliably and to achieve a controlled arc brazing process.

As shown in the measurement record in figure 3, the low melting point of the zinc-based brazing filler metals caused increasingly uncontrolled droplet detachment during the arc burning phase. Compared with a material transfer during short-circuiting, this entailed high heat input into the joining zone.

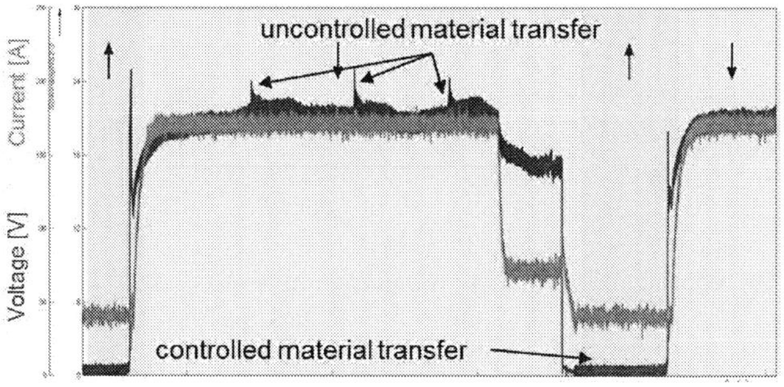

Figure 3: Uncontrolled material transfer during the arc phase of the CMT short arc process

For this reason, high-speed recordings and time-synchronous current and voltage measurement records have been carried out and analysed. Subsequently, appropriate measures for the adaptation of the given characteristics of the brazing process to the zinc-based materials which are to be processed were derived from the gained results. This way it has been possible to achieve a stable and reproducible brazing process.

Since, for the time being, experience about the interactions and dependencies of the selected process parameters and the joining results was not available, a surrogate model has been prepared. A surrogate model is an engineering method to illustrate a relationship between input and output parameters, when an outcome of interest cannot be easily directly measured. The aim of the surrogate modelling was the determination of the influence which the different process parameters exert on the developing joining results and the accompanying mechanical-technological properties. Figure 4 depicts the required steps for the execution of the surrogate modelling.

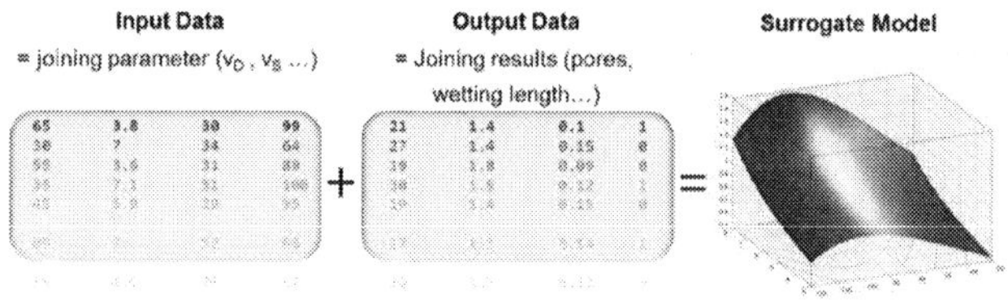

Figure 4: Steps of surrogate modelling

Surrogate modelling delivers the connections between the process parameters (input parameters) and the joining results (output parameters) as a result. The input parameters comprise not only the electric parameters of the brazing process but also the geometrical variables, such as the torch positioning. The joining results are characterized and quantified by the developing weld geometries, the formation of the weld surfaces and the pore formation which, as the case may be, will also appear.

Different surrogate models were created using the "Surrogate Modeling (SUMO) Toolbox" in MATLAB. Figure 5 depicts exemplarily the connection between arc correction and the developing wetting length when a zinc- and aluminum-based filler wire has been used. The arc correction of the CMT-process allows the setting of the spatial dimension of the arc plasma column. In the CMT process, a reduction of the arc length is achieved by a shorter wire retraction time which increases the process frequency. In the reverse case, a positive arc correction will increase the time for the retraction which causes the development of a longer arc and the reduction of the process frequency. A short arc has, in general, positive effects on the process stability while a long arc entails positive properties with regard to expanded wetting to side walls [14].

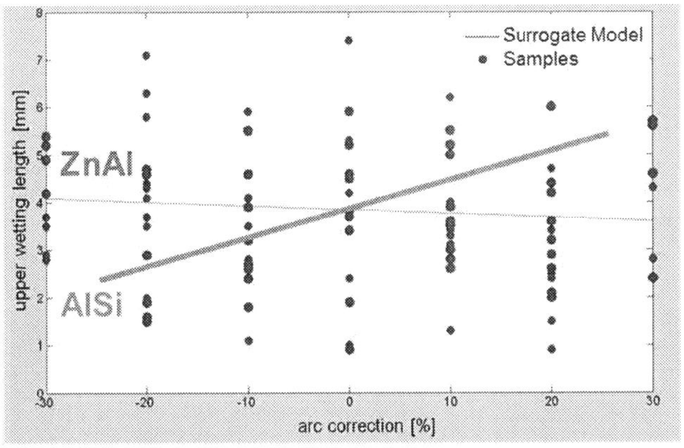

Figure 5: Surrogate model about the relationship between the input parameter "arc correction" and the wetting length using the wire filler materials ZnAl4 and AlSi5

From earlier investigations about the joining of steel-aluminum dissimilar material joints with aluminum wire filler materials it is known that the arc correction is an essential parameter for influencing the developing length of wetting. The selection of a short arc (negative arc correction) allowed the realization of a stable joining process. The developed wetting length was, however, too short which had a negative effect on the joint strength. Although the selection of a positive arc correction allowed the enlargement of the wetting length, it had a negative influence on the process stability.

During the processing of zinc-based brazing filler metals it was possible to demonstrate by means of the surrogate modelling that this parameter exerts almost no influence on the developing wetting length due to the favourable flow behaviour of the zinc-based alloys. Even the selection of a strongly negative arc correction allowed to achieve weld geometries with long wetting length. Since a short arc, due to the shorter wire stick-out of the softer zinc-based brazing material, allows a more stable joining process, a negative arc correction has been selected for the following brazing tests.

Based on the knowledge gained by the surrogate modelling, mechanical-technologically favourable weld geometries with large wetting length and small wetting angles have been produced. Within the framework of the investigations, the filling of a set gap between the joining partners with molten zinc has, among other things, been pursued, figure 6.

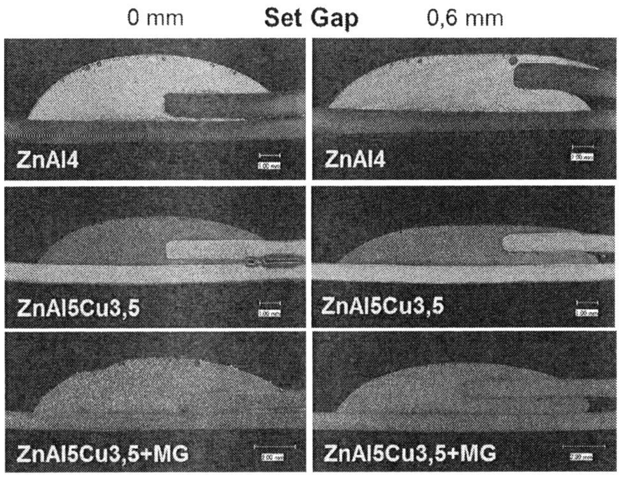

Figure 6: Results of the joining of DX56D/DX56D with setting of a defined gap

The filling of the gap served the purpose of considering manufacturing tolerances which occur in vehicle production; this was also a main requirement from the above-mentioned analysis of demand. Particularly during the joining of DX56D/DX56D, the filling of the gap with zinc brazing material helped to achieve the enlargement of the wetting length and thus the increase of the joint strength properties. With the combination of DX56D/EN AW-6016T4, figure 7, and also with DX56D/DX56D it had been possible to fill a gap width of more than 1 mm with high process reliability.

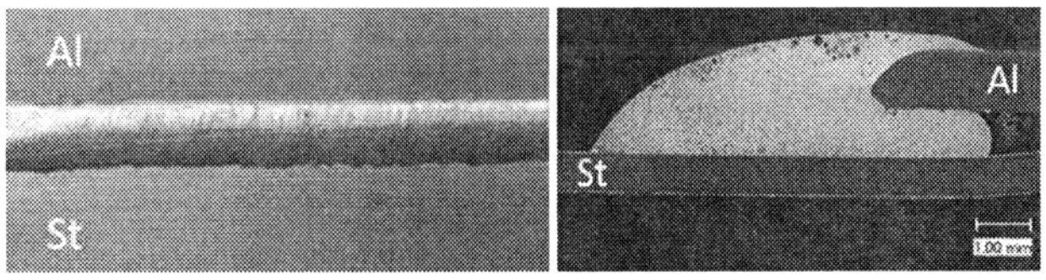

Figure 7: Results of the joining of DX56D/EN AW-6016T4 with setting of a defined gap

Tensile and fatigue tests have been conducted for the combinations DX56D/DX56D and DX56D/EN AW-6016T4 to determine the static and cyclic joint strength. Figure 8 illustrates the tensile joint strength of the combination of DX56D/DX56D by using different zinc-based brazing filler metals.

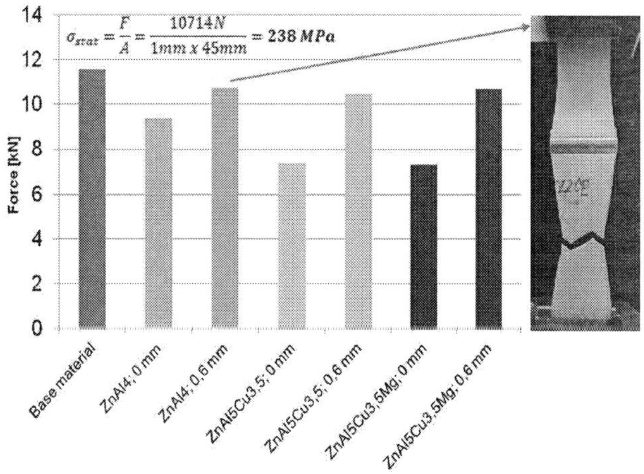

Figure 8: Tensile strength of the joints DX56D/DX56D

Due to the larger wetting length, it is obvious, that a gap between the joining parts, 0.6 mm in this case, increases the tensile strength. This conclusion is not dependent on the type of solder which is used, it is valid for all tested alloys.
The cyclic strength has been ascertained by a servo-hydraulic high-frequency fatigue testing machine with a fatigue limit of $N_G = 2x10^6$. A short extract of the results is shown in figure 9.

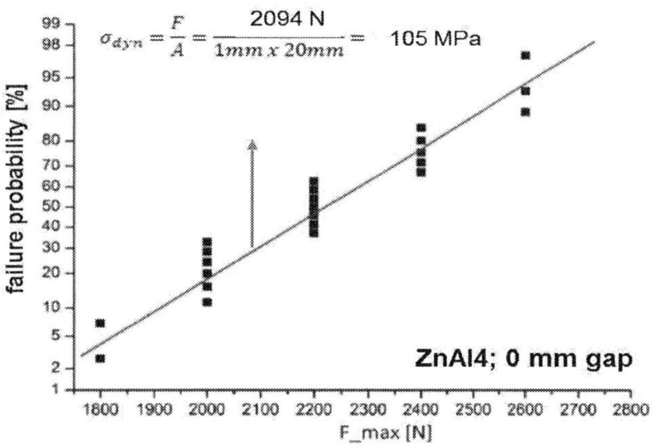

Figure 9: Cyclic strength of the joints DX56D/DX56D

For example, fatigue strength can be reached for a deflection stress of 105 MPa and the probability of a technical failure of 30 %.
The results of the tensile strength tests of the combination DX56D/EN AW-6016T4 are shown in figure 10. All these samples are brazed without flux.

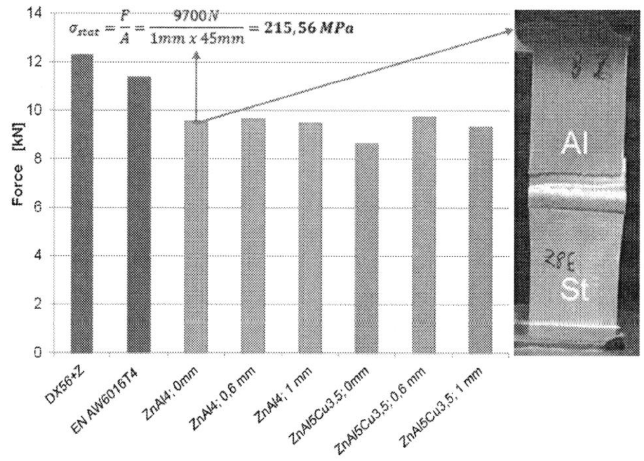

Figure 10: Cyclic strength of the joints DX56D/ENAW 6016T4

The results illustrate that the tensile strength of the aluminum base material in the heat affected zone can be reached applying all tested alloys and gaps.
Cyclic strength was determined with the same method as the steel-steel-samples, figure 11. For a deflection stress of 100 MPa and the probability of a technical failure of 30 % fatigue strength could be reached.

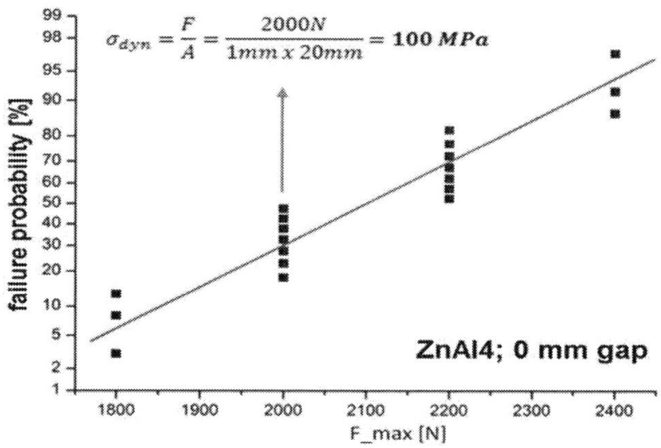

Figure 11: Cyclic strength of the joints DX56D/EN AW-6016T4

Metallographic studies of zinc brazed steel-aluminum-hybrid-joints

To determine the influence of the process parameters on the formation of intermetallic phases, comprehensive metallographic tests have been conducted for the combination DX56D/EN AW-6016T4. In addition, these studies mainly served as an appropriation for the effects of the intermetallic phases on the mechanical and technical properties of flux-free joints. As a result of the studies, it is observed that the intermetallic phases often only appear in the areas which are directly affected by the arc, figure 12.

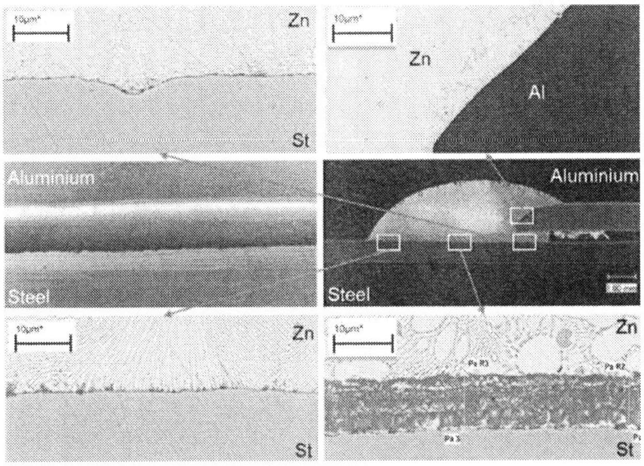

Figure 12: Macro- and micro-sections of the joints DX56D/EN AW-6016T4

By using a digitally controlled short-arc-process and a low melting zinc-based brazing material it is possible to realize an accurate heat input during the joining of steel-aluminum-hybrid-joints. Thereby, on one hand, the formation of intermetallic phases can be widely controlled and, on the second hand, the intermetallic layer can be kept under the critical level of 10 µm. Caused by the different expansion coefficients of the steel base material and the brittle intermetallic phases and the consequent thermal strain there is a way to detach the intermetallic phases off the base material accurately during the joining process. Subsequently, it is possible that a further formation of an intermetallic layer takes place which is possibly also detaching off the steel base material, figure 13, left top.

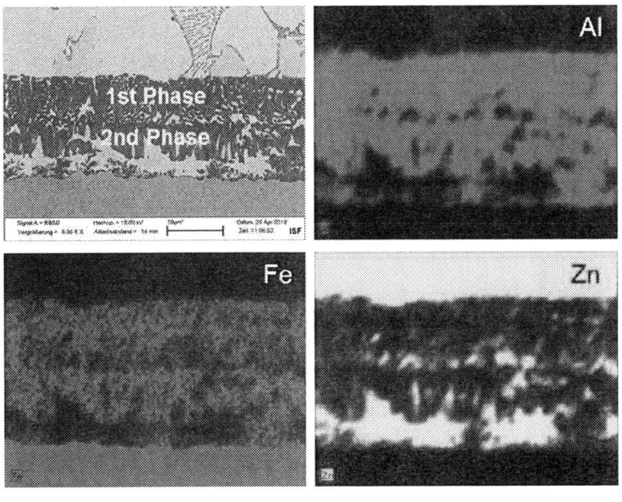

Figure 13: Mapping of the intermetallic phase seam of the joint DX56D/EN AW-6016T4

There is just a very small intermetallic phase (<< 1 µm) at the interface between the molten solder and the base material, which is the most influential spot for the mechanical and technological properties of the joint. The detached and congealed intermetallic phases are circumvented and penetrated by the liquid zinc-bearing melting. Thus the brittle intermetallic phases are embedded in a ductile zinc-matrix. Frequently appearing ruptures during thermal joining of steel-aluminum-hybrid-joints can be closed or even avoided altogether. Due to this effect, the negative influence of the intermetallic phases on the mechanical and technological properties of steel-aluminum-hybrid-joints can be reduced. So the spontaneous and brittle collapse of hybrid-joints resulting from uncontrolled formation of intermetallic phases can be mostly prevented.
Another advantage of using zinc-based solders is the, in comparison with aluminum-based solders, distinctly larger process window which is indicating a high level of reproducibility. Furthermore, there is no need for destructive tests to validate the level of the formation of intermetallic phases.

Realization of a development model

In cooperation with partners of the automotive industry a development model has been evolved to illustrate the field of application of low-melting zinc-based alloys in addition to a low-energy short arc process. During the realization of this model, the joining results, which have been achieved using small-sized test sheets, was successfully adapted to application-related sheets with a consequent major heat dissipation. Figure 14 illustrates the development model, representing a part of an automotive floor assembly. There were both sheets and profiles used.

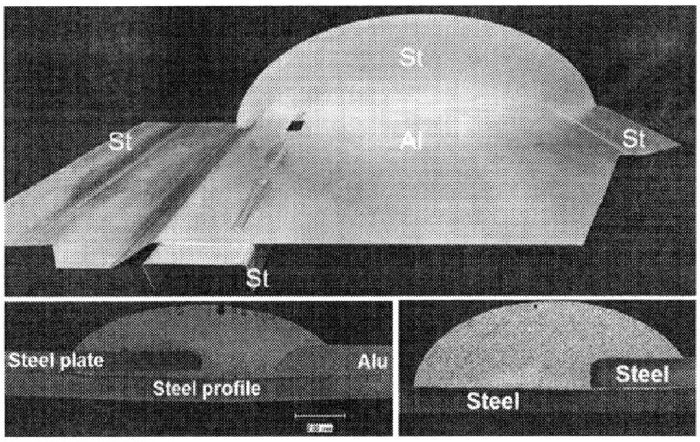

Figure 14: completed development model using the wire material ZnAl4

Apart from the base metal combinations DX56D/DX56D and DX56D/EN AW-6016T4 a combined butt and lap joint has been realized. In doing so three components (steel sheet, aluminum sheet, steel profile) have been joined by only one joining process, figure 14, bottom left. It has to be noted that all joints of the development model have been conducted using exactly the same type of zinc-based filling wire, ZnAl4. Hot-dip galvanized sheets as well as electro galvanized sheets have been used to show the wide field of application of this technology concerning different types of surface coatings. There was no destruction of the surface coating, neither on the front nor on the backside of the joint. The component distortion was quite minor thanks to the low-melting brazing filler metals and the low-energy short-arc process. So it is not necessary to design complex tension devices any more.

Joining of high-strength steels utilising copper-based braze materials

High tensile strength steel is increasingly used in the automotive unibody construction. These sheets which are thermal-mechanically formed by a rolling process to enable their complex properties and structural composition, have both a high tensile strength and a good formability. Especially low-energy short-arc processes in combination with copper solders are well suited to join this kind of steel sheets. Due to the low energy input thermal stressing and a consequent alteration of the structural composition can be kept on a low level.
There are already short-arc processes in use in the automotive industry. Main fields of application for this process are the joining of components focusing on the optic and the corrosive reliability and not strength oriented structures. The reason for this is a phenomenon called undermatching. Undermatching means that the used solder has a lower tensile strength than the base material. For example, the brazing process is used for the rooftop zero gap joint and the connection between rooftop and the rear. In this case, the seams have to be reworked to guarantee non-visible joining zones after the painting process. In the area around the side sill, the requirements on the seams are totally different. The components are joined with slotted hole braze seams in order to attain a larger mounting area. The most important property of this joint is the corrosive reliability because there is no possibility to lacquer or coat the hollow profiles afterwards. Areas being subjected to static and cyclic stress are not yet joined with that type of brazing [13].
A reason against the wide spread application of brazed high tensile strength steel joints is the lack of knowledge regarding the static and cyclic resilience of those. It is a lack of understanding of the different input parameters and the interaction between those and the seam geometry, the type of solder, the process reliability, the reproducibility and the components' properties caused by strain. Another problem in joining high tensile components is the low strength of well braze-able copper-based brazing material.
Brazed joints are mainly qualified by static tensile strength today. Cyclic strength is only tested randomly during the process of qualification. It is nearly impossible to extract conclusions about the cyclic resilience of high tensile materials from common testing methods. The common thread is the lack of knowledge regarding the influencing factors and their correlation.
To improve the knowledge concerning the tensile strength of brazed steel sheet structures, it is necessary to investigate both the notch effect and the properties of the solder material. The focus has to be on the different structural zones of brazed joints concerning their dimensions and properties under cyclic and static stress.

Caused by the huge gap between the possibilities of application and the actual application of brazed high tensile strength joints, there is a current research project called "Load capable arc-solder-joint of high strength steel sheets under metallurgical and geometrical aspects" realized by the ISF, RWTH Aachen and the ISAF, TU Clausthal. The aim is to acquire a basic comprehension with regard to the coherences of geometrical features of the seam, especially of the lap joint, the local metallurgical composition and their influence on the properties of the joint. Furthermore the main influencing factors on the static and cyclic strength are evaluated in detail in order to enable a reliable statement referring to the causes of collapse.
Within the scope of the studies a basic simulation model is developed to validate the structural composition and internal stress of several arc-brazed joints under strain. On the leading point of a load-capable design a fine-tuning of the numeric simulation model is made utilising several geometrical and metallurgical models. The simulation serves to evaluate the static and cyclic strength of metallurgical and geometrical different brazed joints in an effective and efficient way and to identify the reasons of collapse.

Summary

In the interest of lightweight construction in the modern automotive industry there is an increasing usage of both high tensile strength steel and steel-aluminum-hybrid-joints.
Above all digitally controlled short-arc processes in combination with zinc-based brazing material have proved to be suited for the joining of these materials. Due to the results of the surrogate modelling it is now possible to realize seam geometries for both base metal combinations DX56D/DX56D and DX56D/EN AW-6016T4 with proper mechanical and technological properties and suitable wetting length and angle. Furthermore, manufacturing caused tolerances have been considered. The molten zinc fills the adjusted gap and guarantees a reliable joining. The larger wetting length resulting from the filling of the gap increases the static tensile strength of the joint.
Thanks to the lower energy input of this joining strategy there is a way to reduce the negative influence of the brittle intermetallic phases, which unavoidably emerge during the joining of steel and aluminum. This is a result of the embedding of the brittle phases in a ductile zinc-matrix.
The results of static and cyclic tensile strength studies show that tensile strength of the base material in the heat affected zone can be reached for both combinations, DX56D/DX56D and DX56D/EN AW-6016T4.
To illustrate the wide field of application of zinc-based brazing filler metals in combination with a low-energy short-arc process a development model has been made in cooperation with partners of the automotive industry. The same filler wire was used for all combinations (steel-steel, steel-aluminum, steel-steel-aluminum).
Currently a research project is underway at the ISF, RWTH Aachen and the ISAF, TU Clausthal. The focus of this project is on the gap between the possibilities of application and the actual application of arc-brazing with copper-based filler wire. Within the scope of the studies a simulation model has been developed to evaluate static and cyclic strength of metallurgically and geometrically different brazed joints and to identify the reasons of collapse.

Acknowledgments

The authors would like to thank the German Research Foundation DFG for the support of the research work which has been carried out in the research project "Load-capable arc soldered joint of high strength steel sheets under metallurgical and geometrical aspects" (RE 2755/29-1).

References

[1] Ducker Worldwide, EAA, "Aluminium penetration in cars", Report (2012)
[2] Frost, Sullivan, "Strategic Analysis of the European Markets for Body-in-White," Report (2008)
[3] Lesemann, M., Bröckerhoff, M., Urban, P., "The Prospects of Multi-Material Design for a Compact-Class Front Section," ATZautotechnology (07/2008), pp. 16-20.
[4] Volkswagen AG, „Prozess- und Werkstoffinnovationen in der Pulvermetallurgie," Werkstoffsymposium 2012
[5] Feng, J.C., He, P., Hackl, H., "Distribution of Zn and interfacial microstructure of braze-welding CMT joints between aluminium and galvanized steel sheets," Solid State Phenomena, Vol. 127 (2007), pp. 43-48.
[6] Tsai, C.L., Park, S.C., Cheng, W.T., "Welding distortion of a thin-plate panel structure," Welding Research Supplement (Mai 1999), pp. 156-165.
[7] Satoh, K., Terasaki, T., "Effect of welding conditions on residual stress distributions and welding deformation in welded structures materials," Journal of the Japan Welding Society, Vol. 45, No. 1 (1976), pp. 42-53.
[8] Rathod, M.J., Kutsuna, M., "Joining of Aluminum Alloy 5052 and Low-Carbon Steel by Laser Roll Welding," Welding Journal, Vol. 83 (2004), pp. 16–26.
[9] Güngör, Ö.E., Gerritsen, C., "Effect of filler wire composition and metallic coating on the joint performance of aluminium/steel braze-welds," Welding and Cutting, Vol. 7 (2008), pp. 303–312.
[10] Tanaka, T., Morishige, T., Hirata, T., "Enhanced mechanical properties of friction stir welded dissimilar Al–Cu joint by intermetallic compounds", Scripta Mater, Vol. 61 (2009), pp. 756–759.
[11] Yongchao Su, Xueming Hua, Yixiong Wu, "Effect of input current modes on inter metallic layer and mechanical property of aluminum–steel lap joint obtained by gas metal arc welding," Materials Science & Engineering A: Structural Materials: Properties, Microstructure and Processing, Vol. 578 (2013), pp. 340–345.
[12] Kreimeyer, M., Sepold, G., "Laser steel joined aluminium-hybrid structures", Proceedings of ICALEO *'02,* Jacksonville, FL, USA (2002)
[13] Agah Meric Sevim, Möglichkeiten zur Erhöhung der Festigkeitseigenschaften Metall-Schutzgas-gelöteter Überlappstöße hochfester Stähle, Shaker Verlag (Aachen, 2010)
[14] Fronius, Handbuch zu MSG-Schweißstromquelle TransPuls Synergic 4000 (2014)

Electron Beam Joining of γ-Titanium Aluminide

Uwe Reisgen, Simon Olschok, Stefan Gach
RWTH AACHEN University, Aachen, Germany

Abstract

From the economic and also from the production point of view, the development of suitable joining techniques is a decisive factor for industrial application of γ-TiAl. The excellent properties of TiAl such as high specific strength, high temperature loads and creep resistance shall be maintained in the joining point. The brazing process by electron beam offers the advantages of the use of vacuum instead of atmosphere, the precise energy input and the low cycle times. Materials' high brittleness and thus susceptibility to cracking are a challenge for welding and brazing with high cooling rates. The adapted beam deflection for process-integrated post- or pre-heating is one solution to achieve a high strength in the joints. The selection of the brazing filler metal is important for high strength especially in dissimilar material joints like titanium aluminide with steel or titanium aluminide with nickel base alloy. In this presentation, suitable joining strategies for modern γ-TiAl (TMN and TNB V5) via electron beam are presented. Especially dissimilar material joints like TiAl with steel or TiAl with nickel based alloys are presented. Different joints are analysed in detailed for example by cross sections, scanning electron microscope pictures and tensile tests.

Introduction

Nickel based super alloys which have, so far, been used for high-temperature applications are, in the near future, to be substituted by intermetallic materials, for example by the considerably lighter titanium aluminides. Potentials for significant weight saving are up to 40 %. Such weight saving has a strong impact on engine fuel efficiency in aero-engine applications and turbochargers. This applies particularly to parts which are subjected to high temperature and highly dynamic loads. These intermetallic materials are characterised by a high corrosion resistance and also by their high specific strength. A key technology for the application of γ-titanium aluminides will be the joining process. Titanium aluminides are joined by welding or by brazing. Both processes offer specific advantages and disadvantages. In brazing, the melting of the base material is avoided which results in less thermal influence [1]. In welding, the material is molten. As a rule, the results are stronger joints than those achieved by brazing, the thermal influence on the material is, however, higher. This paper is concentrating on brazed titanium aluminide joints. Industrial applications require dissimilar material combinations. For that reason, this paper focuses on dissimilar joints of nickel based alloys or steel and titanium aluminide, which are, for example, required for aero-engine applications.

Materials and experimental details

Brazings have been carried out using the titanium aluminide alloy TMN (Ti-43.5Al-4.0Nb-1.0Mo-0.1B). This alloy belongs to the new β solidifying titanium aluminides. The material has been separated from the ingot and afterwards subjected to remelting, investment casting and hot isostatic pressing. The basic structure is a homogenised lamellar α2/γ-structure, surrounded by ductile β-phase and embedded with γ-phase [2]. Before the brazing started, surface milling and cleaning with alcohol have been carried out. The examined specimens had a size of 80 mm x 15 mm x 4 mm. For brazing dissimilar joints a nickel based alloy is chosen. The specimens of alloy 617 (NiCr23Co12Mo) and steel (42CrMo4) has the size 50 mm x 12 mm x 4 mm.

Overlap joints with an area of 4 x 12 mm in position PA (vertical beam) have been carried out. The electron beam has, for the mentioned tests, been defocused and been moved fast over the workpieces between which the brazing filler metal had been positioned. The brazing filler metal is heated by heat transfer from the basic material. The basis material is heated by the beam. By this method, the workpieces are heating the braze material evenly up to the brazing temperature within the interval of 1000, 1050 and 1100 °C. The brazing time has been varied in the range of 15 to 60 seconds. The beam power has been set with a maximum of 330 W in order to keep a constant workpiece temperature. During the joining process with an electron beam welding machine (pro-beam K7; maximum accelerating voltage: 120kV), the workpiece is in a vacuum with a pressure of approx. $5*10^{-4}$ mbar.

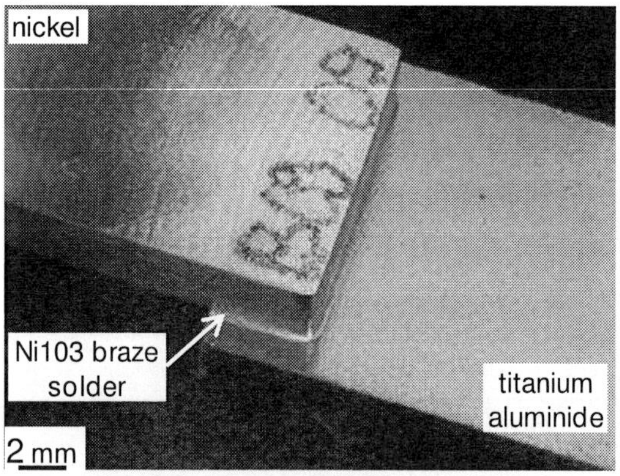

Figure 1. Nickel-titanium aluminide dissimilar joint with brazing filler metal Ni103 at 1050 °C brazing temperature and 30 s brazing time.

The used brazing filler metals were three different materials on nickel-, titanium and copper basis. A standard, high-temperature brazing filler metal for increased application temperature with a good oxidation and corrosion resistance is the nickel based brazing filler metal Ni103. It has a nominal composition of Ni4.5Si3.1B (weight %) and a liquidus temperature of 1054 °C. Ni103 has good joining properties for titanium aluminide [3]. The titanium-based brazing filler metal is composed of 70Ti15Cu15Ni (weight %) with a liquidus temperature of 960 °C. The suitability of this brazing filler metal for titanium aluminides has been demonstrated in [1]. The third investigated brazing filler metal is a pure copper foil with a liquidus temperature of 1083 °C. This material has a high ductility but a low strength. Foil thicknesses are 75 µm for the Ni103 and TiCuNi foil and 100 µm for the copper foil.

Several devices have been used for the evaluation of the joints. For the determination of the brazing temperature, thermocouples of the type K had been inserted into bores in the base metal. The temperature has been set via temperature recordings made by a digital oscilloscope (Yokogawa). After the brazing has been finished, the brazed seams have been subjected to visual examination. Moreover, chemical etching (100 ml H_2O, 3 ml hydrofluoric acid, 6 ml nitric acid) has been applied for the preparation of macro- and microsections. Hardness measurement series in accordance with Vickers have been carried out. Scanning electron micrographs have been performed using a Leo "Gemini 1530".

Results and Discussion

For an estimation of the suitability of the three brazing filler metals for the dissimilar material joint nickel with titanium aluminide, brazings are carried out at a temperature of 1050° C and a brazing time of 30 s. All three brazing filler metals are flowing out of the joining gap and are wetting the base materials. The three brazings have a different appearance, the hardness, the strength and the microstructure.

Brazing with brazing filler metal 70Ti15Cu15Ni

Shear test specimens of the TiCuNi brazed joints are breaking with 77 MPa at the joining surface of the nickel. This points to unsatisfactory fusion at this position. SEM micrographs substantiate this assumption.

By means of a hardness scale in the brazing gap, the brittleness and the strength of the brazing filler metals are estimated, Figure 2. With a hardness of 864 HV0,1, the joint made with brazing filler metal TiCuNi has the highest hardness of the examined specimens. The hardening of the comparable brazings of similar titanium aluminide with titanium aluminide joints is less strong.

The diffusion zone between TiCuNi brazing filler metal and nickel varies between a width of 50 and 100 µm. This seems to be quite wide for a brazing. The diffusion region shows brittle intermetallic phase formation caused by this effect. The surface of the nickel has, moreover, dissolved in the brazing zone. A brazing zone with a width of 120 up to 190 µm has developed from there. Since the brazing foil had a thickness of originally 75 µm, a high number of base material constituents in the brazing filler metal have dissolved.

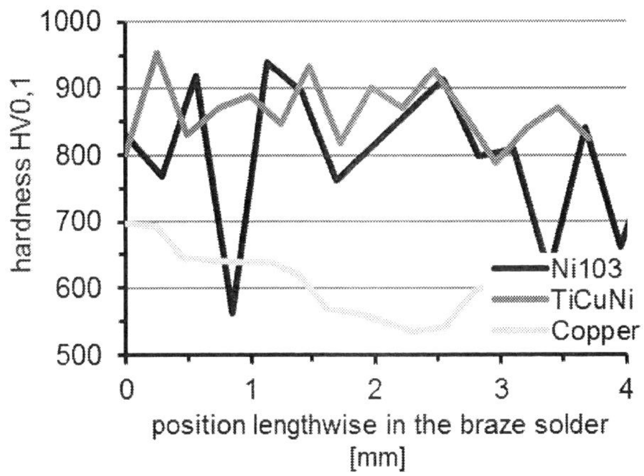

Figure 2. Hardness in the brazing zone of three different brazing filler metals of the dissimilar joint nickel-titanium aluminide at 1050 °C brazing temperature and 30 s brazing time.

The brazing zone is too wide for a high-temperature brazing. The fusion of the nickel base material is inadmissible in brazing. During solidification, brittle intermetallic phases are developing which are hardening the structure. The shear test specimens are breaking at the transition between nickel base material and brazing filler metal. Wetting at this position is poor and the diffusion zone is very wide. A reduced brazing time or brazing temperature does not increase the strength of the TiCuNi brazing filler metal for the dissimilar material joint. The TiCuNi brazing filler metal is, thus, not suitable for a nickel-titanium aluminide dissimilar material joint.

Brazing with brazing filler metal Copper

The copper brazing filler metal is porous and brittle. Shear test specimens are breaking with 49 MPa. The copper brazing is the softest braze material under the examined alloys and thus probably the most ductile brazing joint with a hardness of approximately 607 HV0,1. This brazing joint also shows hardening. A comparable dissimilar brazing with heat treatable steel and titanium aluminide has an average hardness of 379 HV0,1. In connection with the brittle exterior appearance of the copper brazing, this is a sign for intermetallic phase formation in the brazing filler metal.

In the cross-section of the joint with the copper brazing filler metal, a fusion of the titanium aluminide base material is observed, Figure 3. The original edge of the titanium aluminide has been completely molten by the brazing filler metal. The remaining surface of the titanium aluminide is roughened. The brazing gap has widened up to approx. 100 µm.

The diffusion zone between brazing filler metal and nickel base material is well developed and is, with approximately 3 µm, neither too narrow nor too wide. Via diffusion the brazing filler metal has penetrated the titanium aluminide which resulted in the decreasing of the melting point of the titanium aluminide copper diffusion zone. The diffusion zone was thus dissolved. The brazing filler metal is, subsequently, penetrating into deeper layers of the base material.

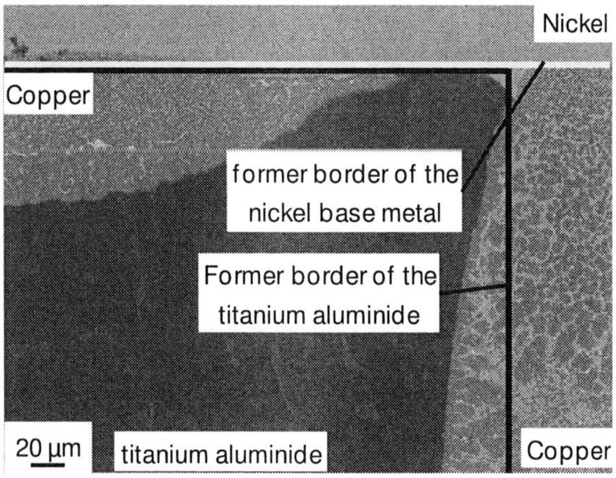

Figure 3. SEM-picture of a copper brazed joint of the dissimilar joint nickel-titanium aluminide at 1050 °C brazing temperature and 30 s brazing time.

The brazings made with a copper brazing filler metal at a reduced brazing temperature of 1000 °C have a lower strength since the brazing temperature is below the solidus temperature of the brazing filler metal. The melting of the base material is much lower in a brazing made at 1050 °C and a reduced brazing time of 15 s. The result is decreased strength since the diffusion rate is too low.

Brazing temperature and brazing time exert influence on the strength of the brazed joint. Tests made with higher and lower brazing temperature and/or shorter or longer brazing time which is decelerating and/or promoting diffusion of the TiCuNi and/or copper brazing joint show that the strength is not increased. It becomes thus apparent that brazings made with the dissimilar material joint nickel-titanium aluminide with the brazing filler metals TiCuNi and/or copper are not suitable due to the mechano-technological characteristics.

Brazing of the dissimilar joint nickel-titanium aluminide with brazing filler metal Ni103

Tests have been made on the basis of the Ni103 brazing filler metal. The minor flow of the Ni103 on both base materials is remarkable. Visual inspection of Figure 1 shows a good appearance of the joint.

The Ni103 brazing filler metal has an average hardness of almost 800 HV0,1. The brazing filler metal has a similar hardness, also in industrially tested brazings of similar titanium aluminide - titanium aluminide joints and dissimilar joints between titanium aluminide and heat treatable steel (42CrMo4) dissimilar material joints. A further embrittlement of this brazing joint has, therefore, not developed in the dissimilar material joint nickel titanium aluminide.

Due to the highest shear strength of 87 MPa, the joint nickel with titanium aluminide with Ni103 brazing filler with 30 s brazing time seems to be better suited for the dissimilar material joint than the other tested brazing times. For this joint brazing temperature and brazing time exert influence on the strength and microstructure of the brazed joint. In a comparison of different brazing temperatures, the nickel-titanium aluminide dissimilar material joint with a Ni103 brazing filler metal has the highest strength at a brazing temperature of 1050 °C. A brazing temperature of 1100 °C and a brazing time of 30 s result in the overheating of the joining zone which is so high that a very strong diffusion between titanium aluminide - base material and brazing filler metal occurs. At this temperature, the diffusion is approximately two up to four times faster than it is at a temperature of 1050 °C, Figure 4.

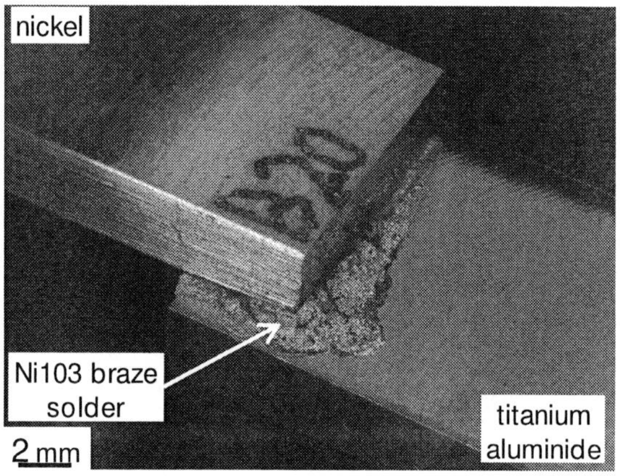

Figure 4. Nickel-titanium aluminide dissimilar joint with brazing filler metal Ni103 at 1100 °C brazing temperature and 30 s brazing time.

A brazing made with a brazing temperature of 1000 °C and a brazing time of 30 s has a shear strength of 68 MPa. In this case, the brazing temperature is too low for a sufficient wetting of the base material.

The effects which brazing times within a range of 15 and 60 s will have on the microstructure have been examined at a brazing temperature of 1050 °C, Figure 5. Here, with a brazing time of 15 s, wetting and diffusion are too low, while the diffusion is too strong with a brazing time of 60 s. This is identical with a brazing temperature of 1100 °C and a brazing time of 30 s. Brazings which were made with a brazing time within the range of 30 and 45 s have the best shear strength values.

The dissimilar material joint nickel with titanium aluminide with Ni103 brazing filler metal, a brazing temperature of 1050 °C and a brazing time of 30 s are, in the following, evaluated with regard to their microstructure. The brazing filler metal has a dendritically grown microstructure, Figure 6. The dendrites are starting their growth at the nickel base material. The ingredient of the dendrites is a nickel mixed crystal (cubic face-centered) with inclusions of chromium and aluminium, Table 1.

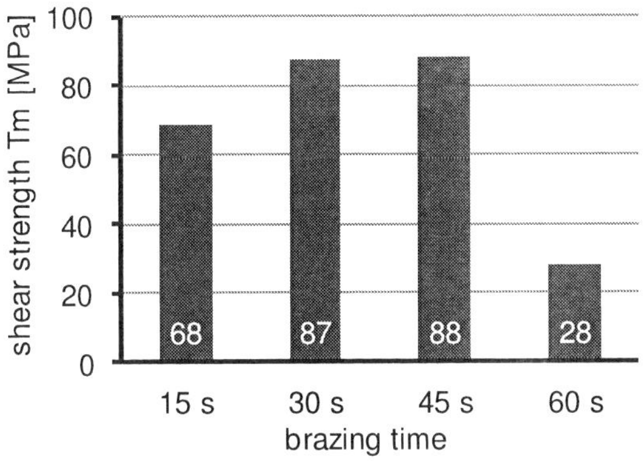

Figure 5. Comparison of the shear strength of different holding times for the dissimilar joint nickel-titanium aluminide with the brazing filler metal Ni103 at 1050 °C brazing temperature.

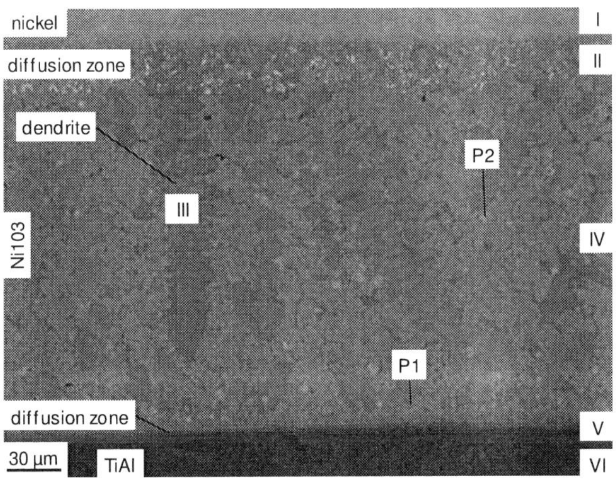

Figure 6. SEM-picture of a brazed joint of the dissimilar material joint nickel-titanium aluminide with the brazing filler metal Ni103 at 1050 °C brazing temperature and 30 s brazing time.

Table 1. Chemical composition (at.-%) of selected points from Figure 6.

Point	Ni	Cr	Mo	Ti	Al	Si	rest
Dendrite	78,6	8,1	1,0	2,8	4,6	3,7	1,2
P1	83,6	4,4	0,0	6,3	1,6	2,0	2,1
P2	23,1	44,7	26,9	2,3	0,1	0,2	2,7

Around the dendrites, another nickel mixed crystal (cubic face-centered) is arranged, which has a similar composition as the dendrites (point P1 in Figure 6). A third constituent in the brazing zone is a phase which is depicted as light points (point P2 in Figure 6). This phase consists of a chromium molybdenum nickel mixed crystal (Cr0,4Mo0,3Ni0,3).

Diffusion zones in the brazing are existing in the direction of both base materials. On the side of the titanium aluminide, a diffusion zone with a width of approximately 5 µm is observed. This zone can be divided into three regions. On the side of the nickel base material which has enriched with constituents from the brazing filler metal, the diffusion zone is very wide, approximately 30 µm. This zone is divisible into three diffusion regions with different compositions. It is remarkable that the brazing zone is, with approximately 120 µm, wider than the brazing foil. The brazing filler metal has been enriched with chromium and molybdenum which have diffused as constituents of the nickel base material into the brazing filler metal. Those have to an increased degree enriched in the light phases (point P2 in Figure 6).

The schematic process of such a brazing starts during the brazing by a strong diffusion between brazing filler metal and nickel base material, Figure 7. Chromium and molybdenum are diffusing from the nickel base material (zone I) into the brazing filler metal. Nickel and silicon are diffusing into the diffusion zone in the opposite direction (zone II). At this point, three phases are developing. These are two nickel mixed crystals with smaller amounts of substitution atoms made of titanium, chromium and/or chromium and silicon. On the side of the titanium aluminide, titanium and aluminium are diffusing into the brazing filler metal and nickel diffuses into the base material. This type of diffusion occurs slower at the point of the nickel base material. Within this diffusion zone at the titanium aluminide, three characteristical regions in transverse to the diffusion vector are developing. In the vicinity of the titanium aluminide, the TiAl/Ti3Al-lamellae are dissolving. At this point, the intermetallic phase Ti3Al is maintained, Figure 8.

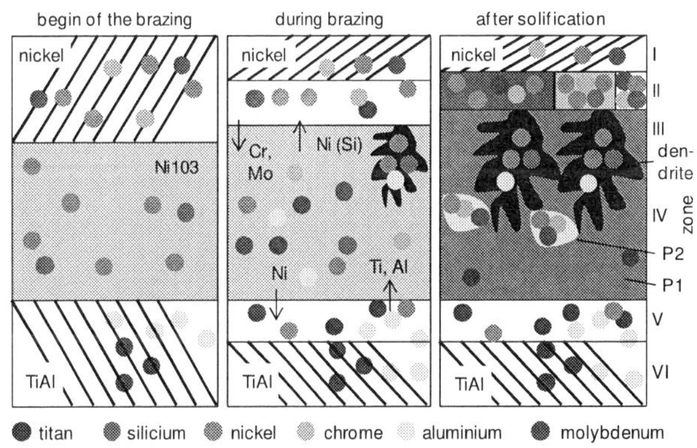

Figure 7. Schematic cycle of a brazed joint of the dissimilar material joint nickel-titanium aluminide with brazing filler metal Ni103 at 1050 °C brazing temperature and 30 s brazing time.

Adjacent to the brazing filler metal, the faster diffusion results in a further phase transformation which is explained by means of Figure 8. With the help of the ternary Al-Ni-Ti phase diagram, the diffusion processes between brazing filler metal and titanium aluminide and the developing phases are specified. At the upper corner, the alloy of the brazing filler metal is depicted. The titanium aluminide base material is positioned below in the centre. The diffusion of aluminium out of the base material results in a diffusion zone with the chemical composition Ti_3Al. With the ever increasing diffusion, nickel is enriching from the brazing filler metal in this region. Two phases which are chemically similar to a $TiNi_2Al$-mixed crystal, are further developing in the diffusion zone.

After the energy input has been concluded, the joining zone is cooling down and the brazing filler metal solidifies. The brazing zone is, at this point in time, very wide. First, the dendrites are solidifying as nickel mixed crystal with silicon and low quantities of aluminium. Enriched with titanium and aluminium, the rest of the liquid phase solidifies with further cooling. The developing phase is characterised by point P1. In this matrix, a further phase is embedded which is characterised by point P2. Point P1 is a nickel mixed crystal and point P2 is a chromium molybdenum nickel mixed crystal.

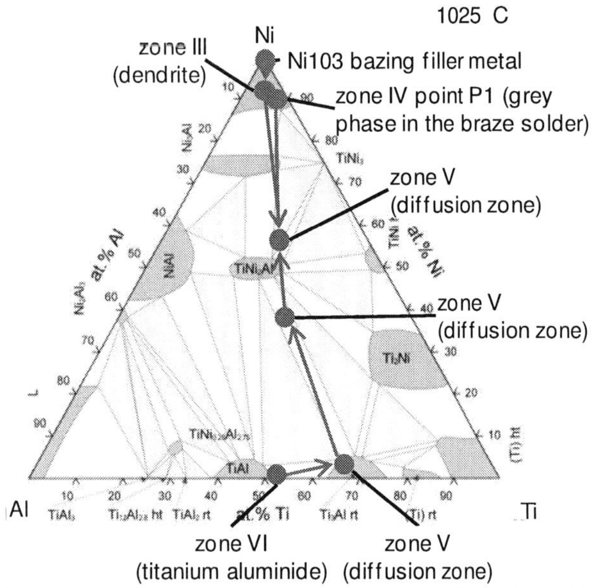

Figure 8. Phase formation of a brazed joint of the dissimilar material joint nickel-titanium aluminide with brazing filler metal Ni103 at 1050 °C brazing temperature and 30 s brazing time [4].

Brazing of the dissimilar joint steel-titanium aluminide with brazing filler metal Ni103

Further experiments have been conducted on the basis of the Ni103 brazing filler metal and the base metal material combination steel-titanium aluminide. Visual inspection show a good appearance of the joint.

The brazing filler metal has a similar hardness as those of the nickel-titanium aluminide dissimilar material joints. A further embrittlement of this brazing joint has, therefore, not developed in the dissimilar material joint steel-titanium aluminide.

Comparing different brazing times at the constant brazing temperature of 1050 °C shows the highest shear strength of 118 MPa at a brazing time of 30 s, Figure 9. For this joint brazing temperature and brazing time exert influence on the strength and microstructure of the brazed joint.

In a comparison of different brazing temperatures in the range of 1000 to 1100 °C, the steel-titanium aluminide dissimilar material joint with a Ni103 brazing filler metal has the highest strength at a brazing temperature of 1050 °C. A brazing temperature of 1100 °C and a brazing time of 30 s result in the overheating of the joining zone which is so high that a very strong diffusion between titanium aluminide - base material and brazing filler metal occurs. However in this case the diffusion zone between steel and brazing filler metal is wider and has a higher strength than joints brazed at 1050 °C. At this temperature, the diffusion is approximately two up to four times faster than it is at a temperature of 1050 °C.

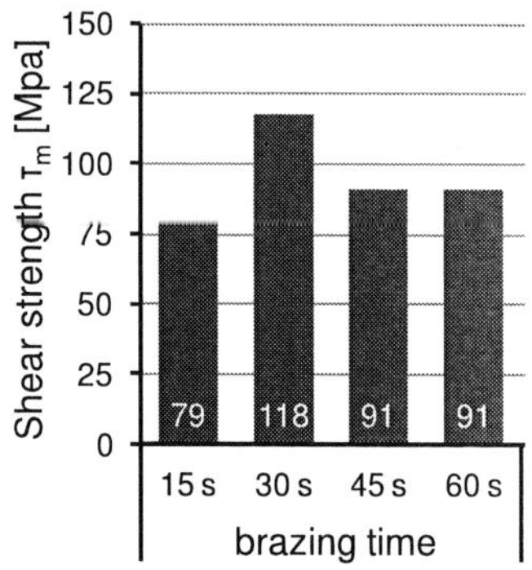

Figure 9. Comparison of the shear strength of different holding times for the dissimilar joint steel-titanium aluminide with the brazing filler metal Ni103 at 1050 °C brazing temperature.

Brazings with 1000 °C brazing temperature have a small diffusion zones between brazing filler metal and base metal. Specimens made at 1050 °C brazing temperature are breaking in the brazing filler metal or in the steel diffusion zone. However in this case the diffusion between steel and brazing filler metal is about 2 µm wide. Specimens made at 1100 °C brazing temperature are breaking mainly in the diffusion zone between the titanium aluminide and the brazing filler metal. In this case the diffusion zone between steel and brazing filler metal is about 9 µm wide. Both diffusion zones growth because the diffusion speed rise by the factor of three when the temperature increases from 1050 to 1100 °C.

In order to get a higher strength of the whole brazing system it could be an advantage to braze with two different brazing temperatures at the different diffusion zones in one joint at one time. For the joint brazing temperatures of 1050 °C and at the titanium aluminide diffusion zone and 1100 °C brazing temperature at the steel diffusion zone could be optimal. This system can ideally be set up by brazing with the electron beam as a heat source. The beam can be fast and precise controlled. This offers a flexible heat input on both specimens.

A joint brazed with the parameters described above has diffusion zones to both base metals, Figure 10. Especially the diffusion zone between steel and brazing filler metal has in this case a sound good width with 6 µm. Due to optimal diffusion on both sides of the joint the tensile strength is with 140 MPa higher than the joints made with similar temperatures on both diffusion zones (the tensile strength R_m = 118 MPa of joints brazed at 1050 °C brazing temperature; R_m = 100 MPa of joints brazed at 1100 °C brazing temperature).

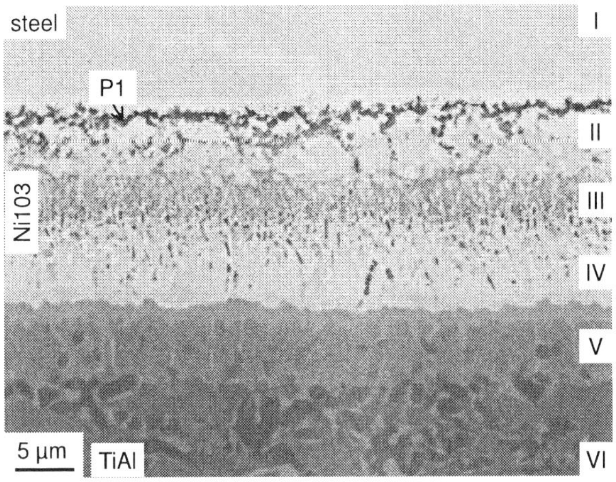

Figure 10. SEM-picture of a brazed joint of the dissimilar material joint steel-titanium aluminide with the brazing filler metal Ni103 at 1050 °C brazing temperature at the titanium aluminide sheet and 1100 °C brazing temperature at the steel sheet and 30 s brazing time.

Conclusions

The joining of dissimilar metal welds is a great challenge for the joining technology. The selection of the brazing filler metal is always a compromise between the suitability of the brazing filler metal for both base materials. Brazing temperature and brazing time have a great influence on the micro-structure and thus on the strength of the joint. What means an optimal brazing temperature and brazing time for one base material may bring along unexpected and unwanted interactions with the second base material. Through indirect energy input via electron beam, specimens have been brazed in vacuum at different temperatures and brazing times.

The Ni103 brazing filler metal is a highly promising brazing filler metal for the production of nickel-titanium aluminide dissimilar material joints as well as steel-titanium aluminide dissimilar material joints. Here, the energy input and the brazing time are important parameters for the control of the diffusion speed. If the diffusion is too low, the adhesion between brazing filler metal and base material is insufficient. If the diffusion is too high, a brittle intermetallic phase develops in the brazing filler metal. With a brazing temperature of 1050 °C and a brazing time of between 30 and 45 s the shear strength value is 87 MPa in the case of nickel-titanium aluminide joints. This joint consists of two diffusion zones and of a dendritically grown brazing filler metal where two mixed crystals are embedded. Complex microstructures and morphology of most interesting specimens have been analysed. In doing so, characteristic zones in the transition zone between base materials and brazing filler metal have been identified and differentiated.

The highest tensile strength can be achieved for dissimilar materials joints if both diffusion zones are brazed with their individual optimal brazing temperatures. The brazing temperature in each diffusion zone must be determined by considering the interaction between brazing filler metal and individual base metal.
In order to braze with individual brazing temperatures at each diffusion zone the brazing process requires a precise energy source. The electron beam with beam deflection offers such precise energy input.

References

[1] Lugscheider, E.: Löten von Nickel- und Titanaluminiden; final report DFG projekt LU 232/82 (2004).
[2] Clemens H., Mayer S.: Intermetallic γ-Titanium Aluminide Based Alloys from a Metallographic Point of View -a Continuation, Practical Metallography, 48 (2011), pp. 64-100.
[3] Reisgen, U., Olschok, S., Backhaus, A.: Microstructure Analysis of Electron-Beam Brazed γ-Titanium Aluminide, Materials Science Forum 690 (2011), pp. 153-156.
[4] Budberg P.B.: Aluminum-Nickel-Titanium, Ternary Alloys, 8 (1993) pp. 7–21.

Metal Leaching of Brazed Stainless Steel Joints into Drinking Water

T. Hartmann
Vacuumschmelze, Hanau, Germany
Thomas.Hartmann@Vacuumschmelze.com

Abstract

Due to hygienic considerations and improved corrosion resistance, the use of stainless steel in heat exchangers and plumbing systems for domestic drinking water installations is increasing. In Europe, and especially in Germany, stainless steel is replacing copper alloys in these applications. The hygienic suitability of stainless steel for drinking water applications is well investigated, documented and approved. However, only a few research papers can be found dealing with hygienic aspects of brazed joints made on stainless steel components, such as heat exchangers.

This paper presents an investigation made on 316L stainless steel joints where plate heat exchanger samples are brazed using a range of copper, nickel and iron based brazing alloys in form of amorphous foils. The metal leaching of these samples into drinking water was examined and the results disclose a distinct improvement in the hygienic suitability of modern VITROBRAZE® compositions over other state of the art brazing alloys.

Introduction

The primary purpose of national and international drinking water standards and guidelines [1-4] is the protection of public health. Drinking water is required for all usual domestic purposes, including drinking, food preparation and personal hygiene. A safe drinking water should not represent any significant risk to health over a lifetime of consumption. To ensure a high level of health protection, drinking water shall meet minimum requirements in the field of appearance, microbiology, radioactivity and chemistry. These values are available based on the scientific knowledge.

Drinking water contains a certain amount of heavy metals due to natural occurrence of these elements or as result of metal leaching from contact with domestic drinking water installations. The maximum concentrations for some brazing related metallic elements are listed in Table 1. Since the precautionary principle has also been taken into account, the German drinking water ordinance [3] demands the minimization principle. Only materials or components should be used for drinking water applications which release a minimum of any substances into the water and should not emit higher values of substances than would be unavoidable in compliance with the state of the art and technology. Due to these hygienic considerations, corrosion resistant stainless steel is replacing copper alloys and zinc coated steel in heat exchangers and plumbing systems for domestic drinking water installations. The hygienic suitability of stainless steel for food processing and drinking water applications is well investigated, documented and approved.

Table 1: Chemical elements of brazing alloys and their maximum concentration limit in drinking water according to the U.S. EPA regulations, the European drinking water directive and the WHO-guideline. Values are given in µg/l.

Element	U.S. EPA	EG council directive 98/83	WHO-guideline value 2011
Fe	-	200	-
Cr	100	50	50
Cu	1300	2000	2000
Ni	-	20	70
Mn	-	50	-
B	-	1000	2400

However, only a few research papers can be found dealing with the hygienic aspects of brazed joints made on stainless steel components, such as heat exchangers [10,12,13]. It is well known that brazed stainless steel structures release a certain amount of involved elements when placed in contact with drinking water [5-13]. Some statements in technical literature conclude that brazed joints on stainless steel components made with Ag-, Cu- or Ni-based brazing alloys do not meet the corrosive or hygienic requirements of drinking water installations, irrespective of the type of brazing alloy used [5,6]. Such statements caused uncertainty to the industry and end users of such components.

Germany's Federal Institute for Material Testing examined several cases of damaged copper brazed heat exchangers after contact with drinking water [7-9]. These investigations disclosed that copper brazed stainless steel heat exchanger exhibit significant corrosion of the brazed joints as well as high copper leaching rates into the drinking water. Although both copper and stainless steel are classified as resistant towards drinking water, the brazed joint may generate corrosive problems if exposed to specific water qualities. Corrosion was always found in the drinking water circuit of such components and did not occur on the heating water circuit. Virgin heat exchangers especially, exhibit high copper leaching rates exceeding the allowable limits during the first weeks of service.

Surprisingly, other investigations show that copper brazed stainless steel heat exchangers may also exhibit significant leaching of nickel into the drinking water. Furthermore, it was discovered that copper brazed heat exchangers exhibit significantly higher nickel leaching rates to the drinking water than components brazed with an nickel brazing alloy [10,11].

Due to the increasing occurrence of nickel hypersensitivity in humans as well as the permanent tightening of the general toxicity level for nickel, strict limitations can be found in some governmental drinking water regulations. In Europe especially, where the nickel limit is set as low as 20 µg/l [2] there is still some confusion as to whether brazed stainless steel components meet the European hygienic requirements. Some research work [12,13] was already done to investigate the emission of nickel ions out of various nickel- and iron based filler metal surfaces into drinking water. These investigations were carried out as a dynamic rig test according to European and German standards, using tube sections coated internally with filler metal as specimen [14,15]. The results show that the hygienic suitability of all tested brazing alloys is comparably as good as the unbrazed 316L surface. However, the specific testing method was configured to investigate individual materials, for example a tube material, where a stable material and surface condition over the whole sample was given. This test method did not reflect the different material conditions occurring on a brazed joint where we have a base material, a filler material and a reaction zone with a material mix of all involved elements. It is well known that such a reaction zone tends to be the weakest area of a brazed component in terms of corrosion resistance.

The target of this study is to provide a baseline investigation for the metal leaching of individual nickel containing brazing materials into drinking water and compare them with the metal leaching of brazed stainless steel structures with surface conditions equivalent to those of a real heat exchanger. One goal was to carry out this investigation based on heat exchanger like samples and with testing methods less complex and expensive than a dynamic rig test according to [14]. Several brazing alloys were tested with the intention of finding suitable brazing alloys where the leaching of undesired metals is minimized. The aim was to identify optimized material combinations leading to a reduced water contamination on one hand and to improved corrosion resistance and service life of the brazed component on the other.

Experimental

Today the majority of plate type, stainless steel heat exchangers are brazed in a vacuum furnace using filler metals in the form of thin metallic foils [11,16,17]. Several nickel and iron based amorphous brazing foils (ABF) as well as a crystalline Cu102 foil are considered in this investigation (see Table 2). The outstanding brazeability of the amorphous VITROBRAZE foils, the microstructural compatibility of the joints and the good mechanical durability of the brazed components are reported elsewhere [16-20]. These foils were prepared to form shims with dimensions of 25 mm x 40 mm x 0.05 mm. The compositions of the brazing foils are listed in Table 2. The austenitic stainless steel grade 316L was chosen as base material.

Heat exchanger plates with a thickness of 0.3 mm and the typical herringbone embossed pattern were provided by a heat exchanger manufacturer. Small samples of these plates, 25 mm x 40 mm were precisely machined using a wire cut EDM machine (Figure 1). To form a structure similar to a heat exchanger, the corrugated 316L samples were stacked with alternating change in direction of the embossed pattern. This results in an open structured 3-dimensional specimen in which a fluid can freely pass. Each specimen consisted of 7 steel layers with a total height of about 16 mm and a total weight of about 18.4 g. All parts are cleaned with Isopropanol in an ultrasonic bath for 15 minutes prior to assembly. One brazing foil shim was placed between each steel layer. A picture of the brazed specimen is shown in Figure 2. Each brazed specimen has 144 punctual brazed joints and a total surface of about 165 cm^2, of which 140 cm^2 is in contact with the brazing alloy. An Al_2O_3 fixture was used to compress the stack during the brazing process and to keep the plates in correct position. Brazing was carried out in a vacuum furnace with a vacuum level of about $1x10^{-4}$ mbar. Individual brazing parameters were set depending on the foil composition (see Table 2).

Table 2: Chemical composition of brazing foils and base material, brazing temperature - T_b and duration at brazing temperature - t_b

Material		Chemical composition in weight %								Brazing conditions	
ISO 17672	VITROBRAZE® designation	Ni	Fe	C	Cu	Cr	Mo	Si	B	T_b [°C]	t_b [min]
Ni610	VZ2111	75.5	4.2	≤ 0.06	-	13	-	4.5	2.8	1130	30
Ni620	VZ2120	82.3	3	≤ 0.06	-	7	-	4.5	3.1	1080	30
-	VZ2150	73.3	-	≤ 0.1	-	18.2	-	7.3	1.2	1190	30
Ni661	-	76.3	-	≤ 0.06	-	15	-	7.3	1.4	1190	30
-	VZ2106	44	35	≤ 0.1	1	11	1.5	6.4	1.5	1190	30
Cu102	-	-	-	-	99.95	-	-	-	-	1130	30
Stainless steel 316L		12	68	≤ 0.03	-	17.5	2	≤ 1	-	-	-

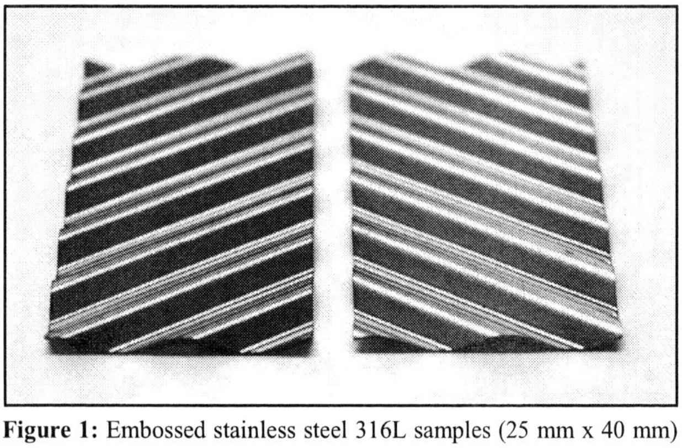

Figure 1: Embossed stainless steel 316L samples (25 mm x 40 mm) prior to assembling and brazing

Figure 2: Vacuum brazed specimen using VZ2150 brazing foil as filler material.

To investigate the metal leaching behavior of the plain metals as well as of the brazed specimens, an immersion test using the local drinking water quality was carried out. A detailed analysis of the local water quality is given in Table 3. Due to the elevated values of hardness and electrical conductivity, this water quality may become problematic under corrosive considerations.

After cleaning and activation of the metal surface of the brazed specimens by rinsing them for 24 h with drinking water, each specimen was completely submerged for 360 h in a closed plastic beaker (Ø 68 mm, h = 80 mm) containing 60 ml of fresh local drinking water. After this test cycle, the specimen was removed from the beaker and the water samples analyzed by means of inductive coupled plasma emission spectrometry to determine the metal concentration leached into the water. For this analysis, a standard analytical instrument, an Agilent Technologies ICP-OES 720 Series Axial, was used. The accuracy of the measurement is within 1-5 % when measuring the amount of heavy metals in aqueous solutions. The lower detection limit for nickel and iron is

Table 3: Quality data of the local drinking water used for this investigation (Water treatment plant II, Hanau, Germany).

Compound		Unit	Averaged value
Temperature		°C	14
pH-value			7.9
Electrical conductivity		µs/cm	637
General hardness		mmol/l	2.55
Hydrogen carbonate	HCO_3^-	mmol/l	3.06
Carbon dioxide	CO_2	mmol/l	0.2
Dissolved oxygen	O_2	mmol/l	0.3
Calcium	Ca	mg/l	75.2
Chloride	Cl	mg/l	66.5
Chlorine dioxide	ClO_2	mg/l	0.06
Iron	Fe	mg/l	0.017
Nickel	Ni	mg/l	< 0.005
Fluoride	F	mg/l	0.1
Potassium	K	mg/l	7.1
Magnesium	Mg	mg/l	15.7
Sodium	Na	mg/l	35.2
Nitrate	NO_3	mg/l	7.4
Phosphate	P_2O_5	mg/l	1.2
Sulfate	SO_4	mg/l	69.9
Total organic carbon	TOC	mg/l	1.8
Uranium	U	mg/l	0.0003

5 µg/l, chromium 2 µg/l and for copper it is 3 µg/l. Values below these limits will be displayed as n.n. - not noticeable. To determine the time dependent metal leaching behavior of the brazed samples a long-term investigation over several test cycles was carried out, because it is well known that due to a passivation process the metal leaching rate of brazed components will reduce after some weeks in service [6,10,13].

To obtain the metal leaching of the individual plain materials the immersion test, following the same testing routine, was carried out using material samples with a total surface area of 140 cm^2. Due to the absence of grain boundaries and their better chemical homogeneity, amorphous metals offer much better corrosion resistance than their crystalline counterparts. As this will certainly influence the metal leaching rate, the amorphous VITROBRAZE® foils were converted to the crystalline state by applying a heat treatment at 500°C for 1 h, which is above the crystallization temperature – T_x of the alloys. The Cu102 foil was tested without heat treatment.

To avoid contamination, the brazed specimens as well the water sample were not in contact with any other metallic component during the entire test procedure. The water sampling was always done after rinsing the water line for 15 minutes. For each test cycle, one plastic beaker with water, but without metal specimen, was processed and analyzed for reference. The analyzed metal content of the reference water was subtracted from values of the metal samples.

Results and Discussion

Plain brazing alloys

In contact with the local drinking water, all tested brazing foils show significant metal leaching for some elements, whereas the plain stainless steel 316L did not release any measurable amount of metal ions. The obtained values for one test cycle are shown in Table 4. Dependent of the chromium and boron content, all nickel foils release different amounts of nickel into the water. The nickel values decrease from Ni620 (777.2 µg/l) → Ni610 (463.5 µg/l) → Ni661 (174 µg/l) → VZ2150 (24.4 µg/l). Metal ions of chromium or iron could not be detected. For the copper foil Cu102, a copper leaching of 754 µg/l could be measured. With the release of only 17.4 µg/l nickel, the lowest metal leaching value is obtained for the Fe/Ni-Cr foil VZ2106.

Brazed stainless steel specimens

In figure 3 a comparison of the measured metal leaching values for the first test cycle of the brazed specimens is shown. Significantly different metal leaching rates were obtained for the brazed specimens when compared to the metal leaching of the plain brazing alloys. The type, as well as the quantity of the released elements differs from alloy to alloy. The variation of the brazing alloy in the brazed specimens leads to significant differences of their metal leaching behavior. The results obtained from the first test cycle showing the highest values for all tested material combinations.

The nickel brazed specimens using the brazing alloys Ni620 and Ni610, containing a relatively high boron amount of approximately 3 wt.%, exhibit the highest metal leaching within this test, whereas the modern nickel brazing alloys containing lower boron content (≤ 1.5 %) and higher chromium concentration release substantially less metal ions into the drinking water. In addition to nickel, a fairly substantial release of iron could be measured for all nickel brazed specimens.

The specimen using Cu102 as brazing alloy released a copper concentration of 1074 µg/l into the water. Furthermore, nickel concentration of 116 µg/l and iron concentration of 5 µg/l was measured. The release of nickel out of austenitic stainless steel joints brazed with copper are in accordance with the results obtained by [9,10]. The microstructural analysis (Figure 4) of the Cu102 brazed specimen exhibits a selective corrosive attack along the grain boundaries of the copper crystals even if the exposure time in drinking water was only 360 h (one test cycle).

The lowest amount of 6.7 µg/l metal ions could be detected for the specimen brazed with the Fe/Ni-Cr brazing foil VZ2106.

Long term investigation of brazed stainless steel specimens

At the initial test cycle the specimens using the brazing alloys Ni661, VZ2150 and VZ2106 release substantially lower amounts of metal ions into drinking water than obtained for the Cu102 brazed specimen. To determine the time dependent metal leaching behavior of these promising specimens a long

Table 4: Metal leaching of the individual plain alloys into drinking water. Total sample surface was 140 cm^2, immersion time 360 h. ABF's are tested in crystalline state (after heat treatment of 500°C for 1 h). Stainless steel 316L was tested in a heat treated condition at 1190°C for 0.5 h in a vacuum furnace. Values are given in µg/l.

Alloy	Cu	Ni	Fe
Cu102	754.0	n.n.	n.n.
Ni610	n.n.	463.5	n.n.
Ni620	n.n.	777.2	n.n.
Ni661	n.n.	174.0	n.n.
VZ2150	n.n.	24.4	n.n.
VZ2106	n.n.	17.4	n.n.
SS316 L	n.n.	n.n.	n.n.

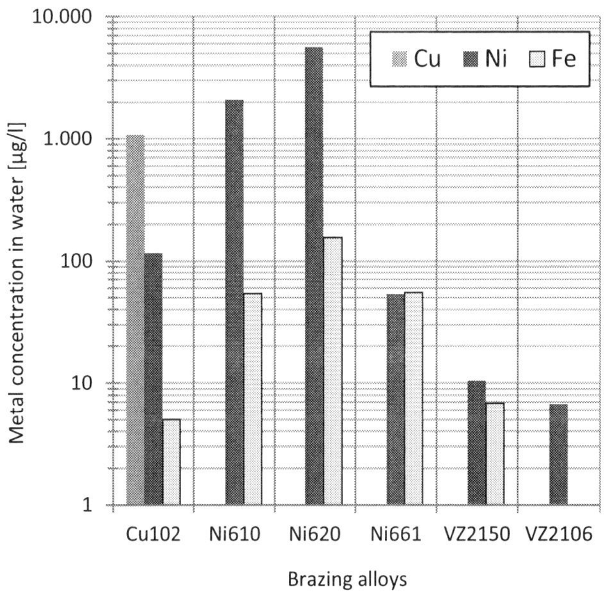

Figure 3: Metal leaching into drinking water of brazed 316L specimens joined using various brazing alloys. Values are given for the 1st immersion cycle (360 h).

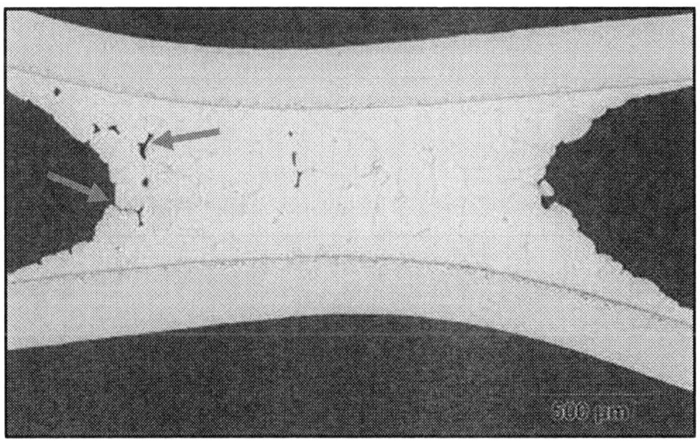

Figure 4: Microstructure of a 316L specimen brazed with Cu102 after 360 h immersion in drinking water. Selective corrosive attack could be observed at the grain boundaries of the copper crystals (red arrows).

term investigation was carried out. Figure 5 compares the nickel leaching of these samples over a period of 6 test cycles. The brazed specimen using VZ2106 show a rapid passivation and did not release any detectable level of metal ions into the drinking water for all further test cycles. The metal leaching for the brazed VZ2150 specimen dropped below the detection limit at the 6th test cycle, and for the Ni661 specimen at the 5th test cycle. For the Cu102 brazed specimen, a distinct release of copper and nickel ions could be measured for all test cycles.

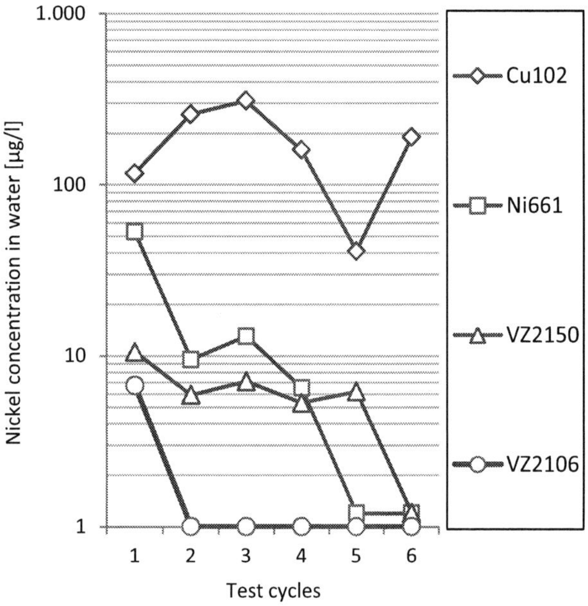

Figure 5: Nickel leaching of brazed specimens into drinking water using various brazing alloys measured over a period of 3 months (360 h per test cycle).

In table 5, the results for the 1st and 6th test cycle are compared and lead to the conclusion that the tested Ni-Cr and Fe/Ni-Cr brazing foils are promising candidates offering a lower susceptibility to corrosion and the associated metal ion leaching. With these compositions it might be possible to obtain brazed stainless steel components offering an improved hygienic suitability over the state of the art copper brazing alloys for drinking water applications.

The metal leaching of brazed stainless steel structures into an aqueous solution is a complicated process affected by many chemical, electrochemical and metallurgical variables. Unexplained today are the mechanisms leading to the significant nickel leaching of the copper brazed specimens. There is no doubt that it originates from the austenitic stainless steel, but it was unexpected that no comparable amounts of Iron and Chromium could be detected since a corrosion process of the stainless steel was assumed. Explanations for this phenomenon may be found on the metallurgical side of braze joint formation. Nickel is soluble in Copper at all concentrations, whereas Iron and Chromium are insoluble in Copper at room temperature. During brazing operation at about 1120°C a partial dissolution of the stainless steel into the liquid Copper will occur. During the following cooling and solidification of the joint, a precipitation of Iron and Chromium crystals will take place at the grain boundaries of the Cu(-Ni) solid solution. Thus, a local corrosion cell is formed leading to an accelerated selective corrosive attack at the grain boundaries of the affected Cu(-Ni) solid solution if they are in contact with a conductive aqueous solution. This may explain our findings of the selective corrosive attack at the copper joint visible in Figure 4. However, further studies are necessary to investigate the involved corrosion mechanisms.

The corrosion resistance of stainless steel is mainly influenced by the presence and the quality of a protective chromium oxide surface layer. Against this, the positive results of brazed VZ2106 joints are surprising. Although the Chromium content of this Fe/Ni-Cr brazing foil is lower than for the Ni-Cr foil VZ2150 (11 % vs. 18 %), the metal leaching into drinking water is on an outstanding low level. The underlying corrosion mechanisms are not clear yet and will be object for further studies. Clearly, the Molybdenum and Copper additions of the VZ2106 alloy will have a beneficial effect to the quality of the protective oxide layer. But it is assumed that also electrochemical aspects will lead to the superior corrosion resistance of this alloy. An investigation of the current density potential of these brazed structures is necessary to complete this research work.

All obtained results can vary with the quality of the individual drinking water. For other water qualities, different results may be obtained.

Table 5: Total amount of Cu+Ni+Fe metal ions leached from brazed 316L specimens into drinking water.

Brazing alloy	Total metal leaching of Cu+Ni+Fe [µg/l]	
	Test cycle 1	Test cycle 6
Cu102	1195.0	820.0
Ni661	108.0	n.n.
VZ2150	17.3	n.n.
VZ2106	6.7	n.n.

Summary

The metal leaching of brazed stainless steel specimens into drinking water was examined. As part of this investigation it appears that drinking water is a corrosive environment demanding a high corrosion resistance of the brazing alloy to obtain an non-critical metallurgical bond in terms of hygienic considerations. The obtained results confirmed that it is not possible to characterize corrosive or hygienic aspects of a braze joint by studying the individual materials. It is imperative to analyze the metallurgical bond and the activities of the involved materials with respect to their thermal processing. A general statement as to whether a pure brazing alloy will release more or less metal ions into water than a brazed joint cannot be given. It was found that not all nickel brazing alloys are comparably suitable for drinking water applications. It must be assumed that the nickel brazing alloys Ni610 and Ni620, containing a high boron content of about 3 wt.%, are not applicable because they release high values of nickel into the water. The specimens brazed with the Ni-Cr alloys Ni661, VZ2150 and the Fe/Ni-Cr alloy VZ2106, containing a lower boron content of ≤ 1.5 wt.%, exhibit the lowest metal leaching rates within our tests and show them to be much more convenient for drinking water applications.

Our obtained results confirmed the assumption that copper brazed 316L joints release a significant amount of nickel in addition to copper and iron. A clear root cause for this specific characteristic could not yet be determined. For the brazed specimens using the Ni661, VZ2150 and VZ2106 brazing foils we have obtained a significantly lower nickel leaching rate than for the copper brazed specimen. The general leaching of metal ions (Cu+Ni+Fe) for the Cu102 brazed specimen is orders of magnitude higher than for the specimens brazed with the above mentioned Ni-Cr and Fe/Ni-Cr brazing foils.

The investigation of the corrosion mechanisms leading to the comparably low metal leaching of some Ni-Cr and Fe/Ni-Cr brazed stainless steel structures will be object of further research work.

With respect to the minimization principle, only materials or components releasing a minimum of substances into the water should be used for drinking water applications. It was shown that brazed stainless steel components using chromium and nickel containing brazing alloys exhibit significantly better hygienic suitability than state of the art copper brazed components.

Acknowledgments

I would like to thank the following colleagues of VAC for their valuable assistance during this investigation: Barbara Plutta, Bernd Hain, Michael Koehler and Harald Staubach.

Literature

[1] N.N. (2011) World Health Organization, Guidelines for drinking-water quality (GDWQ), fourth edition, ISBN: 978 92 4 154815 1

[2] N.N. (1998) EU Council Directive 98/83/EC, On the quality of water intended for human consumption

[3] N.N. (2001) German drinking water ordinance – TrinkwV

[4] N.N. (2009) United States Environmental Protection Agency, National Primary Drinking Water Regulations

[5] G. Herbsleb (1989) Corrosion of joints for stainless steel tubes in water, Materials and Corrosion, Volume 40, Issue 11, pp. 651–660

[6] G. Pajonk (2004) Korrosionsschäden an gelöteten Plattenwärmetauschern, Sonderbände der praktischen Metallographie Vol. 36, pp.371-378

[7] J. Erning (2003) Korrosion von kupfergelöteten Plattenwärmeübertragern – die Folgen, VDI-Berichte Nr. 1765, pp. 85-95

[8] W. Stichel (1998) Corrosion damage in connection with brazed plate heat exchangers made of stainless steel, Materials and Corrosion 49, pp. 189-194

[9] S. Koehler (1999) Damage on brazed plate heat exchangers, Material and Corrosion 50, pp.227-323

[10] N.N. (2009) Migration von Schwermetallen aus Bauteilen der Trinkwassererwärmung, DVGW research projekt W10/01/05

[11] A. Rabinkin (2002) US Patent 6,749,104 B2

[12] B. Wielage (2010) Development of iron filler metals for high-temperature brazing of materials in contact with drinking water, DVS Berichte 263, pp. 189-194

[13] I. Hoyer (2013) Nickel ions in potable water and their migration out of brazing fillers and brazed joints, DVS Berichte 293, pp.176-181

[14] DIN EN 15664: Influence of metallic materials on water intended for human consumption – dynamic rig test for assessment of metal release

[15] DIN 50930-6: Corrosion of metallic materials under corrosion load by water inside of pipes, tanks and apparatus

[16] C. Shu (2012) Developing corrosion-resistant Joints applied in the Plate Heat Exchanger, Advanced Material Research Vol. 410, pp 191-195

[17] T. Hartmann, D. Nuetzel (2009) New amorphous brazing foils for exhaust gas applications; Proceedings of the 4th International Brazing and Soldering Conference, IBSC 2009, Orlando, USA, pp. 110-118

[18] T. Hartmann, D. Nuetzel (2007) Iron containing brazing foils for joining of stainless steels; Proceedings of the 8th international Conference on Brazing, High Temperature Brazing and Diffusion Bonding, LÖT 2007, Aachen; DVS Berichte 243, pp. 140-145

[19] W. S. Chen (2012) Brazing Inconel 625 using two Ni/(Fe)-based amorphous filler foils, Metallurgical and Materials Transactions A, Volume 43A, July 2012, pp 2177-2182

[20] W. S. Chen (2012) Microstructural evolution of brazing 304 stainless steel using an amorphous (Ni,Fe)-based foil, ISIJ International, Vol. 52, No. 5, pp 939-941

Corrosion of α-Al_2O_3/Ag-Cu-Ti/ STS Brazed Joints by Molten Sodium for Sodium Sulfur Batteries

Hoe-jun Heo, Chung-yun Kang
Division of Material Science and Engineering, Pusan National University, Busan, Republic of Korea

Keeyoung Jung
RIST Research Institute of Industrial Science and Technology, Pohang, Republic of Korea

Abstract

One of several joints in sodium sulfur (NaS) battery, ceramic-metal joints were mainly applied with thermal compression bonding (TCB). Al-based insert metal is mainly used to braze with α-Al_2O_3 and Al bracket during TCB process. However, if β"-alumina was broken in emergency, inner parts of battery will grow up to 500℃ due to rapid reactions with Na and S. With Al bracket and Al insert metal, it cannot certify the safety of joints due to their low melting point, so high temperature resistance filler metals are needed. When the new insert metal for TCB process was applied, corrosion resistance should be studied due to loss of bonding area by molten sodium.

The present study seeks to observe corrosion characteristics of α-Al_2O_3 /Ag-Cu-Ti/ SS 430 brazed joints by the molten sodium. The results showed that the Ag-rich matrix in the brazed joints were disappearing after immersing the brazed joints into molten sodium, whereas other phases, Ti-rich continuous reacted layers and Cu-rich beta phases, were not affected. The corroded depth by sodium corrosion increases with exposure time. With Ag-Cu-Ti filler alloys, it cannot satisfy the lifetime of the TCB joints due to fast corrosion kinetics.

Introduction

Sodium sulfur (NaS) battery has become one of the most promising electrochemical devices for large scale energy storage systems (ESS) applications [1]. It utilizes molten sodium and sulfur as electrodes, separating them by a β"-alumina solid electrolyte, which allows sodium ion (Na^+) conduction at high temperatures between 350°C [2~4]. The battery has a lot of attractive features such as high capacity of electronics (up to 230kWh per 1kg), low expense cost (up to 300$ per kWh), high rate charge-discharge (85~90%) and long life cycle (up to 15years). Many practical developments were accomplished for this second battery in Japan and U.S.A. The research workers in SICCAS of China have been developing variable sodium sulfur battery technologies [5].

NaS cell (500~1,200Wh in capacity) is composed with several heterogeneous joints such as ceramic-ceramic, metal-metal, and metal-ceramic joints. Among several joints in the battery, ceramic-metal joints are mainly achieved by thermal compression bonding (TCB) technology. TCB is to bond between ceramic and metal components using an active filler metal which allows chemical bonding at the interface.

In aluminum based structure for NaS cell, Al-Mn alloy such as Al 3003 (eutectic temperature about 653 °C) was utilized as Al bracket and Al-Si-Mg alloys such as Al 4004, Al 4104 and Al 4032 (eutectic temperature in range of 550-580°C) are generally used as filler alloys for TCB in sodium sulfur cells. However, if β"-alumina was broken in emergency during operation, liquid states of sodium and sulfur in the battery will directly react and inner cell temperature will increase rapidly up to 500 °C [6]. With Al bracket and Al filler metal, it cannot certify the safety of joints, so high temperature resistance filler metals are needed.

To select the alternative filler metals, melting temperature should be considered. In the present study, Ag-Cu based active filler metals, which have an active element (Ti, Zr, V, Mg) promoting wetting on the ceramic surface, were tested due to their eutectic temperature in range of 780 °C and maximum application temperature (about 600°C) [7]. Also, aluminum bracket was changed to stainless steel (SS 430), which has melting point about 1510 °C, as metallic part.

In order to ensure long lifetime of the TCB heterogeneous joints in tubular sodium sulfur battery, various issues such as loss of bonding area by molten sodium corrosion, mechanical failure and low He leakage have been studied previously [8-9]. Especially, it is important to sustain a sufficient bonding strength over the lifetime against decrease in bonding area derived from corrosion of filler alloy by molten sodium.

Therefore, reduction of corrosion kinetics of the filler metal by molten sodium is critical. In this study, corrosion characteristics of the α-Al_2O_3 /Ag-Cu-Ti/ SS 430 brazed joints in molten sodium was studied measuring corrosion kinetics.

Experimental Details

Bonding experiment

Table 1 shows commercial Ag-Cu based active filler metals, Nominal compositions in weight percent and eutectic and liquidus temperatures are shown in Table 1. Among these brazing metals, Ag-26.7Cu-4.5Ti was chosen because it has more Ti contents than others in this study. Ag-26.7Cu-4.5Ti sheet form (10mm diameter, 200μm thickness) was applied to the brazing experiment. Cylindrical rod of α-Al_2O_3 (99.7% purity) and SS 430 (10mm diameter, 5mm thickness) were used as the substrates in this study.

Brazing experiments were carried out in a graphite furnace under vacuum lower than $1.0 * 10^{-5}$ torr and heated up to 800 °C, followed by an isothermal holding for 10minutes. Cooling was conducted in rate of 1.7 °C/min. Heat cycle of the brazing experiment is shown in Figure 1.

Corrosion experiment

In order to observe the corrosion characteristics, a number of bonding specimens were exposed to high purity liquid sodium in alumina container. The containers were placed in vacuum furnace after purging argon atmosphere. Corrosion tests were conducted at 350°C up to 50hour. After finishing the tests, the alumina container was cooled down to room temperature in the furnace. The sodium attached on the specimens was cleaned in alcohol.

Table 1 Chemical composition and melting point of commercial silver-copper based active brazing alloys.

Chemical composition (wt. %)				Eutectic (°C)	Liquidus (°C)
Ag	Cu	Ti	Sn		
68.8	26.7	4.5	-	780	900
63.0	35.2	1.75	-	780	815
63.0	34.2	1.75	1.0	775	715

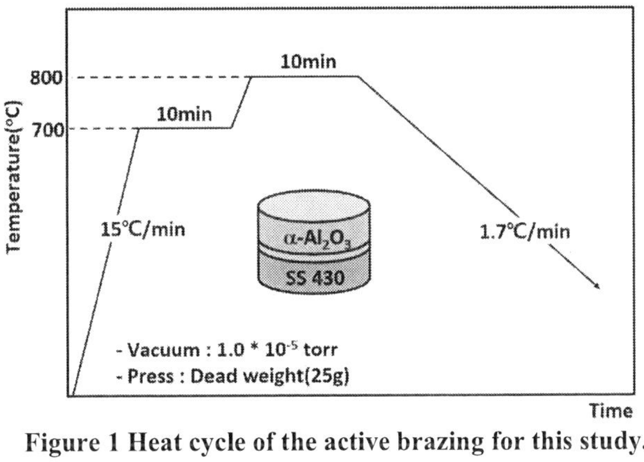

Figure 1 Heat cycle of the active brazing for this study.

Characterization of the joints

The brazed specimens were divided perpendicular to the interface and microstructure of the braze joints. Polished sample was carbon coated and characterized by scanning electron microscope (SEM) and energy-dispersive spectrometer (EDS).

Results

Microstructure of brazed joints

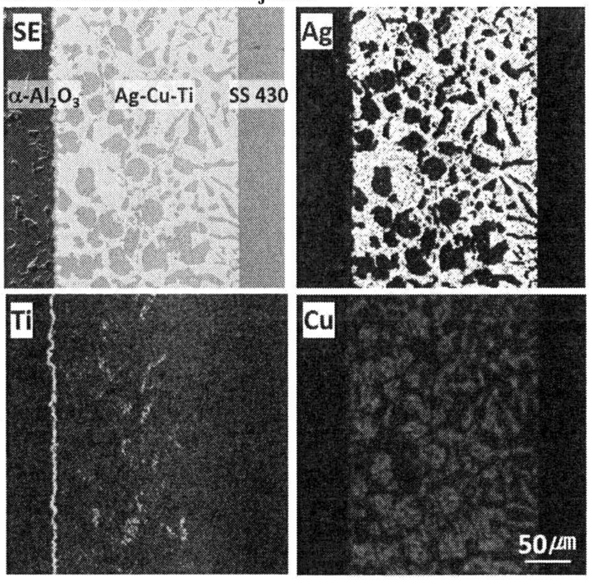

Figure 2 Cross sectional micrographs of brazed joints (EDS elemental mapping)

Figure 2 shows EDS compositional mapping of the joints brazed at 800°C for 10minutes to determine the elemental spatial distribution. Ti-rich continuous layers were formed along the α-Al_2O_3/Ag-Cu-Ti interface. In the brazed metal, lamellar structural phases (alternating Ag-rich alpha and Cu-rich beta phases) and primary beta phases (Cu-rich) are finely distributed in the Ag-rich alpha matrix. In Cu-rich primary beta phases, Cu-Ti phases are present.

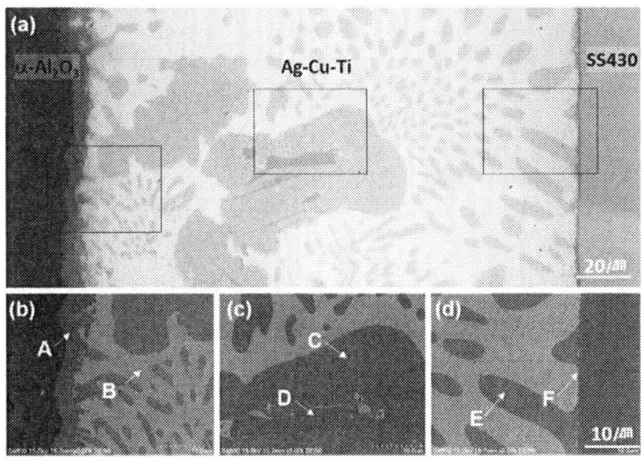

Figure 3 OM and SEM micrographs of joining area brazed at 800°C for 10 minutes.

In Figure 3, magnified images of the joints are shown with optical microscope (OM) and secondary electron (SE). Table 2 shows the EDS analysis results for the phases denoted as A~F in Figure 3. The EDS results indicates that the presence of Cu, Al, Ti and O in layer A and Fe and Ti in layer F. In primary beta phases (labelled C), Ti contents are higher than eutectic Cu-rich beta phases (labelled E) and darker seems to be intermetallic compound (labelled D) due to similar amount of Cu and Ti contents.

Table 2 EDX analysis of points in Figure 3 (at. %)

	Al	Ag	Ti	Cu	O	Fe
A	4.0	-	35.5	34.5	26.0	-
B	1.6	90.9	-	7.5	-	-
C	-	-	34.2	65.8	-	-
D	-	1.3	16.3	82.4	-	-
E	-	2.7	-	97.3	-	-
F	-	-	68.1	-	-	31.9

Corrosion characteristics by molten sodium.

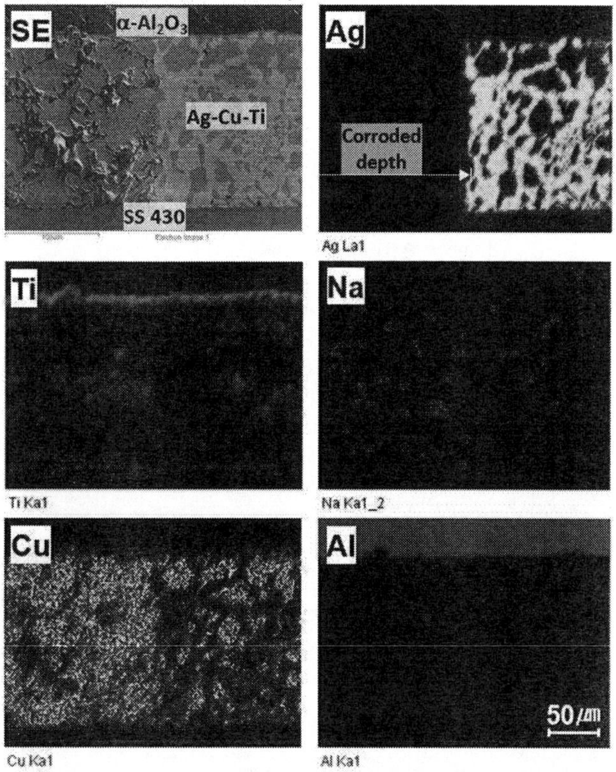

Figure 4 EDX elemental cross-sectional images of inner middle of corroded specimen in molten sodium at 350 °C for 50hr

Cross-sectional images of inner middle of corroded specimen in molten sodium at 350 °C for 50hr and the X-ray maps of the various elements (Ag, Na, Ti, Cu, Al) are shown in Figure 4. With X-ray maps, Ag element is clearly distinguishable from right and left side of the joints. Ag was not detected in the left side, but Ag matrix had existed in the left region. Sodium cannot be detected due to wash in alcohol and other elements such as Ti, Cu and Al are generally distributed on the joints regardless of molten sodium attack. Hence, only Ag-rich alpha matrix was removed due to molten sodium, whereas other phases are not affected.

Ag-Na phase diagram is shown in Figure 5 [10]. It seems that removal phenomenon of Ag matrix by sodium attack was related to liquid state at 350 °C in Na-rich region due to pure molten sodium. Further studies will be needed to demonstrate that conclusively.

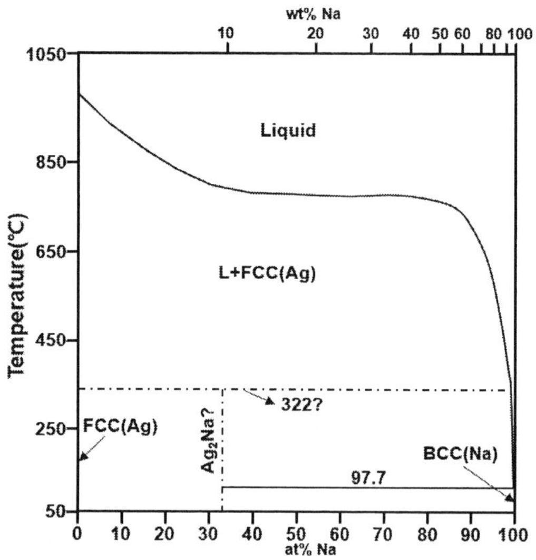

Figure 5 Ag–Na system phase diagram. [10]

Corrosion kinetics by molten sodium

Brazed specimens were exposed to molten sodium for 5, 10, 25 and 50 hours to investigate corrosion kinetics. Corrosion depth was measured with removed length of silver from the surface as denoted in Figure 4(Ag). The square of corrosion depth of corroded joint by molten sodium depending on the time is shown in Figure 7. It means that the corroded depth by molten sodium increases with exposure time. Corrosion kinetics by molten sodium on the brazing joints follows the parabolic rate law ($x^2 = k \cdot t$). The parabolic rate constant was obtained as $k = 3.0 \cdot 10^{-8}$ cm^2/sec. Calculated with the rate constants, corrosion depth from the surface for 15 years is about 47 mm at 350℃. So if Ag-Cu-Ti filler metal was utilized in sodium sulfur battery, it would take 37 days to corrode to the bonding width (~4 mm). Hence Ag-Cu-Ti filler alloy tested in this experiments was not proper to be adopted in sodium sulfur batteries.

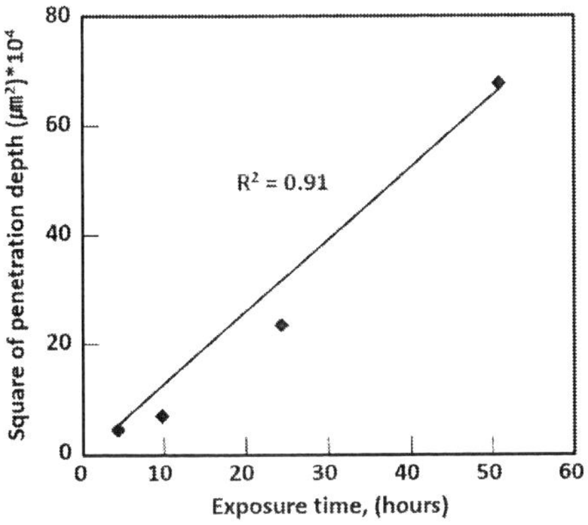

Figure 6 Plot of the square of corrosion depth of corroded joint by molten sodium depending on the time.

Conclusions

α-Al_2O_3/Ag-Cu-Ti/STS 430 brazed joints were exposed to molten sodium at 350°C for 50h to investigate the corrosion characteristics the heterogeneous joints in sodium sulfur cells. Ti-rich continuous layers were formed along α-Al_2O_3/Ag-Cu-Ti and Ag-Cu-Ti/SS 430 interfaces respectively. In the brazed metal, lamellar structural phases (alternating Ag-rich alpha and Cu-rich beta phases) and primary beta phases (Cu-rich) are finely distributed in the Ag-rich alpha matrix. After immersing the brazed joints into molten sodium, only Ag-rich alpha matrix was attacked. The corroded depth by molten sodium increases with exposure time. Corrosion kinetics by molten sodium on the brazed joints follows the parabolic rate law and parabolic rate constant was obtained as $k = 3.0 \cdot 10^{-8}$ cm^2/sec. Ag-Cu-Ti filler alloy tested in this experiments was not proper to be adopted in sodium sulfur batteries due to short application time(37days).

References

[1] J.A. De Gruson, in: The 1991 NASA Aerospace Battery Workshop, 1991, pp. 583-605.
[2] T. Oshima, M. Kjita, A. Okuno, Int. J. Appl. Ceram. Technol. 1 (2004) 269-276.
[3] R. Bauer, W. Haar, H. Kleinschmager, G. Weddigen, W. Fischer, J. Power Sources 1 (1976/77) 109-126.
[4] J.L. Sudworth, A.R. Tilley, The Sodium Sulfur Battery, Chapman and Hall, London, 1985.
[5] Z. Wen, J. Cao, Z. Gu, X. Xu, F. Zhang, Z. Lin, Solid State Ionics 179 (2008) 1697–1701
[6] Keeyoung J, Yoon-Cheol P, Chang-Soo, K.S. Patent 10-1309727
[7] Toshi Oyama and Edgar Vanegas., "Evaluation of Au-Based Active Brazing Filler Metals to Joint Alumina," Proc IBSC 5th, Las Vegas, Nevada, USA, April. 2012, pp. 93-100.
[8] Keeyoung Jung, Journal of Power Sources 250 (2014) 1-14
[9] Keeyoung Jung, Corrosion Science (under review, 2014)
[10] A.D. Pelton, Bulletin of Alloy Phase Diagrams Vol. 7 No. 2 1986

Corrosion Behavior of Contact Reaction Brazed Joints of Magnesium alloy to Steel

Hong Li , Hai-jie Nie , Zhuo Chen
College of Materials Science and Engineering, Beijing University of Technology, Beijing 100124.CHINA

Wei-min Long , Su-juan Zhong
State Key Laboratory of New Brazing Material and Technology, Institute, Zhengzhou 450001.CHINA

Abstract

In recent years, there are has been wide application of magnesium alloys in the industrial field. Corrosion behavior of magnesium alloy especially bonded joints of magnesium alloys and dissimilar materials become more severe. The corrosion behavior of contact reaction brazed joints of magnesium alloy and steel using different coating interlayers (Cu, Ni) was investigated by means of electrochemical techniques. The results show that the magnesium compounds were formed by interfacial reaction in joints. The corrosion current density of brazed seam was increased by nearly an order of magnitude compared with the magnesium alloy. Thus the corrosion resistance of brazed joints deteriorates. However, both the electrode potential and the current density of brazed joints with different metal interlayer were relatively close. The corrosion resistance of the brazed joint using nickel interlayer is slightly superior to those using copper interlayer. Galvanic corrosion is the main mode of corrosion in dissimilar metal joints and the most severely damaged area occurred in the vicinity of the interface of high potential. In addition, using metal interlayer of high potential will accelerate corrosion in brazed interface.

Introduction

Magnesium alloys are called "21st century green engineering of environmentally friendly materials" and have recently been receiving great interest for automotive, electronics, transportation, aerospace, and IT industry[1-3], because of its low density (1.74g/cm^3), specific strength and specific stiffness, shock resistance, anti-magnetic, easy machinability and dimensional stability. However, welding magnesium alloys to dissimilar metals (such as steel, Al...) has become a challenge[4]. Magnesium and steel exhibit poor metallurgical compatibility as the maximum solid solubility of Fe in Mg is 0.00041 at% according to the Mg-Fe binary alloy phase diagram, which brings difficulties to join magnesium alloys to Fe [5]. Although laser welding [6], gas tungsten arc (TIG) welding [7], friction stir welding (FSW)[8], transient liquid phase diffusion bonding (TLP) [9] and hybrid methods [10-11] have been used to weld magnesium alloys, the difficulty of joining of magnesium to steel could be relieved by using brazing techniques. One main problem for brazing magnesium and dissimilar metals is the oxide layer that formed on the surface and prevents wetting and flow of filler metals. Usually fluxes are used to remove this oxide layer but fluxes may also cause corrosion problems. Fluxless soldering with ultrasonic-assisted had been performed to join Mg alloy to steel using Sn based filler metals [12]. Vacuum diffusion brazing of Mg to steel was also carried out by using PVD coating method to deposit an interlayer as filler metals [13].
But problems arise from both brazing technologies and the corrosion behaviour of joined parts. Intermetallic layers formed at the interfaces might degrade the joint strength. In addition, the corrosion rate and corrosion potential difference between base metals and brazed joint affect corrosion resistance. Therefore, the corrosion resistance of brazed joints is not only affected by dissimilar metals corrosion, but also by the complex role of galvanic corrosion between the joint regions [14].
The objective of the present study was focused on the examination of a contact reaction brazing method and study of the corrosion properties of the Mg/steel joint for evaluating the reliability of brazed joints. This study also aims to explore the reasons for the change in corrosion resistance of the brazed joints.

Experimental

The base materials used for brazing were commercial magnesium alloy AZ31B and low carbon steel AISI 1008 plates with their chemical compositions and mechanical properties listed in **Table 1**. Plates were cut into sheets of 30mm×10mm×3mm chips. To prevent the volatilization of magnesium alloy in vacuum furnace, micro-arc oxidation layer was preset on the surface of magnesium alloy AZ31B except the surface to be brazed. Two kinds of intermediate layers (pure copper and pure nickel) were chosen as filler metal to join magnesium and steel. The copper coatings were deposited on the surface of magnesium to be brazed by Cold Gas Spraying (CGS), while the Ni coatings were deposited by Arc-spraying respectively. The bond surfaces of magnesium alloy AZ31B were prepared, using grinding papers with different stages up to 1000 grit, and were subsequently cleaned ultrasonically in alcohol before spraying. The AISI

1008 steel plates were ground using SiC paper to 600 grit and were subsequently cleaned ultrasonically in alcohol before brazing. The bonding process was performed in a chamber under a vacuum pressure of 5 × 10-3Pa. In order to enhance surface contact the joint assembly was pressed under a uniaxial pressure of 0.0375MPa. According to the Mg-Cu and Mg-Ni binary phase diagrams shown in **Figure 1**, eutectic reactions occur between magnesium and steel and nickel at a lower temperature. The samples were heated up to 450°C and kept at this temperature for a dwell time of 10 minutes in order to achieve the thermal equilibrium of the joining couple. Next, the samples with Cu coating were continuously heated up to brazing temperature of 530°C and held for 60min while the samples with Ni coating were heated up to 540°C and held for 35 and 45min respectively.

Thus contact reaction diffusion brazing process between Mg alloy and copper or nickel was carried out. Butt joint was designed and the samples were fixed in a fixture, as shown in **Figure 2(a)**. Prior to the electrochemical measurements, the samples were embedded in cold cured epoxy resin, ground with successive finer SiC abrasive paper, lubricated with water and finally polished with non-aqueous 0.5μm diamond slurry lubricated with glycerol and ethanol. The samples were cleaned ultrasonically in ethanol, rinsing with ethanol and then dried. The electrochemical corrosion properties of both brazed joint of magnesium alloy to steel using different coating interlayers(Cu and Ni) and base metals(magnesium AZ31B and AISI 1008 steel) were investigated by means of current density-potential measurements with PARSTAT 4000 electrochemical workstation, as shown in Figure2(b). The electrolytes used were 3.5% NaCl solution. Additionally, these curves were acquired with a polarization rate of 1mV/s. The microstructures of the brazed joints Mg/Cu(Ni)/AISI 1008 steel were examined with a stereo microscope and a scanning

electron microscope (SEM) with energy dispersive X-ray (EDS).

Table 1 Nominal chemical composition and mechanical properties of the AZ31B and AISI1008 steel plates

Materials	**Elements, Wt.%**								**UTS (MPa)**	**YS (MPa)**
AISI1008	**C** **0.05~0.11**	**Cr** **≤0.10**	**Mn** **0.25~0.5**	**Ni** **≤0.03**	**Si** **≤0.03**	**P** **≤0.035**	**S** **≤0.035**	**Fe** **Bal.**	**386**	**284**
AZ31B	**Al** **3.1**	**Mn** **0.335**	**Zn** **0.82**	**Si** **0.03**	**Fe** **0.005**	**Ca** **0.05**	**Be** **0.1**	**Mg** **Bal.**	**239**	**222.7**

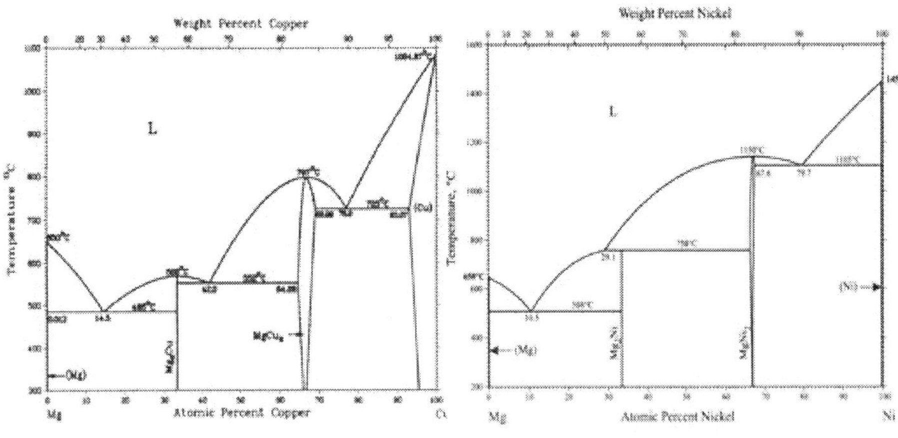

Figure 1 Mg-Cu phase diagram and Mg-Ni phase diagram[10]

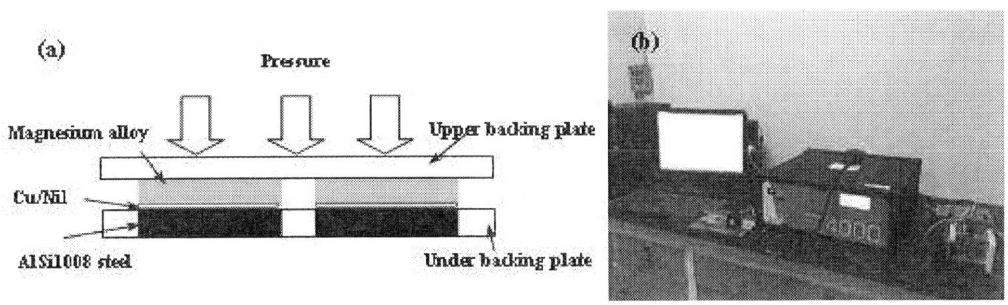

Figure 2 Fixture for contact reaction brazing (a) and PARSTAT 4000 electrochemical workstation (b)

Results and Discussion

Corrosion behavior of AZ31B/Cu coating/AlSi1008 brazed joint

The cross-section of magnesium alloy AZ31B with copper coating deposited by CGS is shown in **Figure 3(a)**. The average thickness of copper coating is about 50μm. At a brazing temperature of 530°C and a 60min holding time, sound joints were obtained since a homogeneous microstructure without voids or cracks was observed along the joint, as it is shown in Figure 3(b). Brazed joints between specimens of magnesium alloy AZ31 and AISI 1008 steel results in the formation of magnesium solid solution (A) and intermetallic phases forming an eutectic microstructure. The intermetallic blocky Mg-Cu-Al ternary phases (B) were embedded in light grey Mg_2Cu (C) phases and magnesium solid solution respectively.

As shown in **Figure 4**, pitting corrosion of the magnesium alloy AZ31B occurs first. With the increase of etching time, pitting gradually expanded to the scope which forms localized corrosion morphology by multiple pitting connected together. From surface topography before and after the corrosion of the steel can be found that the steel in contact with the electrolyte occurs uniform corrosion; Due to the different potentials of copper, magnesium and iron, magnesium acts as the anode which corrodes first in the electrolyte. Then, with corrosion time increasing, significant corrosion did not occur on the surface of the steel and the seam of brazed joints while serious corrosion occurred in the magnesium alloy. It is found that there formed stepped shape morphology between the seam and the magnesium and corrosion even results in significant form of pits.

To obtain more detailed information of the corrosion process at the interfaces, potentiodynamic polarization scans were performed. There is obviously a difference between base materials and brazed joint in electrochemical potentials, as shown in Figure4 and Table2. Due to the higher activity of dissolved base materials and the joints, the corrosion current density could be responsible for the corrosion property. In the cathodic branch of the brazed joint curve, it is obvious that the current density is at least two orders of magnitude higher than the corresponding current densities of the two base materials. The corrosion potential of copper is 337mV which is higher than AISI 1008 (-406mV) and AZ31B (-1422mV). In addition, magnesium alloy AZ31B acted as the anode in the electrolyte which resulted in the dissolution of AZ31B. The Mg_2Cu eutectic phase and Mg-Cu-Al ternary phases were formed by eutectic reaction during brazing process and the existence of Cu element can strengthen the galvanic corrosion of the AZ31B/Cu coating/AISI 1008 brazed joint. In addition, the brazing joint with Cu coating was not able to enhance corrosion performance of the base metals.

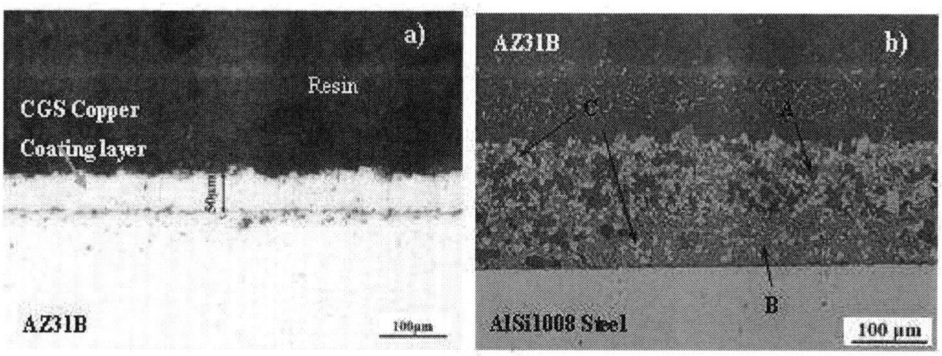

Figure 3 Cross-section morphology of CGS copper coating on AZ31B alloy (a) and SEM micrograph of brazed AZ31B/Cu coating/AISI1008 steel joint interface (b)

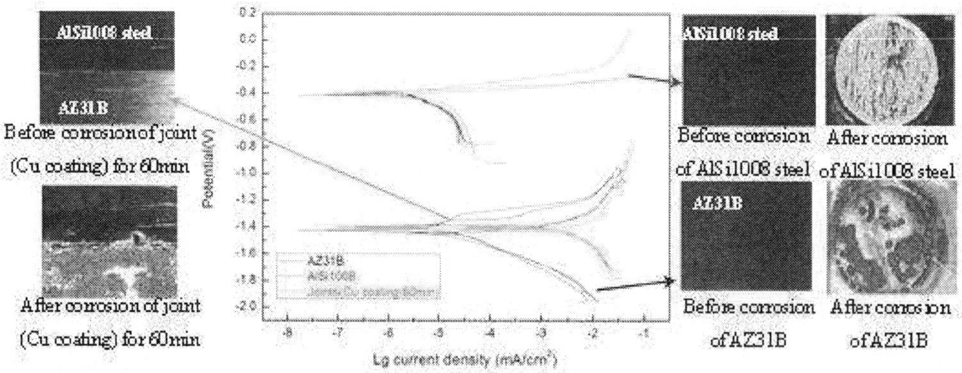

Figure 4 Current density-potential diagrams and corrosion morphology for base materials and brazed joint

Corrosion behavior of AZ31B/Ni coating/AISI 1008 brazed joints

The cross-section of magnesium alloy AZ31B with nickel coating deposited by arc-spraying is shown in **Figure 5(a)**. The average thickness of nickel coating is about 90μm. Then nickel coating was polished to 50μm with SiC abrasive paper of 1000 grit and cleaned by alcohol before brazing. The microstructure of brazed joints with 35min dwell time is shown in Figure 5(b). The interface of brazed joints contains the phase Ni (A), magnesium solid solution (B), the Mg2Ni phase (C) and Mg-Ni-Al ternary phases (D). Because of short holding time, nickel coating did not react completely. Figure 5(c) illustrates the complete dissolution of nickel with increasing the bonding time up to 45 min. The other phases formed in the brazing seam were similar to that of the joint holding for 35 min. While the difference between Figure 5(b) and (c) is that the width of mixing zone (Mg_2Ni phase and Mg-Ni-Al ternary phase) was expanded at dwell time of 45 min.

Figure 6 shows the surface morphology before and after corrosion test. Magnesium, the lowest potential metal, corrodes first in the electrolyte and the higher potential metal (Fe, Ni) can strengthen the corrosion resistance of the magnesium AZ31B surface. Compared to the AZ31B surface of brazed joint with holding time for 45 min, the surface of magnesium alloy corroded more seriously with the dwell time for 35 min.

The potentiodynamic polarization scans were also performed to investigative the corrosion behavior of brazed joint. Compare to the current density curves of AZ31B/Cu coating/ AISI 1008 brazed joint, the AZ31B/Ni coating/ AISI 1008 with different holding time are almost identical as shows in **Figure 7** and **Table. 2**. The difference is that the potential of different brazed joints are not the same, the order from high to low is AZ31B/Ni coating/ AISI 1008 with a dwell time for 35 min, AZ31B/Ni coating/ AISI 1008 for 45 min and AZ31B/Cu/ AISI 1008 for 60 min. So, corrosion resistance of the brazed joint of AZ31B/Ni coating/ AISI 1008 with holding time for 35 min is the best one.

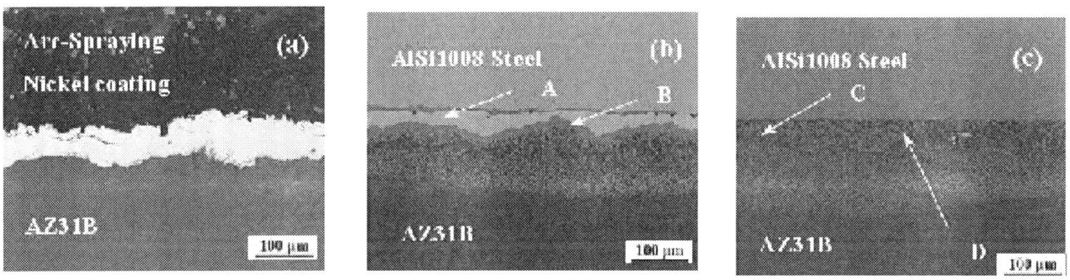

Figure 5 Cross-section morphology of arc-sprayed Ni coating on AZ31B alloy (a) and SEM micrograph of brazed AZ31B/Ni coating/AISI1008 steel joint interface for holding time 35 min(b) and 45 min(c)

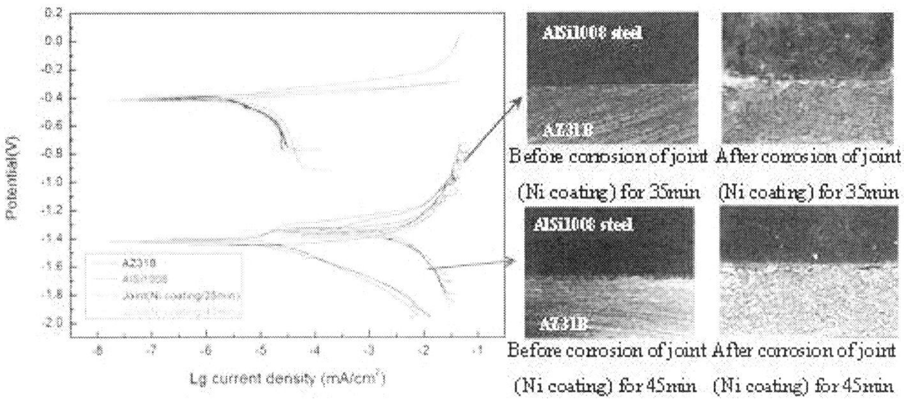

Figure 6 Current density-potential diagrams and corrosion morphology for brazed joints used nickel coating For holding time 35 min and 45 min

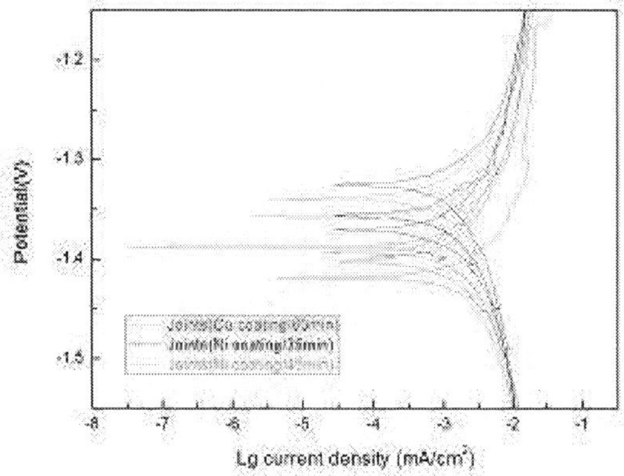

Figure 7 Current density-potential diagrams for brazed joints with different inter metal coating

Table 2 Corrosion current densities and corrosion potentials of base materials and brazed joints measured in 3.5%NaCl solution

		LOG(Corrosion current density)/(mA/cm^2)	Corrosion potential/mA
Base Materials	AZ31B	-4.75±0.15	-1422±20
	AlSi1008 Steel	-4.97±0.10	-406±8
Brazed Joints	AZ31B/Ni/ AlSi1008(35min)	-2.43±0.07	-1373±15
	AZ31B/Ni/ AlSi1008(45min)	-2.33±0.15	-1363±25
	AZ31B/Cu/ AlSi1008(60min)	-2.31±0.15	-1304±10

Conclusions

It has been shown that contact reaction brazing of magnesium AZ31B alloy to AISI 1008 steel was successfully achieved with different metal coating as interlayer. The interface phases of the AZ31B/Cu coating/AISI 1008 steel joint with a bonding temperature of 530°C for 60 min were mainly Mg-Cu-Al ternary phase, eutectic phase Mg_2Cu and Mg solid solution. The seam of AZ31B/Ni coating/AISI 1008 steel joint held for 35 min contains the un-completed dissolution Ni coating, magnesium solid solution, the Mg_2Ni phase and Mg-Ni-Al ternary phases. When the bonding time increased to 45 min, the Ni phase completed dissolved and the other phases formed in the brazed seam was similar to the joint held for 35 min. The corrosion current density of brazed seam is at least two orders of magnitude compared to the base materials. Thus the corrosion resistance of brazed joints deteriorates. The current density of brazed joints with different metal interlayers was relatively close. While the corrosion resistance of brazed joint of AZ31B/Ni coating/AISI 1008 with holding time for 35 min is slightly superior to the other joints. Galvanic corrosion is the main mode of corrosion in dissimilar metal joints and the most severely damaged area occurred in the vicinity of the interface of high potential. In addition, using metal interlayer of high potential could accelerate corrosion in brazed interface.

Acknowledgements

The authors are thankful to the financial support by State Key Laboratory of New Brazing Material and Technology, Zhengzhou Research Institute of Mechanical Engineering (Project SKLABFMT201303).

References

[1] Campbell F.C., Manufacturing Technology for Aerospace Structural Materials, Elsevier Ltd. (London, 2006), pp. 93-118.
[2] Dahle A.K. *et al.*, "Developments and Challenges in Utilization of Magnesium alloys", *Materials Forum*, Vol. 24, (2000), pp. 167-182.
[3] Heakal, E.F. et al. "Influence of halides on the dissolution and passivation behavior of AZ91D magnesium alloy in aqueous solutions," *Electrochimica Acta*, Vol. 54, No. 5(2009), pp. 1545–1557.
[4] Chu, Y.J. *et al.* "Research progress of welding technology of magnesium alloy". *Materials for Mechanical Engineering*. Vol. 6(2010), pp. 10-12, 46.
[5] Elrefaey, A. and Tillman, W. "Application of novel brazing process for joining AZ31 magnesium alloy," *Proc. LÖT2010, Brazing, High Temperature Brazing and Diffusion Bonding*. Aachen, Germany, June. 2010, pp. 361-365.
[6] Wahba, M. and Katayama, S. "Laser welding of AZ31B magnesium alloy to Zn-coated steel," *Materials and Design*. Vol. 35(2012), pp. 701–706.
[7] Tan, C.W. *et al.* "Single and dual beam laser welding-brazing characteristics of magnesium/zinc-coated steel dissimilar alloys," *Chinese Journal of Nonferrous Metals*, Vol. 22, No. 6(2012), pp. 1577-1585.
[8] Jana, S. *et al.* "Friction Stir Lap Welding of Magnesium Alloy to Steel: A Preliminary Investigation". *Metallurgical and Materials Transactions A*, Vol. 41, No. 12(2010), pp. 3173-3182.
[9] Elthalabawy, W.M. and Khan, T.I. "Eutectic bonding of austenitic stainless steel 316L to magnesium alloy AZ31 using copper". *International Journal of Advanced Manufacturing Technology*, Vol. 55(2011), pp. 235-241.
[10] Liu, L. and Qi, X. "Strengthening effect of nickel and copper interlayers on hybrid laser-TIG welded joints between magnesium alloy and mild steel," *Materials & Design*, Vol. 31, No. 8(2010), pp. 3960-3963.
[11] Qi, X. and Liu, L. "Comparative study on characteristics of hybrid laser-TIG welded AZ61/Q235 lap joints with and without interlayers," *Journal of materials science*, Vol. 45, No. 14(2010), pp. 3912-3920.
[12] Wielage, B. *et al.* "Corrosion behaviour of soldered joints of magnesium alloys and dissimilar materials," *Microchemical Acta*. Vol. 5(2006), pp. 151-157.
[13] Tillman, W. *et al*, "New Approaches for Brazing Temperature Sensitive Materials," *Proc 2014 International Conference on Brazing, Soldering and Special Joining Technologies*, Beijing, China, June. 2014, pp. 13-17.

[14] Wloka, J. *et al.* "Corrosion properties of laser beam joints of aluminum with zinc-coated steel," *Corrosion Science*, Vol. 49(2007), pp. 4243-4258.

Study of Atmospheric Corrosion by Urban Pollutants on Refrigeration System Silver Brazed Joints

Nestor Santillan-Soto, Astrid Calderas-Ochoa, Benjamin Valdez-Salas, Rafael Garcia-Cueto, Sara Ojeda-Benitez
Instituto de Ingenieria, Universidad Autonoma de Baja California, Mexicali, Baja California, Mexico
nsantillan@uabc.edu.mx, +52 (686) 566 4150

Samuel Colton
Arizona Western College, Yuma, Arizona, USA
Sam.Colton@azwestern.edu, (928) 344-7570

Abstract

Atmospheric urban corrosion (AUC) is a term that includes environmental contamination from motor vehicles, industrial activity and other domestic emissions i.e. vented gases associated with modern urban dwellings and anthropogenic activity. Gaseous and particulate urban pollutants combine to form chemical compounds which as a consequence act as corrosive agents on silver brazed joints on air-conditioning and refrigeration equipment (ACE). The environmental and economic impacts of silver brazed joint failure have serious consequences due to the release of damaging greenhouse gases (GHG). Analysis of various silver brazed joint samples of different alloy composition (0, 5 and 15 percent silver) that were located in four locations in the city of Mexicali, México were made. The analysis reveals that urban pollutants cause severe corrosion on silver brazed joints. Further analysis was performed to better understand the implications that a corrosion phenomenon has on the urban microclimate studied and the conditions that promote corrosion failures on refrigerant line brazed joints.

Introduction

The climate of Mexicali

Mexicali city is located at 32.55° N, 115.47° W and 4 meters above sea level; represents the Baja California State Capital in Mexico's Northwest region. At the north, it borders with California, USA and comes to be the final part of Sonora desert at the east. This city has a dry and semi-arid climate, and is one of the most extreme in Mexico. July's high temperatures average is 42.2° C and January's high average is 21.1° C. A high solar potential of 90% daylight hours reach each year around 75 mm of rain annually. As a historic data, on July 28, 1995, 52° C was reached, and on January 13, 1963 its all-time lower temperature of -7.0° C [1].

Pollution influences

H_2S is a corrosive and pollutant agent, its main source is the Cerro Prieto geothermal plant located about 40 Km southeast from the city of Mexicali [2]. Gaseous pollutants such as SO_X, NO_X, CO, and O_3 are generated from the diverse industrial activities; important to note is the concentration of industrial parks that are located in the southeast area. Moreover, the presence of municipal sewage ponds adds corrosive contaminants on the west side of the city [3].

GHG emissions due to failures in brazed

Emissions of refrigerant gases in Mexicali, Baja California, are mostly concentrated in specific areas of the city (Fig. 1) according to different geographic and environmental factors. The amount of gas released into the atmosphere by cooling and / or air conditioning equipment installations that have corrosion related failures is between 75 to 100 percent of the total maximum load on each piece of equipment in a period of a year, and sometimes leaks are recurrent in that time period on the same system [4].

Brazing for cooling systems

Ideally, the air condition and refrigeration equipment (ACE) are designed and fabricated to provide comfortable conditions for humans, but service conditions (in polluted cities) its performance decreases due to atmospheric corrosion attacking the braze system components. Different alloys have been developed to obtain better mechanical and anti-corrosion and properties. The AWS classification for brazing filler metal classifications Cu-Ag-P and Ag-Cu-Zn-Sn are widely recommended for copper and high-copper alloys in a variety of applications particularly copper joints including copper piping in cooling systems [5-8].

For the present study the following two hypotheses were established:

- The urban climatic conditions causes an increase in the effects of corrosion in the southeast and west side of the city.
- The city of Mexicali should establish mandatory filler metal requirements and training standards for use by technicians and contractors to minimize the resulting problems associated with urban corrosion and emissions in new equipment and repair services.

Materials and Methods

Based in the preliminary advance published by Calderas [9], Table 1 shows a summary of materials and equipment utilized. All specimens were cut, clean, and afterwards, installed in the different locations described in the Figure 1.

Table 1. Summary of materials

Materials	Description
Base material	Copper type "L" UNS-C12200
Filler metal	BCuP-2 (0% Ag)
Filler metal	BCuP-3 (5% Ag)
Filler metal	BCuP-5 (15% Ag)
Torch	Turbotorch
Gases	MapGas
Flux	-----

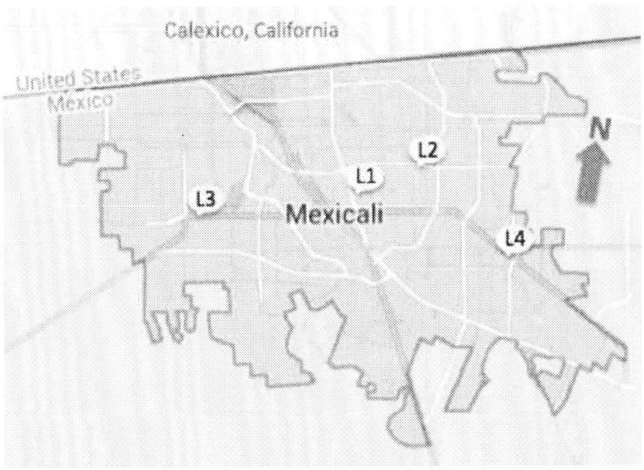

Figure 1. Location of specimens in Mexicali, Baja California, México.

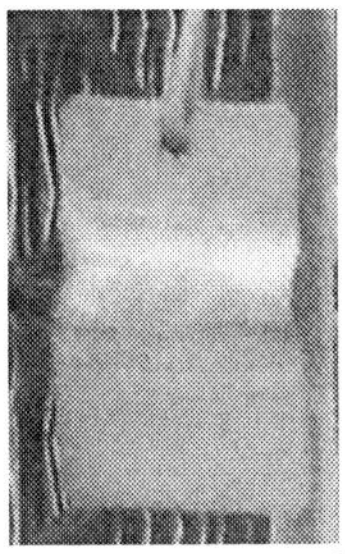

Figure 2. Specimen appearance at L1 site installation

Figure 2 shows the appearance of the specimens at the time they were installed, as can be observed; a clean surfaces is exposed. Additionally, Table 2 presents a brief description of the four locations; all specimens were installed at select supermarket roof top facilities except specimens located at Baja California University (L1 in Fig. 1). The test coupons were exposed during 6 months to environmental conditions close to the condenser coil on the roof, starting in January 2011 and recollected in June 2011. After this collection, visual inspection was performed on base material surfaces, later other advanced techniques (optical microscope and SEM-EDX) allowed further analysis of the corroded surfaces.

Table 2. Experimental site description and instalation date

Site	Description	Date
L1	Urban-residential-commercial	January 10 2011
L2	Urban-residential-commercial-industrial	January 11 2011
L3	Urban-residential-commercial, proximity to the oxidation pond	January 12 2011
L4	Urban-residential-commercial-industrial, proximity to geothermal plant and industrial complex	January 13 2011

Climatic Variables

Wind roses. The influence of the surrounding area involved in the proposed four locations require knowledge of the direction and wind speed at surface level. The prevailing wind in a time range can be represented by a wind rose, which indicates the percentage of time that the wind blows from different directions. The graph uses bars or extensions ranging from the center of a circle to a particular point illustrating the wind direction, the length of each extension will indicate the percentage of time that the wind headed in that direction [10]. WRPLOT View is a program through which wind roses are generated for several meteorological data formats. A wind rose shows the frequency of occurrence of winds in each one of the wind directions provided and speed classes for specific time and place [11].

Temperature and relative humidity; these two fundamental variables are very important because they are the catalysts that promote the velocity and behavior of the corrosion phenomena. Climate variables were provided by the Meteorological and Climatological department of the Institute of Engineering.

Results and Discussions

In this section some brazed specimens are presented. After six months of exposure the coupons changed their appearance, most show black areas as Figure 3 to 6 shows. General and pitting corrosion are present on the surfaces.

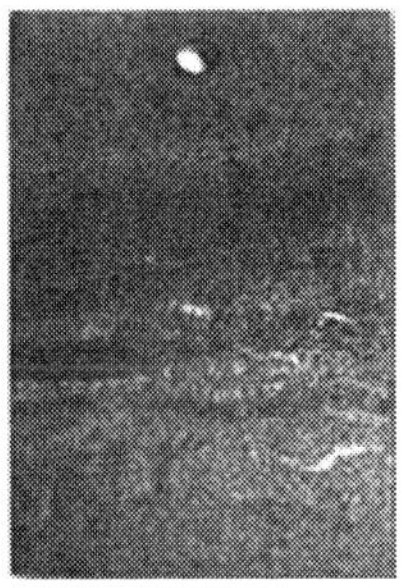

Figure 3. Coupon evaluated in L1 location, 15% w/w Ag.

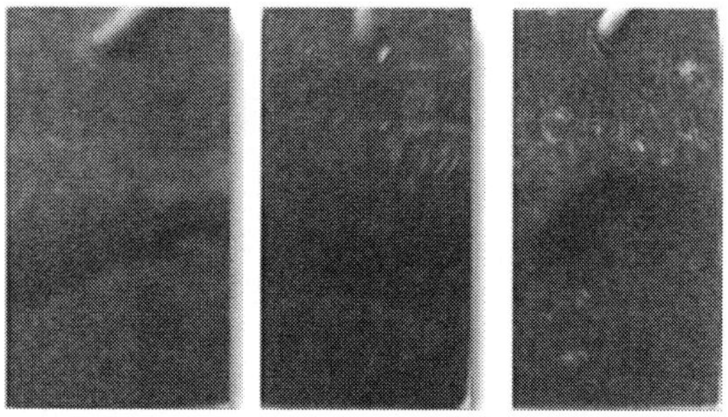

Figure 4. Coupons evaluated in L2 location, 0, 5 and 15% w/w Ag respectively from left to right

Figure 5. Coupons evaluated in L3 location, 0, 5 and 15% w/w Ag respectively from left to right.

Another test location that showed aggressive corrosion formation was location four (L4).

Figure 6. Coupons evaluated in L4 location, 0, 5 and 15% w/w Ag respectively from left to right.

Specimens experience a change in color that occurs at the base metal, (Cu_2S) that is present on the coupons, resulting in an opaque or darkened (Ag_2S and Ag_2O), with other indications of localized corrosion (Figure 5). These black colorations are seen as a product of silver sulfide on the entire surface and base metal.

As can be observed in Figure 5, L1 location shows that the areas of silver braze was not exempt from corrosion attack. A hole and some cracks were created by the environmental conditions. This phenomenon was found repeatedly on the coupon's surfaces at all locations, these morphologies of corrosion and the presence of elemental sulfur (S) were common in most Scanning Elevtron Microscopy (SEM) and Energy-dispersive X-ray spectroscopy analyzes (EDX).

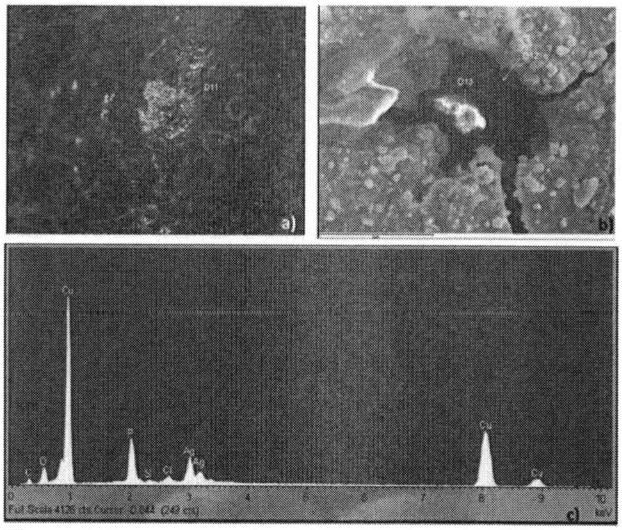

Figure 5. Present a specimen 5% w/w Ag in location L1, a) microscope 50X, b) SEM 1500X and c) EDX analysis.

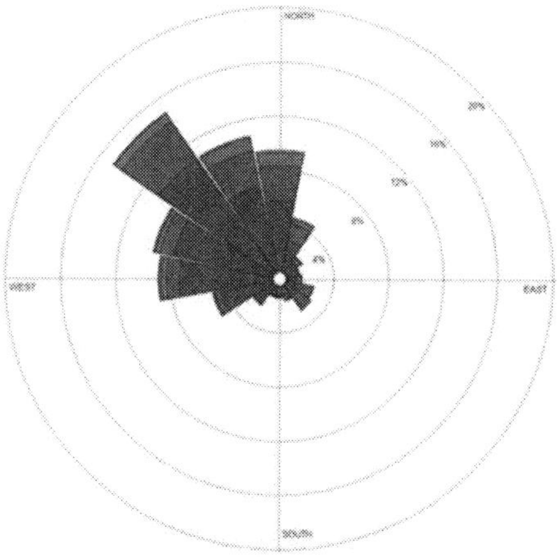

Figure 7. Wind rose of January 2011 at L1 location.

Figure 7 shows the behavior in the prevailing wind direction from January to May. The prevalent winds are from the west, northwest and north. Figure 8 shows the behavior of the wind direction in June. The winds show two prevalent directions (northwest and southeast) due to seasonal changes from spring to summer.

Considering Figure 9, the average values of relative humidity were accentuated in the months from January to April, and average temperatures were increasing gradually, therefore, we can associate the observed results, since these climatic conditions promote corrosion reactions, additionally the presence of sulphidric acid (H_2S) coming from diverse sources will increase the corrosivity of the environment and susceptibility of silver alloys. It is also important to consider the prevailing winds from the west, northwest and north during the winter to the spring season and its change to the south, southeast and east for early summer.

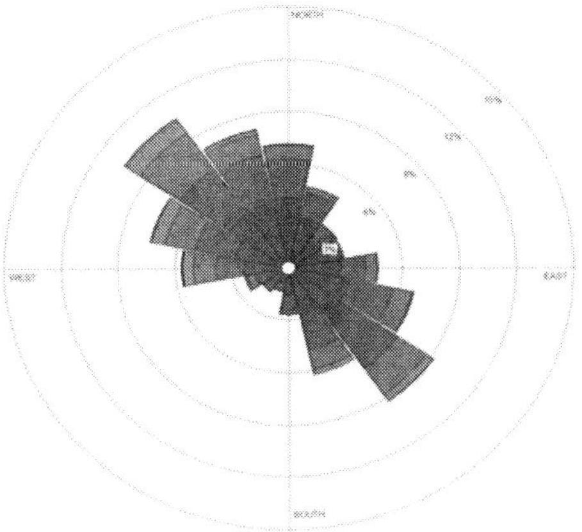

Figure 8. Wind rose of June 2011 at L1 location.

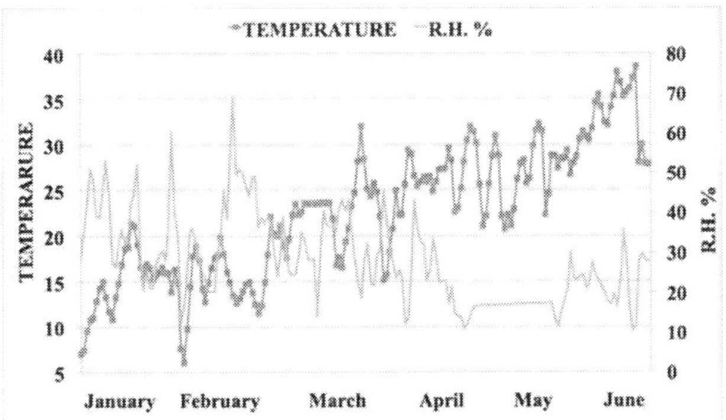

Figure 9. Temperature and relative humidity at L1 location. Monthly Average from January to June.

It is necessary to mention that the equipment (condenser units) are for air-conditioned services and only work during the summer season, the rest of the year they remain unused. Under these conditions, the effects of corrosion becomes more aggressive because the airflow of the condensing units is turned off, and the coupons were exposed to "stagnation conditions" promoting the concentration of the reactive environmental substances on the surfaces of the base metal.

As shown in Table 3 and Figure 9, wind velocities higher than 2.1 m/s have a tendency to increase as the months advance and the lower intensity decreases in the same timeline. Then, the ability to transport corrosive compounds increases and may be associated to wind transport from the oxidation ponds located in the west side of the city as well as vehicle emissions which affect the corrosion process on the surfaces of the specimens.

Table 3. Calms and Wind monthly percentage Jan-Jun 2011.

Month	Calms (%)	0.5-2.1 m/s (%)	>2.1 m/s (%)
January	19.5	59.3	21.2
February	16.9	49.9	33.2
March	15.6	45.4	39
April	10.8	37.7	51.5
May	12.9	37.6	49.5
June	15.2	42.1	42.7

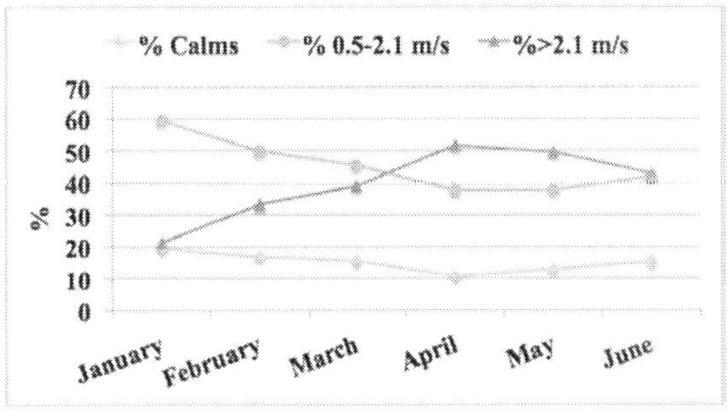

Figure 9. Wind range and calms averages behavior January to June 2011.

Other phenomenon observed was an increase in temperatures, combination of moisture present in the atmosphere (Fig. 9). Do not lose sight that environmental factors such as H_2S and SO_2 concentrations, and weather (high temperature, relative humidity and winds) attack the metal Cu and Ag and alter their condition. All filler metals BCuP analyzed contains silver metal in addition to copper, which in turn is highly susceptible to sulfur compounds (sulfides). This supports the suggestions in the literature, which states that BCuP filler metals are highly anti-corrosive, except in environments rich in sulfur or sulfur compounds [12].

Conclusions

The failures on surfaces impact two aspects: economical, that means repairs and environmental impacts due to the release of damaging greenhouse gases (GHG). We can establish that all locations showed evidence of aggressive corrosive attack on brazed joints and metallic surfaces; climatic variables such as temperature, RH, wind direction and velocity also contribute to accelerate the corrosion process.

According to the results and in agreement with Singler [13] we recommend the following corrective actions: 1) Replace in the short time all the old refrigeration units by modern and more efficient which are free of hazardous refrigerant gases noxious to the, 2) selection of more corrosion resistant filler metals capable to resist urban environments containing sulphide.

Finally, the growth trend in urban areas cannot be stopped, but knowing this, fact based decisions must be taken to remediate the use of outdated refrigeration technology that has supported the use of large scale commercial installations in countries like Mexico. The need to replace outdated technology with new environmentally friendly systems represents an initiative that could be promoted by the manufacturers of refrigeration equipment. The second phase of the research will be to analyze the effects of corrosion due to temperature, relative humidity, wind speed and direction for the months of July to December since the meso-scale condition changes drastically during the summer and fall seasons, then after a year individual results of each study point will show if exist a most corrosive area in the city. Additional to this field study, electrochemical techniques will be applied for Cu-Ag-P and Ag-Cu-Zn-Sn phosphorus-free coupons, in order to compare their potential (Ecorr), current density (Icorr) and corrosion rate (mils per year).

References

[1] García-Cueto, O.R., and Santillán-Soto, N. Modeling Extreme Climate Events: Two Case Studies in México, INTECH (Croatia, 2012), pp. 137-160. L.M. Druyan (Ed.), Climate Models.

[2] G. Lopez, B. Valdez, and M. Schorr, Spectroscopy analysis of corrosion in the electronics industry influenced by Santa Ana Winds in marine environments of Mexico, in Indoor and Outdoor Air Pollution, INTECH (Croatia, 2011), pp. 61-76. J.A. Orosa (Ed.), Indoor and Outdoor Pollution.

[3] Valdez-Salas, B. *et al*, "Copper Corrosion by Atmospheric Pollutants in the Electronics Industry", *Hindawi Publishing Corporation ISRN Corrosion* Vol. 2013, Article ID 846405, (2013), pp. 1-7.

[4] Calderas-Ochoa, A.M. Desarrollo de una metodología para la estimación de emisiones de gases refrigerantes en la ciudad de Mexicali Baja California". Thesis for master's degree, Mexicali, México, (2009).

[5] ASM Handbook. Welding brazing and soldering (Ohio, 1993), ASM International, Vol. 6, pp. 931–932.

[6] Schwartz, M. M. Brazing. (Ohio, 2003), ASM International 2nd Ed., pp. 79–80,

[7] Brazing Handbook. (Miami, 1991) American Welding Society 4th Ed., pp. 284–287.

[8] Harris. Lincoln Electric Co. (2014) www.harrisproductsgroup.com/en/Products/Alloys/Brazing.aspx

[9] Calderas, A., Santillán, N., Valdez, B. and Schorr, M.. "Corrosion of Copper Coils in Air Conditioning Equipment". *Materials Performance*, Vol. 52, No. 9, (2013), pp. 60-63.

[10] Donald, A. C. Essential of meteorology: an invitation to the atmosphere. (Unites States, 1998) Wadworth Publishing Company 3rd edition, pp. 159.

[11] WRPLOT View, Lakes Environmental, (2014). http://www.weblakes.com/products/wrplot/index.html

[12] Welding Handbook. Welding processes- arc gas welding and cutting, brazing and soldering. (Miami, 1978) American Welding Society 7th ed., Vol. 2.

[13] Sigler, D. R. *et al*, "Sulfide-Induced Corrosion of Copper-Silver-Phosphorus Brazed Joints in Welding Transformers". *Welding Journal*, Vol. 86, No. 11 (2007), pp. 340-348.

Characterization of Intermetallic Compound Formation and Growth Kinetics in Lead-Free Solder SAC305

Pedro Henrique Marques Spinola De Andrade, Stephen Liu, Juan Carlos Madeni
Center for Welding, Joining and Coatings Research – Colorado School of Mines, Golden, Colorado, United States of America
pmarques@mines.edu, sliu@mines.edu, jc.madeni@gmail.com

Abstract

This paper describes the experimental techniques necessary for studying intermetallic layer formation and growth throughout the melting and solidification, and thermal aging of the solder alloy. To observe intermetallic compound (IMC) growth during the liquid state of soldering, an experimental setup that quickly raises the temperature, followed by fast cooling (comparable to the actual joint conditions) was required. Temperatures at increments of 25°C above the soldering temperature and times of two, five and eight seconds for growth were carefully controlled. Thermal aging was performed for 20, 50, and 100 hours at 70, 100, and 130°C. The copper substrate/solder alloy interface was examined using light microscopy and SEM-EDS in the as soldered condition and after thermal aging. As expected, intermetallic phases such as Cu_6Sn_5, Cu_3Sn, and Ag_3Sn were observed, with Cu_6Sn_5 being the major phase in as-soldered and thermally aged conditions. Statistical analysis was performed in the IMC thickness data obtained from the long profile analysis. Using Arrhenius analysis, the activation energy for IMC growth was calculated to be approximately 8kCal/mol in the as-soldered condition, and 20kcal/mol for thermal aging, in agreement with reported values in literature, that are in the range of 18 to 27kcal/mol for thermally aged specimens.

Introduction

Due to restrictions applied to the use of lead in industrial applications various lead-free and tin-based binary and ternary alloys solder emerged as an alternative solution [1, 2]. The most widely used lead-free solder alloys are in the SAC family (Sn-Ag-Cu) especially due to their good creep and thermal fatigue resistances [3-5].

The effect of thermal aging over solder joints, simulating in-service conditions, has been extensively studied [6-9]. Intermetallic layers may form and grow early during service time of joined components at the substrate/solder interface [10]. However, formation and growth of intermetallic compounds is not restricted to solid state. In some solder systems, intermetallic compounds are formed at the solder/substrate interface while the solder is still in liquid state [11]. Thus, investigating the kinetics of formation and growth of intermetallic compounds with respect to the period of time the solder is in the liquid state during soldering becomes relevant. Findings as growth rate and activation energy of the intermetallic layer may be used as criteria to define soldering cycles. In fact, this work was conducted to explore the possibility of using those factors as quality control strategies in soldering.

The first portion of the present work was focused in carefully controlling soldering cycles to promote different extents of intermetallic formation and growth while in the liquid state. Melting of the solder alloy, holding time and temperature in liquid state, and solidification were key points to control and compare the different thermal profiles experienced by the produced solder joints. The results obtained include the intermetallic layer growth rate and activation energy related to the reactions that took place in liquid state.

The second part of the work was related to thermal aging of solder samples. New solder joints were produced using a soldering cycle which would surely result in a sizeable intermetallic layer even within the as-soldered condition, and these were subjected to different thermal aging cycles. The evaluation of growth rates and activation energy for solid-state aging accounted specifically for the variation in the thickness of intermetallic layers, since samples were characterized before and after aging.

Experimental Program

The solder alloy selected for all experiments was SAC305. The chemical composition of SAC305 in weight percent is: 96.5% Sn – 3.0% Ag – 0.5% Cu, and its melting temperature range is 217°C to 219°C.

Copper coupons of 25 x 12.5 x 0.2 mm were ground with 1200 grit grinding paper to achieve an oxide-free surface and used as substrate. The solder alloy wire was sectioned to produce several samples of 0.3g that would be applied individually on substrates to produce samples.

To investigate intermetallic compound formation in liquid state three soldering temperatures were selected: T_1=220°C, T_2=275°C, and T_3=300°C, and also three holding times during which the solder alloy would be kept liquid: 2, 5, and 8 seconds. Combining time and temperature parameters, nine different soldering conditions were defined, and three solder samples would be produced for each of those conditions.

The selection of the three temperatures was made based on differential scanning calorimetry (DSC) data [12] for the formation of Cu_6Sn_5, which is shown in Fig. 1, and on the binary phase diagram of Cu-Sn, which is shown in Fig. 2. Even though the soldering system contains silver, the content is low enough to allow an approximate description of the solidification behavior of the solder alloy as a binary alloy. The DSC curve indicates that Cu_6Sn_5 is formed approximately at 275°C. Therefore, T_1 was selected just below the eutectic temperature of 227°C, T_2 is right at the formation temperature of Cu_6Sn_5, and T_3 is that temperature to guarantee the formation of Cu_6Sn_5.

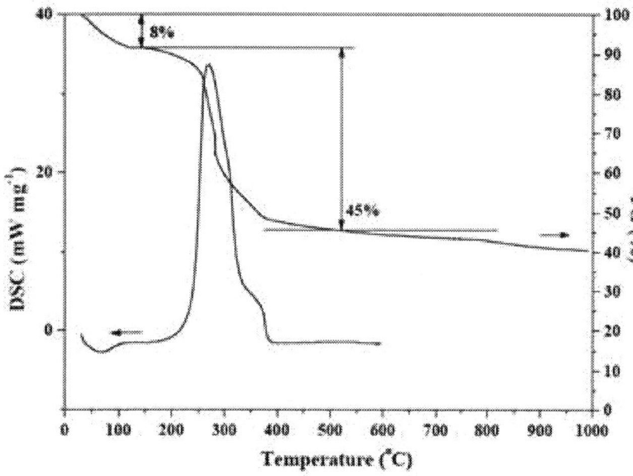

Figure 1. DSC data of formation of Cu_6Sn_5 indicating that the intermetallic compound is formed at 275°C.

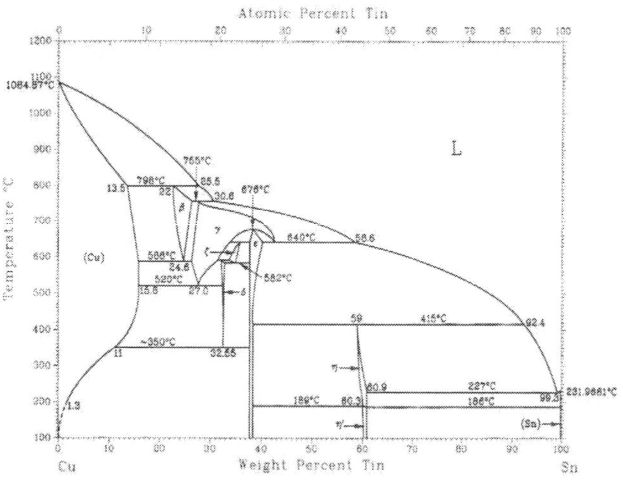

Figure 2. Cu-Sn binary phase diagram. The eutectic point is at 227°C and 60.9% Sn.

During soldering, the copper coupon temperature would be brought up to one of the selected values using a soldering iron that allowed pre-setting and holding a target temperature. The selected holding times refer to the interval between melting of the applied solder alloy and the time during which it was kept in liquid state. Immediately after fulfilling the designated holding time, the heat source would be removed, allowing cooling. Temperature measurements were made using a thermocouple placed at the immediate adjacency of the molten solder to determine the actual temperature of the system, and variations of ±5°C would be accepted. Figure 3 shows the experimental set-up, and Fig. 4 presents a close look on the time-temperature profile experienced by the solder alloy when in liquid state.

Figure 3. Experimental set-up showing the copper coupon used as substrate, the soldering iron used as heat source, and thermocouple position.

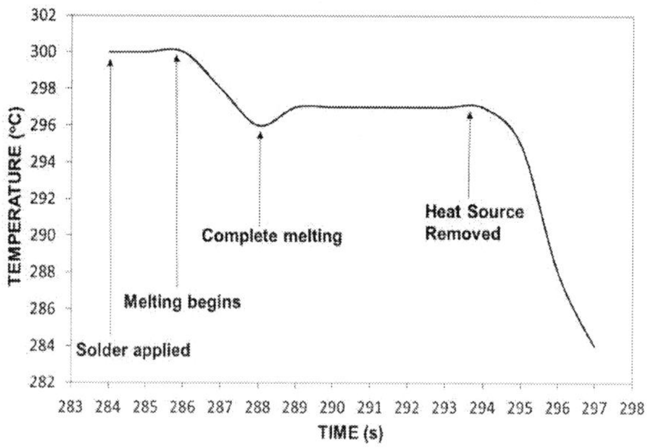

Figure 4. Zoom of a time-temperature profile experienced by one of the produced samples during soldering at 300°C, and with an 8 seconds holding time.

All the metallographic preparation steps that followed sample production were done carefully to avoid high temperatures that could promote further growth of the intermetallic compound that would not be related to the liquid state reactions. All samples were cross-sectioned using a low-speed cutting machine, and cold-mounted in epoxy resin. Grinding, diamond polishing, followed by one hour vibratory polishing using colloidal silica (0.04 μm) were performed. Characterization of the intermetallic layer formed at the solder/substrate interface was done using light microscopy, and SEM-EDS.

Optical micrographs were taken from all 27 produced samples at 500X or 1000X magnification, depending on the resulting thickness of the intermetallic layer, covering lengths of approximately 500 μm. The thickness of the intermetallic layer was measured every 3μm throughout the length. Figure 5 exemplifies how measurements were taken.

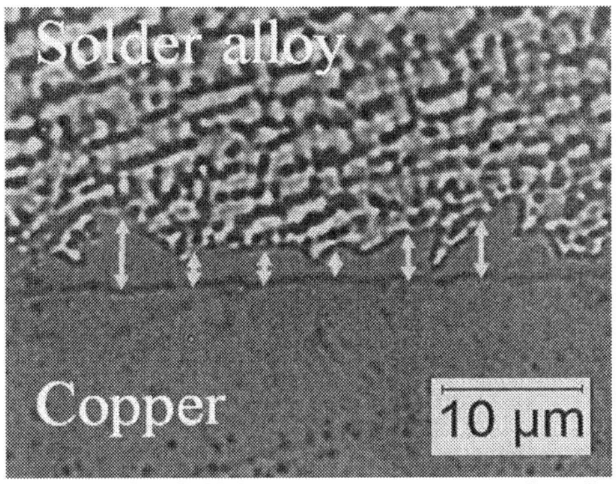

Figure 5. Example of thickness measurements taken from the intermetallic layer.

The following part of this work consisted in investigating solid state-aging growth kinetics. At first, aging temperatures and times were selected: 70, 100 and 130°C, and 20, 50, and 100 hours. Combining the conditions, a total of nine solid-state aging cycles were performed.

Two samples were produced and submitted to each of those aging cycles, giving a total of 18 aged samples. The production of those new samples followed the same procedures outlined earlier for controlling intermetallic formation in liquid state, but only two soldering conditions were used: one half of the samples was produced at 275°C with a holding time of 8 seconds, and the other half at 275°C with a holding time of 2 seconds. Metallographic preparations of samples also followed the same procedures described previously.

All solder samples had their intermetallic layer measured using optical microscopy before being subjected to the respective thermal aging cycles, so that pre-thermal aging and post-thermal aging microstructures could be compared, and allowing for a more accurate evaluation of the resultant growth.

Aging was conducted using furnaces that were able to produce a temperature accuracy of ±0.5°C. Aging times have a negligible error, less than one minute variation from planned times. Samples were water quenched immediately after removal from the furnace.

Results and Discussion

The results related to intermetallic compound formation and growth in liquid state include time-temperature profiles of tests, optical micrographs, SEM images, and EDS scans of the SAC305/Cu interface.

One of the produced samples is shown in Fig. 6, the copper substrate surface surrounding the solidified solder shows signs of uniform heating, which suggests uniform temperature distribution along the SAC305/Cu interface. This observation, associated with temperature-time (T-t) profiles obtained from thermocouple measurements, indicate a proper temperature control. Figure 7 presents the T-t curves of each of the soldering conditions. All plots have the same scales in the X and Y axis; from top to bottom the temperatures are respectively 300, 275, and 220°C, and from left to right holding times are 8, 5, and 2 seconds.

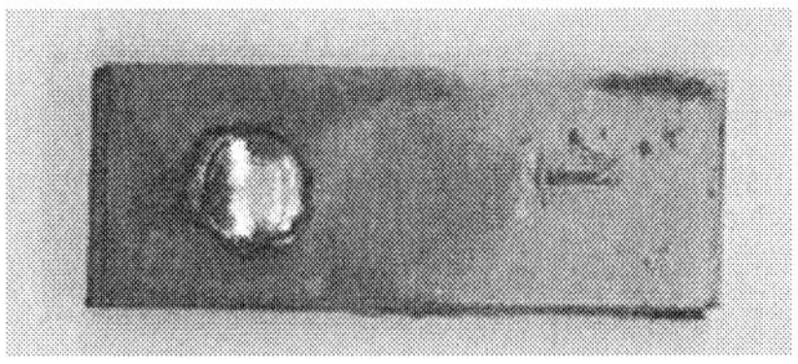
Figure 6. SAC305/Cu joint sample. The uniform coloration surrounding the solder button indicates uniform temperature distribution.

Scanning electron microscope images from samples corresponding to each soldering condition are shown in Fig. 8. It is noticeable that at the lowest temperature (220 °C) the intermetallic layer morphology is uniform, and increasing holding time does not have any a considerable effect on growth. However, as temperature is increased to 275 and 300°C, the effect of holding time on the overall growth becomes more pronounced, and the morphology of the intermetallic layer becomes non-uniform, exhibiting needle-like features with preferred growth directions. Further investigation regarding crystallographic orientation associated with preferred growth was not included in the scope of this work.

The overall impression of thickening and more pronounced non-uniformity suggested by the intermetallic layer images was further examined by the statistical assessment of IMC layer thickness measurements. Figure 9 shows frequency distributions of measurements made for each of the soldering conditions. At 220°C, the distributions are narrow, indicating little variability, and have the smallest mean values. As test temperatures increase, the measurements distributions become wider. Mean values gradually shift to larger values as holding times are increased, demonstrating the extent of growth experienced, and the greater shift are noted at the highest temperature (300°C).

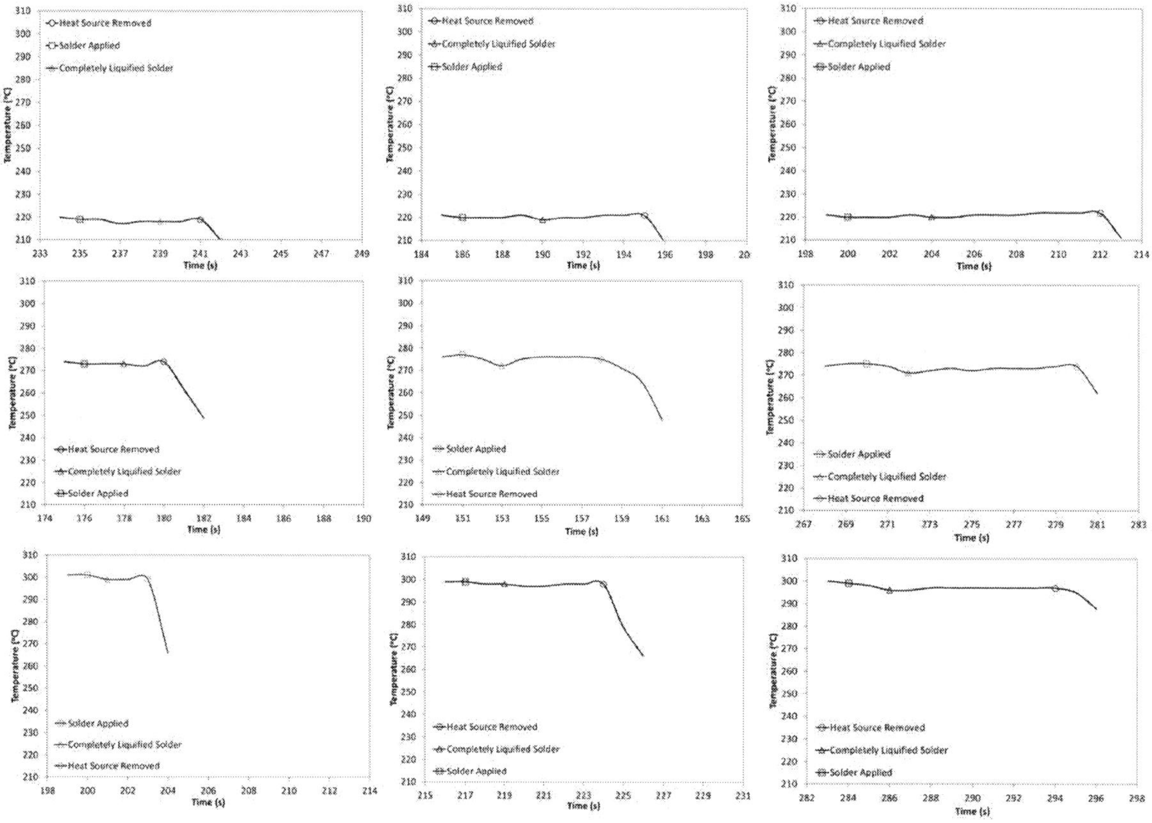

Figure 7. Time-temperature profiles or each soldering condition defined in the experimental program.

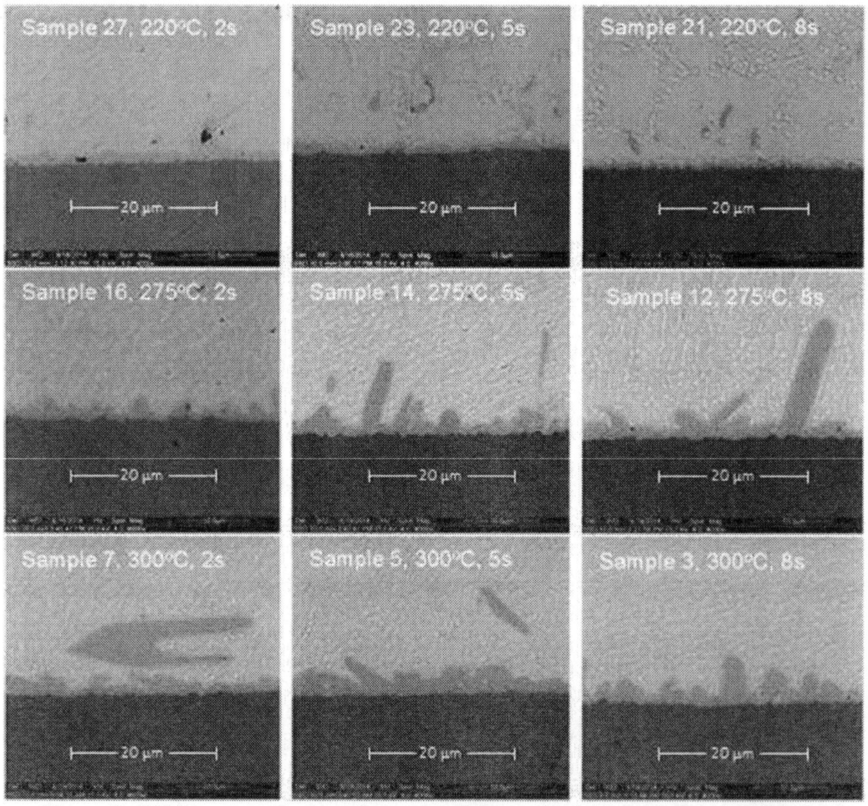

Figure 8. SEM images of selected samples from each of the soldering conditions.

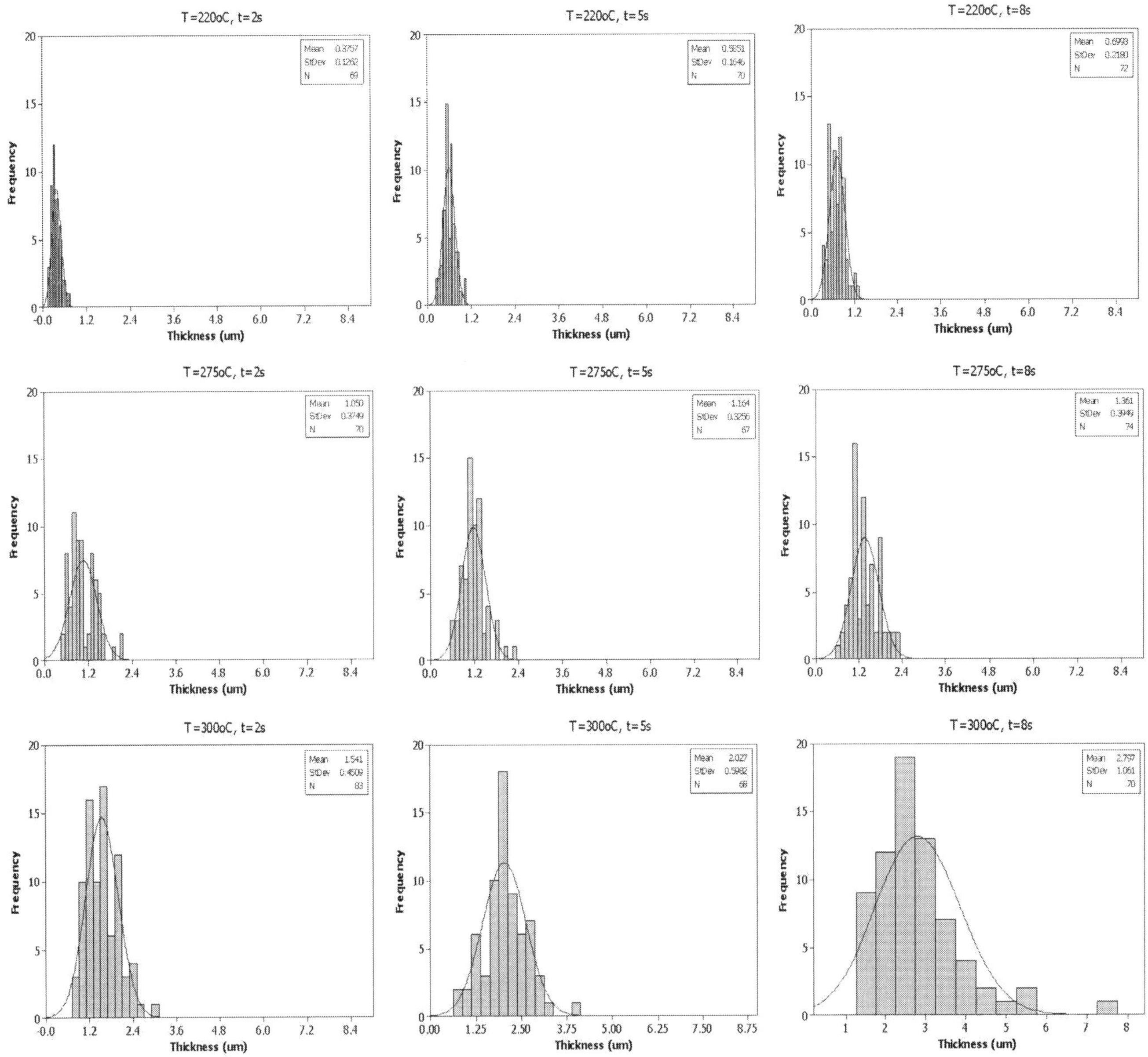

Figure 9. Frequency distributions of the intermetallic layer measurements of individual specimens associated to all soldering conditions.

The mean IMC thickness values are 0.38μm, 0.57μm, and 0.67μm for the samples processed at the temperature of 220°C and times of 2s, 5s, and 8s, respectively. For 275°C, the mean IMC thicknesses are 1.06 μm, 1.21 μm, and 1.33μm for the times of 2s, 5s, and 8s. As expected, for the soldering experiments conducted at 300°C, larger thicknesses of 1.51μm, 2.07 μm, and 2.65μm were observed for 2s, 5s, and 8s. Additional samples were produced at 275°C at 3s, 4s, 6s, and 7s, the IMC thicknesses measured were 1.12μm, 1.24μm, 1.54μm, and 1.63μm, respectively, and those were included in the growth rate calculations that will be presented further on. In this analysis, the IMC layer formed in the as-soldered condition is considered to be composed only by η-Cu_6Sn_5. This assumption was supported by EDS scans that indicated a composition for the IMC layer that is within the expected range expected for Cu_6Sn_5. There is a small transition interval in which Cu_3Sn may be present, but its size is negligible and was not treated separately.

Figure 10 presents one of the EDS line scan performed over the IMC layer, the composition is indicated in atomic percentage. An average value was taken from the portion which overlaps with the IMC layer. The precise atomic composition associated with the compound Cu_6Sn_5 is 45.46%

Sn, and 54.54% Cu. Even though silver is also present in the system, only Cu and Sn were analyzed and the correspondent atomic percentages were normalized to 100%.

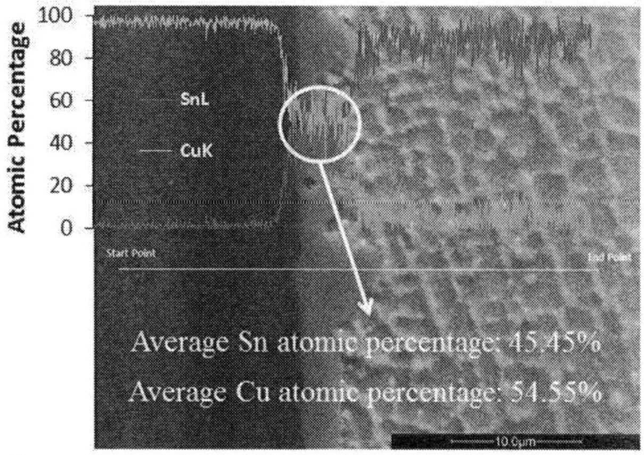

Figure 10. EDS line scan of IMC layer indicating an average atomic composition that matches Cu_6Sn_5.

A typical diffusion-controlled behavior, as observed by other authors in solid state growth of intermetallic compounds, is described by a parabolic-type equation:

$$x = (k_T t)^{1/2} \quad (1)$$

Where x is the mean IMC thickness, k is the growth constant, and t is the dwell time. However, the actual mechanism of IMC for growth in liquid state is influenced by liquid metal flow, substrate erosion, reaction rate between Cu and Sn in addition to the diffusion that takes place. Consequently, the parabolic behavior may not be explicit, and also, for very short times, as those used in the experiments, the growth rate assumes an approximately linear behavior

$$x = A + kt \quad (2)$$

Where x is the mean IMC thickness, A is the intercept, k is the growth constant, and t is the liquid state holding time. In Fig. 11, thickness results corresponding to each temperature conditions were plotted versus time. In fact, a linear correlation adequately describes the growth trend. As temperature increased, the growth rate, represented by the linear slopes, also increased.

The slopes (k_T) of each data set refer to a thermal activated process, expressing thermodynamic and kinetic information. These can be related by a Boltzmann distribution:

$$k_T = k_o exp\left(-\frac{Q}{RT}\right) \quad (3)$$

Where k_T is the growth rate constant for a given temperature T, Q is the activation energy of the process, and R is the ideal gas constant. This expression can be linearized by taking the natural logarithm of both sides, and so k_T becomes an inverse function of T. This graphic representation is shown in Fig. 12, and the slope corresponds to Q/R.

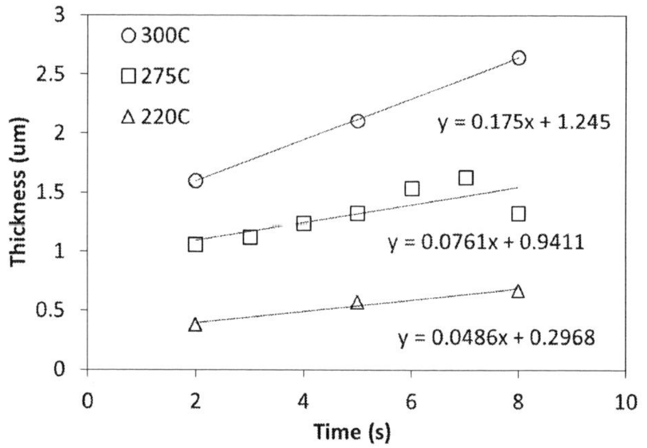

Figure 11. Mean IMC thickness for the three tested temperatures as a function of time.

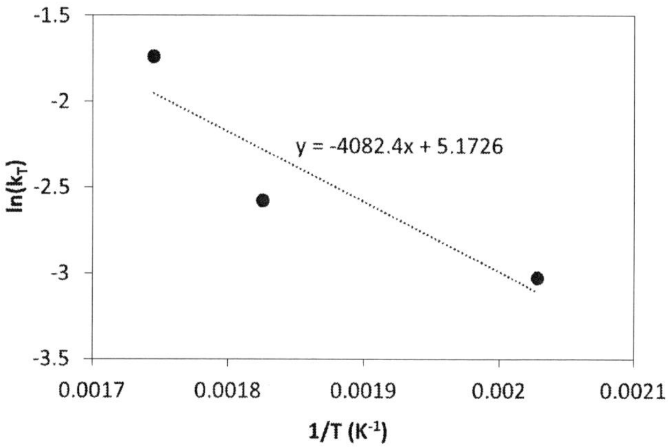

Figure 12. Arrhenius analysis of IMC growth rates in liquid state.

Considering the slope obtained through the Arrhenius plot in Fig. 12, the resultant activation energy is 8.1 kcal/mol (0.35 eV/atom).

Most literature data report values regarding IMC growth during solid-state aging, and activation energy values range from 16 to 27 kcal/mol (0.61 to 1.20 eV/atom). Information regarding activation energies is summarized in Table 1. The solder alloy to which the activation energies refer has a slightly different composition than the one used in the experiments.

The value derived from this work is considerably smaller than all presented references. However, as mentioned previously, reported activation energies refer to solid-state transformations where the controlling mechanism is the diffusion of Cu and Sn atoms, while the presented results are associated with liquid-state processing. Faster diffusion and greater availability of the reacting elements in liquid state appear to strongly impact in lowering the process activation energy, indicating that diffusion might no longer be the rate controlling mechanism of the process. Also, the linear growth behavior observed in Figure 11 suggests that the chemical reaction rate of intermetallic compound formation is the controlling mechanism in liquid state [13].

As proposed by Vianco et al. [8], the mechanism of formation of Cu_6Sn_5 and Cu_3Sn is initiated by immediate covalent bonding between Cu and Sn atoms when liquid solder touches the copper substrate. Cu_6Sn_5 is the first intermetallic to be nucleated, and, as nuclei grow, coalescence occurs and a continuous intermetallic layer is formed. In sequence, Cu_3Sn is formed in the copper-rich side of the interface, but its growth is limited by diffusion of Sn through the Cu_6Sn_5 layer that lies between it and the tin-rich solder. Subsequent growth of IMCs is controlled by reactions that take place in the interfaces between phases, consequently depending also on reactants diffusion to the reaction fronts. This mechanism is illustrated in Fig. 13.

Table 1. Activation energies of $Cu_6Sn_5 + Cu_3Sn$ growth reported in literature.

System	Q (eV/atom)	Reference
Cu/Sn-3.5Ag	1.20	Choi et al.
	0.85	Madeni and Liu
	0.80	Harris et al.
	0.67	Lee et al.
	0.61	Vianco et al.

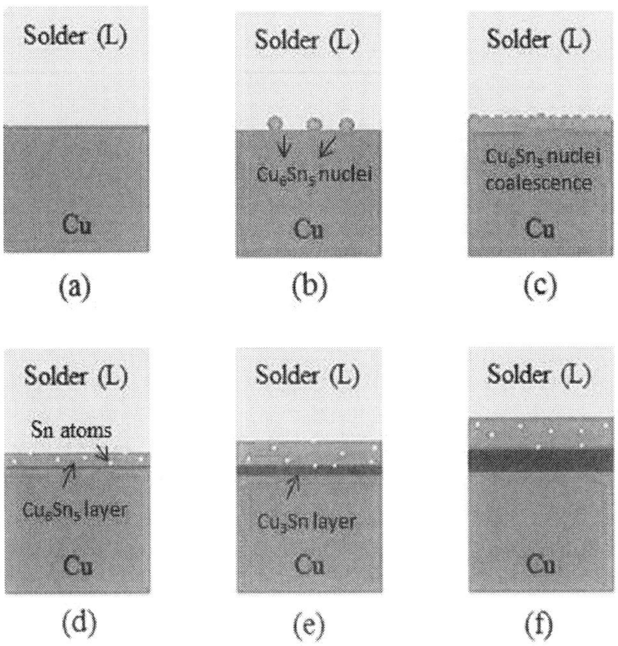

Figure 13. Intermetallic formation mechanism in lead-free Cu-Sn systems[13]. (a) Initial wetting, (b) Cu_6Sn_5 nuclei are formed, (c) coalescence of IMC nuclei, (d) growth of Cu_6Sn_5 layer and diffusion of Sn atoms, (e) formation of Cu_3Sn layer, (f) growth of both IMC layers.

While the system temperature is high enough to sustain diffusion, IMC growth continues, but as the joint reaches ambient temperature it becomes insignificant. In the as-soldered condition, as identified previously in this work, Cu_6Sn_5 was formed in substantially greater amounts, which indicates that diffusion through this layer is indeed a limiting stage for Cu_3Sn formation, and that short reaction times are not sufficient for providing enough reactants to form an identifiable layer. Cu_3Sn was reported to form and grow more noticeably during aging above 70°C [14], and even though, requiring several hours.

To complete the analysis of stages related to formation and growth of IMC and verify that these phenomena evolve through the solid-state aging, samples were subjected to thermal aging cycles.

Samples for thermal aging were produced by the two soldering cycles mentioned earlier in the experimental program, and one specimen corresponding to each soldering condition was thermally aged. Metallographic characterization was done before and after aging, and two examples of micrographs illustrating both stages of characterization are shown in Fig 14.

After thermal aging, Cu_3Sn was clearly identified within the IMC layer. At lower temperature aging the Cu_3Sn layer was no thicker than 0.20 μm, but at the highest aging temperature its mean thickness grew to 0.52 μm, and also became more non-uniform. As Cu_3Sn was not distinguished in the as-soldered condition, the correspondent final thicknesses also represent its growth.

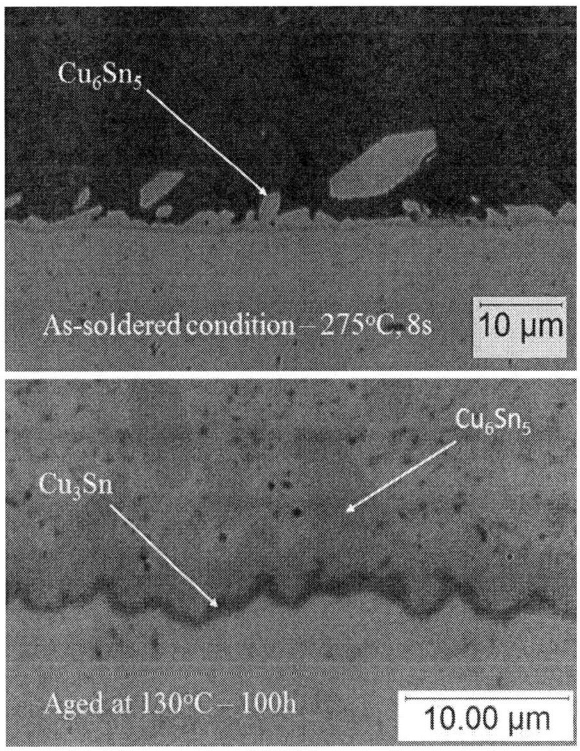

Figure 14. Micrographs of IMC layer before and after aging. The top image was taken at 500X magnification and the bottom at 1000X.

The growth experienced by Cu_3Sn accounts for a significant portion of the total IMC growth. In fact, at 70 and 100°C, for 20 and 50 hours aging, the Cu_6Sn_5 layer exhibited shrinkage in thickness, but the overall growth was compensated by Cu_3Sn. This observation leads to the hypothesis that the Cu_3Sn layer also grew at the expense of Cu_6Sn_5. This conclusion was also reported by Zeng at al. [15]. Cu_3Sn may form by dissolution of Cu_6Sn_5 in the copper substrate, as represented by Equation 4.

$$Cu_6Sn_5 + 9\,Cu \rightarrow 5\,Cu_3Sn \qquad (4)$$

Variations in thickness of both IMCs are summarized in Table 2. At low temperatures Cu_6Sn_5 shrank, what indicates that at early states of aging there is a large growth rate associated with Cu_3Sn that cause dissolution of Cu_6Sn_5. As the transformation advanced, especially at longer times and higher temperatures, Cu_3Sn was formed in larger extents, causing the driving force that promotes its growth and Cu_6Sn_5 dissolution rate to be reduced.

Table 2. Summary of prior and post in thicknesses of IMCs and respective solid state aging conditions.

#	Initial Thickness (μm)	Final Thickness (μm)	Aged at	Variation in Cu3Sn Thickness (μm)	Variation in Cu6Sn5 Thickness (μm)	Total Variation in Thickness (μm)
Produced at 275C - 8s						
1	1.585	1.624	70C/20h	0.183	-0.144	0.039
2	1.586	1.655	70C/50h	0.187	-0.118	0.069
3	1.531		70C/100h			
4	1.541	1.671	100C/20h	0.196	-0.066	0.13
5	1.584	1.769	100C/50h	0.247	-0.062	0.185
6	1.671	1.937	100C/100h	0.207	0.059	0.266
7	1.615	1.916	130C/20h	0.248	0.053	0.301
8	1.653	2.378	130C/50h	0.415	0.310	0.725
9	1.694	2.536	130C/100h	0.526	0.316	0.842
Produced at 275C - 2s						
#	Initial Thickness (μm)	Final Thickness (μm)	Aged at	Variation in Cu3Sn Thickness (μm)	Variation in Cu6Sn5 Thickness (μm)	Total Variation in Thickness (μm)
10	1.307	1.332	70C/20h	0.171	-0.146	0.025
11	1.302	1.354	70C/50h	0.186	-0.134	0.052
12	1.379		70C/100h			
13	1.372	1.565	100C/20h	0.167	0.026	0.193
14	1.353	1.659	100C/50h	0.276	0.030	0.306
15	1.367	1.801	100C/100h	0.212	0.222	0.434
16	1.287	1.672	130C/20h	0.252	0.133	0.385
17	1.335	2.009	130C/50h	0.404	0.270	0.674
18	1.145	2.121	130C/100h	0.518	0.458	0.976

Keeping track of mean thickness values from prior and post aging conditions, the actual variation in thickness, which corresponds to the growth experienced in solid state, was determined. These results account for the combined growth of Cu_6Sn_5 and Cu_3Sn, and are plotted in Figures 15 and 16 as functions of the square root of aging time.

For this portion of the work, as the time provided for the diffusion-based transformations was several orders of magnitude greater than those used in the liquid state experiments, the parabolic behavior presented in Equation 1 could be observed. However, the measured variation in IMC thickness was plotted with respect to the square root of aging times so that the correlation presented graphically would be linear. As expected, increasing aging temperatures, the rate of IMC layer growth, represented by the linear slopes, also increased.

The slopes from Figures 15 and 16 are equal to $K_T^{1/2}$ and the value that possesses thermodynamic meaning is K_T, which represents the growth rate.

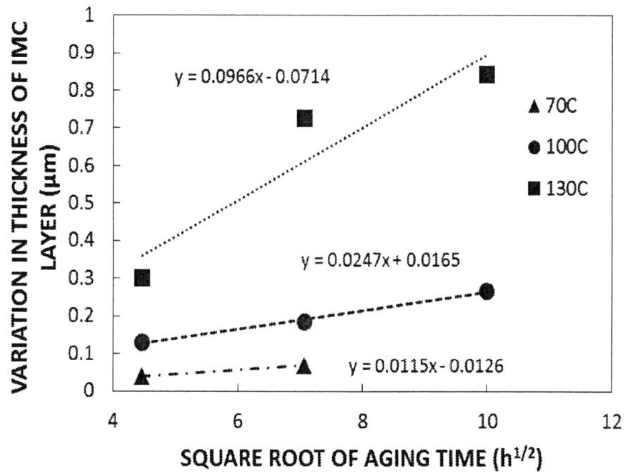

Figure 15. Variation in the IMC thickness of samples produced at 275°C with 8 seconds holding time.

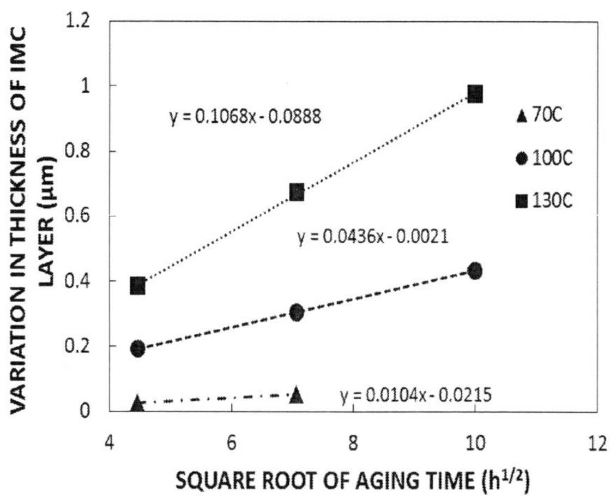

Figure 16. Variation in the IMC thickness of samples produced at 275°C with 2 seconds holding time.

According to slope obtained in Fig. 17, the activation energy for IMC growth during solid state aging was calculated to be 20.3 kcal/mol (0.88eV/atom).

The derived activation energy for solid-state aging of the system Cu/SAC305 is within the range of reported values for a similar system, Cu/Sn-3.5Ag (see Table 1). The differing value determined for transformations in liquid state (0.35eV/atom), which is significantly smaller, must represent alterations in the mechanism of IMC growth that facilitate occurrence of formation and growth reactions.

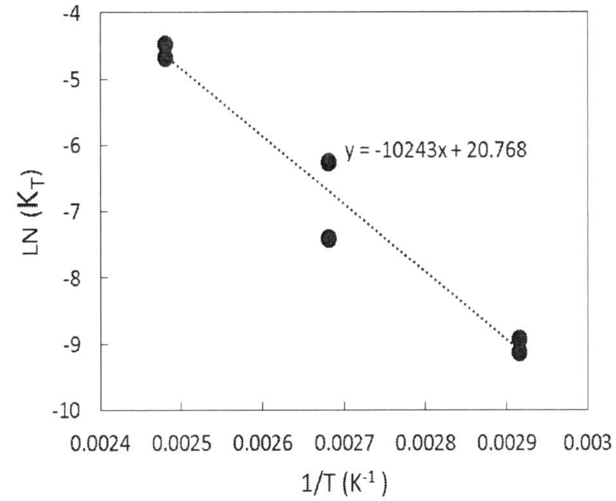

Figure 17. Arrhenius analysis of IMCs growth rate during solid state aging.

Conclusions

The present experiments confirmed that intermetallic compounds are formed and experience growth when the solder is still in liquid state. Cu_6Sn_5 was identified in the interface between copper and SAC305 solder alloy in the as-soldered condition correspondent to three different soldering temperatures. The IMC layer growth ranged from 0.37µm at T=220°C and t=2s to 2.65µm at T=300°C and t=8s. Also, the non-uniform intermetallic layer morphology became more pronounced as soldering temperatures increased, thicknesses were distributed over 0.16-0.71µm to 1.26-7.50µm respectively for T=220°C and t=2s, and at T=300°C and t=8s. The wide range of thickness distributions identified, specially at higher soldering temperatures, imply that mean values shall be used carefully in case of predicting growth behavior and mechanical properties, in view of the statistical nature of IMC growth. The empirical growth kinetics equations developed can be used as one controlling criterion for soldering parameters. The activation energy for IMC layer growth in liquid state for the system SAC305/Cu was calculated to be 8.1 kcal/mol (0.35 eV/atom). The lower activation energy value and linear growth behavior of intermetallic compounds during liquid state processing suggests that the chemical reaction rate became the controlling mechanism as diffusion and reactant availability is strongly enhanced in liquid state.

The correlation between IMC growth and solid-state aging time is parabolic, indicating that it is mostly a diffusion-controlled process. Both, Cu_6Sn_5 and Cu_3Sn were identified in the solder/substrate interface after all aging cycles. After aging at 70 and 100°C, for 20 and 50 hours aging, Cu_6Sn_5 thickness shrank, however, the IMC layer grew due to increase in Cu_3Sn thickness, this finding suggests that Cu_3Sn may also be formed at the expense of Cu_6Sn_5. With increasing aging time and temperature, both IMCs experienced growth, indicating that the rate of dissolution of Cu_6Sn_5 into Cu_3Sn was gradually reduced as transformations progressed. The activation energy for IMC layer growth during solid-state aging was found to be 20.3 kcal/mol (0.88eV/atom), which is in agreement with reported values from literature.

Acknowledgments

The authors acknowledge Hoa Nguyen and Ken Marino from OK International for the financial support and for assistance in the research progress.

References

[1] Abtew, M., Selvaduray, G., "Lead-Free Solders in Microelectronics", Materials Science and Engineering, Vol. 27 (2000), pp. 95-141.

[2] Siewert, T., et al, Database for Solder Properties with emphasis on new lead-free solders, National Institute of Standards and Technology & Colorado School of Mines, 2002.

[3] Terashima, S., et al, "Effect of Silver Content on Thermal Fatigue Life of Sn-xAg-0.5Cu Flip-Chips Interconnects", Journal of Electronic Materials, Vol. 32, No. 12 (2003), pp. 1527-1533.

[4] Kerr, M., Chawla, N., "Creep Deformation Behavior of Sn-3.5Ag solder/Cu Couple at Small Length Scales", Acta Materialia, Vol. 52 (2004), pp. 4527-4535.

[5] Pang, J. H. L., et al., "Creep and Fatigue Characterization of Lead-Free 95.5Sn-3.8Ag-0.7Cu Solder", Electronic Components and Technology Conference, June 2004, pp. 1333-1337.

[6] G. T. Lim, et al., "Temperature Effect on Intermetallic Compound Growth Kinetics of Cu Pillar/Sn Bumps", Journal of Electronic Materials, Vol. 38, No. 11, 2009, p. 2228-2233.

[7] E. Hodulova, et al., "Kinetics of intermetallic phase formation at the interface of Sn-Ag-Cu-X (X = Bi, In) solders of Cu substrate", Journal of Alloys and Compounds, 509 (2011), p. 7052-7059.

[8] P. T. Vianco, A. C. Kilgo, R. Grant, "Intermetallic Compound Layer Growth Kinetics in Non-Lead Bearing Solders", Sandia National Laboratories, Albuquerque, NM.

[9] H. P. Stevens, "A Test Comparison of SAC and Non-SAC Lead Free Solders", Metallic Resources, Inc. Twinsburg, OH.
[10] G. Humpston and Jacobson, D.M. Principles of Soldering and Brazing, 1st Ed. ASM International (1993).
[11] A.C. Harman, "Rapid Tin-Nickel Intermetallic Growth: Some Effects of Solderability", Proceedings, InterNepcon, Brighton, U.K., p.92, (1978).
[12] Y. S. Jang, et al., "Electrochemical Properties of Cu_5Sn_6-C Composite Powders with Mixture of Cu_5Sn_6-Void-C Yolk-Shell, Cu_5Sn_6 Alloy, and Hollow Carbon", Int. J. Electrochem. Science, Vol. 7 (2012) p. 12531 – 1254
[13] J. C. Madeni, et al., "Characterization and growth kinetics of the formation of intermetallic compounds in the liquid state during soldering with lead-free solders", Welding in the World, The International Journal of Materials Joining, (2014).
[14] J. C. Madeni, "Mechanical Properties and Kinetics of Intermetallics Growth in Four Lead-Free Solder Alloys", Sn-3.2Ag-0.8Cu, Sn-3.5Ag, Sn-0.7Cu, and Sn-8.9Zn, Applied to Cupper Substrates" MSc. Thesis, Colorado School of Mines, (2002).
[15] G. Zeng, et al., "A review on the interfacial intermetallic compounds between Sn-Ag-Cu based solders and substrates", Journal of Materials Science, vol. 21 (2010), p. 421-440.

Computational Thermodynamics and Kinetics in Modern Lead-free Solders Development

Shan Jin, Henrik Larsson, Qing Chen, Johan Bratberg
Thermo-Calc Software AB, Norra Stationsgata 93, SE-113 64 Stockholm, Sweden

Paul Mason
Thermo-Calc Software Inc., 4160 Washington Road, McMurray, PA 15317, USA

Abstract

The melting and solidification, as well as the interfacial reactions, highly influence the soldering process and the performance and durability of solder joints. An understanding of these phenomena is essential for the modern electronics industry. This work aims to develop a thermodynamic database for solder/substrate systems and to demonstrate the utility of the database in exploring new solders and understanding and improving existing alloys and processes. A self-consistent solder database within a framework of 16-elements for the calculation of Sn-/Au-/Bi-/Zn-based solder systems has been developed using the CALPHAD method. The database enables the predictions of phase equilibria, phase transformations, solidus/liquidus temperatures, solidification paths, heat evolutions, non-equilibrium micro-segregations etc. Selected applied examples will be presented as well, e.g., the melting behavior of the solders as a function of alloying elements or compositions, the solidification process of the alloys, and the potential interfacial reactions and formation of intermetallic compounds.

Introduction

Electronic devices are becoming smaller, faster, lighter and smarter with each passing year. The development of more functional electronic equipment relies on the development of advanced solder materials and soldering techniques. Soldering is a process in which two or more metal items are joined together by melting and flowing a filler metal, the solder, into the joint. The soldering process can be generally described by the following sequences: melting of the solder; wetting of the substrate by the molten solder; creation of the bonding, i.e., through intermetallic compound (IMC) formation and finally solidification of the solder. In these procedures phase transformations are involved. Computational thermodynamics such as CALPHAD (CALculation of PHAse Diagram) [1] fundamentally links the phases and thus the microstructures to the processing variants i.e., alloy composition, soldering temperature etc. As a result, the properties of the solder joint as a function of chemical composition and temperature can be predicted with the help of computational thermodynamics.

Another important processing parameter in soldering is time, especially in term of dissolution of the substrate and the growth of interfacial intermetallic compounds. The issue of substrate dissolution is becoming more and more critical due to ever increasing miniaturization of the electronic components [2]. On the other hand, the formation of a thin IMC layer is necessary for a successful soldering operation since in most cases it is the IMC layer that bonds the solder to the substrate [3]. However, an excessive growth of the IMC layer often deteriorates the mechanical properties of the joint. All these factors make time control very crucial. DICTRA [4], a software tool for simulations of diffusion controlled phase transformations, can be used for kinetic modeling of the soldering processes where time is a parameter.

Computational thermodynamic and kinetics have been proven to be powerful tools for materials development which greatly shorten the development period and reduces the cost. Nevertheless, the successful use of the thermodynamic and kinetic modeling relies on the development of reliable and self-consistent multi-component databases. In this work, the development of thermodynamic and kinetic databases for solder alloy systems is presented and examples of the applications of the databases to the melting behavior of solders, the reaction with substrates and solidification are demonstrated. The calculations have been performed by using the Thermo-Calc and DICTRA software packages [4].

Thermodynamic and kinetic database

Thermodynamic database

A solder alloy solutions thermodynamic database, TCSLD (current version 2.0), has been developed by Thermo-Calc Software. It contains most of the important solder alloy systems, e.g., Sn-/Au-/Bi-/Zn-based solders and Cu, Ni, Pd etc. relevant substrate or metallization materials. In total 16 elements are included: Ag, Al, Au, Bi, Co, Cu, Ge, In, Ni, Pb, Pd, Pt, Sb, Si, Sn, Zn. TCSLD has been developed using the CALPHAD methodology [1], based on the critical evaluation of various types of experimental data and theoretical information. The TCSLD database enables pre-screening of potential solder candidate compositions, predictions of various thermodynamic properties and phase equilibria, and providing information on possible interfacial reactions. The database has been validated where possible against commercial solder alloys and available experimental information. Some selected examples of the application of TCSLD in solder development are demonstrated in the following sections.

Kinetic database

The development of a mobility database, MOBSLD, for kinetic simulations of soldering processes is in progress. It aims to predict the diffusion behavior, i.e., the dissolution of substrate and the growth of IMCs during the reflow processes. Preliminary results will be shown in the section 'Interface reaction'.

Melting behavior

Melting behavior is usually the first consideration in designing new solder materials. In general, solder alloys are materials with low melting temperatures. The typical melting point of solders is between 90 to 450 ºC. The melting point can determine the actual soldering temperature. The pasty range, or the melting range, is another important characteristic of solder alloys and is defined as the temperature difference between solidus and liquidus temperatures, where the alloy is part solid and part liquid. It is preferable to have a narrow pasty range since alloys with wide pasty ranges are more susceptible to fillet lifting and tombstoning [5]. In addition, the solidus temperature should be significantly higher than the solder joint's maximum operating temperature [6].

A number of promising candidates, e.g., Sn-Ag, Sn-Cu, Sn-Bi, Sn-Zn, Sn-Ag-Cu, Sn-Ag-Bi alloys, have been developed as lead-free solder materials. The addition of Ag, Al, Bi, In, Cu etc. elements is found to lower the melting temperature of Sn-based alloys [7]. Other elements, i.e., Ni, Sb etc., are also commonly used as addition elements in order to improve the mechanical properties. However, they do not reduce the melting temperature. Knowledge on the effect of alloying elements on the melting behavior of lead-free solders is of great importance to the development of new solder alloys and the improvement of existing alloys. This section presents the application of the TCSLD database in predicting the melting behavior of different solder alloy systems. Liquidus temperatures and solidus temperatures reported from different commercial solder companies have been collected and compared to the calculated results.

Melting behavior of commercial solder alloys

In ternary or even higher-order systems, liquidus projection allows easy identification of the low liquidus temperature over the full composition range of each component of the system. Figure 1 shows the calculated liquidus projections at the Sn-rich corner of three typical commercial solder alloy systems, Sn-Ag-Cu, Sn-Ag-Bi and Sn-Ag-In. The liquidus monovariant lines which are shown with thick blue curves indicate the conditions in which two solid phases coexist with the liquid phase. The green curves in the diagram represent constant temperature contours. The compositions of commercial alloys are shown as circles and the corresponding melting points reported by different solder companies [7-14] are marked next to the circles. As can be seen, the temperature difference between the calculated and experimental data is within 1 to 2 degrees in most cases.

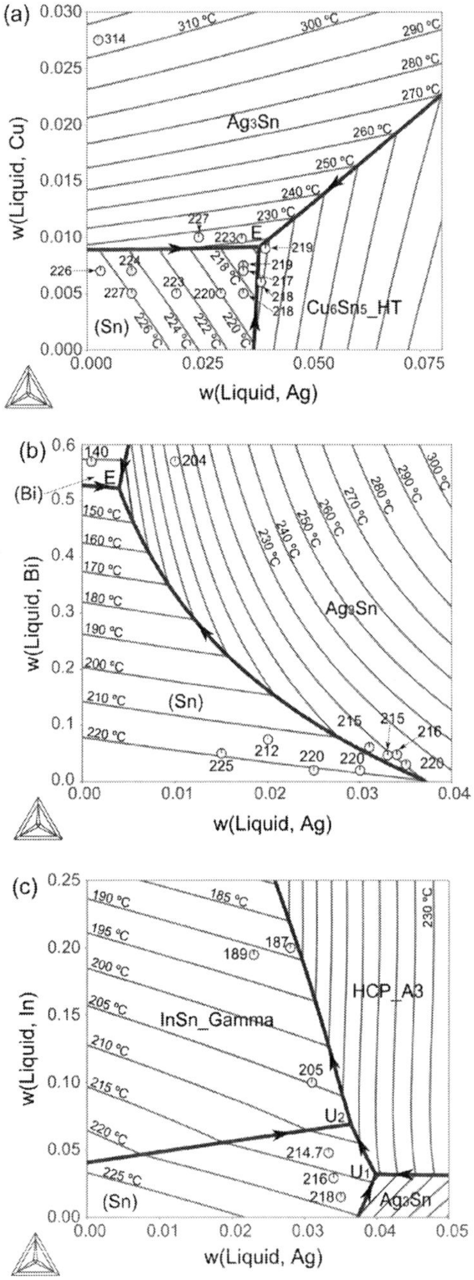

Figure 1 Calculated liquidus projection of a) Sn-Ag-Cu b) Sn-Ag-Bi c) Sn-Ag-In at the Sn-rich corner.

Furthermore, the liquidus and solidus temperatures of a large number of commercial solders have been collected and compared to the calculated temperatures using TCSLD, as shown in Fig. 2. "Critical temperatures" in the axis titles of Fig. 2 refers to either the liquidus or the solidus temperatures. Good agreement between calculations and experiments is observed as well.

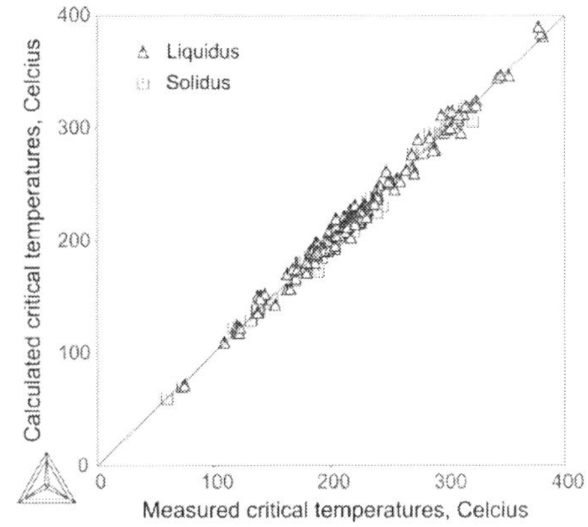

Figure 2 Comparison of experimental and calculated liquidus and solidus temperatures (referred as critical temperatures in the axis titles) for various solder alloys.

Effect of alloying element

With the CALPHAD method, the effect of elements addition on the melting behavior of alloys can also be calculated. As an example, Fig. 3 shows the addition of Bi and Sb on the melting point of the Sn-0.5Ag-0.5Cu alloy. The calculation suggests that Bi presents obvious effects on depressing the melting point and meanwhile broadening the melting range while the addition of Sb has little effect on the melting temperatures. This is supported by the experimental observation by Lu et al. [15]. In addition, the microstructure and properties of the Sn-Ag-Cu alloy system are strongly dependent on alloying elements and impurities, such as Cu, Ag, Sb, Zn, Bi, In and others. Lu et al. [15] found that the addition of Bi and Sb both lead to a gradual decrease of the primary (Sn) phase, which is in consistent with the calculations, as shown in Fig. 4.

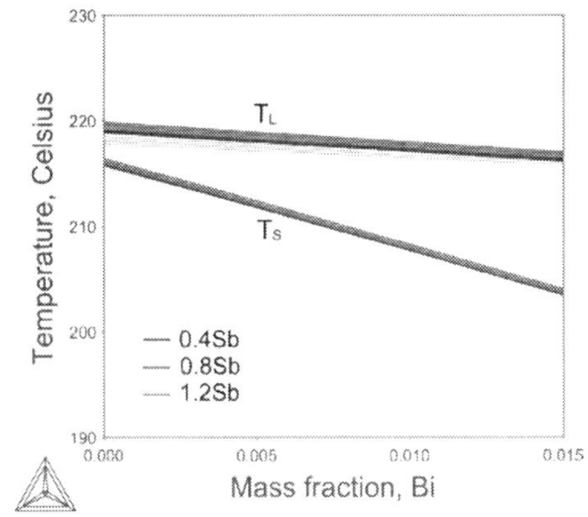

Figure 3 Calculated effect of Bi and Sb addition on melting point of Sn-3.5Ag-0.5Cu.

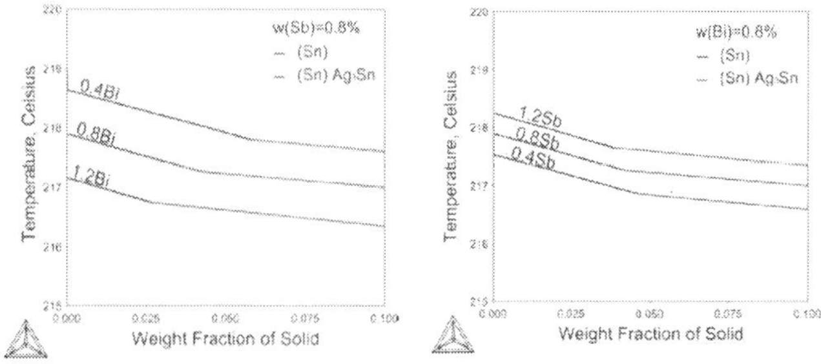

Figure 4 Calculated effect of Bi and Sb addition on the amount of primary phase of Sn-3.5Ag-0.5Cu.

Interface reaction

The interfacial reaction between solder and metallization is an essential process in modern micro-electronic packaging technology for interconnections. The reliability of the joints in the functionality and reliability of electronic components are strongly dependent on the interface reactions that occur between solders and substrate materials. The interface reaction products have a great effect on the mechanical properties of the substrate/solder joint. The understanding and interpretation the solder/substrate interactions and their microstructure evolution at the interface during the soldering process and in applications are crucial for new technology developments. Meanwhile, accurate control of these reactions can have significant impact on the optimization of the soldering process [16].

Formation of interfacial products

The intermetallic compounds at the interface, especially the one that forms first, is critical to the mechanical performance of the soldered joints. The formation of intermetallic compound during the soldering process also plays an important role in the wettability of the solder [17]. Predictability of the interfacial reaction product is thus very important in terms of tailoring the microstructure of the joint through controlling composition or processing parameters.

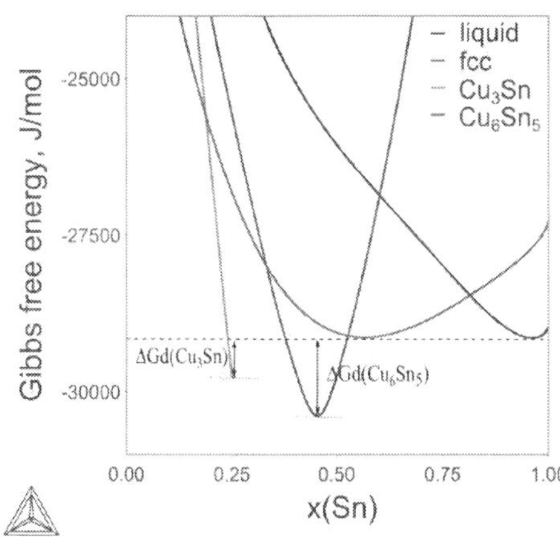

Figure 5 Illustration of driving forces of formation for Cu_3Sn and Cu_6Sn_5 under metastable equilibrium between liquid and fcc at 250 °C in Cu-Sn binary system.

In the present work, the intermetallic compound that forms first at the interface during soldering is predicted through calculating and comparing the driving force of formation. This scheme for predicting the first formed compound was proposed by Lee [17, 18] who suggested that during soldering, the phase which forms first at the interface is the phase with the highest driving force of nucleation from the solder liquid in metastable equilibrium with the substrate. The driving force is thus the Gibbs free energy difference between the set equilibria, i.e., metastable local equilibrium at the interface between the substrate and liquid solder, and the possible forming phases. Figure 5 illustrates the driving forces of formation of Cu_6Sn_5 and Cu_3Sn under the metastable equilibrium between Cu substrate (fcc) and liquid Sn in the Cu-Sn binary system. The dashed line is the common tangent for the metastable equilibrium. In ternary system this common tangent will be a plane. The driving forces of formation of intermetallic phases will then correspond to the vertical distances between the common tangent plane and the Gibbs energy of the phases.

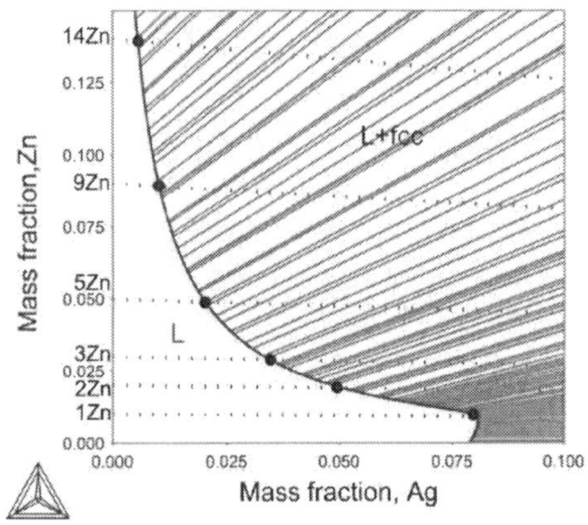

Figure 6 Calculated metastable isothermal section at Sn-rich corner at 260 °C. The dotted lines connect Ag and initial Sn-xZn alloy composition. The cross point (marked with full circle) between the dotted line and the metastable phase boundary is estimated as the interface composition.

The experimental observations by Jao et al. [19] and Jee et al. [20] are taken as examples for illustrating the application of the TCSLD thermodynamic database in interface reactions. Jao et al. [19] and Jee et al. [20] investigated the interface microstructures of Sn-Zn/Ag couples at 260 °C. They found that the interfacial products are different when the Zn contents in Sn-Zn alloys are different. For example, one layer of Ag_3Sn formed at the interface in the Sn-1wt.%Zn/Ag couple while two-layer microstructure was observed in the Sn-2wt.%Zn/Ag and Sn-3wt.%Zn/Ag couples. When the Zn content further increased to 9wt.% or 14 wt.%, three IMC layers, $AgZn_3$, Ag_5Zn_8 and AgZn, were formed. The change of interface microstructure due to different Zn content can be well explained by thermodynamic analysis based on the scheme proposed by Lee [17, 18]. The experimentally observed compound which is adjacent to the solidified solder is considered as the first formed compound and used for comparison with the prediction from thermodynamic calculations.

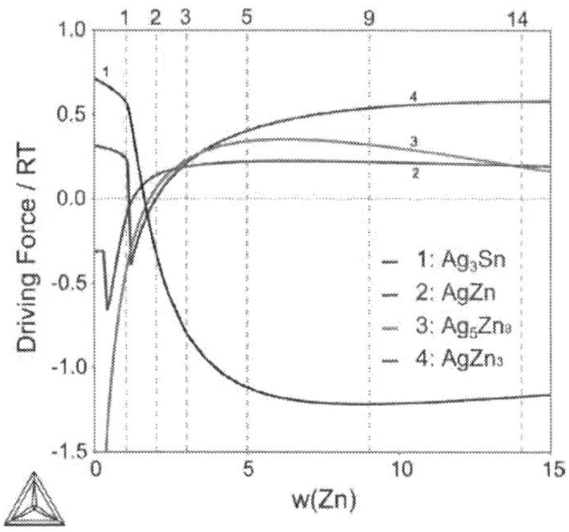

Figure 7 Calculated driving force of formation.

The calculation of the driving force requires the interface composition. The composition at the solder/substrate interface is estimated by calculating the metastable equilibrium between solid substrate and liquid solder phases. Fig. 6 shows the calculated metastable isothermal section of Ag-Sn-Zn at the Sn-rich corner at the reaction temperature, 260 °C, where only the liquid solder phase and solid (Ag) phase (the fcc phase) are considered in calculation. The cross point between the straight line that connects the substrate and the initial composition of liquid solder and the metastable substrate+liquid/liquid phase boundary is estimated as the interface composition, as marked with a full circle in Fig. 6. The driving forces of all the other phases in the system can then be calculated under the two-phase equilibrium between liquid and (Ag) using the developed TCSLD database.

Figure 7 presents the calculated driving force of each phase in the Ag-Sn-Zn system. For the Sn-1wt.%Zn/Ag couple, the Ag_3Sn phase has the highest driving force. According to the scheme, Ag_3Sn should form first at the Sn-1wt.%Zn/Ag interface. This is in accordance with the interface microstructure reported by Jao et al. [19]. The driving force of Ag_3Sn decreases with increasing Zn content. When there is 2wt.%Zn, the phase with the highest driving force is the AgZn phase. For the Sn-5wt.%Zn, Sn-9wt.%Zn and Sn-14wt.%Zn alloys the $AgZn_3$ phase has higher driving force compared to all the other phases in the system. Table 1 lists the predicted first formed phase compared to the experimental observations. As can be seen, the calculations successfully predict for all compositions except for the Sn-3 wt.%Zn alloy. The predicted first formed compound is Ag_5Zn_8 while AgZn was observed at the Sn-3 wt.%Zn/Ag interface during experiment. However, it should be noted that at w(Zn)=3%, the driving forces of Ag_5Zn_8 and AgZn are very close to each other, as shown in Fig. 7. In such a case, other effects, e.g., interfacial energy, misfit strain energy etc., should also be considered for a more accurate prediction of the interface reaction product. Nevertheless, the satisfactory agreement shows the ability of the TCSLD database in predicting interface reactions products. In summary, thermodynamic calculations are a very good starting point to obtain information on possible IMC formations and to understand the interface reactions.

Table 1 List of predicted first formed interface reaction product compared with experiments.

Sn-xZn/Ag couple, 260 °C		
Alloy	Calculation	Experiment [19,20]
Sn-1 wt.%Zn	Ag_3Sn	Ag_3Sn
Sn-2 wt.%Zn	AgZn	AgZn
Sn-3 wt.%Zn	Ag_5Zn_8	AgZn
Sn-5 wt.%Zn	$AgZn_3$	$AgZn_3$
Sn-9 wt.%Zn	$AgZn_3$	$AgZn_3$
Sn-14 wt.%Zn	$AgZn_3$	$AgZn_3$

Kinetics of interfacial reactions

Attempts were made to perform 1D simulations of the growth of the intermetallic compounds between a copper substrate and liquid tin using the DICTRA software [4, 21]. The interdiffusion coefficient in Cu_3Sn was determined by Dreyer et al. [22] in a study of $Cu\text{-}Cu_6Sn_5$ diffusion couples. The interdiffusion coefficient in Cu_3Sn and Cu_6Sn_5 was determined by Mei et al. [23] using Cu-Sn diffusion couples. Gagliano and Fine [24] did a detailed experimental study of liquid tin – solid copper reaction couples at 275 °C.

There are numerous difficulties in trying to capture this complex reaction in a 1D simulation. The Cu_3Sn phase was not experimentally observed for reaction times shorter than 5 minutes. Furthermore, a continuous layer of Cu_3Sn between the substrate and a continuous layer of Cu_6Sn_5 will slow the growth rate of the Cu_6Sn_5 to values far lower than that experimentally observed. If the Cu_3Sn phase is left out of the simulation, the simulated growth rate of a continuous Cu_6Sn_5 layer is still lower than that observed by Gagliano and Fine [24]. This can partly be explained by their observation that open channels between the copper substrate and the liquid tin are retained for approximately 10 minutes. However, the simulations indicate that short-circuit diffusion paths play an important role even after the channels appear to be closed.

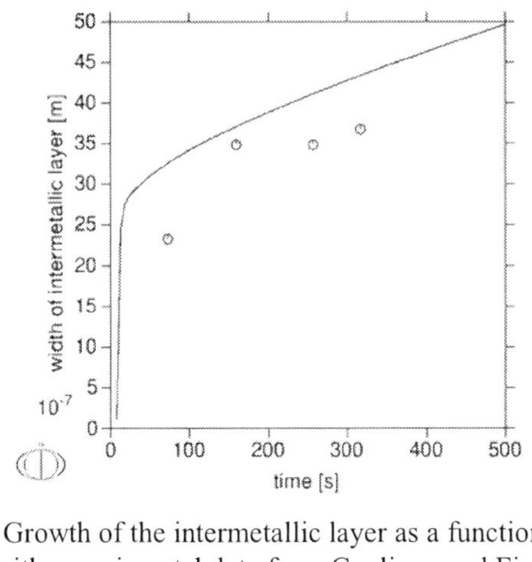

Figure 8 Growth of the intermetallic layer as a function of time along with experimental data from Gagliano and Fine [24].

To get a rough estimate of an "effective diffusivity" through the Cu_3Sn and Cu_6Sn_5 phases, various mobility values were tried. In Fig. 8 is shown the width of the intermetallic layer as a function of time along with experimental data from Gagliano and Fine [24]. In this simulation a mobility value of $3 \cdot 10^{-19}$ $[m^2 \cdot s^{-1} \cdot J^{-1} \cdot mol]$ was used for both Cu and Sn in both Cu_3Sn and Cu_6Sn_5. Both intermetallic phases were entered in a single region in which it is assumed that their relative phase fractions are locally equilibrated. The intermetallic region is separated from the copper substrate and the tin melt by interfaces at which local equilibrium is assumed to hold. It can be seen that there is a dramatic change in growth rate when the tin melt become saturated with copper; the initial stage of the simulation is thus very sensitive to the size of the tin region and the mobility in the liquid.

Solidification

The final step of the reflow process is the solidification of the solder. Though the solidification behavior of solder is also influenced by the consuming of elements during interface reactions, the analysis of the solidification of solder itself can still provide insight on the actual solidification process of the solder. With solidification simulations, the phase formation sequence, phase fraction, phase transformation temperatures etc., can be predicted.

In industry application, non-equilibrium solidification rather than equilibrium solidification is usually observed. Experiences have shown that Scheil simulations, which assume that the diffusion in liquid is very fast while the diffusion in the solid phases is so slow that it can be ignored, often provides predictions close to practical solidification processes. In this section, the solidification processes of Sn-3.7Ag-4Bi-0.9Zn and Sn-Zn-Ag solder alloys are modeled using the Scheil simulation based on the TCSLD database.

Sn-3.7Ag-4Bi-0.9Zn alloy

It is reported that a small addition of Zn to the Sn-Ag solders improves the mechanical strength while maintaining the same level of ductility [25, 26]. Adding a low-melting-point element such as Bi in Sn-Ag-Zn further decreases the solder melting temperature. Therefore Sn-Ag-Bi-Zn alloys have become promising candidates as lead-free solder materials [27]. The solidification behavior of the Sn-3.7Ag-4Bi-0.9Zn (wt.%) alloy is presented in Fig. 9(a). The dotted lines represent the equilibrium solidification and the solid lines represent the Scheil solidification. The phase transformation temperatures modeled by Scheil simulation are listed in Table 2 and compared with experimental data. As can be seen, the phase transformation temperatures are well predicted by the Scheil simulation.

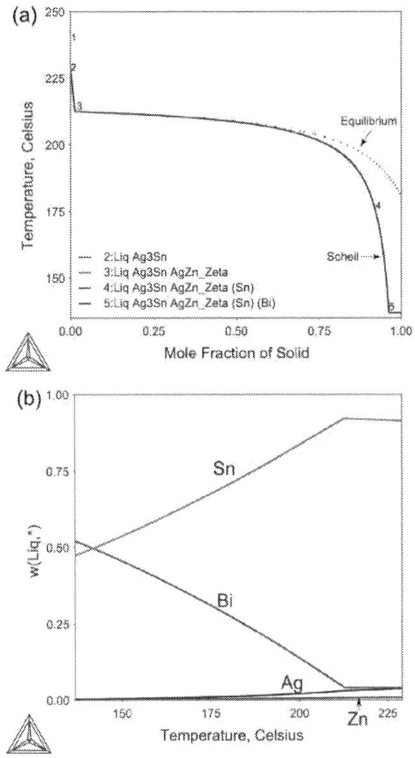

Figure 9 The Sn-3.7Ag-4Bi-0.9Zn (wt.%) alloy: (a) equilibrium solidification (dotted lines) and Scheil solidification (solid lines) simulations; (b) concentration change of Ag, Bi, Sn, Zn in liquid.

In addition, it is noted that the equilibrium solidification ends at T = 180 °C while in the non-equilibirum case the solidification ends at 137 °C. The calculated concentration change of the elements in the liquid phase is given in Fig. 9(b). The concentration of Bi in liquid increases as the solidification continues. Such an enrichment of Bi in liquid makes the temperature fall to near the eutectic temperature of the Sn-Bi binary system [28]. The tendency of precipitation of brittle Bi is connected with the fillet-lifting phenomena at the interface [29]. It is thus important to control the cooling rate in order to avoid the excess formation of Bi precipitates.

Table 2 List of phase transformation temperatures of Sn-3.7Ag-4Bi-0.9Zn (wt.%).

T_{exp} (°C) [27]	220	202	—	139
T_{cal} (°C)	228	212	212	137

Sn-Zn-xAg alloys

Sn-Zn-based alloys are potential candidates as lead-free solder alloys because of their low melting point, good mechanical properties and low cost [30]. It is found that the addition of Ag to Sn-Zn alloys improves the wettability [31] as well as ductility [32, 33]. Tsai and Hwang [34] studied the effect of Ag content on the solidification behavior of Sn-9Zn-xAg (wt.%) alloys. The pasty ranges of Sn-9Zn and Sn-9Zn-xAg (wt.%) alloys calculated using Scheil are shown in Fig. 10(a) with blue color. The experimental results reported by Tsai and Hwang [34] are superimposed for comparison. The agreement is quite satisfactory with a temperature difference of a few degrees. Figure 10(b) presents the calculated mole fractions of the phases using Scheil solidification. The Scheil calculation reproduces well the experimentally measured fractions of the $AgZn_3$ phase [34].

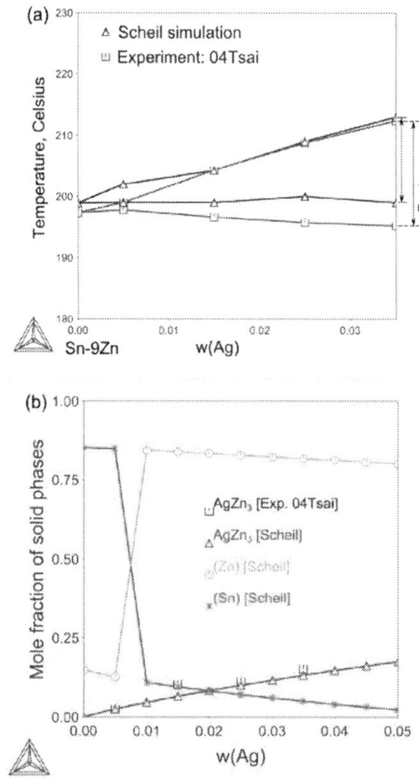

Figure 10 The Sn-9Zn-xAg (wt.%) alloys: (a) calculated pasty range using Scheil simulation compared with experimental data [34]; (b) calculated mole fraction of solid phases using Scheil simulation compared with experimental data [34].

Conclusions

A thermodynamic database for solder alloy systems, TCSLD (version 2), has been developed based on the critical evaluation of experimental and theoretical information using the CALPAHD methodology. Examples of the application of TCSLD on the most relevant processes during soldering, such as melting, interface reaction and solidification, are demonstrated and discussed. Good agreement between calculations and experiments shows the ability of TCSLD in the development of new solder materials and advanced soldering technology.

References

[1] Lukas, H. L., Fries, S. G., Sundman, B., Computational Thermodynamics - the Calphad Method, Cambridge University Press (UK, 2007).

[2] Faizan, M., "On the Mathematical Modeling of Copper Dissolution and IMC Growth in Liquid Solder,"

International Journal of Mechanical Engineering, Vol. 3, No. 2 (2014), pp. 17-26.

[3] Lee, C. C. et al., "Are Intermetallics in Solder Joints Really Brittle," Proc 57th Electronic Components and Technology Conference, Reno, NY, May 29 to June 1. 2007, pp. 648-652.

[4] Andersson, J-O., Helander, T., Höglund, L., Shi, P-F., "Thermo-Calc & DICTRA, Computational Tools for Materials Science", Calphad, Vol. 26, No. 2 (2002), pp. 273-312.

[5] Lee N.-C., Reflow Soldering Processes and Troubleshooting: SMT, BGA, CSP and Flip Chip Technologies, Newnes (2002).

[6] Bath, J., Lead-Free Soldering, Chapter 2: Fundamental Properties of Pb-Free Solder Alloys, Springer (USA, 2007), pp. 33.

[7] Puttlitz, K. J., Stalter, K. A., Handbook of Lead-Free Solder Technology for Microelectronic Assemblies, Marcel Dekker, Inc., (NY, 2004).

[8] http://www.senju-m.co.jp/en/product/ecosolder/.

[9] http://www.genma.co.jp/en/products/pb-free01/index.html.

[10] http://www.coiningllc.com/alloy_list.asp.

[11] http://www.hirschmetals.com/pdfs/SolderAlloyChart.pdf.

[12] Siewert, T., Liu, S., Smith, D.R., Madeni, J.C., Properties of Lead-Free Solders, Release 4.0, NIST and Colorado School of Mines, (2002)

[13] http://www.indium.com/technical-documents/product-data-sheets/.

[14] Vassilev, G. P., Dobrev, E. S., Tedenac, J-C., "Experimental Study of the Ag-Sn-In Phase Diagram," Journal of Alloys and Compounds, Vol. 399, No. 1-2 (2005), pp. 118-125.

[15] Lu, S., et al., "Microstructure and Solderability of Sn-3.5Ag-0.5Cu-xBi-ySb Solders," Proc 11th International Conference on Electronic Packaging Technology & High Density Packaging, Xi'an, Aug. 2010, pp. 410-412.

[16] Bader, S., Gust, W., Hieber, H., "Rapid Formation of Intermetallic Compounds by Interdiffusion in the Cu-Sn and Ni-Sn systems," Acta Metallurgica et Materialia, Vol. 43 (1995) pp. 329-337.

[17] Lee, B. –J., Hwang N. M., Lee, H. M., "Prediction of Interface Reaction Products Between Cu and Various Solder Alloys by Thermodynamic Calculation," Acta Metar., Vol. 45, No. 5 (1997) pp. 1867-1874.

[18] Lee, B. –J., "Thermodynamic Analysis of Solid-State Metal/Si Interfacial Reactions," Journal of Materials Research, Vol. 14, No. 3 (1999) pp. 1002-1017.

[19] Jao, C.-C., et al., "Phase Equilibria of the Sn-Zn-Ag System and Interfacial Reactions in Sn-Zn/Ag Couples," Intermetallics, Vol. 16 (2008) pp. 463-469.

[20] Jee, Y. K., Yu, J., "Interfacial Reactions and Joint Strength of Sn-xZn Solders with Immersion Ag UBM," Journal of Electronic Materials, Vol. 39, No. 10 (2010) pp. 2286-2291.

[21] Larsson H., "A Model for 1D Multiphase Moving Phase Boundary Simulations under Local Equilibrium Conditions", Calphad, Vol. 47 (2014), pp. 1-8.

[22] Dreyer, K.F., Neils, W.K., Chromik, R.R., Grosman, D., Cotts, E.J., "Calorimetric Study of the Energetics and Kinetics of Interdiffusion in Cu/Cu_6Sn_5 Thinfilm Diffusion Couples", Applied Physics Letters, Vol. 67, No. 19 (1995), pp. 2795-2797.

[23] Mei, Z., Sunwoo, A.J., Morris, J.W. Jr, "Analysis of Low-Temperature Intermetallic Growth in Copper-Tin Diffusion Couples", Metallurgical Transactions A, Vol. 23A, No. 3 (1992), pp. 857-864.

[24] Gagliano, R.A., Fine, M.E., "Growth of η Phase Scallops and Whiskers in Liquid Tin-Solid Copper Reaction Couples", JOM, Vol. 53, No. 6 (2001), pp. 33-38.

[25] McCormack, M., Jin, S., "New, Lead-free Solders," Journal of Electronic Materials, Vol. 23, No. 7 (1994) pp. 635-640.

[26] McCormack, M., Jin, S., Kammlott, G.W., Chen, H.S., "New Pb-free Solder Alloy with Superior Mechanical Properties," Applied Physics Letters, Vol. 63 (1993) pp. 15-17.

[27] Wang, X., Liu, Y. C. et al., "Development of A Lead-free Solder: Sn-4.0Bi-3.7Ag-0.9Zn," Soldering & Surface Mount Technology, Vol. 23, No. 1 (2011) pp. 15-21.

[28] Ohtani, H., Ono, Ss., Doi, K., Hasebe, M., "Thermodynamic Study of Phase Equilibria in the Sn-Ag-Bi-Zn Quaternary System," Materials Transactions, Vol. 45, No. 3 (2004) pp. 614-624.

[29] Takao, H., Hasegawa, H., "Influence of Alloy Composition on Fillet-Lifting Phenomenon in Sn-Ag-Bi Alloys," Journal of Electronic Materials, Vol. 30 (2001) pp 1060-1070.

[30] Mavoori, H., Chin, J., Vaynman, S., Moran, B., Keer, L. and Fine, M., "Creep, Stress Relaxation, and Plastic Deformation in Sn-Ag and Sn-Zn Eutectic Solders", Journal of Electronic Materials, Vol. 26, No. 7 (1997) pp. 783-790.

[31] Takemoto, T., Funaki, T., "Role of Electrode Potential Difference between Lead-Free Solder and Copper Base Metal in Wetting," Materials Transactions, Vol. 43, No. 8 (2002) pp. 1784-1790.

[32] Mccormack, M. Jin, S., "Improved Mechanical Properties in New, Pb-Free Solder Alloys," Journal of Electronic Materials, Vol. 23, No. 8 (1994) pp. 715-720.

[33] Ahmed, M., Fouzder, T., Sharif, A., Gain, A. K., Chan, Y. C., "Influence of Ag Micro-Particle Additions on the Microstructure, Hardness, and Tensile Properties of Sn-9Zn Binary Eutectic Solder Alloy," Microelectronics Reliability, Vol. 50, No. 8 (2010) pp. 1134-1141.

[34] Tsai, Y.-L., Hwang, W.-S., "Pasty Ranges and Latent Heat Release Modes for Sn-9Zn-xAg Lead-free Solder Alloys," Materials Transactions, Vol. 45, No. 6 (2004) pp. 1949-1957.

Effects of Adding Porous Cu Interlayer With Different Porosities To Sn-3.0Ag-0.5Cu Lead Free Solder Joint

Nashrah Hani Jamadon, Farazila Yusof, and Mohd Hamdi
Centre of Advanced Manufacturing and Material Processing (AMMP),
Department of Mechanical Engineering,
University of Malaya, Kuala Lumpur, Malaysia
Email :nashrahhani@gmail.com, Phone no :+603-7976 7633

Ariga Tadashi
Department of Metallurgical Engineering, Tokai University, Hiratsuka 259-1292, Japan
Email: ttariga@keyaki.cc.u-tokai.ac.jp

Miyashita Yukio
Department of Mechanical Engineering, Nagaoka University of Technology, Nagaoka 940-2188, Japan
Email: miyayuki@mech.nagaokaut.ac.jp

Abstract

The effects of adding porous Cu interlayer on the mechanical properties and microstructure of Sn-3.0Ag-0.5Cu (SAC305) lead-free solder joints were investigated. The porous Cu interlayer was placed in a sandwich-like layer between Sn-3.0Ag-0.5Cu solder alloy and rod Cu as the substrate metal. Two different pores size approximately (ϕ1.7 mm) and P25 (ϕ1.0 mm) of porous Cu were used in this study. The soldering process of solder joint was carried out in three different soldering temperatures of 267, 287 and 307°C while the holding time was set constant at 300 s. Tensile test was performed to evaluate the joining strength of the solder alloy with loading rate of 0.5 mm/min. The result shows that the joint strength tends to increase proportionally with increasing soldering temperature. This is due to the penetration of molten solder alloy into the porous Cu interlayer. The IMC elemental analysis was identified by energy dispersive X-ray spectroscopy (EDS). The results revealed that during the soldering process, Cu_6Sn_5 compound with scallop like morphology was formed at the interface of the solder alloy and Cu substrate, and solder alloy and porous Cu interlayer.

Introduction

The conventional high temperature solder used in electronics and electrical industries is Pb-Sn which contains high proportion of lead with approximately 90-97 wt% Pb [1],[2]. However, due to the toxicity of lead, research has been focusing on developing lead-free solder to replace this common high temperature lead-based solders [3],[4]. High temperature lead-free solders from gold-based alloy namely, AuSn and AuGe were initially proposed given its favorable high melting temperature. Although, the melting point is suitable for high temperature application, the cost to produce these solders are high [5],[6]. The search for other alternative high temperature lead-free solders thus remains. It was later found that among the potential candidates include Sn-Sb, Bi-Ag, Zn-Sn and Zn-Al solders. However, each of these combinations has its own drawbacks [7]. For example, the solidus temperature of Sn-Sb and Zn-Sn alloys are too low. Zn-Al alloys are highly corrosive and easy to oxidize, while the Bi-Ag alloys have brittleness and low thermal/electrical conductivity issues [8].

Meanwhile, Sn-Ag-Cu solder alloy has also been extensively used as a lead-free solder because of its excellent mechanical properties and economical cost [9]. However its melting point of 220°C does not reach the accepted melting temperature of 250°C to 400°C in the latter category range [10]. The safe operating environment for this alloy is taken between 150°C and 200°C depending on the applications as mentioned elsewhere [1],[5]. George et al. recommended the use of Sn-Ag-Cu solders in applications that operate below 185°C [11]. Hence, modification for this alloy is needed.

A number of studies were carried out to improve the reliability of this Sn-Ag-Cu solder alloy in order to fulfill the requirement of elevated temperature application. The approach involved adding other elements into the solder alloy [12]. Zhao et. al studied the effect of adding RE element-Cerium (Ce) to Sn-Ag-Cu solder joints. It was found that the creep-rupture life of solders increases 9 times or more longer than the solder joints without Ce [13]. Despite this finding, its use remains limited to middle temperature application. It also has been reported that the addition of small amount of Co to Sn-Ag-Cu solder alloys had depleted the phase transformation of IMC and hence, improving its joint solderability [14]. However, the mechanical effect of Co to solder joint has not been discussed in details. This has encouraged other studies to modify Sn-Ag-Cu lead-free solder for high temperature use.

One of the methods to increase the performance of Sn-Ag-Cu solder alloy is by adding porous metal into the material joint as proven elsewhere [15]. It appears that porous metal has excellent thermal conductivity and thus can be applied as an interconnection [16]. Tuan Zaharinie et al. have successfully utilized the porous metal for high-temperature dissimilar material joining. They utilized a porous Cu/Ni interlayer in metal and ceramic joint in brazing processes. They reported that the porous composite of Cu/Ni in an inconel-sapphire joint have modified the thermodynamic activity near the ceramic portion of sapphire site to enhance its solderability [17]. Presently, the use of porous metal is only limited to braze joining and it has not been tested for high temperature solder joining. Only few studies have investigated on the application of porous Cu into solder joining.

This present study examines the utilization of porous Cu interlayer in soldering process to join Cu substrate with common Sn-3.0Ag-0.5Cu Pb-free solder. Two types of porous Cu interlayer with different pore sizes were selected for use at high soldering temperature regime. The interlayer reaction and tensile strength of the joints are discussed in this paper.

Experimental Details

Sn-3.0Ag-0.5Cu (wt%) of lead free solder paste of 0.1 g was used to join the Cu rod. The dimension of Cu substrate (99.99%) is 8 mm in diameter and 30 mm long. The porous Cu interlayer, measured as pore per inch (ppi), was added in the middle of the joining section. Different pore size of porous Cu interlayer namely at 15 ppi (P15) and 25 ppi (P25) were used. Solder joint without the porous Cu interlayer was also prepared as the control sample. The porous Cu interlayer was rolled from its initial thickness to obtain a uniform layer of 0.1 mm. The configuration of Sn-3.0Ag-0.5Cu solder alloy with porous Cu interlayer joined with Cu rod is shown in Figure 1.

During the soldering process, 234 N of load was imposed at the upper jig in order to obtain uniform solder layer (Figure 2). The soldering process was conducted in the standard JISZ 3191 furnace supplied with an Argon gas flow. The furnace is set up as shown in the schematic drawing of Figure 3. Three different temperatures of 267°C, 287°C and 307°C with a constant holding time of 300 s was set. Solder joint without porous Cu interlayer was also prepared as the control sample. After the sample was soldered, it was cut perpendicular to the joint and the cut surface was polished to observe the interface of the joining part. Scanning electron micrograph (SEM) was used to observe the IMC thickness at the interfaces between the solder and Cu rod as well as between the solder and porous Cu. Energy dispersive X-ray spectroscopy (EDS) was utilized to determine the elemental composition of IMC layer.

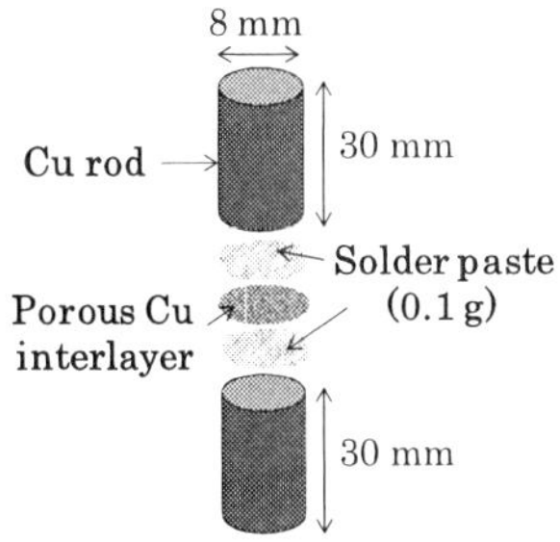

Figure 1 Configuration of solder joint.

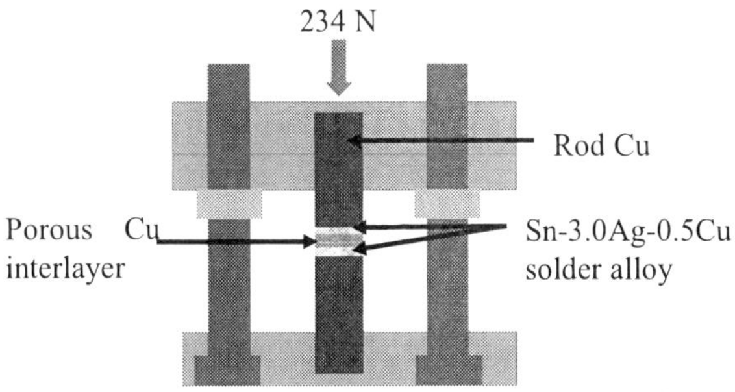

Figure 2 Jig for solder joint holder inside furnace during soldering process.

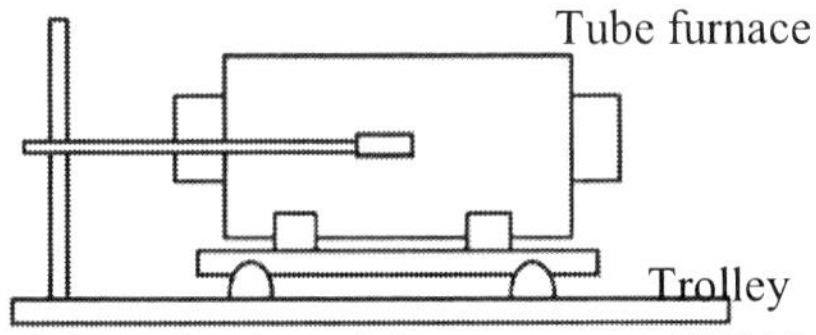

Figure 3 Schematic view of furnace set up (JISZ 3191).

Results and Discussion

Prior to insertion of the solder joint, the porous Cu interlayer was manually rolled. Figure 4 shows the structure of porous Cu interlayer of P15 and P25 before and after rolling. The solid part of the porous Cu has been defined as solid cell walls, (SCW) [16]. The average thickness of SCW for P15 and P25 after rolling are 0.26 mm and 0.06 mm, while the corresponding pore diameter are 0.3 mm and 0.1 mm, respectively.

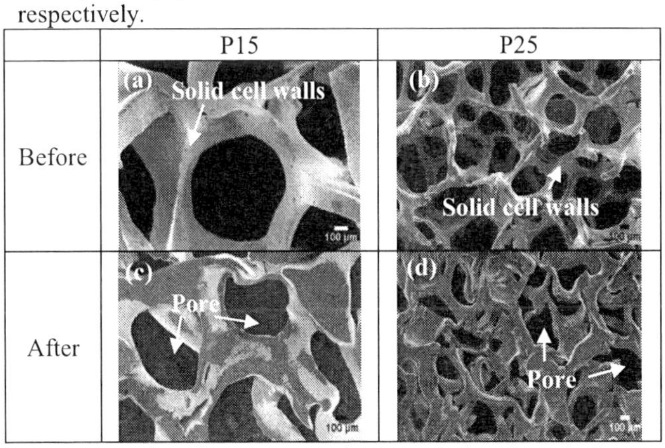

Figure 4 The microstructure of porous Cu interlayer before roll for (a) P15, (b) P25 and after roll (c) P15, (d) P25.

The growth of intermetallic compound (IMC)

It is expected that the IMC layer will develop during the soldering process. Since similar results of SEM backscattered electron micrographs of intermetallic compound (IMC) layers were obtained for all the three soldering temperatures, only the results for 307°C are shown here. Figure 5 and 6 show the SEM backscattered electron micrographs of intermetallic compound (IMC) layers formed at the SAC305/Cu substrate and SAC305/porous Cu (P15 and P25) interface respectively. The elemental percentage measured from EDX analysis is identified from Sn-Cu phase diagram, listed in Table 1 [18].

In Figures 5 (a), (b) and (c), the IMC morphology appear as continuous scallop-like structures of Cu_6Sn_5. This is typical of IMC layer when Sn based solder alloy are joined to a Cu substrate [19]. The formation of IMC layer at SAC305/Cu substrate for solder joint with and without porous Cu are similar given the shape and width of scallops. A very thin layer of Cu_3Sn, was also observed at the interface between Cu_6Sn_5 and the Cu substrate. The formation of a Cu_3Sn layer at at the vicinity of Cu substrate was found to be result of limited diffusion of Sn. It has been reported that the residual quantity of Cu for the reaction comes from solder alloy and not from the substrate. This is because the reactivity of Cu in the solder is more than that of Cu in the substrate at elevated temperature [20]. These lead to micro-voids nucleation at some parts of the interface between Cu_6Sn_5 and Cu_3Sn.

On the other hand, Figure 6 (a) and (b) show the IMC layer at the interface of the SAC305/porous Cu for solder joint with P15 and P25 of porous Cu interlayer, respectively. For IMC layer at SAC305/P15 interface with the size of solid cell wall of 0.26 μm (Figure 4), the IMC layer formed comprised of a similar scallop shape of Cu_6Sn_5 as shown in Figure 6 (a). However, the width of each scallop was narrower to 5% compared to that at SAC305/Cu substrate interface. A very thin layer of Cu_3Sn was also noticable at the interface of SAC305/Cu_6Sn_5 since the temperature is sufficient for formation of Cu_3Sn as mentioned by Lejuste et. al [21]. In contrast, the IMC layer was more uniform at the SAC305/P25 interface as in Figure 6 (b). The scallop-like structures are less pronounced than the one observed at IMC layer in P15. As the SCW for P25 was smaller (0.06 mm), the contact reaction between Cu from porous Cu and molten solder was increased. As a result, the formation of IMC layer is more uniform at the SAC305/P25 interface than that at SAC305/Cu substrate and SAC305/porous Cu interface.

Mechanism of IMC formation

Figure 7 shows the mechanism of IMC formation that occurred at SAC305/Cu substrate and SAC305/porous Cu. The IMC formation at SAC305/P15 interface was quite different than at SAC305/P25 interface. Normally during the soldering process, as molten solder alloy spreads inside the porous Cu, the Cu atom from porous Cu begin to dissolve and subsequently diffuse into the solder matrix. The porous Cu with larger sizes of vacuoles, P15 has reduced the contact area of SCW with molten solder. Therefore, the IMC layer formed is typical of scallop-like structures of the Cu_6Sn_5. In contrast, the reaction of molten solder with Cu atom from porous Cu increased with smaller SCW of P25. The concentration of Sn in the solder matrix is higher than that at the interface of reaction. The diffusion of Cu into the solder led to the first phase of formation of Cu_6Sn_5 scallops. This growth increases via the small spaces between the scallops since more contact area of reaction for P25 of SCW to form less pronounced scallops of IMC. Therefore, the IMC layer is more uniform at the SAC/P25 interface compared to SAC305/Cu substrate interface.

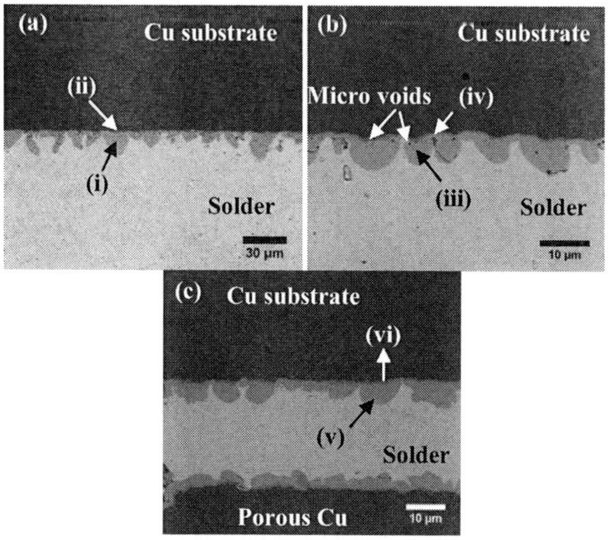

Figure 5 Formation of IMC layer at solder/Cu substrate for solder joint soldered with (a) no porous, (b) P15 and (c)P25 at 307°C with a holding time of 300 s.

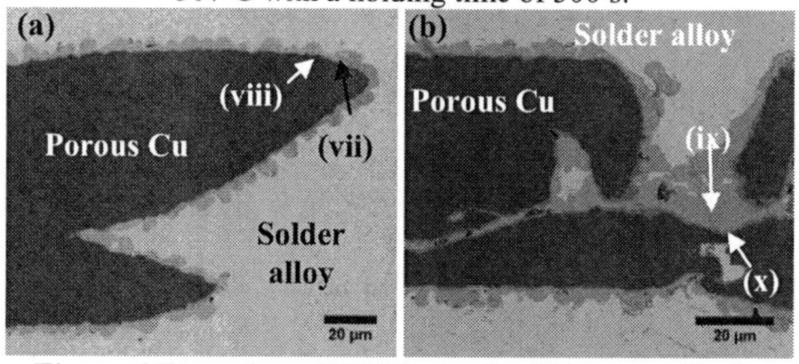

Figure 6 Formation of IMC layer at solder/porous Cu for solder joint soldered with (a) P15 and (b) P25 at 307°C with a holding time of 300 s.

Table 1 EDX analysis in Figure 5 and Figure 6

Spot	Element (at. %)			
	Cu	Ag	Sn	Phase
i	55.17	(0.23)	44.60	Cu_6Sn_5
ii	73.68	0.02	26.30	Cu_3Sn
iii	55.41	(0.18)	44.41	Cu_6Sn_5
iv	67.19	0.04	32.77	Cu_3Sn
v	56.71	(0.13)	43.16	Cu_6Sn_5
vi	72.33	0.00	27.67	Cu_3Sn
vii	55.42	0.08	44.50	Cu_6Sn_5
viii	63.40	(0.28)	36.32	Cu_3Sn
ix	56.35	(0.53)	43.12	Cu_6Sn_5
x	66.25	(0.25)	33.50	Cu_3Sn

The thickness of IMC

Table 2 shows the thickness average of the IMC layer (Cu_6Sn_5+Cu_3Sn) for solder joint with and without porous Cu. The thickness of IMC at the interface of SAC305/Cu substrate with no porous increased 1 μm with added soldering temperature. For solder joint with porous Cu, it was necessary to measure at both interface of SAC305/Cu substrate and SAC305/porous Cu. There was not much difference of IMC thickness at interface of SAC305/Cu substrate and SAC305/P15 interface except at a soldering time of 307°C, in which the thickness at SAC305/Cu was 2 μm thicker than at SAC305/porous Cu interface. Similar observation were noted for the IMC thickness at solder joint with P25 of porous Cu. It can be speculated that the IMC thickness at SAC305/porous Cu reached the limit of deformation around 3-5 μm. Meanwhile, the IMC thickness at SAC305/Cu substrate for solder joint with and without porous Cu has a similar pattern of increment, 4-6.5 μm. This thickness is common with the typical IMC layer at interface of SAC305/Cu substrate as reported elsewhere [22].

Porous	IMC formation	Schematic illustration
No porous	SAC305/Cu substrate	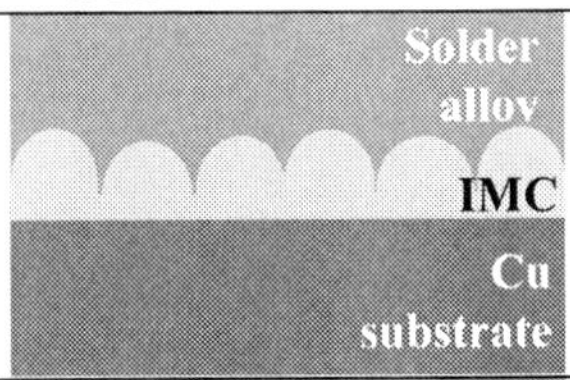
P15	SAC305/porous Cu (thick solid cell wall)	
P25	SAC305/porous Cu (thin solid cell wall)	

Figure 7 Schematic diagram for IMC formation at interface of SAC/Cu substrate and SAC/porous Cu.

Table 2 Average thickness of IMC layer

Soldering temperature, °C	IMC thickness, μm				
	No porous	P15		P25	
		SAC/Cu	SAC/ Porous	SAC/Cu	SAC/ Porous
267	4.23	3.63	3.26	4.58	3.86
287	5.02	4.83	4.11	5.53	4.37
307	6.44	6.54	4.85	6.82	4.52

Effects of adding porous Cu interlayer into solder joint on joint strength

Figure 8 shows the relationship between soldering temperature and joint strength of solder joint with and without porous Cu interlayer at 300 s of soldering time. The error bars shows the standard deviation values of three measured tensile strength for each soldering treatment. The strength of the joints soldered with P15 of porous Cu interlayer at three different temperatures all had similar values of about 23 MPa. In contrast, no clear patterns for P25 were observed. The strength of joint sample soldered with P25 of porous Cu interlayer initially fell from 43 MPa at 267 °C to 20MPa at 287 °C but markedly increased to 64 MPa at 307 °C. It can also be seen that for solder joint without porous interlayer, the strength slightly increased from 45 MPa at 267°C to 48 MPa at 287 °C before it reached 70 MPa at 307 °C. From this studies it would appear that solder joining at elevated temperatures exceeding 300 °C, the addition of porous Cu P25 interlayer to Sn-Ag-Cu solder alloy with 300 s holding time may be recommended.

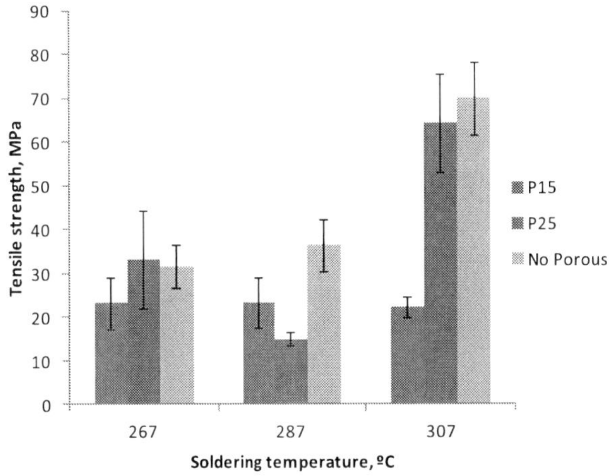

Figure 8 Effect of soldering temperature and pore size of porous Cu interlayer on tensile strength.

Conclusions

In this paper the effect of porous Cu on the tensile strength at different solder joints were investigated. The results from this research can be summarized as follows:

(1) The precipitations of IMC layer at SAC/Cu substrate interface have a pronounced scallop-like structure for solder joint with and without porous Cu.
(2) For solder joint with porous Cu addition, the size of solid cell wall structure for porous Cu influences the growth shape of IMC layer.
(3) The IMC at SAC/P25 interface are distributed more evenly along the interface reaction leading to a less defined scallop-like structure compared to the IMC layer at SAC/P15 interface.
(4) The highest joint strength recorded at 64 MPa showed soldering with addition of P25 porous Cu interlayer at 307°C and 300 s holding time has been identified as the potential soldering process for high temperature solder.

Acknowledgements

The authors greatly appreciate the financial support of this research by High Impact Research Joining Technology Fund, HIR-MOHE-D000001-16001 and Fundamental Research Grant Scheme, FRGS (FP024-2012A). The authors would also like to thank University of Malaya as well as Tokai University and Nagaoka University of Technology, Japan for their contribution s and support on this study.

References

[1] Suganuma, K., Kim, S. -J., and Kim, K. -S., "High-Temperature Lead-Free Solders: Properties and Possibilities,"J. Minerals, Metals & Materials Society, Vol. 61, No. 1 (2009), pp.64-71.

[2] Amalu, E. H., and Ekere, N. N., "High temperature reliability of lead-free solder joints in a flip chip assembly," J. Mater. Process. Technol, Vol. 212, No. 2 (2012), pp. 471–483.

[3] Ogunseitan, O. A., "Public health and environmental benefits of adopting lead-free solders," J. Minerals, Metals & Materials Society, Vol. 59, No. 7(2007), pp. 12–17.

[4] Kim, S. -J., Kim, K. -S., Kim, S. -S., Kang, C. -Y., and Suganuma, K., "Characteristics of Zn-Al-Cu Alloys for High Temperature Solder Application," J. Mater. Trans, Vol. 49, No. 7 (2008), pp. 1531–1536.

[5] Chidambaram, V., Hattel, J., and Hald, J., "High-temperature lead-free solder alternatives," Microelectron. Eng, Vol. 88, No. 6 (2011), pp. 981–989.

[6] Liu, X., Song, K., Davis, R. W., Hughes, L. C., and Hu, M. H., "A Metallization Scheme for Junction-Down Bonding of High-Power Semiconductor Lasers," IEEE Transactions on Advanced Packaging, Vol. 29, No. 3 (2006), pp. 533–541.

[7] Liu, W., Lee, N. -C., and Bachorik, P., "An innovative composite solder preform for TLP bonding — Microstructure and properties of die attach joints," 2013 IEEE 15th Electron. Packag. Technol. Conf. (EPTC 2013), Singapore, Dec. 2013, pp. 635–640.

[8] Zeng, G., Mcdonald, S., and Nogita, K., "Microelectronics Reliability Development of high-temperature solders: Review," Microelectron. Reliab, Vol. 52, No. 7 (2012), pp. 1306–1322.

[9] Guo-kui, J., Xi-cheng, W., Peng, S., and Liu, J., "Tensile Fracture Behavior of Sn-3.0Ag-0.5Cu Solder Joints on Copper," International Symp. High Packaging & Microsystem Integration (HDP'07), Shanghai, China, June. 2007, pp. 5–9.

[10] El-Daly, A. A., Fawzy, A., Mohamad, A. Z., and El-Taher, A. M., "Microstructural evolution and tensile properties of Sn–5Sb solder alloy containing small amount of Ag and Cu," J. Alloys Compd, Vol. 509, No. 13 (2011), pp. 4574–4582.

[11] George, E., Das, D., Osterman, M., and Pecht, M., "Thermal cycling reliability of lead-free solders (SAC305 and Sn3.5Ag) for high-temperature applications," IEEE Trans. Device Mater. Reliab, Vol. 11, No. 2 (2011), pp. 328–338.

[12] Che, F. X., Zhu, W. H., Poh, E. S. W., Zhang, X. W., and Zhang, X. R., "The study of mechanical properties of Sn–Ag–Cu lead-free solders with different Ag contents and Ni doping under different strain rates and temperatures," J. Alloys Compd, Vol. 507, No. 1(2010), pp. 215–224.

[13] Zhao, N., Pan, X., Ma, H., Dong, C., Guo, S., Lu, W., and Wang, L., "The liquid structure of Sn-based lead-free solders and the correlative effect in liquid-solid interfacial reaction," J. Phys. Conf. Ser, Vol. 98, No. 1 (2008), pp. 012029.

[14] Cheng, F., Gao, F., Nishikawa, H., and Takemoto, T., "Interaction behavior between the additives and Sn in Sn–3.0Ag–0.5Cu-based solder alloys and the relevant joint solderability," J. Alloys Compd, Vol. 472, No. 1–2 (2009), pp. 530–534.

[15] Jamadon, N.H., Yusof, F., Hamdi, M., Ariga, T., "Effect of Adding Porous Cu on the Microstructure and Mechanical Properties of Pb-Free Solder Joint," Proc 35th Electronic Manufacturing Technology Symposium (IEMT), Perak, Malaysia, Nov. 2012, pp. 1–5.

[16] Thewsey, D. J., and Zhao, Y. Y., "Thermal conductivity of porous copper manufactured by the lost carbonate sintering process," Phys. Status Solidi, Vol. 205, No. 5 (2008), pp. 1126–1131.

[17] Farazila, Y., Hamdi, M., Ariga, T., and Fadzil, M., "Microstructure Analysis of Brazed Sapphire to Inconel ® 600 Using Porous Interlayer," Welding Journal, Vol. 92, No. 11 (2013), pp. 332–335

[18] Kattner, U. R., "Phase diagrams for lead-free solder alloys," J. Minerals, Metals & Materials Society, Vol. 54, No. 12 (2002), pp. 45–51.

[19] Huangl, X., Lee, S. W. R., Li, M., and Chen, W. T., "Effect of High Temperature Storage on Reliability of Sn-Ag-Cu Flip Chip Solder Bumps," International Conf on Electronic Mater. & Packaging (EMAP), Kowloon, Hong Kong, Dec. 2006, pp. 7–12.

[20] Kar, A., Ghosh, M., Ghosh, R. N., Majumdar, B. S., and Ray, A. K., "Evolution of mechanical and electrical properties of tin–lead and lead free solder to

copper joint interface," Mater. Lett, Vol. 62, No. 1 (2008), pp. 151–154.

[21] Lejuste, C., Hodaj, F., and Petit, L., "Solid state interaction between a Sn–Ag–Cu–In solder alloy and Cu substrate," Intermetallics, Vol. 36, (2013), pp. 102–108.

[22] Wu, C. M. L., Yu, D. Q., Law, C. M. T., and Wang, L., "Properties of lead-free solder alloys with rare earth element additions," Mater. Sci. Eng. R Reports, Vol. 44, No. 1(2004), pp. 1–44.

Effect of Ga on the Inoxidizability and Wettability of Sn-0.5Ag-0.7Cu-xGa Solder

Luo Dongxue, Xue Songbai*, Xue Peng, Xu Jiachen

College of Materials Science and Technology,

Nanjing University of Aeronautics and Astronautics, Nanjing, China

E-mail addresses: xuesb@nuaa.edu.cn (Xue Songbai), Tel.: +86 2584896070

Wang Jianxin

School of Materials Science and Engineering, Jiangsu University of Science and Technology, Zhenjiang, China

Abstract

The wetting and oxidation resistance properties of Sn-0.5Ag-0.7Cu solders with different amount of Ga were investigated in this work. Results show that with proper Ga addition, the microstructure was refined and the IMCs in the matrix of Sn-0.5Ag-0.7Cu solder became small and uniform. It was also found that both the wettability and the oxidation resistance properties were better than those of the Sn-0.5Ag-0.7Cu solder alloy. In addition, the relationship between wettability and oxidation resistance was also studied which it was confirmed that as a surface-active element, Ga could enrich on the surface of the solder, so that the composition and structure of the liquid solder surface were changed remarkably, which benefited the properties correspondingly. These results show that as a modifier to the solder, Ga plays an important role in improving the properties of low-Ag Sn-Ag-Cu solders.

Introduction

At present, with the promulgation of the WEEE and RoHS directives, electronic manufacturing has shifted to green manufacturing system of "lead-free assembly" [1]. As the extensions of Sn-Ag/Sn-Cu-based solders, Sn-Ag-Cu solder is considered as the best choice to replace the traditional Sn-Pb solder. Compared with the other binary and ternary lead-free solders, Sn-Ag-Cu solder has good wettability and reliability, and its tensile strength is 1.8 times better than the conventional solders as well. In addition, it has excellent thermal fatigue performance and fine pitch soldering of electronic components and the bridging phenomenon can hardly happen.

The existing Sn-3.8Ag-Cu lead-free solder cannot meet the new trend in the field of electronic packing due to the severe interface reaction, the negative effects of the brittle compound in the solder joints and the high cost of Ag as well. Therefore, it is necessary to explore the Sn-Ag-Cu lead-free solder with low-Ag content. However, with the reduction of Ag content, the properties such as wetting behavior and thermal fatigue performance become worse [2-5].

As well known, Ga can be easily enriched on the surface of the Sn-Pb solder liquid surface and form a dense protective film to improve the oxidation resistance of the solder [6-7]. Similarly, in this paper, Ga is added into the solder to ameliorate such problems. This paper aims to study the new low-Ag lead-free solder with a good prospect. By adding proper amount of Ga, the microstructure and properties of the solder are optimized, which can be close to those of the Sn-3.8Ag-Cu solder, thereby reduced the manufacturing costs. The results in this paper possess important practical values for the development of the electronic industry.

Experimental procedure

Alloy Design and Preparation

Sn, Ag, Cu and Ga of 99.9% purity were used to prepare the solder alloys, and the compositions of element Ga were shown in table1. The melting process was conducted under nitrogen atmosphere at 350℃ for several minutes while mechanical stirring was carried out every minute. After melting, the solder was cooled to room temperature and cut into bars of 40mm length. The compositions of the solders were investigated by inductively coupled plasma emission spectrometer (ICP-AES).

Table 1 Composition of solder alloy (wt.%)

	Solder alloys					
	1	2	3	4	5	6
Ga (wt.%)	0	0.05	0.1	0.25	0.5	1
Initial alloy	Sn-0.5Ag-0.7Cu(SAC0507)					

Solderability Test

The solderability test was carried out with an SAT-5100 Solder Checker. Using the wetting balance method, the wetting force and wetting time were recorded. For the advance preparation, the Cu foils (30mm×5mm×0.3mm) were cleaned in ethanol and dried. The temperatures were set as 245, 255, 265 and 275℃ and water-soluble flux was chosen. The stress situation of the Cu foil in the wetting process was shown in figure1. Each test was conducted 5 times and the average result was obtained for analyzing.

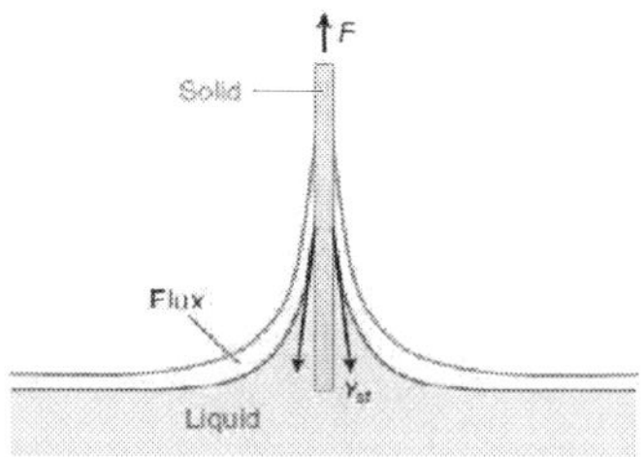

Figure1 The stress situation of the Cu foil

Microstructure

The microstructure was examined by optical microscope and scanning electron microscope (SEM). A solution of 5% HNO_3 and 95% alcohol was configured to etch the samples. Energy Dispersive X-ray Spectrometry (EDS) was adopted to identify the phase structure of the solder.

Oxidation resistance test

The surface oxide film of the solder after a certain time of heat preservation can visually estimate the degree of the solder oxidation. Different components of the solders were melted on the solderability tester and kept at 245℃ for 30 min, recording its surface oxide film morphology.

To further investigate the oxidation resistance of the solder, high-temperature oxidation weight gain method was used. Solders of different components were rolled into solder sheets of about 0.15mm. After grinding and polishing to remove the oxide film, the solder was cleaned with ethanol and dried to produce small square pieces of 0.1mm×10mm×10mm. The samples were put into a furnace at 245±5℃, and removed to cool every 10 hours till 60 hours. Analytical balance was used to weigh the total weight of the solder and the crucible. The resulting data was analyzed to show the growth of the sample mass.

Results and Discussion

Microstructure

Figure2 showed the microstructure of as-soldered SAC0507-xGa solders (x=0, 0.05, 0.1, 0.25, 0.5, 1). The microstructure of the SAC0507 solder was shown in Figure2(a). β-Sn, eutectic region and intermetallics were notified separately and it's easy to discover that β-Sn took the largest proportion and the IMC particles and eutectic were dispersed within Sn-rich matrix. According to the previous researches [8-9], the IMC particles were mainly Cu_6Sn_5 and Ag_3Sn. Compared with the high-silver solder, the content of the brittle Ag_3Sn in the matrix was reduced. These large particles and β-Sn grain always appeared in low-Ag SAC solder, which might be a threat to the premature failure [10-13].

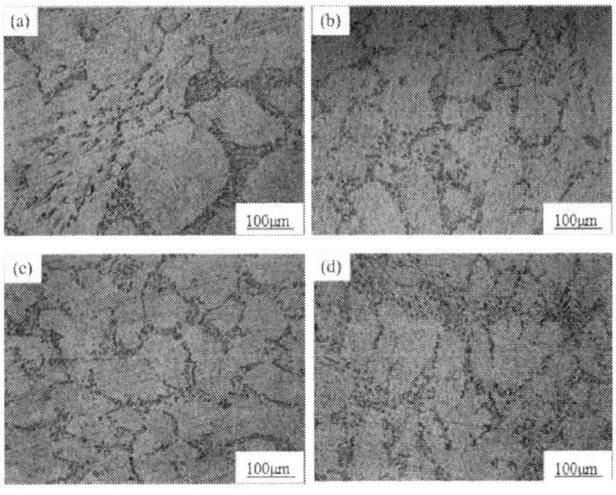

Figure2 The microstructure of SAC0507-xGa solders: (a)SAC-0Ga, (b)SAC-0.05Ga, (c)SAC-0.1Ga, (d)SAC-0.25Ga, (e)SAC-0.5Ga, (f)SAC-1Ga

As the content of Ga element gradually increased, the matrix structure refined, the coarse IMC matrix particles became smaller and distributed more uniform, and showed ellipsoidal-shaped instead of long-strip-shaped. When the Ga content was 0.5wt.%, the microstructure reached the most uniform and fine and the grain reached the highest degree of refinement. The IMC particles in the SAC-0.5Ga solder were smaller, with respect to the SAC solder. On the basis of the strengthening principle, the mechanical properties of the solder could also be enhanced. However, as the amount of Ga increased to 1 wt.%, some new phases with large size and non-uniform distribution could be found. The compositions were analyzed by EDX, the result was shown in figure3. It could be seen that the phase was mainly composed of Sn, Cu and Ga.

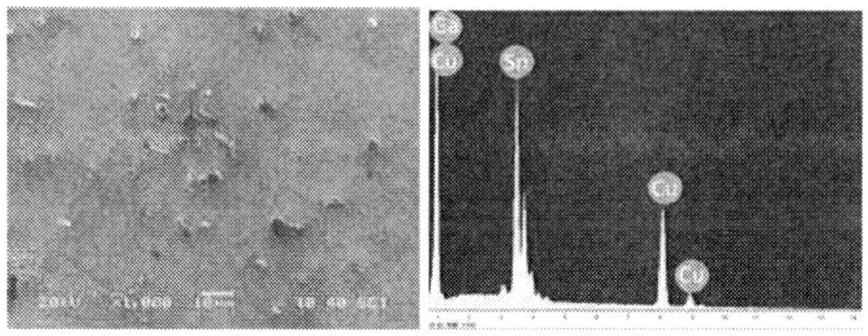

Figure3 The EDX result of SAC0507-1Ga solder

As shown in figure4 of the Ga-Sn binary phase diagram [14], as the content of Ga element is less than its maximum solubility in Sn which is 6.4%, β-Sn mainly exists in the form of solid solution at room temperature.

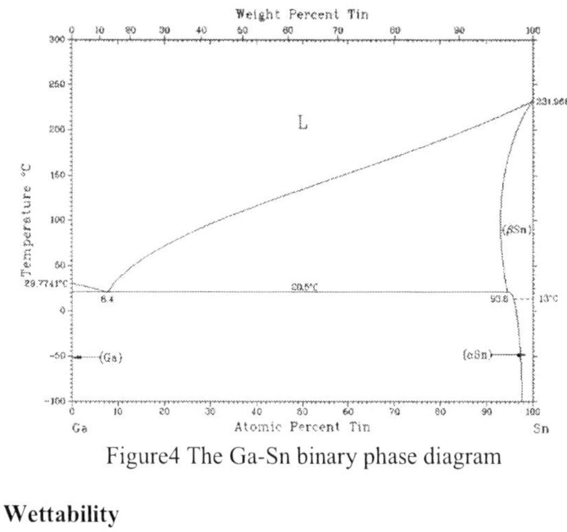

Figure4 The Ga-Sn binary phase diagram

Wettability

It is reported that Ga element can decrease the surface tension of alloys [15]. In this paper, the improvement to the wetting ability due to Ga addition was also found. Wetting balance method was used and the wetting forces and wetting time of solders with different content of Ga at different temperatures were tested and the results were shown in figure5. At certain temperature, according to the test results, with the Ga content increasing, the wetting force significantly increased, while the wetting time showed a relatively-large reducing tendency. And the best wettability was achieved when the content of Ga was 0.5%, and then the wettability showed a slight downward trend when the addition kept increasing, which suggesting the proper addition of Ga was about 0.5wt.%.

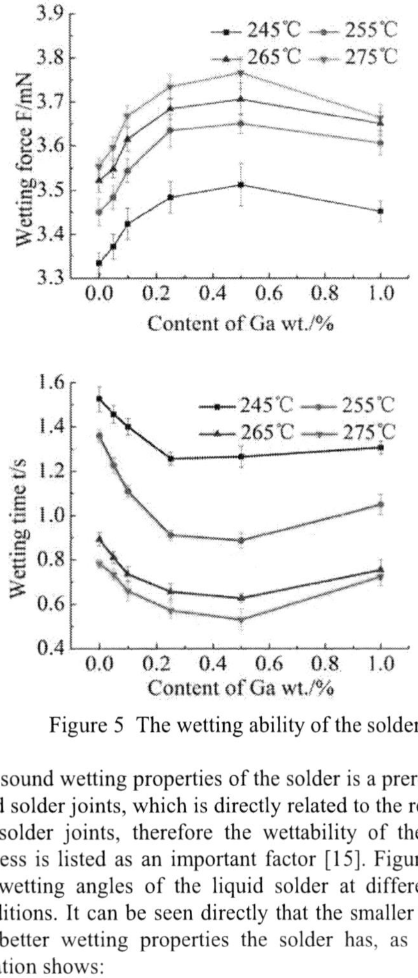

Figure 5 The wetting ability of the solders

The sound wetting properties of the solder is a prerequisite for good solder joints, which is directly related to the reliability of the solder joints, therefore the wettability of the soldering process is listed as an important factor [15]. Figure6 showed the wetting angles of the liquid solder at different wetting conditions. It can be seen directly that the smaller value of θ, the better wetting properties the solder has, as the Young equation shows:

$$\sigma_{gs} = \sigma_{ls} + \sigma_{gl} \cos\theta$$

θ represents the wetting angle, and σgs, σgl, σls represent the interfacial tensions between the solid–gas, liquid–gas, liquid–solid interfaces separately.

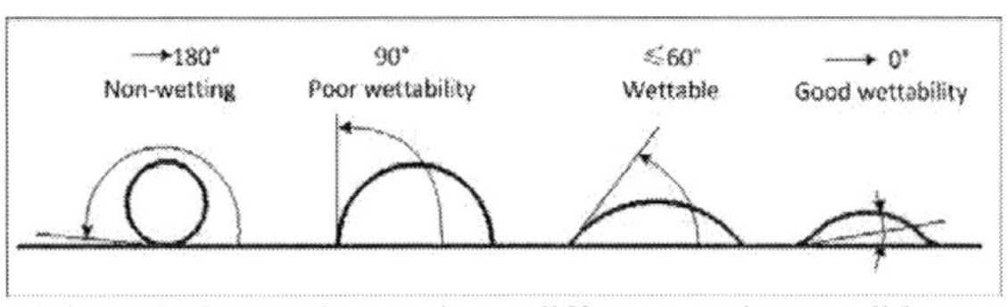

Figure6 The wetting angles at different wetting conditions

Typically, the compositions and the surface tension of the solder and the base metal are two important factors that affect the solder's wettability. Under certain circumstances that the compositions are identified, the surface tension of the solder and base metal plays a key role in the wettability of the solder, and its influence factors are temperature, flux, metal surface oxides and surface active substances, etc. The copper used were deoxygenated, and the flux was the same, so it could be estimated that the surface active substances were added to improve the solder's wettability by way of reducing the surface tension.

As mentioned above, Ga is a surface active element, so a strong positive-adsorption action can take place. As a result, the composition and structure of the liquid solder surface are changed, the interfacial tension is reduced and the flowability is significantly improved, which caused the refinement of wetting process of the liquid solder. In this paper, when the content of Ga exceeded 0.5%, the surface enrichment got relatively stable, so the wettability no long rose. However, the generation of the Ga-rich phase limited the active effect of Ga element, decreasing its improvement of the solder's wetting properties.

Oxidation Resistance

The oxidation of solder in molten state is a trouble to be faced during soldering, which the generated oxides have adverse factors in the wettability and reliability of the solder joints in the service process. Under the protection of N_2 atmosphere, the oxidation resistance ability can be improved, while in actual industrial production, it requires optimization in the complexity and cost of the equipment and technology, which is not feasible[16]. Early studies shows that Ga is highly enriched on the surface of Sn-Pb solders and the degree of Ga enrichment can rapidly reach 34000-fold for a nominal additive amount [17], the enriched Ga can form a protective film to prevent the oxidation of the solder alloy. In this chapter, the improvement about oxidation resistance at high temperature, which added Ga into SAC, has been studied.

1. Surface morphology after high-temperature oxidation

To simulate the reflow process condition, the temperature was set 245℃. The Sn-0.5Ag-0.7Cu-xGa solder was in molten state and the surface atoms were prone to react with oxygen in the air so that the quality of the solder changed. Therefore, difference in the change in the solder's quality could be a factor to characterize the difference between the oxidation resistance properties of the solders. The surface morphology of several SAC0507-xGa solders after incubation at 245℃ for 30min showed in Figure7. As can be seen, the surface of SAC0507 solder was covered with a thick layer of gray dense oxide slag. With the increase of the content of Ga, the thickness of the surface oxide film became thinner and thinner, which for SAC-0.5%Ga solder, the surface was brighter and we could see little oxide layer, which showed excellent oxidation resistance. However, when the content of Ga reached about 1wt.%, some colored oil slick existed on the surface of the Sn-0.5Ag-0.7Cu-1Ga solder, which we can see a little oxide layer on the surface of the Sn-0.5Ag-0.7Cu-1Ga solder(see figure7d).

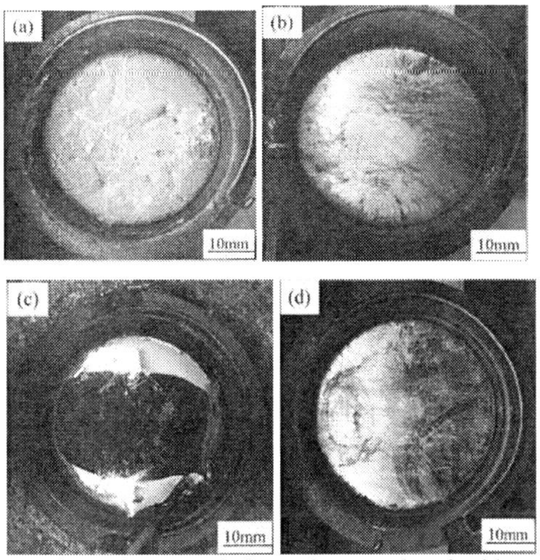

Figure7 The surface morphology photographs of solders after incubation at 245℃ for 30min: (a)SAC-0Ga, (b)SAC-0.1Ga, (c)SAC-0.5Ga, (d)SAC-1Ga

2. Weight gain after high-temperature oxidation

High-temperature oxidation weight gain method is used to quantitatively analysis the oxidation resistance of the solders. The mass ratio of the oxidized solder weight gain was calculated to study the impact of law of element Ga, the results were shown in figure8. From the comparison, for solders of certain Ga content, the oxidation degree of the solder increased in the oxidation process, but when the incubation time exceeded 30 hours, the trend of increase declined. In the same oxidation period, the weight gain of solder excluding Ga was the highest, indicating the oxidation resistance was the worst and an appropriate amount of Ga could improve the oxidation resistance of the solder. The oxidation rate of the solder decreased with increasing time and this was because a protective oxide was formed on the solder surface, which inhibiting the further oxidation of the metal matrix efficiently.

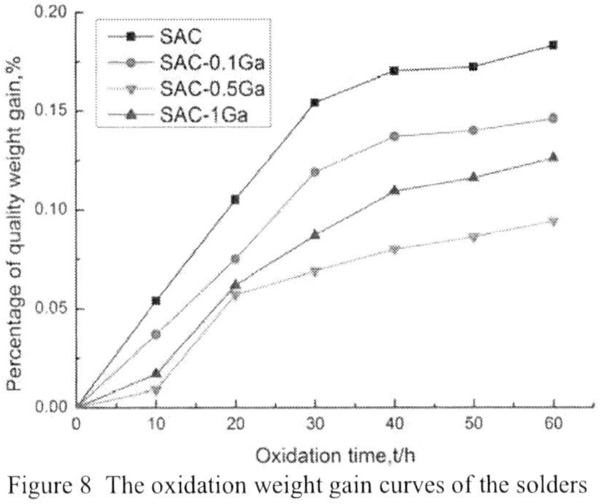

Figure 8 The oxidation weight gain curves of the solders

With the increase of Ga element in the solder, the oxidation resistance was improved, and the SAC0507-0.5Ga had the minimal weight gain, illustrating that its improvement in the oxidation resistance properties was the most significant. It was worthwhile that the best ingredient point was in agreement with the wettability, further proving the relationship between the oxidation resistance and the wettability of the solder. The addition of Ga could not only increase the oxidation resistance ability, but also reduce the surface tension of the solders. As a result, after adding proper amount of Ga, the wettability improved. However, when there was too much Ga, the Ga-rich phase would lead to the uneven distribution of Ga in the surface, so that the O atoms could be in contact with the solder, which worsening the oxidation resistance properties.

Conclusions

1. A small amount of Ga could effectively improve the microstructure of the solder matrix in Sn-0.5Ag-0.7Cu lead-free solder. The structure gradually refined and the IMC in coarse particles became small and evenly distributed. When the addition amount was about 0.5%, the effect was the most obvious.

2. The wettability of the solder was improved with the increasing of the content of Ga, and reached the maximum when the amount was 0.5%. Then with the increase of Ga, the wetting properties showed a slight downward trend. This was mainly because Ga could be enriched in the solder surface so as to reduce the surface tension of the solder. However, with excess Ga, the solder wettability was reduced owing to the formed Ga-rich phase.

3. The influence of Ga in the oxidation resistance was similar to that in wettability, namely, the molten solder became relatively bright with proper Ga addition, the oxidation resistance increased and achieved the best at 0.5% Ga. But as the content of Ga continued to increase, there was oxide film on the solder surface. The oxidation rate with time slowed down due to the blocking role of the formed oxidation layer.

Acknowledgments

This work was supported by the Key Laboratory of Advanced Welding Technology of Jiangsu Province, China (JSAWT-14-04). This work was also supported by the Priority Academic Program Development of Jiangsu Higher Education Institutions.

References

[1] Xiao Z X, Xue S B, Hu Y, *et al*, "Properties and microstructure of Sn-9Zn lead-free solder alloy bearing Pr", *Journal of Materials Science: Materials in Electronics*, Vol. 22, No.6 (2011), pp. 659-665.

[2] Zhao J, Qi L, Wang X, *et al*, "Influence of Bi on microstructures evolution and mechanical properties in Sn–Ag–Cu lead-free solder", *Journal of Alloys and Compounds*, Vol. 375, No. 1 (2004), pp. 196-201.

[3] Kim K S, Huh S H, Suganuma K., "Effects of intermetallic compounds on properties of Sn–Ag–Cu lead-free soldered joints", *Journal of Alloys and Compounds*, Vol. 352, No. 1 (2003), pp. 226-236.

[4] Chuang T H, Wu M W, Chang S Y, *et al*, "Strengthening mechanism of nano-Al_2O_3 particles reinforced Sn3. 5Ag0. 5Cu lead-free solder", *Journal of Materials Science: Materials in Electronics*, Vol. 22, No. 8 (2011), pp. 1021-1027.

[5] Shnawah D A A, Said S B M, Sabri M F M, *et al*, "Novel Fe-containing Sn–1Ag–0.5 Cu lead-free solder alloy with further enhanced elastic compliance and plastic energy dissipation ability for mobile products", *Microelectronics Reliability*, Vol. 52, No. 11 (2012), pp. 2701-2708.

[6] Dong-xue Luo, Song-bai Xue, Zai-qian Li., "Effects of Ga addition on microstructure and properties of Sn-0.5Ag-0.7Cu solder", *Journal of Materials Science: Materials in Electronics*, Vol. 25, No. 8 (2014), pp. 3566-3571.

[7] Dong-xue Luo, Song-bai Xue, Shuang Liu., "Investigation on the intermetallic compound layer growth of Sn-0.5Ag-0.7Cu-xGa/Cu solder joints during isothermal aging", *Journal of Materials Science: Materials in Electronics*, Vol. 25, No. 12 (2014), pp. 5195-5200.

[8] Hammad A E., "Investigation of microstructure and mechanical properties of novel Sn-0.5Ag-0.7Cu solders containing small amount of Ni", *Materials & Design*, Vol. 50 (2013), pp. 108-116.

[9] Glazer J., "Microstructure and mechanical properties of Pb-free solder alloys for low-cost electronic assembly: a review", *Journal of Electronic Materials*, Vol. 23, No. 8 (1994), pp. 693-700.

[10] Zhang L, Xue S B, Gao L L, *et al*, "Microstructure characterization of SnAgCu solder bearing Ce for electronic packaging", *Microelectronic Engineering*, Vol. 88, No. 9 (2011), pp. 2848-2851.

[11] El-Daly A A, El-Taher A M., "Evolution of thermal property and creep resistance of Ni and Zn-doped Sn–2.0Ag–0.5Cu lead-free solders", *Materials & Design*, Vol. 51 (2013), pp. 789-796.

[12] Gao P, Lin J, Lei Y, *et al*, "Drop performance research of silver content on the SnAgCu based lead-free solder of joint level", *Proc 14th International Conference on Electronic Packaging Technology of IEEE*, Dalian, China, Aug. 2013, pp. 809-813.

[13] Yang Liu, Fenglian Sun, Yang Liu, Xuemei Li., "Effect of Ni, Bi concentration on the microstructure and shear behavior of low-Ag SAC–Bi–Ni/Cu solder joints", *Journal of Materials Science: Materials in Electronics*, Vol. 25, No. 6 (2014), pp. 2627-2633.

[14] T. J. Anderson and I. Ansara, "The Ga-Sn (gallium-tin) system", *Journal of Phase Equilibria*, Vol. 13, No. 2 (1992), pp. 181-189.

[15] Wenxue Chen, Songbai Xue, Hui Wang, Jianxin Wang, Zongjie Han., "Investigation on properties of Ga to Sn–9Zn lead-free solder", *Journal of Materials Science: Materials in Electronics*, Vol. 21, No. 5 (2010), pp. 496-502.

[16] Ren Xiao-Xue, Li Ming, Mao Da-Li.,"Effect of Alloying Elements on the High-temperature Oxidation Resistance of Sn-Zn Based Lead-free Solder", *Electronic Components and Materials*, Vol. 23, No. 11 (2004), pp. 40-44.

[17] Peng Xue, Songbai Xue, Yifu Shen, Fei Long, Hong Zhu. "Mechanism of Reaction between Nd and Ga in Sn-Zn-0.5 Ga-xNd Solder", *Journal of Electronic Materials*, Vol. 43, No. 9 (2014), pp. 3404-3410.

Joining of Copper at Low Temperatures Using Ag Nanoparticles: Influence of Process Parameters on Mechanical Strength

S. Hausner, S. Weis, B. Wielage and G. Wagner
Technische Universitaet Chemnitz, Institute of Materials Science and Engineering, Chemnitz, Germany
susann.hausner@mb.tu-chemnitz.de

Abstract

Nanoparticles exhibit a decrease in sintering and melting temperature with decreasing particle size in comparison to the corresponding bulk material. After melting or sintering of the nanoparticles, the material behaves like the bulk material. Therefore, high-strength and temperature-resistant joints can be produced at low temperatures, which is of big interest for various joining tasks.

Joints (substrate: Cu) were prepared with an Ag nanoparticles-containing paste. The influence of the adjustable process parameters joining pressure, joining temperature, holding time, heating rate, thickness of paste application, surface treatment, pre-drying process and subsequent heat treatment on the strength behavior of the joints was investigated. It is shown that in particular the joining pressure exerts an essential influence on the achievable strengths. In addition, temperature, holding time and thickness of paste application have a significant effect on strength behavior. In contrast, the pre-drying process, heating rate, surface pre-treatment and subsequent heat treatment possess hardly any influence on joint strength.

Introduction

Due to the large specific surface area (high surface-to-volume ratio), nanoparticles exhibit a decrease in sintering and melting temperature with decreasing particle size, Fig. 1.

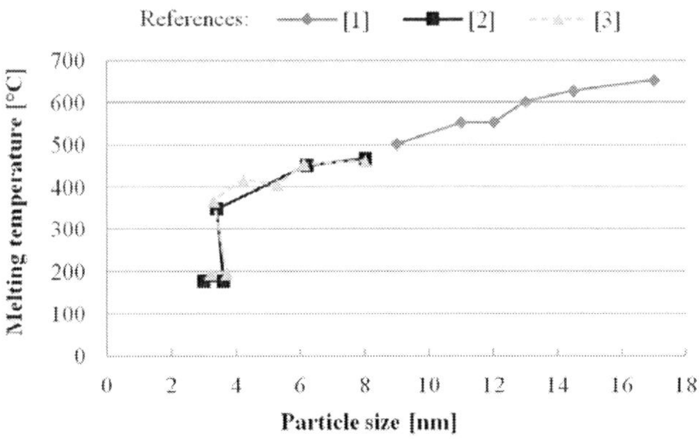

Figure 1: Literature values for the experimental determination of the melting point of silver nanoparticles depending on the particle size [1–3]

This effect, which had been theoretically predicted already 1909 by Pawlow [4–5] and experimentally verified since the 1950s [6–8], was technically implemented only a few years ago and is now the subject of intense research. Especially in the area of microelectronics, many research papers deal with printing and sintering of electrically conductive structures with nanoparticle inks [9–12], but the utilization of this effect is also of great interest for joining technologies, because after melting of the nanoparticles, the material behaves like the bulk material. Thus, high-strength and temperature-resistant joints can be produced at low temperatures. This results in a high potential for various joining processes. Previous publications only concern with joining of components in power electronics. Especially, joining with nanoparticles as a substitute for soldered joints, which tend to form cracks during thermal cycling, was investigated [13–24]. Nevertheless, there is a much greater potential for joining with nanoparticles. A possible application is the joining of high-performance materials respectively materials with adapted and optimized microstructure. With low joining temperatures, structural damage, e.g. abnormal grain growth or undesirable phase transformation, can be avoided. The utilization of this effect would be also of great interest for the joining of materials with different coefficients of thermal expansion such as carbide-metal joints and ceramic-metal joints, to reduce the often critical thermally induced residual stresses of the joints. There is also an increasing demand for novel hybrid compound joints, for example between fiber-reinforced composites and metals, where low joining temperatures are required. The list of possible applications could be extended further. This demonstrates the great potential of joining with nanoparticles for highly stressed structural components as an alternative for brazing processes.

The principal aim of previous studies in the field of power electronics was to reduce the temperature and the necessary pressure, because electronic components have only a limited thermal and mechanical loading capacity. However, with regard to a possible alternative for brazing processes, these parameters are less critical. For brazed joints, especially the resulting strength is of crucial importance. Therefore, in the present study, significantly higher pressures and temperatures are used to investigate whether joining with nanoparticles could be a possible alternative for brazing processes. In addition, in previous publications, only the parameters temperature, pressure, holding time and particle size were varied and investigated. However, the quality of joints is influenced by a complex interplay of much more process parameters, most of which are systematically examined in the present work. The objective is to provide a basis for a fundamental understanding of this novel joining process. Besides temperature, pressure and holding time, heating rate, thickness of paste application, different surface pre-treatments of the substrates, a pre-drying process, as well as a subsequent heat treatment are investigated.

Experimental

A commercially available Ag nanopaste of the company Harima Chemicals, Inc. (Japan) was used for the experiments. Conventionally this nanopaste is used in microelectronics for printing and sintering of electrically conductive structures [25–26]. The composition and properties of the nanopaste with the company name NPS are shown in Tab. 1. With this spreadable paste, an application similar to conventional solder- or brazing-pastes is possible. The paste has a very high particle content of 81.8 wt%, which is advantageous for joints with low porosity. The nanoparticles are suspended in a solvent and through the addition of a dispersing agent surrounded with an organic shell, Fig. 2. This organic shell leads to strong repulsive forces between the particles, so that agglomeration can be avoided [13, 27–28].

Table 1: Properties of the nanopaste (manufacturer's specification)

Particle size	8 – 15 nm (mean diameter: 12 nm)
Metal content	81.8 wt.%
Organic substances	1-Decanol: 1 – 2 wt. % Petroleum hydrocarbon: 2.5 – 3.5 wt.% Dispersant A: 6 – 10 wt.% Dispersant B: 1 – 3 wt.% Additives A: 2 – 3 wt.% Additives B: 0.8 – 1.5 wt.%

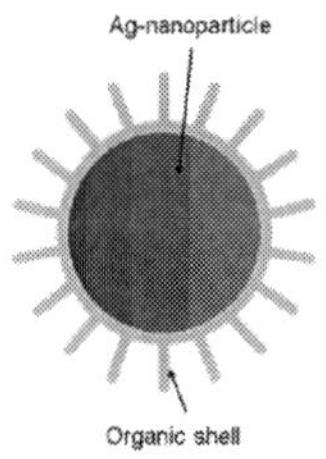

Figure 2: Schematic figure of a nano-particle with organic coating

The shape, distribution and size of the nanoparticles were determined using transmission electron microscopy (TEM, Hitachi H8100). The thermal behavior of the nanopaste was characterized by simultaneous thermal analysis (STA 409C, Netzsch Gerätebau GmbH). The analyses were carried out in air and argon with a heating rate of 10 K/min. The heating was performed from room temperature to 500 °C, followed by cooling (10 K/min) back to room temperature (Tab. 2a). In order to exclude possible post-reactions, further analyses were carried out with a second run up to 1000 °C (Tab. 2b).

The determination of the tensile shear strength was performed using the sample geometry shown in Fig. 3. Pure copper (EN: Cu-ETP, DIN material number 2.0060) was used as substrate with a joining area of 9.0 mm x 7.5 mm. When joining with nanoparticle-containing suspensions, the application of pressure during the joining process is necessary to achieve a dense microstructure of the joints with good strengths [29]. Therefore, a modified hydraulic press was used for the joining experiments, Fig. 4. The heating of the samples was carried out in air by means of induction. The desired pressure was applied on the entire joining surface of the sample with a punch, Fig. 4. Tab. 3 shows the process parameters that were varied during the experiments. Reference samples with defined parameters were prepared. Based on these reference samples, one parameter was changed, while the others were held constant, Tab. 3. The surface pre-treatment of the samples consisted of grinding, up to grain size 400, and degreasing in acetone. Further variations of surface pre-treatment included polishing, up to 1200, blasting with Al_2O_3, d = 0.5 mm, 1 bar, pickling using ammonium persulphate, 50 °C, 30 s and the deposition of a Ag coating by means of physical vapor deposition (PVD). For pre-drying, the samples with the applied nanopaste were dried in air for 4 min at a temperature of 100 °C prior to the joining process. For the subsequent heat treatment, the samples were heat treated without pressure for 20 min at a temperature of 300 °C in a muffle furnace in air after the joining process.

The examination of all tensile specimens – at least 5 individual samples per varied parameter – was performed with a universal tensile testing machine Zwick/Roell Z020 at room temperature and a strain rate of $10^{-3} \cdot s^{-1}$.

Table 2: Analysis conditions of the STA studies

a) Standard Analysis		b) Analysis with two runs	
		1st run	2nd run
Heating rate	10 K/min	10 K/min	10 K/min
Maximum temperature	500 °C	500 °C	1000 °C
Cooling rate	10 K/min	10 K/min	10 K/min
Atmosphere	Air and Argon	Air	Air

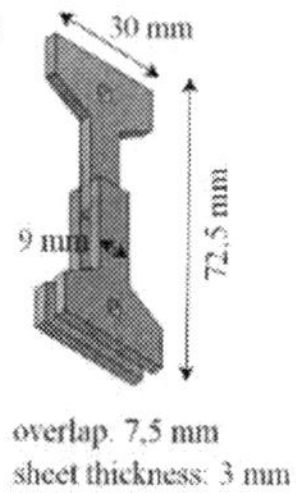

Figure 3: Sample Geometry

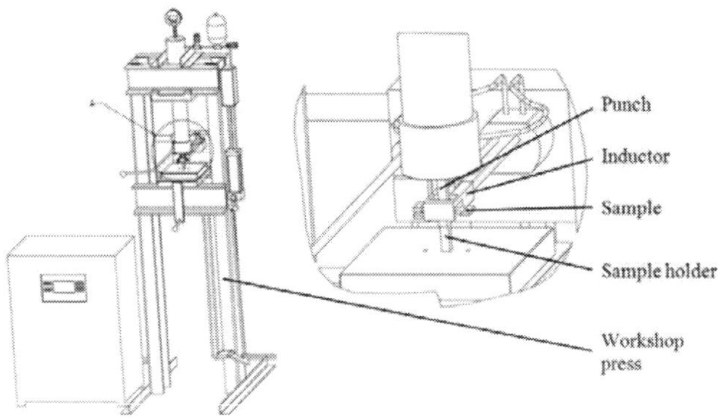

Figure 4: Schematic figure of the joining system

Results and Discussion

TEM Investigations

Fig. 5 shows transmission electron micrographs of the nanoparticles and the particle size distribution. The nanoparticles are homogeneously distributed. The size of the spherical particles was determined by measuring 1000 particles of different micrographs. The particle size distribution has a maximum at 6–7 nm, Fig. 5.

Table 3: Varied process parameters

	p	T	t	υ	s	ST	Extra
Reference	40 MPa	300 °C	10 min	150 K/min	20 µm	Ground	-
VARIATION	0 MPa 5 MPa 10 MPa 20 MPa 40 MPa 60 MPa 80 MPa	✓	✓	✓	✓	✓	-
	✓	250 °C 300 °C 350 °C 400 °C 450 °C	✓	✓	✓	✓	-
	✓	✓	2 min 5 min 10 min 30 min	✓	✓	✓	-
	✓	✓	✓	25 K/min 150 K/min 500 K/min	✓	✓	-
	✓	✓	✓	✓	10 µm 20 µm 50 µm 100 µm	✓	-
	✓	✓	✓	✓	✓	Ground Polished Blasted Pickled PVD-Ag	-
	✓	✓	✓	✓	✓	✓	Pre-drying
	✓	✓	✓	✓	✓	✓	heat treatment

✓ Used parameter
p Pressure
T Temperature
t Holding time
υ Heating rate
s Thickness of paste application

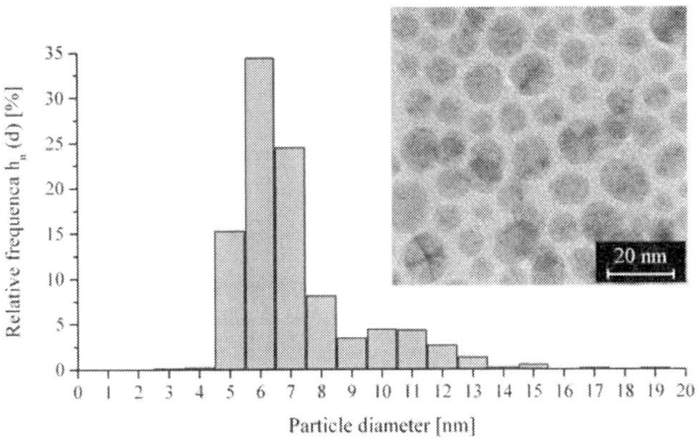

Figure 5: TEM image of the nanoparticles and particle size distribution of the nanopaste

STA Investigations

The thermal behavior of the Ag nanopaste was investigated by simultaneous thermal analysis (STA). Fig. 6 shows a typical differential scanning calorimetry (DSC) and thermogravimetry (TG) curve of the paste. The analysis was carried out in air. Between 250 °C and 410 °C, the DSC curve shows two exothermic peaks, which can be assigned to the gradual decomposition and later oxidation of the stabilizer molecules attached to the silver surface [30]. The TG curve first shows a strong mass loss up to about 275 °C. This area can be assigned predominantly to the evaporation of the solvent of the paste. In the area of the exothermic peaks, the TG curve shows a linear decrease, which corresponds to the removal of the organic shell of the nanoparticles. After the oxidative decomposition of the organic shell, the nanoparticles almost spontaneously sinter by flow processes due to the high specific surface energy [13, 28, 30]. Subsequent SEM analyses of the DSC samples (Fig. 6) show, that after the exothermic reaction, a sintered structure with an average grain size of about 1.3 µm has developed.

Fig. 7 shows the DSC and TG curves of an STA analysis, in which, after the first heating in air, the sample was cooled down to room temperature, before a second run up was performed. The second heating procedure was carried out up to a temperature of 1000 °C, also in air. With the second STA run, it was intended to investigate whether post-reactions take place such as further evaporation of solvents or thermal reactions of the nanoparticles. The DSC curve of the second run shows that no suchlike effects take place at low temperatures. Also a further mass loss was not observed. The DSC curve of the heating process shows a clear endothermic melting peak at a temperature of 962 °C, which corresponds to the values of the melting temperature of silver given in the literature. Thus, it is shown that the reactions of the first STA run have completely occurred. The sintered material possesses now the thermal properties of the bulk material.

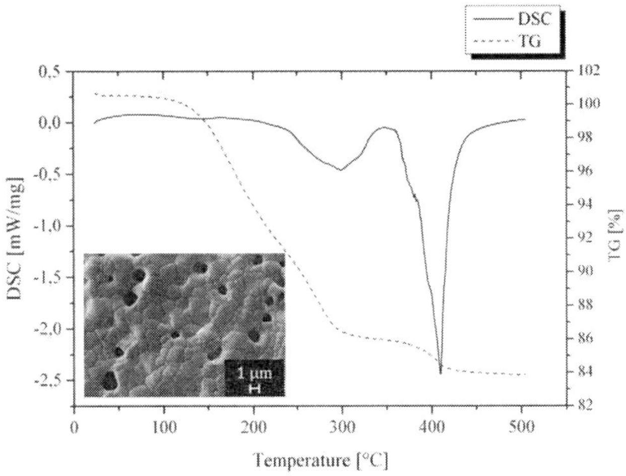

Figure 6: DSC and TG curves of the Ag nanopaste, analysis in air

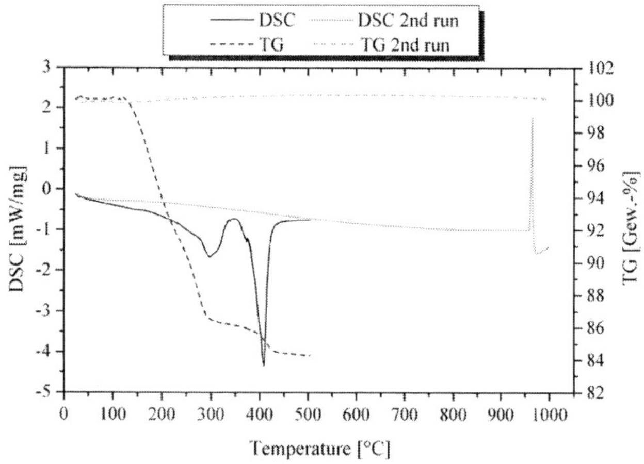

Figure 7: DSC and TG curves with two runs

STA investigations were also performed in argon, Fig. 8. In comparison to the correspondent analysis in air, the TG curve shows in argon up to a temperature of about 275 °C a similarly strong mass loss, which can be attributed to the evaporation of the solvents. In the further course of the TG curve in argon, no significant mass loss is observed, which results in a higher residual mass content for the analysis in argon, Fig. 8. In the temperature range between 350°C and 450°C, the TG curves for air and argon show a different progression. In the inert atmosphere, there is no mass loss, suggesting that the degradation of the organic shell does not occur. This is proved by the DSC curve in argon. Similar to the analysis in air, although less pronounced, the DSC curve in argon possesses a small first exothermic peak at about 300 °C. However, at the second exothermic peak for the DCS course in air, no further reaction is observed in argon. This leads to the conclusion, that the organic shell of the particles is not removed when using the inert gas argon. This is consistent with previous studies [13, 14, 16, 24, 30], which show that oxygen is necessary for the decomposition of the organic shell (oxidation reaction).

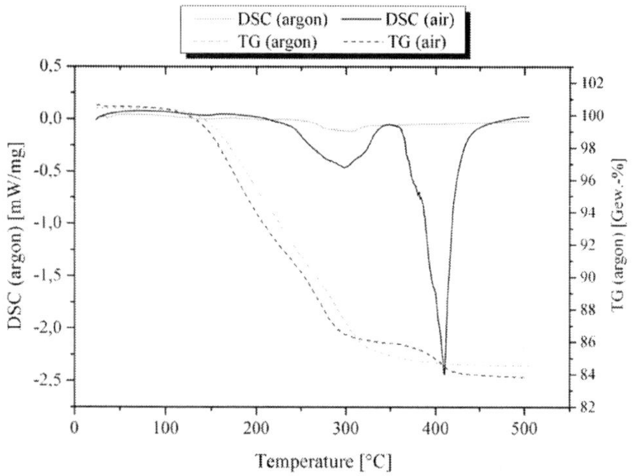

Figure 8: Comparison of the DSC and TG curves in air and argon

Strength Behavior

As a function of pressure

To achieve a dense microstructure, it is necessary to apply pressure while joining. First investigations on the mechanical properties were done to characterize the influence of joining pressure on joining strength. Fig. 9 shows the achievable tensile shear strengths as a function of joining pressure at a constant joining temperature of 300 °C and a holding time of 10 min and a paste application of 20 µm. It appears that the joining pressure has an essential influence on the strength behavior. A significant increase in strength from 5 MPa up to 95 MPa is observed with increasing joining pressure from 20 MPa to 80 MPa. The increase in strength is a result of a significant densification of the sintered structure caused by the increasing pressure, Fig. 9.

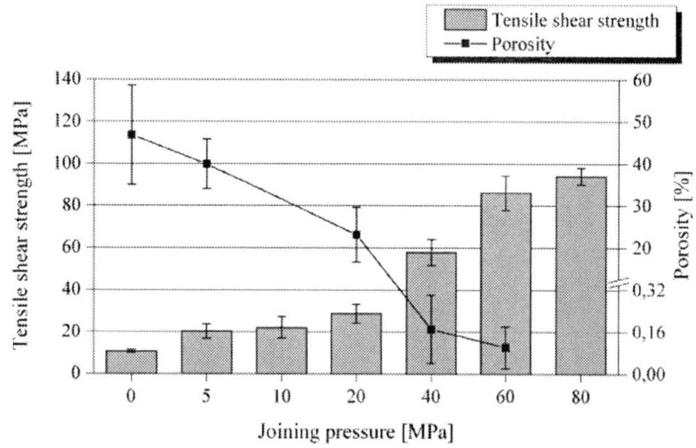

Figure 9: Tensile shear strength and porosity as a function of the joining pressure (temperature: 300 °C, holding time: 10 min, paste application: 20 µm)

As a function of temperature

The joining temperature influences the tensile shear strength in a lesser extent compared to the joining pressure, Fig. 10. The investigations were performed with a constant joining pressure of 40 MPa, a holding time of 10 min and a paste application of 20 µm. At a temperature of 250 °C, the strength is still relatively low at around 41 MPa. For 300 °C strength increases up to 57 MPa and does not change significantly up to 350 °C. Up to a temperature of 400 °C, it rises slightly to 64 MPa and decreases for higher temperatures, finally. The increase in strength up to a temperature of 400 °C correlates with the results of the DSC and TG investigations: At a temperature of about 400 °C the decomposition of the organic shell and the sintering process are completed. With further increase of the temperature, a decrease in strength is observed. That is a result of the scaling of the copper surface, which can be observed macroscopically.

As a function of holding time

Holding time has a significant influence on the strength behavior, Fig. 11. Joining was performed with a constant joining pressure of 40 MPa, a temperature of 300 °C and a paste application of 20 µm. Up to a holding time of 30 min, strength increases up to a maximum of 77 MPa. But at the same time the standard deviation increases considerably. It might be reasonably assumed that due to the longer holding time, an improved adhesion respectively bonding to the base material is achieved, however, this still needs to be confirmed by fracture surface characterization.

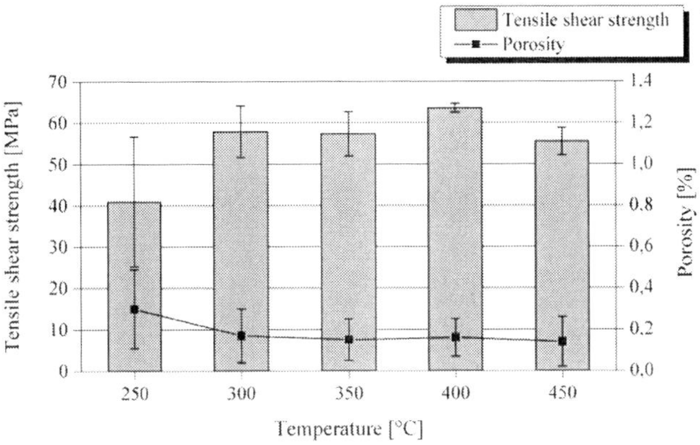

Figure 10: Tensile shear strength as a function of the joining temperature (pressure: 40 MPa, holding time: 10 min, paste application: 20 µm)

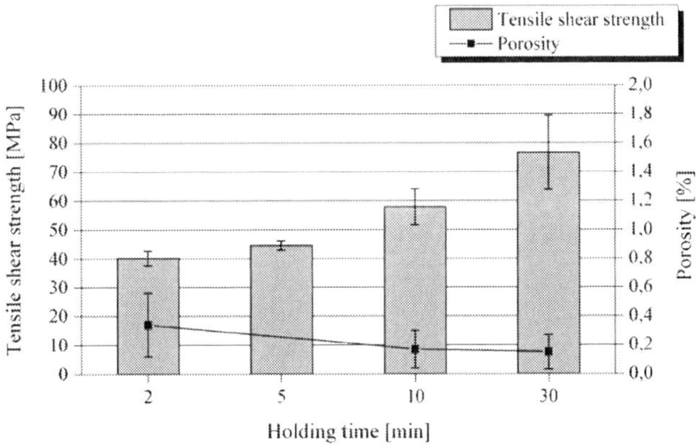

Figure 11: Tensile shear strength as a function of the holding time (temperature: 300 °C, pressure: 40 MPa, paste application: 20 µm)

As a function of heating rate

Furthermore, the influence of the heating rate on the strength behavior was examined, because studies in the field of sintering [31–33] found that the heating rate influences the sintering progress. At a constant temperature of 300 °C, a pressure of 40 MPa and a holding time of 10 min, three different heating rates (25 K/min, 150 K/min and 500 K/min) were tested. The diagram in Fig. 12 shows that the heating rate has no influence on the achievable strength. Only at a heating rate of 500 K/min, a slightly reduced strength is observed, but the marginal deviation is in the range of standard deviations of the other heating rates. However, it must be noted that the investigated heating rates are relatively high due to the use of induction heating. It cannot be excluded that slower heating rates, e.g. in furnace, have another influence.

As a function of paste application

At a temperature of 300 °C, for the joining pressures 5 MPa and 40 MPa and a holding time of 10 min, the influence of the thickness of paste application on the strength behavior was investigated, Fig. 13. It is shown that the thickness of the paste application has a significant influence on the achievable strength. An optimum thickness of paste application of 20 µm was determined, which leads to the highest achievable strengths. Thinner or thicker paste application results in lower strengths, regardless of the pressure. For a thinner application (10 µm), the surface roughness of the base materials cannot be compensated. Non-adherent areas between joint and base material are observed. With a thicker paste application, an incomplete sintering of the particles can be noticed, which results in lower strength values.

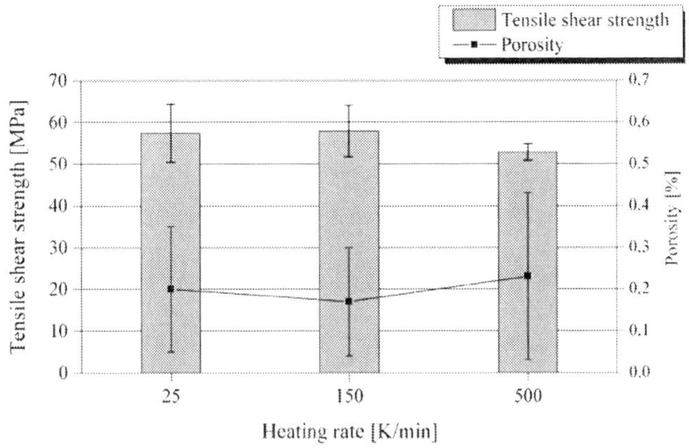

Figure 12: Tensile shear strength as a function of the heating rate (temperature: 300 °C, pressure: 40 MPa, holding time: 10 min, paste application: 20 µm)

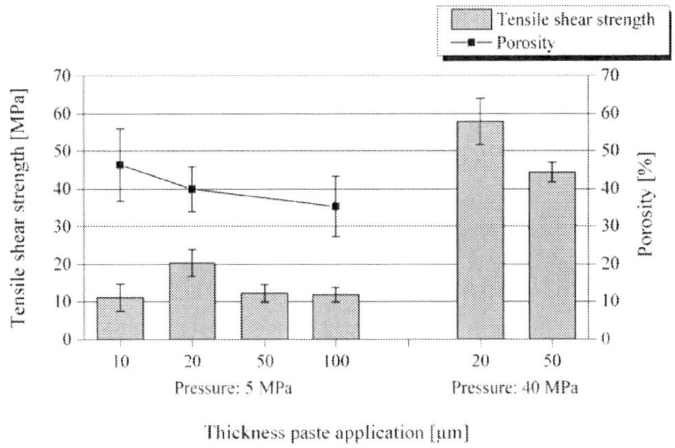

Figure 13: Tensile shear strength depending on thickness of paste application (temperature: 300 °C, holding time: 10 min, paste application: 20 µm)

As a function of surface pre-treatment

In joining processes, the surface treatment of the substrate decisively influences the quality of the joints. Therefore, various surface pre-treatments were investigated at a temperature of 300 °C, a pressure of 40 MPa and a holding time of 10 min, Fig. 14. While the tensile shear strength is almost the same when using polished samples in comparison to ground samples, the strength slightly increases with blasted samples, which is most likely caused by an improved adhesion due to the larger surface of the blasted samples. The slightly increased strength of PVD Ag-coated samples can also be attributed to an improved adhesion, which corresponds to previous studies [14, 34–35]. The pickled samples show a more pronounced increase of strength. Most likely, this is due to an improved removal of the copper surface oxide, which leads to an enhanced adhesion.

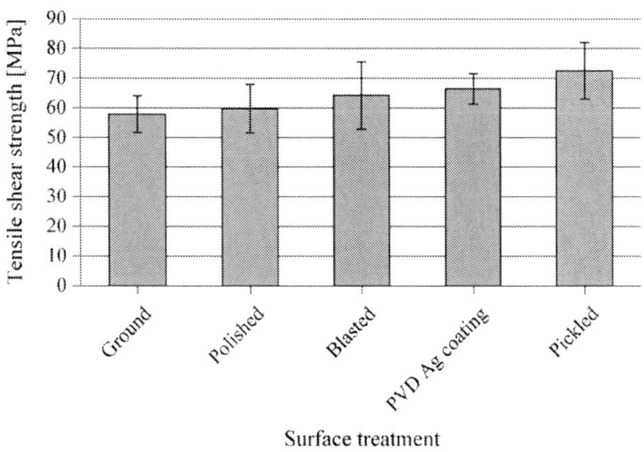

Figure 14: Tensile shear strength depending on the surface treatment (temperature: 300 °C, pressure: 40 MPa, holding time: 10 min, paste application: 20 µm)

As a function of a pre-drying process

For the investigation of a pre-drying process, a temperature of 300 °C, a joining pressure of 40 MPa and a holding time of 10 min was used. The upstream drying process, 4 min at 100 °C, was carried out in order to analyze, whether the strength can be increased, if a certain amount of the organic solvent is removed before the joining process and therefore the porosity can be reduced in the joint. Fig. 15, left part, shows the resulting strengths. It appears that, with the use of a high joining pressure of 40 MPa, the pre-drying process has no significant influence on the achievable strength.

As a function of a subsequent heat treatment

In further experiments, the effect of a subsequent heat treatment was investigated. Again, the samples were joined at a temperature of 300 °C, a pressure of 40 MPa and a holding time of 10 min. After joining, the samples were heat treated in a muffle furnace at a temperature of 300 °C for 20 min in air without pressure. This results in a total holding time of 30 min at a temperature of 300 °C (Fig. 15, a). This allows a direct comparison with the previously outlined analyzed samples, which were joined with a holding time of 30 min under pressure (Fig.15, c). The aim of this study was to investigate, whether the pressure has to be applied for the entire holding time, or if the increase in strength can be achieved by longer holding times, even without the application of pressure. However, Fig. 15 shows that the subsequent heat treatment does not influence the strength. The strength of the heat-treated samples is comparable to the strength of the samples, which were joined with a holding time of only 10 min. An increase in strength as for the samples, which were joined for 30 min under pressure, cannot be observed. However, it may be possible that higher temperatures and longer holding times, as those are used in conventional heat treat-treatment processes, lead to higher strengths also without pressure.

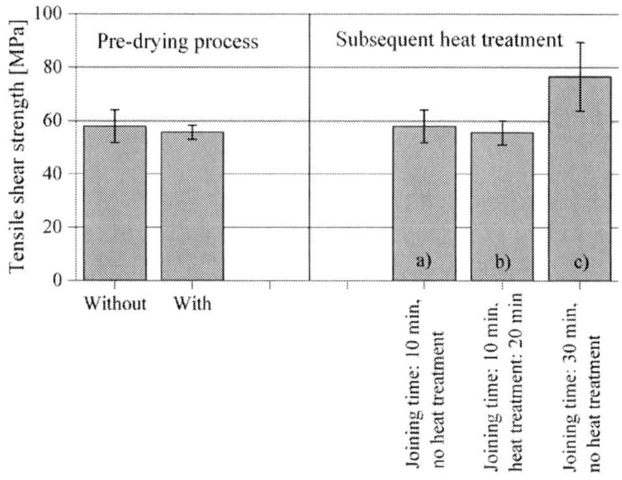

Figure 15: Influence of pre-drying and subsequent heat treatment on tensile shear strength (temperature: 300°°C, pressure 40 MPa, paste application: 20 µm)

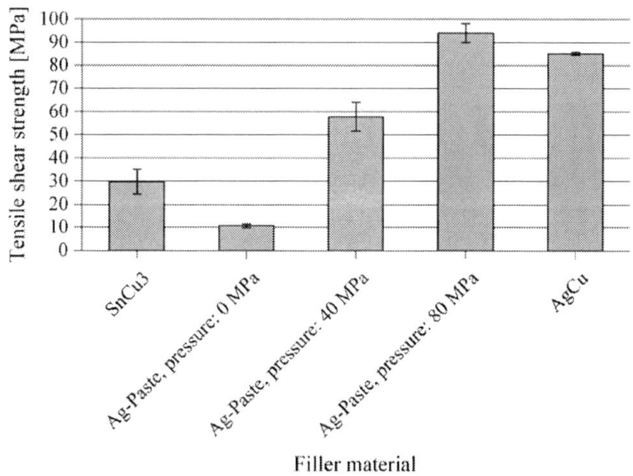

Figure 16: Tensile shear strength of soldered and brazed joints in comparison to samples joined with Ag nanopaste at different pressures (temperature: 300 °C, holding time: 10 min)

Comparison with soldering and brazing

In Fig. 16, the strength values of samples with different pressures, which exerted the greatest influence on the strength properties, are compared to conventional soldered and brazed samples. For soldering, SnCu3 was used as filler metal and soldered at 300 °C for 10 min without pressure. For brazing, an AgCu brazing filler metal (72 wt. % Ag, 28 wt. % Cu) was used und brazed at 780 °C for 10 min also without pressure. The results clarify that the application of pressure is necessary to produce high quality joints. Without pressure, the strengths are even lower than the strengths of soldered joints. With increasing pressure, the strength increases significantly and at high pressures, the strengths of conventional brazed joints are exceeded. This demonstrates that joining with nanoparticles is also an alternative for brazing processes. However, this requires the application of a high joining pressure, which is critical for a practical application. But the results show that for a sufficient quality of the joints, the joining pressure is essential. Thus, in practical applications, the sample geometry has to be adjusted to ensure the application of pressure.

Conclusions

The investigations prove that the nanopaste offers a great potential for joining at low temperatures. The DSC analyses showed that the organic shell of the nanoparticles decomposes at temperatures of about 410 °C followed by a sintering process. The sintered structure possesses the thermal properties of bulk silver. The investigations of the strength behavior show, that it is possible to produce joints even at lower temperatures (of approximately 300 °C) with good strength properties and high temperature stability.

The variation of the process parameters reveals that in particular the joining pressure exerts an essential influence on the achievable strengths. Also temperature, holding time and thickness of paste application have a considerable effect on strength behavior. In contrast, the pre-drying process, heating rate, surface pre-treatment and subsequent heat treatment exhibit hardly any influence on joint strength.

With high pressures, the strength values of conventional brazed joints can be improved. It can therefore be stated that joining with nanoparticles is also a high interesting alternative for brazing processes.

Acknowledgments

This work was performed within the Federal Cluster of Excellence EXC 1075 "MERGE Technologies for Multifunctional Lightweight Structures" and supported by the German Research Foundation (DFG). Financial support is gratefully acknowledged.

References

[1] Shyjumon, I. et al, "Structural deformation, melting point and lattice parameter studies of size selected silver clusters", Eur. Phys. J. D, Vol. 37 (2006), 409-415.

[2] Jiang. Q. et al, "Size-dependent melting point of noble metals", Materials Chemistry and Physics, Vol. 82 (2003), 225-227.

[3] Castro, T. et al, "Size-dependent melting temperature of individual nanometer-sized metallic clusters", Physical Review B, Vol. 42 (1990), 8548-8556.

[4] Pawlow, P., " Über die Abhängigkeit des Schmelzpunktes von der Oberflächenenergie eines festen Körpers", Z. Phys. Chem., Vol. 65 (1909), 1-35.

[5] Pawlow, P., " Über die Abhängigkeit des Schmelzpunktes von der Oberflächenenergie eines festen Körpers (Zusatz)", Z. Phys. Chem., Vol. 65 (1909), 545-548.

[6] Takagi, M., "Electron-Diffraction Study of Liquid-Solid Transition of Thin Metal Films", J. Phys. Soc. Jpn., Vol. 9 (1954), 359-363.

[7] Wronski, C.R.M., "The size dependence of the melting point of small particles of tin", Br. J. Appl. Phys., Vol. 18 (1967), 1731-1737.

[8] Meyer, K., Physikalisch-chemische Kristallographie, Deutscher Verlag für Grundstoffindustrie (Leipzig, 1968).

[9] Chou, K.-S. et al, "Fabrication and sintering effect on the morphologies and conductivity of nano Ag particle films by the spin coating method", Nanotechnology, Vol. 16 (2005), 779-784.

[10] Park, S. H. et al, "Electrical properties of silver paste prepared from nanoparticles and lead-free frit", Journal of Nanoscience and Nanotechnology, Vol. 7 (2007), 3917-3919.

[11] Kim, J.-W. et al, „Electrical characterization of screen-printed conductive circuit with silver nanopaste", Japanese Journal of Applied Physics, Part 1, Vol. 48 (2009), 06FD14/1-6.

[12] Mathews, N. et al, "Printing materials for electronic devices", International Journal of Materials Research, Vol. 101 (2010), 236–250.

[13] Ide, E. et al, "Metal-metal bonding process using Ag metallo-organic nanoparticles", Acta Materialia, Vol. 53 (2005), 2385-2393.

[14] Ide, E. et al, "Bonding of Various Metals Using Ag Metallo-Organic Nanoparticles-A Novel Bonding Process Using Ag Metallo-Organic Nanoparticles", Mater. Sci. Forum, Vol. 512 (2006), 383-388.

[15] Tatsumi, H. et al, "Sintering mechanism of composite Ag nanoparticles and its application to bonding process - Effects of Ag_2CO_3 contents on bondability to Cu-", Advanced Materials Research, Vol. 24-25 (2007), 499-502.

[16] Akada, Y. et al, „ Interfacial Bonding Mechanism Using Silver Metallo-Organic Nanoparticles to Bulk Metals and Observation of Sintering Behavior", Materials Transactions, Vol. 49 (2008), 1537-1545.

[17] Bai, J. G. et al, "Thermomechanical reliability of low-temperature sintered silver die attached SiC power device assembly", IEEE Transactions on Device and Materials Reliability, Vol. 6 (2006), 436-441.

[18] Wang, T. et al, „ Low-Temperature Sintering with Nano-Silver Paste in Die-Attached Interconnection", Journal of Electronic Materials, Vol. 36 (2007), 1333-1340.

[19] Kun, Q. et al, "Effect of interconnection area on shear strength of sintered joint with nano-silver paste", Soldering & Surface Mount Technology, Vol. 20 (2008), 8-12.

[20] Calata, J. N. et al, "Sintered nanosilver paste for high-temperature power semiconductor device attachment", International Journal of Materials and Product Technology, Vol. 34 (2009), 95-110.

[21] Lei, T. G. et al, "Low-Temperature Sintering of Nanoscale Silver Paste for Attaching Large-Area (> 100 mm^2) Chips", IEEE Trans. Compon. Package Tech., Vol. 33 (2010), 98-104.

[22] Alarifi, H. et al, "Bonding of Cu wires by solid state sintering of Ag nanoparticles at low temperatures", Materials Research Society Symposium Proceedings, Vol. 1207 (2010), 7-12.

[23] Hu, A. et al, "Low temperature sintering of Ag nanoparticles for flexible electronics packaging", Applied Physics Letters, Vol. 97 (2010), art. no. 153117.
[24] Knörr, M., Verbinden von Leistungshalbleiterbau-elementen durch Sintern von nanoskaligen Silberpartikeln, Dissertation Universität Erlangen-Nürnberg (2011).
[25] Mathews, N. et al, "Printing materials for electronic devices", International Journal of Materials Research, Vol. 101 (2010), 236-250.
[26] Kim, J.-W. et al, "Electrical characterization of screen-printed conductive circuit with silver nanopaste", Japanese Journal of Applied Physics, Part 1, Vol. 48 (2009), 06FD14/1-6.
[27] Nagasawa, H., "Physical Characteristics of Stabilized Silver Nanoparticles Formed Using a New Thermal-Decomposition Method", Phys. Stat. Sol. (A), Vol. 191 (2002), 67-76.
[28] Jiang, H., "Conductivity Enhancement of Nano Silver-Filled Conductive Adhesives by Particle Surface Functionalization", Journal of Electronic Materials, Vol. 34 (2005), 1432-1439.
[29] Hausner, S. et al, "Low temperature joining of copper by Ag nanopaste: Studies on the strength behavior", Adv. Mater. Res., Vol. 925 (2014), 420–427.
[30] Lu, D. et al, "Characterization of Silver Flake Lubricants", Journal of Thermal Analysis and Calorimetry, Vol. 59 (2000), 729-740.
[31] Wang, S. et al, "Densification of Al_2O_3 powder using spark plasma sintering", Journal of Materials Research, Vol. 15 (2000), 982-987.
[32] Freim, J. et al, "Microwave sintering of nanocrystalline y-Al_2O_3", Nanostructured Materials, Vol. 4 (1994), 371-385.
[33] Groza, J. R. et al, "Nanoparticulate materials densification", Nanostructured Materials, Vol. 7 (1996), 749-768.
[34] Siow, K. S., „Mechanical properties of nano-silver joints as die attach materials", Journal of Alloys and Compounds, Vol. 514 (2012), 6-19.
[35] Khazaka, R. et al, "Review on Joint Shear Strength of Nano-Silver Paste and Its Long-Term High Temperature Reliability", Journal of Electronic Materials, Vol. 43 (2014), 2459–2466.

Mechanical Properties of Mo-Ni Brazing Filler Metal Caused by Additions of SiC Nano-Particle Powder (I)

Dusan P. Sekulic
Harbin Institute of Technology, Harbin, China
College of Engineering, University of Kentucky, Lexington, KY 40506, U.S.A
E-mail: sekulicd@engr.uky.edu
Hai Fu
College of Engineering, University of Kentucky, Lexington, KY 40506, U.S.A
Anatol Rabinkin
Brazing and Joining Consultant, LLC, Watertown, MA 02472, USA
Daniel Busbaher
Ceradyne, Inc., a 3M Company, Lexington, KY 40511, USA
Edgar Leone and Tom Preuss
Exel Laboratory Services, Dover, NJ 07801, USA

Abstract

Addition of SiC nano-particles to the eutectic Mo-Ni brazing filler metal changes both the microstructure of the solidified Mo/Mo-Ni joints and the mechanical properties of the material. A set of brazed specimens was manufactured from pure solid Mo base metal and Mo-Ni powder having eutectic composition as the brazing filler metal. The powder was doped with 1 wt. %, 3 wt. % and 5 wt. % of the nano-SiC powder. Mechanical testing was made using a pure shear mode method developed earlier. This study offers evidence of a non-linear change of the joint fracture shear stress value vs. that of the non-doped braze sample. The maximum joint fracture stress level is observed in the specimen containing 3 wt. % SiC additive. For the first time nano-indentation experiments were carried out using atomic force microscopy (AFM) to measure the hardness of sub-micrometer joint phases with resolution < 50 nm. Results of these measurements on a nanometer scale, combined with detailed SEM and EDX information on phase composition and morphology, along with mechanical property testing, provide a basis for explaining high joint strength in relationships with microstructures achieved due to Mo-Ni filler metal doping with SiC.

Introduction

Motivation

The present work was initiated first in order to alter, decrease the liquid $Mo_{.40}$-$Ni_{.653}$ alloy spreading characteristics over pure Mo during joining. It was made adding to micro-sized $Mo_{.40}$-$Ni_{.653}$ eutectic alloy powder small amounts of SiC nano-particles via doping made by mechanical mixture of both powders [1-4]. Not only spreading characteristics were indeed affected but refining of the post-solidification eutectic microstructure was discovered. Such effect, never observed earlier in eutectic alloys, served as a stimulus to carry out wide scope studies of how molybdenum brazing may be affected in general via brazing filler metal doping.

Detailed description of the experimental procedures and sample preparation for brazing runs will be given in this paper. Particular focus is on mechanical testing (macro and nano) of the Mo joints brazed using MoNi filler metals with various concentrations of SiC nano particles.
Detailed metallographic studies of the induced changes in joints microstructure phase composition and their morphology are elaborated in an accompanying twin paper, [5], and will not be presented here. The twin paper [5] also contains discussion and explanation of the fundamental origins of doping effects observed in these joints.

Specimens and Procedures

Specimens for testing mechanical properties of Mo/MoNi + SiC/Mo brazements

Mechanical testing was made using the two-dimensional shear mode method developed in [6]. Standard AWS C#2 lap-shear method with 3-mm thick and about 100-130 mm long sample was rejected as inappropriate because in the standard testing a tri-axial stress situation develops in specimens under load. In addition the specimens used there are prohibitively expensive to manufacture. In our case we prepared specimens from two blocks with rectangular geometry and 10 x 10 x 50 mm dimensions which were machined from solid Mo rods. The blocks were placed in a special device in which they were tack-welded to the supporting plates in the lap-type configuration with fixed 50 μm gap (Fig. 1). Before brazing a filler metal powder was deposited not only in this gap but a small amount of it was also place outside the gap to fill completely the brazing gap upon melting. The brazing heating history, controlled atmosphere conditions and quench were identical to the conditions described in [5]. After brazing operation the tack-welded plates, i.e. fixers, and an excessive amount of filler metal in the fillets were machined off the brazed samples in order to create a perfect rectangular geometry of the specimens, including brazements as well. After this machining operation one could measure accurately

both brazements dimensions and determine the critical stress parameters during joint mechanical testing. Figure 2 shows samples after brazing and machining operations together with appearance of brazed samples after mechanical testing.
To compare the strength of the joint with virgin Mo base metal, the step-like specimens, Fig. 3, so-called Z-specimens, were machined from the solid Mo block to have their geometry similar to that of the brazed ones. These Z-specimens were placed in the heat treating furnace together with specimens to be brazed during heat treating operation to get the same state of the virgin Mo as in the brazed samples.

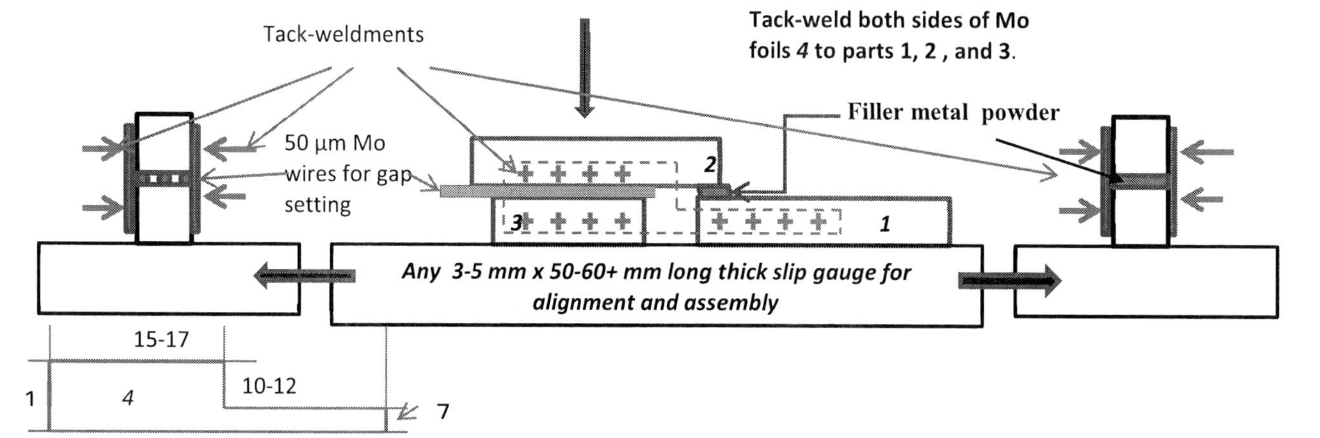

Figure 1 Schematic diagram of the brazement assembly (slip gauges for setting precise brazing gap, fitting parts, the filler metal placement location, and tack-welding fixing operations marked).

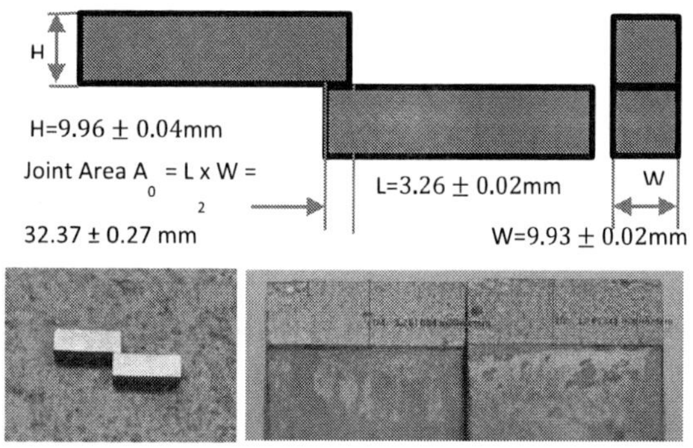

Figure 2 Brazed specimen dimensions and their view after failure under mechanical testing.

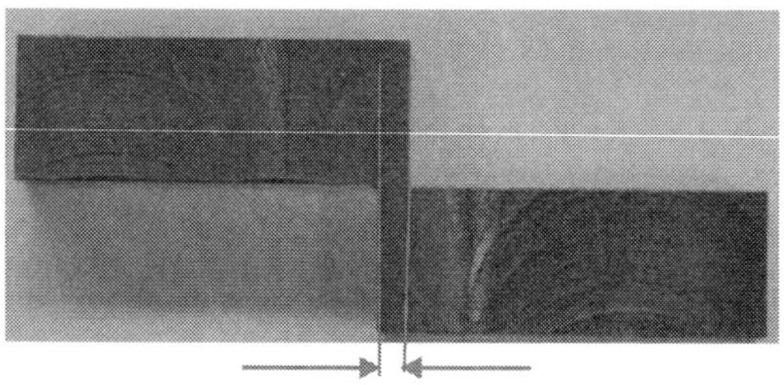

Figure 3 Pure Mo solid specimen, Z block

Mechanical testing procedure

Figure 4 a shows schematics of a special guide which was placed on the Instron table during testing. This guide creates lateral support to specimens in the lap-type configuration thus avoiding formation of the three-dimensional stress field during vertical loading. The general view of the guide with assembled pure Mo specimen and cemented carbide pushers before inserting into the guide is shown in Fig. 4 b. Whereas the total view of the testing arrangement in an Instron 5582 machine is given in Fig. 5. The loading of all tested specimens was applied with 1.25 mm/min rate.

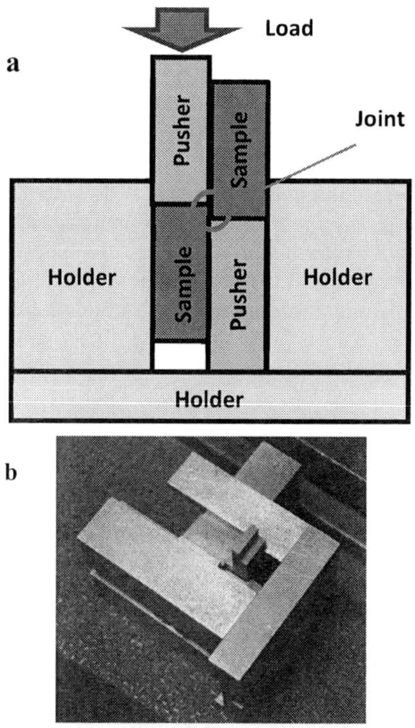

Figure 4 The schematics and a view of the supporting guide for keeping specimens and WC pushers aligned during testing

Figure 5 The general view of assembly placed in the Instron 5582 instrumentation.

Nano-indentation hardness testing procedure

Nano-indentaion experiments were performed using atomic force microscopy (AFM) in order to characterize the relative hardness of phases in Mo/Ni braze joints that were modified by the addition of SiC nano particles. Because silicon and carbon were found to be non-uniformly distributed among the Mo/Ni phases the AFM nano-indentation measurements would characterize the role of the silicon and carbon as active factors changing the joint microstructure morphology and as potentially hardening components in the joints' phases.

Previously AFM was used not only for specimen's surface characterization but also for assessment of elastic modulus of some polymer fibers and creep evaluation [7]. To the best of our knowledge the AFM nano-indentation application as a phase hardness measurement method in metallographic studies was not suggested previously or even used in studying sub μm phase formation in nanoparticle strengthened Mo/Ni alloys.

Nano indentation AFM utilizes a fine tip probe mounted at the end of a cantilever to measure the surface morphology. In our case for hardness assessment we used Veeco (Digital Instruments) Dimension 3100 AFM instrumentation. In this instrument a special diamond tip is mounted to a stainless steel cantilever that can be pressed into the surface of a sample with a known force to create a nano indentation. The spring constant (k) of the probe is 249 N/m. The resident frequency of the diamond indentation probe is about 64 kHz, considerably less than the typical Si probes (which are about 300 kHz). An indent is made by forcing the tip into the surface until the required cantilever deflection is reached. The cantilever force used in this study was 1.73×10^{-5} N.

For each indent a plot of the cantilever deflection versus the Z displacement (called a force plot) was created. An image of the indent was recorded by SEM. Afterwards the depth of the indent was determined by performing a cross-section scanning analysis through the center of the indent (see black insert in Fig. 7). The diamond tip movement was used to perform this scanning.

Samples analyzed in the study were as following:

- Sapphire (Al_2O_3)
- Gold film on Cr
- Molybdenum crucible in location far from the joint interface
- Phases in the Control 4-2A Mo/Ni braze joint (no SiC additive)
- Phases in the brazed joint containing 5% SiC nano particles (9-2A Mo/Ni braze joint).

The comparative hardness of phases analyzed was determined in relationship with that of sapphire and gold film on pure Cr.

Results

Macro scale mechanical properties

The load/displacement curves of all brazed and virgin Mo specimens upon testing are shown in Fig. 6. As is seen in Fig. 6 all dopped specimens are stronger than undopped one. The best result, shear strength increase on about 26%, is obtained in the brazement with 3 % SiC. The nonlinear shear strength concentration dependence with maximum value at 3 % SiC dopping may probably occur because this alloy has the finest mixture of the joint eutectic phases and advantageous morphology specifics in the joint interface zone. The view of a typical for all joints fracture face is presented in Fig. 2. The failure of all brazed specimens is occurred in the brazement zone and proceeds in a brittle mode-no plastic deformation stage is present on loading curves in Fig. 6.

In the molybdenum Z-specimen case failure occurred at about 436 MPa that is close to data on upper shear strength under torsion testing which is in [9]. A good coincidence of results for pure Mo,-strain rate 1.25 mm/min,-with that of the ASM standard data proves that our measurements method is rather reliable to apply in brazement testing for characterizing the joint strength. The pure Mo specimen failure was also initiated in the ends of the overlap "neck" area, -i.e. simulating brazement,-with failure cracks propagating in the specimen body at about 45° upon failure, Fig. 7.

The strength of brazements manufactured in current work is equal or even exceeding 50 % of that of pure Mo, i.e. twice as higher than that obtained in [1]. Therefore, application of Ni-Mo alloy should be considered as a rather encouraging stimulus for using it as a brazing filler metal for molybdenum joining.

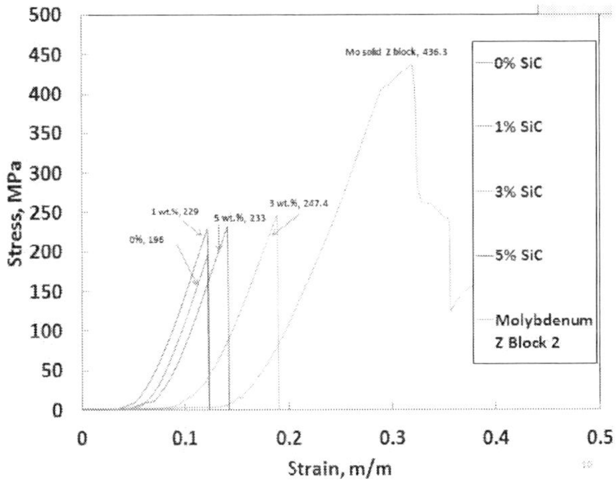

Figure 6 Stress/strain curves of tested brazed samples and the solid Mo Z block.

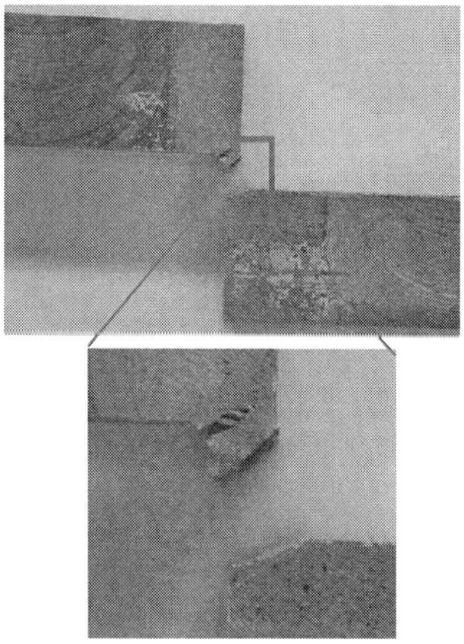

Figure 7 Solid Mo Z block after shear testing.

Nano scale mechanical properties

A semi-quantitative summary of results will be given herewith to help elucidating the doping effects on brazement microstructure and mechanical properties. Fig. 8 presents the SEM image of indents made in the Ni matrix phase of the 5 wt. % SiC specimen as an example of data sets obtained on all tested specimens. It shows a few indentations made for better statistics and the depth profile of the indent determined by the diamond tip scanning.

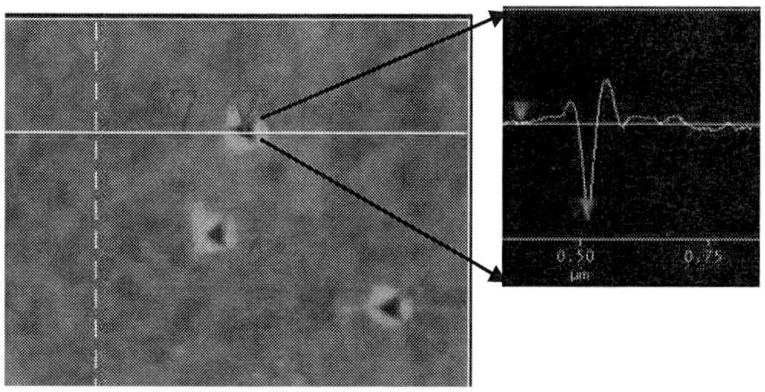

Figure 8 AFM indents made in Ni-rich phase of the 9-2A specimen of 5 wt. % SiC alloy [5]. The indent depth is ~ 4.2 nm.

The AFM nano hardness analysis data are presented in Fig. 9. They depict the average value of indent depths on various materials and phases in our samples for comparative purposes. It is worth to underline that the smaller the indent the higher directly corresponding microhardness. The standard Moh's hardness values for metallic gold, molybdenum and aluminum are also given for comparing the phase's properties in relationships with that of well-known metals.

We are currently working on developing a procedure which will convert the indent depth data directly into the microhardness values given in standard scale(s) such as the Vickers hardness testing (VH). The results will be published later.

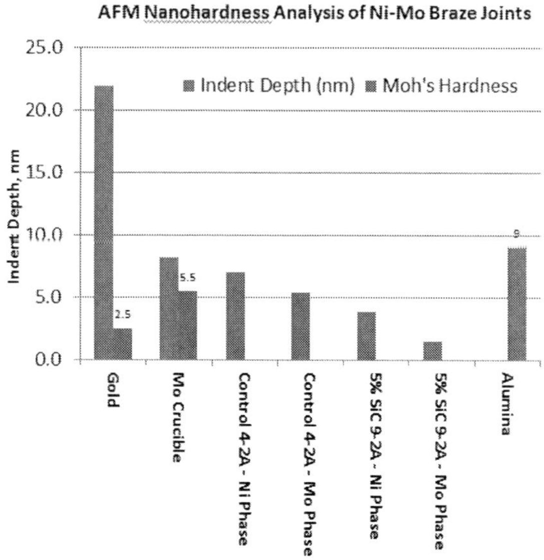

Figure 9 Nano hardness of a pure gold and aluminum specimens, virgin Mo crucible and Mo-and Ni-rich phases in Mo-Ni and Mo-Ni+5 wt. % joints.

As follows from data in Fig. 9 the hardness of brazement phases is ranked in the following order (soft to hard):

[Gold] << [Mo crucible] < [Control 4-2A (0 % SiC) – Ni-rich phase] < [Control 4-2A – Mo-rich phase] < [5% SiC 9-2A – Ni-rich phase] < [5% SiC 9-2A – Mo-rich phase] < [Sapphire (Al_2O_3)].

The AFM "nano" hardness analysis correlates well in the general way with the EDX and WDX microprobe analysis results on silicon and carbon concentration in joints phases [5]. Both these analyses revealed that silicon and carbon are dissolved as solutes in basic MoNi and Ni(Mo) joint phases without changes in their crystal structure. It was also found that silicon and carbon concentration higher in the Mo-rich phases in the samples with "nano" SiC additions.

The observed increase of hardness in both Ni- and Mo-rich phases is caused by well-known solid solution strengthening effect: It is due to the substitutional mechanism for Si, which has atomic radius not strongly different from those of Mo and Ni, and the interstitial one for carbon atoms. The strengthening effect relative to non-doped one is the most pronounced, about 3 times increase, in Mo-based phase of the alloy with 5 % SiC.

Conclusions

The strength of brazements manufactured in current work which are doped with "nano" SiC powder is equal or even exceeding 50 % of that of pure Mo. It is twice as higher than that obtained previously in [1]. Therefore, application of Ni-Mo alloys should be considered as a rather encouraging stimulus for using it as a brazing filler metal for molybdenum joining.

Acknowledgments

The work of the lead author (DPS) and HF have been supported in part through US NSF Grant CBET # 1234581. Authors acknowledge an in-kind support of Ceradyne, Inc. (formerly Semicon Associates), a 3M Company. The first author acknowledges support by the 1000 Plan Foreign Expert Professor Talent Program at the Harbin Institute of Technology, Harbin, China. Dr. Liu Wen currently of Delphi Advanced Thermal Systems, Lockport, NY, contributed to the study at its early stage at the University of Kentucky Brazing Laboratory.

References

[1] Sene F. and Motta, C. "Synthesis and characterization of Ni-Mo filler brazing alloy for Mo-W joining for microwave tube technology, *Mat. Res.* vol. 16 no. 2, 2013, pp. 417-423.

[2] Busbaher D., Liu, W., and Sekulic, D.P. "High Temperature Brazing of Mo/Mo-Re with a Nano-Composite Mo-Ni Filler," 2010, *DVS-Berichte*, Vol. 263, pp. 211-214.

[3] Busbaher D., Liu, W., and Sekulic, D.P. "High Temperature Brazing Using Nano-Particles Doped Filler Metal for Dispenser Cathode Applications," 2010 IEEE International Vacuum Electronics Conference, IVEC-2010, pp. 151-152.

[4] Busbaher, D., Liu, W., Moecher, D., and Sekulic, D.P. "High Temperature Brazing of Porous Tungsten with Nano Structured Mo-Ni for a Dispenser Cathode Application," 2011 IEEE International Vacuum Electronics Conference, IVEC-2011, pp. 323-324.

[5] Sekulic, D.P., Fu, H., Rabinkin, A., Busbaher, D., Leone, E., and Preuss, T. "Positive changes in the microstructure morphology of Mo-Ni eutectic brazing filler metal caused by additions of SiC nano-particle powder (II)," Proceedings of 6th IBSC, 2015, Long Beach, CA (to be published).

[6] Rabinkin A. and Pounds S. "Effects of Load on Brazing with METGLAS® MBF-2005 Filler Metal", *Weld. J.*, (1988) v. 67, N5, p. 33.

[7] Oliver W. and Phar G., "Measurement of hardness and elastic modulus by instrumented indentation: Advances in understanding and refinements to methodology", *J. Mater. Res.*, (2004), Vol. 19, No. 1, pp 3-20.

[8] Croft, D., Shed G. and Devasia S., "Creep, Hysteresis, and Vibration Compensation for Piezoactuators: Atomic Force Microscopy Application" , *J. Dyn. Sys., Meas., Control (*2001) 123(1), pp. 35-43.

[9] ASM Handbook, 1990, vol. 2, p. 1142, Fig. 76.

Positive Changes in the Microstructure Morphology of Mo-Ni Eutectic Brazing Filler Metal Caused by Additions of SiC Nano-particle Powder (II)

Dusan P. Sekulic
Harbin Institute of Technology, Harbin, China
College of Engineering, University of Kentucky, Lexington, KY 40506, U.S.A
E-mail: sekulicd@engr.uky.edu
Hai Fu
College of Engineering, University of Kentucky, Lexington, KY 40506, U.S.A
Anatol Rabinkin
Brazing and Joining Consultant, LLC, Watertown, MA 02472, USA
Daniel Busbaher
Ceradyne, Inc., a 3M Company, Lexington, KY 40511, USA
Edgar Leone and Tom Preuss
Exel Laboratory Services, Dover, NJ 07801, USA

Abstract

A doping of the micro-sized eutectic alloy powder by SiC nano-particles has been applied in order to decrease the liquid $Mo_{.40}$-$Ni_{.653}$ alloy spreading over pure Mo. The pure solid Mo was brazed with Mo-Ni powder having eutectic composition. The later powder was doped with 1 wt. %, 3 wt. %, and 5 wt. % of the nano-SiC powder. The initial (0% SiC) typical eutectic microstructure consisting of large hypo-eutectic NiMo dendrite crystals and a mixture of MoNi and Ni(Mo) eutectic intermetallic phases was observed in the joint area. The microstructure of the doped samples depends strongly on the concentration of the SiC power. Both metalloids, Si and C, are dissolved upon filler metal melting, and are found as solid solution components in Mo- and Ni-base intermetallic eutectic joint phase components. The bulk of the joint having 3 wt.% SiC addition has fine and uniform mixture of eutectic phases which is most favorable from a mechanical strength point of view. The phase compositions of observed microstructures are discussed in relationship with phases existing in all parental binary and ternary phase diagrams known for all possible combinations between all four elements, -Mo, Ni, Si and C,- participating in joint formation.

Introduction

The state of the art and research objective

Brazing is applied in many molybdenum applications [1-3]. Unfortunately, the low melting brazing fillers metals such as BAg-series alloys cannot be used at high temperatures at which Mo is utilized predominantly. As a response to Mo service conditions, high temperature brazing filler metals such as V-35Nb, V-Mo [4], Ni-14Cr-6Fe (wt.%) [2], Ti-based alloys [1, 3], Fe-15Mo-5Ge-4C-1B, and Ni-Mo eutectic alloys were developed and are used. The Ni-14Cr-6Fe (wt. %) filler metal yield joints with a rather low, 132 MPa shear strength [2]. TLP-type processing in which thin layer of Ni was used for diffusion welding requires a long temperature annealing that could bring a negative embrittlement near the recrystallization temperature [4]. Recently, Ni-Mo filler metal with eutectic composition was exploited in joining solid molybdenum with porous tungsten for microwave technology [6]. The filler metal powder was produced using elemental powders milling. The brazing operations were carried out at 1400°C for 15 min. The resulting joint tensile strength was rather low, about 88 MPa. No detailed description of joint microstructure was presented.

Experimental

Specimen Material Preparation

All specimen materials used in this work were manufactured with specially prepared binary Mo-Ni alloy micron-scale powder as the major base component. The powder had 34.71 at. % Mo and 65.28 at. % Ni (46.5/53.5 wt. %) composition and < 350 mesh dimensions (40-150 μm particle size). The first specimen material designated as 4-2a, Table 1, consisted only from Mo-Ni powder and served as "the base line" for consideration of metallurgical effects on joint properties via additions of SiC powder to the binary Mo-Ni powder. The SiC powder particle dimensions were in "nano" range, i.e., < 50 nm. Three doped powder mixtures were prepared via addition of 1, 3 and 5 wt. (2.44, 6.96, and 10.88 at.) of % SiC to Mo-Ni micro-scale size powder. The doping was accomplished by mixing binary Mo-Ni powder and SiC powder together in a Fisher-Kendall mixer during an extensive period of time.

To calculate the atomic fraction of all 4 elements, i.e, Mo, Ni, Si, and C, in each specimen material , we considered first the Mo-Ni base weight as 100 wt. %. Then, the total weight of each mixture was 101, 103, and 105 wt. % of the base weight, correspondingly. The elemental atomic fractions were recalculated relative to this each individual combined weight considered as the new 100 %. The Table 1 shows the concentration, at. %, of Mo, Ni and SiC in each tested composition and specimen's material designation.

Table 1 Concentration, at. %, of Mo, Ni and SiC in each tested specimen material and their designation.

Specimen Designation	Formula for powder loading (1:1 atomic fraction)	Powder mixture composition*, at. %			
		Mo	Ni	Si	C
4-2a	Mo/Ni without SiC;	34.71	65.28	-	-
7-2a	Mo/Ni + 1 wt. % SiC;	33.86	63.69	1.22	1.22
8-3a	Mo/Ni + 3 wt. % SiC;	32.29	60.73	3.48	3.48
9-2a	Mo/Ni + 5 wt. % SiC.	30.94	58.19	5.44	5.44

*Calculated relative to the real powder composition. It is slightly off, to the Ni side, of the eutectic composition indicated on the Mo-Ni binary alloy phase diagram.

Specimens Preparation and Thermal Treatment

The powders were loaded in small caps made of pure molybdenum. Afterwards, the loaded caps were placed together in a controlled atmosphere furnace (75%H_2+25%N_2) and heated until melting the powders within the caps. During heating, the specimens were at temperatures above 1319°C for about 1 min with the temperature peak achieving 1362°C. A relatively fast cooling to the room temperature was followed. The temperature/time of the complete thermal cycle is given in Fig. 1.

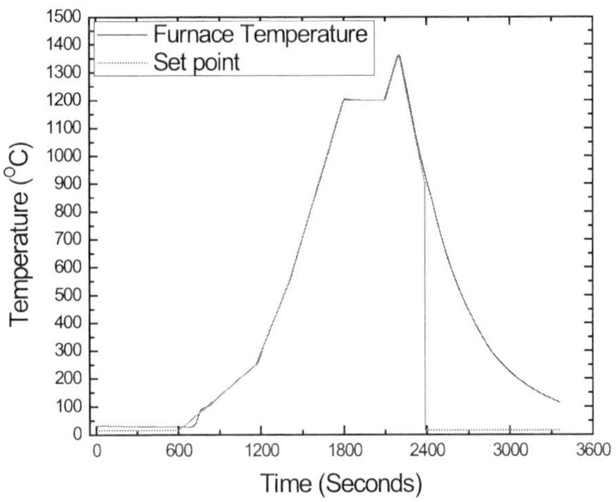

Figure 1 Temperature/time profile of the heat treatment applied to the specimens.

As a result of this procedure the powder was melted, reacted with Mo, solidified, and created the joints between each of the alloys and solid Mo (Fig. 2, general view). The solidified alloys had filled about a half of the cap volume.

Thus formed specimens imitate one half of a virtual brazing joint with "infinitely thick" brazing filler metal contacting solid Mo. All four specimen materials were then used to study metallurgical reactions between Mo and Mo-Ni+SiC at the joint interface and microstructure of a brazing filler metal at the interface vicinity and beyond. For this they were cut in half, imbedded in the standard 31-mm diameter plastic cylinders and subjected to the following multi-stage polishing procedure: First, specimens were polished using different silica sand papers following by the diamond polishing. Lastly, Stuers OPU paste was applied to finish samples preparation.

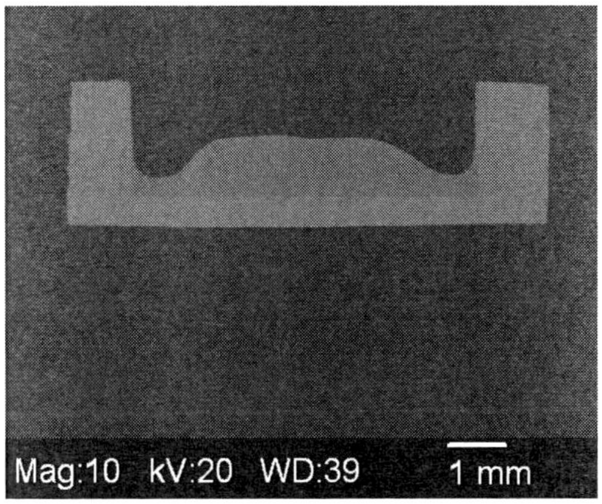

Figure 2 The cross-section of a typical Mo container with crystallized filler metal (FM).

Metallographic and DTA analysis

Metallographic studies were made using optical microscopy (OM), and scanning electron microscopy (SEM), performed on a JEOL JXA 840 electron probe microanalyzer. Both secondary electron imaging (SEI), and backscattered electron imaging (BEI) techniques were used. Part of SEM analysis was performed with both energy dispersive X-ray (EDX) spectroscopy, and wavelength dispersive X-ray (WDX) spectroscopy analytical methods. WDX analysis was used to provide additional information not obtainable from the EDX analysis.

Table 2 DTA Data

Sample	Formula for powder loading	Composition of powder mixture, at. %				Temperature, °C	
		Mo	Ni	Si	C	T_{liq}	T_{crys}
4-2a	Mo/Ni (1:1 atomic fraction) without SiC	34.1	65.3	-	-	1320	1310
7-2a	Mo/Ni (1:1 atomic fraction) + 1 wt. %SiC	33.9	63.7	1.2	1.2	1320	1300
8-3a	Mo/Ni (1:1 atomic fraction) + 3 wt. %SiC	32.3	60.7	3.5	3.5	1320	1290
9-2a	Mo/Ni (1:1 atomic fraction) + 5 wt. % SiC	31.0	58.2	5.4	5.4	1350 (?)	1285

DTA was carried out on NETZSCH STA 449C instrumentation. The consideration of DTA data of all four alloys showed that liquidus, T_{liq} determined during heating is more or less the same for all alloys. However T_{liq} cannot be considered as an alloys objective melting characteristic due to too slow powder melting during the transient procedure of DTA during the specimen heating. Therefore, the crystallization temperatures of alloys in completely liquid state indeed characterizes effect of doping by metalloids such as Si and C on alloy melting characteristics. As it is expected and observed both the melting temperature and consequently surface tension of MoNi liquid are decreasing with SiC addition

Microstructural observations

Mo-Ni Filler Metal Joint (Sample material 4-2a)

$Mo_{0.34}Ni_{0.65}$ powder reacts well with pure Mo forming a good joint: The interface between Mo and Mo-Ni is well developed showing a strong bonding of the crystallized Mo-Ni with Mo (Figs. 3a and 3b). No any visual defects at the interface such as pores, etc. are seen (Figs. 3a and 3 b). The joint crystallizations starts at the interface with formation of a thin layer of the high-melting phase of this eutectic, namely, $Mo_{0.34}$ $Ni_{0.65}$ (Fig. 3b, thick arrow). This continuous, 5-10-µm thick layer consists of rounded crystals of Mo-rich phase. The area in vicinity of the Mo BM surface, outlined by a dashed line in Fig. 4, is definitely enriched in Mo as a result of its dissolution in the liquid FM and not sufficient compositional equilibration of the FM liquid phase during brazing, see Tables 3 and 4. The excessive, relative to the pure eutectic composition, amount of Mo is present in MoNi dendrites formed in the interface area.
In addition, the excessive Mo amount yields formation of large dendrites of hypo-eutectic NiMo crystals grown in the eutectic mixture which has the equilibrium composition (Fig. 3a, dashed circle and Fig. 3b). Because the average alloy composition is close to the eutectic one, no MoNi dendrite should form. Thus the presence of these hypo-MoNi crystals confirms non-equilibrium distribution of Mo and Ni in the joint interface vicinity. Whereas a more or less uniformed typical eutectic mixture is formed in the rest of the specimen (Fig. 3 b, dashed circle and Fig. 3 c).
All crystals formed in this specimen have rounded morphology. No any faceted crystals typical for MoNi intermetallic crystals and Mo-based carbides and silicides are observed.
In the area outstanding on 200-300 µm from the joint interface the microstructure consists of only very fine mixture of MoNi and Ni(Mo) eutectic phases (Figs. 3b, dashed circle and 3c). Their composition is close to eutectic one which can be found in the Mo-Ni phase paternal diagram.

Mo-Ni Filler Metal + 1.22 at% SiC Joint (Sample material 7-2a, 1 wt. % SiC)

An SEM image of this joint is given in Fig. 5. A substantial number of well-faceted Mo-enriched crystals are present in eutectic crystal mixture (Figs. 5, 6, and 7 (point 1) beyond of the interface area. Optical analysis revealed that these large, dozens-micron-sized Mo-rich crystals appeared to be much harder than the other phase structures based on variation in polishing edge retentions (see inset in Fig. 5). Their composition (Table 5) is very close to that of the major MoNi intermetallic phase existing in Mo-Ni alloy system according to the Mo-Ni binary phase diagram. They also contain 5.8 at. % Si. This silicon concentration is higher than the average concentration of Si in the specimen. Therefore, it is obvious that Si prefers associating with Mo-rich phase.

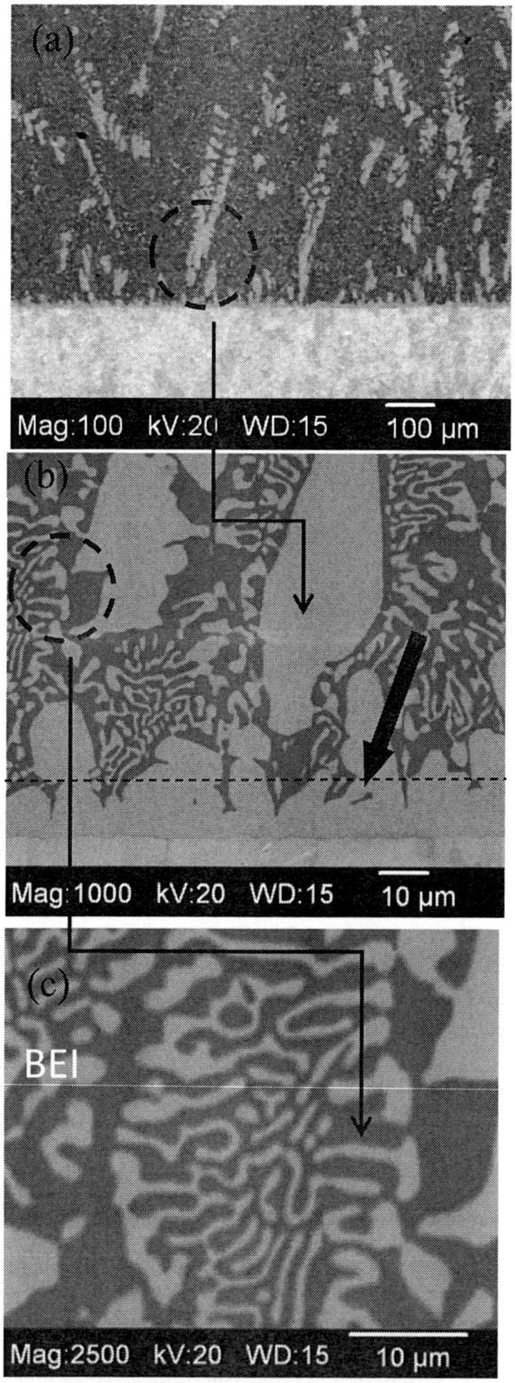

Figure 3 SEM micrographs of the 4-2a Mo-Ni control specimen containing no SiC.

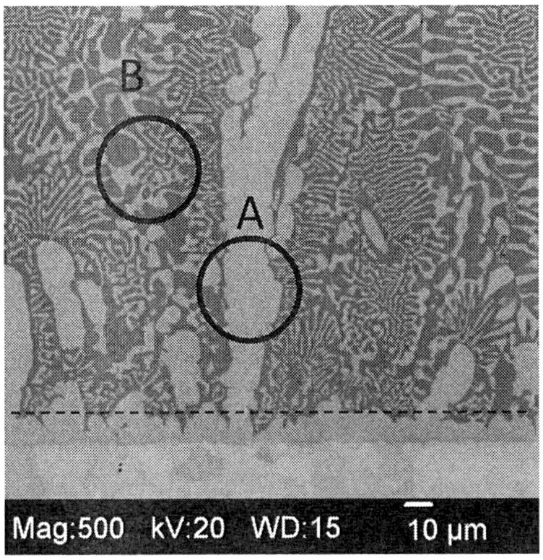

Figure 4 Microstructure, and phase composition of Mo-Ni 4-2a control specimen without SiC (See Tables 3 and 4 for the phase composition at the locations A and B)

Table 3 Phase composition at the location A, Fig. 4

Elements:	WT%	AT%	K_A	K_F	K_Z	Intensity	P/bkg
MoL	65.09	53.29	0.916	1	0.955	3037.409	0
NiK	34.91	46.71	0.974	1	1.079	1945.676	0

Table 4 Phase composition at the location B, Fig. 4

Elements:	WT%	AT%	K_A	K_F	K_Z	Intensity	P/bkg
MoL	45.16	33.51	0.872	1	0.932	1863.818	0
NiK	54.84	66.49	0.982	1	1.053	2862.735	0

A large fraction of MoNi crystals is present which don't have strong faceting (Fig. 7, points 2 and 3). These areas are definitely similar to eutectic areas in non-doped alloy. They also contain less both, Mo and Si. Still, variations of Mo, Ni, and Si concentrations in various MoNi crystals observed in this alloy are insubstantial.

Some MoNiSi crystals have well distinguished small spherical particles located in their centers.

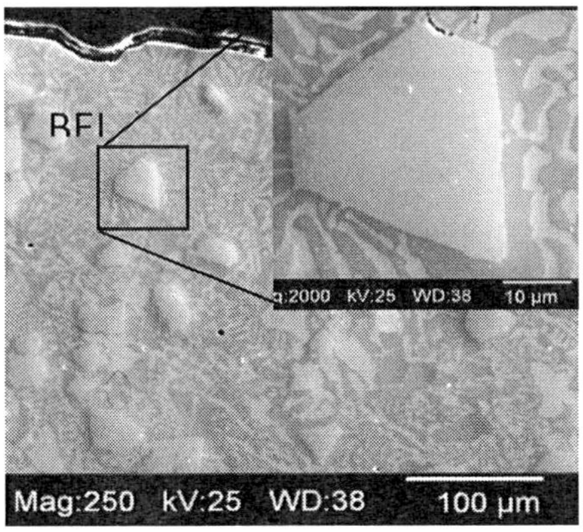

Figure 5 Microstructure of Mo-rich phase observed in Mo-Ni 7-2a specimen containing 1 wt.% SiC

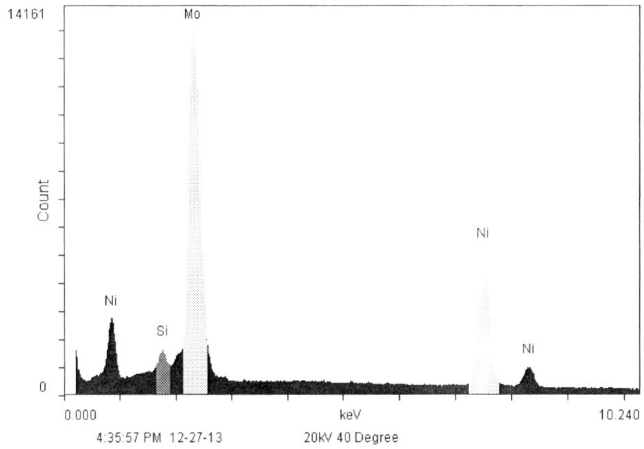

Figure 6 Composition of Mo-rich phase observed in Mo-Ni 7-2a specimen containing 1 wt.% SiC

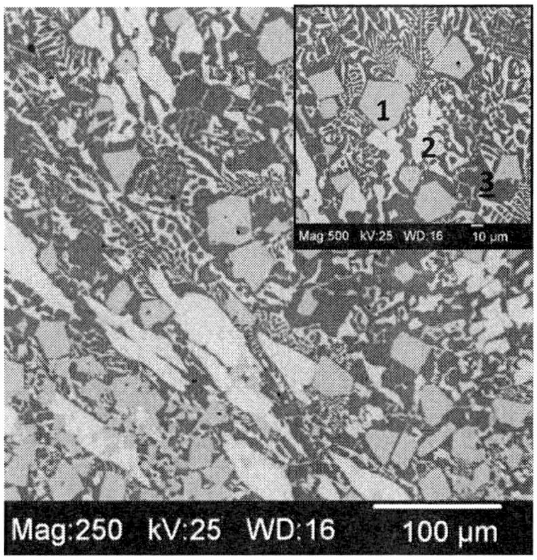

Figure 7 Microstructure of Mo-Ni specimen containing 1 wt.% SiC (4-2A). Note that there are two types of Mo-rich phases containing different concentrations of Mo and having different morphology.

Table 5 SEM/EDX composition of the Mo rich phase (Fig. 7, point 1)

Elements:	WT%	AT%	K_A	K_F	K_Z	Intensity	P/bkg
SiK	2.11	5.81	0.588	1.016	1.165	377.906	0
MoL	67.71	54.5	0.919	1	0.955	3188.172	0
NiK	30.17	39.69	0.972	1	1.079	1690.323	0

MoNi Filler Metal + 3.48 at. % SiC Joint (Sample material 8-3a, 3 wt. % SiC)

The 3 wt. % SiC sample has a substantially different microstructure compared with 1% sample. Its SEM micrograph is given in Fig. 8.

Figure 8 SEM micrograph of the Mo/MoNi+3 wt.% SiC interface.

As seen in Fig. 8, a narrow Mo-enriched layer is attached to Mo in a similar way as in a sample with 0 and 1 % SiC (arrows). The following interface region between the Mo crucible and the bulk of the brazing filler metal has a number of bands of Mo and Ni rich phases seen in Fig.8. These bands are alternating layers of large, fine and intermediate sized Mo/Ni phases. The bulk of the joint material has a relatively fine MoNi phase located in Ni-rich matrix as shown in Figs. 8. EDX analysis clearly showed that the Si content is much higher in the bright Mo-rich phases. An interesting result of the study is that the Ni is higher in the large Mo rich phases in Band 1 adjacent to the Mo interface, compared to the finer Mo rich structures in Band 2. Also, Si is lower in the large phase structures in Band 1, adjacent to the Mo crucible. Still, again the concentration differences are not great to pay a strong attention to these variations.

MoNi Filler Metal + 5.44 at. % SiC Joint (Sample material 8-3a, 5 wt. % SiC)

The microstructure of this joint consists of two distinguished zones: 1. The narrow interface zone seen in a SEM image and presented in Fig. 9. This zone is located between Mo base metal and white dash outline and it contains Bands #1, 2 and 3 (Fig. 9). 2. The bulk of the joint outside of the dashed line.

As follows from interface observations depicted in Fig. 8 the interface zone is more or less similar in appearance, phase locations and their composition to that in the specimen containing only 1 wt.% SiC. Indeed, a narrow Mo-enriched layer (arrows) is attached to Mo, followed by thin Ni-rich (point 1). A rather wide band of Mo-enriched crystals (point 2) separating interface zone from the rest of the joints and it is attached to the latter. The rest of the joint microstructure consists of a fine, uniform mixture of two Mo-rich (point3) and Ni-based (point 4) phases which constitute the bulk of eutectic. The phase dimensions in this area are smaller than 10 μm. Neither large MoNi crystals, similar to those observed in 1 wt. % SiC specimen, nor carbides nor silicides crystals are found there.

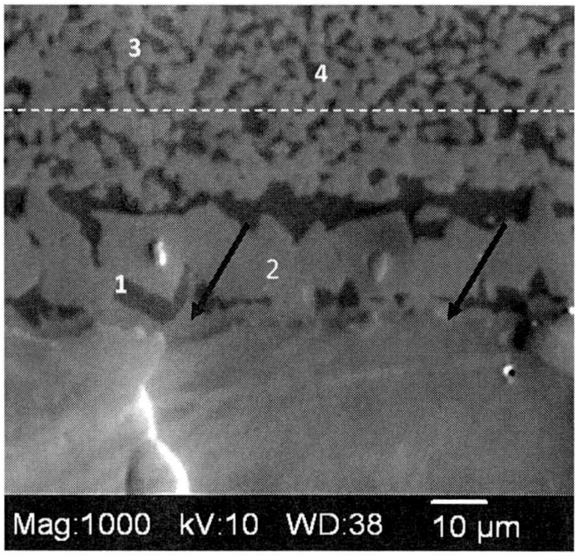

Figure 9 Areas in the joint interface of alloy with 5 wt.% SiC in which WDX Intensity were measured.

It is important to underline a drastic difference in morphology of the eutectic in the bulk of the joints having 0 vs. 3 and 5 wt. % Si. In the first case the microstructure is a classical, "hand book", eutectic build of dendrite like network with rounded shapes, whereas in joints containing 3 and 5 wt. % of SiC it is an assembly of very fine cuboidal, trapezoidal particles of Mo-rich phase segregated in the Ni-rich matrix. All these particles have flat facets outlined by straight lines. The particles sizes are in order of a few micrometers. On the other hand the morphology of the alloy with 1 wt.% SiC is transitional: It is a mixture of the "classical" plate–like eutectic structure observed in the binary Mo-Ni alloy and newly observed combination of faceted crystals of Mo-rich phase distributed rather evenly in Ni-rich matrix in alloy with 5 wt.% SiC.

A review of parental phase diagram of Mo-C, Mo-Ni, Mo-Si, Ni-Si, Si-C, Ni-C, C-Ni-Si, C-Mo-Ni, Mo-Ni-Si alloy systems was made. From this review one can expect the following probable changes in composition of the liquid Mo-Ni-C-Si phase as well as eutectic MoNi and Ni(Mo) phases as a result of SiC alloy doping:

- SiC should dissolve in liquid Mo-Ni on separate, elemental components which exist as such in forming solid phases upon crystallization. This dissolution was confirmed in present work applying both energy dispersive X-ray (EDX) spectroscopy and wavelength dispersive X-ray (WDX) spectroscopy [7] analytical methods.
- Melting and crystallization temperatures of Mo-Ni alloys should be lower with addition of SiC. Indeed the DTA data during crystallization prove this suggestion.

- Both MoNi and Ni(Mo) eutectic phases may contain Si and C as solid solution components without changes in the crystal structure of these phases in Si and C concentration ranges studied in the present work.

Addition of Si and C seems to be moving liquid composition closer to the quaterrnal eutectic point. This thus decreases likelihood for segregation of prime hypo-eutectic dendrites in the middle of solidifying alloys. Indeed no dendrites were observed at all doped compositions in contrast to what was observed in the non-doped alloy.

In short conclusion of this paper it may be noted that change in MoNi morphology from spherical to faceted can be reasonably explained by decreasing the surface energy of liquid and solid phases when Si is added. This is the consequence of the well-known effect of Si additions to transition metals surface tensions [8]. The surface tension decreasing, in turn, leads to a lower driving force for crystal rounding in lamellae formation. The latter phenomenon is always an expression of high surface energy contribution to the total crystal free energy and thus its substantial effect on phase morphology. Dramatic increase of number of crystals formed in the bulk of Mo-Ni+SiC alloy upon crystallization yields more and more finer eutectic mixture. This phenomenon is also result of very strong decrease of surface tension of MoNi liquid and solid phases due to addition of Si. Detailed discussion and explanation of doping on microstructure and mechanical properties of these Mo/MoNi+SiC joints are given in the second paper of the same authors presented herewith at IBSC 2015 and given in [9].

Conclusions

1. Novel advantageous metallurgical technique, - i.e., addition of small amounts of SiC "nano" powders to regular micron scale-sized MoNi powder which is used as a brazing filler metal, - is proposed. The mixture of MoNi + "nanno" SiC powders is successfully used in wetting and joining pure molybdenum.
2. The Mo/MoNi + SiC joint microstructures were studied by multiple analytical methods yielding details of its phase component compositions in various joint locations.
3. The major beneficial changes of the brazed joints morphology due to alloying of Mo + Ni alloy with SiC moderate additions take place in the joint microstructure. These changes are expressed in transformation, replacement of typical eutectic morphology consisting of dendrites imbedded within mixture of lamellae into a new type of refined microstructure consisting of a mixture of very small evenly distributes eutectic crystals having cuboidal and trapezoidal form. This transition expresses itself by changes in forms and dimensions of crystallized eutectic phase components from rounded lamellae crystals to facetted ones.
4. Si and C in moderate amounts can dissolve completely in Mo-Ni liquid and exist as solvent elemental species in MoNi and Ni(Mo) solid phases forming upon crystallization of Mo-Ni-Si-C alloys. The preferential phase for Si location is Mo-rich phases.
5. No formation of separate crystalline carbides and silicide phases is noticed in all joints under the present study.

Acknowledgments

The work of the lead author (DPS) and HF have been supported in part through US NSF Grant CBET # 1234581. Authors acknowledge an in-kind support of Ceradyne, Inc. (formerly Semicon Associates), a 3M Company. The first author acknowledges support by the 1000 Plan Foreign Expert Professor Talent Program at the Harbin Institute of Technology, Harbin, China. Dr. Liu Wen currently of Delphi Advanced Thermal Systems, Lockport, NY, contributed to the study at its early stage at the University of Kentucky Brazing Laboratory.

References

[1] Brazing Handbook, (2007), 5th Ed., AWS, CH. 31, pp. 559-561.

[2] Spravochnik po paike, Russ., Moscow Machinostroenie, Edit. Petrunin I., (1984), pp. 256-258.

[3] Metals Handbook, Vol. 6, Welding, Brazing and Soldering", ASM, (1993), pp. 941-943.

[4] Lundberg G. *et al,* "Brazing Molybdenum and Tungsten for High Temperature Service", *Weld. J.* Supp. (1978), No.10, p. 311s.

[5] Lin C. *et al*, "Brazing Mo using Ti–Ni–Nb braze alloys", *Int. J. Refractory Metals and Hard Materials*, v.29, N5, Spt. 2011, pp. 641-644.

[6] Sene, F.F. and Motta, C.C., "Synthesis and Characterization of Ni-Mo Filler Brazing Alloy for Mo-W Joining for Microwave Tube Technology," *Materials Research,* (2013) Vol. 16, pp. 417-423.

[7] Leone E., "Electron Microprobe X-ray Chemical Shift and Line Profile Changes in Silicon Containing Materials," *Scanning Electron Microscopy/ 1984/III*, SEM Inc., AMF O'Hare (Chicago), IL, 60666, pp. 1023-32, 1984.

[8] Allen B. in "Liquid Metals", Chap. 4, Marcel Dekker, NY, (1972), p. 186.

[9] Sekulic D. P., Fu, H., Rabinkin, A., Bushbaher, D., Leone, E., and Preuss, T., "Mechanical Properties of Mo Joints brazed using Mo-Ni brazing filler metal powder doped by SiC nano-particles (I)," Proceedings of 6th IBSC , 2015, Long Beach, CA, (to be published).

Pressureless and Low-temperature Silver-Sintering using Solvent-free Nanoparticle Pastes

A. Oestreicher, M. Lerch, and T. Röhrich
Technische Universität Berlin, Berlin, Germany

Abstract

In the last decade low-temperature silver sintering has emerged as an attractive alternative for conventional soldering. In this study a novel metal-organic silver bulk material enriched with organic-free, agglomerated silver nanoparticles was used for silver paste preparation. During an intermediate heating step excess organics are volatilized and solvent-free silver nanopastes are obtained. Due to the presence of a sufficient amount of small silver nanoparticles and increased contact areas owing to the pasty consistency of the material up to the sintering step, pressureless sintering is possible. Well adhered silver layers on copper were obtained.

Introduction

Silver nanoparticles allow sintering at temperatures of 250 °C and less and are discussed as promising potential candidates to form highly conductive silver layers. Once sintered, they assume the melting temperature of the bulk material (961 °C). Sintered silver joints are therefore expected to withstand high-temperature environments in high-performance next-generation power devices and microelectronic packaging and to have high joint reliability and durability. Low-temperature silver sintering reflects the urgent need for new bonding techniques and -materials because conventional solders are no longer able to meet advanced requirements.

Early methods of solid-phase nanosintering at low temperatures necessitated high pressures up to 40 MPa. The application of pressure involved densification control during sintering through increased contact surface area and neck growth of neighboring silver particles and enhanced diffusion between silver nanoparticles and substrate. Later developments led to low-pressure assisted applications [1-3]. But there has been a desire for entire elimination of pressure during the sintering stage and pressureless approaches have been introduced [4-8]. However, sintering under pressure and pressureless sintering may have different degrees of densification resulting in significant differences in their properties, performance and reliability.

The substrates have a crucial influence on the interface to the sintered silver, too. Frequently used substrates are gold and silver metallized surfaces because good adherence of the sintered silver can be expected [9]. Bare copper as substrate material has also been an object of interest. Copper is less costly than noble metals like silver or gold and is highly conductive, too.

One approach to achieve denser silver layers in a pressureless bonding process is to use different-sized silver particles anticipating more efficient particle packing density [10]. However, the common way is the use of silver particles in combination with additives like solvents to adjust the viscosity of the paste. They are usually removed in a preheating step before sintering because rapid volatilization of solvents during a sintering step produces voids. The percentage of the silver in the paste is further increased in the evaporation process leading to more contacts between neighboring silver particles.

Silver sintering material is commercially available as inks intended for printing techniques and only in very few cases for pressureless joining applications. In this study silver nanomaterial has been synthesized in standard one-pot reactions in air from purchased chemicals [11]. The obtained metal-organic silver complex deviates in its properties from those of the commonly known derived from fatty acids. The new solder material combines several advantageous features: Nanoparticles are only generated upon heating. Different-sized silver particles are present during sintering. Furthermore, pressureless sintering is promoted through the presence of pasty material until the sintering step. This way, increased contact areas between the particles to be sintered are possible. Paste formation is achieved due to the properties of the material and the pastes contain no solvents or other additives. It is shown, that although pressure is completely excluded, dense silver layers may be obtained with good adherence to a copper substrate surface.

Experimental

Silver pastes were made from a bulk metal-organic complex and additional organic-free, agglomerated silver nanoparticles [11]. Mixtures in the ratios of 1:1 to 1:3 were heated to 190 °C in an oxygen-free atmosphere and submitted to a predefined dwell time during which excess organics are released. After cooling the samples to room temperature, the remaining silver material was pasty. Sintering was performed in ambient atmosphere in a furnace at temperatures of 180 °C to 240 °C for coating purposes and at 220 °C to 240 °C for joining conditions. The sintering temperatures and holding times used do not represent the minimum values. Bare copper was used as substrate material, cleaned with citric acid before usage.

Cross-sections of the sintered samples were made and investigated by optical microscopy. The ratio of pores in sintered silver layers were compared by means of graphical analysis of cross-sectional images of optical microscope images.

Results and Discussion

The used silver complex is solid after preparation. Upon heating it shows different stages [11]. It melts in a yellow liquid at 110 °C, followed by the generation of >2 nm small silver nanoparticles in this liquid. Further heat treatment leads to the formation of a molten metal-like silver phase. After decomposition of the organic complex and degassing of the constituents a sintered silver layer is left.

Once the nanoparticles have formed, the heating process is irreversible. When cooling down the molten metal-like silver phase to room temperature, waxy material is obtained, consisting of small silver nanoparticles capped by the complexing material. If heat treatment proceeds, the organic matter is gradually reduced through decomposition and volatilization. Cooling down the samples after different times of isothermal heating, the silver nanomaterial shows different viscous stages. As long as all organics are still present, waxy material is obtained. When excess organics is burned off, the nanomaterial again assumes the metal-like silver phase. Depending on the amount of added agglomerated silver nanoparticles and the time of isothermal heating accordingly a rather broad variation of viscous stages may be produced. These findings were used for paste preparation in an intermediate heating step.

Melting the solids in a liquid and subjecting them to a heating step means also that the material undergoes compaction. It should be noted that sintering occurs between neighboring particles. Therefore the resulting micro structure strongly depends on the way the particles are arranged before sintering. Compacted material may provide more contacts between adjacent particles.

Figure 1 demonstrates the effect when a small amount of waxy silver nanomaterial was loosely placed on a copper plate and directly submitted to a sintering temperature of 250 °C. The obtained silver framework did not change its shape during sintering (Fig. 1, above). However, when one drop of ethanol was put on top of the nanomaterial, the liquid caused further compaction leading to more contacts especially to the substrate (Fig. 1, below). Therefore, the use of pasty material may provide a maximum of contacts between the particles involved in a pressureless sintering process.

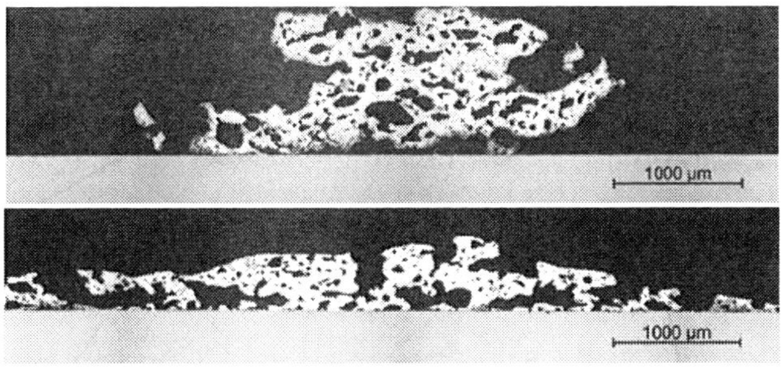

Figure 1: Nanosilver of waxy consistency sintered on copper at 250 °C. Above: Onto the copper surface as obtained; Below: Only one drop of ethanol on top of the waxy material may ensure more contacts to the substrate and between the silver particles

Figure 2 shows the sintered structures obtained when waxy nanosilver material was directly subjected to different sintering temperatures. As equally seen in Fig. 1 (above), only poor adhesion could be achieved for lack of close contacts to the substrate. It is quite evident that a decrease in sintering temperature produced less voids in the sintered structure. This effect might be attributed to the more violent outgassing of the organics at elevated temperatures. Therefore, the heating profile has a crucial impact on the resulting silver structure.

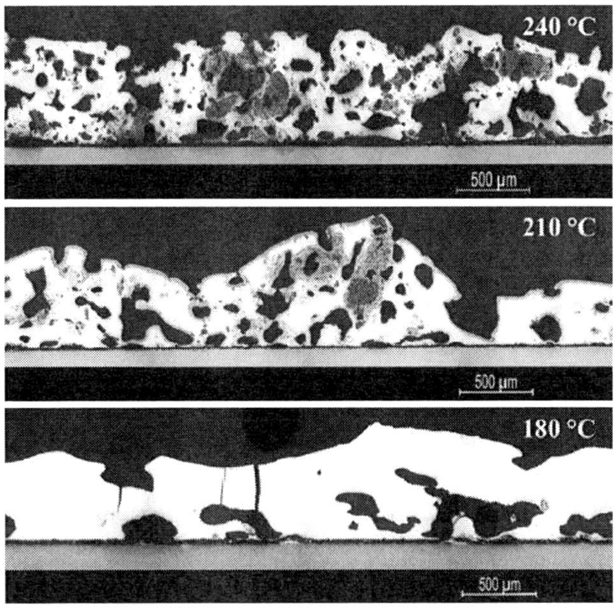

Figure 2: Nanosilver of waxy consistency sintered on copper at different temperatures. The number of voids is clearly reduced at lower sintering temperatures

In consideration of the effects described above void-less and dense sintered silver layers may be obtained. Figure 3 shows a >600 µm thick sintered layer on copper using pasty material with organic contents of 10 wt%. This sample was slowly heated to 160 °C, followed by fast heating to 220 °C and a holding time of 10 minutes and a final fast heating ramp to 240 °C with a holding time of 40 minutes. The resulting micro structure was not completely homogeneous. Graphical analysis on cross-sectional images revealed that there was no measurable quantity of porosity within the silver layer with bright appearance in the image. In the darker fields were porosities found, but could not be precisely determined.

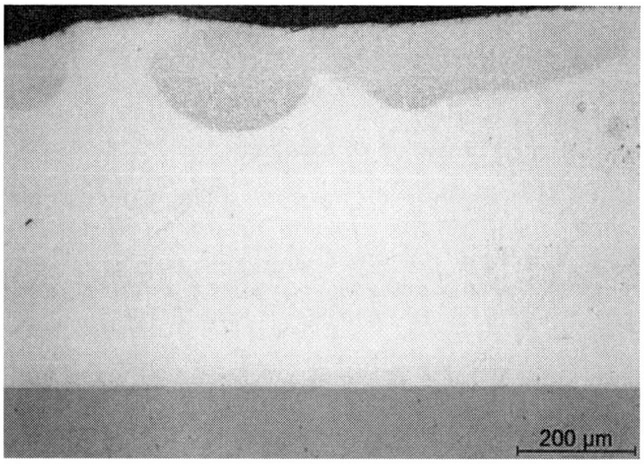

Figure 3: Pasty nanomaterial sintered on copper at 240 °C resulting in a predominantly dense microstructure with good adherence to the substrate

A thick silver layer on copper consisting of four layers with varying porosities sintered on top of each other is shown in Fig. 4. The different porosities in the structures are caused by their different compositions during paste preparation. Good adherence to the substrate was found as well as between the silver layers. For this sample, the porosity analysis gave values of 16% for layer 2, 36% for layer 3 and 15% for layer 4, respectively.

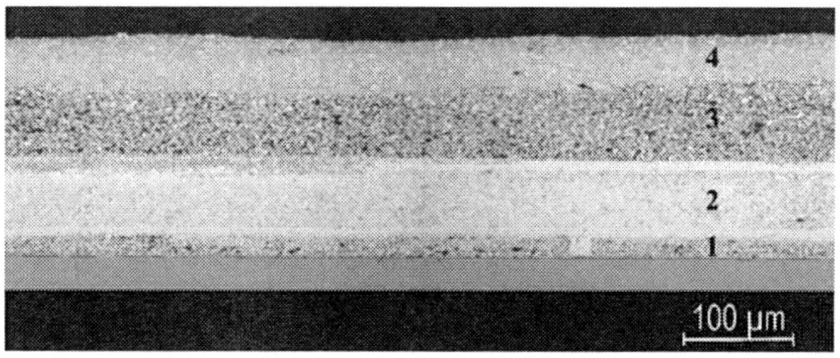

Figure 4: Four sintered silver layers on top of each other with different porosities

The application under joining conditions is far more challenging. There may occur capillary effects. Most notably, the presence of oxygen during sintering is of vital importance. However, this is limited under joining conditions because gaseous products have simultaneously to find out their way between the substrates. Producing defined porosities of the silver microstructure is difficult.

For joining purposes silver pastes with low organic contents to be volatilized would be most preferable. Using the method described above, low-organic silver pastes were prepared. For the joint of Fig. 5, a paste was used of only 4 wt% organic residues. The sample was heated to 250 °C within 30 minutes with a holding time of 20 minutes. As can be seen in Fig. 5, tendencies to structures of high porosities and partial poor adhesion to the copper substrate were found.

Figure 5: Copper-silver-copper joint using low-organic silver paste; sintering temperature 250°C, layer thickness 25 µm

While for the paste of Fig. 5 no alcohol was used for the formulation, a small amount of a long-chain alcohol (Butylglycol) was added to the paste used in Fig. 6 before application. The result was a well distributed sintered silver layer over the cross-sectional length (Fig. 6, above). Yet, the higher magnification (below) reveals that no adherence to copper was achieved. The silver particles sintered only to each other and not to the substrate.

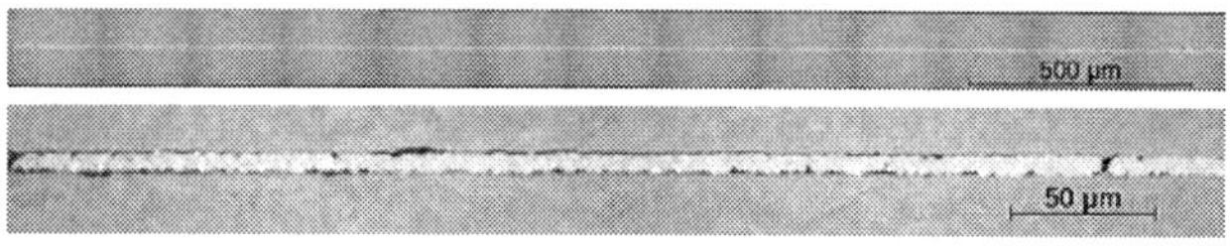

Figure 6: Copper-silver-copper joint using low-organic silver paste; sintering temperature 240°C, layer thickness 5 µm; Below: section with higher magnification

The used solder material contained small, with organic ligands stabilized silver nanoparticles and larger agglomerated bare silver particles. The latter are expected to only diffuse closer during sintering and not to contribute to adherence in the absence of pressure. This is different for the smaller and with organic capping material stabilized nanoparticles, provided by the organic silver complex. After bond cleavage between silver and the carboxylate group, the bare and highly reactive small silver clusters may either merge with the agglomerated nanoparticles or diffuse to the substrate surface.

Therefore, to achieve adherence of sintered silver to copper, the presence of enough small nanoparticles which are closely related with a sufficient amount of stabilizing organic content is crucial.

The organic constituents of the samples of Fig. 5 and Fig. 6 were greatly reduced during an intermedium heating step to reach a value of 4 wt%. One such sample is shown in the TEM image of Fig. 7. When stabilization of the small nanoparticles was no longer possible due to excessive volatilization of organics, the small particles fused with the larger agglomerated silver particles. These are not able to promote adhesion to a substrate in a sintering step.

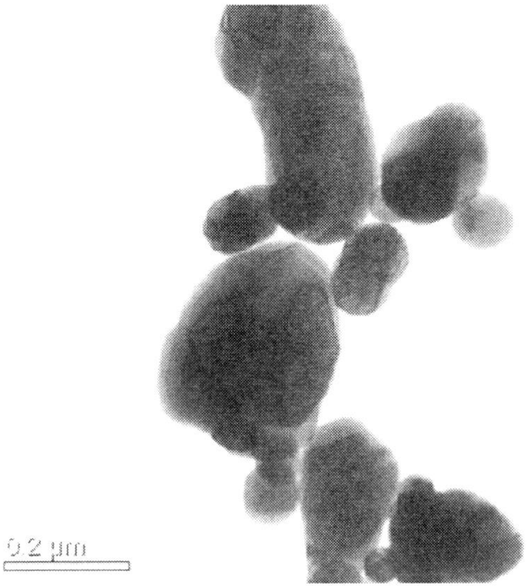

Figure 7: Diffusion of smaller nanoparticles onto the larger ones after volatilizing a part of stabilizing organics

Accordingly, within the experiments of this study, 10 wt% and more organics in the paste led to joints with good adherence between the interfaces of silver and copper. Figure 8 shows a well adhered sintered layer between two copper plates. The sample was slowly heated to 160 °C with a holding time of 30 minutes followed by fast heating to 220 °C and a holding time of 30 minutes.

Figure 8: Copper-silver-copper joint using a silver paste with 18 wt% organic residues; sintering temperature 220°C, layer thickness 5 µm

It is discussed in the current research that nanosilver joints with 100% density are not only not practical nor desirable because of the need to balance the elastic modulus and other mechanical properties [12]. It might be therefore acceptable when the joints exhibit small voids within a rather dense silver structure (Fig. 9). The sample of Fig. 9 was obtained under the same conditions as that of Fig. 8.

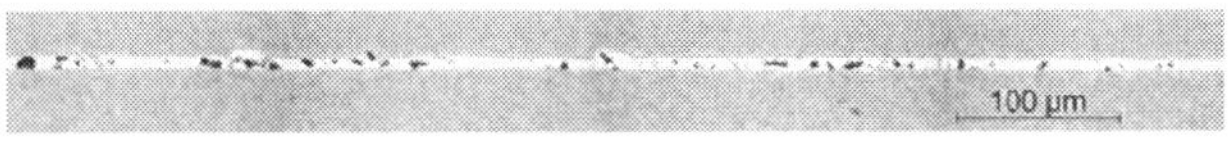

Figure 9: Copper-silver-copper joint using a silver paste with 18 wt% organic residues; sintering temperature 220°C, layer thickness 5 µm

Figures 10 and Fig. 11 show other examples of sintered silver layers, well adhered to copper. The layer of Fig. 10 has a thickness of 40 µm. The sample was heated to 210 °C within 40 minutes, followed by fast heating to 250 °C within 5 minutes and a holding time of 20 minutes.

Figure 10: Copper-silver-copper joint using a silver paste with 17 wt% organic residues; sintering temperature 250°C, layer thickness 40 µm

Figure 11: Copper-silver-copper joint using a silver paste with 10 wt% organic residues; sintering temperature 250°C, layer thickness 30 µm

The well-adhered silver layer of Fig. 11 was achieved with a paste containing a minimum organic share of 10 wt%. The heating ramp was 30 minutes until 250 °C, followed by a holding time of 20 minutes.

There are still some challenges remaining. Determination of shear strength gave values of 35 MPa that exceeded those of the conventional Sn-based solders. Further measurements are in progress.

Conclusions

It has been shown that pressureless silver sintering at low temperatures is not only feasible, but may also lead to well adhered silver layers on copper. It has been confirmed by several experiments, that a minimum of small nanoparticles are necessary to achieve adhesion. Therefore, a minimum of organic content which serves as capping material for the small nanoparticles, is also a prerequisite. Optical microscope of cross-sectional images and porosity analysis of the sintered silver indicate that dense microstructures may be produced.

Acknowledgments

The authors would like to acknowledge the Federal Ministry of Education and Research BMBF for financial support.

References

[1] Ogura, H. *et al*, "Carboxylate-Passivated Silver Nanoparticles and Their Application to Sintered Interconnection: A Replacement for High Temperature Lead-Rich Solders", *J. Electron. Mater.*, Vol. 39, No. 8 (2010), pp. 1233-1240.

[2] Chen, G. *et al*, "Pressure-Assisted Low-Temperature Sintering of Nanosilver Paste for 5x5 mm^2 Chip Attachment," *IEEE Trans. Comp. Packag. Manuf. Technol.*, Vol. 2, No. 11 (2012), pp. 1759-1767.

[3] Mei, Y. *et al*, "Simplification of Low-Temperature Sintering Nanosilver for Power Electronics Packaging," *J. Electron. Mater.*, Vol. 42, No. 6 (2013), pp. 1209-1218.

[4] Calata J. N. *et al*, "Sintered nanosilver paste for high-temperature power semiconductor device attachment", *Int. J. Materials and Product Technol.*, Vol. 34, No. 1/2 (2009), pp. 95-110.

[5] Lei, T. G. *et al*, "Low-Temperature Sintering of Nanoscale Silver Paste for Attaching Large-Area (>100 mm^2) Chips," *IEEE Trans. Comp. Packag. Technol.*, Vol. 33, No. 1 (2010), pp. 98-104

[6] Wang, S. *et al*, "Fabrication of interconnects using pressureless low temperature sintered Ag nanoparticles," *Mater. Letters*, Vol. 85 (2012), pp. 61-63.

[7] Yan, J. *et al*, "Pressureless bonding process using Ag nanoparticle paste for flexible electronics packaging," *Scripta Mater.* Vol. 66 (2012), pp. 582-585.

[8] Zheng, H. *et al*, "Chip-Bonding on Copper by Pressureless Sintering of Nanosilver Paste Under Controlled Atmosphere", *IEEE Trans. Comp. Packag. Manuf. Technol.*, Vol. 4, No. 3 (2014), pp. 377-384.

[9] Ide, E. *et al*, "Bonding of Various Metals Using Ag Metallo-organic Nanoparticles – A Novel Bonding Process Using Ag Metallo-organic Nanoparticles - ," *Mater. Sci. Forum*, Vol 512 (2006), pp. 383-388.

[10] Morisada, Y. *et al*, "A Low-Temperature Pressureless Bonding Process Using a Trimodal Mixture System of Ag Nanoparticles," *J. Electron. Mater.*, Vol. 40, No. 12 (2011), pp. 2398-2402.

[11] Oestreicher, A. *et al*, "An innovative method for joining materials at low temperature using silver (nano)particles derived from [$AgO_2C(CH_2OCH_2)_3H$}", *Appl. Surf. Sci.*, Vol. 265 (2013), pp. 239-244.

[12] Siow, K. S.; "Mechanical properties of nano-silver joints as die attach materials", *J. Alloys and Compounds*, Vol. 514 (2012), pp. 6-19.

Influence of Brazing Parameters on Microstructure and Strength of PCD-Cemented Carbide Joints

Christian Leinenbach, Nico Weyrich, Rémi Longtin
Empa-Swiss Federal Laboratories for Materials Science and Technology, Duebendorf, Switzerland
christian.leinenbach@empa.ch

Leif Dahl
Sandvik Coromant, Stockholm, Sweden

Abstract

In this work poly-crystalline diamonds (PCD) were brazed to cemented tungsten-carbide (WC) by an active braze alloy based on the Ag-Cu-In-Ti system (Incusil-ABA). Different process conditions and their influence on graphitization of the PCD, the microstructure and the mechanical properties of the joints were studied. To investigate the extent of graphitization at different brazing temperatures RAMAN spectroscopy was carried out. It showed only limited graphitization at brazing temperatures up to 800 °C. The microstructural investigation was done by means of scanning electron microscopy (SEM) and energy dispersive spectroscopy (EDS). With increasing process temperature the degree of (interface) reactions increased in combination with the formation of Co-Ti-rich particles. For the evaluation of the mechanical properties shear test were performed. The highest shear strength (281 MPa) could be achieved for samples brazed at 740 °C.

Introduction

The application of diamond as an abrasive component in material removal tools, e.g. for grinding and cutting, is evident due to its outstanding hardness and high strength, directly linked to its covalent structure. Different ways exist to attach diamond onto a substrate; either mechanical bonding between diamond and substrate, i.e. resin-bonding, sintering or electroplating, or chemical bonding, i.e. active brazing with Cu-, Ni- and Ag-based filler alloys [1-4].

Poly-crystalline diamond (PCD) is an excellent material as cutting tool for machining of non-ferrous light-weight materials such as aluminum, MMC (metal matrix composites, e.g SiC/Aluminum), CFRP (carbon fiber reinforced plastics), GFRP (glass fiber reinforced plastics) and their layer combinations with Aluminum or Titanium [5+6]. Large application areas of such materials are found in Aerospace and Automotive industries where the use of light-weight materials is constantly increasing. Figure 1 illustrates a typical insert and tool for milling operations. PCD as inserts and brazed round tools are also frequently used in turning and drilling applications.

The PCD is typically manufactured from diamond powder, allowing cobalt (as catalyst) to infiltrate the compacted powder under high pressure and high temperature [7+8]. Thus, the single diamond crystals are sintered into PCD discs. Tips manufactured out of such discs are usually brazed to a carbide base and finally ground into inserts. Since, the customers require inserts which are smaller and of more complex geometry, the process requirements of brazing are increasing. Brazing of pure PCD to carbide is very challenging, because sufficient wetting of both substrates and subsequent reaction of the braze alloy with the base materials is required to achieve adequate joint strength. In this context residual cobalt in the PCD must also be taken into account as it influences the interface reactions. In addition to that cobalt may act as a catalyst for graphitization at high brazing temperatures (higher than 700 °C) which leads to a degradation of the PCD [9+10]. In this work different brazing conditions (temperature / time) were studied with respect to the degree of graphitization as well as the evolving microstructure and strength of the joints. The optimal brazing conditions for sufficiently high bond strength and maintained PCD grade properties were deduced.

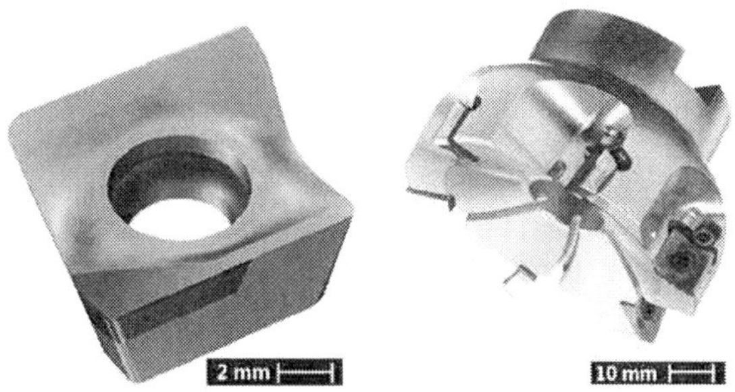

Figure 1: Insert and tool for milling applications

Testing materials and experimental setup

For this study, cemented tungsten-carbide (WC) platelets of 12.5 by 12.5 mm and a thickness of 5 mm were used (Sandvik Coromant, Stockholm, Sweden). The Co-sintered PCD specimens had a dimension of 3x3x1 mm and a diamond grain size of 5 μm (Sandvik Coromant, Stockholm, Sweden). As filler metal Incusil-ABA powder was applied with a maximum grain size of 44 μm (Wesgo Metals, Hayward-CA, USA). The braze alloy exhibits a solidus temperature of 605 °C and a liquidus temperature of 715 °C. For the joining experiments the filler metal powder was mixed with a self-made octyl-acetate binder to produce a homogeneous paste. Afterwards a braze alloy foil was made out of the paste by spreading it out on a Teflon covered plate and drying it at 150 °C in an airflow furnace. Almost identical filler alloy thickness of (110 ± 10) μm was achieved for all specimens after drying.

Prior to the brazing process the substrates were ultrasonically cleaned with acetone and subsequently degassed at a base pressure of $p = 10^{-5} - 10^{-6}$ mbar for 45 min at 800 °C in a Torvac high-vacuum furnace (Cambridge Vacuum Engineering LTD, Cambridge, United Kingdom) in order to remove organic residues in the near-surface layers of the substrate due to machining. The brazing foil and the diamond were then placed onto the WC substrate. This design was chosen allowing the performance of all necessary investigations in a thorough way at the same specimen.

Brazing was carried out in the above mentioned high vacuum furnace ($p = 10^{-5}$ to 10^{-6} mbar) at 650°C, 690°C, 710°C, 740°C and 800°C with a dwell time between 10 min and 45 min, using a special brazing jig. Ten specimens were brazed for each set of process parameters. By applying a small pressure (0.25 to 2 MPa) on the samples floating and rotating of the PCD during brazing was prevented and sufficient wetting was ensured. After cooling in the high vacuum furnace the specimens were used for RAMAN spectroscopy, metallographic investigation and shear strength measurement. RAMAN spectroscopy was done with a WITec Confocal Raman Microscope 200 (WITed, Ulm, Germany). As a light source a continuous blue laser was used with a wavelength of 532 nm and a power of 1 mW. For metallographic investigations cross sections were prepared by ion milling (Hitachi IM4000, Hitachi High-Technologies Europe GmbH, Krefeld, Germany) and analyzed by scanning electron microscopy (SEM; ESEM XL 30, FEI, Eindhoven, Netherlands). Shear tests were carried out using a table top shear testing device (STM-20A, Walter+Bai AG, Loehningen, Switzerland). Samples were adjusted in the machine in a way that the load was introduced approximately 50 μm above the brazing gap. The shear strength of the joints was determined at the maximum force ($\tau = F_{max} / A$; $A = 16mm^2$).

Results

Raman spectroscopy was used to detect the occurrence and extent of graphitization in the PCD due to heating above 700 °C in the presence of cobalt. The Raman spectra of the 5 μm un-sintered diamond powder, a reference PCD which was not brazed and brazed PCD samples are shown in Figure 2. In most spectra, there is a strong background due to photoluminescence presumably due to the presence of cobalt in the PCD. The characteristic diamond peak centered at 1332 cm^{-1} which is seen most clearly in the diamond powder, is present in all spectra although weakens substantially in intensity at 740 °C and above . The appearance of the peak at 1580 cm^{-1}, corresponding to so-called G-peak of graphite, is most evident at 740 °C and above although it does appear to exist as a wide peak when the PCD are brazed at 650 °C. The reduction in intensity of the diamond peak and the appearance of the G-peak qualitatively indicate that the PCD samples underwent noticeable graphitization when brazed at 740 °C and above. It is not possible to accurately quantify the graphitization due to the noise in the spectrum. However, the extent of graphitization can be described as limited due the fact that the diamond peak can still be resolved at 800 °C indicating the presence of both sp^3 and sp^2 bonding characteristic of diamond and graphite respectively.

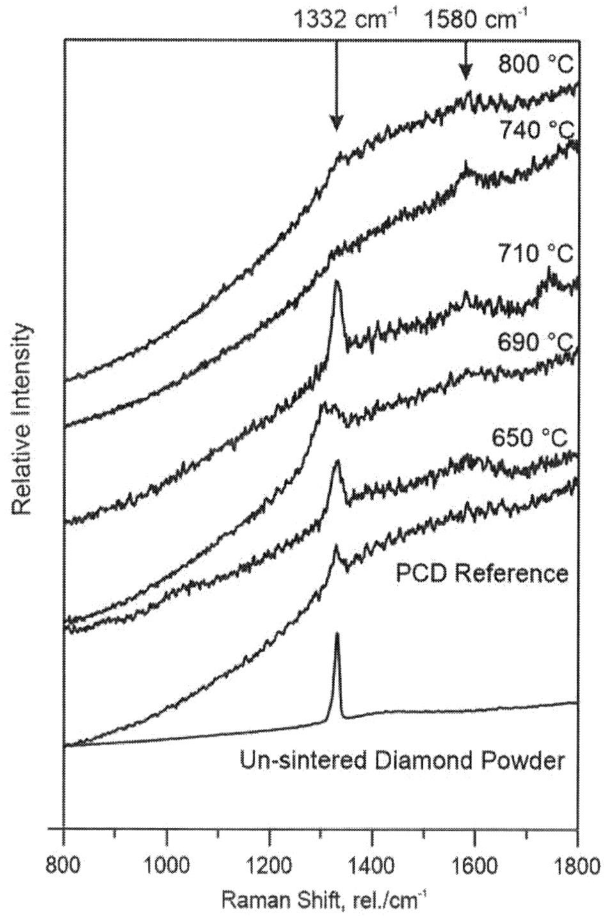

Figure 2: Raman spectra showing PCD graphitization as a function of brazing temperature

Three brazing parameter combinations were chosen for comparison and microstructural investigation (740°C - 10 min, 690°C - 20 min and 650°C - 45min). A process temperature of 740 °C is well above the liquidus temperature of the braze alloy and within the brazing temperature range recommended by the manufacturer of the alloy. During brazing at this temperature the whole alloy is in the liquid state. As can be seen in Figure 3a the braze alloy flows on the carbide surface while simultaneously wetting the PCD and the WC. The samples showed the formation of pronounced fillets. After reducing the brazing temperature to 690 °C the fillets were smaller and less pronounced (Figure 3b). The alloy did not flow on the surface of the WC and as a result the area covered with braze alloy did not change during brazing. This is due to the fact that the process temperature of 690 °C is just below the liquidus temperature of Incusil-ABA. Thus, the braze alloy is only partly liquid and much more viscous compared to 740 °C. At a brazing temperature of 650 °C, which is far below the liquidus temperature of the alloy, no spreading and fillets formation was observed (Figure 3c). The braze foil remained almost unchanged in dimension and appearance before and after brazing. However, a sound joint between PCD and WC could be achieved.

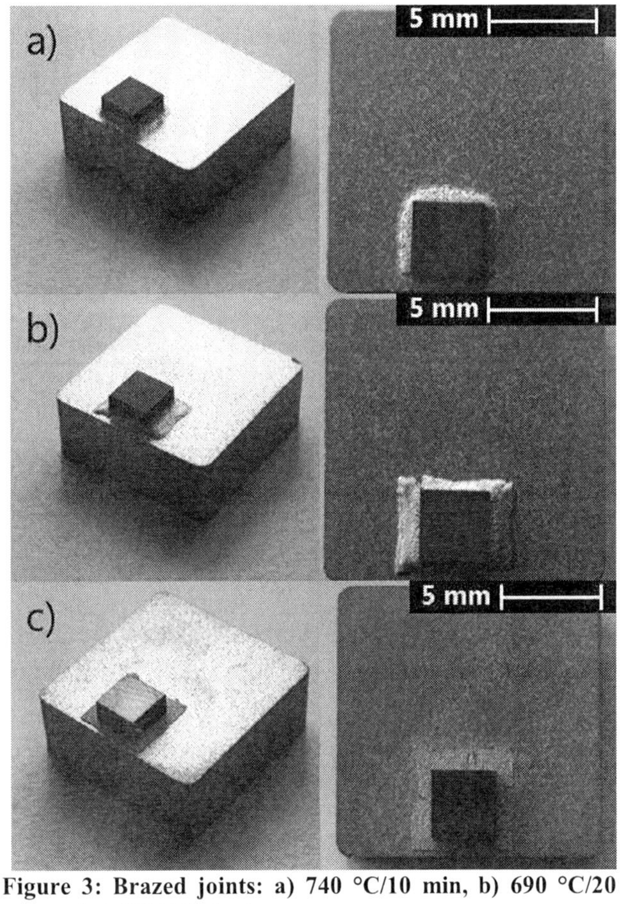

Figure 3: Brazed joints: a) 740 °C/10 min, b) 690 °C/20 min and c) 650 °C/45 min.

In addition to the macroscopic differences resulting from the varying brazing parameters also microstructural change were found. Figure 4 shows the microstructure obtained after joining PCD to WC at 740 °C for 10 minutes. The WC platelet is located at the bottom of the picture while the PCD is visible on the top. During the joining process the braze alloy penetrated the PCD to a certain extent which led to a roughening of the interface. Within the brazing gap two main phases are present which are Cu-rich or Ag-rich, respectively. The Cu-rich phase has a composition of 90 at.-% Cu and some additions of Ag, In, Ti and Co. This phase appears in big and almost continuous grains, which are mainly oriented towards the braze-substrate interfaces. Around the centerline of the joint the Ag-rich phase was mainly found with a composition of 76Ag-15In (at.-%) and some dissolved Cu and Ti. In addition to that, particles with a darker contrast appeared in the micrograph of Figure 4. These particles were too small to be measured via EDS precisely; nevertheless the EDS suggested that they are Ti and Co-rich. For a brazing temperature of 740 °C the particles are distributed over the whole brazing gap width. In contrast to that the Ti/Co-rich phase was only found close to the WC-braze interface when the joining temperature was reduced to 690 °C as can be seen in Figure 5. Compared to the microstructure of the joint performed at 740 °C the grain growth of the Cu-rich phase is also less pronounced. Distinct grains of different size were detected in the micrograph of Figure 5 and showed the tendency to accumulate at the interfaces. The grains of Cu solid solution were embedded in a matrix of a Ag-rich compound. This Ag-rich compound had a comparable composition (78Ag-12In-Cu at.-%) as already known from the analysis of the joint done at 740 °C.

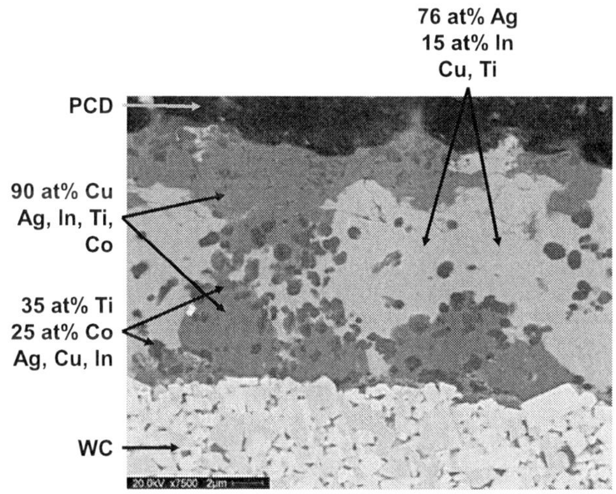

Figure 4: Microstructure of PCD brazed to WC at 740 °C for 10 minutes

No extensive reactions of the braze alloy with the base materials could be found in the microstructural analysis of the joints produced at 650 °C for 45 minutes (Figure 6). A small reaction layer was observed at the WC-braze interface, which is presumably Ti-Co rich. As Figure 6 shows no Ti/Co particles formed during the joining process. The brazing gap consisted of a Cu-rich phase (96Cu-Ag-In) and a Ag-rich phase (78Ag-13In-Cu) with a composition comparable to the phases already known form the joints at 690 and 740 °C. The Cu solid solution appeared as distinct grains with no indications for accumulation.

In addition to the microstructural analysis shear test were performed to evaluate the mechanical properties of the joints. The results of the shear tests are presented in Figure 7 and Table 1. Besides the brazing temperatures mentioned above

(650 °C, 690 °C and 740 °C) joints for the mechanical tests were also prepared at 710 °C and 800 °C, respectively. Overall the highest shear strength could be achieved for the joints which were brazed at 740 °C for 10 minutes. The average was determined to 281 MPa with single values between 200 and 329 MPa. Subsequent fracture surface analysis revealed that fracture occurred partly in the filler alloy and partly at the filler alloy PCD interface. Thus, it was concluded that the fracture mode in this case was cohesive-adhesive.

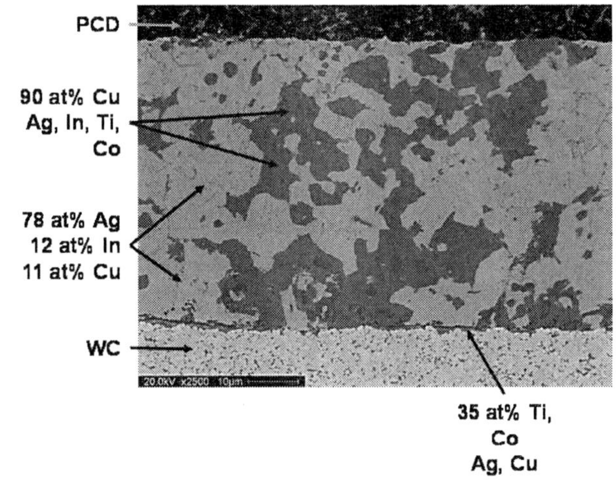

Figure 5: Microstructure of PCD brazed to WC at 690 °C for 20 minutes

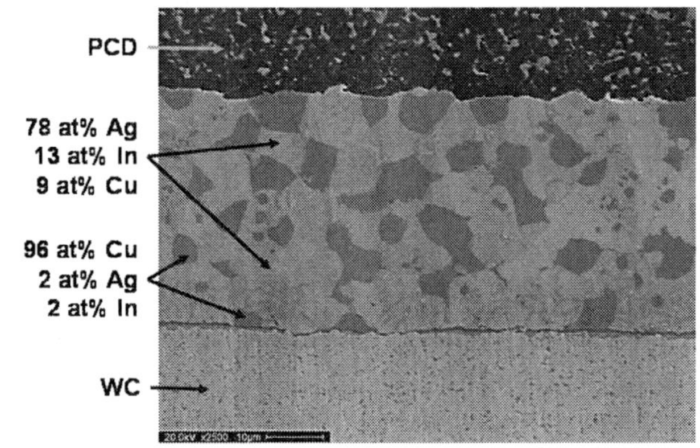

Figure 6: Microstructure of PCD brazed to WC at 650 °C for 45 minutes

Shear strength values in a comparable range could be achieved with specimens brazed at 690 °C and 710 °C, respectively. For the brazing process at 710 °C the mean shear strength was 269 MPa which is 12 MPa lower compared to the value achieved at 740 °C. A further decrease of the brazing temperature to 690 °C led to a slight decrease of the shear strength as well. The average strength of the joints was 258 MPa with a standard deviation of 47 MPa. Figure 8 shows a typical fracture surface for the joints at 690 and 710 °C. Fracture occurred close to the braze PCD interface in a cohesive-adhesive manner. However, most of the filler alloy was found on the WC platelets and the adhesive fracture was predominant.

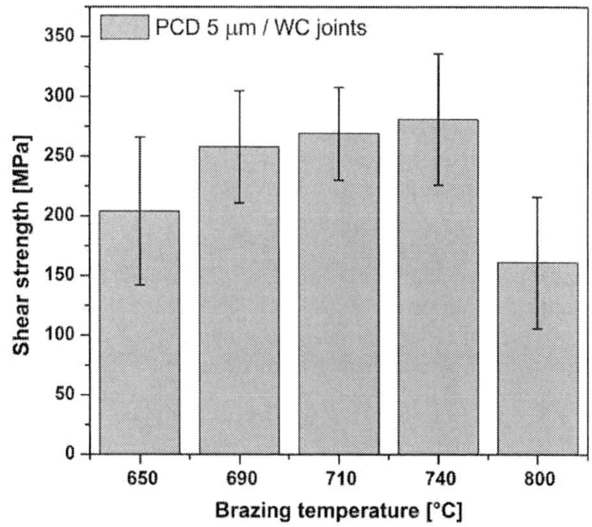

Figure 7: Shear test results of joints produced at different process conditions

Table 1: Shear test results – average and deviation

Temperature - Time	Avg. Strength MPa ± SDEV	Min. [MPa]	Max. [MPa]
800°C - 10 min.	161 ± 55	114	233
740°C - 10 min.	281 ± 55	200	329
710C° - 10 min.	269 ± 39	204	305
690°C - 20 min.	258 ± 47	224	336
650°C - 45 min.	204 ± 62	135	282

When the brazing temperature was set to 650 °C, far below the liquidus temperature of Incusil-ABA, an average of 204 MPa was achieved, which is 71% of the maximum bond strength of the specimens brazed at 740 °C. Single test values were in the range between 135 and 282 MPa. The majority of the joints broke cohesively in the filler alloy. Figure 9 shows typical fracture surfaces where both substrate materials are covered with braze after the mechanical test.

An increase in brazing temperature over 740 °C did not show a positive effect on the mechanical properties of the joints. Increasing the process temperature to 800 °C led to a decrease of shear strength by 120 MPa. The average strength was 161 MPa with a standard deviation of 55 MPa. This is the lowest value among the tested process conditions. For this type of samples the predominant fracture mode was the PCD fracture. Either the PCD splintered during testing or fracture occurred in the PCD close to the PCD-braze interface.

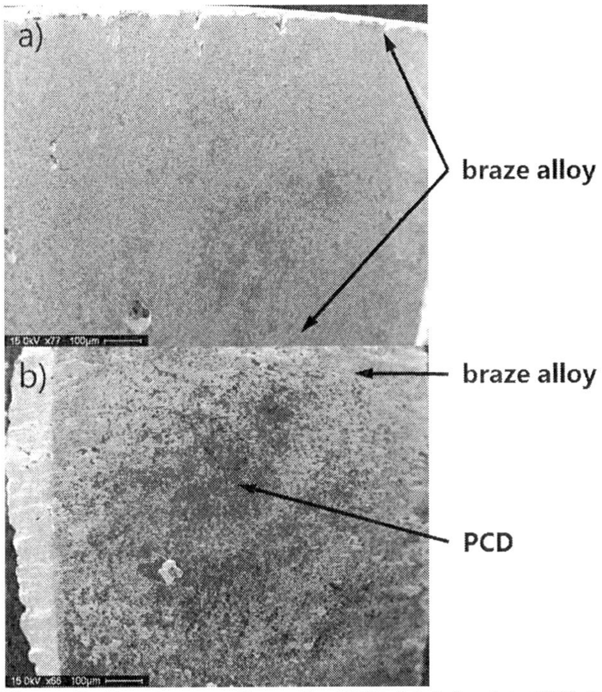

Figure 8: Fracture surface of a PCD (a) joined to WC (b) at 690 °C for 20 minutes

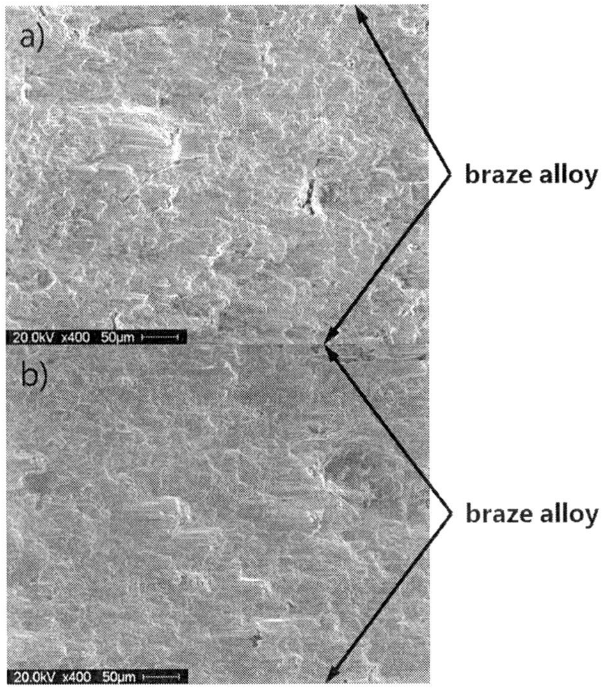

Figure 9: Fracture surface of a PCD (a) joined to WC (b) at 650 °C for 45 minutes

Discussion

The lowest shear strength was detected for samples brazed at 800 °C for 10 minutes. Since the parts mainly fractured in the PCD it was concluded that the PCD is damaged at this brazing temperature. The integrity of the PCD can harmed at high process temperatures by an agglomeration of the Co binder phase. Consequently, the cohesion of the sintered diamond crystals is no longer sufficient. In addition to that an infiltration of the PCD by the braze alloy and a strong reaction may negatively influence the stability of the PCD and therefore of the joint. A combination of the mentioned effects led to the lowest shear strength compared to the other brazing conditions.

On the contrary, the highest strength was achieved for the specimens brazed at 740 °C. In this case the microstructural analysis already showed a reaction of the filler alloy with that PCD, also indicated by the formation of Co-Ti particles. Good bonding of the filler to the substrates could be obtained leading to a mixed (cohesive-adhesive) fracture behavior. Moreover, the formation of big almost continuous Cu-rich grains was beneficial for the mechanical properties of the joints. Since the Cu-phase is ductile it is able to compensate local peak shear stresses.

For the samples joined at 710 °C and 690 °C the shear strength value were in the same range, but a bit lower compared to a brazing temperature of 740 °C. From the microstructure it was already known that the reaction of the braze alloy with the substrate was less pronounced at lower brazing temperature. Less Co-Ti particles were found and also the Cu-rich grains were smaller and more distinct. The weaker interface reaction consequently led to fracture behavior dominated by adhesive fracture. In addition to that the smaller Cu-rich grains cause less stress compensation leading to the lower shear strength.

The microstructural investigation of the samples brazed at 650 °C for 45 minutes did not show any obvious indication for interface reactions as well as for the formation of Co-Ti particles. Nevertheless, reasonable shear strength values were measured superior to the ones achieved at a brazing temperature of 800 °C. Fracture did not occur at one of the filler-substrate interfaces but within the filler alloy. This result shows that sufficient interface reactions have taken place which provided a good bonding of the braze alloy to the substrates. Since the alloy did not melt completely at 650 °C the cohesion of the metal particles in the braze layer was weaker than the adhesion to the substrates leading to the described fracture behavior and the lower shear strength.

Conclusions

It was shown that it is possible to braze PCD to cemented WC using Incusil-ABA braze alloy. Different process conditions and their influence on graphitization, microstructure and mechanical properties were studied. Only limited graphitization was detected by RAMAN spectroscopy for at brazing temperatures of 740 °C and above. An increase in brazing temperature caused stronger (interface) reactions and a more pronounced formation of Co-Ti particles. With increasing process temperature also the shear strength increased up to a threshold at 740 °C. Exceeding 740 °C led to a significant decrease of shear strength due to a damage of the PCD. Although reasonable bond strength could be achieved via brazing below the liquidus temperature of the braze alloy, overall 740 °C gave the best results and was recommended as a brazing temperature for brazing PCD to WC with Incusil-ABA filler alloy. Overall, it was demonstrated that under specific brazing conditions high strength joints can be achieved without harming the PCD grade properties.

References

[1] Buhl, S. *et al*, "Influence of the brazing parameters on microstructure, residual stresses and shear strength of diamond-metal joints," *Journal of Materials Science*, Vol. 45 (2010), pp. 4358-4368.

[2] Li, W.C. *et al*, "Interfacial segregation of Ti in the brazing of diamond grits onto a steel substrate using a Cu-Sn-Ti brazing alloy," *Metallurgical and Materials Transactions A*, Vol. 33, No. 7 (2002), pp. 2163-2172.

[3] Chattopadhyay, A.K. *et al*, "Induction brazing of diamond with Ni-Cr hardfacing alloy under argon atmosphere," *Surface and Coating Technology*, Vol. 45, No. 1-3, pp. 293-298.

[4] Yamazaki, T. *et al*, "Role of the reaction product in the solidification of Ag-Cu-Ti filler for brazing diamond," *Journal of Materials Science*, Vol. 33, No. 5, pp. 1379-1384.

[5] Hung, N.P. *et al*, "Cutting tools for metal matrix composites," *Key Engineering Materials*, Vol. 138-140 (1998), pp. 289-325.

[6] Teli, R, "Machining of Composite Materials," *CIRO Annals – Manufacturing Technology*, Vol. 51, No. 2 (2002), pp. 611-634.

[7] Katzman, K. *et al*, "Sintered diamond compacts with a cobalt binder," *Science*, Vol. 172, No. 3998 (1971), pp. 1132-1134.

[8] Akaishi *et al*, "Synthesis of fine-grained polycrystalline diamond compact and its microstructure," *Journal of the American Ceramic Society*, Vol. 74, No. 1 (1991), pp. 5-10.

[9] Mehan, R.L. *et al*, "Thermal degradation of sintered diamond compacts," *Journal of Materials Science*, Vol. 24, No. 3 (1989), pp. 942-950.

[10] Westraadt, J.E. *et al*, "Characterization of thermally degraded polycrystalline diamond," *International Journal of Refractory Metals and Hard Materials*, Vol.48 (2015), pp. 286-292.

Packaging of the Fiber Bragg Grating Sensors Using Brazing Methods

Yulong Li, Changjin Wen, Xiaowu Hu and Ronghua Hu
Key Lab for Robot & Welding Automation of Jiangxi Province, Mechanical & Electrical Engineering School, Nanchang University, Nanchang city, Jiangxi province, PR China
Corresponding author: Yulong Li, Email: liyulong1112ster@gmail.com, Telephone: 13657091860

Abstract

A fiber Bragg grating (FBG) is a distributed Bragg reflector constructed in a short segment of optical fiber that reflects particular wavelengths of light and transmits all others. It can be used as a sensor for monitoring the change of the temperature, stress and strain of a structure. However, there are some difficulties for the applications of bare FBGs in a harsh environment due to the fragile characteristics.

In this paper, bare fiber and FBG sensors were chemical plated and electroplated firstly, and then packaged on the surface of metal substrates by using brazing methods in the controlled atmosphere. Carbon steel and AlSi eutectic alloy was used as the substrate and filler, respectively. A finite element model on the FBG smart structure was used to analyze and determine the response of the surface brazed FBG to the thermal load during the brazing process and subjected to load after metallization. Simulated results show: the Bragg wavelength of surface brazed FBG shifted down ~ 10 nm during the brazing process; when one side of the structure subjected to uniformly distributed load, the Bragg wavelength shifted accordingly, and the strain sensitivity was 1.3 pm/με.

Introduction

Smart structures have the characteristics of inherent response to the internal and external changes, which consist of sensing element, driving element, control system and substrates [1]. Fiber Bragg grating (FBG) sensors are regarded as one of the potential ideal carrier for information transmission and sensing because of their advantages of immunity to electromagnetic interference, long-term stability, high multiplexing capacity, small size and light weight [2-5]. However, there are some difficulties for the applications of the bare FBGs in a harsh environment due to the fragile characteristics. Therefore, packaging FBGs onto the surface or into the inside of the engineering parts is difficult. After the packaging, FBG based smart structures with the ability of detecting changes and immunity to the harsh environment are obtained.

At present, the FBG smart structures in applications are mainly based on the composite materials [6-9], although a few are based on the metal materials. FBG smart metal structures got a few researches may be attributed to the reasons as follows: FBGs bonded on the metal structures have the ability of detecting temperature and strain, while the sensing suffers from the lifetime of the packaging materials/adhesives; during embedding into metal parts, the FBGs are affected by the high temperature and thermal stresses. Despite that, Marcus Perry et al. [10] embedded FBGs into ferrules with high-temperature fillers by using induction brazing method. However, they showed little concentration on the wavelength change of FBGs during the packaging process. Mathias S. Müller et al. [11] had the metallic coated FBGs surface bonded on a cantilever beam with an adhesive solder. They investigated wavelength change of the FBG during cooling process of the adhesive with a high Young's modulus, while the processing temperature of the soldering in their work was as low as 200 °C.

In our previous work [12], FBGs have been satisfactorily metallic encapsulated by a two-step method, after that, the metallized FBGs have been successfully embedded into 42CrMo steel and 1100 aluminum foils by using soldering and ultrasonic welding methods, respectively. In this paper, FBG sensor was chemical plated and electroplated firstly, and then packaged on the metal substrate by using brazing methods. An attempt to investigate the Bragg wavelength change of FBGs during the brazing process, a finite element model was established. The model was used to analyze and determine the response of the surface brazed FBG to the thermal load during the brazing process and subjected to load after metallization.

Packaging of FBGs

The sensitivity of FBG should be enhanced; also, the FBG should be protected encapsulation before application. In the previous work, a chemical plating and electroplating method for metallization packaging of FBG was presented. The chemical-plating coating on the FBG was so thin that the sensitivity of the FBG was only enhanced a little; further, the FBG with a thin coating could not bear the high temperature during the brazing embedding or surface mounting process. However, the chemical plating coating could be used as the conductive layer for the following electroplating process. After electroplating, the FBG was well protected and the sensitivity of the FBG was markedly enhanced due to the thick metal coat.

In this study, FBGs were mounted on the surface of the metal parts by using brazing methods in a controlled atmosphere.

Figure 1 shows the macroscopic cross-section of the surface mounted FBG, a steel substrate and AlSi filler were used.

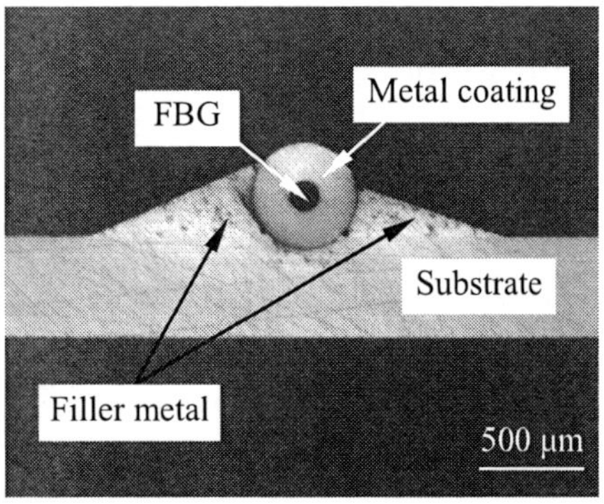

Figure 1 Macroscopic cross-section of the surface mounted FBG

Simulations

Simulation on FBG Smart Structure During the Brazing Process

An attempt to investigate the Bragg wavelength change of FBG during the brazing process, a finite element model was established in Cartesian coordinate, as shown in Figure 2. In the model, the materials for simulation analysis were listed as follows: silica glass of FBG, nickel plating coating, AlSi12 eutectic alloy, 316L stainless steel. The simulation analysis was based on the experimental assumption that the brazing was conducted in the controlled atmosphere, the temperature of which decreases from 600 °C to 25 °C with the cooling rate of 25 °C/s, as well as the convective heat transfer coefficient of 2×10^{-5} W/(mm^2·°C).

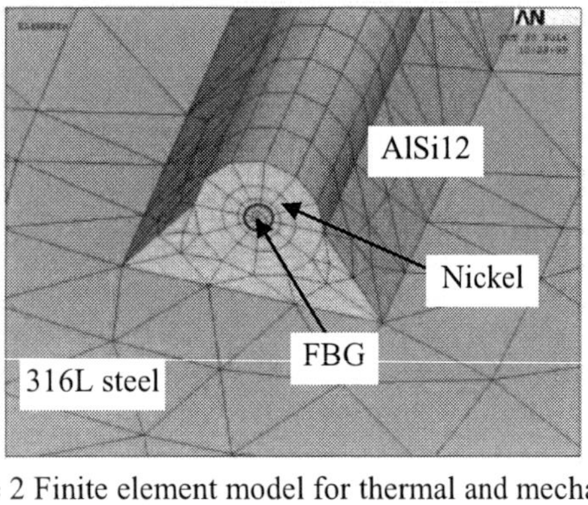

Figure 2 Finite element model for thermal and mechanical simulation

The sequential thermal and mechanical analysis method was required in the simulation. For the thermal analysis, the 3D thermal SOLID 87 and SOLID 90 elements were applied. After the switch from the thermal elements to mechanical elements, the corresponding 3D mechanical SOLID 187 and SOLID 186 elements were applied to determine the strain. Figure 3 presents the simulated von Mises strain diagram of the 20 times magnified deformation. Obviously, the contraction of AlSi12 eutectic alloy shows larger than that of other materials. Figure 4 shows strain distributions diagram along the axial direction of FBG, and the axial and radial strain are represented by $\varepsilon_z = -9.244\times10^{-3}$ and $\varepsilon_r = \sqrt{\varepsilon_{r1}^2 + \varepsilon_{r2}^2} = -5.02\times10^{-3}$, respectively.

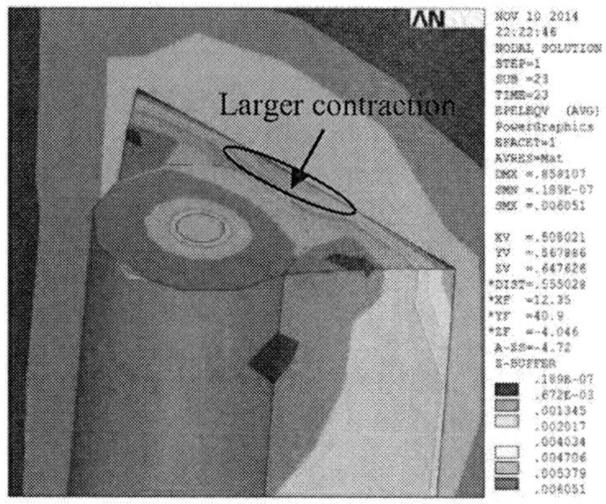

Figure 3 Simulated deformed von Mises strain diagram

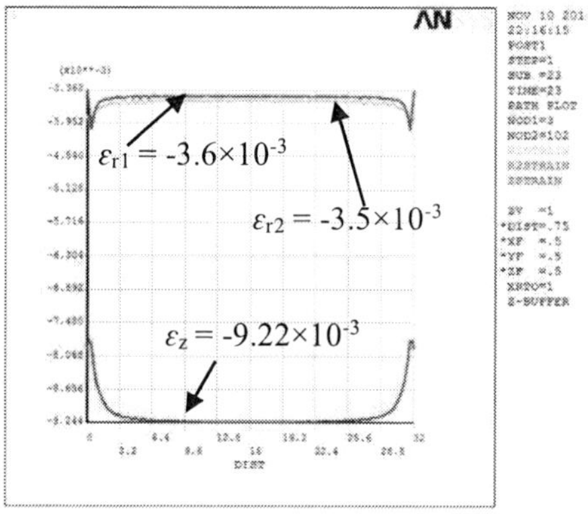

Figure 4 Strain distributions along the axis of FBG after brazing

The refractive spectral shift of the Bragg reflection wavelength due to elastic strain is given as [5]:

$$\Delta\lambda_B = \lambda_B\{\varepsilon_z - n_{eff}^2/2[(p_{11}+p_{12})\varepsilon_r + p_{12}\varepsilon_z]\} \quad (1)$$

where n_{eff} is the effective refractive index, p_{11} and p_{12} are the elastic-optic coefficients, ε_z and ε_r represent the axial and radial strain, respectively.

According to the Equation (1), with the parameters values of λ_B = 1550 nm, n_{eff} = 1.456, p_{11} = 0.121, p_{12} = 0.27, ε_r =-5.02×10^{-3} and $\varepsilon_z = -9.244\times10^{-3}$ applied, the wavelength value shifted, $\Delta\lambda_B$ = -7 nm.

Simulation on FBG Smart Structure Subjected to Load

In order to analyze and determine the response of the FBG smart structure to load, the strains and the corresponding Bragg wavelength change of FBG subjected to load was investigated assisted by the established model simulation. The one side of the structure was fully-constrained, and the opposite side was under uniformly distributed load of 1 MPa, as illustrated in Figure 5. Accordingly, Figure 6 presents strain distributions diagram along the axial direction of FBG subjected to load, and the axial and radial strain are represented by $\varepsilon_z = 8.5\times10^{-7}$ and $\varepsilon_r = \sqrt{\varepsilon_{r1}^{\,2} + \varepsilon_{r2}^{\,2}} = -5.2\times10^{-7}$, respectively. Correspondingly, the total strain is calculated, $\varepsilon = \sqrt{\varepsilon_z^{\,2} + \varepsilon_r^{\,2}} = 9.96\times10^{-7}$.

According to the Equation (1), with applying the parameters values shown above, the wavelength value shifted, $\Delta\lambda_B = 1.3$ pm. Correspondingly, the strain sensitivity is calculated, $K_\varepsilon = \Delta\lambda_B / \varepsilon = 1.3$ pm/με. Therefore, the simulation study presents the feasibility of the structure for determining of the load and strain change.

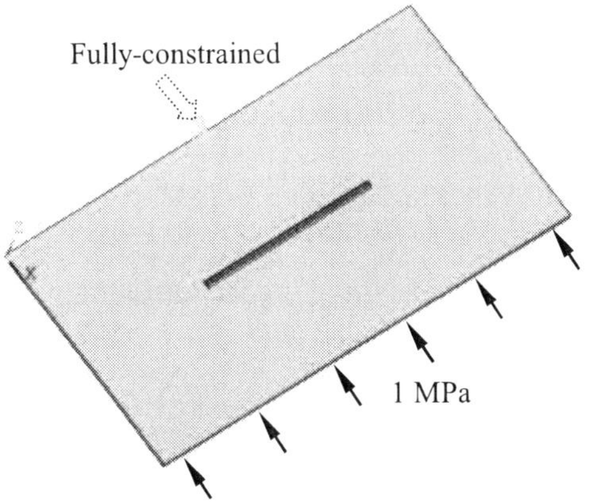

Figure 5 Schematic of model subjected to load

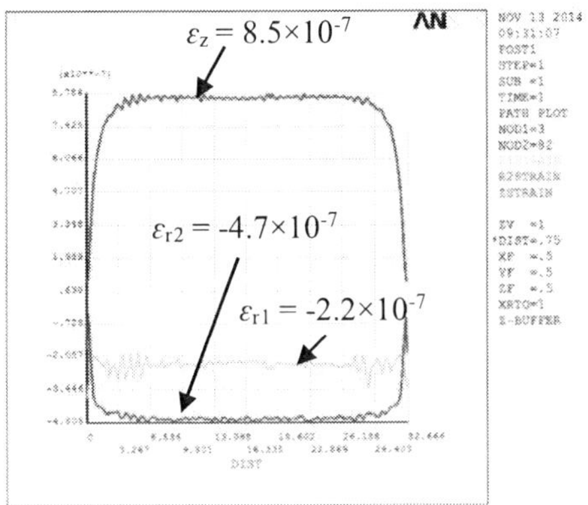

Figure 6 Strain distributions along the axis of FBG subjected to load

Conclusions

The bare fiber and FBG sensors were chemical plated and electroplated firstly, and then packaged on the metal substrate by using brazing methods in the controlled atmosphere. A finite element model of FBG smart structure was established, which was used to analyze and determine the Bragg wavelength change of FBG during the brazing process as well as subjected to load after metallization. Simulated results show: during the brazing process, the Bragg wavelength shifted down ~10 nm; when one side of the structure subjected to uniformly distributed load, the Bragg wavelength shifted accordingly, and the strain sensitivity was 1.3 pm/με.

Acknowledgments

This work was supported by National Natural Science Foundation of China (No.51265035), Scientific Research Foundation for the Returned Overseas Chinese Scholars, State Education Ministry, Natural Science Foundation of JiangXi province (20132BAB206024).

References

[1] Robison, R., "Smart Structures," Civil Eng, Vol. 62, No. 11 (1992), pp. 66-68.

[2] Rao, Y. J., "Recent Progress in Fiber-optic Extrinsic Fabry-Perot Interferometric Sensors," Opt Fiber Technol, Vol. 12, No. 3 (2006), pp. 227-237.

[3] Prakash, O. et al, "Effect of the Spatial Coherence of Ultraviolet Radiation (255 nm) on the Fabrication Efficiency of Phase Mask Based Fiber Bragg Gratings," Opt Commun, Vol. 263, No. 1 (2006), pp. 65-70.

[4] Colpo, F. et al, "Characterisation of Residual Stresses in a Single Fibre Composite with FBG Sensor," Compos Sci Technol, Vol. 67, No. 9 (2007), pp. 1830-1841.

[5] Majumder, M. et al, "Fibre Bragg Gratings in Structural Health Monitoring-Present Status and Applications," Sensor Actuat A-Phys, Vol. 147, No. 1 (2008), pp. 150-164.

[6] Wnuk, V. P. et al, "Process for Mounting and Packaging of Fiber Bragg Grating Strain Sensors for Use in Harsh Environment Applications," Proc. of SPIE, San Diego, CA, Mar. 2005, pp. 46-53.

[7] Seongmin, J. et al, "Enhanced Sensitivity of the FBG Temperature Sensor Based on the PbO-GeO2-SiO2 Glass Optical Fiber," J Lightwave technol, Vol. 28, No. 18 (2010), pp. 2697-2700.

[8] Wen, Q. Z. et al, "Design and Synthesis of a Packaging Polymer Enhancing the Sensitivity of Fiber Grating Pressure Sensor," Prog Nat Sci-Mater, Vol. 18, No. 2 (2008), pp. 914-917.

[9] Mizunami, T. et al, "High-sensitivity Cryogenic Fibre-Bragg-grating Temperature Sensors Using Teflon Substrates," Meas Sci Technol, Vol. 18, No. 2 (2008), pp. 197-200.

[10] Perry, M. et al, "Induction Brazing of Type-I Fiber Bragg Gratings into Kovar Ferrules Exploiting Curie Transition," IEEE Sens J, Vol. 12, No. 7 (2001), pp. 816-823.

[11] Müller, M. S. et al, "Soldering Fiber Bragg Grating Sensors for Strain Measurement," Proc. of SPIE, Perth, Australia, Apr. 2008, pp. 70040B-70040B-4.

[12] Li, Y. L. et al, "Metal Coating of Fiber Bragg Grating and the Temperature Sensing Character after Metallization," Opt Fiber Technol, Vol. 15, No. 4 (2009), pp. 391-397.

Evaluation of Low-temperature Amorphous Foils for Brazing Titanium and Ceramics

Alexander E. Shapiro
Titanium Brazing, Inc., Columbus, OH
ashapiro@titanium-brazing.com

Yury Flom
Goddard Space Flight Center, NASA, Greenbelt, MD
yury.a.flom@nasa.gov

Abstract

New brazing filler metals in the form of amorphous foils 50 microns thick, and 75 mm wide, were evaluated for vacuum brazing of titanium alloys, graphite, and ceramics. The following amorphous foils were tested: (a) TiBraze®200 (Ti-20Zr-20Cu-20Ni wt.%), having the melting range 845-863°C (1553-1585°F), (b) TiBraze®590 (Zr-17Ti-20Ni-1Hf wt.%), having the melting range 796-813°C (1465-1495°F), and (c) TiBraze®800 (Zr-14.7Ti-12.6Ni-7Cu-1Hf wt.%) having the melting range 772-786°C (1422-1447°F).

Shear strength of brazed joints of CP titanium and Ti-6Al-4V alloy and microstructures of joints studied after brazing at different temperatures are reported. Brazing temperatures of titanium joints with amorphous foils are significantly lower than that of standard filler metal AWS BTi-1 (Ticuni®).

Brazing of titanium-to-ceramics and graphite with new foils should be done at temperatures greater by 50-60°C than that of brazing titanium-to-titanium. If compared to traditional active brazing filler metals that do not spread on ceramic surfaces, the tested Ti- and Zr-based amorphous foils revealed excellent wetting, and sometimes spreading, along ceramic surfaces of alumina and boron nitride, as well as graphite. Also, all tested amorphous foils provided good quality of brazed samples of fin-plate titanium heat exchangers and honeycomb panels.

Introduction

Three titanium-based filler metals dominate today in the brazing industry for joining titanium heat exchangers, honeycomb panels, tooling, spacecraft parts, etc. : BTi-1 (Ti-15Cu-15Ni wt.%) in the form of cladded thin foil, BTi-5 (Ti-20Zr-20Cu-20Ni wt.%) in the form of powder, and BTi-3 (Ti-37.5Zr-10Cu-15Ni wt.%) in the form of powder or sintered thick foil [Ref. 1-5]. BTi-1 provides a high strength of brazed joints but it requires a brazing temperature above 950°C (1742°F) and that significantly exceeds the $\alpha \rightarrow \beta$ transus of all titanium base materials, while brazing alloys BTi-3 and BTi-5 have brazing temperatures below or equal to $\alpha \rightarrow \beta$ transus of titanium base alloys but they are not supplied in the form of a thin foil [Ref. 2-4].

The BTi-5 (TiBraze®200) and two new zirconium-based filler metals TiBraze®590 and TiBraze®800 are now manufactured and supplied in the form of amorphous foils having thickness in the range of 40-50 microns and width up to 75 mm (3 inches). Both zirconium-based filler metals provide brazing below the $\alpha \rightarrow \beta$ transus of α- and ($\alpha+\beta$)-titanium base alloys, and TiBraze®800 has a melting range even below near-β and β-titanium base alloys. This presents an opportunity to manufacture brazed parts from such alloys as Timetal 21S and Ti-3Al-10V-2Fe without affecting mechanical properties of base materials. Also, these foils resolve the problem of furnace brazing of titanium to copper, because brazing temperatures are much lower than the titanium-copper eutectic point 885°C (1625°F).

These homogeneous and ductile foils have excellent wetting and melting characteristics (Table 1) and good compatibility not only with titanium or zirconium alloys but also with ceramics, graphite, and carbon-carbon composites. When used in a preplaced preform, these brazing materials supply a minimum and accurate amount of filler metal in the joint clearances, resulting in joint mechanical properties unattainable for filler metals in powder or sintered strip forms.

Table 1 Amorphous foils for brazing titanium and ceramics

Brazing Filler Metal	Chemical Composition, wt.%	Melting temperature range,	
		°C	°F
TiBraze®200	Ti-(19-21)Zr-(19-21)Cu-(19-21)Ni	845-863	1553-1585
TiBraze®590	Zr-(16-18)Ti-(19-21)Ni-<1.5Hf	796-813	1465-1495
TiBraze®800	Zr-(14-16)Ti-(12-14)Ni-(6.5-7.5)Cu-<1.5Hf	772-786	1422-1447

These amorphous foils were experimentally evaluated as brazing filler metals in combinations of base materials representing Titanium Grade 2, Titanium Grade 5 (Ti-4Al-4V), stainless steel AISI 304, copper, nickel-plated carbon steel 1018, ceramics (alumina, hexagonal boron nitride, silicon carbide) and graphite. Such characteristics as flow and filling the gap at different brazing temperatures, shear strength and microstructure of brazed joints were analyzed and reported in this paper.

Experimental Procedure

Standard lap steel specimens were used for brazing and testing according to AWS C3.2M/C3.2:2008, *Standard Methods for Evaluating the Strength of Brazed Joints*. The base metal thickness was 0.125" (3.175 mm) and width was 0.5" (12.7 mm). Three to five specimens were fabricated for every combination of the base and braze metals that were subjected to tensile testing to determine shear strengths and evaluate their ductility. Shims of amorphous filler metals were placed between base material parts and brazed in vacuum furnace at 10^{-5}-10^{-6}Torr. Samples were loaded during heating and cooling by dead weights, which provided compression ~250 g/cm^2. Brazing temperatures and times are presented in Table 2.

Model samples of titanium heat exchanger and honeycomb panel were also manufactured by brazing at the same process parameters. Titanium Grade 2 foils were used for plates and Ti-3.5Al-2.5V alloy foil - for the honeycomb core and fins of the heat exchanger samples.

Brazed joints were cut, mounted in KonductoMet® and polished for studying the microstructure of resulting brazed joints. Polished cross-sections of brazed joint were etched for 1-2 min with a water solution containing hydrofluoric acid (3-4 mL) and nitric acid (3-4 mL). Finally, the samples were washed with ethanol.

Using an optical microscope, micrographs of each sample were studied in order to determine the microstructure and joint quality. Images included macro-sections, fillets, interfaces, joint metals, and defects. ESM with EDX analysis for the first-order approximation of phase compositions was used for studying fine structure and phase composition of the joint metal and diffusion zones.

Table 2 Brazing process parameters

Brazing Filler Metal	<u>Metal-to-metal:</u> CP Titanium, Ti-4Al-4V, Ti-3.5Al-2.5V, stainless steel 304, copper	<u>Ceramics:</u> alumina, hBN, SiC, graphite
TiBraze®200	**890-900°C (1634-1652°F)**	**960-970°C (1760-1770°F)**
TiBraze®590	**860-870°C (1580-1598°F)**	**910-920°C (1670-1688°F)**
TiBraze®800	**820-830°C (1508-1526°F)**	**860-870°C (1580-1598°F)**
Holding time	**10 min**	**40 min**

Results and discussion

Brazed joints of titanium-to-titanium, titanium-to-SS304, and titanium-to-copper are shown in Figures 1-3, titanium-to-alumina and titanium-to-SiC, in Figures 4-5, and brazed joint of graphite in Figure 6. Brazed samples of titanium honeycomb panel and a fin-plate heat exchanger are shown in Figures 7-8. Average shear strengths of metal-to-metal brazed joints is presented in Table 3. All three brazing filler metals exhibited good wetting of titanium, graphite, and ceramics base materials, and even flow along ceramic surfaces (Figure 4) that is not typical for traditional active filler metals. Overheating and longer holding time when brazing ceramic and graphite (in comparison with metal-to-metal joining) gave better results both in quality and strength consideration of ceramic and graphite brazed joints. Recommended parameters of brazing processes are presented in Table 2.

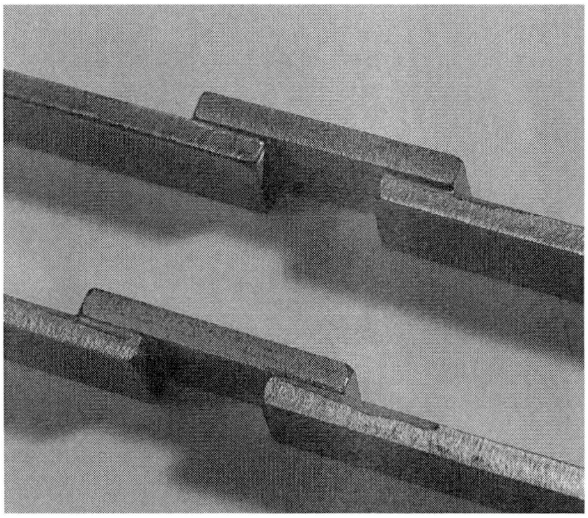

Figure 1 Ti-6Al-4V alloy brazed by TiBraze200

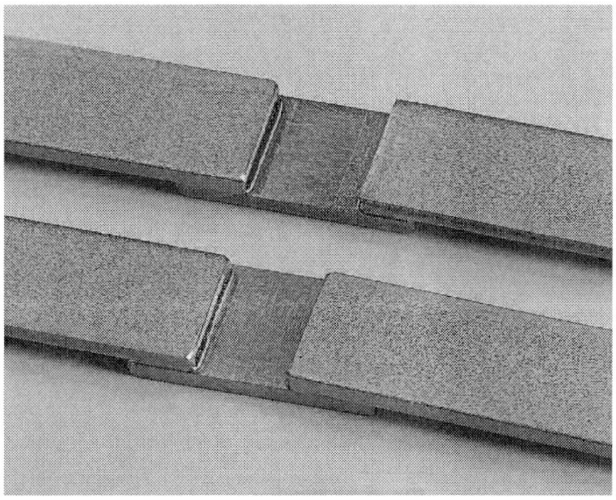

Figure 2 Ti6Al4V alloy brazed to SS304 by TiBraze590

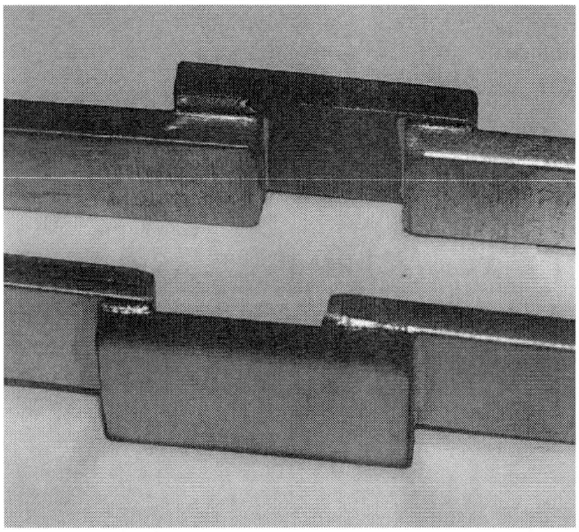

Figure 3 Copper brazed to CP Titanium by TiBraze800

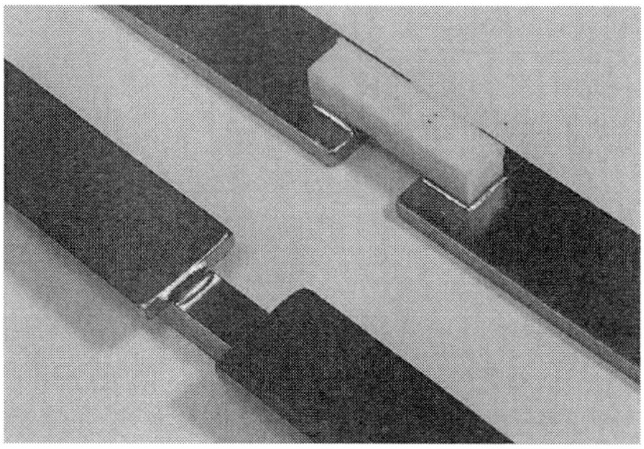

Figure 4 Alumina brazed to Ti-6Al-4V alloy by TiBraze200

Figure 5 Silicon carbide brazed to CP Titanium by TiBraze800

Figure 6 Graphite brazed by TB200

Figure 7 Titanium honeycomb core brazed by TiBraze200

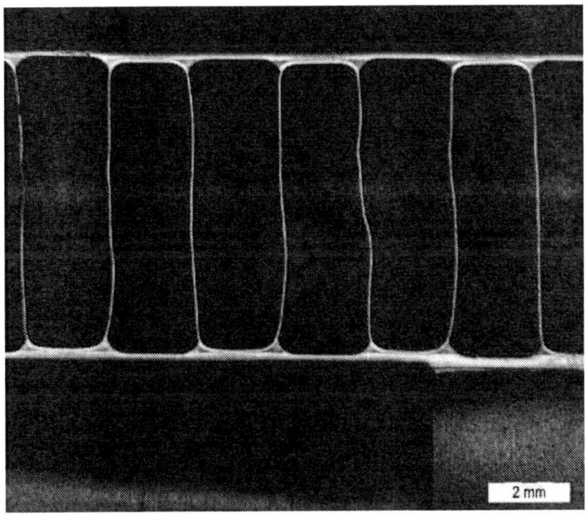

Figure 8 A section of titanium heat exchanger brazed by TiBraze200

We presume that significant differences in strength of brazed joints manufactured with a titanium-based filler metal TiBraze200 and zirconium-based filler metals TiBraze590 and TiBraze800 results from a significant difference in joint metal microstructures formed by these brazing alloys. Structure of the TiBraze200 joint metal is a matrix of 90Ti-4.5Al-2.5Zr wt.% alloy, with small content of other elements (black phase in Fig. 9) well-reinforced with needle-like crystals containing 75Ti-3Al-2Zr-5V-7Ni-5.5Cu wt.% (white needles in Figure 9). As a reinforced matrix, this structure resists crack propagation during mechanical loading of the brazed joint, and the crack occurring in the fillet propagates to the base metal (Figure 10).

Table 3 Shear strength of metal-to-metal brazed joints, MPa

Base metals	TiBraze200	TiBraze590	TiBraze800
CP Titanium (Titanium Grade 2)	216.5-271.2 244.0	200.1-235.4 217.8	166.5-214.4 190.4
Ti-6Al-4V alloy (Titanium Grade 5)	259.4-307.7 276.7	197.7-218.6 208.2	188.3-227.2 207.8
Ti-6Al-4V alloy + SS304	122.5-152.2 137.4	62.5-97.5 80.0	-
CP Titanium + Copper	-	83.8-110.7 97.2	55.9-69.0 62.5
CP Titanium + Ni-plated carbon steel	-	95.6-124.3 110.0	69.9-83.4 76.6

Appearance of significant amounts of vanadium and aluminum in the joint metal, on the one hand, and relatively low content of zirconium, copper and nickel in both matrix and needle-like phases, on the other hand, suggests a very active diffusion exchange of elements between base and braze alloys. Zirconium-based filler metal TiBraze590 forms a two-phase, non-reinforced microstructure of the brazed joint (Figure 11).

This cast structure is typical for brazed joints and it cannot resist crack propagation. The matrix phase is represented by large grains of the 49Zr-25Ti-22Ni wt.% alloy slightly alloyed with aluminum, vanadium and hafnium. Grains of a dark phase are uniformly distributed in the matrix and constitute a high-temperature alloy 72Zr-24Ti-2.5Hf wt.%.

At a critical tensile load, multiple cracks are initiated and propagated along grain boundaries of the matrix phase, between the hard "islands" of the high-temperature phase. Further, it is very likely that crack propagation immediately causes, in the joint metal, a shear strength in the structure that lower in comparison with TiBraze200.

Brazed joints made with zirconium-based amorphous foils also have a wide diffusion zone clearly seen in Figure 11. This zone of Ti-6Al-4V alloy is transformed by zirconium and nickel that diffused into the base metal from the joint to form the composition 72Ti-16Zr-4Ni-3Al-4V wt.%. Just behind this diffusion zone, the base metal contains only 2 wt.% of zirconium and about 1 wt.% of nickel. At the distance about 50 microns, the base metal contains only traces of the brazing alloy components. No intermatallic layers were found at the interface of the base and joint metals that makes titanium brazed joints resistant to brittle failure. Also: wide diffusion zones and the lack of intermetallics characterize titanium brazed joints made with zirconium-based filler metals as the joints having low structural stress concentration, which makes them potentially resistant to dynamic loading.

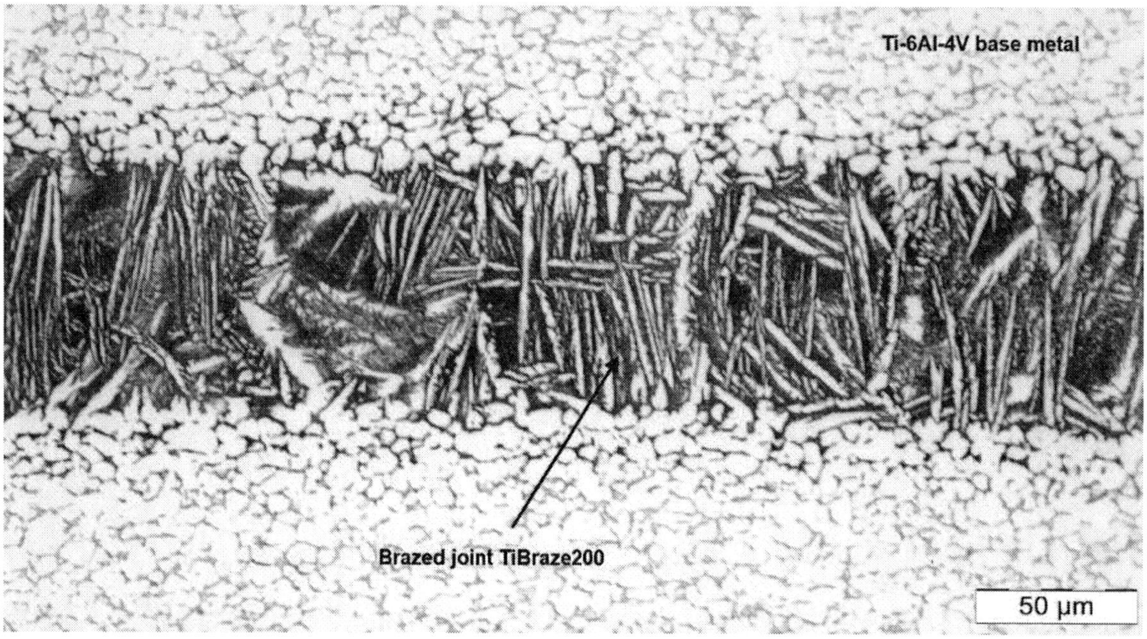

Figure 9 Microstructure of brazed joint of Ti-6Al-4V alloy (Titanium Grade 5) made using TiBraze200 amorphous foil as a filler metal, x500

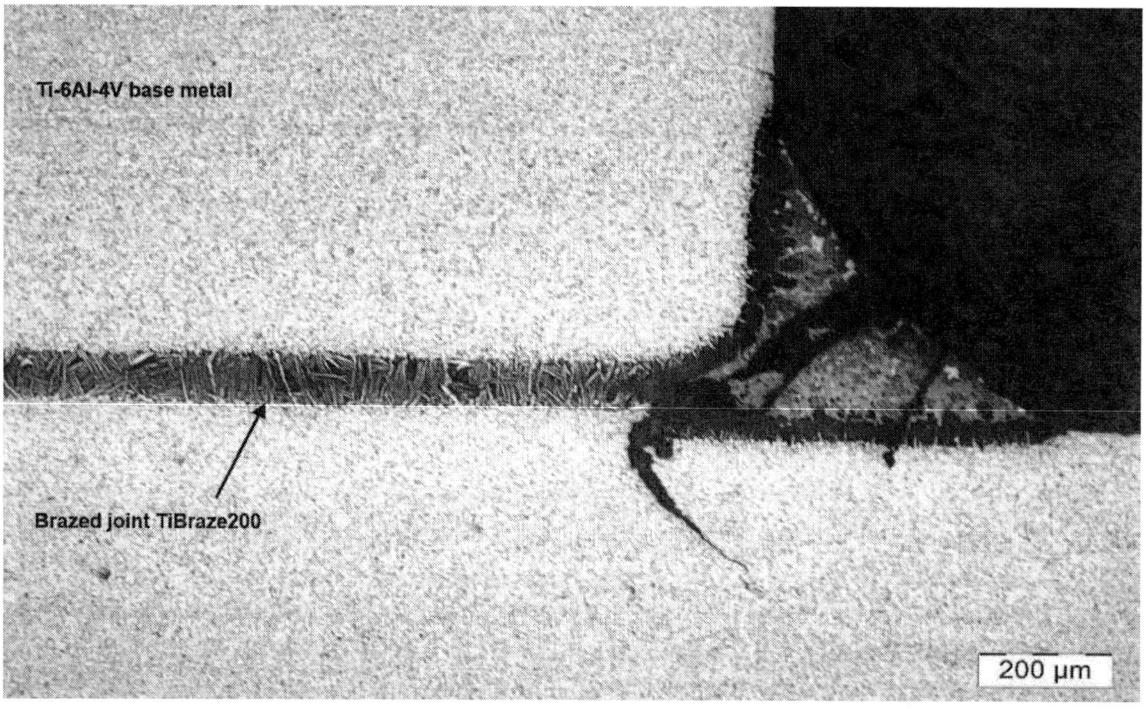

Figure 10 Macrostructure and crack propagation in brazed joint of Ti-6Al-4V alloy (see Figure 9) made using TiBraze200 amorphous foil as a filler metal: x200.

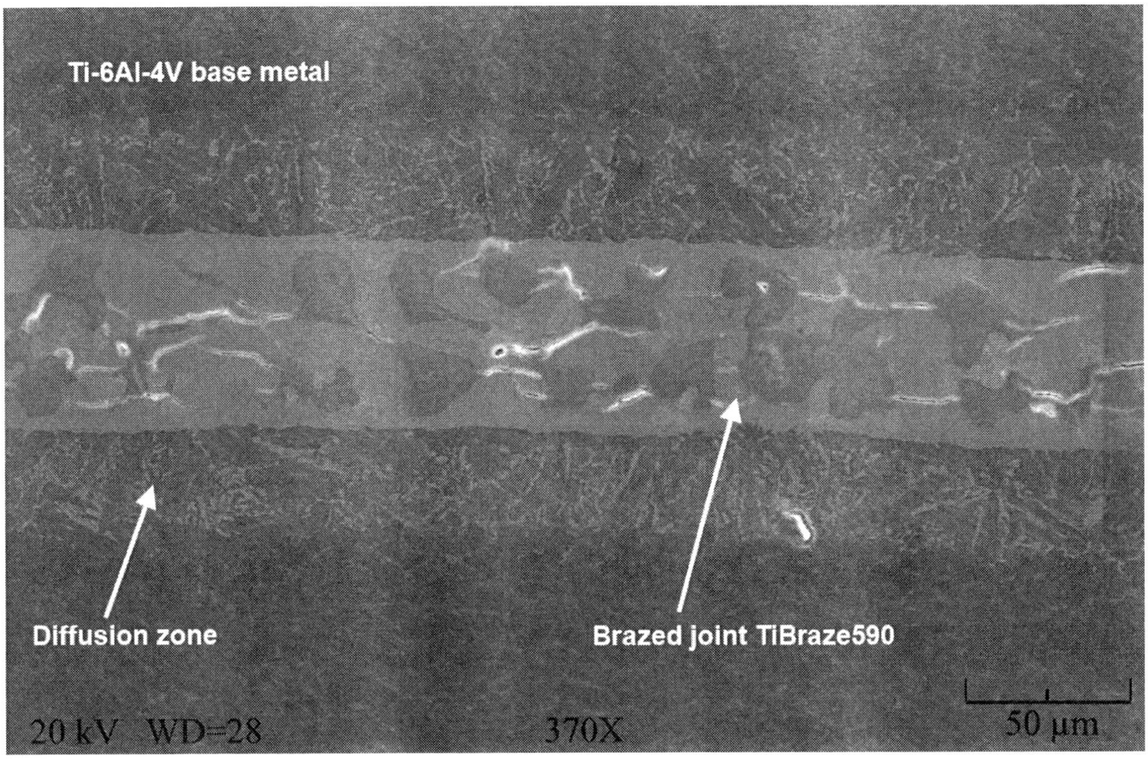

Figure 11 Microstructure of brazed joint of Ti-6Al-4V alloy (Titanium Grade 5) made using TiBraze590 amorphous foil as a filler metal: x370.

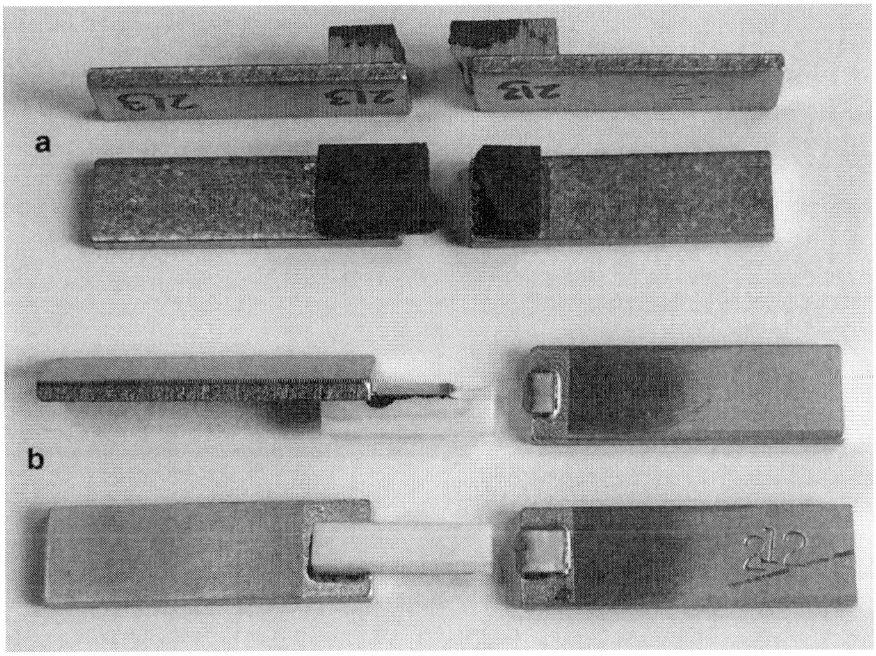

Figure 12 Brazed joints of titanium-to-graphite (a) and Ti6Al4V-to-alumina (b) failed along the graphite and ceramic bodies. Brazing filler metal is TiBraze200.

Shear strength of graphite-to-titanium brazed joints wasn't measured: because all the samples failed in graphite body (Figure 12a), even at the overlap of one-two thickness of the graphite part. Shear strength of ceramic-to-titanium brazed joints also was tested using non-standard "bridge-like" samples (Figures 4 and 5). Most of alumina-titanium, SiC-titanium, and hBN-titanium joints failed along the ceramic body (Figure 12b), not at the joint, which confirms that brittleness of ceramic body played crucial role in testing:---this means that the geometrical design of the samples should be changed. However, these test results at least demonstrate that adhesive of the braze materials tested, to ceramics and graphite, is sufficiently strong to resist significant shear loads, and titanium-to-ceramics brazed joints can respond to the requirements of reliability at larger overlaps. Mechanical testing of ceramic-to-metals brazed joints manufactured with new amorphous foils as brazing filler metals shall continue using modified sample design.

Conclusions

1. New titanium- and zirconium-based amorphous foils provide reliable vacuum brazing in a wide range of temperatures below $\alpha \rightarrow \beta$ transus not only of α - and $(\alpha+\beta)$-titanium base alloys but also near-β and β-titanium base alloys. Where used they result in satisfactory strength in brazed joints of various combinations of base materials and have good compatibility with base materials: specifically, with CP titanium or Ti-6Al-4V alloy, and in joints of dissimilar base material titanium-to-copper, titanium-to stainless steel or to nickel-plated carbon steel, ceramics (alumina, silicon carbide and boron nitride), and graphite. Joints made with the amorphous foils had low erosion of the base metal and high brazed joint quality in titanium heat exchangers and honeycomb panels.

2. Different brazing temperatures are recommended for joining ceramic or graphite and metal-to-metal: --brazing ceramics and graphite should be done at temperatures greater by 50-70°C than that of brazing metal-to-metal.

3. Brazed joints formed with all amorphous foils are characterized by complex multiphase microstructure of the joint metal and wide recrystallized diffusion zones. Some structures (TiBraze200) work as reinforced composite materials and resist crack propagation during mechanical loading. No intermatallic layers were found at the interface of the base and joint metals and that makes titanium brazed joints resistant to brittle failure. Effect of microstructures on mechanical behavior of titanium and ceramic brazed joints requires further study aiming at the optimization of brazing process parameters suitable for joining different combinations of base materials.

References

[1] Rabinkin, A., Liebermann, H., Pounds, S., Taylor, J., Reidinger, F., and Siu-Ching, L. 1991. Amorphous Ti-Zr-base Metglas®, brazing filler metals. *Scripta Met. et Mater.* 25, 399–404.

[2] Onzawa T., Suzumura, and Ko M. W. 1990. Brazing of titanium using low-melting-point Ti-based filler metals, *Welding Journal*, vol. 69, No. 12, 462s-467s.

[3] Shapiro A. E. and Rabinkin A. 2003. State of the art of titanium-based brazing filler metals, *Welding Journal*, vol. 82 (10), 36-43.

[4] Matsu K., Miyazawa Y., Ariga T. 2006. Titanium brazing for manufacturing titanium heat exchangers, *Proc. of 3rd Int. Brazing and Soldering Conf., April 24-26, 2006, San Antonio, TX, USA,* 307-309.

[5] Shapiro A. E., Flom Y. A., Moxson V. S., and Duz V. A. 2013. Cost-effective sintered brazing foils for joining titanium alloys and ceramics, Proc. of 10th Int. Conf. on Brazing, High-temperature Brazing and Diffusion Bonding, LÖT-2013, *DVS-Berichte,* vol. 293, 7-10.

Wettability of Co–Ni–Nb–V Alloy on SiC Ceramic Matrix Composites and Interfacial Reactions

Wen-Wen LI, Bo CHEN, Hua-Ping XIONG * *and Wen-Jiang ZOU*
Welding and Plastic Forming Division, Beijing Institute of Aeronautical Materials, Beijing, China
**Corresponding author. Tel: +86-10-62496680; fax: +86-10-62456925. E-mail address: xionghuaping69@sina.cn*

Abstract

A novel composition of Co–Ni–Nb–V alloy was designed for the joining of fiber-reinforced silicon carbide matrix composites, including Cf/SiC and SiCf/SiC composites. In the present paper, the wettability of Co–Ni–Nb–V alloy on silicon carbide matrix composites was studied by the sessile drop method. After heating at 1573 K for 20 min in vacuum the Co–Ni–Nb–V alloy exhibited a low contact angle on the SiC composites. The interfacial microstructure was investigated by SEM equipped with EDS. A multi-layer reaction structure was found at the Cf/SiC–brazing alloy interface and the interfacial microstructure can be described as the follows: VC band/NbC band/discontinuous VC band/Co(Ni, Si) solid solution/residual brazing alloy. In the case of SiCf/SiC, only VC band was formed at interface, signifying that the activity of Nb was lower than V. The elements of Co and Ni mainly existed in the form of Co–Si or (Co,Ni)–Si. The interfacial reaction mechanism was also discussed in this paper.

Introduction

Silicon carbide (SiC) shows the superior high temperature oxidation-resistance and mechanical properties [1]. The fiber reinforced silicon carbide ceramic matrix composites have the property combination of low density, low coefficient of thermal expansion (CTE), good high-temperature strength and thermal shock resistance as well as enhanced toughness owing to the embedded carbide fibers or silicon carbide fibers [2]. As a result, the fiber reinforced SiC composites have great potential of being used as new lightweight structural materials for high-temperature aerospace applications [3-4]. In many cases the joining of SiC ceramic composite to itself or to metals is needed.

Brazing, due to its simplicity, lower cost investment and potential as a mass production process, is an appropriate joining technology for SiC ceramic matrix composites. So far, Cu–Ti, Ag–Cu and Ag–Cu–Ti brazing alloys have been the base of SiC ceramic composite joining [5-8]. But the service temperature of the joint brazed with Ag–Cu–Ti or Cu–Ti filler metal is limited to 500°C [9]. To make full use of high-temperature property advantages of SiC ceramic composite, it is necessary to search for new brazing alloys with stable high-temperature performance. B. Riccardi et al. developed a brazing alloy composed of Si–44Cr at. % eutectic and the intermetallic $CrSi_2$ for SiC_f/SiC composite, and the corresponding joint showed a shear strength of about 60 MPa [10]. H. P. Xiong et al. designed a Cu–Pd–V system filler alloy used for joining C_f/SiC. $VC_{0.75}$ reaction band was formed at interface, and the central part of the joint was composed of (Cu, Pd) solid solution and eutectic-like phase of Pd_2Si+Cu_3Pd. The room-temperature three-point bend strength was 128 MPa, the joint strength at temperatures of 873 K~1073 K was even higher than the room-temperature strength [11]. However, the research on high-temperature active filler metals is far insufficient.

In the present study, a Co–Ni–Nb–V brazing filler alloy was proposed for the joining of SiC composites. Design of Co–Ni–Nb–V alloy was a result of the following considerations: Firstly, it has been recognized that V and Nb are widely used in ceramic joining as active elements [12-14], and they are strong carbide-forming elements. Therefore the addition of Nb and V into the brazing alloy should be favorable to its wettability on the SiC ceramic composites. Secondly, elements Co and Ni are commonly-used primary constituents in commercial superalloys. Correspondingly the Co(Ni)-based brazing alloy matrix should have superior high-temperature properties. Thirdly, element Nb has a large affinity with element Si, and thus it is worth investigating the possibility of forming Nb–Si refractory compounds at the joining interface.

In this paper, a preliminary wetting experiment of Co–Ni–Nb–V alloy on the C_f/SiC and SiC_f/SiC was conducted. The reaction mechanism at the interface was also discussed. The purpose of this study is to provide a basis for the future research on development of new high-temperature brazing filler metals for SiC ceramic composites.

Experimental procedure

The CoNi–(10~20)Nb–(8~16)V(wt.%) alloy used for wetting experiment was prepared by the following procedure: First, the high-purity powders (200~300 mesh) of Co, Ni, Nb and V were weighed out in the desired ratio and mixed mechanically until homogenization. And then some mixture powder of 80 mg~100 mg was compacted into a cylindrical billet with a diameter of 4 mm. The size of wetting sample was Φ4mm×2mm. The ceramic composites used in wetting experiment were 3D-C_f/SiC and 3D-SiC_f/SiC ceramic composites. The ceramic composite samples were ground by silicon carbide papers and ultrasonically cleaned in the acetone solvent before the experiment. The contact angles of the brazing alloy on SiC composites were measured with the sessile drop method.

The wetting experiment was conducted in vacuum furnace at 1573 K for 20 min. The vacuum at 1573 K was better than 3.0×10^{-2} Pa. During the heating process, the powder billet melted into liquid drop in situ, and then spread on the composite surface. The contact angles of two samples were measured using an optical microscope after cooling down to room temperature.

A JEOL-5600LV type scanning electron microscope (SEM) and an Oxford Inca type energy-dispersive spectrometer (EDS) were used for microstructure observation of the wetting interface and semi-quantitative composition analysis of some typical micro zones.

Results and discussion

After being held at 1573 K for 20 min, the Co–Ni–Nb–V showed a low contact angle of 10° and 2° respectively on C_f/SiC and SiC_f/SiC ceramic composite, as shown in Figure 1. It can be seen that good bonding was achieved at the interface and the liquid metal partially infiltrated into the SiC_f/SiC substrate.

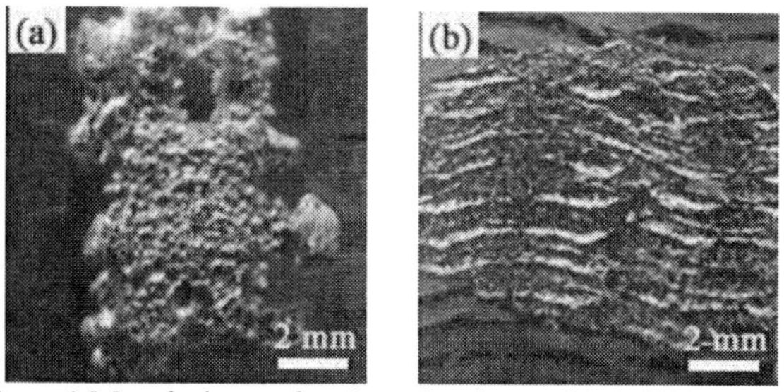

Figure1 Morphology of molten droplet on (a) C_f/SiC and (b) SiC_f/SiC after being held at 1573K for 20 min.

Figure 2 presents the back-scattered electron image of microstructure of the wetting interface between Co–Ni–Nb–V alloy and C_f/SiC composite and the area distribution maps of elements V, C, Si, Co, Nb and Ni. The EDS results of some typical micro zones marked in Figure 2 (a) are shown in Table 1.

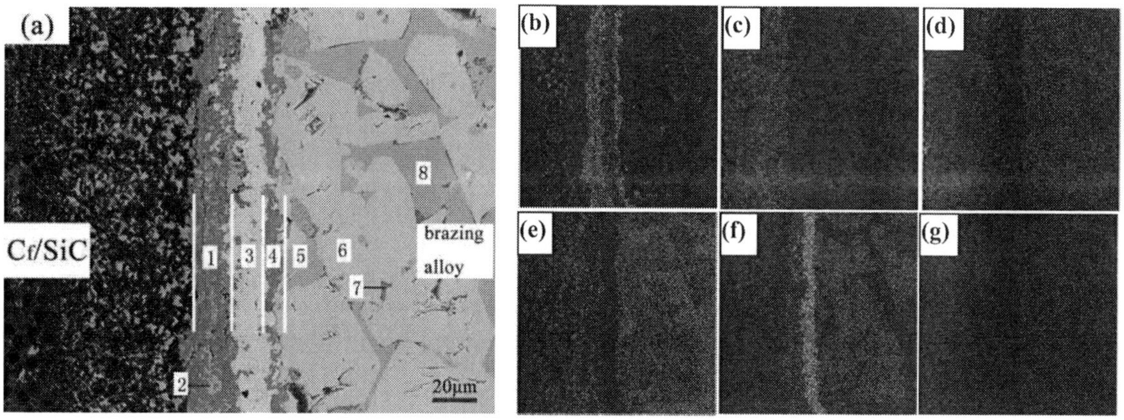

Figure 2 Back-scattered electron image of microstructure of the wetting interface between Co–Ni–Nb–V alloy and C_f/SiC composite and the area distribution maps of elements V(b), C(c), Si(d), Co(e), Nb(f) and Ni(g).

Table 1 EDS results of typical regions marked in Figure 2 (a)

Typical regions	Composition (at %)							Deduced phases
	Au	C	Nb	Co	Si	V	Ni	
1	-	40.8512	8.8261	0.8275	0.2945	48.9473	0.2534	VC
2	0.7344	3.6746	0.0591	67.7921	13.6856	2.6882	11.366	Co(Ni,Si) solid solution
3	-	41.054	44.6669	0.4473	0.0763	13.6229	0.1326	NbC
4	-	41.8304	8.6169	0.9731	0.2345	48.0709	0.2742	VC
5	0.714	2.4571	0.0344	68.0195	14.5332	1.9128	12.329	Co(Ni,Si) solid solution
6	-	5.9057	16.5736	40.4109	23.3024	2.086	11.7213	residual filler metal with V depleted and Si dissolved
7	-	43.0095	7.8178	2.2084	0.5292	45.9143	0.5208	VC
8	0.6764	3.6453	0.0288	65.4143	12.8668	4.5124	12.8540	residual filler metal with Nb depleted and Si dissolved

It is clear that sufficient interfacial reactions occurred and the liquid metal infiltrated into the ceramic matrix. Closest to the C_f/SiC surface, a dark-grey reaction band (labeled "1" in Figure 2 (a)) was first formed, with a thickness of about 25 μm. It's well known that EDS was one of semi-quantitative chemical analysis method, and it was widely used for the approximate speculation of the formed phases in the analyzed microzones. Here, according to the composition analysis, it can be identified as VC compound. Interestingly, there existed some alloy micro zones (labeled "2") inside the reaction band "1", mainly Co(Ni,Si) solid solutions. The next reaction band (grey-white region "3") was a new carbide layer, enriched Nb and C, with a thickness of 18 μm, as shown in Table 1. The atom ratio of Nb and C for the reaction band "3" was about 1:1, and so it should be NbC compound. At the interface between the NbC band and the residual alloy, one discontinuous VC reaction layer "4" appeared, about 10 μm in width. The sandwiched-structure of VC/NbC/VC can be confirmed by element distribution maps (b) and (f) in Figure 2. The microstructure of residual alloy was composed of Co(Ni,Si) solid solution marked as "5", residual filler metal "6" with V depleted and Si dissolved, and residual filler metal "8" with Nb depleted and Si dissolved. To be noticed, a very small amount of VC phases marked as "7" existed among the brazing alloy matrix, which can signify that C atoms decomposed from SiC matrix diffused into filler metal and reacted with residual V elements.

From a thermodynamic point of view, VC was stable than SiC, for example, at 1573 K, the free energy of formation of VC was -87.02KJ·mol^{-1}, far lower than that of SiC, -67.35 KJ·mol^{-1} [15]. Therefore, the VC compound can be generated spontaneously at the high temperature for the V-containing alloy/SiC system. Similarly, Nb had strong affinity with C to form NbC compound. It can be seen from the Figure 2, the VC reaction band preferentially formed at the interface closest to SiC composite surface and NbC band formed subsequently, indicating that the element of V presented higher activity than Nb in this reaction system. In the meantime, Si atoms decomposed from SiC would diffuse towards the brazing alloy and react with Co and Ni intensively to form Co(Ni,Si) solid solutions [16].

On the basis of dissolution enthalpy, the enthalpy of Nb in liquid Si was -173 KJ·mol^{-1} while the enthalpy of Co, Ni and V in liquid Si was -78 KJ·mol^{-1}, -85 KJ·mol^{-1} and -121 KJ·mol^{-1} [17]. This means that Nb should have the strongest affinity with Si among the four elements in the brazing alloy. However, no evidence was found for the formation of Nb–Si compound at the interface. It is believed that the formation and growth of Nb–Si compound were not only controlled by the thermodynamic condition but also the kinetic factors.

Figure 3 shows the back-scattered electron image of microstructure of the wetting interface between Co–Ni–Nb–V alloy and SiC_f/SiC composite and the area distribution maps of elements V, C, Si, Co, Nb and Ni. Compared with Figure 2, the microstructure of wetting interface changed distinctly. Furthermore, some micro-cracks were visible at the interface.

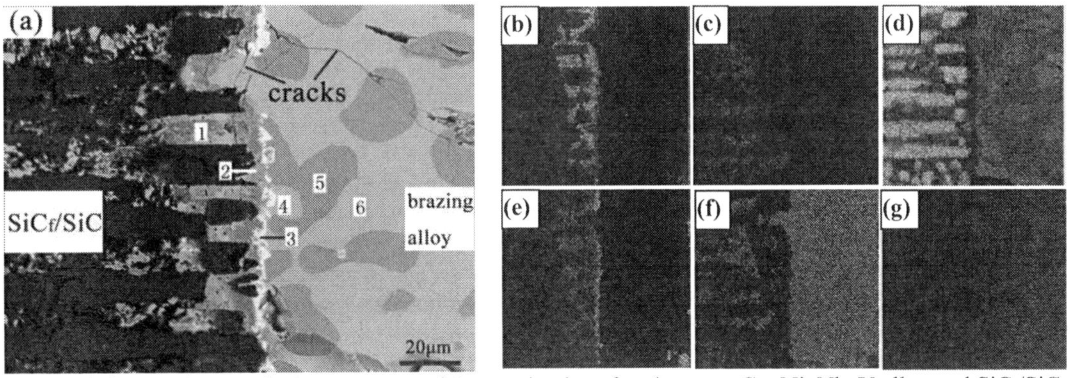

Figure 3 Back-scattered electron image of microstructure of the wetting interface between Co–Ni–Nb–V alloy and SiC_f/SiC composite and the area distribution maps of elements V(b), C(c), Si(d), Nb(e), Co(f) and Ni(g).

Table 2 EDS results of typical regions marked in Figure 3

Typical regions	Composition (at %)						Deduced phases
	C	Nb	Co	Si	V	Ni	
1	46.52	7.37	9.02	8.93	25.54	2.61	VC
2	45.72	6.50	4.99	17.56	24.38	0.84	VC dissolved with Si
3	54.80	10.35	2.01	1.08	31.18	0.59	(V,Nb)–C
4	49.88	16.07	9.61	5.51	15.80	3.13	(Nb,V)–C
5	15.87	0.98	38.97	41.76	0.16	2.25	Co–Si
6	17.92	1.03	39.61	27.93	1.44	12.07	(Co,Ni)–Si

It can be seen from the Figure 3 (a) that an evident reaction between alloy and SiC_f/SiC composite also occurred. In front of the wetting interface, the active element V diffused into SiC matrix as shown in Figure 3 (b), according to the EDS analysis listed in Table 2, the region in SiC matrix marked as "1" can be denoted as VC. The composition of micro-zone "2" near the interface was similar to region "1", meaning that VC compound was also formed here. And then, there appeared a discontinuous reaction band marked as "3", where both V and Nb elements segregated at the interface close to SiC_f/SiC composite, as presented in Figure 3 (b) and Figure 3 (e). The EDS analysis results confirmed the formation of (V,Nb)–C compounds in areas "3" and "4", signifying that interfacial reaction between brazing alloy and SiC_f/SiC was attributed to the active role of elements Nb and V. The microstructure of residual alloy was characterized by isolated Co–Si phase labeled "5" and continuous (Co,Ni)–Si phase labeled "6" in Figure 3 (a).

It should be noted that, at the wetting interface related to the SiC_f/SiC composite, no independent NbC reaction band was formed, indicating again that the Nb activity was inferior to V. On the one hand, in terms of thermodynamics, the dissolution enthalpies of element Nb, V in liquid metal Co and Ni were listed in Table 3. It can be deduced that the affinity of V with Co and Ni should be remarkably lower than that of element Nb. On this basis, the activity of element Nb in liquid Co(Ni)–based alloy should be lower than that of element V. On the other hand, the formation of NbC would become difficult when free carbon fibers were absent in the SiC_f/SiC composite.

Table 3 Dissolution enthalpies of Nb and V in liquid metals Ni and Co[17]

Liquid Metal	ΔH_{Nb} (KJ·mol^{-1})	ΔH_V (KJ·mol^{-1})
Ni	-143	-75
Co	-118	-58

Similar to the interface between the Co–Ni–Nb–V alloy and the C_f/SiC composite, no Nb-Si compound was formed at the reaction interface of Co–Ni–Nb–V alloy and SiC_f/SiC composite.

Conclusions

In this paper, the wettability of a novel alloy of CoNi–(10~20)Nb–(8~16)V(wt.%) on C_f/SiC and SiC_f/SiC composites was studied. The conclusions can be summarized as the follows:

(1) At 1573 K for 20 min, the new Co–Ni–Nb–V alloy exhibited a low contact angle of 10° and 2° respectively on C_f/SiC and SiC_f/SiC ceramic composites. Good bonding was achieved at the interface between C_f/SiC composite and the brazing alloy.

(2) The microstructure at the interface between Co–Ni–Nb–V alloy and C_f/SiC composite was VC band/NbC band/discontinuous VC band/Co(Ni,Si) solution/residual brazing alloy. The elements V and Nb exhibited high activity to react with C_f/SiC.

(3) Under the same condition, the interfacial microstructure of new alloy and SiC_f/SiC composite was different, compared with the system of new alloy and C_f/SiC composite. Only V played active role in interfacial reaction, whereas NbC reaction band was absent. It appeared that the Nb activity was inferior to V.

Acknowledgments

This research work is sponsored by National Natural Science Foundation of China (Contract Nos.59905022, 50475160 and 51275497). We also would like to thank Aeronautical Science Foundation of China (Grant 2008ZE21005).

References

[1] Snead, L. L. *et al*, "Silicon carbide composites as fusion power reactor structural materials," *Journal of Nuclear Materials*, Vol. 417, No. 1–3 (2011), pp. 330–339.
[2] Xiong, J. *et al*, "Brazing of carbon fiber reinforced SiC composite and TC4 using Ag-Cu-Ti active brazing alloy," *Materials Science and Engineering A*, Vol. 527, No. 4–5 (2010), pp. 1096-1101.
[3] Riccardi, B. *et al*, "Issues and advances in SiC_f/SiC composites development for fusion reactors," *Journal of Nuclear Materials*, Vol. 329-333 (2004), pp. 56–65.
[4] Riccardi, B. *et al*, "Low activation brazing materials and techniques for SiC_f/SiC composites," *Journal of Nuclear Materials*, Vol. 307-311 (2002), pp. 1237-1241.
[5]Yang, Z. *et al*, "Microstructural evolution and mechanical properties of the joint of TiAl alloys and C/SiC composites vacuum brazed with Ag–Cu filler metal," *Materials Characterization,* Vol. 62, No. 9 (2011), pp. 825–832.
[6] Chen, B. *et al*, "Microstructure and strength of C_f/SiC joints with Ag–Cu–Ti brazing fillers," *Journal of Materials Engineering,* Vol. 329, No. 10 (2010), pp. 27–31 (in Chinese).
[7] Cui, B. *et al*, "Reaction-composite brazing of carbon fiber reinforced SiC composite and TC4 alloy using Ag–Cu–Ti–(Ti+C) mixed powder," *Materials Science and Engineering A* Vol. 562, No. 2 (2013), pp. 203–210.
[8] Guo, W. et al, "Microstructure evolution and mechanical properties of vacuum-brazed C/C composite with AgCuTi foil," *Materials Science and Engineering A,* Vol. 564, No. 3 (2013), pp. 192–198.
[9] Qin, Y. and Feng, J. "Active brazing carbon/carbon composite to TC4 with Cu and Mo composite interlayers," *Materials Science and Engineering A*, Vol. 525, No. 1–2 (2009), pp. 181–185.
[10] Riccardi, B. *et al*, "Issues of low activation brazing of SiC_f/SiC composites by using alloys without free silicon,"

Journal of Nuclear Materials, Vol. 329–333 (2004), pp. 562–566.
[11] Xiong, H. *et al*, "Interfacial reactions and joining characteristics of a Cu–Pd–V system filler alloy with C_f/SiC. composite," *Ceramics International,* Vol. 40, No. 6 (2014), pp. 7857–7863.
[12] Nakao, Y. *et al*, "Bonding of Si_3N_4 to metals with active filler metals," *Transaction of Japan Welding Society,* Vol. 20 (1989), pp. 66–76.
[13] Ito, Y. *et al*, "Joint of silicon nitride and molybdenum with vanadium foil, and its high-temperature strength," *Journal of Materials Science,* Vol. 28, No. 18 (1993), pp. 5014–5018.
[14] Fukai, T. *et al*, "Interfacial structure and reaction mechanism of SiC/V joints," *Transactions of JWRI*, Vol. 26 (1997), pp. 93–98.
[15] Park, J. S. *et al*, "Kinetic control of silicon carbide/metal reactions," *Materials Science and Engineering A,* Vol. 259, No. 2 (1999), pp. 279–286.
[16] Xiong, H. et al, "Wetting behavior of Co based active brazing alloys on SiC and the interfacial reactions", *Materials Letters*, Vol. 57, No. 22–23 (2003), pp. 3417– 3421.
[17] Miedema, A. R. *et al*, "On the normal Hall effect in nickel alloys," *Journal of Physics F: Metal Physics,* Vol. 7, No. 1 (1977), pp. 353–359.

Understanding the Run-out Behavior of a Ag-Cu-Zr Braze Alloy When Used to Join Alumina to an Fe-Ni-Co Alloy

P.T. Vianco, C.A. Walker, D. De Smet, A. Kilgo, B.M. McKenzie, P.M. Kotula, and R.L. Grant*
Sandia National Laboratories
Albuquerque, NM USA
**505-844-3429; ptvianc@sandia.gov*

Abstract

The run-out phenomenon occurs when a significant portion of the molten filler metal exits the gap region and accumulates on an exterior, base material surface. Run-out can cause excessive void formation due to the loss of filler metal from gap volume. It can also create stress concentrations that lead to base material cracking, especially when the latter is a brittle material – e.g., ceramics and glasses. At first, run-out appears to result from simply the physical displacement of too-much filler metal for the final gap dimensions (thickness and footprint). On the other hand, an active braze application was investigated that showed run-out to resulting from an instability in the wetting and spreading process. The filler metal, 97Ag-1Cu-2Zr (wt.%), was used to join an alumina (Al_2O_3) ceramic to a Kovar™ base material. The driving force for run-out was the metallurgical reaction between the Fe, Ni, and Co constituents of Kovar™ with the elemental Al released by the reduction-oxidation reaction between Zr and Al_2O_3. A constraining factor is the surface tension of the molten filler metal in conjunction with the brazement geometry. Based on this study, a modification to the filler metal composition, or use of a barrier coating on the Kovar™ surface, would provide the most promising mitigation strategies against run-out.

Introduction

Run-out Phenomenon
The optimization of a braze joint requires that close control be maintained of the amount of filler metal supplied to the gap. Too little filler metal results in voids and non-filled regions of the brazement that can jeopardize both strength and hermeticity. On the other hand, having an excess quantity of filler metal causes the molten braze alloy to be ejected from the gap and collect on an exterior surface as *run-out*. The run-out filler metal may wet to ancillary structures beyond the joint or freeze to non-wettable surfaces. An example of the adhesion of filler metal run-out to base material surfaces is shown in Fig. 1a where the 97Ag-1Cu-2Zr (wt.%, abbreviated Ag-Cu-Zr) filler metal ran out from the joint made between Kovar™ (trademark of Carpenter Technologies, Reading, PA) and an alumina (Al_2O_3) ceramic. The scanning electron microscope (SEM) image in Fig. 1b shows a cross section view of the excess filler metal.

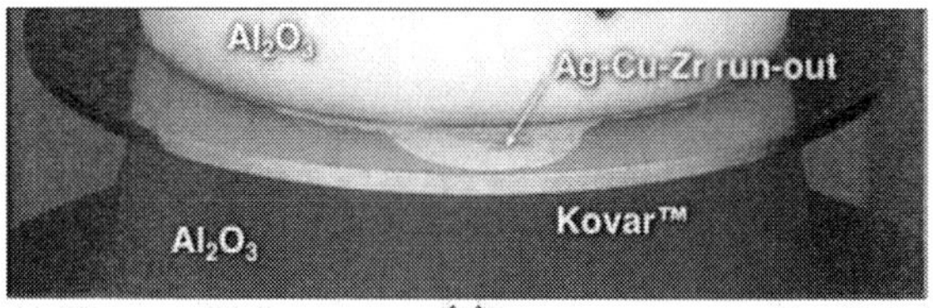

(a)

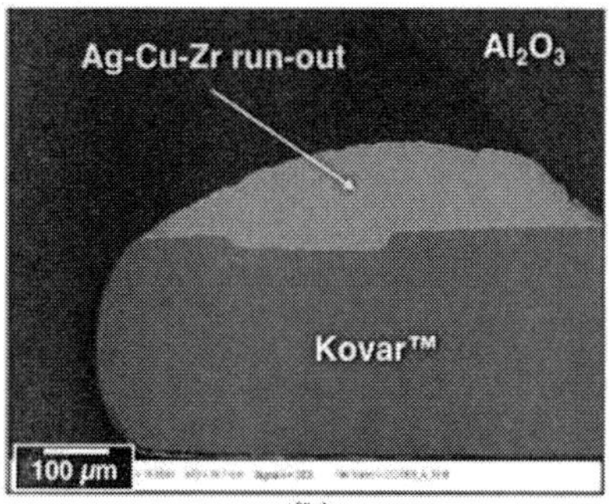

(b)

Figure 1 (a) Photograph shows run-out by a Ag-Cu-Zr active braze alloy being used to join two Al_2O_3base materials to a Kovar™ spacer. (b) An SEM photograph was taken of run-out made visible by the metallographic cross section.

There can be significant consequences to a run-out event in terms of braze joint performance. Run-out "blobs" or "lobes" can pose unsightly cosmetic defects or serve as an impediment to visual and x-ray inspection. When the excess filler metal collects outside of the gap upon solidification, it can interfere with next-assembly fit-up and mechanical functions. Run-out removes filler metal from the joint that can lead to reduced mechanical strength and a potential loss of hermeticity. Poor strength and non-hermeticity may be further aggravated by excessive solidification shrinkage, which contributes to an undersupply of filler metal within the confined geometry of the gap. The latter circumstance is illustrated in Fig. 2a. The inset image shows, what appears to be, excessive void development in the braze joint. However, upon closer analysis, the voids were actually caused by solidification shrinkage enhanced through the loss of filler metal due to run-out.

(b)

Figure 2 (a) SEM photograph illustrates the voids observed in the x-ray radiograph (inset) created by the combination of Ag-Cu-Zr filler metal lost to run-out and solidification shrinkage of the remaining braze alloy. (b) Optical micrograph shows a crack in the ceramic base material due to the run-out of the Au-Cu-Ni-Ti filler metal.

Perhaps the most problematic consequence of run-out is the generation of residual stresses in the base material. This phenomenon is illustrated by the optical micrograph in Fig. 2b. The run-out of a Au-Cu-Ni-Ti filler metal caused cracking in the ceramic member of the assembly. The high stress condition created by the run-out was confirmed by computational modeling, the results of which, are shown in Fig. 3. The run-out and no-run-out braze joints are shown by the optical micrographs in Figs. 3a and 3b, respectively. Finite element models were built from those images; they are shown in Figs. 3c and 3d, respectively. The stress contours predicted by the constitutive model are show in Figs. 3e and 3f for the run-out and no-run-out conditions, respectively. When run-out is present, a peak stress of 42,000 psi is generated in the ceramic, which is more than enough to cause a crack. That stress decreases to the relatively benign 4,400 psi when run-out was absent.

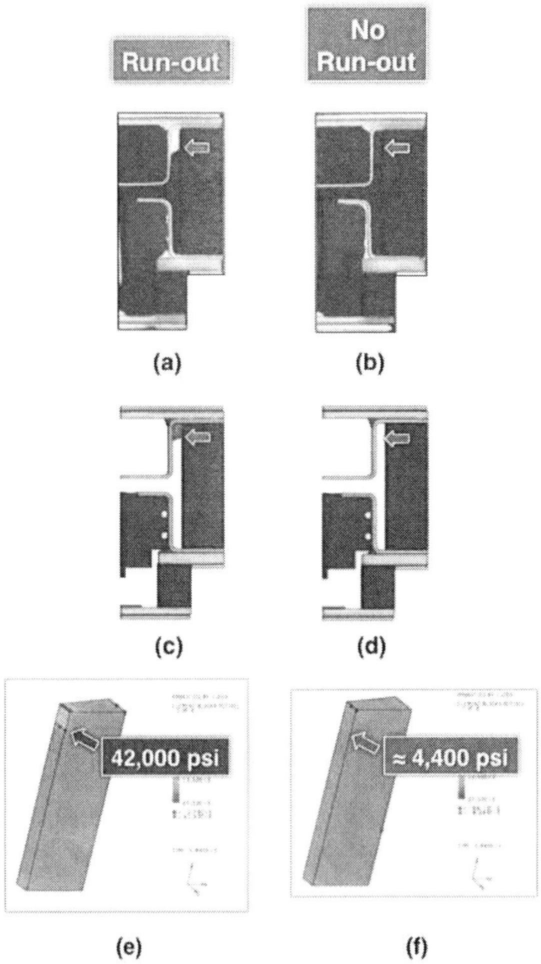

Figure 3 (a, b) Optical micrographs of Au-Cu-Ti-Ni braze joints between Nb and ceramic for which run-out was present or absent. (c, d) Finite element models of the run-out and no-run-out cases used to predict the residual stresses. (e, f) Tensile stress contours and maximum values predicted by the constitutive model for the two conditions.

The aforementioned consequences of run-out have long been recognized by the brazing industry. Several mitigation approaches have been explored to control excessive filler metal flow. One approach is to reduce the peak brazing temperature in order to increase the viscosity of the molten filler metal. The only filler metals for which this approach has merit, are those alloys that have an extended pasty range. When the filler metal has a significant difference between the solidus and liquidus temperatures, $T_l - T_s > 10°C$, there is sufficient margin in a typical furnace process to control the filler metal flow properties by strategically selecting a peak temperature that is within the alloy pasty zone. Unfortunately, molten metal viscosity, alone, is not very sensitive to temperatures above the liquidus point. Thus, in the case of eutectic or near-eutectic filler metals ($T_l - T_s < 10°C$), altering the brazing temperature offers little control over the flow behavior of the liquid filler metal.

A second approach is to add one or more alloying elements to the filler metal. The goal is to create a wider pasty range in order to allow the process temperature to provide a measure of control on filler metal flow. However, altering the filler metal composition comes with it, many other, often synergistic effects that can degrade braze joint manufacturability, performance, and reliability.

A third strategy is to modify the *geometry* of the base material surface. Adding corners to the base material surface would be expected to slow or halt the flow of molten filler metal. In fact, such a feature is observed in Fig. 1b. The trough was added to the Kovar™ base material to "capture" the filler metal and minimize further run-out. Clearly, the trough technique was not successful.

All of these measures have met with little or no success to control filler metal flow. Their lack of effectiveness, however, suggested that other driving forces were responsible for the run-out behavior, not simply a physical displacement of molten filler metal.

As a summary, there are significant consequences to the run-out phenomenon that extend beyond simply "wasted" filler metal. Run-out can interfere with next-assembly function as well as lead to the damage of adjoining base materials. The inability of early mitigations steps to control run-out provides an indication that this behavior is not simply the physical displacement ("squishing out") of excess molten filler metal from the brazement.

Characterization of Run-out for the Target Application

The Ag-Cu-Zr active braze alloy was considered for an application that required that a Kovar™ structure be joined to an Al_2O_3 ceramic part. The tensile button configuration was used to initially assess braze joint strength, hermeticity, and microstructure prior to the development of actual hardware [1]. A photograph of the tensile button specimen is shown in Fig. 4. Two Al_2O_3 "buttons" were brazed to a Kovar™ "spacer." The spacer had 0.0015 in. dimples that controlled the thicknesses of both gaps. The Ag-Cu-Zr filler metal was in the form of an annular ring. Its footprint was the same as that of the Al_2O_3 faying surface: 0.400 in. inner diameter (ID) and 0.625 in. outer diameter (OD). The preform was 0.002 in. thick. There were two varieties of Al_2O_3 ceramics that are distinguished by their different colors. The small compositional differences responsible for the color variations did not have an effect on any properties investigated in this study.

Shown in each of Figs. 5a and 5b is one-half of two, post-tested tensile buttons. Both samples were brazed at 965°C for 20 min in a 600 torr Ar atmosphere. The yellow dashed circles indicate the boundaries of the faying surface area. The specimens exhibited significant filler metal run-out as indicated by the arrows. Although two run-out "lobes" formed in Fig. 5a, most often, there was only one per side of the Kovar™ spacer. The run-out lobe on one side did not exhibit a particular correlation with the lobe on the other side.

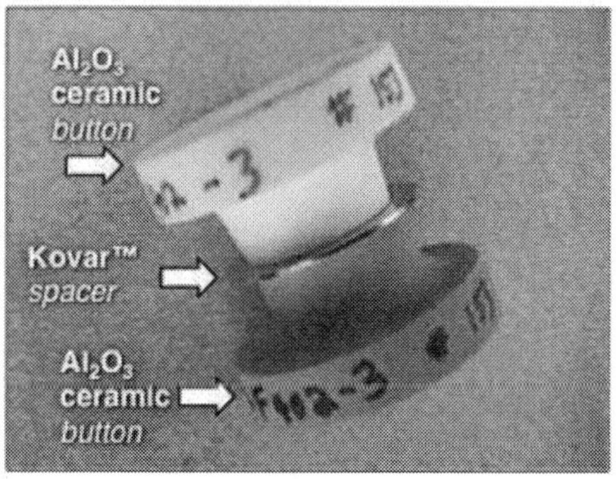

Figure 4 Photograph shows the tensile button configuration. There are two Al_2O_3 ceramic buttons brazed to each of the two sides of the Kovar™ spacer.

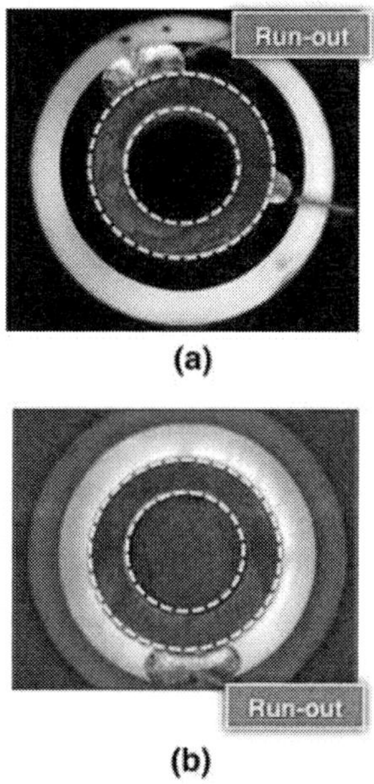

Figure 5 (a, b) Optical photographs show run-out (arrows) experienced by the Kovar™ spacer/Ag-Cu-Zr/Al_2O_3 braze joint. The tensile buttons were brazed at 965°C for 20 min in 600 torr Ar prior to being pull tested; the faying surfaces are between the dashed circles.

An experiment was performed to explore the possibility that run-out was not simply the ejection of excess filler metal from the joint. A smaller footprint of filler metal was placed in the same tensile button braze joint, by reducing the OD of the Ag-Cu-Zr preform from 0.625 in. to 0.500 in. One tensile button was assembled with a single such preform in the joint. The smaller OD reduced the filler metal volume by 60% so that any occurrence of run-out could not be attributed to a physical displacement effect.

A second tensile button sample was made with *two* of these preforms. The preforms were stacked upon one-another, which in effect, returned to overall braze alloy volume to approximately equal to that of the single preform of nominal OD (0.625 in.). The purpose of the latter specimen was to determine if run-out was encouraged by a strong degree of capillary flow by the molten filler metal.

Both tensile buttons were brazed at 985°C for 5 min (600 torr Ar). The temperature and time were higher and shorter, respectively, than the nominal conditions. Nevertheless, they were still well within the process window for this filler metal. The tensile buttons were subsequently pull tested to facilitate the documenting of the run-out behavior.

Figure 6 shows optical micrographs that were taken of post-pull tested samples. The run-out is identified by the arrows. The tensile button in Fig. 6a, which had one preform, showed three primary run-out lobes and several lesser wetting fronts. The fact is that run-out took place under this circumstance of limited filler metal decouples the phenomenon from a physical displacement of molten filler metal.

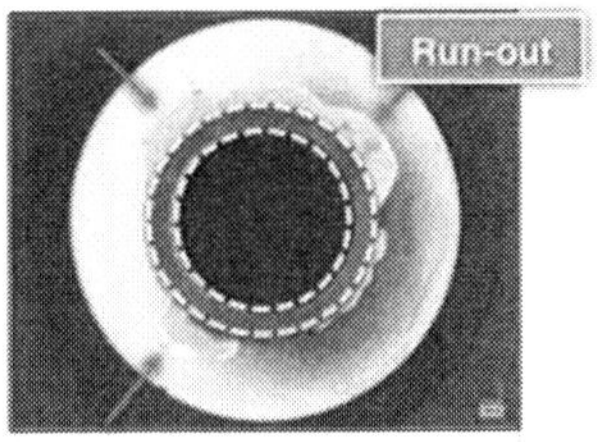

(a)

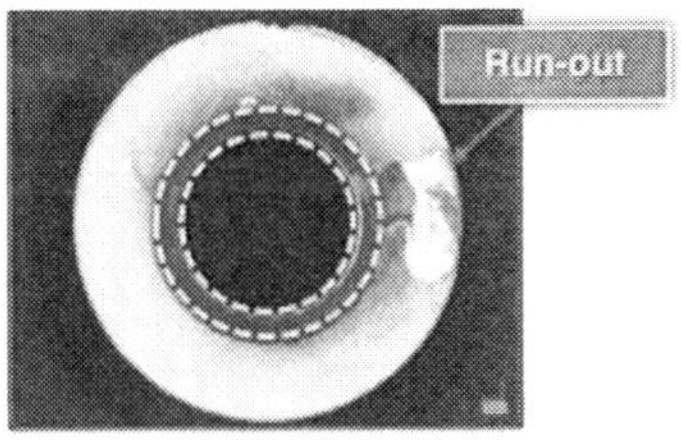

(b)

Figure 6 Optical photographs show run-out (arrows) experienced by the Kovar™ spacer/Ag-Cu-Zr/Al_2O_3 braze joint. The process conditions were 985°C for 5 min in a 600 torr Ar atmosphere. (a) This sample was fabricated with a single Ag-Cu-Zr preform having a reduced OD of 0.500 in. (b) This specimen was fabricated with the same, smaller preform footprint, but having two preforms stacked upon each other. The run-out is identified by the arrows.

The tensile button shown in Fig. 6b, which had two preforms, exhibited only a single run-out lobe. That run-out lobe was relatively voluminous, indicating that it carried the excess filler metal of the two preforms. Nevertheless, it occurred in a *single* location indicative of run-out, as opposed to there being a general flow of excess filler metal squished out over a larger area of the Kovar™ spacer surface.

In summary, the experimental evidence showed that the molten filler metal did not respond to changes of volume in a manner that would indicated that run-out was a physical displacement of excessive filler metal. Rather, run-out appeared to result from an instability in the wetting and spreading behavior of molten Ag-Cu-Zr on the Kovar™ base material. The goal of the study described in this report was to collect evidence that could lead to a root-cause determination for run-out in the system of Ag-Cu-Zr active filler metal and the base materials, Kovar™ and Al_2O_3. A longer term objective is to use the root-cause knowledge to develop a first-principles methodology to mitigate against run-out.

Material Systems

The materials systems is the same as that used to exemplify the run-out phenomenon in the previous section. The filler metal is the active braze alloy, 97Ag-1Cu-2Zr (wt.%, abbreviated Ag-Cu-Zr). The T_s is 940 – 950°C and T_l is 960 – 970°C. The filler metal preforms had these dimensions: 0.400 in. ID; 0.625 in. OD, and 0.002 in. thickness. The base materials were (a) the Kovar™ alloy (b) the 94% Al_2O_3 ceramic. The nominal braze process had a peak temperature of 985°C; time duration of 5 min at peak temperature; and a 600 torr Ar atmosphere. The braze joint microstructure is exemplified by the SEM image in Fig. 7. Besides the filler metal and base materials, the brazements were characterized by distinct reaction structures at the two interfaces. Because those structures are critical to the analysis, they are described in greater detail in the next section.

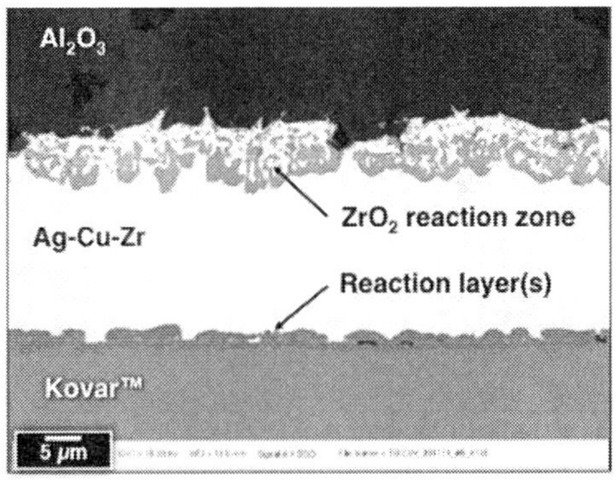

Figure 7 SEM photograph shows the general microstructure of the braze joint that formed between the Kovar™ and Al_2O_3 ceramic base materials using the Ag-Cu-Zr active filler metal. The process conditions were 985°C; 5 min; and 600 torr Ar atmosphere.

Interface Reactions

Cross Section Microstructure

A detailed discussion is presented here of the reaction layers that develop at the Ag-Cu-Zr/Al_2O_3 and Ag-Cu-Zr/Kovar™ interfaces. These reactions form the basis for understanding the run-out phenomenon. The reaction zone created at the Ag-Cu-Zr/Al_2O_3 interface is shown in Fig. 8a. The reaction zone is a coalescence of ZrO_2 particles that develop, initially, at the immediate Ag-Cu-Zr/Al_2O_3 interface as shown in Fig. 8b. The particles separate from the interface and migrate into the near-interface filler metal. The ZrO_2 particles are a product of the reduction-oxidation (redox) reaction between Zr and Al_2O_3 ceramic below:

$$3Zr + 2Al_2O_3 = 4Al + 3ZrO_2 \qquad (1)$$

The reaction is *not* spontaneous, having a calculated Gibb's free energy of +610 cal. This magnitude of free energy is well within the error of literature data so that the reaction may very well be spontaneous (negative), albeit, perhaps only slightly so. Regardless, the redox reaction does, in fact, take place as is evidenced by the appearance of the ZrO_2 zone. The Cu component of the filler metal does not have a role in this reaction; rather, it remains homogeneously-distributed throughout the bulk Ag-Cu-Zr alloy.

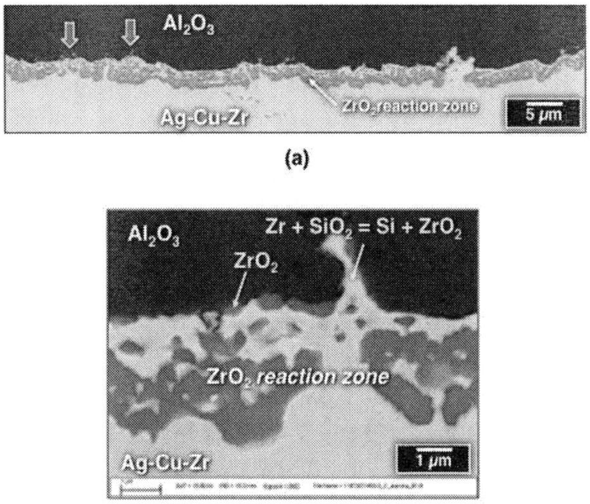

Figure 8 (a) Low magnification SEM photograph shows the ZrO_2 reaction zone at the Ag-Cu-Zr/Al_2O_3 interface. The magenta arrows indicate locations where the Zr/SiO_2 redox reaction removed SiO_2 grain boundary glassy phase, thereby allowing filler metal to infiltrate between the Al_2O_3 grains. (b) High magnification view of the immediate Ag-Cu-Zr/Al_2O_3 interface showing formation of the ZrO_2 particles and infiltration of the inter-grain region by the filler metal as SiO_2 is lost to the Zr/SiO_2 redox reaction.

The ceramic has grain boundary phase based on the SiO_2 glass that bonds the Al_2O_3 grains, together. There is a corresponding redox reaction between the SiO_2 phase and Zr:

$$Zr + SiO_2 = Si + ZrO_2 \qquad (2)$$

This reaction is predicted to be spontaneous because it has a negative Gibb's free energy, -46,100 cal. The evidence of this redox process is replacement of glass material from between the Al_2O_3 grains with filler metal. This effect occurred in Fig. 8a at sites indicated by the magenta arrows and is shown at high magnification in Fig. 8b.

Aside from the fidelity of the reference information, there are three physical phenomena that could drive the Zr/Al_2O_3 redox reaction into spontaneity:

a. Zr/SiO_2 redox reaction,
b. Free energy of solution as elemental Al enters the molten Ag-Cu-Zr filler metal, and
c. Reactions at the Kovar™/Ag-Cu-Zr interface.

It is unlikely that (a) would pose a significant effect given the limited quantity of SiO_2 present in the ceramic. Scenario (b) could potentially contribute to the driving force. The Ag-Al binary alloy phase diagram exhibits a eutectic composition of 71Ag-29Al (wt.%) that has a eutectic temperature of 567°C, which is well below the brazing temperature [2]. That property, together with the extensive composition range in the δ phase (Ag terminal phase) favors a low energy barrier for the dissolution of Al into the molten filler metal. If this scenario did not contribute to the driving force for the Zr/Al_2O_3 redox reaction, it would be *unlikely* to impede it.

The scenario (c) could provide a significant contribution of driving force in support of the Zr/Al_2O_3 redox reaction. Although there is not a particularly strong driving force for Ag to react with the Fe, Ni, or Co constituents of Kovar™, there is an affinity between those elements and Al because they readily form covalent (intermetallic) compounds [3].

The reaction that takes place at the Kovar™/Ag-Cu-Zr interface is illustrated in Fig. 9. The SEM image in Fig. 9a shows the interface microstructure. The sub-layers were identified using energy dispersive x-ray (EDX) as well as elemental spectral analysis. The latter method, the results of which shown in Fig. 9b, uses a thin section formed by focused ion beam (FIB) milling and then analyzed by transmission electron microscopy (TEM) techniques. (Note the magnification marker.) Neither technique included the standards and calibrations necessary to quantify the compositions. Clearly, the Al released by the redox reaction at the Ag-Cu-Zr/Al_2O_3 interface was integral to the Kovar™/Ag-Cu-Zr interface microstructure.

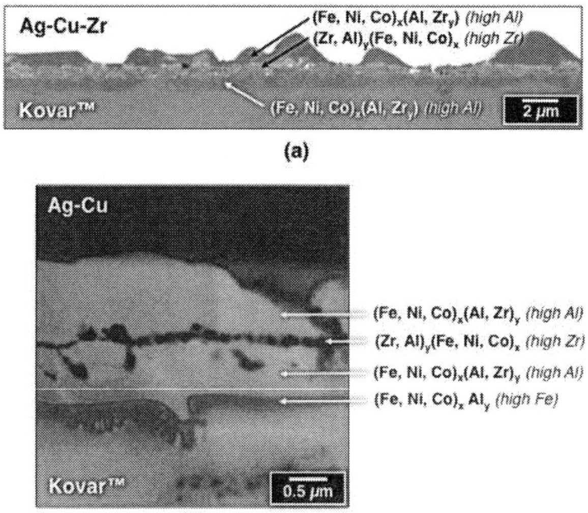

Figure 9 (a) SEM image shows the reaction sub-layers that form at the Kovar™/Ag-Cu-Zr interface in the braze joint fabricated under nominal brazing conditions (985°C, 5min, 600 torr Ar). (b) Elemental spectral analysis shows a map of elements that comprise the reaction sub-layers at the Kovar™/Ag-Cu-Zr interface.

A qualitative accounting was made of the sub-layers as follows: There is the high-Al, $(Fe, Ni, Co)_x(Al, Zr)_y$ layer that borders with the bulk filler metal (Ag-Cu). Below it is the Zr-rich, $(Zr, Al)_y(Fe, Ni, Co)_x$ layer and underneath it, a second, Al-rich reaction sub-layer also denoted as $(Fe, Ni, Co)_x(Al, Zr)_y$. The fourth and final layer, which is next to the Kovar™ base material, is high in Fe and has a dendritic morphology. That layer was designated as $(Fe, Ni, Co)_xAl_y$. Lastly, it was observed that Cu was not detected in any of the reaction sub-layers; it remained dispersed in the filler metal.

The presence and morphologies of the sub-layers varied somewhat, sample-to-sample, as well as location-to-location along the same interface. The two high-Al reaction sub-layers were generally present, but to varying thicknesses. The Fe-rich, $(Fe, Ni, Co)_xAl_y$ layer was observed more infrequently. The high-Zr layer was also typically observed, but having different morphologies, depending on the local availability of Zr. The somewhat diffuse structure, which is exemplified in Fig. 9a, was the norm. In the event that more Zr was available from the filler metal – e.g., often at the fillet edges – then, a thick, continuous reaction layer was observed in place of the diffuse structure at the interface.

Based upon observations compiled from countless cross section analyses, a scenario was formulated that describes the development of the reaction layers at the two interfaces. Experimental data indicate that this process is relatively insensitive to brazing parameters within the nominal process window of [965°C, 995°C] and (5 min, 20 min] for this filler metal. In fact, these reactions occur so rapidly that they cannot be controlled within these parameter ranges. Certainly, a contributing factor to the quickness of the reactions was the fact that all diffusion takes place across the *molten* filler metal.

That sequence of events leading up to the reaction layer formation is described using Fig. 10. Captions are provided underneath the pictures. The analysis will assume that the sub-layers that form at the Ag-Cu-Zr/Kovar™ interface, do so sequentially. It is plausible that a sub-layer

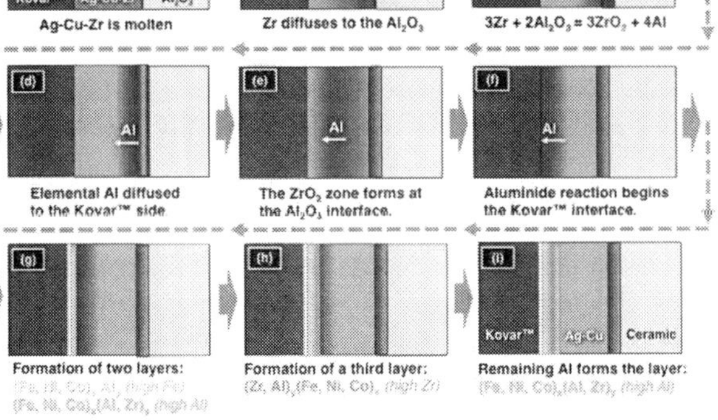

Figure 10 Diagram that depicts the proposed sequence whereby the reaction layers are formed at the Ag-Cu-Zr/Kovar™ and Ag-Cu-Zr/Al_2O_3 interfaces.

could develop from a reaction between two pre-existing sub-layers. A validation of this scenario was beyond the scope of this particular study.

The proposed scenario is described, below:

(a) The molten Ag-Cu-Zr alloy contacts the Kovar™ and Al_2O_3 base materials.

(b) Immediately, Zr diffuses to the latter.

(c) At the Ag-Cu-Zr/Al_2O_3 interface, the redox reaction releases elemental Al into the filler metal.

(d-f) The elemental Al diffuses to the Ag-Cu-Zr/Kovar™ interface and reacts with the Fe-, Ni-, and Co-components of Kovar™ to begin formation of the reaction layers described earlier in Fig. 9.

(g) The high Fe, $(Fe, Ni, Co)_xAl_y$ layer forms adjacent to the Kovar™ base material. Between that layer and the molten Ag-Cu-Zr alloy grows the high Al, $(Fe, Ni, Co)_x(Al, Zr)_y$ reaction layer.

At this point, Zr participates in the Ag-Cu-Zr/Kovar™ interface reactions. This observation implies one of two scenarios: (1) There is an excess of Zr in the molten filler metal because the redox reaction did not consume it all at the Ag-Cu-Zr/Al_2O_3 interface or (2) the presence of the $(Fe, Ni, Co)_xAl_y$ drives the $(Fe, Ni, Co)_x(Al, Zr)_y$ reaction that competes with the redox reaction for Zr in the filler metal.

(h) The driving force for Zr to react at the Ag-Cu-Zr/Kovar™ interface increases as is evidenced by formation of the Zr-rich, $(Zr, Al)_y(Fe, Ni, Co)_x$ reaction layer.

(i) The Zr reaction slows and the remaining Al in the filler metal is used to complete the Ag-Cu-Zr/Kovar™ interface reaction. The same high Al, $(Fe, Ni, Co)_x(Al, Zr)_y$ reaction layer forms at the filler metal interface. The EMPA technique confirmed that Al was absent from the filler metal to within the detection limits of that method (±0.5 wt.%).

The Cu constituent of the Ag-Cu-Zr alloys does not have a detectable presence in *any* of the interface reaction stochiometries. The EPMA determined that the remaining filler metal had retained all of the nominal 2 wt.% Cu concentration as well as confirmed the absence of Zr and Al to within the detection limit.

In summary, the Kovar™/Ag-Cu-Zr/Al_2O_3 materials system was investigated for the reactions that take place at the mutual interfaces. The Zr/Al_2O_3 redox reaction occurred at the Ag-Cu-Zr/Al_2O_3 interface as evidenced by the formation of ZrO_2 reaction zone and the release of elemental Al. The proposed reaction sequence was also predicted at the Ag-Cu-Zr/Kovar™ interface. The layers are a product of reactions between Zr, Al, Fe, Ni, and Co. The completeness of these reactions was confirmed by the absence of both elemental Al and Zr from the filler metal.

Wetting Front Metallurgy

The above discussion demonstrated that aluminide and zirconide reactions, which have very strong driving forces, take place at the Ag-Cu-Zr/Kovar™ interface. The role that those reactions have in the wetting and spreading behavior of this system was examined in order to find a correlation, if any, between them and the run-out phenomenon. The analysis began by examining the wetting edge of the Ag-Cu-Zr alloy on the Kovar™ base material.

Shown in Fig. 11a is a tensile button that was pull tested to reveal the run-out of the Ag-Cu-Zr filler metal. The yellow box indicates the location of run-out and the microanalysis. Figure 11b is a secondary electron (SE), SEM image of the edge of the run-out. The bulk filler metal formed a lip by lifting from the surface. Because the surface under the lip does not appeared fractured, the lip was formed by a retraction of the bulk Ag-Cu-Zr during solidification.

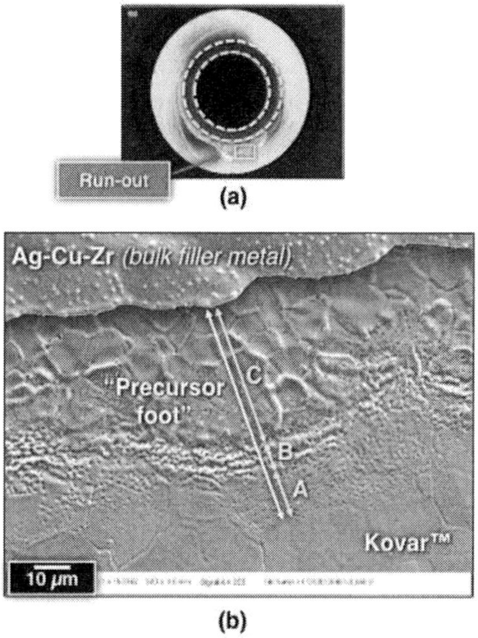

Figure 11 (a) Photograph shows a tensile button that was pull tested to reveal run-out on the Kovar™ spacer. The process conditions were 985°C, 5 min, and 600 torr Ar. (b) SEM (SE) image of the edge of the run-out taken in the yellow box of (a). A precursor foot developed that had three zones "A," "B," and "C" as distinguished by surface topography.

At the edge of the Ag-Cu-Zr filler metal is a "precursor foot" that is made visible by the high magnification, SEM image in Fig. 11b. The precursor foot, which extend approximately 40 μm from the bulk filler metal edge, has three zones labeled "A", "B", and "C" that are distinguished by the surface topography. Surface and near-surface (limited bulk) diffusion, together with the reaction, were the expressed mechanisms responsible for the precursor phenomenon.

The composition of the precursor foot was analyzed by the EDX analysis. The back scattered electron (BSE), SEM image, together with the Fe, Ni, and Al maps are shown in Fig. 12. The regions "B" and "C" exhibited a very strong presence of Al and an enrichment of Ni versus Fe when compared to unreacted Kovar™ at the lower, right-hand region of the image. Metallographic cross sections determined that the displaced Fe had concentrated in the reaction layer under the Ni-rich, aluminide layer observed at the surface. The "A" region has a reduced amount of Al that causes it to be, chemically, nearly indistinguishable from unreacted Kovar™ alloy. Zirconium was absent from the precursor foot with the exception of a one-or-two isolated particles. The extent of the precursor foot away from the bulk filler metal demonstrated the strong driving force for the diffusion/reaction process between Al, Ni, and, Fe.

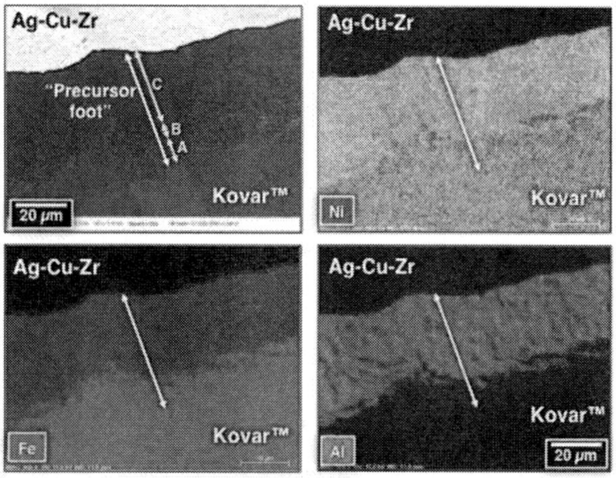

Figure 12 SEM (BSE) image and EDX analysis maps of Ni, Fe, and Al taken at the edge of the run-out analyzed in Fig. 11. The precursor foot and three surface topographies have been labeled in the SEM image.

A comparison was made between the precursor foot ahead of the run-out region as shown in Figs. 11 and 12, and the edge of the Ag-Cu-Zr fillet where run-out was absent. The latter structure was taken from the same sample as displayed in Fig. 11a, but at a different location that is shown by the yellow box in Fig. 13a. (The image in Fig. 13a image was rotated with respect to Fig 11a.) The SEM (SE) image is shown of the fillet edge in Fig.13b, which highlights the surface topography. The edge of the is comprised of an array of particles. The raised "plateau" in the upper right-hand corner is the fracture surface. The edge of the filler metal is very abrupt when compared to the gradual decrease of fillet thickness at the edge of the run-out in Fig. 11b.

The corresponding SEM (BSE) image is provided in Fig. 13c, which shows the compositional footprint of the fillet edge. There appear to be three distinct regions: two phases comprising the particle region and the filler metal matrix. The black contrast of the raised plateau is Al_2O_3 that was pulled out of the surface of the mating Al_2O_3 button by the pull test.

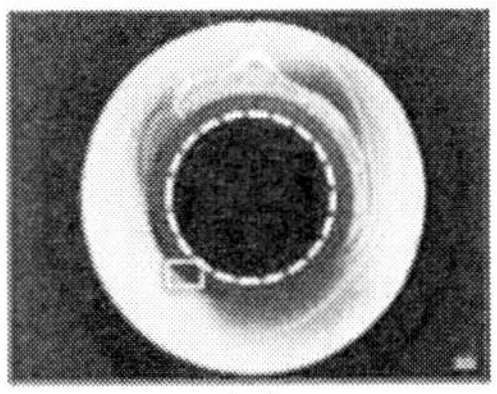

(a)

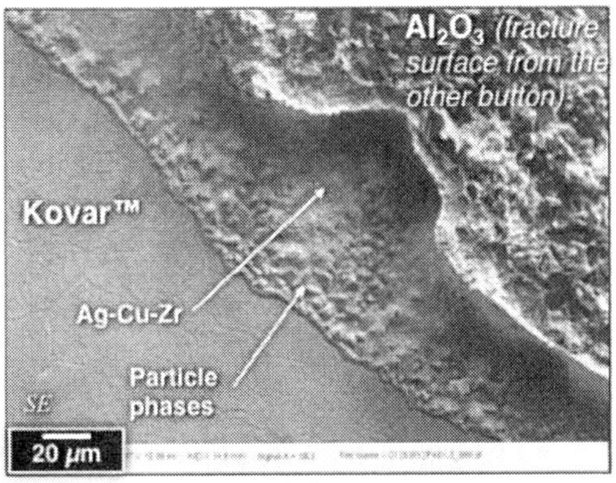

(b)

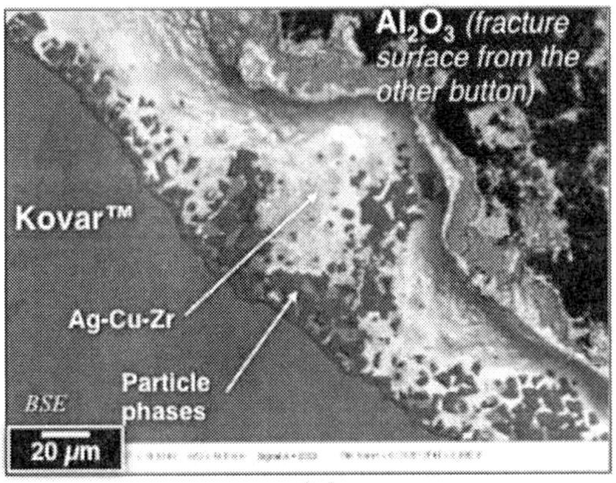

(c)

Figure 13 (a) Photograph shows a tensile button that was pull tested to reveal run-out on the Kovar™ spacer. The process conditions were 985°C, 5 min, and 600 torr Ar. (b) SEM (SE) image shows the edge of the wetted filler metal (yellow box) where run-out did *not* occur during brazing. (c) SEM (BSE) image was made of the same location as (b).

The EDX analysis was used to obtain a qualitative assessment of the phase phases present in Fig. 13. The x-ray maps are shown in Fig. 14 of the same, non-run-out region. The particle phases were comprised Fe and Ni as well as Zr and Al; the latter two elements appear to spatially compliment each-other. The particles of brighter gray tone were rich in Zr while the other particle phase was rich in Al; they represent the exposed zirconide and aluminide reactions, respectively. The matrix phase was the remaining Ag-Cu-Zr filler metal that was devoid of its Zr constituent.

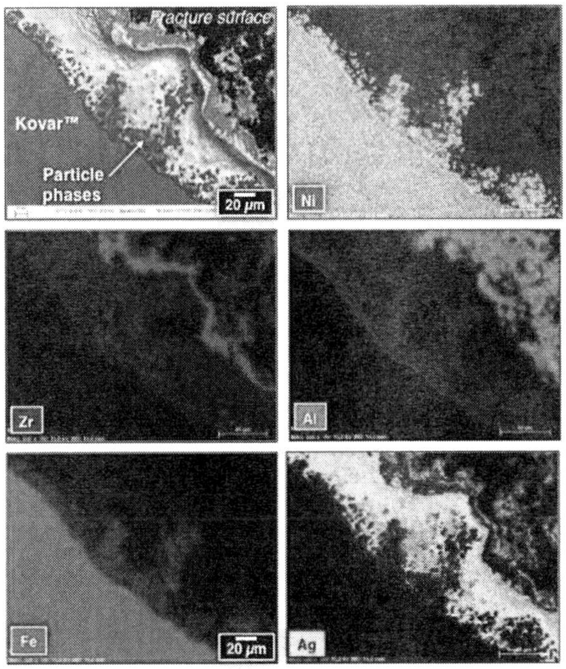

Figure 14 SEM/BSE image as well as Zr, Al, Fe, Ni, and Ag EDX maps are shown of the fillet edge where run-out was not observed on the Kovar™ spacer. The process conditions were 985°C, 5 min, and 600 torr Ar, pull test sample.

The EDX analysis in Fig. 14 allowed for labeling the likely compositions of the particle phases, which was performed in Fig. 15. Figure 15a repeats the same region of fillet edge without run-out. The high magnification image is provided in Fig. 15b that shows the particle phases and their proposed compositions. The dark particles predominate, which are the high-Al, $(Fe, Ni, Co)_x(Al, Zr)_y$ phase. This phase also appears to form a contiguous boundary along the perimeter of the fillet edge. The light-gray material is the high-Zr, $(Zr, Al)_y(Fe, Ni, Co)_x$ phase; it surrounds the high-Al particles. The filler metal was identified as Ag-Cu-Zr although the Zr component was lost to the interface and redox reactions.

The final analysis of the interface reactions examined the fillet edges by metallographic cross section. Shown in Fig. 16 are the SEM image as well as the Al and Zr EDX analysis maps that originated from a run-out region of the specimen shown in Figs. 11 and 12. The SEM image shows the raised edge that formed during solidification of the filler metal (black arrow). There was an extensive diffusion/reaction between Al and the Kovar™ constituents to a distance of about 50 μm ahead of the sessile drop edge (orange arrow), which corresponds to the surface morphology in Fig. 11b. It is presumed that the layer is the high-Al, $(Fe, Ni, Co)_x(Al, Zr)_y$ phase. It was not possible to discern specific microstructural features in Fig. 16 that were responsible for the individual zones "C," "B," and "A" belonging to the precursor foot in Fig. 11b other than they reflected gradually decreased thicknesses of the $(Fe, Ni, Co)_x(Al, Zr)_y$ layer.

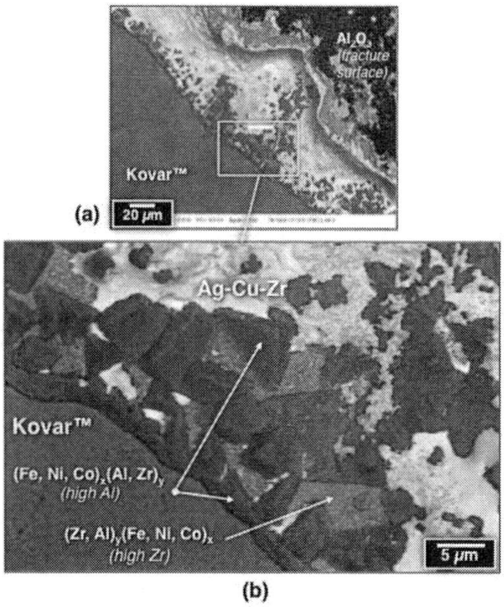

Figure 15 (a) Low magnification SEM (BSE) image shows the fillet edge of the pull test sample where run-out did not occur during brazing. The yellow box indicates the region of interest, which was the particle phases. (b) High magnification SEM (BSE) image shows the fillet edge. The particle phases have been identified with the most likely compositions.

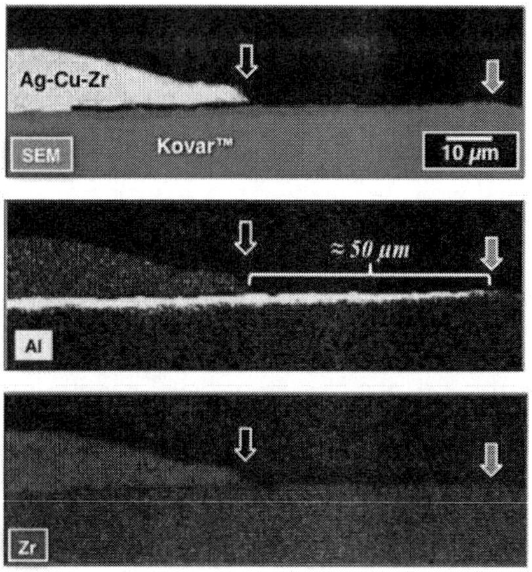

Figure 16 SEM image shows the edge of the Ag-Cu-Zr sessile drip at a *run-out location*. The process conditions were 985°C, 5 min, and 600 torr Ar. EDX maps of Al and Zr show the extent of the precursor foot (between the black and orange arrows) and that it was comprised of the high-Al, (Fe, Ni, $Co)_x(Al, Zr)_y$ phase that gradually decreased in thickness away from the filler metal edge.

A similar cross section was made of the Ag-Cu-Zr fillet edge at a location where run-out was absent. The corresponding images are shown in Fig. 17 that include an SEM photograph together with the Al and Zr EDX maps. The aluminide reaction layer extended approximately 16 µm from the fillet edge (black arrow) into the Kovar™ (orange arrow). The layer was comprised of the same large, high-Al, (Fe, Ni, $Co)_x(Al, Zr)_y$ phase particles ("A") that were observed in Fig. 15b. That phase also comprised the perimeter structure in Fig. 15b, that is now observed in-profile at location "B" of the Al map of Fig. 17. Unlike the run-out region, the high-Zr, (Zr, $Al)_y(Fe, Ni, Co)_x$ phase layer accompanied the high-Al phase to the same distance, being between the latter and the Kovar™ base material. Its surface morphology was shown in Fig. 15b.

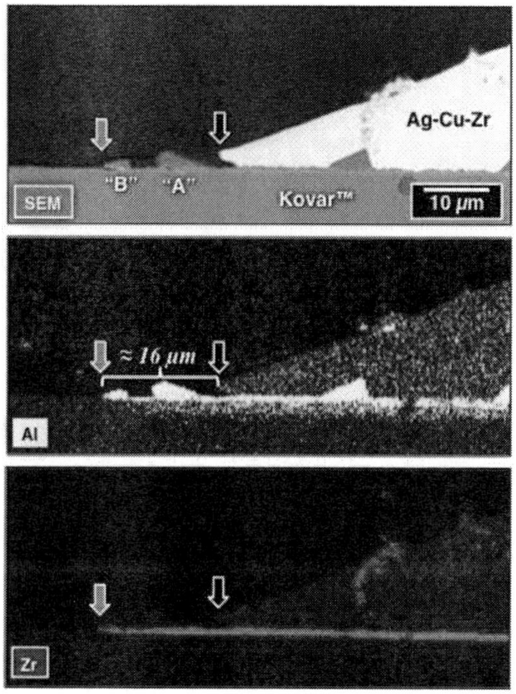

Figure 17 SEM image shows the edge of the Ag-Cu-Zr sessile drip at a *non-run-out location*. The process conditions were 985°C, 5 min, and 600 torr Ar. The EDX maps of Al and Zr show the extent of the diffusion/reaction by the high-Al and high-Zr phases (between the black and orange arrows).

A comparison was made between the images in Fig. 16 (run-out) and 17 (no run-out). The analysis indicates that co-formation of the high-Al, (Fe, Ni, $Co)_x(Zr, Al)_y$ *and* the high-Zr, (Zr, $Al)_y(Fe, Ni, Co)_x$ phase underneath it, impeded the diffusion/reaction of the high-Al phase from extending further from the filler metal edge as the precursor to run-out. The fact that the high-Zr layer was underneath the high-Al layer suggests that the former grew prior to the arrival of elemental Al from the redox reaction at the Ag-Cu-Zr/Al_2O_3 interface. This scenario implies that the source of the instability is formation of the high-Al, (Fe, Ni, $Co)_x(Al, Zr)_y$ phase when it is unchecked by the co-development of the high-Zr, (Zr, $Al)_y(Fe, Ni, Co)_x$ phase.

The above scenario indicates that run-out is stochastic in nature, similar to corrosion behavior. Run-out begins with a local perturbation in the system – e.g., in the case of the molten filler metal, either an excess of Al that drives the high-Al, (Fe, Ni, $Co)_x(Al, Zr)_y$ diffusion/reaction or a deficiency of Zr that fails to create the high-Zr, (Zr, $Al)_y(Fe, Ni, Co)_x$ phase.

At that point, the high-Al diffusion/reaction predominates, generating the precursor foot that locally accelerates wetting and spreading by the molten filler metal out, onto the Kovar™ base material surface. It cannot be ruled out that the system perturbation was caused by a local anomaly on the Kovar™ surface, although there was no explicit evidence to this effect. Clearly, there is a factor that limits run-out to one or two locations. One potential explanation is a drop in overall system free energy resulting from (a) the run-out diffusion/reaction, itself, and (b) a change to the surface energy of the molten filler metal. The latter contribution is a collaborative effect between the molten alloy surface tension and the joint geometry. The significance of the surface energy is investigated in the next section.

Base Material Geometry

The above analysis points to the aluminide reaction, when unchecked by the high-Zr reaction, as the underlying factor in the wetting and spreading instability responsible for run-out. The next analysis was performed to determine the extent to which, surface energy, which is controlled by the surface tension of the molten braze alloy and joint geometry, has a role in the run-out phenomenon.

Test Specimens

Duplicate tensile button samples were fabricated that represented one of four variants constructed of different spacer and base materials. Those variants are represented schematically by the cells in Fig. 18. The upper left-hand cell represents the "All Kovar™" joint that has the individual buttons and the spacer fabricated from the alloy. To its right is the baseline variant – Al_2O_3 buttons and a Kovar™ spacer. The lower left-hand cell has Kovar™ buttons and an Al_2O_3 spacer, which is reverse to the traditional configuration. Lastly, there is the "All ceramic" sample that has all of the components constructed of Al_2O_3 ceramic base material.

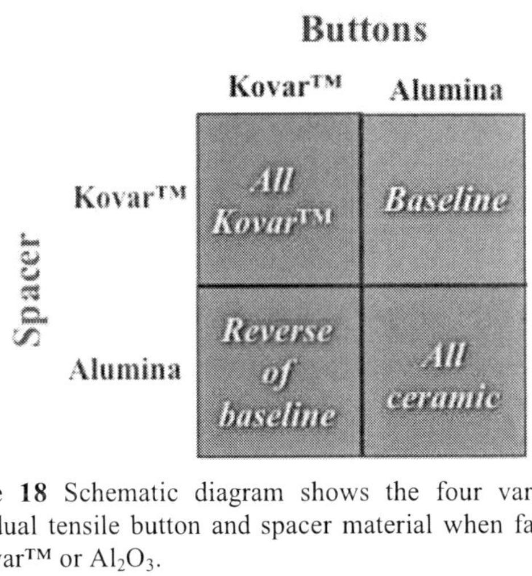

Figure 18 Schematic diagram shows the four variants of individual tensile button and spacer material when fabricated of Kovar™ or Al_2O_3.

A photograph of the test samples is shown in Fig. 19. The buttons and spacers were brazed together using a 0.002 in. thick preform of the Ag-Cu-Zr filler metal. The process conditions were 985°C, 5 min, and 600 torr Ar. The braze joint gap was controlled by dimples in the case of the Kovar™ spacer, and 0.001 in. ribbons when the Al_2O_3 spacer was in place.

Figure 19 Photograph shows the variants of the tensile button samples.

Each sample was inspected for the presence of run-out. Those results are summarized in Fig. 20. The baseline sample exhibited the expected run-out behavior. There was an absence of run-out in the all-Kovar™ braze joints. This finding confirms that the elemental Al released by the Zr/Al_2O_3 redox reaction is required for the run-out phenomenon to take place; it is not simply a physical displacement ("squishing out") of excess molten filler metal.

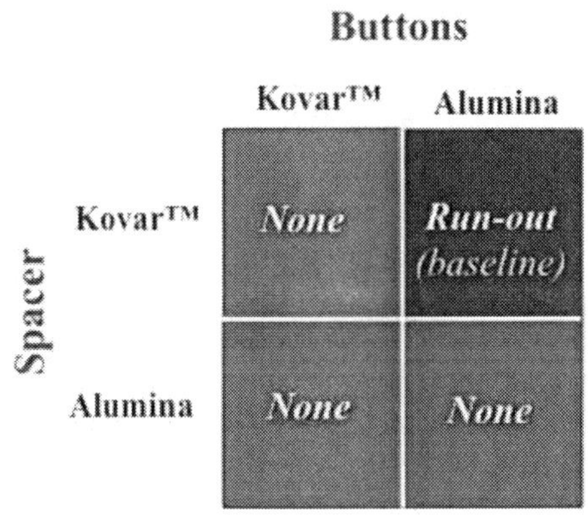

Figure 20 Schematic diagram shows the four variants between the spacer and button base materials.

The all-Kovar™ samples were examined by metallographic cross sections in order to determine the fate of the Zr component of the Ag-Cu-Zr filler metal. The microstructure of these braze joints is represented by the SEM image in Fig. 21. Reaction layers formed at both Ag-Cu-Zr/Kovar™ interfaces between Zr and the Fe, Ni, and Co constituents of the base material. Those reactions consumed all of the Zr content in the filler metal. The remaining filler metal contained Ag and the Cu constituent; the latter did not participate in the interface reaction.

There was a noticeable difference between the microstructures of the two reaction layers. That difference became less significant with decreasing gap thickness. There were no obvious reasons for the different layer morphologies. The EDX maps indicated that, qualitatively, the two layers had the same chemistry. The reaction layer compounds were heavier in Ni than in the Fe, despite the opposite proportionality of the two elements in the Kovar™ base material. The consequence was a slightly higher concentration of Fe in the Kovar™ immediately adjacent to the interface. Cobalt was also present in the reaction layers.

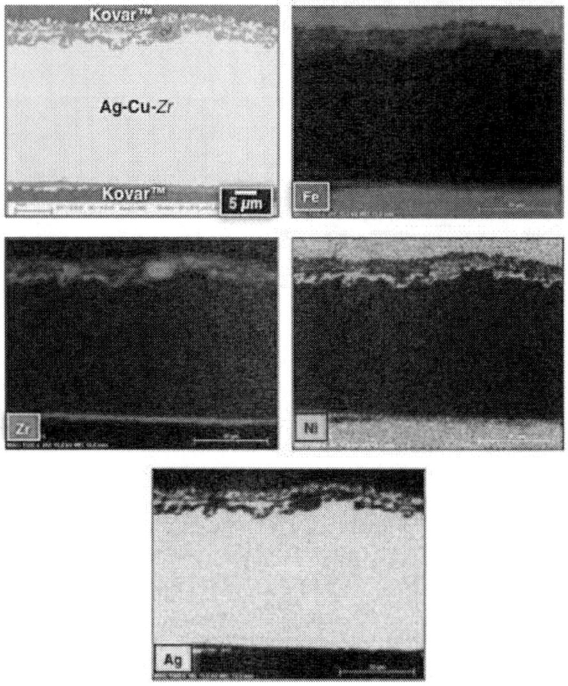

Figure 21 SEM image and EDX maps show the physical metallurgy and composition, respectively, of the Ag-Cu-Zr braze joint fabricated between the Kovar™ buttons and Kovar™ spacer.

This sample established, conclusively, that there is a significant driving force for Zr to react with the constituents of the Kovar™ base material. However, those reactions do not provide the driving force for the run-out phenomenon. The evidence to this effect is supported by the SEM image in Fig. 22. The cross section was taken of the fillet, just below the red dot in the inset photograph. The observable reaction layers extended as far as indicated by the yellow arrows. The filler metal wet and spread further onto, and up, the spacer and button, respectively, but not in a manner characteristic of run-out. Therefore, it is the aluminide reaction, not the zirconide reaction, that drives the run-out behavior when both can develop in the metal-ceramic braze joint.

The analysis turns to the samples having Kovar™ buttons, but a ceramic spacer. The SEM photographs in Fig 23 confirmed formation of the ZrO_2 zone that results from the redox reaction. It was expected that run-out would be observed, since the redox reaction generated elemental Al. But, such was not the case. Attention turned to the Ag-Cu-Zr/Kovar™ interface, which is shown in Figs 24a and 24b. This microstructure can be compared to Fig. 9. The high Al, $(Fe, Ni, Co)_x(Al, Zr)_y$ layer was absent that was located adjacent to the Kovar™ base material. The high-Zr, $(Zr, Al)_y(Fe, Ni, Co)_x$ alloy was observed along the entire interface and it was very thick. The second, high-Al layer occurred intermittently along the interface, being present as either isolated particles (Fig. 24b) or as a discontinuous layer. Although this Ag-Cu-Zr/Kovar™ interface structure was not typical as is shown Fig. 9, it was also not unusual amongst the many test specimens.

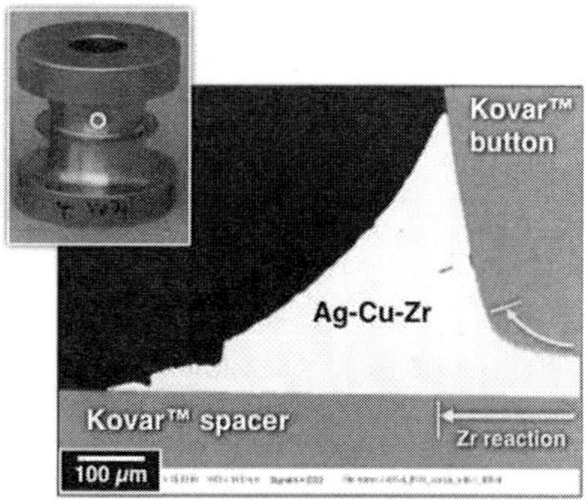

Figure 22 SEM image shows the cross section of the fillet that formed under the red dot of the inset photograph for the all-Kovar™ test sample made with the Ag-Cu-Zr filler metal. The yellow arrows indicate the extent of the zirconide reaction.

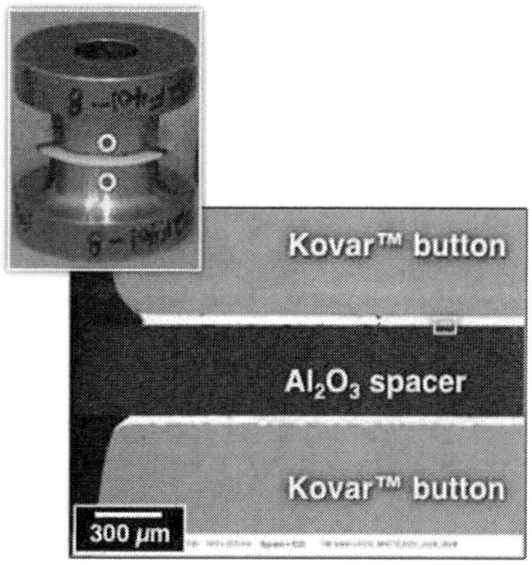

(a)

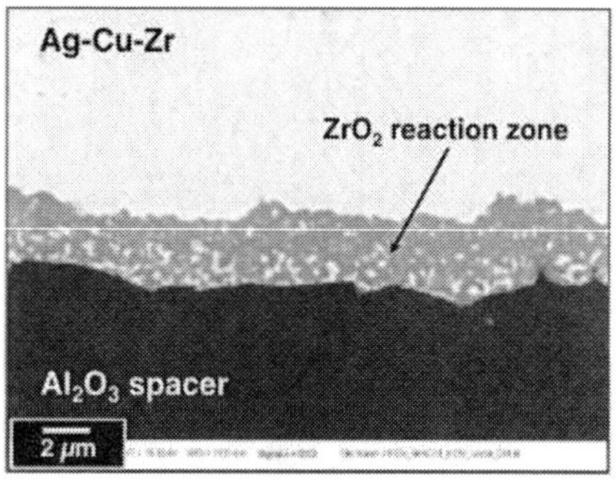

(b)

Figure 23 (a) SEM image shows the cross section of the Ag-Cu-Zr braze joint made between the Al_2O_3 spacer and the Kovar™ buttons. The inset photograph shows the sample; the cross section was made between the red dots. (b) High magnification SEM photograph shows the ZrO_2 reaction zone formed at the Ag-Cu-Zr/Al_2O_3 interface.

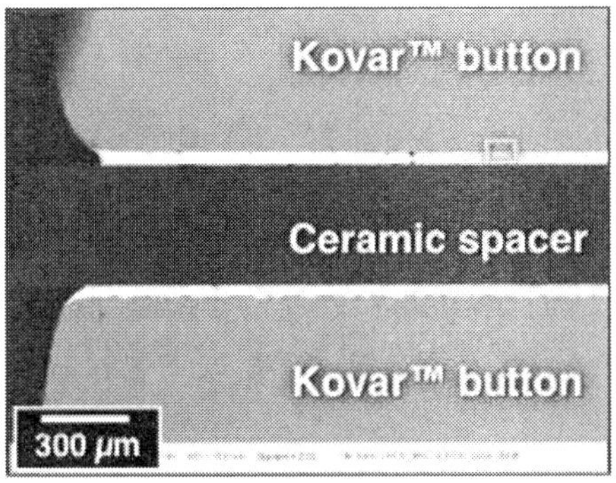

(a)

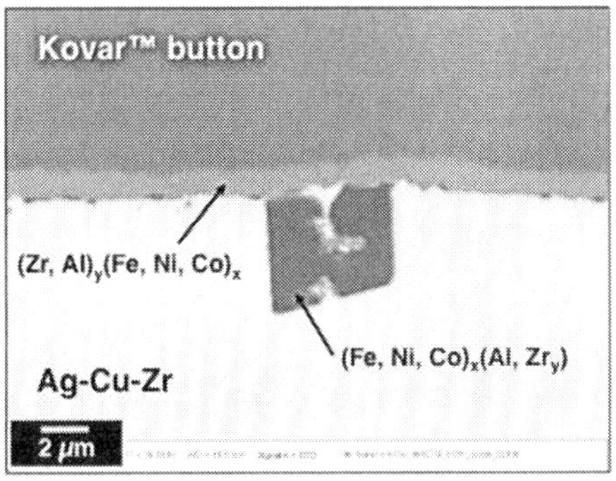

(b)

Figure 24 (a) SEM image shows the cross section of the Ag-Cu-Zr braze joint between the red dots of the inset photograph that illustrates the tensile button made with the Al_2O_3 spacer and Kovar™ buttons. (b) The SEM photograph shows the reaction layers that formed at the Ag-Cu-Zr/Kovar™ interface.

All of the physical metallurgy was in place to support a run-out event. The aluminide layer developed at the Ag-Cu-Zr/Kovar™ interface in the gap and extended through the fillets and well up the wall of the button in a manner indicative of the run-out phenomenon. This observation is illustrated in Fig. 25. The cross section of the Kovar™/Al_2O_3/ Kovar™ braze joint is repeated in Fig. 25a. The picture in Fig. 25b highlights the region in front of the fillet edge, which is shown in higher magnification in Fig. 25c. The corresponding EDX Al map in Fig. 25d, confirmed that the aluminide diffusion/reaction occurred well beyond the fillet edge. But, it could not be determined if the diffusion/reaction occurred locally or around the entire button circumference.

The aluminide diffusion/reaction observed in Fig. 25 should have led to run-out up the wall of the button. The absence of the phenomenon was due to the geometry of the fillet region and the contributing factor of molten filler metal surface tension.[1] This point is described with the assistance of Fig. 26. The solid blue line marks the limit of the filler metal fillet. The fillet shape that would be required in the event of run-out up the button, is described by the dashed trace. The latter fillet configuration would require the filler metal to wet an additional distance "A" further out, on the Al_2O_3 surface. However, the Ag-Cu-Zr active braze alloy cannot spontaneously spread on the ceramic surface. Therefore, although the aluminide diffusion/reaction can progress up the sidewall of the Kovar™ button uninhibited, the surface tension of the molten filler metal cannot permit the latter to follow the reaction layer to create a run-out lobe.

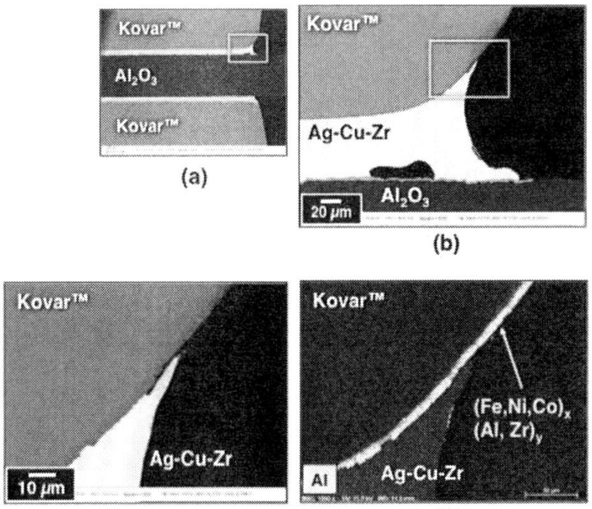

(c) (d)

Figure 25 (a) SEM image shows the cross section of the Ag-Cu-Zr braze joint made between the Al_2O_3 spacer and the Kovar™ buttons. (b) The medium magnification image shows the top of the filler metal fillet. (c, d) SEM image and EDX map of Al illustrate the extent of the aluminide diffusion/reaction beyond the fillet edge.

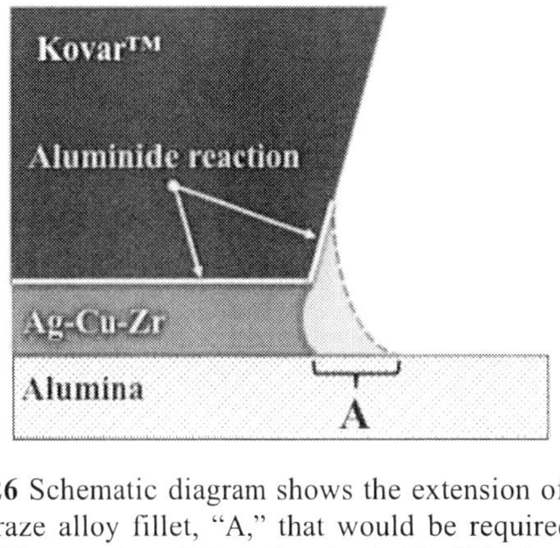

Figure 26 Schematic diagram shows the extension of the Ag-Cu-Zr braze alloy fillet, "A," that would be required for the molten filler metal to follow the aluminide diffusion/reaction up the sidewall of the Kovar™ button as run-out.

The final configuration in this experiment was the all-Al_2O_3 specimen. As expected, the redox reaction took place at both Ag-Cu-Zr/Al_2O_3 interfaces. Run-out was not observed on the Al_2O_3 surfaces, thereby confirming that the Kovar™ base material and, specifically, the aluminide reactions, were required for run-out to take place.

[1] Strictly speaking, interfacial tensions, of which the surface tension of the fillet is one part; joint geometry; and gravity control wetting and spreading by the molten filler metal.

In the absence of the Kovar™ base material, the Al_2O_3/Ag-Cu-Zr/Al_2O_3 provided the opportunity to measure the quantity of elemental Al generated by the redox reaction. Electron probe microanalysis (EPMA) was used to measure the concentrations of Al and Zr present in the filler metal field. One of the replicate traces is shown in Fig. 27.

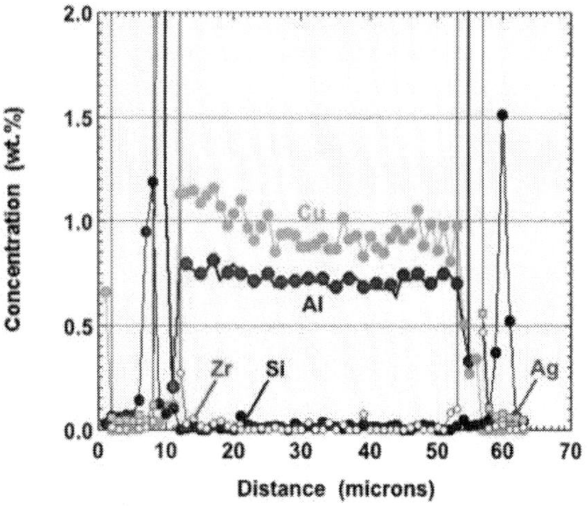

Figure 27 EPMA trace across the Al_2O_3/Ag-Cu-Zr/Al_2O_3 braze joint. The concentration scale was magnified in order to show the concentrations of Zr and Al in the bulk filler metal field.

The EPMA data confirmed that all of the 2 wt.% Zr component in the filler metal had been consumed by the redox reaction. In addition, the concentration of elemental Al resided was in the range of:

$$0.7 - 0.9 \text{ wt. \% Al.}$$

It is interesting that a relatively small quantity of elemental Al was responsible for the aluminide reactions and the run-out phenomenon observed in this brazement.

Lastly, this sample provide evidence with respect to the driving force behind the Zr/Al_2O_3 redox reaction. The absence of the Kovar™ base material eliminated the aluminide reactions as a contributor to that driving force. Therefore, it is concluded that either the free energy of solution, which results from elemental Al entering the filler metal, or an inaccuracy in the literature values of standard free energies of formation, was responsible for the spontaneity of the Zr/Al_2O_3 redox reaction.

In summary, the analysis of the mixed configuration, tensile buttons yielded important information regarding the run-out phenomenon: (a) The strong reaction between Zr and the constituent elements of the Kovar™ base material does not drive the run-out phenomenon. Rather, run-out is driven, expressly, by the aluminide reaction. (b) The geometry of the braze joint, in conjunction with the surface tension of the molten filler metal, can determine the presence or absence of run-out, even in the presence of the strong aluminide reaction layers.

Conclusions

1. The run-out behavior degraded braze joints made between Kovar™ and Al_20_3 ceramic using the active filler metal, 97Ag-2Zr-1Cu (wt. %). Because it did not respond to different parameters within the acceptable process window or to features added to the Kovar™ surface, a study was conducted to determine the root-cause of the underlying wetting and spreading instability as a first-step towards its mitigation.
2. Aluminum was released into the molten Ag-Cu-Zr filler metal by the redox reaction between Zr and the Al_2O_3 base material. The Al reacted with the Kovar™ constituents, creating an aluminide reaction layer that provided the primary driving force behind the wetting and spreading instability.
3. The surface tension of the molten Ag-Cu-Zr filler metal, in conjunction with the braze joint geometry, had a significant effect on the extent of the run-out.
4. The findings of this study indicate that the strategies having the greatest promise of successfully mitigating run-out are (a) alloy modifications to the filler metal, or (b) the use of coatings on the Kovar™ surface. The objective of either approach is to inhibit the aluminide reactions at the Ag-Cu-Zr/ Kovar™ interface.

Acknowledgments

The authors wish to thank Lisa Deibler for her careful review of the manuscript. Sandia is a multiprogram laboratory operated by Sandia Corporation, a Lockheed Martin Company, for the United States Department of Energy's National Nuclear Security Administration under contract DE-AC04-94AL85000.

References

[1] ASTM F19-11, "Standard Test Method for Tension and Vacuum Testing Metallized Ceramic Seals," (West Conshohocken, PA; 2011).

[2] "The Ag-Al (Silver-Aluminum) System," ed. by A. McAlister, *Bull. of Alloy Phase Diagrams,* Vol. 8, No. 6 (1987), p. 526.

[3] Yue, T., Yang, H., Li, T., and Huang, K., "The Synthesis of Graded Thermal Barrier Coatings on Nickel Substrates by Laser Induced Thermite Reactions," *Mater. Trans. Jap. Inst. of Metals*, Vol. 50, No. 1 (2009), pp. 219-221.

A New Iron-base Filler Metal for Brazing of Stainless Steel

Hai-Sheng Zhao, Hua-Ping Xiong*, Hui Pan, Jun-Feng Huai
Welding and Plastic Forming Division, Beijing Institute of Aeronautical Materials, AVIC, Beijing 100095, China
The corresponding author: xionghuaping69@sina.cn

Abstract

To solve corrosion failure problem of the copper-filler-brazed oil coolers, a new iron-base filler metal was designed, in which elements Si and B as well as Cu-Ti binary alloy were added as the temperature depressants. This brazing alloy was fabricated into filler foils by a rapidly-solidifying technique. The addition of Cu-Ti binary alloy decreased the needed amount of Si and B, thus it had an effect on the improvement in the mechanical properties of the brazed joints. Vacuum brazing of stainless steel was carried out at 1120 ℃ for 15 min. The corresponding joint microstructure was analyzed by means of SEM, XEDS and XRD, and typical joint was mainly composed of solid solution with a small quantity of $(Fe, Cr)_2B$, (Ti, Fe)B and (Fe, Ni, Ti)B compounds. The brazed joints offered an average tensile strength of 375.3 MPa at room temperature, about 15% higher than that of copper-filler-brazed under the identical joining condition. Moreover, compared with nickel-base filler metal, the new iron-base filler metal showed the advantages of lower cost and better mechanical properties.

Introduction

Austenitic stainless steel containing much Cr and Ni elements, has good corrosion resistance in any environment of oxidizing, neutral or light reductive medium. Due to good plasticity and toughness as well as cold-workability, the austenitic stainless steel has been widely used in the fields of automotive industry, food industry, medical apparatus and instruments, petrochemical industry and aerospace, etc. [1]. Brazing technique for stainless steel has been successfully used with silver-based filler metals, manganese filler metals, nickel-based filler metals and gold-based filler metals, and copper filler metal in inert gas atmosphere [2-6]. Due to the fluctuating costs of raw materials such as nickel, new Ni-based filler metals with lower nickel content or Fe-Cr-based filler metals were investigated recently [7-8]. However, the majority of stainless steel brazing is still carried out using nickel-based filler metals in a vacuum environment, but the gap clearances of the brazing joints should be strictly controlled to achieve joints with satisfactory properties. In fact, presently in automotive industry, pure copper filler was widely used for the brazing of oil cooler [9], but the copper-filler-brazed joints were easy to be corroded in oil medium due to the formation of $CuSO_4$, resulting in the failure of oil coolers after a certain life time. Furthermore, the mechanical properties of the copper filler brazed joints remain to be improved.

To solve the above problem, a new type of Fe-Ni-Cr system filler metal was designed and fabricated into filler foils for the subsequent brazing experiments in the present paper. Firstly, considering the corrosion problem, Fe-Ni-Cr system filler metal should be undoubtedly prior to the traditional copper filler. Secondly, elements Si and B were added into the new Fe-Ni-Cr system filler alloy to lower the melting temperature of the filler alloy, and equally important, to ensure corrosion- and oxidation-resistance of the joints. But the amount of Si and B was strictly limited, because they are brittle compound forming elements [6, 10]. Thirdly, according to the binary phase diagram, elements Cu and Ti can react with each other forming various low melting point intermetallics. Therefore, Cu and Ti were added into the filler metal to further lower its melting temperature and accordingly the needed amount of Si and B for the decrease of melting point can be decreased. It is believed that the decrease of the amount of elements Si and B is favorable to the improvement of the mechanical properties of the brazed stainless steel joint.

This paper presents the microstructure of the joints brazed with the new Fe-Ni-Cr system filler metal and the mechanical properties of the corresponding joints. The evolution of the joint phases and the joining mechanism are discussed.

Experiments

The experimental base metal is 1Cr18Ni9Ti stainless steel, and its nominal composition is shown in Table 1.

New designed filler alloys were prepared by vacuum arc-melting method, and the achieved ingots were broken up to pieces. The filler sample was heated to 1120 ℃ for 15 minutes in the ZKH-1 vacuum brazing furnace in order to study their melting behavior and their wettability on the base metal. The melting sample is showed in Figure 1.

In this paper, the selected alloy is named F20 with a nominal composition shown in Table 2. According to a differential thermal analysis (DTA), the solidus temperature and liquidus temperature of this filler are respectively 1042 ℃ and 1052 ℃. The brazing foil of F20 was fabricated in a vacuum and argon-protecting furnace with a rapidly-solidifying technique. According to the previous melting experimental results, vacuum brazing was conducted with these foils under 1120 ℃/15 min. The mechanical properties of the brazed joints were tested and contrasted to the copper-filler-brazed joints at the same brazing and testing conditions. The microstructures of the brazed joints were analyzed by means of scanning electron microscope (SEM) and X-ray energy-dispersive spectrometer (XEDS).

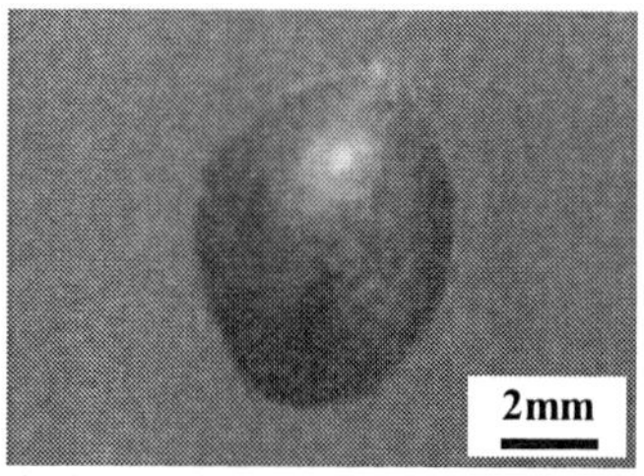

Figure 1 Morphology of the wetting specimen after heated at 1120 ℃ for 15 minutes

Table 1 Nominal composition of 1Cr18Ni9Ti stainless steel

Element	Fe	C	Mn	Si	Cr	Ni	S	P	Ti
wt.%	Bal.	≤0.12	≤2.0	≤0.80	17.00~19.00	8.00~11.00	≤0.025	≤0.035	5(C-0.05)~0.8

Table 2 Nominal composition range of the F20 filler metal

Element	Fe	Ni	Cr	Si	B	Cu+Ti
wt.%	Bal.	12.0~21.0	6.0~15.0	3.5~5.5	2.0~3.0	3.0~11.0

Results and Discussion

Mechanical Properties

Table 3 shows the tensile strength of the brazed joints using F20 filler metal compared with that using copper filler under the same joining conditions. After 1120 ℃/15 min brazing procedure, the average tensile strength of the F20 brazed joints is 375.3 MPa, 15% higher than that of copper-filler-brazed joints, and 64% higher than that of traditional BNi82CrSiB brazed joints [11].

Table 3 Comparison of the tensile strength of the brazed joints using two kinds of filler metals

Filler metal	Tensile strength σ_b, MPa			
	Individual value			Average value
F20 (Fe-Ni-Cr based)	422.1	345.0	358.8	375.3
Copper	406.9	324.1	249.1	326.7

The SEM images of the fracture surfaces of the brazed samples can be seen in Figure 2. On the fracture surface of the F20 brazed joint there exist many tearing edges and a small quantity of dimples, and furthermore it exhibits more homogeneous and fine microstructure. In contrast, the surface of the copper-filler-brazed joint is very coarse and flat. This result is in good agreement with the improved mechanical properties brazed with the F20 filler metal.

Figure 2 Fracture surface of specimen: (a) F20 brazed joint; (b) copper-filler-brazed joint

Analysis of joint Microstructure

Figure 3 shows the microstructure of the typical joint brazed with F20 filler metal. The joint is composed of diffusion-reaction zones (I), solid solution zones (II) and central zone (III). During the brazing procedure, active elements of B and Si diffused into the base metal and reacted with Fe, Cr and Ni forming grayish phase, as point "1" shows in Figure 3b. Of the diffusion-reaction zones, the microstructure change may lead to stress concentration when bearing stress, so it may be the weak zone determining the mechanical properties of the joint.

Near the interface in the base metal, there exist some small black blocks which may be titanium carbides as point "2" shows. There are two solid solution zones between diffusion-reaction and central zone. At the beginning of cooling stage after brazing process, solid solution formed along the interfaces and grew into the seam center, accordingly low melting point liquids were pushed to the seam center and solidified with preformed solution at a relatively low temperature [12]. So the central zone is composed of solid solution and a small quantity of compounds with small size (black particles as point "3" and dark particles as point "4").

To study the phase composition of the joint, XEDS analysis was carried out. According to the results in Table 4, the microzone "5" in the base metal almost has the same chemical composition as nominal, and microzone "6" in II zone and microzone "7" in III zone have similar elements compositions and are both Fe-based solid solutions which have some solubility of Ni, Si, Cu and Cr, but the amount of Si in microzone "7" is a bit higher than that in microzone "6" due to its diffusion into base metal and gathering in the brazing seam center [13]. Combined with Fe-B and Ni-B phase diagrams, the grayish phase in I zone may be $(Fe, Cr)_2B$ in which B is diffused from the filler metal into the base metal. The black block as point "2" in the base metal near the interface contain lots of Ti and should be TiC compounds. In the III zone, the black particles contain lost of Ti, Fe and B, thereby they may be (Ti, Fe)B. And the dark particle as point "4" containing more Fe, Ni and Si may be (Fe, Ni, Ti)B with some element Si dissolved in it.

To find out the diffusion and relative content of the elements in the brazed joint, Element area distribution analysis of the Figure 3c was carried out and the result is presented in the Figure 4. Almost all of element Ti concentrated in the borides of black or dark particles which are lack of Fe, Ni and Si. Other elements, such as Cr, Si and Cu, were uniformly distributed in the whole brazing seam. In F20 filler metal, a small quantity of Ti can form some compounds to lower the melting temperature of the filler metal, then segregated as TiB particles. And these small particles may strengthen the brazed joints to some extent.

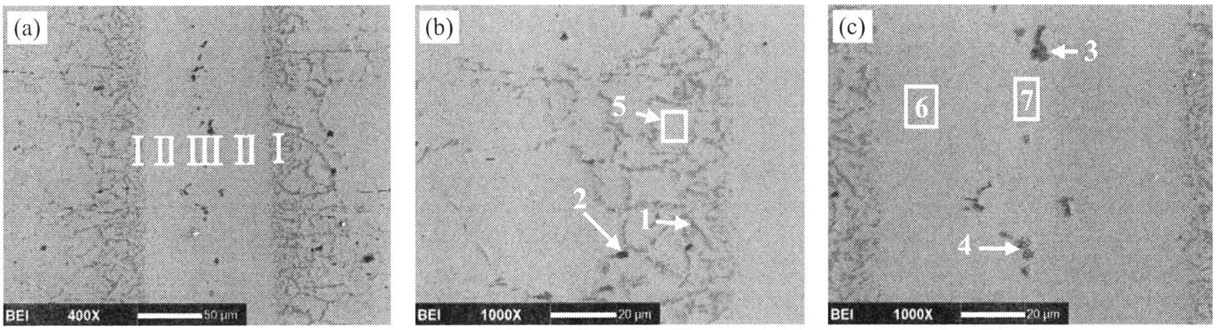

Figure 3 The microstructure of the 1Cr18Ni9Ti brazed joint using F20 filler metal: (a) The whole joint; (b) Diffusion-reaction zone; (c) Central zone

Table 4 The XEDS analysis of the typical phases and regions (at. %)

Microzone/point	Fe	Ni	Cr	Si	B	Cu	Ti	Possible phases
"1" grayish phase	36.16	2.21	33.80	—	26.67	Mn0.94	0.22	$(Fe, Cr)_2B$
"2" black block	3.67	0.41	1.46	—	—	C12.62	81.84	TiC
"3" black particle	13.75	5.74	2.16	3.05	54.98	0.59	19.73	(Ti, Fe)B
"4" dark particle	27.81	11.90	2.04	6.72	43.55	1.35	6.63	(Fe, Ni, Ti)B
"5" solid solution	72.36	9.10	16.45	—	Mn1.45	—	0.64	$(Fe, Cr, Ni)_{SS}$
"6" solid solution	63.51	18.93	5.21	9.68	—	2.47	0.20	$(Fe, Ni, Si, Cr, Cu)_{SS}$
"7" solid solution	58.45	22.57	3.55	12.36	—	2.77	0.30	$(Fe, Ni, Si, Cr, Cu)_{SS}$

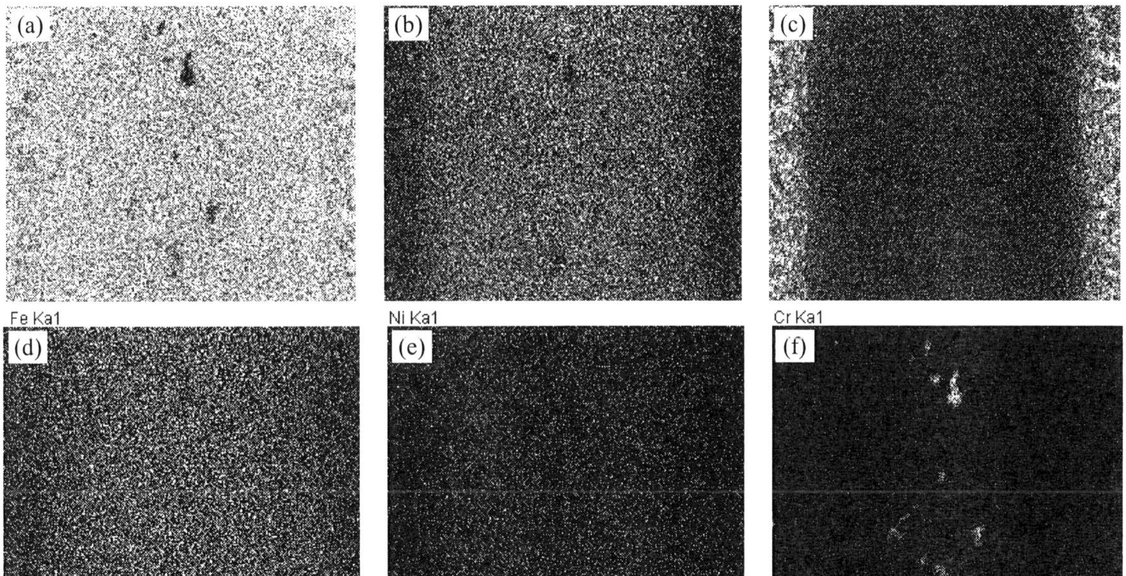

Figure 4 Element area distribution maps of the brazed joint: (a) Fe; (b) Ni; (c) Central zone; (d) Si; (e) Cu; (f) Ti

Similarly, the microstructure of the copper-filler-brazed joint was analyzed, and the backscattering electron images are presented in Figure 5. The microstructure is relatively simple and homogeneous. There is not obvious diffusion-reaction zone between base metal and brazing seam, but diffusion and solution can be seen along the interface. During the brazing procedure, some grains of the base metal dissolved into the melted copper, then solidified and distributed throughout the brazing seam as the grayish particles as point "8" shows. Meanwhile, melted copper diffused into the base metal, especially along the grain boundaries, and weakened the boundary of the base metal accordingly. Combined with Cu-Fe binary phase diagram, the microstructure of the brazing seam is a eutectoid structure based on Cu-base solid solution with a few α-Fe particles segregated on it.

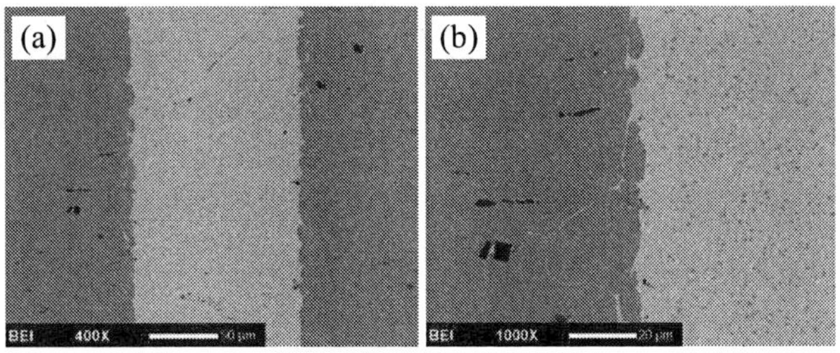

Figure 5 Microstructure of the 1Cr18Ni9Ti brazed joint using copper filler metal: (a) The whole joint; (b) Interface

The phase composition was identified from the fracture surface by X-ray diffraction showed in Figure 6, indicating that the brazed joint using F20 filler mainly consists of Fe-base solid solution, as well as a small amount of compounds of Fe_2B. And the copper-filler-brazed joint is mainly composed of Cu-base solid solution. Thus, the fracture of F20 brazed specimen occurred in the diffusion-reaction zone, whereas the copper-filler-brazed specimen fractured in the brazing seam.

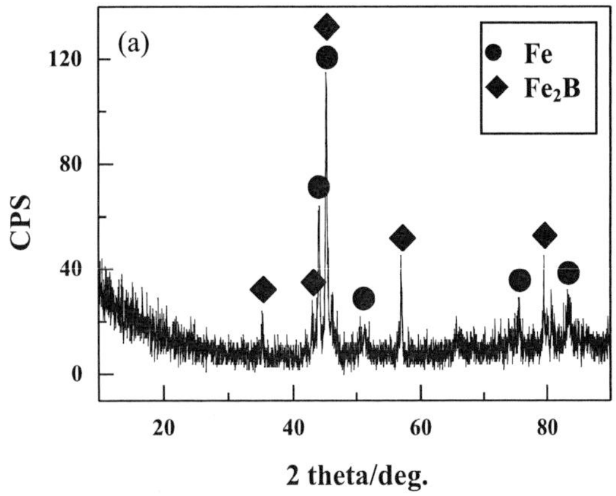

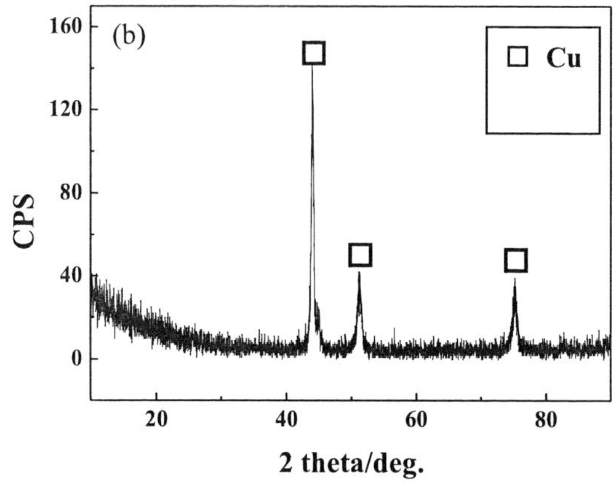

Figure 6 X-ray diffraction analysis of specimen using; (a) F20 and (b) Copper filler metal

Conclusions

In this paper, vacuum brazing of 1Cr18Ni9Ti stainless steel was carried out under the brazing condition of 1120 ℃/15 min using a new designed Fe-Ni-Cr system filler metal with Si and B as well as Cu-Ti binary alloy added in as the temperature depressants. The mechanical properties of the joints using this filler metal were evaluated compared with that using pure copper filler metal. Primary conclusions are summarized in the following:

1. Typical joint is composed of diffusion-reaction zones, solid solution zones and central zone.
2. The brazed joints using the new designed Fe-Ni-Cr system filler metal showed an average tensile strength of 375.3 MPa, 15% higher than that of copper-filler-brazed joints under the identical joining condition.
3. The fracture surface of the F20 brazed joint reveals tearing edges and a small quantity of dimples, and exhibits more homogeneous and fine microstructure, indicating good mechanical properties of the brazed joint. In contrast, the fracture surface of the copper-filler-brazed joint is very coarse and flat.
4. The diffusion-reaction zone bases on primary Fe-based solid solution with a small quantity of $(Fe, Cr)_2B$, and the central zone of the brazed joint is composed of solid solution and small particles of (Ti, Fe)B and (Fe, Ni, Ti)B.

References

[1] Zhang Q. Y., Zhuang H. S., Brazing and Soldering Manual, (Beijing, 2008), pp. 199.
[2] Pan H., Sun J. S., Liu X. F., "Microstructure of Braze Metal of Stainless Steel Brazed with Bag25CuZn alloy," Journal of Aeronautical Materials, Vol. 20, No. 3 (2000), pp. 94-97.
[3] Ou C. L., Liaw D. W., Du Y. C., "Brazing of 422 Stainless Steel Using the AWS Classification BNi-2 Braze Alloy," Journal of Materials Science, Vol. 41, No. 9 (2006), pp. 6353-6361.
[4] Yan G. et al, "Effects of Filler Metal on Microstructures

of Stainless Steel Brazed Joint," Casting · Forging · Welding, Vol. 40, No. 11 (2011), pp. 179-181.
[5] Jing J. N. et al, "Microstructure and Microhardness of 316L Stainless Steel Joint Seam Vacuum Brazed by BNi7 Ni-Based Filler Metal," Materials for Mechanical Engineering, Vol. 37, No. 1 (2013), pp. 10-13.
[6] Fan Y. H., Yan J., "Torch-Brazing of T2 Copper with 1Cr17Ni2 Stainless Steel," Development and Application of Materials, Vol. 24, No. 3 (2009), pp. 42-45.
[7] Ulrika P., "New Iron-Chromium Based Brazing Filler Metal for Demanding Stainless Steel Applications," Proceeding of the 4th International Brazing and Soldering Conference, Orlando, Texas, USA, Apr. 2009, pp. 125-129.
[8] Matsu K. et al, "Mechanical Properties of Fe-Cr Based Brazing Filler Metals," Proceeding of 10th International Conference on Brazing, High Temperature Brazing and Diffusion Bonding Conference, Aachen, Germany, Jun. 2010, pp. 48-51.
[9] Xu R. H. et al, "Corrosion Resistance of Pure Copper-Filler-Brazed Joint of Oil Cooler," Journal of Welding - Production Theme, No. 3 (2011), pp. 35-37.
[10] Tokunaga T., Nishio K., Hasebe M., "Thermodynamic Study of Phase Equilibria in the Ni-Si-B System, Basic and Applied Research: Section Ⅰ," Journal of Phase Equilibria, Vol. 22, No. 3 (2001), pp. 294-298.
[11] Pan H., Zhao H. S., Sun J. S., "Effects of BNi82CrSiB Brazing Alloy in Different Forms on Microstructures and Mechanical Properties of the Brazed Joints," Materials Science and technology, No. 17(Sup. 1, 2009), pp. 9-11.
[12] Yeh M. S., Chuang T. H., "Effects of Applied Pressure on the Brazing of Superplastic INCONEL 718 Superalloy," Metallurgical and Materials Transactions A, Vol. 28, No. 6 (1997), pp. 1367-1376.
[13] Grushko B., Weiss B. Z., "The BNi-5-Inconel 718 "Binary" System," Materials Science and Engineering, Vol. 74, No. 1 (1982), pp. 19-27.

Microstructure and Strength Properties of Austenitic and Ferritic Stainless Steels When Brazed with Ni-Cr-P Amorphous Brazing Foils

Eric Theisen and Will Coughlan
Metglas, Inc., Conway , SC, USA
eric.theisen@metglas.com

Abstract

Austenitic and ferritic stainless steel are important alloys used in the manufacturing of heat exchangers. These alloys are used in heat exchangers in automotive industry applications such as exhaust gas re-circulator and heat recovery units. They are also used in heat exchangers in the power generation industry and oil and gas industry. The following paper describes the microstructure and single-lap shear strength results of both types of stainless steel alloys when joined with Ni-Cr-P amorphous brazing foils.

Introduction

Austenitic stainless steel (SS316) and ferritic stainless steel (SS444) are important alloys used in the manufacturing of heat exchangers. These alloys are used in heat exchangers in automotive industry applications such as exhaust gas re-circulator and heat recovery units. They are also used in heat exchangers in the power generation industry and oil and gas industry. Further details on automotive applications for braze foils can be found elsewhere [1]. High corrosion resistance is often required for heat exchangers and previous studies have focused on the role of high Cr containing foils for preventing corrosion [2, 3, 4]. It has been shown that the corrosion performance of the alloys becomes much worse with high levels of B and low levels of Cr in the braze filler metal and results in chromium boride precipitation. This pulls the Cr from the solid solution state of the base material reducing the corrosion resistance in the braze interface [5].

Substituting the B for P in the foil prevents the chromium boride precipitation in the base material. However, there is little information on the joint strength properties available. The following study describes the microstructure and single-lap shear strength results of both types of stainless steel alloys when joined with Ni-Cr-B based foils as well as Ni-Cr-P based foils. Table 1 shows the chemistry and melting characteristics of the MBF51 and MBF67 foils which are available in commercial form. The MBF67 foil is designed to be a low melting point, high corrosion resistant alloy for applications such as exhaust gas recirculation coolers.

Table 1: Metglas Brazing Foils nominal chemistries and melting characteristics.

Alloy	Nominal Composition, wt%							Melting Temp C	
	Cr	Fe	Si	B	P	Mo	Ni	Sol	Liq
MBF51	15	--	7.3	1.4	--	--	Bal	1030	1126
MBF67	25	<1	1.5	0.5	6	1.5	Bal	890	970

Experiments

Samples of SS316 and SS444 brushed to a surface finish of 0.17 micron average roughness were obtained for brazing trials using MBF51 and MBF67 foil. The brazing process and mechanical testing procedures followed the AWS C3.2M procedure for single-lap shear specimens. Figure 1 shows images of the samples after the braze cycle where the sample thickness is 0.125 " and the braze overlap distance is 0.5 ". Brazement samples joining two 4 " by 1 " SS316 and SS444 sample coupons were also carried out in the vacuum brazing furnace for cross-sectional analysis. The samples of SS316 and SS444 were put in the vacuum furnace together, along with the sample coupons for both of the MBF51 foil and MBF67 foil heating cycles. The vacuum in the furnace was held at 10^{-4} torr minimum during the brazing cycle. The furnace temperature was ramped up and held for 20-45 minutes at dwell temperature 1 (260°C) and then ramped to dwell temperature 2 and held for another 20-45 minutes to ensure temperature and pressure stabilization. The temperature was then ramped to required brazing temperature for each alloy and held for 15 minutes. Table 3 lists the dwell and brazing temperatures for each alloy. The furnace was then back filled with nitrogen and quick cooled to a temperature about 200-300°C below which oxidation of joints would occur.

After the brazing cycles were completed the shear lap samples were analyzed using an Instron machine for the average stress at failure. The results of the tests for each base material and foil combination are given in Table 3. It is seen that the samples using the MBF51 foil have much higher stress at failure and that the failure occurs within the base material for both the SS316 and SS444 alloys. The samples using the MBF67 foil showed lower stress at failure and fractured within the braze layer. When the failure occurs in the braze layer, the maximum stress on the base material is estimated using the maximum load at failure. Similarly, when the failure occurs in the base metal the corresponding joint shear stress is reported and it is implied that the joint strength is greater than the reported value since the failure did not occur in the joint.

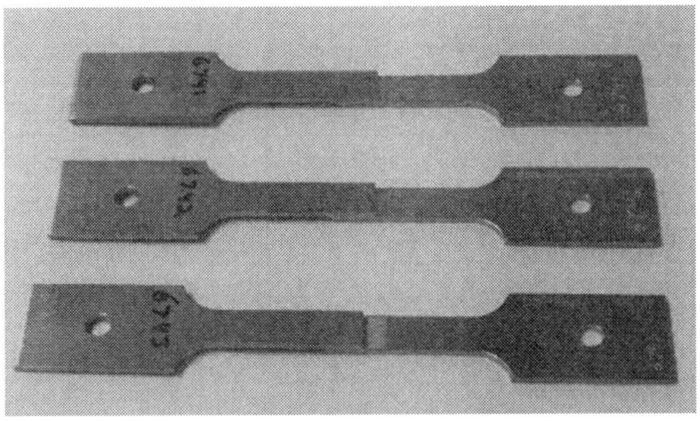

Figure 1. Single-lap shear specimens of SS444 after vacuum brazing with MBF67 foil.

Table 2: Dwell temperatures and brazing temperature for each of the MBF alloys.

Alloy	Dwell Temp. 1	Dwell Temp. 2	Brazing Temp.
MBF 51	260°C	1000°C	1195°C
MBF 67	260°C	870°C	1090°C

Table 3: Joint stress at maximum load for SS316 and SS444 brazed with MBF51 and MBF67 foil.

Alloy	Foil	Failure Location	Base Metal Stress at Max Load (ksi)	Joint Stress at Max Load (ksi)
SS316	MBF51	Base Material	86.1 +/- 0.8	> 21.5
SS316	MBF67	Braze Layer	> 43	10.8 +/- 0.5
SS444	MBF51	Base Material	54.6 +/- 4.6	> 13.6
SS444	MBF67	Braze Layer	> 43	10.7 +/- 1.3

Following the vacuum braze cycle, the brazement sample coupons were cross sectioned and then progressively polished down to a sub-micron colloidal silica solution finish using standard metallographic techniques. The samples where then analyzed on a Hitachi S3400N Scanning Electron Microscope (SEM) and an elemental map using an EDAX Genesis energy dispersive x-ray analysis was performed.

Results and Discussion

The SEM images from the brazement sample coupons for SS316 and SS444 brazed with MBF51 foil are shown in Fig. 2a and Fig. 2b respectively. It is observed that there is a much greater penetration into the base material for the SS316 material relative to the SS444 material and has been observed previously [6].

a)

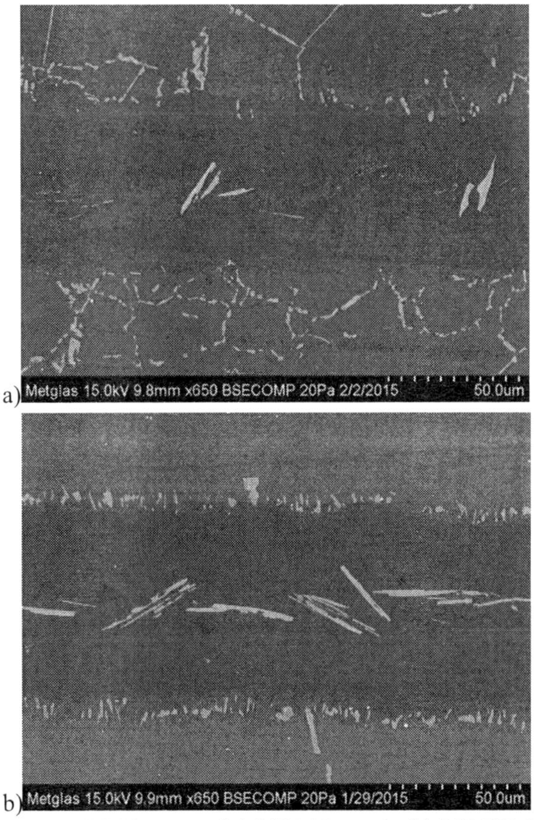

b)

Figure 2. SEM images of a) SS316 brazed with MBF51 foil and b) SS444 brazed with MBF51 foil.

The elemental maps for the SS316 material are given in Fig 3 showing the relative concentrations of Cr, Fe, Ni and Si. Light elements such as B are not easy to identify in this investigation but it is inferred that the B diffuses into the base metal leaving a solid solution of Ni, Cr and Si in the braze joint as discussed elsewhere [6]. This results in high levels of chromium borides in the grain boundaries of the base SS316 and a corresponding absence of Fe and Ni.

Figure 4 shows the elemental maps for the SS444 base material brazed with MBF51 foil. There is not the same base material penetration along the grain boundaries indicating that the B remains mainly within the braze layer. Here it is inferred that the chromium boride phases remain along the interface of

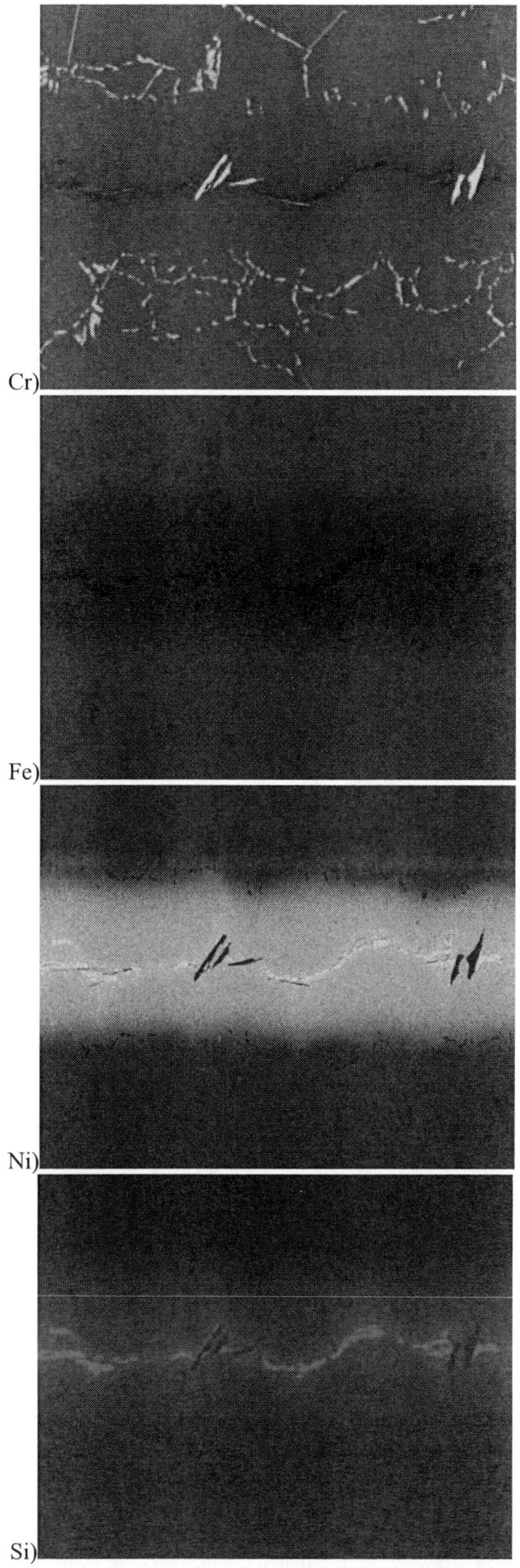

Figure 3. Elemental analysis of SS316 brazed with MBF51 showing the relative concentrations of Cr, Fe, Ni and Si.

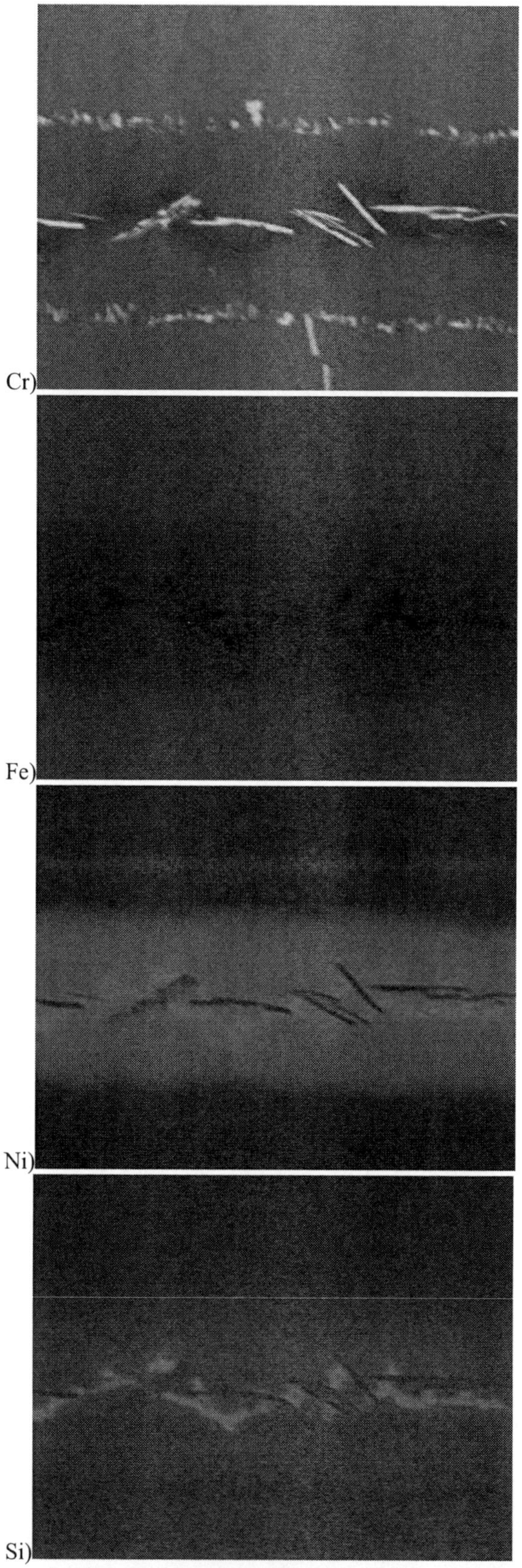

Figure 4. Elemental analysis of SS444 brazed with MBF51 showing the relative concentrations of Cr, Fe, Ni and Si.

the base material and along the centerline of the braze layer where the strong presence of Cr is observed. There are also Ni-Si intermetallic phases along the centerline of the braze layer which is surrounded by a solid solution of Ni-Cr-Si-Fe.

The SEM images from the brazement sample coupons for SS316 and SS444 brazed with MBF67 foil are shown in Fig 5a and Fig 5b respectively. It is observed that there is a very little penetration into the base material through the grain boundaries due to the low B levels of MBF67 foil. The interface between the base material and the braze layer is sharper for the SS444 material compared to the SS316 material.

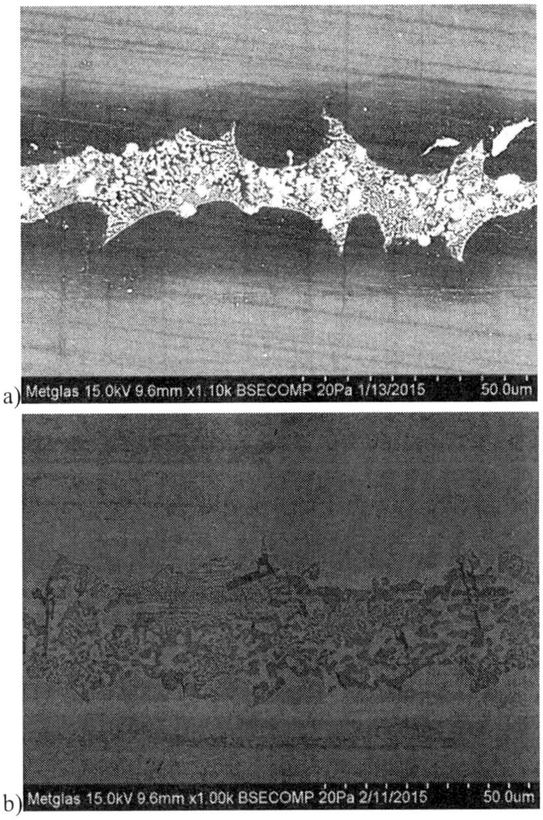

Figure 5. SEM images of a) SS316 brazed with MBF67 foil and b) SS444 brazed with MBF67 foil.

The elemental maps for the SS316 material is given in Fig 6 showing the relative concentrations of Cr, Fe, Ni and P. Here the center line of the braze layer consists of Ni-Cr-P intermetallic phases along with block phases of Cr. The center line is surrounded by a Ni-Cr solid solution. Similar results for the elemental maps SS444 brazed with MBF67 foil are shown in Fig 7. The major difference noted between the elemental maps in Fig. 6 and Fig. 7 is the Cr distribution. It is very uniform throughout the brazement in the SS316 samples, with higher concentrations seen in the blocky chromium precipitates. However, for the SS444 samples there is a lower concentration of Cr seen in the Ni rich interface zone of braze joint.

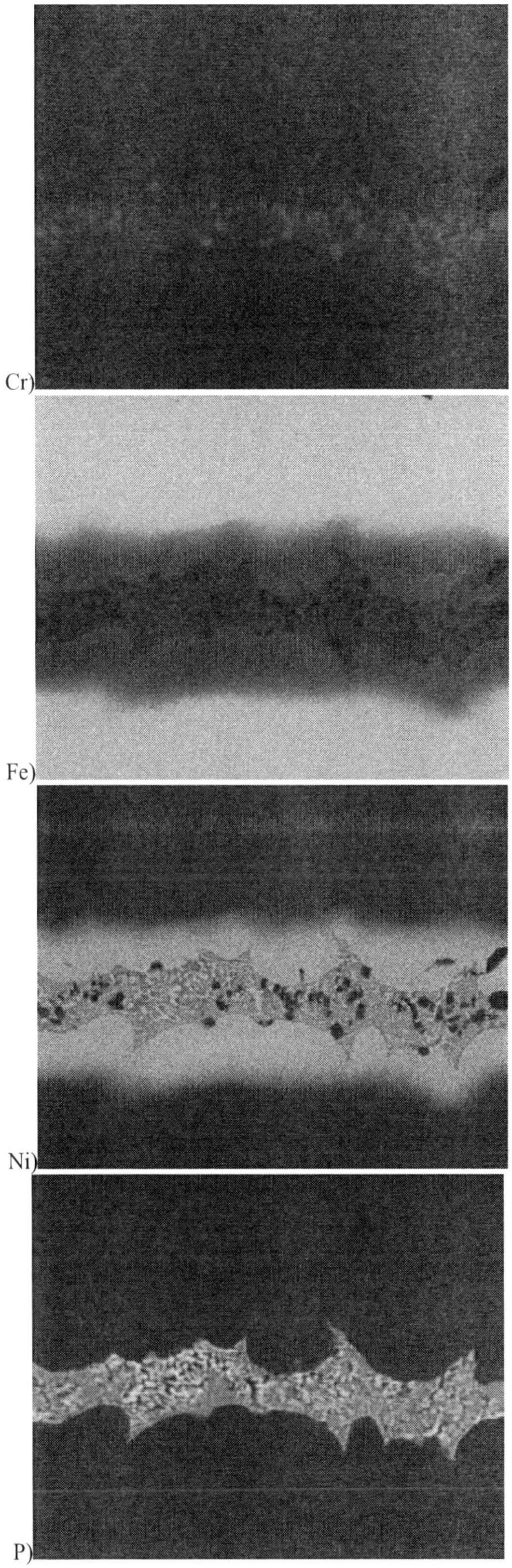

Figure 6. Elemental analysis of SS316 brazed with MBF67 showing the relative concentrations of Cr, Fe, Ni and P.

Figure 7. Elemental analysis of SS444 brazed with MBF67 showing the relative concentrations of Cr, Fe, Ni and P.

The joint strength data given in Table 3 show that the failures occur within the base SS316 and SS444 material when using the MBF51 foil. The failures occur near the ultimate tensile strength of the base material, which are 75 ksi and 40 ksi respectively, and indicate a strong and ductile braze joint. However, the failure location is within the braze joint for each of the MBF67 brazed samples and occur at the same applied stress for both the SS316 and SS444 base materials. The intermetallic phases within the braze layer are brittle and allow for crack propagation as discussed elsewhere [7]. This is likely the cause of the lower shear strength in the material brazed with the MBF67 foil. Analysis of the single-lap shear samples show that the braze layer fractures along the center line of the braze layer.

Conclusions

Newly developed MBF67 foil is a Ni-Cr-P-Si based foil the can be cast with good ductility, high corrosion resistance and low melting points [2, 3 ,4]. Brazing austenitic stainless steel and ferritic stainless steel with MBF67 foil is compared to brazing with MBF51 foil, a Ni-Cr-B-Si based foil, and significant differences are found. The MBF51 brazements show deep B diffusion within the grain boundaries of the base material for austenitic SS316. A solid solution of Ni-Cr-Si remained in the braze layer. However, the ferritic SS444 did not show the B penetration into the base material. Rather B seems to remain in the braze layer and a number in intermetallic phases remained in the braze layer. The MBF67 brazements also show very little penetration into the base metals for both SS316 and SS444 materials. There also tends to be more intermetallic phases contained within the braze layer for the P containing foils. The intermetallic phases in the MBF67 brazement reduces the shear strength of the braze layer and shows fracture along these phases. The stress at failure was almost the same when the failure occurs within the braze layer for the samples brazed with MBF67 foil. The stress at failure was near the ultimate tensile strength of the base materials for the samples brazed with MBF51 foil. While other studies have shown that there is a greater corrosion resistance in the brazement when using MBF67 foil, there seems to be a trade-off where the joint strength is not as large. The choice of brazing foil may depend on which aspect is more important for the application.

References

[1] Rabinkin, A., "New brazing development in the automotive industry," IBSC 2009: Proceedings from the 4th International Brazing and Soldering Conference, (2009), pp. 380-386.

[2] Weinstein, M., and Lee, L., "Further developments in boron free Nickel-Chromium-Phosphorus-Silicon brazing filler metals"," Wall Colmonoy Technical Article, (2012) www.wallcolmonoy.com.

[3] Hartmann, T., and Nuetzel, D., "Chromium containing amorphous brazing foils and their resistance to automotive exhaust gas condensate," IBSC 2012: Proceedings from the 5th International Brazing and Soldering Conference, (2012) pp. 394-400.

[4] Theisen, E., and Coughlan, W., "Recent Developments in Amorphous Brazing Foils," 2014 International Conference on Brazing, Soldering and Special Joining Technologies, (2014) pp. 120-124.

[5] Chen, W., and Shiue, R., "Brazing Inconel 625 Using Two Ni/(Fe) Based Amorphous Filler Foils," Met Trans A, Vol 43 issue 7, (2012), pp. 2177-2182.

[6] Leone,E., et al, "Microstructure of thin gauge austenitic and ferritic stainless steels brazed using Metglas amorphous foils," Weld.World, 50, N1, (2006) pp. 3-15.

[7] Yuan, X., et al, "Microstructures, Mechanical and Chemical Properties of TLP-Bonded Joints in a Duplex Stainless Steel with Amorphous Ni-Based Insert Alloys," Met Trans A, Vol43A, (2012) pp. 1989-2001.

Brazing of Ferrite Stainless Steel with Ni-Based Amorphous Brazing Foils

Yasuyuki Miyazawa, Masato Ishihara, Kei Ota
Tokai University, Hiratsuka-shi, Kanagawa-ken, Japan
Email: ymiyazawa@tokai-u.jp
Phone: +81-463-58-1211

Yoshio Bizen
Hitachi Metals, Ltd., Minato-ku, Tokyo, Japan
Email: yoshio.bizen.bq@hitachi-metals.com
Phone: +81-3-5765-4520

Abstract

Ni-based brazing filler metal is common filler metal for stainless steel brazing. Microstructure at the austenite stainless steel brazed joint with Ni-based filler consists of many kinds of phases. These phases influence strongly the joint properties. So until now there are lots of investigations about brazing of austenite stainless steel with Ni-based brazing filler metals. Recently, ferrite stainless steel has been employed, because of economical reason. In these days, new type Ni-based amorphous brazing filler metal foils containing a little amount of Boron have been developed. So brazing of ferrite stainless steel with Ni-based amorphous foils was investigated. Type 444 and 316L stainless steel were employed as base metals. BNi-2, BNi-5a, BNi-5b and Ni-Cr-Si-B-P type brazing filler foils produced by rapid solidification technology were used in this study. Brazing was done by electric heated furnace under vacuum atmosphere. T-joint and single lap specimens were employed to investigate interfacial reaction during brazing and to estimate mechanical properties of the joints. Elemental distributions at the brazed joint were analyzed by Electron Probe Micro Analyzer. According to observation results and EPMA results, Boron diffused to ferrite stainless steel base metal from brazing filler and B-Cr alloyed phase was formed at the base metal.

Introduction

Ni-based brazing filler metal is common filler metal for stainless steel brazing [1]. It has high corrosion resistance and heat resistance properties. On the other hand, complex cross-sectional microstructures at the brazed joint, which was brazed with Ni-based brazing filler metal, influence strongly mechanical properties of the brazed joint.

So it is required to clarify an interfacial reaction during brazing and control the cross-sectional microstructure at the brazed joint.

Generally the amount of brazing filler metal at the joint influences brazed joint microstructure and mechanical properties. So control of the amount of brazing filler metal is very important and it is needed to control the joint gap. However, it is not possible to keep the recommended joint gap during brazing because of technical and economic reasons.

Amorphous brazing filler metal foil is manufactured by rapid solidification (RS) technology, and so it has not contained an organic binder [2]. Generally it is noted that an amorphous brazing filler metal foil is so expensive as compared with powder type filler metal. When filler metal foil was employed, it is easy to control the amount of filler at the brazing joint, and a volatilization step for organic binder is not need, and a damage at the inside wall of the brazing furnace, which is caused by organic binder, is decreased. Therefore a big cost difference between the conventional brazing process by powder type filler and new process by filler metal foil is decreased.

Amorphous brazing filler metal foil has to contain small amount of B because of RS technology [2]. B is important and requisite element for manufacturing of metallic amorphous foil. It is very difficult to manufacture a ductile amorphous Ni-based brazing foil without B by RS technology.

Recently, instead of austenite stainless steel, ferrite stainless steel has been employed as a base metal for manufacturing field, particularly automobile filed. There are a few researches for investigation of brazing ability of ferrite stainless steel until now.

So in this study, brazing of ferrite stainless steel by amorphous Ni-based brazing filler metal foils was investigated by observation of cross-sectional microstructure and the elemental distributions by EPMA at the brazed joint. Purpose of this study is to reveal behavior of B element during brazing and the interfacial reaction.

Experimental procedures

Base metal and brazing filler metals used in this study
Ferrite stainless steel, SUS444 (JIS), was employed as base metal. Chemical compositions of the base metal were shown

in Table 1. The base metal contained 18.4 mass% Cr and not contained Ni.

Table 1 Chemical composition (mass %) of base metal stainless steel, SUS444

Fe	Cr	Si	Nb	Mn	Mo	Ni	P	C	S	Cu	N
Bal.	18.37	0.29	0.44	1.01	1.86	0.17	0.035	0.01	0.002	0.2	0.014

Amorphous Ni-based brazing filler metal foil, which was manufactured by RS technology, was used in this study. This article reported results in that case of three kinds of amorphous Ni-based brazing filler metal foils.

Chemical compositions of amorphous Ni-based brazing filler metal foil used in this study were shown in Table 2. These compositions were analytical results, not commercial value. These compositions were similar to AWS brazing filler metal standard's chemical compositions [3]. Metglas(R) MBF20 was typical brazing metal foil, which contained 3.01mass%B, and its thickness was 38.1 μm in accordance with the BNi2 standard. Metglas(R) MBF50 was filler metal foil, which contained 1.28mass%B and 18.54mass%Cr, and its thickness was 25.4 μm in accordance with the BNi5a standard. Metglas(R) MBF51 was filler metal foil, which contained 1.35mass%B and 14.82mass%Cr, and its thickness was 25.4μm in accordance with BNi5b. Amorphous Ni-based brazing filler metal foil used in this study was manufactured by Metglas, Inc. and Hitachi Metals, Ltd [4].R

Table 2 Chemical compositions (analytical value) and temperatures of Ni-based amorphous brazing filler metal foils used in this study.

	Chemical composition (mass %)					Temperature(°C)	
	Ni	Cr	Fe	Si	B	Solidus	Liquidus
MBF20	Bal.	7.06	3.43	4.92	3.01	969	1024
MBF50	Bal.	18.543	0.198	7.154	1.28	1052	1144
MBF51	Bal.	14.82	0.4	7.33	1.35	1030	1126

Specimens and brazing conditions

T-joint specimen was employed in this study as shown in Fig.1. T-joint specimen was an applicable specimen for estimation of a brazing-ability of brazing filler metal foil and investigation of an interfacial reaction during brazing [5]. A single 20mm x 40mm preform of brazing filler metal foil was placed between the two stainless steel plates as shown in Fig.1. And they were clumped together by stainless steel jig. Brazing was done with an electric heated conventional type vacuum furnace.

Heating rate was 10K/min. Brazing time was 5min at each brazing temperature. Brazing temperatures were 1050 and 1150 degree of centigrade by consideration of liquidus temperature of each brazing filler metal foil as shown in Tabel3. Brazing steps did not contain a volatilization step for organic binder because of no binder in the brazing filler metal foils.

Estimation and analytical procedure

After brazing, a specimen was cutted along the specimen center line by middle speed cutter. And cross-section of the specimen was polished by emery paper and cloth polishing. And an aqua regia was used as an etching liquid. Brazing-ability was estimated by cross-sectional observation results and elemental distributions analytical results with EPMA. Cross-sectional microstructure at the brazed joint was observed with optical microscope and SEM.

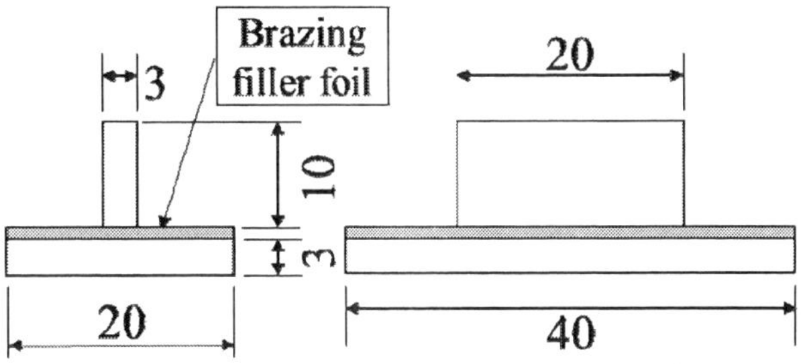

Figure1 Schematic diagram of the T-joint specimen.

Results and Discussions

Typical cross-sectional microstructure at the SUS 444 brazed joint, which was brazed by Metglas(R) MBF20, was shown in Fig.2 Cross-sectional microstructure was observed at the brazed joint including fillet area. Generally cross-sectional fillet shape and microstructure at the brazed joint influenced a performance of manufacturing product such as EGR cooler. On the other hand, cross-sectional microstructure at the parallel area influenced strongly brazed joint strength.

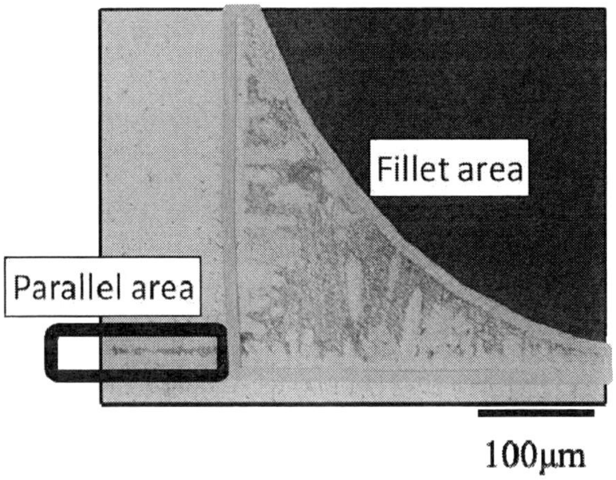

Figure2 Typical cross-sectional microstructure at the SUS 444 brazed joint, which was brazed by Metglas(R) MBF20. Brazing temperature was 1050 °C. Brazing time was 5min.

In the case of brazing filler metal foil, filler foil was placed between base metals preliminarily, so that a sound brazed joint without void and/or clack was obtained. Therefore we separated the brazed joint fillet area into a fillet area and a parallel area as shown in Fig.2 and investigated.

Gap at the parallel area was about 30 μm. This 30μm gap was smaller than recommended gap for Ni-based brazing filler metal [1]. So it was recognized that an excellent brazed joint at the parallel area was obtained. And this 30μm joint gap was similar to the brazing filler metal foil thickness. It was recognized that applicable brazing joint gap was obtained because of relative thin brazing filler metal foil.

Typical elemental distributions by EPMA at the SUS 444 brazed joint, which was brazed with Metglas(R) MBF20, were shown in Fig.3. According to these analysis results by EPMA, the fillet area microstructure was typical brazed joint microstructure [7]. The fillet area microstructure consisted of Ni primary crystal containing Si and Cr as solid solute elements, remained lamellar Ni-Cr-Si-B eutectic and Cr-B intermetallic compound (IMC) phase. This fillet area microstructure was similar to the microstructure at the SUS 304 brazed joint, which was brazed with Ni-based powder type brazing filler metal, BNi2. It was revealed that stainless steel type influenced hardly the fillet area microstructure at the brazed joint according to these results.

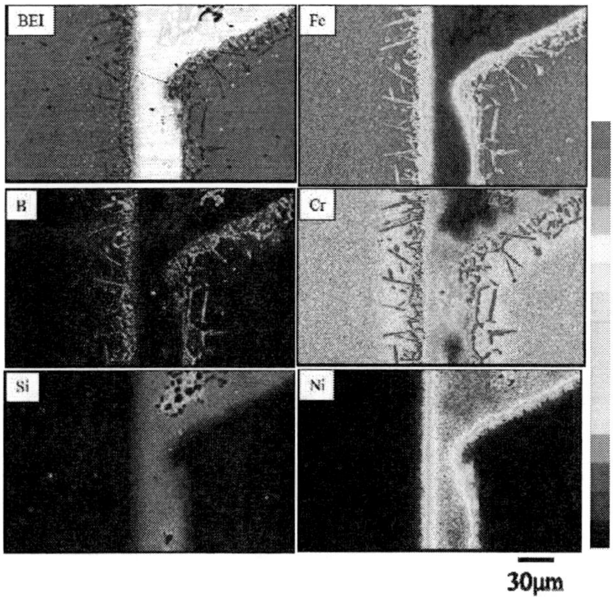

Figure3 Typical elemental distributions by EPMA at the SUS 444 brazed joint, which was brazed with Metglas(R) MBF20. Brazing temperature was 1050 °C. Brazing time was 5min.

Cross-sectional microstructure at the stainless steel base metal near the interface between a fillet area and a base metal was similar to that at the stainless steel base metal near the interface between a parallel area and a base metal. In both cases, B diffused from brazing filler metal to stainless steel base metal, and it reacted Cr and Cr-B IMC phase was formed. Shape of a Cr-B IMC phase at the base metal area near the interface was inordinate shape. And the Cr-B IMC phase at the base metal area near the interface was dispersed randomly. On the other hand, an acicular Cr-B IMC phase was formed at the middle distance base metal from the interface and it grew up vertically to interface.

In that case of austenite stainless steel, boron diffused intergranularly along the base metal grain boundary from brazing filler metal, and Cr-B intermetallic compound phase was formed along the its grain boundary.

On the other hand, in that case of ferrite stainless steel, grain of the base metal coarsened during heating, so that intercrystalline diffusion of B from filler to base metal occurred. It appeared that an acicular Cr-B IMC phase was formed because of a preferential diffusion pass for B at the ferrite phase. Diffusion length of boron in this study was similar to that by calculation by existing data.

Cr depleted zone was formed at the base metal, because Cr was depleted by formation of the Cr-B IMC phase. It appeared that Si could not diffuse from filler metal to base metal as compared with B, because Si atom diameter and solid solubility limit of Si into Ni solid solution were big.

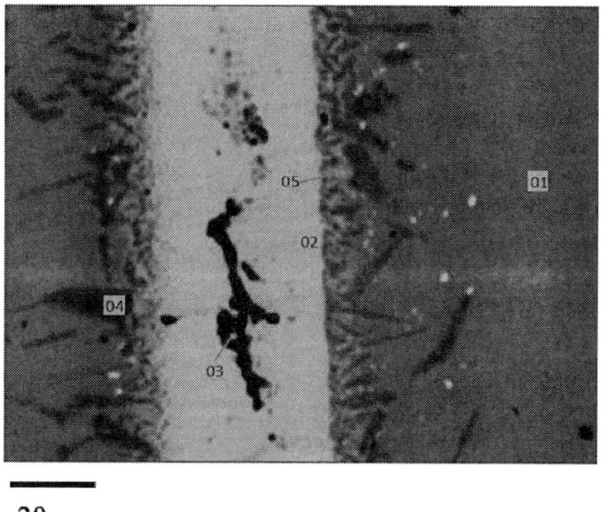

No.	Region	Analytical results (at%)				
		Cr	Ni	Fe	Si	B
1	Base metal	20.4	-	78.8	0.8	-
2	Brazed layer at parallel area	7.2	63.3	22.7	6.8	-
3	Black phase at parallel area	41.0	15.5	3.9	3.4	36.2
4	Black phase at base metal	37.8	-	26.8	-	35.4
5	Base metal near interface	15.4	21.0	62.0	1.6	-

Figure4 Typical point analysis results by EPMA at the parallel area brazed joint brazed by Metglas(R) MBF20. Brazing temperature was 1050 °C. Brazing time was 5min.

It recognized that Ni diffused slightly from filler metal to the base metal from elemental maps. So point analysis by EPMA was done. Typical point analysis results by EPMA were shown in Fig.4. It appeared that alpha phase was transformed into gamma phase according to analytical result of Ni at the base metal near the interface and Fe-Cr-Ni ternary phase diagram. And it recognized that black phase at the brazed layer was Cr_2B phase.

Typical cross-sectional microstructure and elemental distributions by EPMA at the brazed joint, which was brazed with Metglas(R) MBF51, were shown Fig.5. Metglas(R) MBF51 was manufactured according to B-free Ni-based brazing filler metal BNi5. However 1.35mass%B (analytical value) was contained for form-ability of amorphous brazing filler metal foil. Cr-B IMC phase was formed at the fillet area similar to in that case of Metglas(R) MBF20. On the other hand, relative narrow gap at the parallel area was caused by filler foil thickness. Sound brazed joint was obtained. Cr-B IMC phase was formed at the base metal near the interface. B diffused from filler metal to base metal and Cr depleted zone was observed.

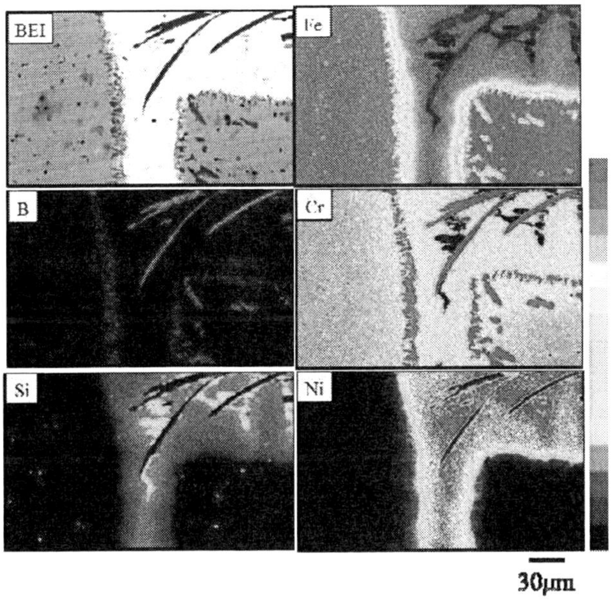

Figure5 Typical cross-sectional microstructure and elemental distributions by EPMA at the brazed joint, which was brazed with Metglas(R) MBF51. Brazing temperature was 1150 °C. Brazing time was 5min.

Then Metglas(R) MBF50 containing more little amount of boron as compared with Metglas(R) MBF51 was employed as filler metal and brazing was done. Boron content in this filler metal was 1.28mass% (analytical value). Typical observation and analysis results were shown in Fig.6. Similar results were obtained.

At fillet area, Cr-B IMC phase was observed. Solid solubility limit of B into Cr solid solution was less than 1.0mass% from Cr-B binary phase diagram. So it appeared that Cr-B IMC phase was formed at the fillet area by minimal B content in filler.

B solid solutes into Ni solid solution slightly, but into Fe solid solution hardly. Therefore Ni primary solid solution at the brazed layer contained a little amount of B after brazing. On the other hand, it appeared that Fe-Cr solid solution contained B as solid solute element hardly, so that Cr-B IMC phase was formed.

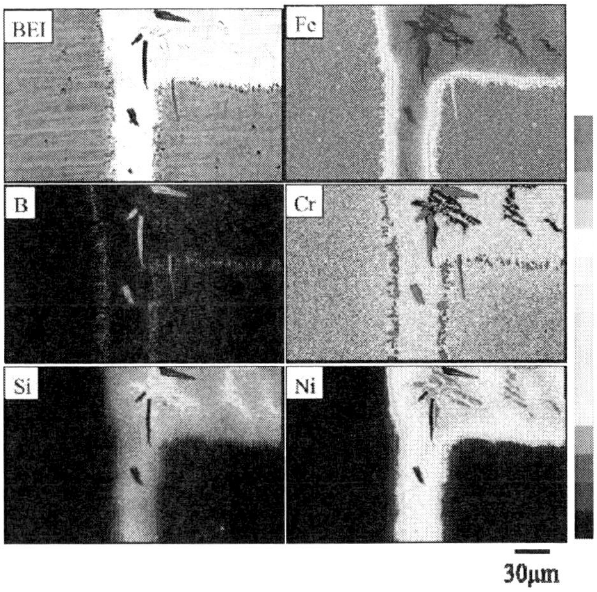

Figure6 Typical observation and analysis results by EPMA at the brazed joint. Brazing temperature was 1150 °C. Brazing time was 5min.

Diffusion time of B from fillet area to the base metal was not enough because of large amount of fillet [6]. So it appeared that Cr-B IMC phase was formed near the remained eutectic region.

Solid solubility limit of B into Ni or Fe-Cr solid solution phase was very little, so it was recognized that boride was formed at the base metal area because of minimal B diffusion from filler metal to stainless steel. And diffusion ability of B was high. So B content in brazing filler metal foil influenced strongly stainless steel brazed joint microstructure.

Then other types of amorphous Ni-based brazing filler metal foil containing more little amount of B as compared with Metglas(R) MBF50 and Metglas(R) MBF51 were employed as filler metal and brazing was done. Cross-sectional microstructures at the brazed joint were observed and the elemental distributions were analyzed by EPMA. Effect of B content (analytical value) of the amorphous brazing filler metal foil on B elemental distribution by EPMA was shown in Fig.7.

When B content of amorphous brazing filler metal foil exceeded 1mass%, B diffused from filler metal to base metal and Cr-B IMC phase was formed at the base metal according to these analytical results.

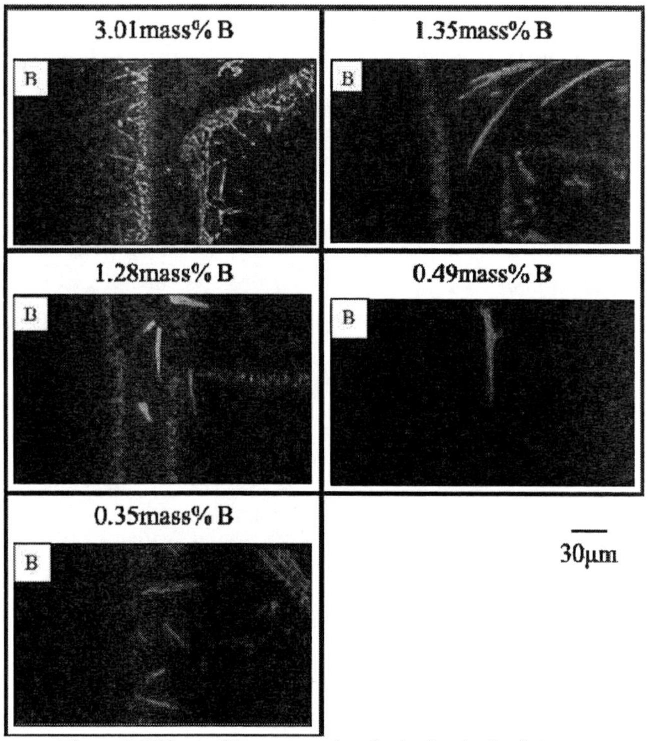

Figure7 Effect of B content (analytical value) of the amorphous brazing filler metal foil on B elemental distribution by EPMA.

Conclusions

In this study, brazing of ferrite stainless steel by amorphous Ni-based brazing filler metal foils was investigated by observation of cross-sectional microstructure and the elemental distributions by EPMA at the brazed joint. Purpose of this study is to reveal behavior of B element during brazing and the interfacial reaction. Following results were obtained.

1. We separated the brazed joint fillet area into a fillet area and a parallel area and investigated.
2. It was recognized that applicable brazing joint gap was obtained because of relative thin brazing filler metal foil.
3. It was revealed that stainless steel type influenced hardly the fillet area microstructure at the brazed joint according to these results.
4. B diffused from brazing filler metal to stainless steel base metal, and it reacted Cr and Cr-B IMC phase was formed.
5. Shape of a Cr-B IMC phase at the base metal area near the interface was inordinate shape. And the Cr-B IMC phase at the base metal area near the interface was dispersed randomly. On the other hand, an acicular Cr-B IMC phase was formed at the middle distance base metal from the interface and it grew up vertically to interface.
6. It appeared that an acicular Cr-B IMC phase was formed because of a preferential diffusion pass for B at the ferrite phase.
7. When B content of amorphous brazing filler metal foil exceeded 1mass%, B diffused from filler metal to base metal and Cr-B IMC phase was formed at the base metal according to these analytical results.

References

[1] AWS C3 Committee, Brazing Handbook 5th edition, AWS (Miami, 2007), pp. 67-100.
[2] Rabinkin,A, "New application for rapidly solidified brazing foils", *Welding Journal*, 68(10), pp. 39-46
[3] AWS C3 Committee, "Specification for Filler Metals for Brazing and Braze Welding 10th edition", *AWS A5.8M/A5.8:2011*, pp. 1-19.
[4] Hitachi Metals, Ltd., "Amorphous brazing filler metal foil", Hitachi Metals technical data sheet
[5] Y. Miyazawa, "Joining ability of Ni-based brazing filler metal manufactured by rapidly solidification technology (in Japanese)," *Tokai University Doctor Thesis*, (1993)
[6] Kinzokugakkai, Kinzoku data book 3rd edition (in Japanese), Maruzen (Tokyo, 1993), pp. 20-25.
[7] Y. Miyazawa, K. Saito, T. Fukikoshi, T. Takizawa, "Brazing of ferritic stainless steel and microstructure at the brazed joint", *IBSC Proceedings of the 5th International Brazing and Soldering Conference*, Las Vegas, NV, April, 2012, pp. 236-239

Induction Joining Of AISI 316L To AISI 304 Stainless Steels With Fe-Based Amorphous Alloy Ribbon Interlayer

J. A. Verduzco, J. Lemus-Ruiz and V. H. Verduzco
Universidad Michoacana de San Nicolás de Hidalgo, Morelia, Michoacán, México
verduzco@umich.mx, +52 443 3223500, ext. 4050

Abstract

In recent years, the production and utilization of metallic glasses for several applications have attracted the interest by many researchers and investigation groups. This is because these materials poses high ductility, chemical homogeneity and high mechanical and corrosion resistances due to their vitreous nature, compared to their crystalline counterparts; in addition, metallic glasses have low melting temperatures and have the capability of being self-fluxed, such characteristics make these materials able to be used as the joining element to bond dissimilar materials, for example stainless steels and ceramic to steel. This work presents a study of the microstructure characterization and interdiffusion of the main elements (Fe, Cr, Ni and P) participating in the bonding process of AISI 316L stainless steel/$Fe_{60}Ni_{12}Cr_8P_{13}B_7$ metallic glass ribbon/AISI 304 stainless steel joints. The joints were performed at a temperature of 1000 °C for different joining times carried out by induction heating into a furnace under an Ar controlled atmosphere. It was found that the interface line tended to disappear as the joining time increase and that the elements that more diffused were Cr and Ni.

Introduction

One of the attractive properties of metallic glasses obtained by rapid solidification is that it is possible to achieve solid solutions exceeding limits of solubility, obtaining new alloys with unique properties and readily to be used in several applications such as element joining of dissimilar materials when they are fabricated in shape of foils; in this form this materials are ductile and flexible and can be used as the joining element without using additional materials like fluxes, as is the case when powder is used as the joining element [1]. For instance, Ni-base metallic glass braze foils have been used to join zirconium diboride (ZrB2)-based ultra-high-temperature ceramic-matrix composites to themselves and to commercially pure Ti [2], and carbon–carbon, carbon–silicon carbide, and silicon carbide–silicon carbide composites have been vacuum brazed to Ti and Hastealloy X [3]. Between various methods to braze materials are laser, resistance furnace chambers and induction heating; the last method has been used because is fast and economical compared to the other. Botstein et al. [4] joined by induction heating the biomaterial Ti-6Al-4V with a 25Ti-25Zr-50Cu metallic glass foil, reporting strong and ductile joints. In addition, some Ni, Mg, Zr, Cu, Pd and more recently Fe- based metallic glass in ribbon form have been used to joint alloys such as Inconel 718 to Inconel X-750 [5], Ti-based alloy for biomedical applications [6] and AISI 316L [7]. Among the many metallic glass alloys used as the filler metals, those based on Fe containing phosphorous have not been explored to study their attractive potential even though it is reported its high corrosion resistance in 1N HCl [8]. This work presents a study of the microstructure characterization and interdiffusion of the main elements (Fe, Cr, Ni and P) participating in the bonding process of AISI 316L stainless steel/$Fe_{60}Ni_{12}Cr_8P_{13}B_7$ metallic glass ribbon/AISI 304 stainless steel joints.

Experimental procedure

Sample Preparation:
Samples of AISI 316L and 304 stainless steel (SS) having dimensions of 10 mm long, 5 mm width and 3 mm thick were joined employing a melt spum $Fe_{60}Ni_{12}Cr_8P_{13}B_7$ metallic glass ribbon, MGR (38 μm thick and 3 mm width) as the filler element. Table 1 presents the chemical composition of the three materials, obtained using a scanning electron microscope (SEM) JEOL JSM-6400J by energy dispersive X-Ray spectroscopy technique (EDX). The melting temperature (1019 °C, see Figure 1) for the glassy alloy was determined in a differential scanning calorimeter TA SDT Q600, for the present work, the joining temperature was established to be 1000 °C, since at this temperature the ribbon shows a readily to join the SS in a transient liquid phase.

Table 1.- EDS chemical microanalyses of the starting materials.

Material	Atomic %							
	Fe	Cr	Ni	P	Mo	C	Si	Mn
AISI 304	70.13	18.34	7.23	-	-	5.03	0.69	-
AISI 316L	66.90	16.80	9.16	-	1.21	5.85	1.21	1.47
MGR	58	9	12	13.2				

Note: Due to the SEM does not detect B, the rest of the composition of the metallic glass ribbon is supposed to be of this element.

Joining Procedure:
The SS surface to be joined (10x3 mm^2) were firstly grounded with silicon carbide papers starting from 800 to 1200 grit; next, the surfaces were polished using a 1 μm diamond paste

to obtain a mirror-like surfaces. Two SS surfaces to be joined were placed face to face in a sandwich type arrangement accommodating the Fe-based ribbon between the SS surfaces. The assembly was then collocated in a boron nitride coated graphite die and subjected to a mechanical torque of 0.5 Nm to keep the assembly static. The assembly was introduced into a chamber of radio frequency induction furnace. A vacuum of $2.13x10^{-2}$ Pa was realized in the chamber and the filled with argon to 133 Pa. This flushing procedure was repeated 2 times in order to assure an inert atmosphere, the system was left with a flowing positive pressure. Finally, the joining process was started by supplying a power of 5kW, and then it was gradually increasing in order to heat the assembly at a heating rate of 52 °C min^{-1}. The temperature was monitored with a k-type thermocouple. Figure 2 shows a schematic equipment set up were the joining experiments were carried out. Once the sample to be joined reached a temperature of 1000 °C, it was held for periods of 4, 8, 12 and 16 minutes. The samples were cooled down at rate of 70 °C min^{-1} to room temperature.

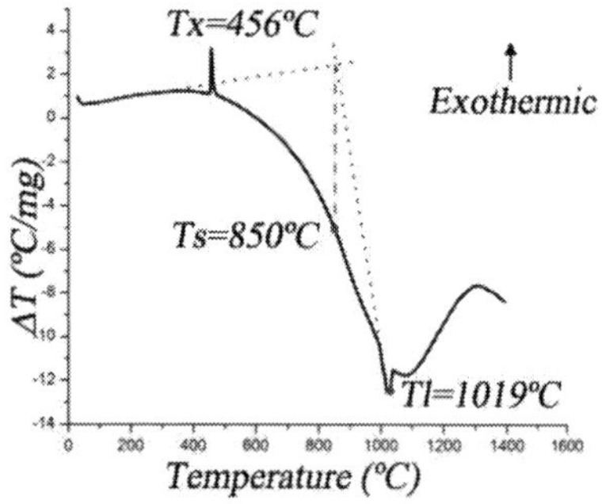

Figure 1.- Differential thermal analysis of the metallic glass ribbon. Tx=Crystallization temperature, Ts=Solidus temperature and Tl=Liquidus temperature.

Microstructural Characterization:

Microstructural analyses of the joined zones were performed on polished cross-sections of the joined samples using SEM. The interfaces and precipitated particles formed in the joining zone were subjected to punctual microanalyses by EDX.

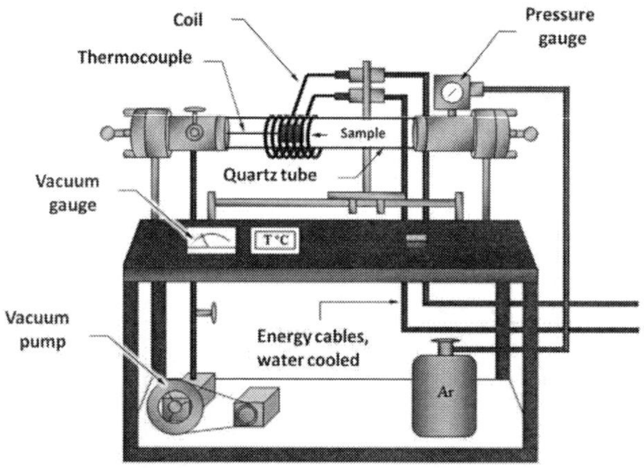

Figure 2.- Schematic set up of the chamber furnace.

Results and Discussion

Microstructural Characterization

It was possible to join the SSs for all joining times. Figure 3 the microstructure obtained for each joining time. It is clearly observed the joining zones between the austenitic steels and the metallic ribbon (interlayer), which does not show a well defined or limited notable interfaces. For joining times of 4, 8 and 12 minutes, it is observed in Figures 3(a-c) a phase along the middle of the interlayer (labeled as "A"). In these joining times, the width of the zone A, did not increase considerably by increasing the joining time; however, for the joining time of 16 minutes, the zone A disappeared (Figure 3(d)). It can be suggested that for this joining time, the phosphorous can diffuse more intensely out the joining zone and be dissolved in the base metal due to an increment in the diffusivity.

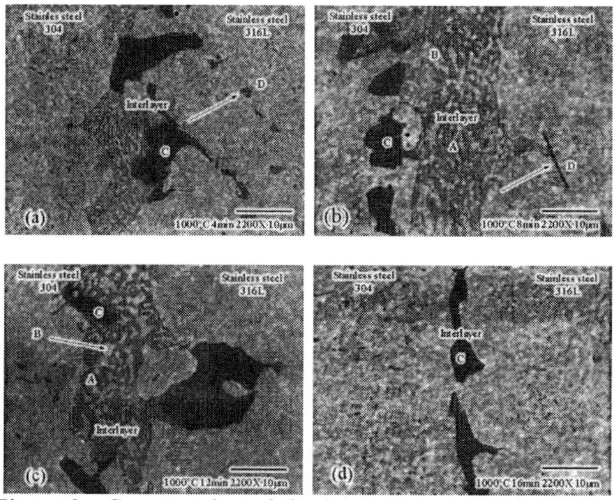

Figure 3.- Cross section of the joining zones for the joining times of (a) 4, (b) 8, (c) 12 and (d) 16 minutes.

In the central line of the joints some precipitated particles were formed (indicated as "C" in the photomicrographs) for all joining times; also, some of the same precipitated particles were formed in both SSs. In Figure 3(a) it is observed that the formation of a precipitated particle of the joining zone divides the zone A, provoking a discontinuity of the interlayer, it is also observed the formation of some precipitated particles outside the interlayer, which have the form of small rounded particles, indicated as "C". For the sample joined for a holding time of 8 minutes (Figure 3(b)) the precipitated particles were formed in the side of the AISI 304 SS. The sample joined for 12 minutes holding time presented an increase in the size of the precipitated particles (Figure 3(c)). Regarding to the formation of grey islands in to the zone A (labeled as "B"), these are long and irregular; their sizes increased as the joining time increased, becoming more visible for the joining time of 12 minutes.

Figure 4 shows a mapping analyses in order to observe the distribution of the elements for the sample joined for 4 minutes. Figure 4(a) presents the photomicrograph of the sample where the mapping was performed, indicating the zones A and C. From Figure 4(b and c) it is observed that the concentration of Fe is higher in the SSs, but in minor amounts in zone A and practically none in zone C. In the case of Cr, it is homogeneously distributed on the SSs, presenting major concentration in zone C. The distribution of P is presented in Figure 4(d), this element is present in zone A and in higher concentration in zone C, this element diffused towards the steels although in minor amounts. Regarding to Ni, it is observed a uniform distribution along the reaction zone and in the SSs as seen in Figure 4(e), however in zone C this element is not present. Finally, this analysis shows that Mo is present in zone C and also diffused to the AISI 304 SS in small amounts (Figure 4(f)).

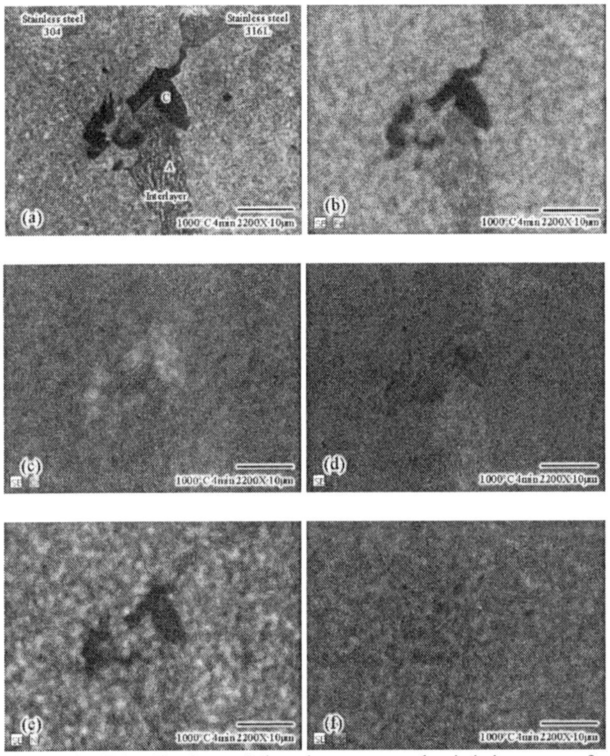

Figure 4.- Distribution of the elements at the joining zone for the sample joined for a holding time of 4 minutes, (a) Image of the mapping analyses, (b-f) Fe, Cr, P, Ni and Mo distribution, respectively.

Identification of Phases

Table 2 presents the punctual chemical microanalyses realized at the zones shown in Figure 3. The predominant phase formed at the interlayer (zone A) is composed mainly for Fe, Cr, Ni and P, with a small amount of Mo. The elongated grey islands (labeled as B) formed inside the zone A, have high concentration of Fe, Cr and Ni. Figure 5 shows the spectrograms for the zones labeled as A, B, C and D for a sample joined for 8 minutes, these are representatives for the rest of the samples joined for the other joining times. The precipitated particles labeled as zone C have high contents of Fe, Cr, Mo and P, which indicate the possible formation of phosphorides and possibly borides of Cr, Ni and/or Mo at the joining zone.

Table 2.- EDS chemical composition of the formed phases shown in Figure 3.

Zone	Joining time	Wt. %							
		Fe	Cr	Ni	P	Mo	C	Si	Mn
A	8 min	55.28	19.49	10.42	10.22	1.50	1.74	-	-
	12 min	55.27	22.46	9.60	9.89	1.27	1.51	-	-
B	4 min	71.70	17.58	7.37	0.09	-	1.39	0.43	-
	8 min	68.98	9.27	16.66	0.69	-	1.765	0.30	-
	12 min	70.17	9.88	17.03	0.88	-	1.83	0.21	-
C	4 min	22.11	60.61	-	8.07	5.80	2.44	-	-
	8 min	22.91	60.21	-	8.33	4.09	2.68	-	-
	12 min	22.42	61.84	-	8.15	3.53	2.30	-	-
	16 min	20.28	62.66	-	7.73	5.57	2.81	-	-
D	8 min	39.47	41.77	6.14	6.90	3.19	1.99	-	-

In order to identify the formed phases in the different joining zones, the isotherms at 650 ºC of the Fe-Cr-Ni [9] and Fe-Cr-Mo [10] ternary equilibrium phase diagrams shown in Figure 6 were used to support the chemical composition found and presented in table 2. Table 3 presents the phases and symbols representing the phases indicated in Figure 6, corresponding to the zones A, B, C and D labeled in Figure 3. Regarding to the Fe-Cr-Ni isotherm diagram (Figure 6(a)), the composition of zone A presents "γ" and "σ" phases; meanwhile zone B corresponds to the austenitic "γ" phase. The precipitated particles labeled as zone C, have Fe, Cr and Mo, observing the Fe-Cr-Mo isotherm diagram of Figure 6(b), it is shown that for this composition a (Cr-Mo)+σ intermetallic phase is formed. The elements found in zone D are Fe, Cr and Ni, where there is a Fe50:Cr50 ratio, which in agreement to the Fe-Cr-Ni isothermal diagram, the corresponding phase is σ.

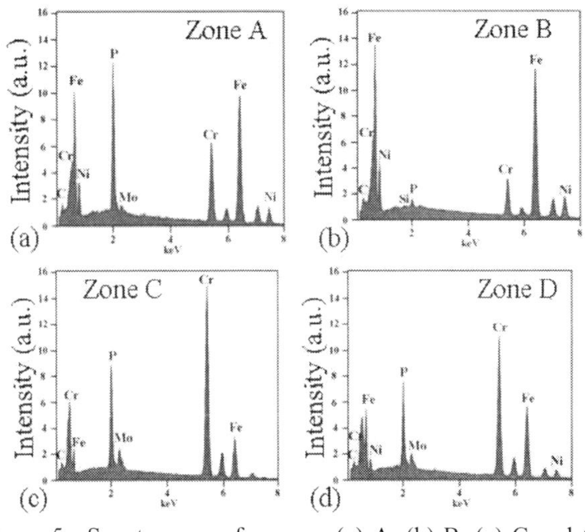

Figure 5.- Spectrograms for zones (a) A, (b) B, (c) C and (d) D, corresponding to Figure 3(b). These are representatives of the same zones for all other joining times.

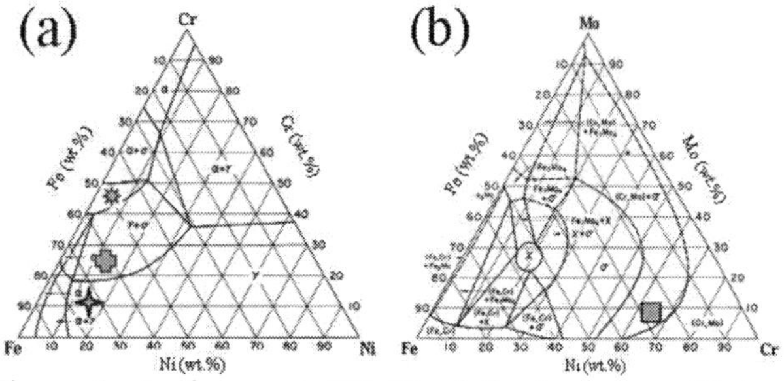

Figure 6.- Isotherms at 650 °C of the ternary equilibrium diagrams (a) Fe-Cr-Ni [9] and (b) Fe-Cr-Mo [10] used to determine the phases formed. Symbols represent the phases:

= γ, = γ + σ, =σ and =(Cr-Mo)+σ.

Table 3.- Determined phases at the different zones indicated in Figure 3, regarding to the isotherms at 650 °C of the ternary diagrams of Figure 6.

Zona	Fase	Símbolo
A	γ + σ	
B	γ	
C	(Cr,Mo) + σ	
D	σ	

Diffusion of Elements

Figure 7 the chemical microanalyses performed at 7 points starting on the AISI 304 SS, crossing the joining zone (point 4 of Figure 7(a)), and finishing on the AISI 316L, in order to observe the diffusion behaviour of Fe, Cr, Ni and P. The results are presented as a function to the position of the points, each set of curves corresponds to its joining time (Figures (b-e)). For all joining times, the Fe concentration at the interphase was lower than in both stainless steels. The Cr concentration in both SSs diminished from a joining time of 8 minutes; however, the amount of this element was higher than 12 at. % for all joining times.

The amount of Ni notably diminished at the interlayer from the joining time of 8 minutes, obtaining near to 8 at. % and 0 at. % for the joining times of 12 and 16 minutes, respectively. Regarding to the P diffusion, this element tended to concentrate at the interlayer, reaching concentrations of 15 at. % for the joining time of 12 minutes, P also diffused towards the SSs, where its concentration was of 1 at. % approximately, from the joining time of 8 minutes.

In order to observe more clearly the diffusivity behaviour of Fe, Cr, Ni and P, their concentrations at points 3 (AISI 304), 4 (interlayer) and 5 (AISI 316L) were plotted as a function of joining time (Figure 8). Fe concentration decreased at the interlayer and lightly increased in both SSs, being more notably after 8 minutes holding time. On the other hand, Cr concentration had an inverse behaviour, since at the interphase its concentration considerably increased (up to 65 at. %), decreasing in both SSs. In the case of Ni, it is observed that its atoms diffused from the interlayer to the austenitic SSs increasing their concentrations of this element, and consequently decreasing at the interlayer. Finally, the amount of P was concentrated at the interlayer, increasing its content as the joining time was increased, it can be supposed that this diffusivity originated the formation of the elongated islands (zone B) into the interlayer, since the amount of P in zone B is near to cero.

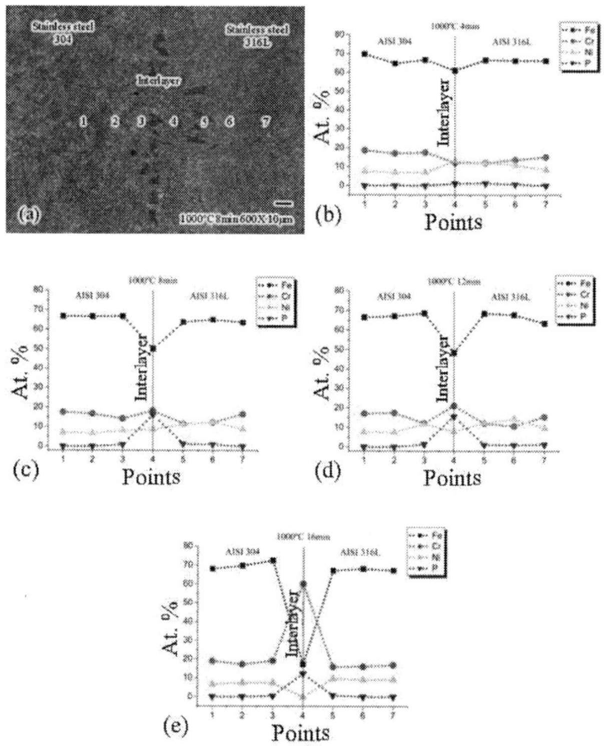

Figure 7.- (a) Joining cross section for the sample joined for a holding time of 8 minutes; (b-e) punctual chemical microanalyses for the joining times of 4, 8, 12 and 16 minutes, respectively.

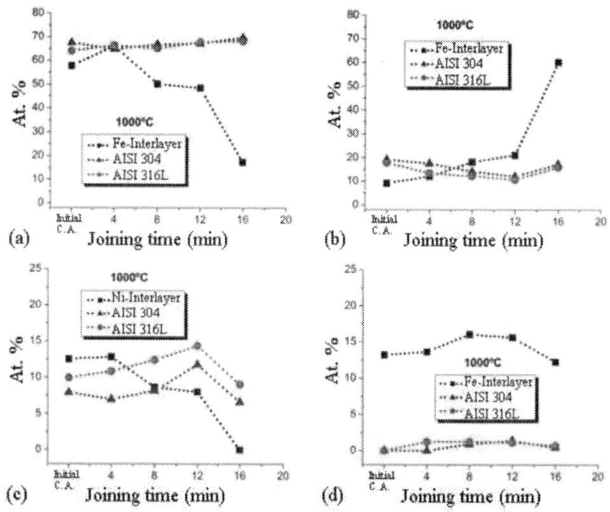

Figure 8.- EDS chemical microanalyses for (a-d) Fe, Cr, Ni and P, respectively, as a function of joining time,

corresponding to points 3, 4 and 5 of Figure 7, regarding to the AISI 304 SS, interlayer and the AISI 316L SS, respectively.

By comparing the diffusivity of Cr and Ni in Figure 8(b and c), it is observed that Cr has more intense change of concentration at the joining zone, this is because the diffusion velocity of Cr in Fe-γ at 1000 ºC is $11.6x10^{-12}$ cm^2s^{-1} [11], meanwhile the diffusion velocity of Ni en Fe-γ at the same temperature is $2.41x10^{-12}cm^2s^{-1}$ [12] (almost one over fifth slower than for Cr).

Conclusions

It was possible to join two different stainless steels (AISI 304 – AISI 316L) with a non commercial Fe60Ni12Cr8P13B7 metallic glass ribbon for the 4 joining times: 4, 8, 12 and 16 minutes.

The joining mechanism was carried out by diffusion in liquid state, which involves the formation of a reaction zone composed by different intermixed elements.

About the formed microstructures, the central phase in the joining zone was mainly composed by a Fe-Cr-Ni-P solid solution, which regarding to the Fe-Cr-Ni ternary diagram, it presents a $\gamma + \sigma$ phase.

The precipitated particles formed inside the interlayer, are composed with high contents of Fe, Cr, Mo and P, which regarding to the Fe-Cr-Mo ternary diagram, it presents a (Cr-Mo) + σ phase.

The precipitated particles formed outside the interlayer, showed high contents of Fe, Cr and Ni, with a Fe50:Cr50 ratio, which regarding to the Fe-Cr-Ni ternary diagram, it corresponds to a σ phase.

Acknowledgments

We acknowledge the financial support received from the Universidad Michoacana de San Nicolás de Hidalgo and CONACYT México.

References

[1] Rabinkin, A., "Brazing with (NiCoCr)-B-Si amorphous brazing filler metals: alloys processing, joint structure, properties, applications", Science and Technology of Welding and Joining, Vol. 9, No. 3 (2004) pp. 181-199.

[2] Singh, M. and Asthana, R., "Joining of zirconium diboride-based ultra high-temperature ceramic composites using metallic glass interlayers", Materials Science and Engineering: A, Vol 460-461 (2007) pp. 153-162.

[3] Singh, M. *et al.*, "Brazing of ceramic-matrix composites to Ti and Hastealloy using Ni-base metallic glass interlayers", Materials Science and Engineering A, Vol. 498 (2008) pp. 19–30.

[4] Botsiein, O. *et al.*, "Induction brazing of Ti-6Al-4V alloy with amorphous 25Ti-25Zr-50Cu brazing filler metal", Materials Science and Engineering A, Vol. 206 (1995) pp. 14–23.

[5] Wu, X. *et al.*, "Induction brazing of Inconel 718 to Inconel X-750 using Ni-Cr-Si-B amorphous foil" J. Mater. Process. Technol. Vol. 104 (2000) pp. 34-43.

[6] Miura, E. *et al.*, "Mechanical property and corrosion resistance evaluations of Ti-6Al-7Nb alloy brazed with bulk metallic glasses", Mater. Trans. JIM, Vol. 48 (2007), pp. 2235-2243.

[7] Verduzco, J. A. et al., "Microstructure and electrochemical properties of the bonding zone of AISI 316L steel joined with a Fe-based amorphous foil", J. Mater. Process. Technol., Vol. 210 (2010), pp. 1051-1060.

[8] Naka, M. *et al.*, "High Corrosion Resistance of Chromium-Bearing Amorphous Iron Alloys in Neutral and Acidic Solutions Containing Chloride", Corrosion, the J. of Sci. and Engng., Vol. 32, No. 4 (1976) pp. 146-152.

[9] Raynor, G. V. and Rivlin, V.G., Phase Equilibria in Iron Ternary Alloys (Phase Diagrams of Ternary Iron Alloys IV. The Institute of Metals (London, 1988) pp. 485.

[10] Wada, T., Iron–Molybdenum–Nickel. Metals Handbook, **Vol. 8**, ASM Metals Park, (OH, 1973) pp. 431–1431.

[11] Bowen, A. W. *et al.*, "Solute diffusion in alpha- and gamma-iron", Metal. and Mater. Trans. B, Vol. 1, No. 6 (1970) pp. 1695-1700.

[12] Hirano, K. *et al.*, "Diffusion of nickel into iron", Acta Metallurgica, Vol. 9, No. 5 (1961) pp. 440-445.

Investigation of the Properties of Brazed Joints under Application-Relevant Loading

Bobzin, K.; Öte, M.; Wiesner, S.; Puidokas, S. M.; Matthes, D.
IOT – Surface Engineering Institute, RWTH Aachen University

Abstract

Generally, only a marginal part of the properties available for brazed joints can be directly transferred to real industrial components. The aim of this study is to achieve a better understanding about the behavior of brazed joints under application-relevant loading conditions. Typical base and filler materials, such as stainless steel, heat-resistant steel and a nickel-based alloy as well as nickel, iron and gold-based filler materials, which are often used in miscellaneous industrial applications, were chosen and tested under different loading conditions. The influence of different filler metals and the characteristics of the brazed seam on the properties of brazed joints were investigated under static and dynamic loads together with thermal or corrosive influences. The results allow a better understanding of the critical factors that must be considered in the dimensioning of such relevant brazed joints and thus contribute to an increased use of the brazing technology in future.

Introduction

The brazing technology gains increasingly importance because of its manifold application possibilities to join different materials, complex geometries and highly loaded components. A wide range of information about the wetting behavior of the base material or about the interactions between base and filler metal exists for widely-used base/filler metal combinations. On the other hand, knowledge about the mechanical properties of the joints under different loading conditions is rare. Because of time and cost related reasons, the mechanical properties of the joint can be tested and determined only for one single case of loading on simple specimens. However, brazed joints are subjected to different mechanical, thermal and corrosive loads simultaneously in real application environment. Additionally, the geometry of the brazing gap, the different base materials and parameters of the brazing process have a great influence on the structure of the braze joints, which result inevitably in a change of the properties of the braze joint.

Another point to be taken into consideration is the changing of mechanical behavior of the joints with respect to different loading conditions. In most cases, only the pure tensile or shear strength of the brazed joints is measured and determined. However, in practice the joints are usually subject to tensile and shear loads simultaneously.

In order to determine the influence of different loading magnitudes and conditions on the mechanical behavior, in this research the brazements were tested mechanically with a varying angle of the joint. With an increasing angle of the joint the proportion of the shear stress to the tensile stress changes during the mechanical loading. Additionally, the influence of corrosion and different temperatures on the mechanical behavior was investigated to ensure practice relevant loading conditions.

Experimental procedure

Materials and brazing process

In this study different filler metals relevant to miscellaneous industrial applications were investigated. For copper filler metals, a method to investigate the properties of the brazements is already presented in [1-3]. In this study different base metal/filler material combinations are considered. The relation between the structure of the brazement, different loading conditions and the mechanical properties of the joints was investigated for nickel, iron and gold based braze materials with steel and nickel based joining partners. Nickel based filler metals allow the production of strong joints, highly resistant against hot gas corrosion. They can be used in high temperature applications and they are often applied to join temperature and corrosion resistant steels and nickel based super alloys. However, these filler metals tend to build brittle hard phases if the thickness of the brazing gap is not sufficiently small. Because of the high material costs of nickel, iron based braze materials were used in some applications. These have a comparable corrosion resistance with lower material costs. Gold filler materials are used in aero-engines to join nickel based super alloys, where high mechanical strength at high temperatures is required. In contrast to the nickel based filler metals, gold filler metals do not have a tendency to build brittle hard phases [4-6]. The base metals and filler metals used, the investigated combinations and the performed experiments are listed in Table 1.

Table 1: Used materials and performed experiments

Base metal	Filler metal		
	Ni 620 [1)]	Amdry 805[2)]	Au 827 [1)]
Stainless steel (1.4301) X5CrNi18-10	T, C, V	T, C	
Heat resistant steel (1.2344) X40CrMoV5-1	T, C		
Inconel 625 (2.4856) NiCr22Mo9Nb	T, C		T, C
T = tensile test, C = corrosion test, V = vibration test			
1) DIN EN ISO 17672 [7] 2) Fe29Cr18Ni7Si6P2Re			

All investigations were performed with tensile test samples after DIN 50125:2009 "Testing of metallic materials - Tensile test pieces" [8]. The samples were cylindrical with a diameter of 12 mm. The angle of the brazing joint α was set to be 0°, 15°, 30°, 45° and 60° perpendicular to the centerline (Figure 1). As a result of this setup, the proportion of tensile and shear stress occurring in the brazement changes with respect to the angle α during tensile tests. A brazing gap of about 50 µm was set for all samples. Before brazing, the joining surfaces were polished and cleaned with ethanol. The filler material was applied as foil (Au 827, Ni 620) or paste (Amdry 805).

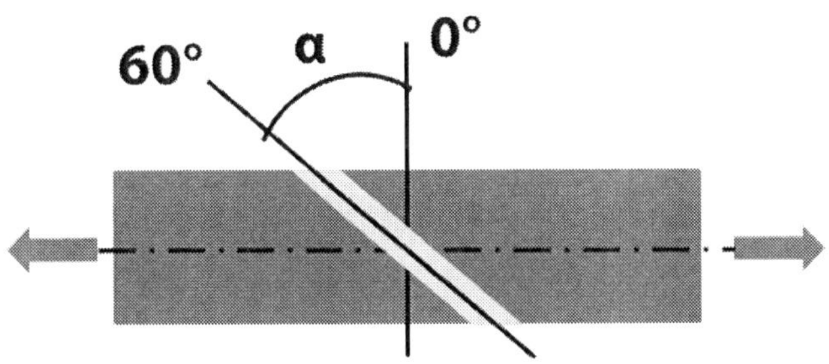

Figure 1: Geometry of the braze samples

The samples were joined in a vacuum furnace. The combination 2.4856/Ni 620, 1.2344/Ni 620 and 1.4301/Ni 620 were joined at 1090 °C for 10 minutes in a vacuum of 9 x 10^{-5} mbar. After joining 1.2344 with Ni 620 an additional heat treatment at 520 °C/2 hours was accomplished to harden the base material. The 2.4856/Au 827 joints were brazed for 10 minutes at 1050 °C. The 1.4301/Amdry 805 samples were brazed at 1205 °C for 5 minutes. An additional thermocouple was used for quality assurance. This thermocouple was assembled to one of the samples to measure the temperatures at the interior region of the sample. After brazing, all samples were brought to final geometry by turning operation.

Test methods

The specimens were tested on a servo-hydraulic tensile test machine with a tensile speed of 0.04 mm/s at room temperature (RT) and at application relevant temperatures of 400 °C and 600 °C. To evaluate the behavior of the brazed joints under corrosive atmospheres the dip-dry-test was used. This is a cyclic test to evaluate the hot-gas corrosion. In this test normed sea water (DIN 50905-4:1987 "Corrosion of metals; corrosion testing; corrosion testing in liquids under laboratory conditions without mechanical stress" [9]) is applied on the samples. After a drying period of 50 hours at 815 °C the procedure is repeated for four times. The corrosion tests were followed by tensile tests with the above mentioned specifications or by a vibration test. These tests were realized at temperatures from room temperature up to 600 °C. The tests are based on DIN 50100:1978 "Testing of Materials; Continuous Vibration Test; Definitions, Symbols, Procedure, Evaluation" [10]. Some samples were examined with light microscope and scanning electron microscope with energy dispersive micro area analysis.

Results and Discussion

Behavior of joints at static loads

In the Figures 2 and 4, results of tensile tests at different temperatures and different angles of the joint are shown. The tensile strength of nearly all samples is lower at higher temperatures in comparison to their strength at room temperature. The average difference between the strength at room temperature and at higher temperatures is approximately 50 - 75 MPa for all samples.

When testing the samples 2.4856/Au 827 and 2.4856/Ni 620 with an increasing angle of the joint the strength of the joint decreases, Figure 2. With an increasing angle, a higher amount of shear stress is induced in the braze leading to lower overall tensile strength of the specimens.

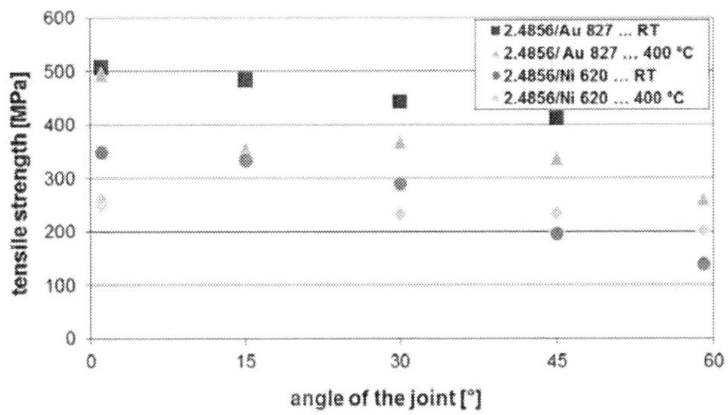

Figure 2: Tensile strength of joints at RT and 400 °C depending on the angle of the brazed seam

By investigating the fractured surfaces of the brazed specimens visually, significant differences can be observed, Figure 3. These differences depend strongly on the angle α and the testing temperature. For example, the fracture surfaces of 2.4856/Au 827 at 0° at RT show the typical surface of a sliding fraction. At 400 °C a more ductile fracture can be observed. At an angle of 30° the specimens show a shear fracture. At 400 °C this is more evident in comparison to RT. Significant deformations of the grains in the form of elongated combs are visible at the samples with a braze angle of 60°. This is obviously the result of the high amount of shear stress. In all specimens the deformation at higher temperatures is more evident than that at RT.

The 1.2344/Ni 620 specimens do not show a clear angle dependency of their strength, Figure 4. A possible reason for this can be gathered from Figure 5. The figure shows two fractured specimens after tensile test. The left picture shows a cross-section of a fractured specimen with a braze seam width of nearly 80 µm. A dark phase occurs in the light-colored matrix. As this phase is notably harder than the matrix, it acts as a notch in the brazed joint leading to higher stresses. Therefore, cracks start to propagate at this phase at relatively low external stresses running through or along the hard phase. The specimen with a brazement width of approx. 50 µm, shown in the right picture, does only have an indistinct ribbon made of dark phase in the middle of the seam reducing the strength of the joint. As it fractured through the diffusion zone between base material and brazement it can be stated that the notch effect due to the brittle phase does not occur in this case and that the strength of the specimens therefore is higher.

Figure 3: Overview of fracture surfaces (left) with detail (right) of 2.4856/Au 827 at different angles and test temperatures

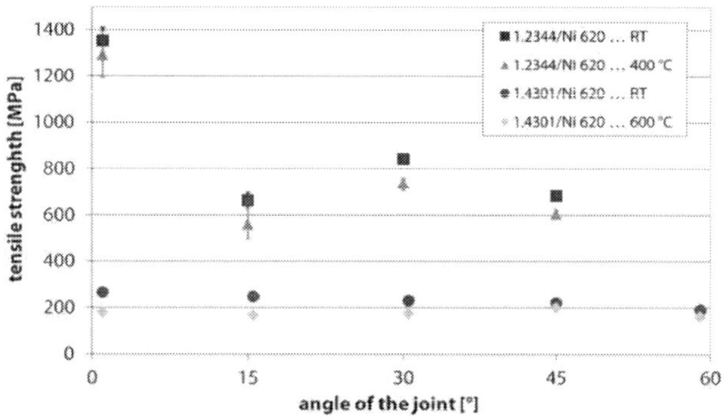

Figure 4: Tensile strength of joints brazed with Ni 620 filler metal at different temperatures depending on the angle of the brazed seam

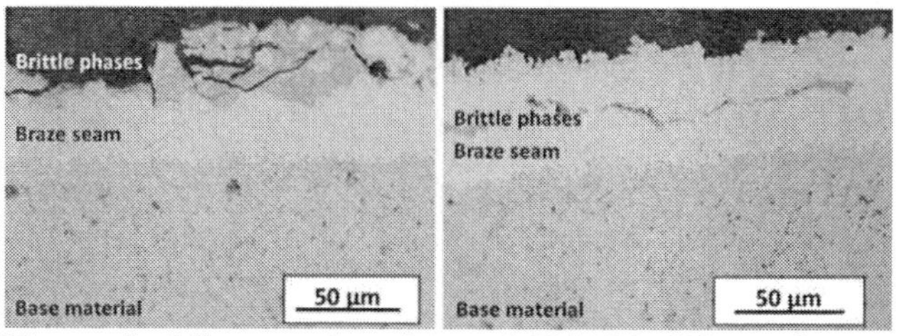

Figure 5: Micrographs of two 1.2344/Ni 620 specimens after tensile test; left: fracture through brittle phases in the middle of the braze seam; right: fracture in diffusion zone

The formation of a line of brittle phases has a significant negative influence on the mechanical behavior of the joint, resulting in a lower strength of the specimens. This formation mainly depends on the size of the brazing gap. Therefore, the strength of the brazed 1.4301/Ni 620 joints is mainly a result of the thickness of the brazing gap rather than being influenced by the angle of joint. Unintended variations of the thickness of the braze gap cannot be avoided in practice.

Behavior of joints at static loads after corrosive attack

During the dip-dry-test the samples were corrosively and thermally loaded. The surface of the base material of the specimens is homogenously attacked by corrosion. The brazed seams show a different behavior under the corrosive attack, Figure 6. The brazements 1.4301/Amdry 805 show a corrosion depth of approx. 40 µm. However, in case of 2.4856/Au 827 joints, a selective corrosion of Au-Si phases to a depth of approx. 150 µm can be observed. The 2.4856/Ni 620 specimens show, in contrast to the brazements shown in Figure 6, no significant difference in the corrosive attack zone for the base and braze materials after dip-dry-test.

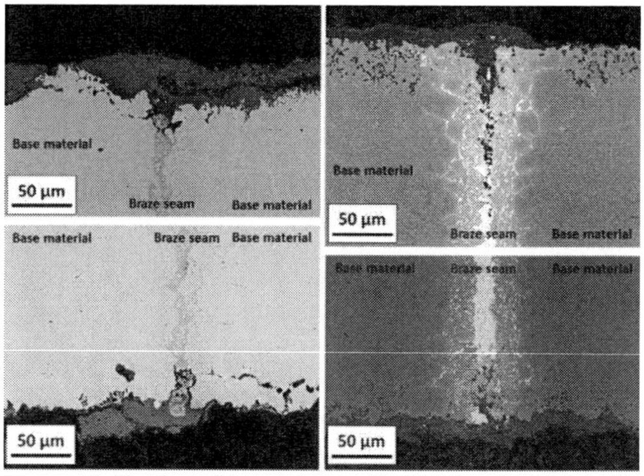

Figure 6: Border of brazed seam 1.4301/Amdry 805 (left) and 2.4856/Au 827 (right) after dip-dry-test

By analyzing the brazed seams in the center of the specimens significant changes in the structure before and after the dip-dry-test can be observed in some cases. This can be related to the thermal loading during the tests. The brazed joints with Ni 620 show a significant change in the brazement, Figure 7. The brittle phases dissolve in the base material based on diffusion processes. Only a few isolated little phases can be observed in the brazed seam after the dip-dry-test. At the center of the 2.4856/Au 827 joints no significant changes in the braze seam before and after dip-dry-test could be detected. The phases in the brazed seam seems to be stable at the test temperatures, so they do not dissolve in base material. Therefore, only an insignificant change in mechanical properties of the brazed joint is expected.

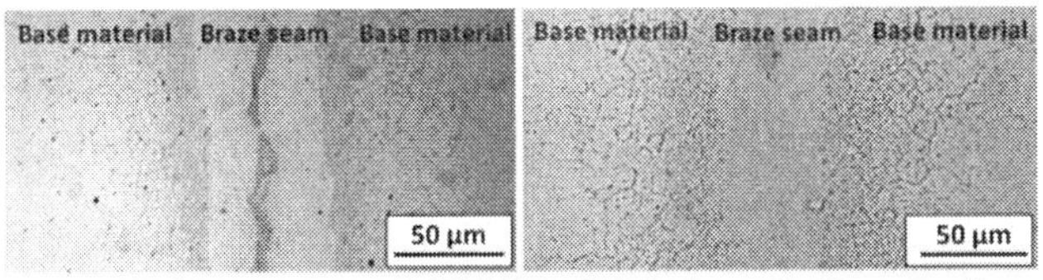

Figure 7: Braze seam 2.4856/Ni 620 before (left) and after (right) dip-dry-test

The influence of corrosion on the tensile strength of the joint is shown in Figure 8. As expected, the corrosion test has no significant influence on the tensile strength of the joint 1.4301/Amdry 805, because there were no changes in the structure of the brazement before and after dip dry test visible. The 2.4856/Au 827 joint shows an increase in strength of about 110 MPa between the specimens before and after corrosion. In the structure of the brazement before and after the dip-dry-test no significant change could be detected in braze seam. An explanation for the increase in the strength of the joint could be fine homogenization processes in the brazement at the high temperatures during the dip-dry-test. The joint 2.4856/Ni 620 shows an increase of the strength after dip-dry-test of about nearly 120 %. This is the result of the aforementioned diffusion of the brittle phase forming elements into the base material. Further information about the behavior of iron and nickel brazements under different static load conditions with and without corrosive load are published in [11, 12].

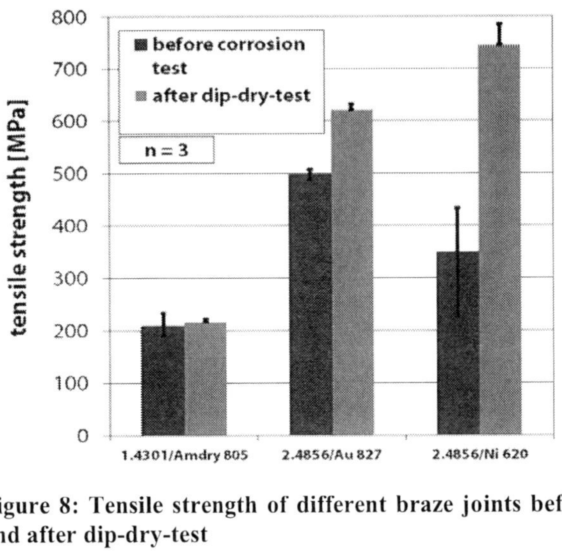

Figure 8: Tensile strength of different braze joints before and after dip-dry-test

Behavior of joints at dynamic loads

The behavior of the 1.4301/Ni 620 joint was investigated at vibration test at different temperatures. Figure 9 shows the comparison between the stress and the fatigue lives of the samples. The test was interrupted after 2,000,000 cycles (fatigue threshold). The fatigue threshold of the 1.4301/Ni 620 specimens at room temperature lies at about 190 MPa, which means at over 70 % of their tensile strength. At 600 °C the 1.4301/Ni 620 specimens show a similar dynamic behavior as at room temperature. However, the strength of the samples at 600 °C amounts to only about 40 % of the strength at room temperature.

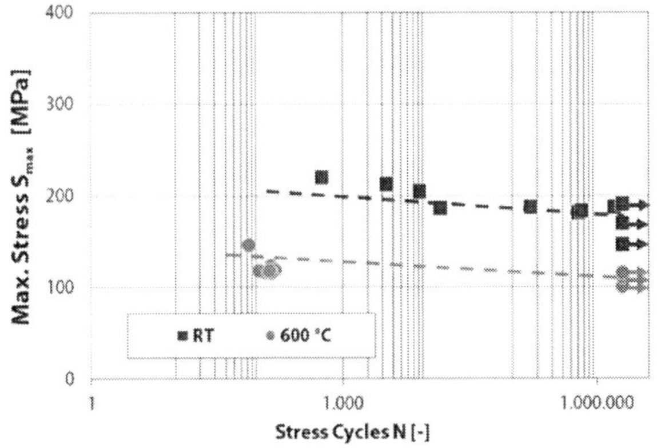

Figure 9: Stress life diagram of brazed 1.4301/Ni 620 specimens at RT and 600 °C

The fraction surfaces of 1.4301/Ni 620 samples were investigated after the vibration test. NiFe-, Si- and Ni_2B-phases are detected at the surfaces. At the surfaces of the specimens, which reached a lower cyclic fatigue at a relative high maximal stress, the visible amount of a hard phase is much higher than in the specimens which reached a much higher number of cycles. This phase is mainly determined as Ni_2B. Figure 10 shows fatigue cracks next to uncovered grains of these phases on the fracture surface of the brazement. These phases are uncovered on the fracture surface because the crack runs along their grain boundary. At lower load, the crack mainly runs around the Ni_2B hard phases through the soft matrix. In the specimens that were tested at a load of over 70 % of the tensile strength the crack runs through the hard phases. So the hard phases in the brazement have a positive influence at lower stress intensities as they decrease the crack velocity and the crack stability. At higher stresses the influence is negative because the phases are overcome by the crack. This results in a faster failure of the joints under dynamic load. This process is similar to the results of the mechanical behavior of particle reinforced materials which are described in [13]. Further information about the design and behavior of highly stressed brazed joints under different load conditions is given in [14].

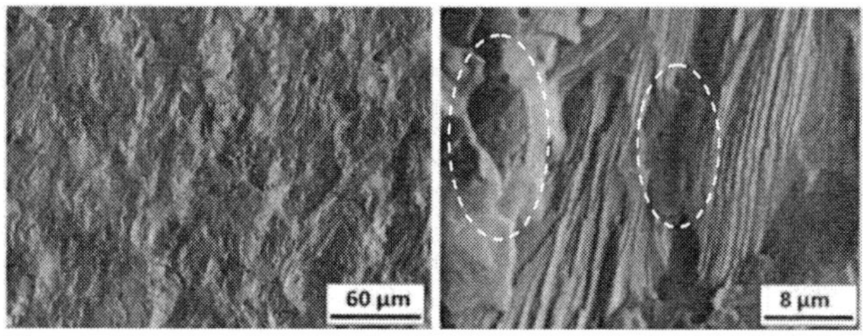

Figure 10: Fracture surface of a specimen tested at RT; left: overview; right: fatigue striations near a grain boundary in detail (hard phases are outlined)

Conclusions

The mechanical behavior of the brazed steel and nickel based specimens using nickel, iron and gold-alloys as filler materials was investigated. The influence of different mechanical strain and shear load conditions, the influence of corrosion and temperature on the strength of the joints was determined. Vibration tests were carried out for chosen joints at room temperature and at 600 °C.

Mostly, the static strength decreases at higher temperatures. Some brazed joints show a significant dependency on the predefined angle of the joint at tensile load. An increasing angle of the joint relative to the strain axis causes a higher shear stress amount. This results in a lower overall strength at the specimen as well as in a more ductile failure of the joints, which can be observed in the fracture surfaces. This effect could not be observed at brazements with a continuous line of brittle phases, which induces a local notch effect in the structure of the braze seam. These specimens do not show a significant angle dependency of their strength.

During the dip-dry-test of the specimens brazed with nickel-filler metal, the brittle phases in the brazement dissolve, resulting in a higher strength of the joint after corrosion test. When using iron or gold filler materials this effect could not be observed, because they do not build as much brittle phases during the brazing process. However, all brazed joints were attacked by corrosion at their surface.

The tested 1.4301/Ni 620 specimens show a fatigue threshold of about 70 % of the tensile strength. The investigations show a significant influence of brittle phases on the mechanical behavior under cyclic loads.

References

[1] Bobzin, K.; Tillmann W. et al., "Influence of dynamic stresses and operating environment on the mechanical properties of copper-based braze joints," Proc 5th International Brazing and Soldering Conf, Las Vegas, NV, April. 2012, pp. 327-331

[2] Bobzin, K.; Tillmann W. et al, "Systematik zur Auslegung und anwendungsrelevanten Prüfung von gelöteten Fügeverbunden," Proc 12th Werkstofftechnisches Kolloquium, Chemnitz, Germany, October. 2009, pp. 169-179

[3] Bobzin, K.; Tillmann, W. et al., "Systematische Untersuchung der Eigenschaften gelöteter Fügeverbunde mit anwendungsrelevanten Prüfverfahren," Schweißen und Schneiden, Vol. 65, No. 5 (2013), pp. 254-261

[4] Müller, W., Metallische Lotwerkstoffe, Arten Eigenschaften, Verwendung, DVS-Verlag (Düsseldorf, 1990)

[5] Dorn, L. et al., Hartlöten, Grundlagen und Anwendungen, Expert Verlag (Sindelfingen, 1985)

[6] Zaremba, P., Hart- und Hochtemperaturlöten, DVS-Verlag (Düsseldorf, 1988)

[7] DIN EN ISO 17672:2010 Brazing - Filler metals, DIN e. V. (Berlin, 2010)

[8] DIN 50125:2009 Testing of metallic materials - Tensile test pieces, DIN e. V. (Berlin, 2009)

[9] DIN 50905-4:1987 Corrosion of metals; corrosion testing; corrosion testing in liquids under laboratory conditions without mechanical stress, DIN e. V. (Berlin, 1987)

[10] DIN 50100:1978 Testing of Materials; Continuous Vibration Test; Definitions, Symbols, Procedure, Evaluation, DIN e. V. (Berlin, 1978)

[11] Bobzin, K.; Tillmann W. et al., "Reliable Design Procedure of Iron-Based Braze Joints," Proc 10th Brazing, high temperature brazing and diffusion bonding, LÖT 2013, Aachen, Germany, June. 2013, pp. 261-266

[12] Bobzin, K.; Tillmann W. et al., "Systematik zur Auslegung und anwendungsrelevanten Prüfung von Nickel-Basis Lötungen," Proc 9th Brazing, high temperature brazing and diffusion bonding, LÖT 2010, Aachen, Germany, June. 2010, pp. 17-22

[13] Suresh, S. et al, Fundamentals of Metal Matrix Composites, Butterworth-Heinemann (Stoneham, 2002)

[14] Puidokas, S. M., Design of highly stressed brazed joints, Shaker-Verlag (Aachen, 2013)

New Brazing Method Used Thermal Gradients in the Conventional Furnace

Mitsuharu Kurata, Satoshi Kogi, Yasuyuki Miyazawa
Tokai University, Hiratsuka-shi, Kanagawa-ken, JAPAN
Email: mthr@mail.com
Phone: +81-463-58-1211

Abstract

In generally speaking, a brazing is good at joining for dissimilar metals. Then, two kinds of different interfaces are formed. The reason is that the filler metals put between dissimilar metal and another dissimilar metal. There are two kinds of interfacial reaction at joint interface. However, in spite of the filler metals into the joining surface, there are formed brittle inter metallic compounds. The brittle intermetallic compound is the cause of the destruction. Therefore, we suggested new type brazing method and did brazing dissimilar metals. The method is different from the conventional brazing. One of the conventional brazing methods is to heat the sample in the uniformly heated furnace and to hold certain period time. But this new type brazing method is able to make thermal gradients in the furnace by three of heaters has been equipped by the furnace. Using this new type brazing method, brazing surface has been controlled two kind of interfacial reaction by thermal gradient in the furnace. Then, we got tough assembly. To inspect this new type method, we made the furnace and did pre experiment by copper to copper brazing. After we confirmed availability of this new type brazing method, we did the experiment by aluminum to copper with thermal gradients on this new type method. The each sample shape is sphere 8mm and vertical 35mm. In this experiment, we had using the two kinds of filler metal. One is Brazing Sheet which is thin cladding material containing aluminum mainly. We put in Brazing Sheet to surface of base metal, aluminum side. Second is MBF2005 which is thin filler metal foil. MBF2005 contain Cu/P mainly. We put in MBF2005 to surface of base metal, copper side. After finishing brazing, we did metallographic observation, electron prove micro analyzer, inter-facial analysis and point analysis. These results showed at the bottom of the result and discussions.

Introduction

Welding is significant basic process. For example, arc welding, friction stir welding and so on. Almost welding process is obtained part of joint as line. On the other hand, brazing process is obtained part of joint for surface. From these backgrounds, it is said that the brazing process are good at joining of dissimilar metals and dissimilar metals brazing process has been needed for industrial fields. Now, Al/Cu, Al/Fe, Ti/SUS and etc., are needed in industrial fields. However, there are no filler metals for brazing dissimilar metals. Then, in this research, we improved brazing process of dissimilar metals. Firstly, we bought the vertical electric furnace such as Figure.1.

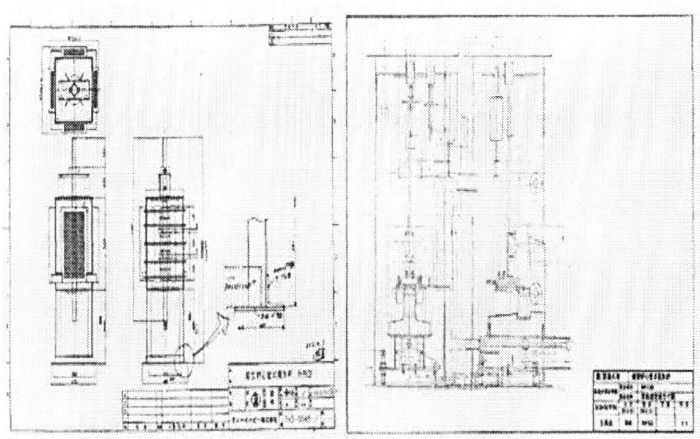

Figure.1 Diagram of vertical furnace.

This furnace has three heaters. There are able to control temperature in parts by three heaters in the furnace. Also, this furnace is able to make thermal gradient into the joining interface positively. Next, we had revamped the furnace. The main revamping is adding lifting and lowering device which using jack of motor cycle and microscope stage showed in Figure.2.

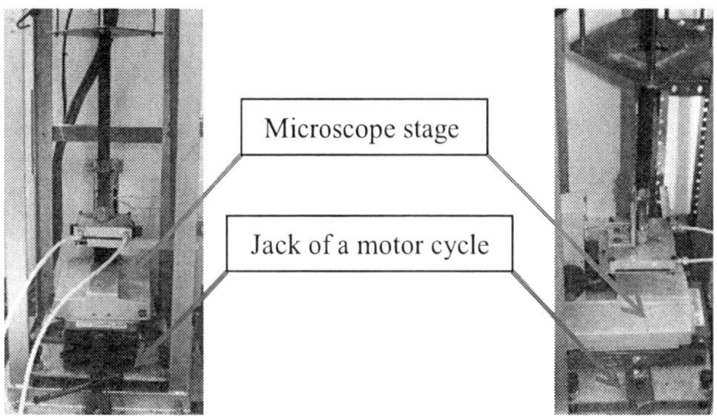

Figure.2 Lifting and lowering device of the vertical furnace

The lifting and lower device can be moved by a few micrometers. The sample is in the original assembly such as Figure.3.

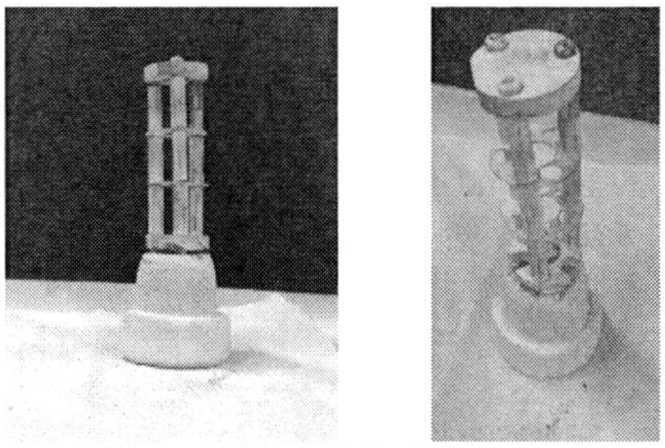

Figure.3 The original assembly

We took Ar gas with 10 percent hydrogen for atmosphere and we took silica gel in to the hose pipe to avoid oxidation.

Experimental procedures

The typical brazing Cu to Cu

We did the typical brazing copper to copper to confirm the new type brazing process. Firstly, we did brazing for copper to copper with non-thermal gradient. We set all three heaters to 700°C. The each samples shape is sphere 8mm, vertical 35mm. We used MBF2005 as filler metal which chemical composition is showed at Table 1. Secondary, we did brazing of copper to copper with thermal gradient. We set top and middle heaters to 700°C and set under heater to 300°C. The sample shape and filler metal are same as brazing of copper to copper with non-thermal gradient.

Table 1 Chemical Composition(mass%) and melting point(°C) of the filler metals.

		Chemical composition(mass%)									Thickness (μm)	Temperature(°C)	
		Al	Si	Fe	Mn	Cu	Zn	Ni	Sn	P		Solidus	Liquidus
Brazing Sheet	Middle	Bal.	0.5	0.17	1.1	0.15	2.5	—	—	—	70 Clad thicness10%	—	—
	Outside	Bal.	7.5	—	—	—	—	—	—	—		577	617
MBF2005		—	—	—	—	Bal.	—	5.5	9	6.5	40	591	641

The brazing Al to Cu

In this experiment, each sample's shape is same the typical brazing of copper to copper. This time, we took aluminum and copper as base metals. Each base metal's degree of purity is more than 99 percent. In addition, we used two kinds of filler metal. We put Brazing Sheet for base metal, aluminum side and MBF2005 for base metal, copper side [1]. Each filler metal's chemical compounds showed at Table1. We had using the two kinds of filler metal. Brazing Sheet is thin cladding material and containing aluminum mainly. MBF2005 is thin filler metal foil for copper brazing generally containing Cu/P mainly to surface of part of joint copper side.

After finishing copper to copper brazing and aluminum to copper brazing, we did metallographic observation, electron prove micro analyzer, inter-facial analysis and point analysis.

Results and Discussions

In this research, before brazing dissimilar metals, we studied the effect of thermal gradients on brazing same metal. Firstly, we took copper to copper for base metal and MBF2005 as filler metal. Then, we did brazing with non-thermal gradient that we set three heaters to 700℃. The typical result showed at Figur.4.

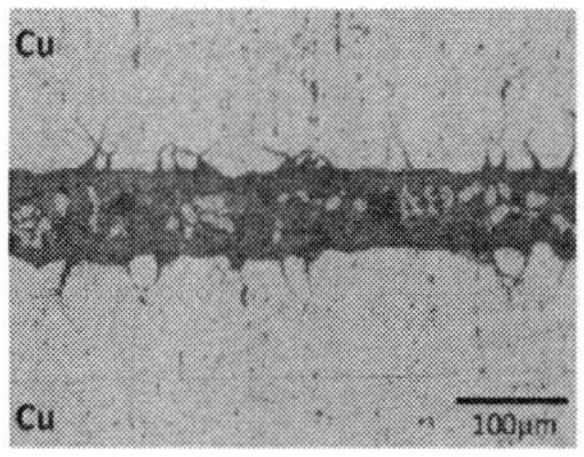

Figure.4 Cross-sectional microstructure at the brazed joint with non-thermal gradient that we set temperature 700℃ at all heaters.

This is typical copper to copper the brazed joint. Next we did brazing with thermal gradients that top and middle of heaters set up to 700°C and bottom heater set up to 300°C. Then we took the result that it is metallic microstructure such showed at Figure.5.

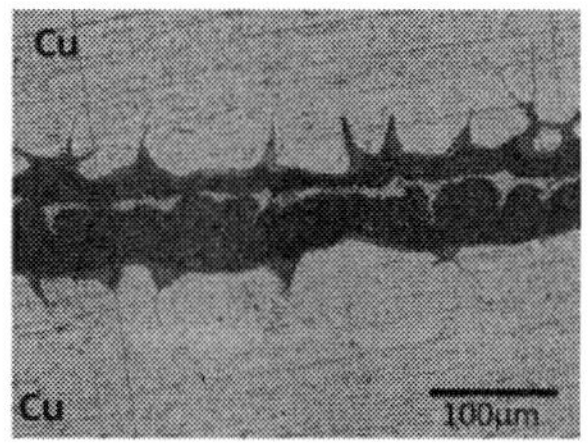

Figure.5 Cross-sectional microstructure at the brazed joint with thermal gradients top and middle of heaters set up to 700°C and bottom heater set up to 300°C.

We compare these results. Penetration grain boundary each interface is confirmed by the sample with non-thermal gradients. On the other hand, the sample with thermal gradients has grain boundary which is different from the sample of non-thermal gradients. It can be seen that the position of the eutectic is shifted to upper side. For the results, we did brazing dissimilar metals with thermal gradients we took the base metal which aluminum and copper .The results of brazing showed at Figure.6. There are different melting point brazing sheet and MBF2005. Also, liquidus temperature and solidus temperature are different. Then, we need changing temperature at each zone of the furnace. We set 300℃ for top heater and 600℃ for middle and bottom heater. Besides, we put the aluminum to the upper side of original assembly and copper to bottom side of original upper side. Finishing brazing, the brazed sample moved to upper side of furnace and cooling nearby entry of Ar gas with 10 percent hydrogen.

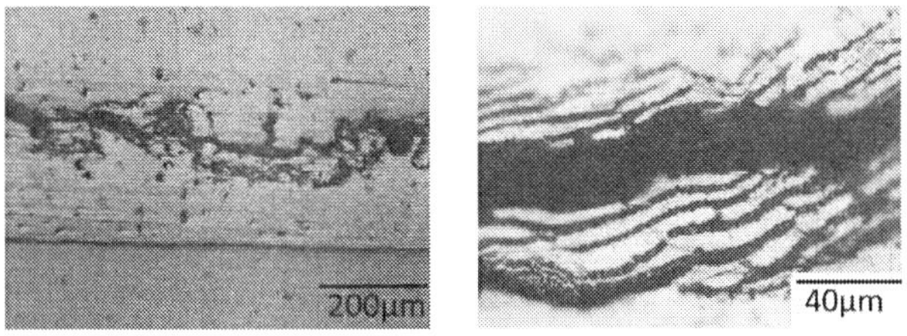

Figure.6 Cross-sectional microstructure at the brazed joint of Al to Cu with thermal gradients.

After cooling, we did metallographic observation and EPMA. The results were shown at Fig.7.

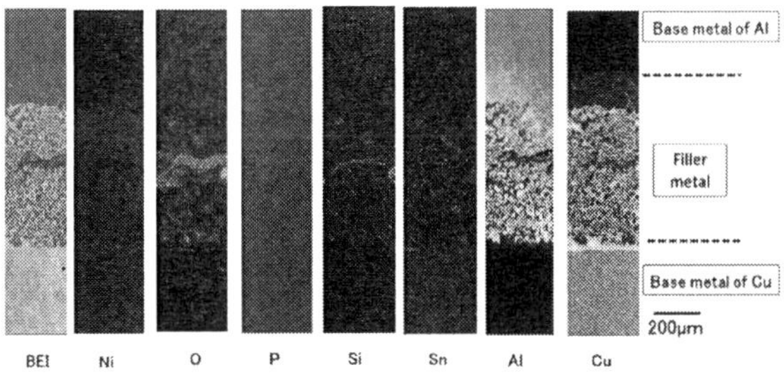

Figure.7 The result of brazing Al/Cu by EPMA

According to metallographic, we confirmed black layer and hole at the brazed surface vicinity. According to the result of EPMA, we confirmed some layer of the filler metal at copper side. We did point analysis and interfacial analysis of the layer is shown in Figure.8 (a), (b), Table 2 and 3.

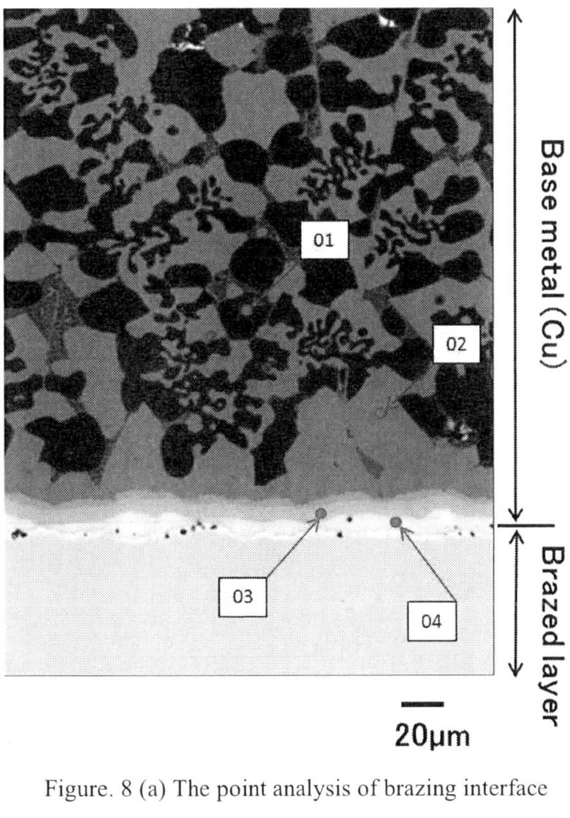

Figure. 8 (a) The point analysis of brazing interface

Table 2 The result and consideration of the point analysis.

No.	at%					Result and Consideration
	Al	Cu	Zn	P	Sn	
1	97.2	2.2	0.5	-	-	Confirmed α phase
2	66.8	33.2	-	-	-	Confirmed $\alpha+\theta$ phase
3	34.1	65.2	-	-	0.7	Confirmed γ phase
4	78.3	18.8	-	2.9	-	Confirmed $\alpha+\theta$ phase

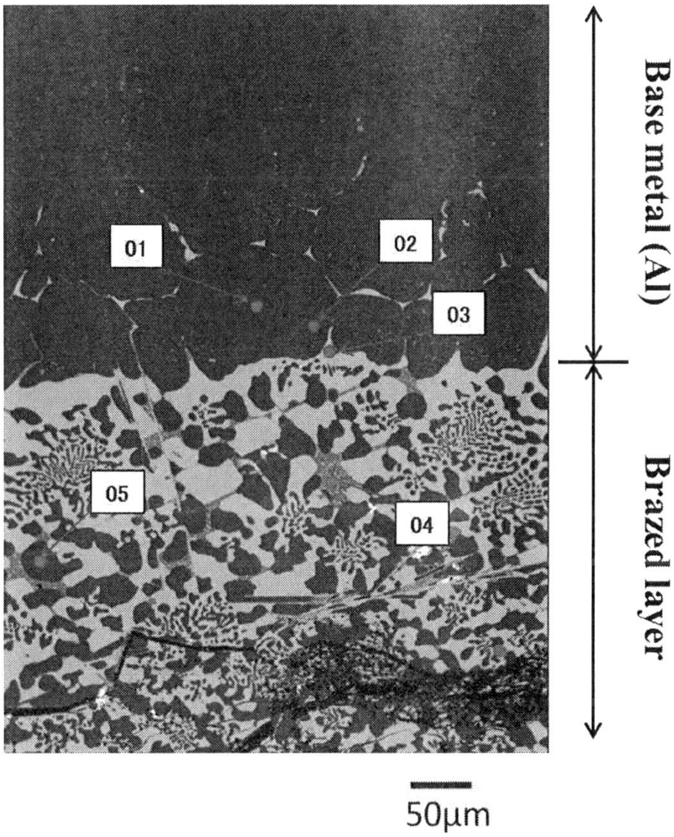

Figure. 8 (b) The point analysis of brazing interface

Table 3 The result and consideration of the inter facial analysis

No.	at%				Explain and Consideration
	Al	Cu	Ni	P	
1	99.2	0.8	-	-	Base metal
2	94.5	5.5	-	-	Base metal. The Cu diffused to Al.
3	33.3	65.7	0.9	0.2	We thinking of γ phase because the brazing filler metal and diffused to Base metal
4	78.8	21.2	-	-	The Cu diffused to after brazing base metal and formed θ phase.
5	97.8	2.0	-	0.2	α phase.

These analyses showed the results that the layer which are showed containing of aluminum and copper interface. In the inter face of copper side, we confirmed element of Al rich from No.4 of Figure.8 (a).

We accepted the Sn from MBF2005 because the chemical composition is containing Sn. In the meanwhile, the generally speaking, solder of copper to copper with the solder is containing Sn and Sn is compatible with copper. Guessingly these reasons, the Sn containing MBF2005 which put on copper side [2] [3]. Next, we confirmed the crack by the metallographic. We have to make another new thermal gradient into the furnace to control the crack which we confirmed this time in the future.

Conclusions

(1) We suggested and established the new type brazing method of dissimilar metal to another dissimilar and we confirmed the validity of the new type brazing method with thermal gradient in the conventional furnace to obtain the following conclusions.
(2) We did pre experiment of brazing copper to copper.
(3) We did experiment of brazing aluminum to copper took the two kinds of filler metal. In the case of brazing copper to copper with MBF2005, there was different the solidification structure of the brazing filler metal layer by furnace position. Therefore, there were controlled solidification processes of the brazed samples.
(4)From these experiments, in order to suppress interface reaction, there are requiring to keep the heating base metal of aluminum for lower temperature and making thermal gradients relatively large.
(5) We got the brazed joint but it had not enough strength. This is said that two kinds of filler metal were melted together and formed brittle layer of intermetallic compounds.

Acknowledgments

Public benefit corporation Tanigawa Thermal Technology Promotion Fund, 2012 fiscal subsidies and, we received the auxiliary of fiscal 2012 faculty of Engineering Research and Education Grants. We represent my deep appreciation to both. In addition, we owe engineer Yasuhiro Tsukamoto belongs to Department of Material Science, School of Engineering, Tokai University our gratitude.

References

[1] IHI foreign news Vol.50 No.4 (2010), pp12-15(in Japanese)

[2] Xia, C Z;Li, Y J;Puchkov, U A;Gerasimov, S A;Wang, J "Materials Crack analysis near vacuum brazing interface of Cu/Al dissimilar materials using Al-Si brazing alloy", Materials Science and Technology; Mar 2009; 25, 3; ProQuest Central, pg 383 (in Japanese)

[3] Takeshi Kobayasli, Toru NIaruyatta and Graduate StudentYuki Sakaguchi, Engineering,Kansai University,Lecturer, Dro Eng. Sn in Cu–Zn Alloy, (Received IV[arch 30,2005;Accepted June 2,2005)

Hump-Back Versus Straight Through Continuous Furnaces ... When and Why?

By

Stephen L. Feldbauer, Ph.D.

Abbott Furnace Company
1068 Trout Run Rd.
St. Marys, PA 15857

Figure 1. Hump-Back Furnace

Background

There are two types of the continuous belt furnace: hump-back and straight through. Either furnace is capable of processing products that require very reducing atmospheres. It is not uncommon to see dew points below -70°F in either type of furnace.

The hump-back furnace has a belt that travels the length of the heated and cooling sections of the furnace and returns along the furnace bottom. The surface upon which the belt rides as it passes through the heated and cooling sections has a slope in the front and rear portion of the furnace that results in a "hump" in the overall pass-line of the furnace.

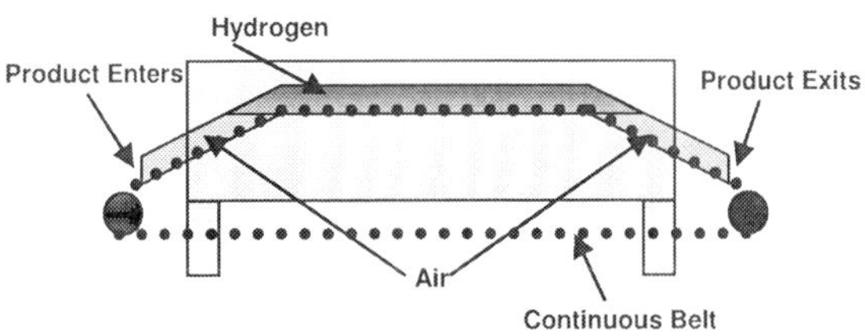

Figure 2. Schematic of a Hump-Back Furnace

The "hump" is designed such that the floor of the elevated portion of the pass-line is physically above the openings at the entrance and exit of the furnace. When an atmosphere such as Hydrogen is used in the furnace, the "hump" takes advantage of the differences in the density between the very "light" Hydrogen and the much heavier Air to form a Hydrogen pocket that is very pure and ideal for processing components that require an atmosphere that is very reducing in nature. The flow rate of the process atmosphere is minimized because there need only be enough fresh atmosphere introduced into the "pocket" to replace the actual gas volume that is used in the process and that which is consumed in the reaction with the air at the interfaces at the front and rear of the "pocket".

In the past, the hump-back furnace was the most common type of furnace that competed with vacuum furnaces. In brazing applications, for example, the hump-back furnace affords the producer the ability to braze larger volumes of product with equivalent quality, higher production rates and higher thermal efficiency than with a vacuum furnace.

Figure 3. Straight Through Continuous Furnace

A straight through continuous furnace has a belt that rides on a horizontal surface throughout the furnace.

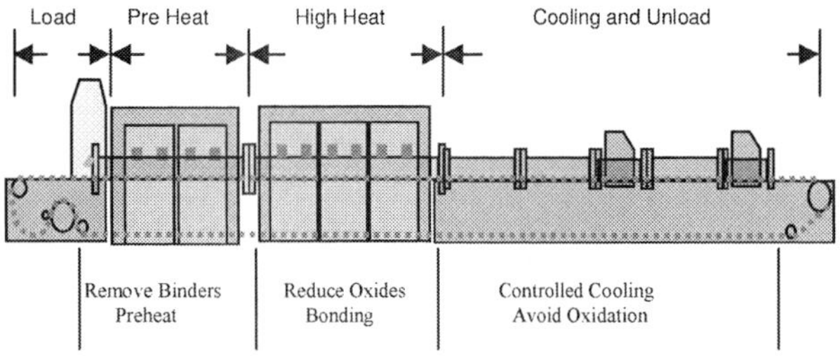

Figure 4. Schematic of a Straight Through Continuous Furnace

Until recent years, it was thought that in a continuous belt furnace the atmosphere control and dew point could not be controlled to a level sufficient to perform processes such as brazing stainless steel without the use of a flux. However, due to advances in atmosphere technology and furnace design, this is now a very common approach for large production rate, critical atmosphere quality processes.

Optimal atmosphere control is achieved through zoning, atmosphere composition, multiple atmosphere injection points, flow rate and overall directionality of the atmosphere flow in the furnace. Approximately 80% to 90% of the total atmosphere that is introduced into the furnace should flow toward the front of the furnace. Forward atmosphere flow provides optimal heating and a minimization of atmosphere usage to flush any volatiles from the product while preventing air from entering the furnace opening.

Advantages and Disadvantages

Although a hump-back furnace has the advantage of reduced atmosphere flows when compared to the straight through furnace, there are some disadvantages that may out weigh the advantages.

The total height and length of the product and fixture must be restricted to ensure that the hump is effective in maintaining the seal through the stratification of the atmosphere. The typical working envelope where a hump-back furnace begins to have advantages over a straight through furnace is at a total part and fixture height above 8 inches and length under 24 inches.

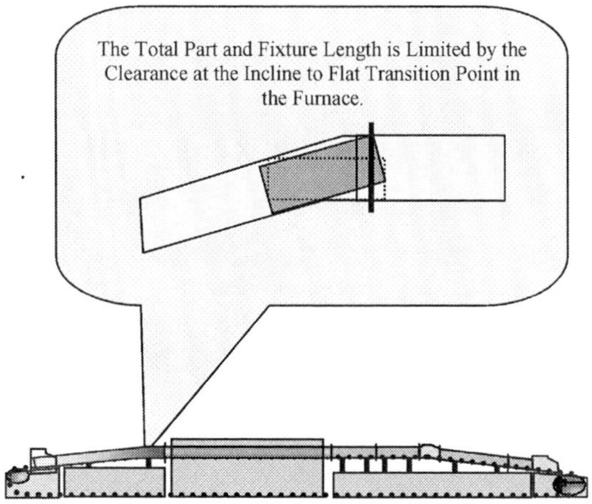

Figure 5. Schematic of Part and Fixture at the Transition of a Hump-Back Furnace

In most cases where the total length exceeds 24 inches, the part and fixture will not have enough clearance to safely transition from the incline to the flat section of the furnace. As well, the hump presents potential issues with respect to product tipping while transitioning from the incline to the flat region of the furnace.

The geometry of the incline also results in an increase of the tension that must be put onto the belt to pull the additional loaded length of belt over the transition. Belt life is inversely proportional to tension; hence, a hump-back furnace will see ~5% reduction in the life of the belt when compared to that of a straight through furnace. Likewise, the slope on the front and rear on the furnace require that more belt be purchased for a hump-back furnace. The net result is an increase in belt costs for a hump-back furnace in excess of 10% over a straight through furnace.

Two other significant disadvantages of a hump-back furnace are a direct result of the increased size of the furnace. The furnace will occupy more floor space and have a larger capital cost. A hump-back furnace is typically 20% more expensive than a straight through furnace that could produce the same product with the same quality and production rate.

A less quantifiable disadvantage of a hump-back furnace is the increased height of the furnace. This makes the overall maintenance of the furnace more difficult than if the elements and electrical components of the heated section were lower and accessible without a ladder or platform.

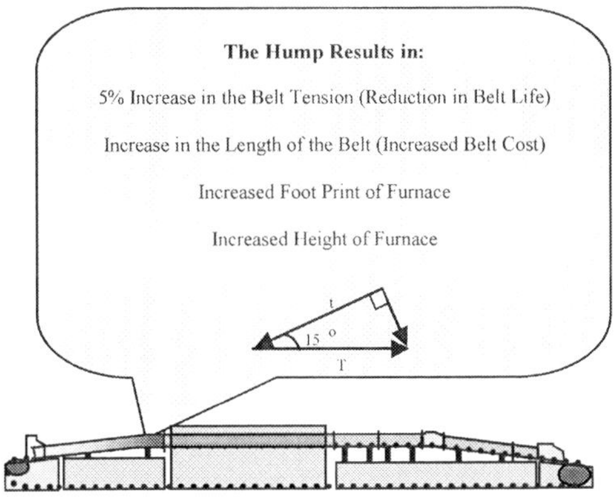

Figure 6. Effect of Slope to Flat Transition in a Hump-Back Furnace

The primary advantage of the hump-back furnace is the savings in the amount of atmosphere necessary to keep the air out of the furnace. The savings is directly proportional to the cost of the gas and the amount used. It then follows that the advantage is a strong function of a.) the composition of the process atmosphere and b.) the opening of the furnace, through which the part and fixture must pass to enter the furnace.

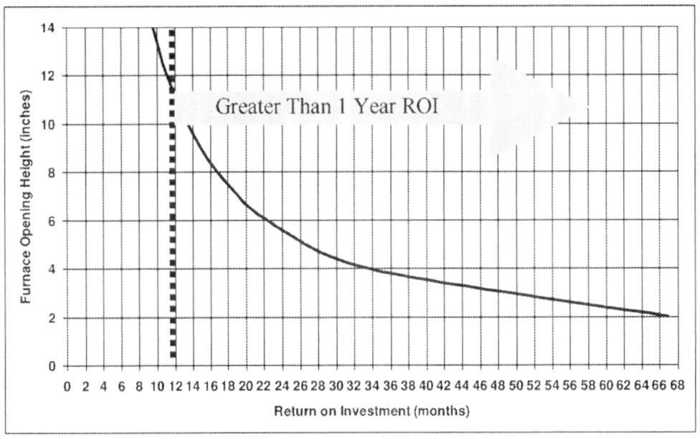

Figure 7. Opening Size Versus Return on Investment for a Hump-Back Furnace using 25% Hydrogen / 75% Nitrogen

So when does it make sense to use a hump-back furnace? Figure 7 illustrates the estimated return on the incremental investment for a hump-back furnace over a straight through furnace when using a relatively low mix of Hydrogen and Nitrogen. Figure 7 is based on \$1.40 per 100 scfh of Hydrogen and \$0.40 for 100 scfh of Nitrogen, an atmosphere composition of 25% Hydrogen and 75% Nitrogen and a 24 inch wide belt furnace. In this case, the furnace opening must be in excess of 11 inches to have a typically desirable return of 1 year or less on the additional investment for the hump-back.

Figure 8 is the same plot for a high Hydrogen atmosphere; in this case 75% Hydrogen and 25% Nitrogen. The higher volume of Hydrogen increases the cost of the atmosphere and moves the curve down. Now smaller parts, requiring an opening of 8 inches or more will have a return of 1 year or less.

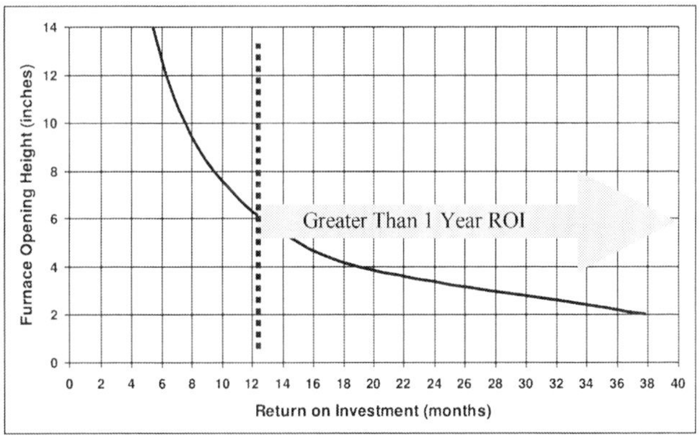

Figure 8. Opening Size Versus Return on Investment for a Hump-Back Furnace using 75% Hydrogen / 25% Nitrogen

Summary

Metallurgically speaking, the same product can be produced in either a hump-back or a straight through furnace. However, the choice of which furnace type to use is somewhat complex. Many variables impact the economics and functionality of these furnaces. The atmosphere composition, part and fixture dimensions, maintenance, belt costs, furnace foot print, and capital cost must all be considered.

Figure 9 summarizes this decision process in a convenient decision tree. Generally speaking, unless the part plus fixture is less than 24 inches long and over 6" tall (requiring an opening of 8 inches or more), a straight through furnace is likely the better choice.

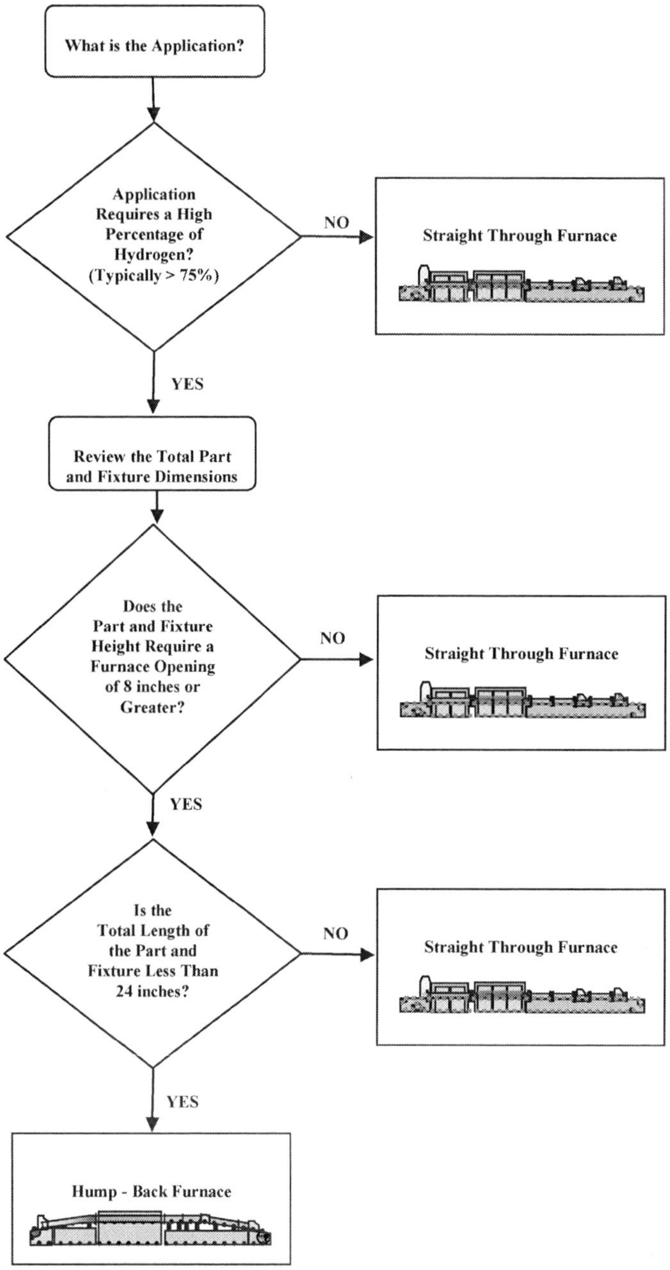

Figure 9. Hump-Back versus Straight Through Furnace Flow Chart

How to Make the Best Use of Your Furnace: Maintenance Tips

Jim Grann
Ipsen, Cherry Valley, Illinois, United States
800-727-7652

Abstract

Today's competitive world and business environment often requires one to do more with less – to make equipment last longer and run more efficiently. Successful operation and maintenance of a vacuum furnace system not only requires strict adherence to a carefully scheduled maintenance program, but also a deeper understanding of preventative and corrective maintenance – such as how to properly detect, identify and correct leaks, repair and replace heating elements and more.

By following these maintenance tips and best practices for troubleshooting your vacuum furnace, you can ensure your furnace equipment has a long and healthy lifespan, as well as enhance braze quality.

Introduction

As businesses invest in expensive capital equipment, knowing when and how to perform maintenance becomes a primary concern. After all, if not cared for properly, equipment can easily break down, thus creating unexpected costs and slowing down production throughput. As a result, one of the main challenges businesses find themselves faced with is how to prolong the life of their furnace systems while also increasing productivity. This is why one of the first – and most essential – steps for protecting your investment is to routinely employ preventative and corrective maintenance in day-to-day operations.

Knowing how to properly maintain and troubleshoot your furnace system is an excellent way to protect your investment, as well as minimize downtime, increase productivity, ensure a long lifespan, achieve a quality braze and cut down on operational costs. However, maintenance is a broad topic; some of the essential preventative and corrective maintenance areas every furnace owner and operator should be familiar with include:

- Creating daily, weekly, monthly and yearly maintenance schedules
- Detecting, identifying and preventing leaks, from properly using a helium mass spectrometer to knowing the probable locations of leaks
- Repairing and replacing heating elements

This paper will cover these key areas for keeping your furnace running at peak performance and, by following these maintenance best practices, one can consistently achieve a quality braze while also cutting down on operational costs and unplanned downtime.

Corrective and Preventative Maintenance

Now that you and/or your company have made an investment in a heat-treating furnace system, it can be daunting to know where to start when it comes to maintenance. Before beginning maintenance on your vacuum furnace equipment, it is essential to understand that there are two basic types of maintenance: corrective and preventative.

Corrective maintenance takes place after a furnace fails to function properly. It typically involves looking at the common symptoms to determine the probable causes of failure and, ultimately, correct them.

Preventative maintenance, on the other hand, involves regular inspection and maintenance of the equipment. This form of maintenance is regarded as the most effective method for protecting your equipment and extending its lifespan, as well as preventing unexpected downtime.

Creating Maintenance Schedules

As one moves forward from an understanding of the two primary forms of maintenance to implementation, it is essential to start by creating accurate records if you are to maintain an efficient maintenance program. After you start up your equipment, it is important to establish a performance log/baseline that records the following:

- Blank-off pressures of the mechanical pumps
- Pumpdown time to a given pressure
- Ultimate vacuum
- Linear leak rate for when the chamber is blanked-off from the pumping system

The items listed above are of special importance in operating a vacuum furnace, which is why one must record every observation or change. Since the equipment's performance will deteriorate with age, being able to make an accurate diagnosis avoids the unnecessary dismantling of components.

For example, if you record the furnace's ultimate vacuum, linear leak rate, etc. from the beginning, as well as any changes that occur over time, you can then use the log to determine if performance in one of those areas has deteriorated. As a result, you are better equipped to narrow down and identify potential causes, as well as the proper corrective actions, without wasting large amounts of time and resources. Overall, maintaining historical logs is invaluable to maintenance personnel who are dealing with both preventative and corrective maintenance.

In addition to maintaining historical logs, another effective method for ensuring proper maintenance and care of one's vacuum furnace is to adhere to a carefully scheduled preventative maintenance program. By utilizing daily, weekly, monthly and yearly maintenance checklists (Figure 1), one is better able to track all the items that should be regularly inspected.

Maintenance Checklist

The following items should be checked according to the schedule outlined:

1. **Daily**

____ Check the condition of the hot zone elements and shielding. Be sure to check the element retaining wires. Are any broken or missing?

____ Check the water lines for leaks and any hoses that appear hotter to the touch than usual; this would indicate a lack of normal flow (closed loop system).

____ Visually inspect all water lines for flow and temperature (open system).

____ Check the oil level and condition of all pumps. The roughing pump oil should be circulating.

____ Check that the roughing pump is operating at 140°F (60°C). Adjust pump water miser as necessary. (160°F/71°C for gas ballasting operation.)

____ Check the temperature and cooling-water profile of the diffusion pump.

Figure 1 – A checklist of preventative maintenance tasks that should be performed on a daily basis.

Some essential maintenance activities that should be performed include changing the oil in the vacuum pump every two to six months, replacing dynamic seals (such as door seals and poppet valve seals) every year and replacing the jack panel (work thermocouples) assembly every year. Replacement of the control thermocouple should follow applicable guidelines. Vacuum sensing gauges also need to be replaced, cleaned or rebuilt as required. While adhering to a set maintenance schedule is important, it is also essential to make sure equipment and parts are properly cleaned during the maintenance process.

Cleaning of Parts and Vacuum Furnace

Now that you have established a maintenance routine, it is important to note that the majority of maintenance time spent on a vacuum brazing furnace is devoted to cleaning metal oxide deposits that form inside the chamber and hot zone during the brazing process. It is important to do this regularly – as well as know the correct cleaning techniques – so as to avoid damaging your equipment.

Mechanical cleaning is the usual method for removing metal oxides on an aluminum-brazing furnace. Scraping of the metal oxides from chamber walls and hot zone shields must be done with nonferrous scrapers so as not to generate a spark that may ignite the fast-burning oxide.

If the metal oxide buildup is too heavy and hard to scrape off, you may have to perform an air cleanup cycle to help dislodge large oxide clusters. After the majority of the oxide is removed from the furnace chamber and hot zone, a normal vacuum cleanup cycle is ran to further condition the furnace prior to placing it back into production.

In addition to maintaining your vacuum furnace, brazing also requires parts to be cleaned properly and handled with care so as not to introduce contamination prior to the brazing cycle. It is essential to remember that what goes into the furnace will come back out – no matter how small or insignificant it originally seemed. Putting dirty parts into the furnace can increase the rate at which the hot zone deteriorates, as well as negatively affect production times and the quality of future brazed parts.

All grease, oil and particulates must be cleaned off the parent and filler metal surfaces, and assemblers need to be careful not to transfer oils from their skin to these surfaces while assembling and fixturing the parts together. Typical cleaning methods include vapor degreasing, hydrocarbon wash, aqueous washing, acid etching and vacuum de-oiling.

Oftentimes, sandblasting will be used to clean fixtures. However, such a cleaning method is generally not recommended as it can cause untold amounts of damage to the furnace chamber. This is because there is often residue on parts, and every time an item is cleaned via sandblasting, particulates are removed and become mixed up in the sandblast material. Not only are these unknown particulates not rated for temperature or vacuum, but they can also cross-contaminate any items that are later sandblasted, thus negatively impacting the process and equipment.

Repairing and Replacing Heating Elements

Sandblasting fixtures and/or parts is not the only thing that can cause unexpected damage to one's furnace; not properly maintaining the heating elements can also negatively impact the furnace and its temperature uniformity. When it comes to the heating elements on a vacuum furnace, maintenance practices can vary depending on whether one is working with graphite or metal. One generally replaces graphite heating elements rather than repairing them; this is because graphite is less brittle than molybdenum, so one doesn't have to worry about accidently breaking surrounding heating elements during the replacement process. Metal heating elements, on the other hand, can have a whole section replaced (just like graphite), or they can be repaired with an element patch.

Element patches allow you to clamp on small sections that bridge the broken element, rather than having to unbolt the heating elements between connection points and replace an entire section. Such repair patches are typically preferred as molybdenum becomes extremely brittle once it has been heated, making it easier to break surrounding elements during replacement. Breaking surrounding elements can quickly lead to additional, unexpected repair costs.

Inspecting the heating elements on the vacuum furnace should be part of every maintenance schedule as broken heating elements can lead to temperature problems, such as inadequate heat and poor uniformity. One should check the heating elements monthly for signs of deterioration. The shields and heating elements should also be checked every two weeks for signs of discoloration. If they begin to show signs of discoloration, a leak check and/or furnace cleanup cycle may be required.

However, if you have an older furnace or one that has been out of production for some time, it is important to know that a vacuum furnace must be taken higher in temperature and soaked long enough to reach a steady-state heat loss condition. For most Ipsen-brand furnaces, it is recommended that average cleanup temperatures should not exceed 2,200 °F (1,204 °C). Always make sure your furnace can withstand this temperature before proceeding with a cleanup cycle.

To determine if a heating element needs to be replaced, one should check the resistance to ground, as well as inspect for signs of arcing, signs of cracking and degradation on the bottom third of each element and signs of low temperature alloys under the elements. It is vital that all low temperature alloys – such as fixtures, grids and baskets – are removed before running a high-temperature cleanup cycle, as the alloys will melt onto the heating elements, as shown in Figure 2.

Figure 2 – Alloys left behind during a cleanup cycle melted onto the hot zone's heating elements.

If you are running a high-temperature cleanup cycle and realize you melted some lower temperature alloys, you should immediately stop the cycle, let the furnace high-vacuum cool and then wait until morning to check the furnace. This allows the melted alloy to solidify before you attempt to remove it.

Finally, it is extremely important that all replacements have the same resistivity (i.e., within eight percent) of the parent material used in the original elements. All too often, the material can differ and have a different resistance depending on the manufacturer. If you end up changing out heating elements and the replacements have a different level of resistivity, it can also negatively affect the furnace's temperature uniformity. By being proactive in replacing and checking heating elements, as well as knowing how to correctly repair or replace them when needed, you can help ensure your equipment continues to run smoothly for decades to come.

Leak Detection, Identification and Prevention

While using continuous maintenance can help ensure peak performance of the furnace, one isn't always able to prevent everything. Sometimes things simply go wrong. Which is why it is essential to not just know how to perform preventative maintenance, but also how to perform corrective maintenance on your furnace (i.e., troubleshoot different problem areas).

A prime example of this is dealing with the pumping systems. Since the pumpdown speed, ultimate vacuum and vacuum rate of rise characterize the performance of the furnace, it is essential that the pumps and chamber be kept clean and in good condition at all times. Any volatile material left inside the chamber will increase pumpdown time and the leak-up rate.

However, one of the major problems that arises in vacuum furnace maintenance is that the furnace will not pump down. To determine the primary cause, one must then move from preventative maintenance to corrective maintenance, thus following a systematic procedure. Typically, loss of vacuum in the chamber during a leak rate test signifies outgassing and/or the presence of a real leak. Other types of leaks include permeation, diffusion, backstreaming, internal leaks and virtual leaks, as illustrated in Figure 3.

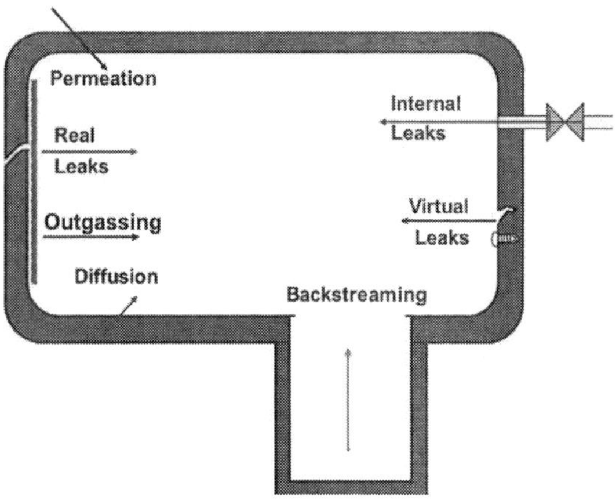

Figure 3 – The different types of leaks and ways they enter the furnace system.

In order to fully understand the maintenance procedures for identifying, correcting and preventing leaks, it is important to know what these different terms mean and how to deal with them, specifically outgassing and real leaks.

Outgassing
Outgassing is the release of absorbed (low vapor pressure) contaminants present in the vacuum system. These contaminants evaporate at various rates during the heat treatment process and the effect is a continuous rise in pressure.

Outgassing can be detected by comparing successive leak-rate values after long pumping cycles. If the leak-rate value improves with each comparison (Figure 4), outgassing can be considered a factor contributing to the total pressure rise.

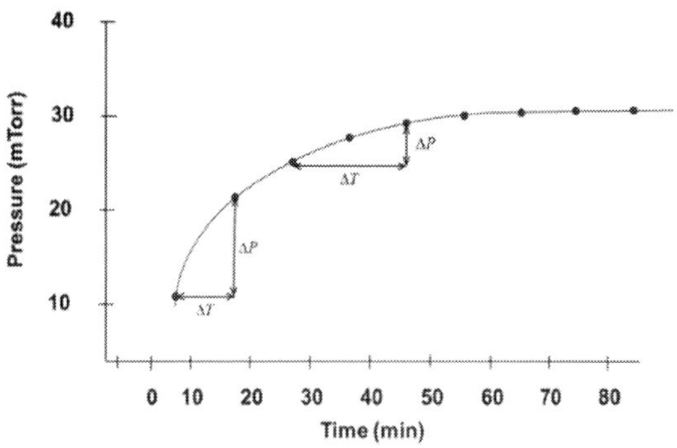

Figure 4 – The leak rate gradually improves over time, meaning the initial rise in pressure can be attributed to outgassing.

If the leak-up rate remains approximately the same (linear) after the successive long cycles (Figure 5), a real leak may be considered the principal cause. The first step here is to check the previous cycle in the furnace to see if it was normal.

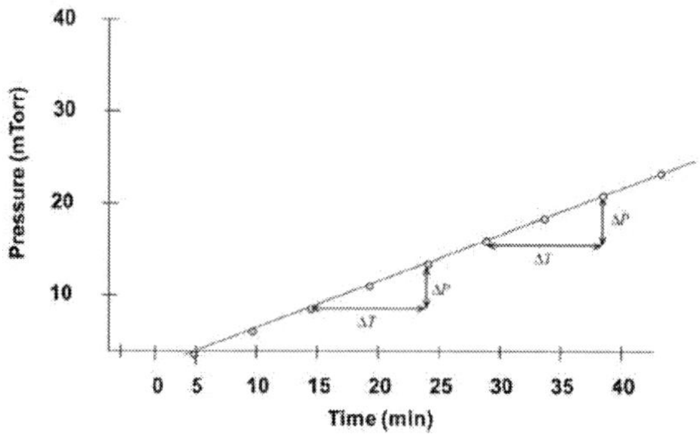

Figure 5 – The leak rate remains the same over time, signifying the presence of a real leak.

It is also important to note that foreign material can contaminate the vacuum gauge tubes. If maintenance was conducted prior to poor performance, there may be material left in the chamber that could cause excessive outgassing. Gauge tubes can be checked by utilizing a reference gauge tube; one can purchase such a reference tube from the system supplier.

Real Leaks
If one has ruled out outgassing and determined a real leak is the culprit, the next step is to identify and detect the leak. Leak detection can be done several ways, depending on the available leak-detecting equipment. The two most common methods for locating the presence of real leaks are utilizing either a solvent or a helium mass spectrometer.

Solvent Technique
Using a solvent, such as acetone, is a simple but effective method for locating intermediate-sized leaks. The procedure involves carefully spraying the suspected area with the solvent (acetone is preferred) and monitoring the most sensitive vacuum instrument of the chamber. If a leak is present in the area sprayed, the solvent will expand in the chamber, showing an increase in chamber pressure via the vacuum gauge.

The pressure increase may take five to ten seconds to appear, so it is necessary to wait this length of time before proceeding to the next area. The sensitivity of this procedure greatly increases with lower pressures. One should observe the usual precautions when using hazardous solvents. Do not spill solvent on painted surfaces as acetone will remove many finishes.

Helium Mass Spectrometer
If the solvent technique is not used, a helium mass spectrometer can also be employed to determine the leak's source. When utilizing a helium mass spectrometer to perform a leak check, it is recommended that you first calibrate the leak detector. To begin leak checking the furnace, the detector should be connected to the inlet side of the mechanical pump. Vacuum systems using a blower have a port for leak testing located between the blower and the mechanical pump. Using the blower and the diffusion pump employs the maximum

sensitivity of the leak detector. It is also important to start leak checking at the top of the vessel and work your way down since helium is lighter than air.

Common Leak Problem Areas

When looking for leaks in the chamber, check the joints or fittings with a rotating or reciprocating movement. These are most susceptible to leaking. The door gasket is the most abused seal in the chamber; as a result, one should inspect it after each cycle and lubricate it daily.

The inert gas system can also be a problem area in furnace maintenance. Leaks in the gas system will cause discoloration of the workpieces being heat treated. The appearance of color on the workpieces after heat treatment is the best indicator of a leak or contamination problem; whereas, bright, clean 17-7 or 17-4 precipitation-hardening (PH) stainless steel is an indication of a clean, tight furnace.

It is also vital to know that the liquid nitrogen/argon systems – which convert the liquid to a gas (with a holding tank at the furnace) – are more reliable than gas systems using bottled nitrogen or argon since bottled gases are usually not as pure as liquid gases. This is because the bottles often become contaminated when they are completely emptied and the valves are left open.

Finally, copper pipes and silver brazed joints are recommended for keeping all of the gas lines tight. This is because threaded pipe fittings may develop leaks and introduce air during quenching which will contaminate the furnace and discolor the workpieces.

If discoloration does persist after checking for inert gas leaks, the common practice is to vacuum cool some sample workpieces (a few pieces of austenitic and martensitic stainless steel specimens) and inspect for appearance. If the workpieces still experience discoloration, the nitrogen/argon system can be ruled out as a source of the problem and a leak in the chamber is evident.

Cold Linear Leak Rate

While knowing how to find and identify leaks is an imperative part of furnace maintenance, it is also crucial to perform a leak rate check and know the linear leak rate before you ever run the furnace. By doing so, one is able to determine if there are any leaks that need to be corrected. In addition, if the furnace's linear leak rate ever gets too high, it signifies the furnace has leaks that need to be found and fixed.

Determining the Linear Leak Rate

In order to determine the true linear leak rate, or the rate of rise, of the vacuum furnace, one would perform a rate of rise test – commonly referred to as the CDE (clean, dry, empty) linear leak rate. To do this, one would:

1. Pump the furnace down
2. Let it sit for a predetermined amount of time (depending on the temperature of the furnace and the processes being ran)
3. Stop the process cycle, which separate the pumps from the furnace
4. Watch the rate of rise over a one-hour time period

However, this is an area where some fail their mandated leak rate, as one cannot necessarily obtain a true linear leak rate in a one-hour time sample. One needs multiple samples to ascertain linearity.

For example, a leak rate of five microns per hour or less is required. The vacuum furnace is pumped down to high vacuum, shut off for an hour and, at the end of the hour, it has gone up to eight microns. This means the leak rate is eight microns/hr. If you looked at just this time sample, you would fail the mandated leak rate.

On the other hand, if you allow the furnace to sit for another hour, one is better able to determine if the leak rate is truly linear. If it is truly linear, the leak rate will double (go up by eight microns) in that second hour. At that point, you can say it is linear and it fails. However if, in that hour, it goes from eight to ten microns instead of doubling, then the furnace would receive a passing leak rate of five microns/hour.

All too often, people give up prematurely and end up leak checking the furnace again, running a cleanup cycle or calling maintenance. As a result, a solid 24 hours of production time is needlessly lost because they did not allow the furnace to sit for one more hour.

By letting the furnace sit for an additional amount of time, one can confirm either:

1. It was linear and truly leaking, or
2. It was not linear yet, but now it passes and you are back in production in two hours rather than 24 hours.

In the end, this extra time not only lets you cut down on unnecessary maintenance and lost production costs, but it also allows you to be proactive in protecting your investment.

Knowing the Acceptable Leak Rates

Knowing the cold linear leak rate is only one small component of the overall picture. When determining the acceptable leak rate for the furnace, there are two factors to consider. First, one should consider the acceptable leak rate for the furnace. Second – and sometimes more importantly – one should consider the acceptable leak rate for the parts being processed.

When working with Ipsen furnaces, it is important to know that Ipsen furnaces with all-metal hot zones historically ship at a linear leak rate of two microns/hour or less, and Ipsen furnaces with graphite hot zones ship at a linear leak rate of five microns/hour or less. The reason for the difference between the leak rates is due to the hygroscopic nature of the hot zone and its ability to maintain gases and water vapors; as a result, it is slightly harder to achieve lower leak rates on a graphite furnace. Nevertheless, these leak rates are acceptable for the furnace and are Ipsen's ISO Standard for Pre-Shipment Qualifications.

Sometimes, though, one has an older, dirty furnace with carbon buildup that is running low-quality production. The question then, is, what is an acceptable leak rate you can run the furnace at without damaging it?

If you want to look at just the furnace, you could consider 15 microns/hour (anything under 15 microns/hour is acceptable for most commercial vacuum heat treating).[1]

In other instances where you are going to run titanium and super high alloys, the parts are now driving the specifications. This is because to run titanium you might need a leak rate of three microns or less in a furnace with a graphite hot zone (even though Ipsen's furnace ships at five microns). In essence: the production often directs or mandates what acceptable leak rates can be. Other times, with less critical parts, the furnace mandates what the leak rates can be.

Why is this the case? Consider what happens if a furnace is ran at too high of a leak rate. It consumes itself – just like a lightbulb. For example, there is a vacuum inside the lightbulb with a filament. When one cracks the lightbulb's glass, air rushes in and hits the hot filament, vaporizing it. That is exactly what happens to the hot zone. If too much air is going inside, the hot zone starts to consume itself and vaporize. Which is why, if one has an excessive leak rate, one can actually destroy a hot zone in a matter of months.

If you just invested in a new furnace, you want to make sure it lasts as long as possible; this is why it is essential to carefully consider the acceptable leak rates for your furnace and your parts.

Understanding Leak Anomalies

In addition to acceptable leak rates, there is often confusion surrounding the process of leak checking. For instance, one needs to remember helium is lighter than air. Therefore, if you are leak checking a furnace, you should start at the top, as mentioned earlier, because helium floats upwards. This is especially important to remember when leak checking a large furnace because it will take longer for the helium to dissipate from surrounding areas.

One must always be cognizant of the surrounding environment when leak checking. If the wind were blowing in your face, you would not start on a 25-foot-long furnace with the wind blowing down its length; you would start at the end where the wind is blowing towards so, as you leak check, the wind blows the helium across areas that have already been checked.

Of course, there is more to leak checking than that. For instance, one also needs to consider if there are multi-pump or single-pump units, how to cycle the pumps, when to have the pumps on, when to have the pumps off, etc.

Let's say you are leak checking a furnace system with four diffusion pumps. While four pumps would give you a much lower vacuum, that does not mean you should end your final leak check with all four pumps on.

Rather, you should shut three of them off and just use one. The reason for this is that, when a leak detector is connected to a pump and there is a small leak, you are asking the helium to go across all of the remaining pumps to reach the final one. However, there may not be enough helium to reach the pump the leak detector is hooked up to. As a result, one needs to close the first three main valves so, if there are any leaks in the furnace, the only route of escape for the helium is the pump with the currently connected leak detector.

Hot Leak Rate

In the pursuit of correctly performing maintenance on your furnace system and protecting it from unnecessary damage, it is essential to remember the furnace also needs to be leak checked when it is at temperature. So many times a furnace can pass a cold leak check and a cold leak rate test but, when it is at temperature, something may start leaking.

For example, if the furnace has been previously leak checked with a helium mass spectrometer and the linear cold leak rate has been verified – but the parts have been exposed to oxygen and are coming out discolored – the furnace might be leaking when it is hot.

This is why, periodically, one needs to:

1. Heat the furnace up to 2,000 °F (1,093 °C)
2. Let the furnace soak out for an hour to degas
3. Do a physical leak check on the furnace with a helium mass spectrometer, while also taking care of the energized feedthroughs

Trying to perform an automated leak test with the software on the furnace (or, in other words, a *hot leak rate check* with the software) yields a pass/fail point of reference only. It does not provide a legitimate linear leak rate or qualification.

Validation of the leak rate is a mechanical (either software or man-interface) manipulation of the furnace controls when the furnace is pumping, stopping, watching and calculating a leak rate of rise. That does not necessarily mean that is the true linear leak rate, but rather – as mentioned before – a pass/fail point of reference.

[1] **NOTE:** While 15 microns might be tolerable for most furnaces and the furnace will not be damaged, 15 microns is **not** a number Ipsen condones. Please follow the specifications outlined in your specific furnace manual.

For instance, one opens the furnace door, closes it, pumps down for thirty minutes and does a leak rate check with the software for 10 minutes, ending up with a leak rate (rate of rise) of 30 microns. However, that is not the true linear leak rate. It simply means that – based on how you designed and performed the experiment – it is 30 microns for that day. If the same thing is done repeatedly every day and, if the furnace is operating correctly and the parts are clean, 30 is a good reference number. But if one day it goes from 30 microns to 300 microns and the parts come out blue, then there's a problem.

Ipsen's software, as demonstrated in Figure 6, is also capable of performing both hot and cold leak rate checks. Many operators of vacuum equipment routinely conduct a leak test before energizing the heating elements. This is particularly useful when the workload consists of very expensive material. In this case, the furnace is pumped down (e.g., to 10^{-4} Torr) for at least one hour (with the vacuum valve connecting the furnace to the pump closed), and the software is programmed to check the rate of rise in internal furnace pressure over a specified time interval. This allows you to get a data point known as either a *hot leak rate point or cold leak rate point.*

Again, this data point is not to be construed as the true linear leak rate; it is also to be considered a point of reference only.

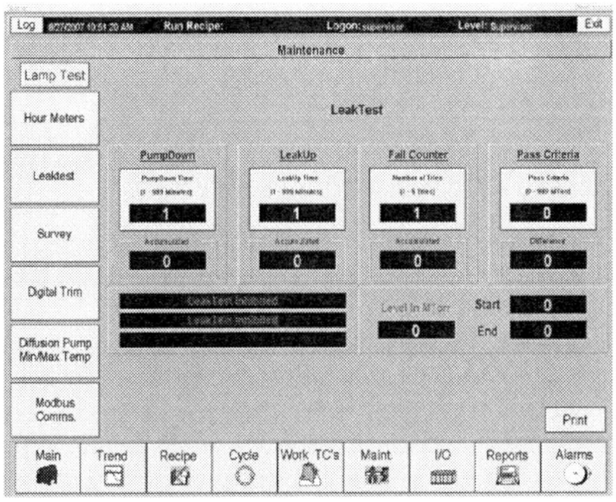

Figure 6 – Ipsen's automated leak check software performs hot and cold leak rate checks, helping signify if there are points of change that should be addressed.

Vacuum Valve Troubleshooting

Sometimes, though, knowing how to leak check a furnace, obtain a linear leak rate, etc. is not enough to solve the problem with the furnace. When performing corrective maintenance, it is also important to consider the vacuum valves as an extension of leaks because they can also cause issues with the furnace. Valves can leak between adjacent systems, which can then affect the leak rate, part colors and quality in the vacuum furnace. To check the seal integrity of the various seals and valves, one should perform the following maintenance steps:

Main Valve

1. Pump down the main vacuum chamber to 1,000 microns or less.
2. Press the *Process Cycle Stop* button to stop the pumping of the furnace.
3. Select the vacuum sensor tube that is reading the vacuum level of the diffusion pump. Normally this will read somewhere in the range of 20 to 200 microns.
4. While you are watching the vacuum level in the diffusion pump, press the *Chamber Release* button.
5. As the inert gas enters the main vacuum chamber, there should be no upward change (loss of vacuum) in the diffusion pump's vacuum reading.
6. If the diffusion pump seems to be backfilling at the same rate as the main chamber, there is a leak at the main valve dish seal.

Roughing Valve

1. Pump down the main vacuum chamber to 1,000 microns or less.
2. Press the *Process Cycle Stop* button to stop the pumping of the furnace.
3. While the vacuum level in the main chamber begins to slowly climb, have someone stop the roughing pump and air release the roughing line.
4. As air enters the roughing line, there should be no upward change (loss in vacuum) in the diffusion pump's vacuum reading.
5. If the chamber rapidly loses vacuum when you are venting the roughing lines, the roughing valve is leaking.

Foreline Valve

1. Pump down the main vacuum chamber to 1,000 microns or less.
2. Press the *Process Cycle Stop* button to stop the pumping of the furnace.
3. Select the vacuum sensor tube that is reading the vacuum level of the diffusion pump. Normally, this will read somewhere in the range of 20 to 200 microns.
4. While you are watching the vacuum level in the diffusion pump, have someone turn off the roughing pump and air release the roughing lines.
5. As air enters the roughing lines, there should be no upward change (loss of vacuum) in the diffusion pump's vacuum reading.
6. If the diffusion pump loses vacuum, the foreline valve is leaking.

Holding Valve

Typically, if the oil keeps disappearing from the holding pump and there are no signs of external oil leakage, the holding valve is leaking across its own seal, subsequently pulling oil backwards into the diffusion pump.

In the end, by knowing the proper procedures for detecting, identifying and preventing leaks in a vacuum system – from using a helium mass spectrometer to troubleshooting the vacuum valves – you can ensure your heat treatment equipment is continuously kept in prime condition.

Conclusions

Overall, it is essential to know how to protect your investment, especially when working with expensive equipment in a highly competitive business world. After all, the regular performance of maintenance is a vital part of equipment upkeep and achieving and maintaining a quality braze.

By understanding how to implement and perform the essential basics of furnace maintenance and upkeep – such as diagnosing and troubleshooting the furnace, performing a leak check or repairing the heating elements – you can begin to extend the lifespan of your equipment, while also minimizing the cost to your company. In the end, with the proper utilization of preventative and corrective maintenance, one can begin to generate higher uptime percentages, reduce unplanned downtime situations and yield greater shop throughput.

Vacuum Levels Influence on the Properties of Various High Temperature Brazing Alloys

Owe Mårs,
Höganäs AB, Höganäs Sweden

Eduard Hryha
Chalmers University of Technology, Gothenburg, Sweden

Abstract

The importance of a good vacuum for brazing Chromium containing alloys on stainless steel is well known. The vacuum level required is depending on the Cr content in the brazing alloy, higher content requires a higher vacuum level. Also the alloying composition of the steel in the substrate from the point of view of elements with high oxygen affinity as e.g. chromium, plays a significant role as well. However, due to the higher Cr-content in the brazing alloy, the dominating factor for what vacuum quality is required is the composition of the brazing alloy.

This paper deals with thermodynamic and experimental investigation of the vacuum level required to achieve a good brazing behavior of different alloys. The vacuum levels required, or being more specific, oxygen partial pressures needed, are not at all enough to reduce the Chromium oxide even at the lower chromium activities present in low Chromium containing compositions at the typically applied brazing temperatures. However, brazing powder is covered by heterogeneous oxide, formed by the thin oxide layer of the main matrix element and oxide islands, rich in stronger oxide forming elements. Hence, the oxides controlling the behavior of the powder during process is strongly determined by the removal of this thermodynamically unstable surface oxide layer of base matrix element. Importance of vacuum level/quality is therefore expressed in terms of its effect on oxide transformation and distribution during brazing process that will determine final properties of the component.

Introduction

High temperature brazing with corrosion resistant alloys is very much focused on the utilization of advantages of the controlled atmosphere due to the fact that both, powder production and brazing process are performed at higher temperatures than the standard heat treatment. Modern powder production method allows manufacturing of the metal powders with very low oxygen content. From the powder activation point of view it means that the surface of the powder is covered by easily reducible oxide of the main (matrix) element – e.g. nickel or iron – with the presence of particulate oxide phases, formed by oxides with higher thermodynamic stability [1-3]. Careful analysis of the powder surface used for pressed and sintered applications indicate that the most of the powder surface is covered by easily reducible oxide of the main elements of the alloy (Fe/Ni) – about 95% of the area. Hence, sintering of such a powder has to be not so different from plain iron or nickel powder. Particulate oxide features are typically formed by alloying elements with higher thermodynamic stability, present in the alloy, as e.g. Cr, Mn, Si, P, etc. which are commonly used in brazing alloys to enhance corrosion resistance and depress the melting point. These oxides are characterized by significantly higher stability than the Fe or Ni oxides, see Fig.1, and their amount is strongly determined by alloy composition and further powder handling [1-3]. Hence, reduction of such thermodynamically stable oxides requires significantly better conditions during the sintering process.

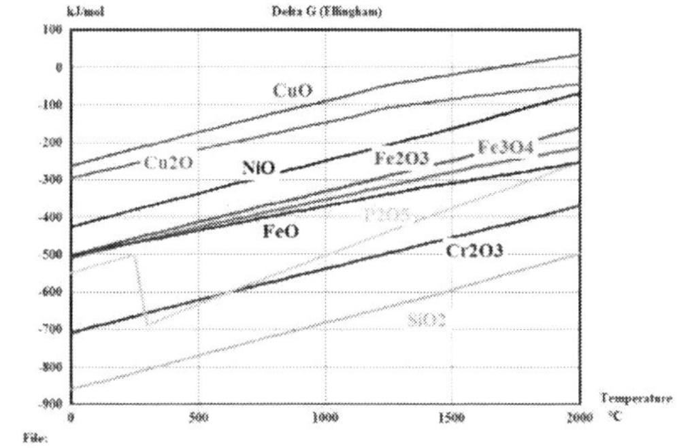

Figure 1. Ellingham–Richardson diagram indicating stability of oxides of interest (HSC Chemistry 7.1).

This means that the atmosphere requirements for brazing stainless steel with Ni or Fe based brazing alloys is significantly more stringent and can be characteristic like for very dry hydrogen atmosphere [4,5] or ultra-high vacuum [6]. Therefore, taking into account high-purity of the starting powder, the most significant part during brazing of such materials is to avoid formation of such oxide phases during heating stage [7, 8].

It is very important to remember that powdered material has surface area that is about 10 000 times larger than the surface area of the same material in a bulk form. This results in significantly higher surface reactivity of the powder and accelerated mass-transfer of the alloying elements at elevated temperatures that makes highly-alloyed powder much more sensitive to atmosphere quality from the oxygen potential point of view [4, 6]. This is especially important during the heating stage, when the temperature is low and hence, in case of the same vacuum level/processing gas purity, reducing potential of the process is rather low. This can lead to the extensive oxide transformation from the iron/nickel based surface oxide into more thermodynamically stable oxides formed by alloying elements, present in the matrix, as e.g. Cr and Si [7, 8]. Due to the significantly higher stability of these oxides, see Fig.1, they will be later seen as inclusions in the brazed joint or even as a shell of oxides on the powder particle or on the surface of the brazing paste, see Fig 6.

To be able to apply brazing powder in an efficient manner it is often turned into a paste. This paste is then used to apply the brazing alloy in a precise and efficient manner. This creates another issue during heating stage which is the removal of the binder phase. Previous studies on the binder/lubricant removal [9, 10] indicate that the binder decomposition during heating stage leads to evolution of the oxidizing gas species as carbon dioxide, water vapor, etc. Presence of such compounds in the porosity around the particles significantly worsens the local oxygen potential in the compact that can lead to significant oxidation of the powder surface and formation of the thermodynamically stable surface oxides [9]. The risk of such oxide formation significantly increases with the increasing of the debinding temperature/ heating rate as this will result in the presence of oxidizing gas species from the binder decomposition at high temperatures. If the temperature is high enough these oxidizing gases will decompose when the metals with high affinity to oxygen is forming oxides. Hence, accurate control of the debinding process has to be performed in order to avoid such oxide formation that are impossible to reduce during the further brazing process and will inevitably result in the presence of defects at the original application point of the braze alloy, inclusions and various oxide shells on the final brazed component.

Previous studies on different alloyed powders [6-8] in PM indicate that the most critical phase during sintering from the oxide reduction/transformation point of view is the heating stage between 800 and 1000°C. At this temperature interval significant risk of oxide transformation into more stable oxides exits that is following be enclosure inside the growing inter-particle neck and can lead to the presence of rather extensive defect sites. The reason is that at this temperature range temperature activation is high enough for effective mass-transfer and there are two processes that are taking place almost in parallel: growth of the inter-particle necks and reduction/transformation of the residuals of surface oxides. Hence, final quality of the inter-particle neck is determined by the difference in thermodynamics of oxides present and kinetics between these two processes. Parallels can be drawn with the behavior of the powdered brazing alloy as this temperature is below the melting point. If the oxides are formed before melting they will not melt together with the alloy but remain as inclusions or slag.

The most important role of the heating stage for pressed and sintered components is to provide efficient conditions for the reduction of surface oxides. Surface oxides comprise diffusion barriers that hamper formation of sinter necks between the metal particles and hence has to be removed at as early stages of sintering as possible [4, 7, 8]. It is very important to emphasize that in case of brazing, process is typically performed in vacuum, and not in rich hydrogen-containing processing atmospheres, typically utilized to reduce stable surface oxides. At the same time, alloys used for brazing do not contain carbon and hence carbothermal reduction of the oxide phases, that is the most effective mechanism for the reduction of stable oxides at elevated temperatures [4, 5], can't be applied. Additionally brazing alloys have a high content of the alloying elements with high oxygen affinity, as Cr and Si, meaning that the risk of formation of the stable surface oxide is really high if brazing process is not controlled properly.

Brazing process is typically performed in vacuum, without presence of any reducing agents, and hence the basic mechanism of oxide removal is through the dissociation of the oxides according to $\frac{2}{y}M_xO_y = \frac{2x}{y}M + O_2$. Thermodynamic evaluation of the Gibbs free energy change for the mentioned above reaction using thermodynamic modeling software HSC Chemistry allows evaluation of the temperature dependence of equilibrium oxygen partial pressure for the oxides of interest, see Fig.2. At an oxygen partial pressure below each curve for the specific oxide in the diagram, the oxide is decomposed, above each curve – metal is oxidized (required conditions for reduction of silicon oxide are marked on the Fig.2 with the arrow, as an example).

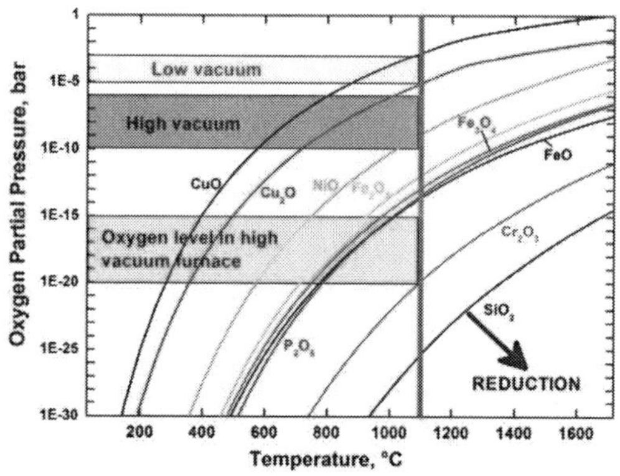

Figure 2. Equilibrium oxygen partial pressure for metal oxides/metal systems of interest in relation to the vacuum level and oxygen partial pressure in the industrial vacuum furnaces.

Analysis of the curves indicate that the decomposition of the even rather thermodynamically unstable oxides of iron and nickel requires lowering of the oxygen partial pressure in the vacuum furnace by several orders of magnitude in comparison with the pressure in the vacuum furnace, see Fig.2Analysis of the curves indicate that even at 1100°C for the reduction of Fe_2O_3 and NiO oxides, that constitute dominant part of the surface oxide in case of powders used for brazing, oxygen partial pressure in the vacuum furnace has to be below $1.2*10^{-12}$ and $1.9*10^{-9}$ bar, respectively. However, it is important to remember that dominant fraction of the residual gas, present in the vacuum furnace during processing, is the inert gas from purging and not oxygen. Hence, vacuum level in the furnace does not represent an actual oxygen potential in the atmosphere, surrounding brazing powder, and actual oxygen potential in the industrial vacuum furnace is in the range of the partial pressures, marked in light blue color in Fig.2. Analysis of the oxygen partial pressure in dependence on temperature and vacuum level indicate that reduction of the iron and nickel oxide layers is possible and takes place during heating stage. However, oxygen partial pressure is too high to allow reduction of the more stable oxides of Cr and Si. Therefore, if Cr and Si oxides are formed in the material during heating stage – they are impossible to reduce later in the heating cycle. Hence, the most important stage during brazing of the alloys with high content of Cr and Si is the heating stage as improper debinding or too poor vacuum level/purity during heating brings a significant risk of formation of stable Cr and Si-based oxides that are impossible to reduce even at brazing temperature in industrial vacuum furnaces at typically used temperature profiles and vacuum level.

Therefore, debinding at low temperature, attaining high vacuum in the furnace at as low temperature as possible and proper load of the furnace (overloading will result in high oxygen partial pressure due to decomposition of larger amount of oxides) is strongly recommended to avoid the risk of defects formation, connected to the presence of oxide phases.

Experimental

A number of experiments have been conducted to demonstrate the importance of a good atmosphere during the heating of brazing paste. The vacuum furnace used was equipped with a pre vacuum pump and a high vacuum pump. The pre vacuum pump was capable of creating a vacuum level of $5*10^{-5}$ bar. The high vacuum pump was capable of creating a vacuum level of $2*10^{-7}$ bar. Different conditions were created by engaging the high vacuum pump at different temperatures. None of the vacuum levels are sufficient to reduce oxides according to the Ellingham diagram but a lower vacuum is definitely a sign of less oxygen available and less oxidization. Apart from the differences in engaging the high vacuum pump all brazing were done identically with the same furnace cycle and brazing temperature.

Figure 3. Brazed T bar of Ni bal , 29% Cr, 7 % P, 4 % Si High vacuum from 400° C. No oxide at paste application point.

Figure 4. Brazed T bar of Ni bal , 29% Cr, 7 % P, 4 % Si, High vacuum from 500° C. Clear oxide residual at paste application point. Good brazing seam

Figure 5. Brazed T bar of Ni bal , 29% Cr, 7 % P, 4 % Si High vacuum from 600° C. Large oxide residual at paste application point. Good brazing seam but less alloy available, smaller fillet.

All samples have been brazed at the same temperature of 1100° C and the same vacuum, $2*10^{-7}$ bar. But the appearance of the sample after brazing is very different. There is a lot of oxidization taking place during the heating of the sample before the alloy melts. The oxide formed then is impossible to be reduced. This oxides is consuming some of the alloy as it is tied into the shell. When the high vacuum pump is turned on it lowers the pressure to a level where evidently not enough oxygen is remaining available to continue forming oxides. At lower temperatures the oxidizing process is not fast enough and the diffusion rate is to slow to form larger amounts of oxides or lead to extensive oxide transformation.

Result

When analyzing the residual it is confirmed that it is made up by the oxidized powder particles. The surface of the particles have received a thin ¨skin of oxides and they prevent the molten alloy from flowing. Further away from the surface the powder particles are less oxidized and more of the molten alloy have been able to flow. When the molten alloy reaches the gap the capillary force is acting on it and the alloys penetrates the gap and forms a solid joint. Enough material is pulled from the braze alloy reservoir at the application point of the paste to fill all the gap and form a nice fillet. Remaining on the site of the alloy application is the oxides that has formed and never where able to melt and flow. The conditions during forming decides the extent of these oxides as the vacuum during brazing have not been sufficiently low to reduce any of them.

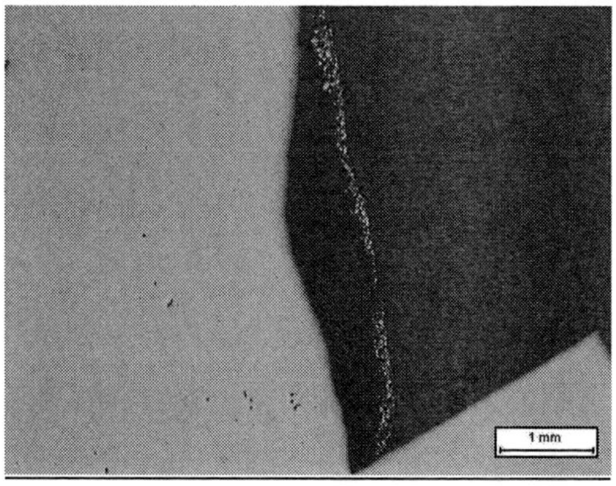

Figure 6. Cross section of the braze residual. Thin shell of the oxidized powders but the bulk of the alloy has flown in to the gap.

Conclusion

Three factors determines the amount of oxides to form, during the heating of the sample to brazing temperature. It is the availability of Oxygen, The availability of elements with high affinity to oxygen and the temperature.

The availability of oxygen is primarily decided by the level of vacuum. A lower pressure makes less oxygen molecules available for reaction with the elements with high affinity to oxygen. But as mentioned in the theory chapter the debinding of the binders in the paste can also release oxidizing gases such as CO2 and H2O. This changes the availability of oxygen around the powder particles. When the binder is decomposed and vaporized is directly depending on the temperature and the level of vacuum. This means if the high vacuum pump is engaged at a lower temperature the binder will vaporize at a lower temperature.

The availability of elements with high affinity to oxygen is depending on the composition the alloy. Larger amounts of Cr and Si will result in an alloy more sensitive to vacuum level. But the presence of these elements in the alloy is determined of what performance the alloy has.

When brazing a component it will be exposed to all temperatures from room temperature up to the brazing temperature. The key to avoid oxidization is to ensure that when the component reaches a temperature were the elements with high affinity to oxygen will start to react with oxygen the availability of oxygen has to be minimized.

References:

[1] L. Nyborg, E. Hryha: "Surface Oxides on Gas and Water Atomized Steel Powders", Advances in Powder Metallurgy & Particulate Materials-2014, Proceedings of the 2014 International Conference on Powder Metallurgy & Particulate Materials, MPIF, May 18-22, Orlando, FL, USA, Vol.2, pp. 153-161, ISBN:978-0-9853397-6-0,

[2] E. Hryha, C. Gierl, L. Nyborg, H. Danninger, E. Dudrova: "Surface Composition of the Steel Powders Pre-Alloyed with Manganese", Applied Surface Science, 2010, Vol. 256, No.12, pp. 3946-3961, ISSN: 0169-4332.

[3] A. Weddeling, K. Zumsande, E. Hryha, S. Huth, L. Nyborg, S. Weber: "Characterization of High-Mn-Cr Austenitic Steel Powder Fe-19Mn-18Cr-C-N", Proc. of EURO PM2011 Congress & Exhibition, 9-12 October 2011, Barcelona, Spain, Vol.2, p.15-20, ISBN: 978-1-899072-23-1.

[4] E. Hryha, E. Dudrova, L. Nyborg: "On-line control of processing atmospheres for proper sintering of oxidation-sensitive PM steels", Journal of Materials Processing Technology, 2012, Vol. 212, No.4, pp. 977-987, ISSN: 0924-0136.

[5] E.Hryha,L.Nyborg: "Thermogravimetry Study of the Effectiveness of Different Reducing Agents during Sintering of Cr-prealloyed PM Steels", Journal of Thermal Analysis and Calorimetry, 2014, Vol.118, No.2, pp. 825-834, ISSN: 1388-6150, DOI 10.1007/s10973-014-3915-z.

[6] K. Zumsande, A. Weddeling, E. Hryha, S. Huth, L. Nyborg, S. Weber, N. Krasokha, W. Theisen: "Characterization of the surface of Fe–19Mn–18Cr–C–N during heat treatment in a high vacuum – An XPS study", Materials Characterization, 2012, Vol. 71, pp.66-76, ISSN: 1044-5803.

[7] E. Hryha: "Application of Fractography for Investigation of Surface Oxide Reduction/Transformation and Inter-Particle Necks Formation During Sintering of Prealloyed with Cr and Mn PM Steels", Powder Metallurgy Progress, 2014, Vol. 14, No.1, pp. 24-31, ISSN: 1335-8987 (print), ISSN:1335-4533(online), http://www.imr.saske.sk/pmp/issue/12014/PMP_Vol14_No1_p_024-031.pdf

[8] E. Hryha, L. Nyborg: "Oxide Transformation in Cr-Mn-Prealloyed Sintered Steels: Thermodynamic and Kinetic Aspects", Metallurgical and Materials Transactions A, 2014, Vol. 45, No. 4, pp. 1736-1747, ISSN: 1073-5623 (print), ISSN: 1543-1940 (online), DOI: 10.1007/s11661-013-1969-3.

[9] E. Hryha, H. Borgström, K. Sterky, L. Nyborg: "Influence of the Steel Powder Type and Processing Parameters on the Debinding of PM Compacts with Gelatin Binder", Journal of Thermal Analysis and Calorimetry, 2014, Vol.118, No.2, pp. 695-704, ISSN: 1388-6150, DOI 10.1007/s10973-014-3839-7.

[10] S. Karamchedu, E. Hryha, L. Nyborg: "Influence of delubrication process on the properties of sintered PM components", Proceedings EURO PM2013 Congress & Exhibition, 15-18 September 2013, Gothenburg, Sweden, Vol.2, pp. 163-138, ISBN: 978-1-1899072-31-1.

Mechanically Supported Solid State Bonding for joining Material Combinations with Large TEC-differences

E. Skiera*, B. Kuhn
ForschungszentrumJülich GmbH,
Institute for Energy and Climate Research (IEK),
52425 Jülich, Germany
*Corresponding Author: Tel.: +49-2461-61-3688
e-Mail: e.skiera@fz-juelich.de

Abstract

The development of high-temperature electrochemical devices like solid oxide fuel and electrolysis cells, oxygen and hydrogen separators and reformers poses a great challenge in joining of metals/metals, metals/ceramics and ceramics/ceramics. To maintain the operability of such equipment, the seals have to be reliable and robust in isothermal high temperature as well as in thermal cycling operation. As a solution "Reactive Air Brazing (RAB)" was successfully applied over the past years for joining a variety of material combinations for high temperature application. To date successful RAB-joining was limited to material combinations with a maximum difference in thermal expansion coefficients of about 2-3 ppmK-1 (i. e. 8YSZ/X1CrWNbTiLa22-2 (German designation: 1.4755) or BSCF/AISI 314 (X15CrNiSi25-21, German designation: 1.4841).
This limitation was overcome by so-called "Mechanically Supported Solid State Bonding (M3SB)" which enables joining of material combinations with larger differences (ΔTEC > 3 ppmK-1, i.e. BSCF/1.4755, maximum ΔTEC = 4 ppmK-1 or 8YSZ/1.4841, maximum ΔTEC = 7.5 ppmK-1) in thermal expansion coefficients. In brazing the final microstructure of the joint is limited to structures which form during solidification of the molten joining material. In M3SB the joint is formed in "solid state" by applying low mechanical stress at high temperature (but below the melting point of Ag) and thus the final microstructure can be closely tailored to the intended application prior to joining.

Introduction

Joining of similar and dissimilar materials is a critical technical challenge for high temperature applications, since hermetic sealing between the joining partners and an economical way of manufacturing complex components has to be provided [1].
Several joining techniques such as glass joining, diffusion bonding and active metal brazing are currently used, but each possesses some form of trade-off or exhibits some penalty in terms of joint properties, ease of processing, and/or cost [1-5].
Recently a simple and economical joining technique referred to as reactive air brazing (RAB) has been developed and applied for joining several substrates [1, 6-13].
Unlike traditional ceramic-to-metal brazing (e.g. active metal brazing), which must be carried out in vacuum or reducing gas environment, the brazing process takes place in ambient air without need of surface reactive fluxing agents or inert cover gases [14].
Up to date successful RAB-joining is limited to material combinations with a maximum difference in thermal expansion coefficients of about 2-3 ppmK-1. Therefore the braze material and the joining partners have to be chosen accordingly to their thermal expansion coefficients.
RAB has been successfully applied for joining BSCF-membrane and SOFC-half-cells and their corresponding metallic joining partners (BSCF/1.4841, 8YSZ/1.4755) [15, 16].
A change of the joining partners leads to failure during cooling down from the RAB-process due to the stresses that arise from the large differences in thermal expansion coefficients.
Therefore an alternative joining process for material combinations with a large mismatch in thermal expansion coefficients has to be applied.
In the 1980s de Bruin et al. [17] presented a method for chemically bonding ceramics to metal in air. A joining temperature below the melting point of the lowest melting component was used.
In this paper we transferred this principle to join ceramics to steel with a silver interlayer in between.
To reduce the arising stresses and to enable a sufficient bonding strength the bonding temperature is reduced below the melting temperature of the interlayer. Additionally the bonding process is supported by a mechanical load and an increased holding time. The right combination of temperature, pressure and time allow the formation of a joint in "solid state".
The microstructure of the joint has a direct influence on mechanical properties like strength, creep strength and thermal cycling stability [18-20].
A closely tailored microstructure not only enables improved strength, but also increases the thermal cycling stability. In RAB the microstructure cannot be controlled and is limited to structures which form during solidification of the molten joining material. Many RAB-brazes present a single crystalline microstructure across the thickness of the joint and are subject

to large stresses at the interfaces during thermal cycling [19]. Failure of these brazes mostly occurs before twinning and recrystallization can accommodate for the stresses. A polycrystalline microstructure on the other hand facilitates stress reduction by recrystallization from the first thermal cycle, and thus increased thermal cycling stability up to more than 10 times [19].

In M3SB the microstructure of interlayers can be defined accordingly to the required properties prior to the joining process. Since the joint is formed in "solid state" the microstructure is not limited to structures formed during solidification.

The bonding process of so-called "Mechanically Supported Solid State Bonding (M3SB)" not only enables joining of up to now not applicable high ΔTEC combinations like 8YSZ/1.4841 and BSCF/1.4755 with sufficient fracture energies and good contour accuracy. M3SB also enables a closely tailored microstructure with superior mechanical properties.

Experimental

Sample preparation

Materials

As a simplified standard interlayer system with a minimum of chemical constituents pure Ag foils (purity: 99.995 %, thickness: 100 µm) produced by Schlenk GmbH, Germany were applied. These are not well suited for commercial application due to poor wettability on the joining partners and possible pore formation during annealing in air [21], but serve well for the examination of basic microstructure / strength correlations.

The joining partner materials utilized for the experiments outlined in this paper were: ferritic steel AISI 314/Crofer® 22 H (1.4755 - X1CrWNbTiLa22), austenitic steel 1.4841 (X5CrNiMo17-12-2) as the metallic joining partners, TZ3Y as the ceramic joining partner in ambient temperature delamination testing, and BSCF as the ceramic joining partner for tensile testing.

Delamination testing specimen

The TZ3Y ceramic foils (thickness: 100 µm, supplier: Nippon Shokubai, Japan) were symmetrically joined in between steel substrates (thickness: 3 mm) manufactured from commercial austenitic steel 1.4841. Specimen geometry and preparation are described in detail in [13, 22, 23].

Compound tensile (CT) specimen

"Compound tensile" specimens (CT) were produced by joining two bar stocks (diameter: 19 mm, height: 20 mm) made from ferritic steel 1.4755 with a BSCF cylinder (diameter: 17 mm, height: 4 mm) in between.

Joining (M3SB)

The interlayer foils were manually positioned between the joining partners.

Joining was performed in air in a TesT 113.5kN electromechanical testing machine with a hold time of 4 hours at maximum temperature while applying a permanent pressure. The joining temperature was chosen to be lower than the melting point of silver to ensure contour accuracy. In RAB Ag is not applied as braze due to its poor wetting properties. In M3SB it was used successfully as an interlayer since the joint is formed in "solid state".

Joining parameters were 800 °C with a permanent pressure of 6 MPa for BSCF/1.4755, and 550 °C with a permanent pressure of 6-12 MPa for 8YSZ/1.4841.

Four point bending test

To characterize the interfacial fracture energy of the joints the method introduced by Charalambides et al. [24] was adopted. Testing set-up and data evaluation are described in detail in [13, 22, 23].

During bending, cracks initiate from the notch tip of the multilayer specimen and propagate into the multilayer system along the path of lowest fracture resistance [21].

Four-point bending tests were performed at ambient temperature using a TesT 113.5kN electromechanical testing machine with ceramic four-point bending fixtures (45 mm specimen length, 40 mm / 20 mm span ratio) applying a constant deflection rate of 0.33 µms-1 [21].

Results and discussion

Brazing of BSCF/1.4755 leads to failure upon cooling from joining due to the large differences in thermal expansion coefficients. As an alternative M3SB with a pressure of 6 MPa at 800 °C was applied. The as-joined microstructure is shown in Fig. 1. At the steel/interlayer interface a thin reaction layer can be observed (Fig. 1, 2b). The ceramic/interlayer interface shows no reaction whatsoever, but it should be noted that the Ag-interlayer follows the surface roughness of the ceramic (Fig. 1, 2a).

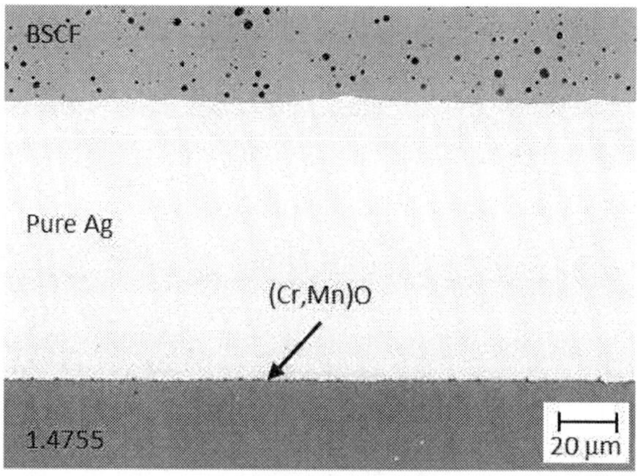

Figure 1: As-joined 1.4755/BSCF (M3SB: T.: 800°C, t.: 4 h, σ.: 6 MPa).

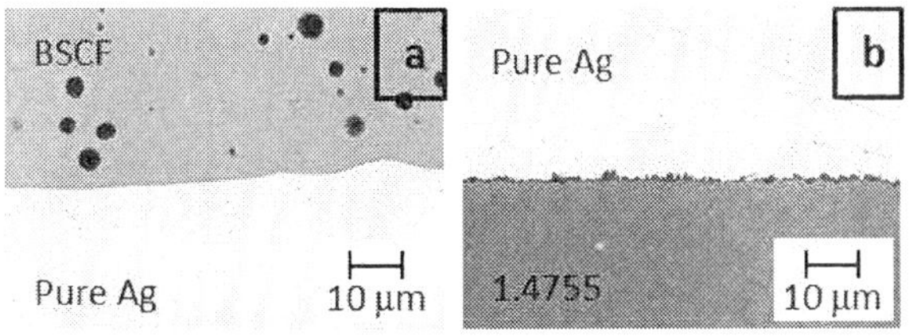

Figure 2: As-joined 1.4755/BSCF (M3SB: T.: 800°C, t.: 4 h, σ.: 6 MPa), a) ceramic/interlayer-interface, b) steel/interlayer-interface.

Reactive Air Brazing of 8YSZ/1.4841 samples either leads to failure upon cooling from joining or at very low fracture energies. Besides the known bad wetting behaviour of pure Ag-braze, pore formation and braze spillage have been observed (Fig. 3). The large differences in thermal expansion coefficients lead to failure at the braze/metal interface.

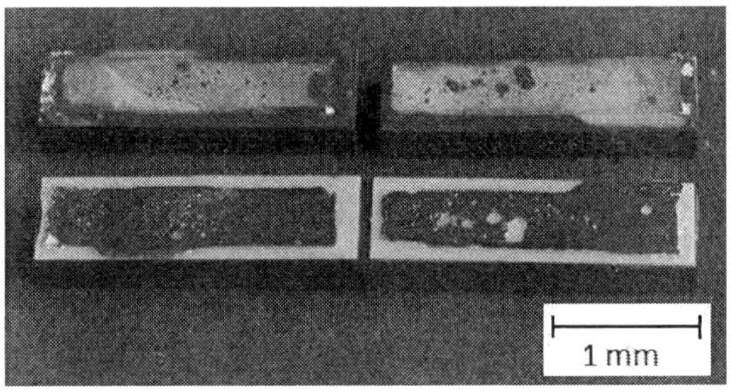

Figure 3: Detached as-brazed 8YSZ/1.4841 (RAB: T_{max} = 1000 °C).

Starting with the parameters that enabled successful joining of BSCF/1.4755, 8YSZ/1.4841 was joined with a permanent pressure of 6 MPa at 800 °C. The resulting fracture energies were as low as after RAB.

To reduce the arising stresses at the interfaces and to limit the oxide growth at the steel/interlayer interface the joining temperature was reduced to 550 °C and the influence of variation in joining pressure was investigated. Joining was successfully achieved for all parameter sets.

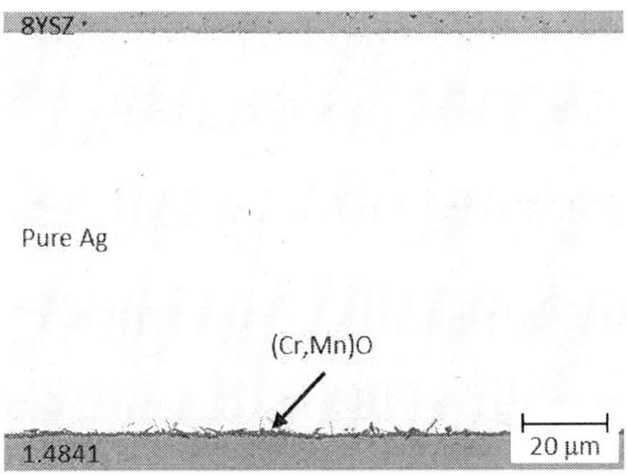

Figure 4: As-Joined 8YSZ/1.4841 (M3SB: T.: 550 °C, t.: 4 h, σ.: 6 Mpa).

The microstructure after joining is shown in Fig. 4 for 8YSZ/1.4841. The interfaces are without visible imperfections after joining. Only at the steel/interlayer interface a thin reaction layer can be observed (Fig. 4).

To evaluate the influence of annealing on the joints 8YSZ/1.4841 specimens were annealed in air at 800 °C for 1000 h (Fig. 5). After annealing the reaction layer was about 10-15 µm thick (Fig. 5, 6). It consisted of (Cr, Mn)O on the steel side and a (Ag, Cr, Mn)O towards the Ag-interlayer. At the steel/reaction layer interface silver pockets can be observed. During annealing diffusion processes take place forming a silver layer in between the reaction layer and the substrate surface.

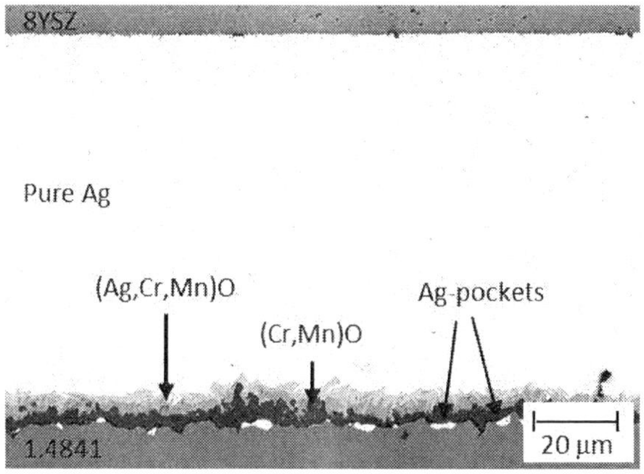

Figure 5: 8YSZ/1.4841 M3SB after 1000 h annealing in air at 800 °C.

The reaction layer is shown in detail in Fig. 6. An element mapping was performed in the highlighted area. The results are shown in Fig. 7. Starting from the steel the first part of the reaction layer is a CrO with minor Mn content, followed by a (Cr, Mn)O. The following layer is made up from needle shaped (Ag, Cr, Mn)O with varying compositions. Cr- and Mn-content are higher closer to the steel, whereas the Ag-content rises towards the pure Ag-interlayer.

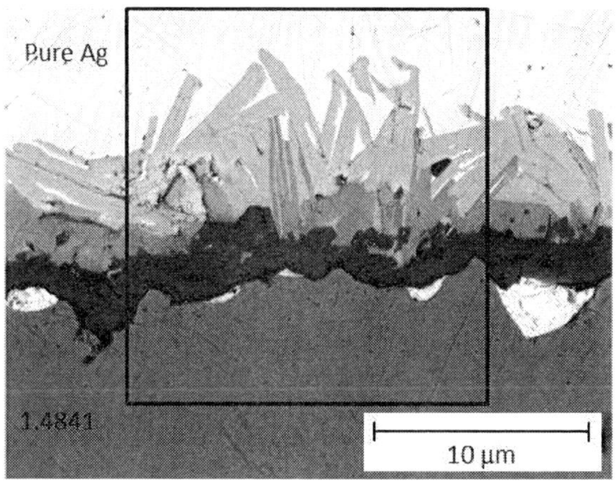

Figure 6: Reaction layer after 1000 h annealing in air.

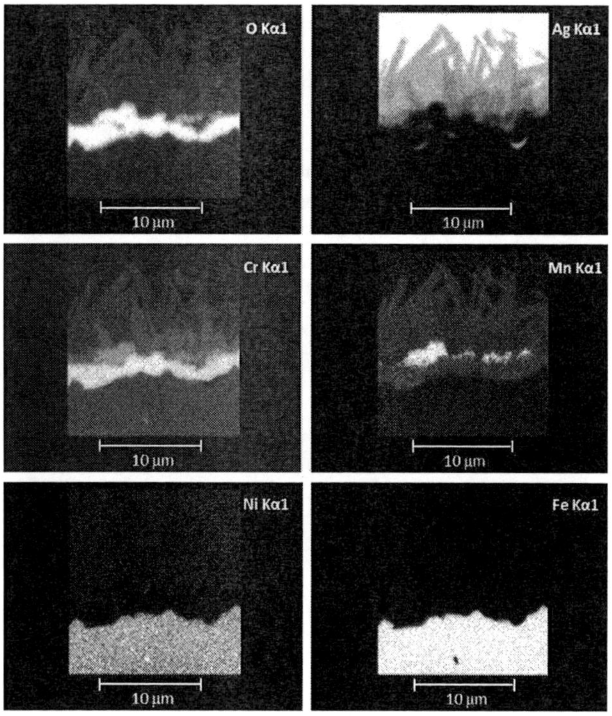

Figure 7: Element mapping after 1000 h annealing in air.

As previously noted joining of 8YSZ/1.4841 has been performed with 3 different permanent pressures. To take into account the reduced temperature the applied pressure was increased step-wise. For the considered pressure values no influence on the fracture energy could be determined (Fig. 8).

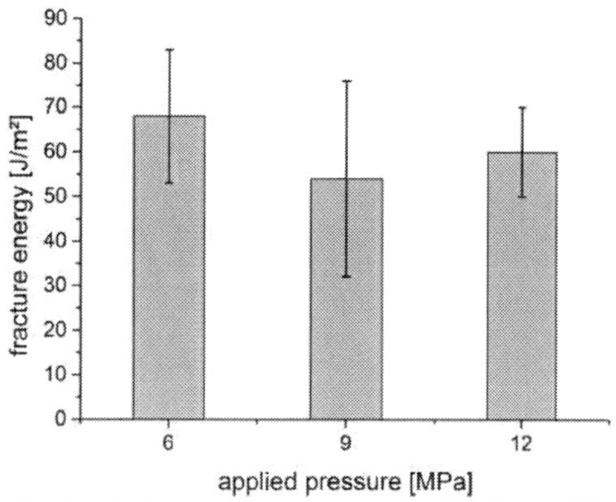

Figure 8: Applied pressure values during M3SB at 550 °C and measured fracture energies.

Due to the number of available specimens only for 8YSZ/1.4841 fracture energies in the as-joined state and after annealing in air were determined (Tab. 1). Compared to the as-brazed state, where failure occurred at low fracture energies (Fig. 3), in the as-joined state after M3SB at 550 °C sufficient fracture energies have been achieved. First annealing experiments in air at 800 °C indicate fracture energies comparable to those of the as-joined state (Tab. 1). For annealing experiments the number of tested specimens was limited and therefore only a lower limit for the fracture energy and no standard deviation are given.

Table 1: Comparison of fracture energy after brazing and M3SB (550 °C) for 8YSZ/1.4841.

	Fracture Energy [Jm^{-2}]
As-brazed (RAB)	12 ± 10
As-joined (M3SB)	69 ± 24
1000 h, 800 °C, Air	>80

Crack propagation and failure in the as-joined state occurs dominantly along the ceramic/interlayer interface. At the steel/interlayer interface a stronger joint has been achieved with M3SB compared to RAB. This change is also reflected in the rise of the fracture energy values.

Conclusions

The concept of M3SB could be successfully applied for joining the material combinations BSCF/1.4755 with a maximum ΔTEC = 4 ppmK-1 and 8YSZ/1.4841 ΔTEC = 7.5 ppmK-1.
Microstructural analysis revealed that by applying M3SB strong joints and interfaces without visible imperfections have been achieved. In the as-joined state only a thin reaction layer at the steel/interlayer interface could be observed.
Annealing of 8YSZ/1.4841 in air at 800 °C led to a growth of the reaction layer thickness, but fracture energies remained constant.
Further investigation will focus on the fracture energy evolution after annealing and the transfer of M3SB to other material combinations.

Acknowledgments

The authors gratefully acknowledge the German Research Foundation (DFG) for supporting this study in the framework of the PAK524 programs and H. Reiners, B. Werner, D. Liebert, D. Grüner and E. Wessel for their support in mechanical testing, metallographic preparation and microstructural investigation.

References

[1] Kim, J. Y. et al., "Ag–Al based air braze for high temperature electrochemical devices," Journal of Hydrogen Energy, Vol. 33, (2007), pp. 3952-3961.
[2] Eichler, K. et al., "Degradation effects at sealing glasses for the SOFC," Proceedings of the fourth European solid oxide fuel cell forum, Oberrohrdorf, Switzerland, 2000, pp. 899–906.

[3] Lewinsohn, C. A. et al., "Joining of silicon carbide composites for fusion energy applications," Journal of Nuclear Materials, Vol. 283, (2000), pp. 1258-1261
[4] Pippel, E. et al., "Structure and composition of interlayers in joints between SiC bodies," Journal of the European Ceramic Society, Vol. 17, (1997), pp. 1259–65.
[5] Kim, J. H. et al., "Bonding of alumina to metals with Ag–Cu–Zr brazing alloy," Journal of Material Science Letters, Vol. 16, (1997), pp. 1212-1215.
[6] Weil, K. S. et al., "Reactive Air Brazing: A Novel Method of Sealing SOFCs and Other Solid-State Electrochemical Devices," Electrochemical and Solid-State Letters, Vol. 8, No. 2, (2005), pp. A133-A136.
[7] Kim, J. Y. et al., "Effects of CuO Content on the Wetting Behavior and Mechanical Properties of a Ag–CuO Braze for Ceramic Joining," Journal of the American Ceramic Society, Vol 88, No. 8, (2005), pp 2521-2527.
[8] Hardy, J.S. et al., "Joining Mixed Conducting Oxides Using an Air-Fired Eletrically Conductive Braze," Journal of the Electrochemical Society, Vol. 151, No. 8, (2004), pp. J43-J49.
[9] Kim, J. Y. et al., "Silver-copper oxide based reactive air braze for joining yttria-stabilized zirconia," Journal of Material Ressource, Vol. 20, No. 3, (2005), pp. 636-643.
[10] Kim, J. Y. et al., "Novel Metal-Ceramic Joining for Planar SOFCs," Journal of The Electrochemical Society, Vol. 152, No. 6, (2005), pp. J52-J58.
[11] Kim, J. Y. et al., "Effects of atmospheres on bonding characteristics of silver and alumina," Journal of Hydrogen Energy, Vol. 33, (2008), pp. 4001-4011.
[12] Schüler, C. C. et al., "Direct silver bonding – an alternative for substrates in power semiconductor packaging," Journal of Materials Science-Materials in Electronics, Vol. 11, No. 5 (2000), pp. 389-396.
[13] Weil, K. S. et al., "The effect of TiO2 on the wetting behavior of silver–copper oxide braze filler metals," Scripta Materialia, Vol. 54, (2006), pp. 1071-1075.
[14] Kuhn, B. et al., "Fracture Mechanics Experiments to Characterize the Failure of Ceramic/Metal Joints in SOFC Stacks," ECS Transactions, Vol. 7, No. 1, (2007), pp. 413-419.
[15] Li et al., "Mechanical behavior of advanced reactive air brazed joints," Brazing and Soldering 2012: IBSC Proceedings of 5th International Conference, Las Vegas, NV, April. 2012, pp. 444-449.
[16] Kaletsch et al., "Joining Oxygen Transport Membranes by Reactive Air Brazing," Brazing and Soldering 2012: IBSC Proceedings of 5th International Conference, Las Vegas, NV, April. 2012, pp. 437-443.
[17] H. J. de Bruin and C. E. Warble, U.S. Patent 4,050,956 A.
[18] Kuhn, B. et al., "Mechanical properties of reactive air brazed (RAB) metal/ceramic joints. Part 1: Viscoplastic deformation of silver based reactive air brazes," Advanced Engineering Materials, (2014), DOI: 10.1002/adem.201400072.
[19] Kuhn, B. et al., "Mechanical properties of reactive air brazed (RAB) metal/ceramic joints. Part 2: Tailored microstructure for thermal cycling resistance," Advanced Engineering Materials, (2014), DOI: 10.1002/adem.201400073.
[20] Brandenberg, J. et al., "Dual-Gas stability of RAB-joints in SOFC applications," Advanced Engineering Materials, (2014), DOI: 10.1002/adem.201400098.
[21] Li, C. et al., "Improving Contour Accuracy and Strength of Reactive Air Brazed (RAB) Ceramic / Metal Joints by Controlling Interface Microstructure," Advanced Engineering Materials, Vol. 14, No. 6, (2012), pp. 394-399.
[22] Kuhn, B. et al., "Mechanical performance of reactive-air-brazed (RAB) ceramic/metal joints for solid oxide fuel cells at ambient temperature," Journal of Power Sources, Vol. 193, No. 1, (2009), pp. 199-202.
[23] Malzbender, J. et al., "Failure probability of solid oxide fuel cells," Ceramic Engineering and Science Proceedings, Vol. 26, No. 4, (2005), pp. 293-297.
[24] Charalambides, P.G. et al., "A Test Specimen for Determining the Fracture Resistance of Bimaterial Interfaces," Journal of Applied Mechanics, Vol. 56, (1989), pp.77-82.

Diffusion Bonding Systems - State of the Art

Simon Jahn, Felix Gemse, Steffen Dahms, Sabine Sändig
ifw - Guenter-Koehler-Institute for joining technologies and material testing, Jena, Germany
sjahn@ifw-jena.de, +49 3641 204 151

Abstract

The solid state joining process "diffusion bonding" is used more often worldwide because of the process properties like areal joint without any additional filler material. Examples of application are optical devices, micro fluidic components or heat transfer elements and heat exchangers. The market for diffusion bonded parts is growing fast, but it is still small today. Despite this, more than 30 companies offer diffusion bonding systems all over the world. In this contribution, different commercially available diffusion bonding systems as well as lab scale solutions are presented as well as design principles and their differences. One of the most important criteria is to achieve a homogenous load distribution in the parts to join. For this, different concepts will be shown, too.

Introduction

Diffusion bonding, also known as diffusion welding, is often called the ideal joining process, because no filler material is necessary and the bond area can often not be detected. Typical for diffusion bonds are high bond strength (comparable to base material properties) as well as multi-material joints. The main process steps are shown as schematic in next figure.

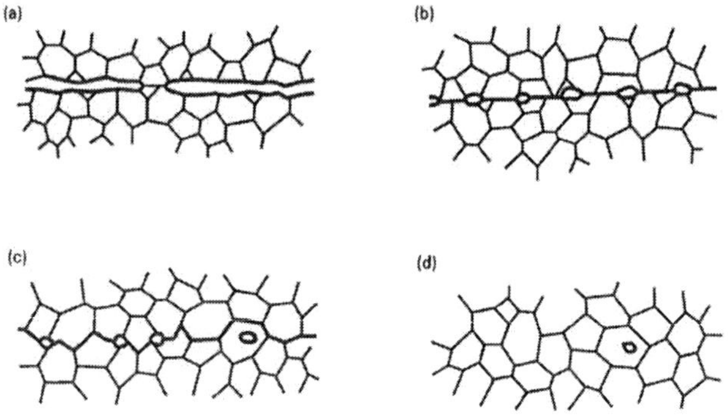

Figure 1: Principle of diffusion bonding [1]

To achieve a diffusion bond, it is needed to press the parts to join together (Fig. 1 a) and enhance the diffusion of atoms by raising the temperature (b – d). Because of the load, faying surfaces get in close contact and allow atoms to "move" through the interface. Diffusion itself can be controlled by temperature, whereas higher temperatures lead to faster diffusion. At the end of the process, the bond line is not visible any more in the case of similar material joining. To protect the parts /material during the bonding at elevated temperatures, a non-reactive atmosphere has to be used for most material, e.g. to avoid oxidation. For all three important elements of a diffusion bonding system (temperature, load application as well as atmosphere) different solutions were developed and commercially available in the market. In the following some special solutions will be explained.

Temperature

In order to achieve diffusion as well as to promote the bond by changing the mechanical properties, the parts have to be heated to a certain material specific temperature. This is typically below the melting point or the solidus temperature.

Heating

The raise of the temperature can be done with direct or indirect heating of the part. The first mentioned group includes resistive and inductive heating of the part. While in resistive heating the part is connected into an electric circuit, the inductor of inductive heating generates heat surface-near. Because of the necessary electrical power, both methods are in use for small parts only. Sometimes resistive heating is combined with external heat generation to reduce the temperature gradient within the part itself. For reduced residual stresses, it is important to keep the temperature gradient within the part small during the bonding cycle. This gradient is depending on the thermal properties of the assembly to join (material properties) as well as on the principle of the heat generation. Further influencing factors explained in the next chapters are the atmosphere and the load application including the pressing plates (size, material). In vacuum the part cannot be heated convective, only by radiation or conductive (by heated pressing plates). In old system designs, it is often essential to use thick pressing plates for a homogeneous load distribution, which raises the mass in the furnace significant. Because of the "dead weight", long cycle times are the result of this procedure. The use of higher heating power is not solution, because it must be avoided to overheat the part (even on the surface). If the surface temperature is above the solidus or near the melting temperature, partial melting can occur. This has to be taken in consideration, especially while using principles, which heat the surface only (e.g. inductive heating). For diffusion bonding in production scale, the heating by resistive heated elements made for example of molybdenum, tantalum or graphite are the most important. Further heating methods comprise beam heating (laser or electron beam) and radiation heating by infrared or halogen lamps [2]. The type of heating (and the necessary power for heating) depends further on material to join and part size. For joining Al-parts only, the maximum temperature is given by the melting/solidus point of the material (around 600° C), whereas for ceramic bonding temperatures up to 2000° C can be necessary. The maximum or so-called design temperature of a diffusion bonding system influences the system costs significant.

Cooling

At the end, after heating and soaking, the joined assembly has to cooled to room temperature respectively to the unload temperature, which can be higher. In the case of phase transformation materials (e.g. tool steels), it is possible to combine the bonding and heat treatment process (same as in use for brazed parts). For this, the cooling rate must be high enough. But too huge temperature gradients can lead to distortion and in the worst case to delamination of the bond. Additional stresses are induced by (wanted) phase transformation (depending on the density skip, e.g. austenite to martensite). The cooling procedure is always a compromise between time/necessary cooling rate and residual stresses.

If there is no forced cooling, the cooling rates, especially while bonding in vacuum are very low. To achieve the necessary cooling rates, additional devices to remove the heat are often integrated into the bonding system. All common is the use of inert gas. The simplest solution is to realize an inert gas flow through the system, sometimes direct after passing the part into the vacuum pump. Disadvantages are the still low cooling rates and the gas consumption. In order to raise the cooling rates, internal fans can be built into the hot zone for the circulation of the "quenching" gas (Fig. 2). A further enhancement is possible by the use of a heat exchanger (internal or external) to cool the circulating gas.

Figure 2: Diffusion bonding system with gas recirculation by blower and external heat exchanger for cooling the gas

The efficiency of the cooling with gas is a function of the gas pressure in the system as well. With higher pressure (and so higher gas mass in the system). For highest cooling rates a combination of a heat exchanger and a high pressure chamber should be used. In this case, the quenching gas pressure can be as high as 30 bar in the system, if the hot zone is constructed suitable for higher pressure.

Atmosphere

Main function of the chosen atmosphere is to protect the material in order to avoid negative reaction between ambient air and part (e.g. certain ceramic). Beside oxygen, the adsorption of nitrogen or hydrogen can lead to a reduction of ductility. Because of this, a protective environment is necessary for the bonding process in most cases.

Vacuum

The "cleanest" atmosphere is vacuum. Depending on the pumping rate(s) of the vacuum system and the resulting pressure, there are different vacuum classes (low: 760 to 25 Torr, medium: 25 to $1\times10{-}3$ Torr and high: $1\times10{-}3$ to $1\times10{-}9$ Torr). High or very high vacuums often need a combination of different pumps, like piston, roots or scroll compressor pump for rough vacuum combined with oil diffusion or turbo pump. For holding the vacuum, rotary vane pumps support the other pumps, too. Vacuum contributes the joining process by enhancing the outgassing and the removal of oxides. But is must be mentioned, that the resulting "gas" composition in the vacuum chamber should be controlled as well, especially with regard to oxygen content. For this, oxygen sensors (only for oxygen) and mass spectrometer (gas composition) are in use. The last listed solution is the more expensive one. Compared to commercially pure argon, the impurity content in a vacuum pressure of 133 Pa is lower.

Protective gas

Main disadvantage of vacuum is the low heat transfer between part and heating elements. If there is a protective gas between the heating elements and the part, the heat transfer can occur convective, too. This is more efficient compared to radiant heating only in vacuum. Depending on the material to bond, a combined heating method can be used as well, e.g. in the lower temperature range within gas (convective and radiant) and at higher temperatures heating in vacuum only. After passing a certain temperature, the atmosphere is switched from gas type to vacuum and radiant heating only. A solution with protective gas can be cheaper compared to the generation of a vacuum (depending on the used gas, e.g. type and purity).

Load

The load application can be subdivided into load application method, how the load is applied, and into concepts for load distribution, how the load is transferred from a single point to the whole faying surface area.

Load application method

There are once again two main principles, the external application and the internal generation. The last mentioned can only be used in the case of dissimilar joints. For this, it is important, that the two materials differ in their coefficient of thermal expansion (CTE) and strength, e.g. joints between titanium alloys and stainless steels. For a joint between Ti2 and Al99.9, the titanium sheet was bended in a u-shape around an Al cube and then heated up to 550° C. With an FEM-simulation it was confirmed, that the resulting pressure between both parts (at joining temperature) was in the range of 15 MPa. Conducted vacuum leak test on Ti-Al6-V4 – stainless steel (1.4301) joints have proved a vacuum tightness of 10-5 (mbar*l)/s [3].

The method of utilization of CTE-differences can also be used for application of an external load. Figure 3 [4] contains a device for joining with thermal expansion. Main advantage of this method is that no pressing equipment is necessary and "conventional" vacuum brazing or heat treatment furnaces are sufficient. In opposite, the maximum load is limited and the tooling expensive.

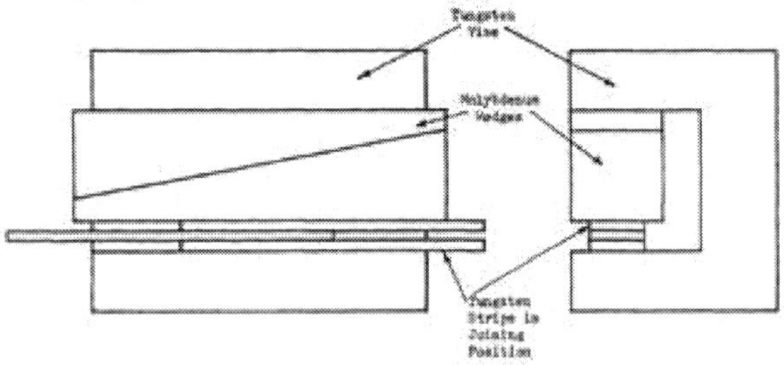

Figure 3: Device for the application of load by different CTE-materials [4]

Similar is the addition of weight on top of an assembly to achieve the load ("gravitation method"). Again the maximum load is limited, but there are lower tooling costs. Because of the disadvantages, presses are often in use for diffusion bonding in the industry. These systems are cheaper in operation and more flexible. In the simplest way, the press is operated by hand (e.g. screw drive, Fig. 4). Because of the thermal expansion of the parts, the applied load raise during heating. In the case of manually operated presses, a lower load has to be set at the beginning of the process or the load must be reduced in order to avoid massive deformation of the part.

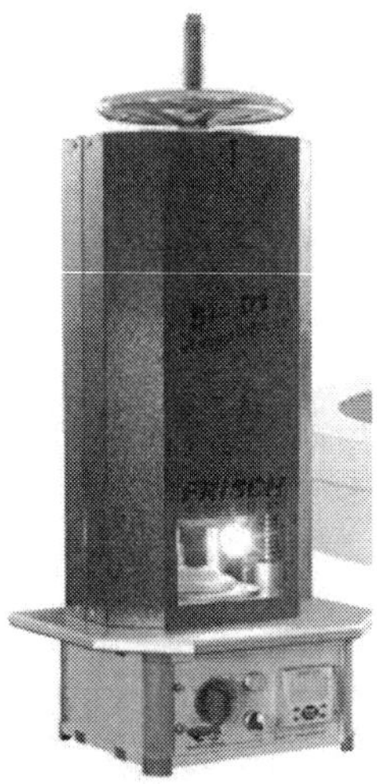

Figure 4: Commercial compact diffusion bonding system (courtesy of Frisch GmbH, Pforzheim, Germany) [5]

In low cost systems, pneumatic cylinders are integrated as another concept for applying load. But air is compressible. Systems with incompressible fluids (hydraulic) can be controlled with more precision. That is one reason for the variety of hydraulic based vacuum hot presses. Disadvantages of hydraulic based systems are the energy consumption of the hydraulic power unit and the maintenance. Electro-mechanical based presses (e.g. electromotor operated screw or worm drive) have significant higher electrical power efficiency. Figure 5 contains a system with an electrically driven press.

Figure 5: Diffusion bonding systems with electro-mechanically operated presses

Another point to mention is the maximum applicable load. Depending on this value, the diffusion bonding system has to be designed with a press frame (high load, Fig. 2) or the vacuum vessel itself can take the load (small load, Fig. 5).

Load distribution

Beside the application of the load, the distribution of the load inside the joining zone / faying surfaces is most critical for the process. The crush of the part (height before and after bonding) should be uniform for the whole part. In the conventional diffusion bonding system design, there is only one support for the pressing plate on both sides, top and bottom. In the case of larger parts, the pressing plates bend and with this the effective load at the edges is reduced. Result of the bending behavior is a different bond quality over the whole part. One method to visualize the load distribution within the faying surfaces is the use of so-called Prescale measurement

film [6]. According to [7], the foils can be used up to a temperature of 180° C.
The typically used technique to avoid the bending is to enhance the pressing plate thickness (and thus reduces the effective usable height between the pressing plates as well). The increase of thickness is necessary, because there are often only single point contacts between pressing plate and the press (Fig. 6). Another thing to consider for this procedure is, that the mass is increased and with this the heating as well as cooling consumes more time (longer cycle time).

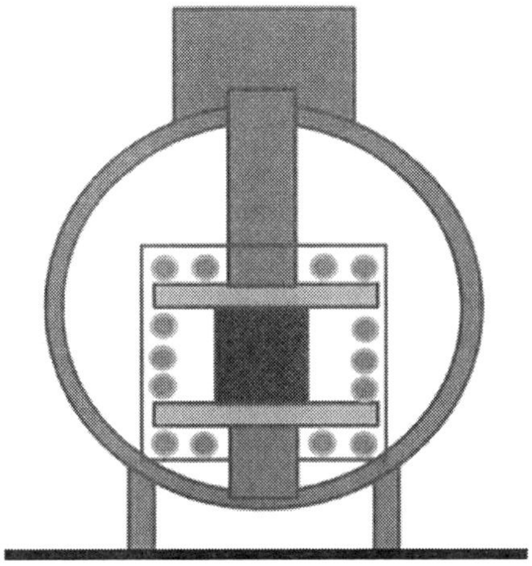

Figure 6: Schematic of conventional diffusion bonding system design

Another approach is shown in Fig. 7. The bottom pressing plates is supported by a multi columns or multiple support setup. Based on this setup, the thickness of the bottom pressing plates can be reduced significant and the size of the hot box can be used effective for part size. Another value results out of the lower weight of the pressing plates: less energy is required to heat the pressing plates themselves.

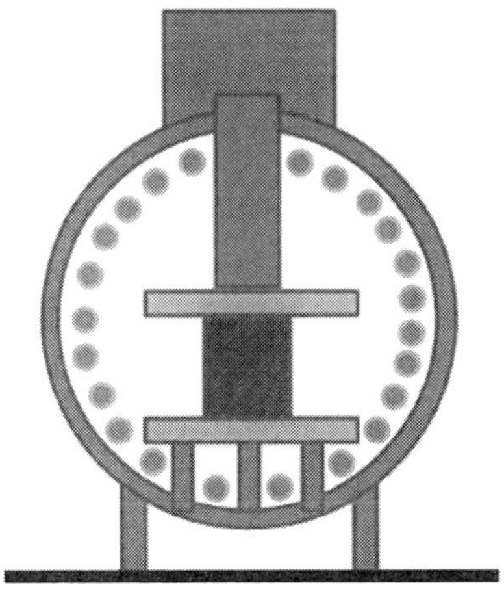

Figure 7: Schematic of diffusion system design for low bending of bottom pressing plate

Figure 8 shows a system with a multiple bottom pressing plate support. Because of the use of standard press design on the top, the system costs for such a setup rises not significant. But the disadvantage of bending is still there for the top pressing plate.

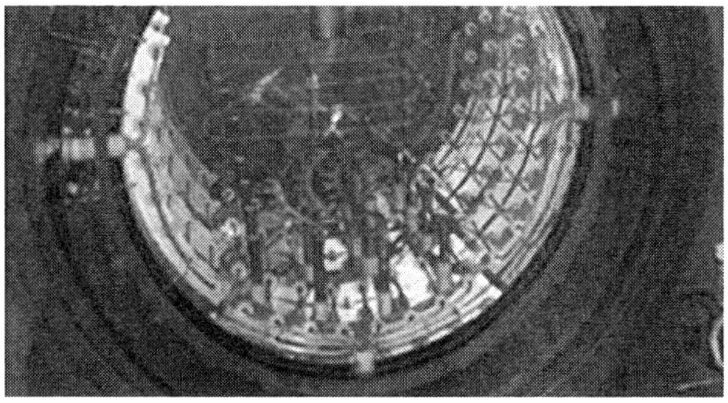

Figure 8: Diffusion bonding system with multi supports for bottom pressing plate and one single support for top pressing plate

In order to avoid the bending of the top pressing plate as well, a multiple support for the top pressing plate is commercially available up to a pressing plate size of 800 x 600 mm² [8] (Fig. 9).
A further design uses instead of one common used hydraulic press, an independent pressing axis for each support. Advantages are that it is possible to join several parts (mass series production), even if the single parts differ in height/tolerances or size of faying surface, in one cycle or to set load gradients within a large part. There are systems in the market with more than 10 independent pressing axes.

Figure 9: Diffusion bonding system with multiple supports for top and bottom pressing plate

Compared to the uniaxial pressing, the effort for the load distribution is less in the case of multi axial pressing (isostatic pressing). For the HIP (hot isostatic pressing) process, it is necessary to build a "container" or "envelope" for every part, which is lost after bonding. During the HIP the part is shielded by the container and the load as well as the heat transfer is done by the pressing medium. Because of the multi axial pressing, complex shaped joining areas can be bonded (e.g. parts requiring bonding on an inner diameter [9, 10, 11]). Because of the additional costs for the envelope production as well as the machining after bonding, HIP diffusion bonding process is more expensive and less in use compared uniaxial pressing processes.

Designs

Two different basic concepts dominate in industrial diffusion bonding, the furnace and hot press based. Furnace based systems provide in most cases a more uniform temperature distribution and control within the work zone as well as higher maximum joining temperatures. To apply the necessary load, presses are integrated into the furnace, e.g. an H-design with a hydraulically operated ramp. This basic concept (furnace with additionally integrated hydraulic press) is the most common. Further, these systems are often combined with single point bushing into the hot area to support the pressing plates only. In order to enhance the output, semi-continuous systems are an option. Often these systems consist out of three chambers, one for (pre-)heating, one for pressing (heated as well) and one for cooling [12, 13].
In furnace design, cold-wall and hot-wall types are used. For joining materials, which tend to outgassing, it is important to remove the exhaust/sublimate without to condense inside the furnace (e.g. at a cold wall). More rugged and resistant to contamination is the hot-wall furnace concept. A combination of the advantages of both types is the so-called ISO-furnace by MUT Advanced heating GmbH (Fig. 10) [14]. Inside of a cold-furnace one hot-wall furnace is installed. Result of the ISO-concept is a very clean furnace, even after several years use with outgassing materials.

Figure 10: Schematic of cold-wall (left), hot-wall (middle) and ISO (right) furnace (courtesy of MTU Advanced Heating GmbH, Jena, Germany)

In opposite to furnace based concepts, in the case of hot press based system, the load applying device (press) is present as well as a heating of the pressing plates (often integrated into the press table). Specialty of the hot press based design (compared to a furnace based) is the heated pressing plates. Because of this, there is conductive heat transfer into the parts and areal heating. For the heating of flat assemblies, this method is particularly recommended. But for tall parts, it is not sufficient to heat the bottom and the top only. Additionally extra heating elements for higher parts need to be attached to the hot press as well as a protective chamber. The hot press based systems stand for higher pressing forces and homogenous load distribution. This can be contributed to the origin use of those hot presses, sintering of ceramics [15] or forming of metal. In Fig. 11 a hot press system is shown without and with thermal insulation [16].

Figure 11: Hot press based diffusion bonding system, press size 700 x 700 mm^3 (courtesy of FormTech GmbH, Weyhe-Dreye, Germany)

Further system setups, which often cannot be classified, can be seen in labs/universities. One possibility to get a low cost diffusion bonding system is to "misuse" material testing machines [17], e.g. tensile test machine with integrated furnace or Gleeble systems [18]. Advantage of the procedure is that the testing machines are often equipped with high resolution length measuring devices. These can be used to monitor the crush of the part, but more important to determine high temperature properties and behavior of the materials to bond, e.g. creep rate at a given temperature and load. The other option are "self-made" systems, e.g. with a floating system as base (Fig. 12) [19].

Figure 12: Self-made diffusion bonding system [17]

Jewelry is, in addition to technical applications, a further market for diffusion bonding [5]. For this, special systems are available.

Diffusion bonding worldwide

Lab scale systems are distributed worldwide and available for R&D as well as series production. Most of existing diffusion bonding systems are designed for small size parts and in use for jewelry, micro-mechanical respectively micro-system devices like mixer or heat exchanger. In the last years, midsize parts like heat exchanger, heat transfer elements and parts for forming industry required larger systems. Because of this, new diffusion bonding systems with joining areas of 500 mm in diameter respectively 300 x 300 mm² have been installed. But there is still a trend for larger systems above the mentioned dimensions. Reasons for this are the demands for large heat exchangers or large area bonding, e.g. metal – ceramic joints for applications in next evolutionary stage of power generation (fusion reactors). In opposite to the allocation of medium and small size systems, which are available worldwide, the large ones are concentrated in the Asian region (Japan, China) and North America. In Europa, there are only a few systems in job shops for joining areas above 400 x 400 mm².

Conclusions

The advantages of the joining process diffusion bonding (joining below melting temperature, high strength, areal bonding, suitable for dissimilar joints) contribute mostly to the high demand for such bonded parts. Typically applications are based in the possibility to realize multi-material combinations in one single part as well as light-weight structure design. With the growing part size, the preferred design (furnace with integrated press) has to be adjusted. Most important criterion for the adjustment in the design is load distribution.
To conclude, it is to mention, that there is no diffusion bonding system for effective universal use. To operate the system effective and economically, it is necessary to adjust/setup/specify the system with regard to the requirements of the parts to bond (size, temperature, load, system concept, etc.).

References

[1] Owszarski, W.A. and Paulonis, D.F.: Application of Diffusion Welding in the USA, Welding Journal, 22, 1981

[2] Wallheinke, H.-D.: Beitrag zur Klärung der Vorgänge beim Diffusionsschweissen am Beispiel der Werkstoffpaarung Kupfer und Nickel, Braunschweig, Germany, 1976

[3] Wilden, J.; Jahn, S.: Entwicklung einer Fügetechnologie zum Herstellen von Mischverbindungen mit Titanwerkstoffen bei niedrigen Temperaturen, project report, DVS-Forschungsvereinigung, Düsseldorf, Germany, 2008

[4] Brophy, J. H. et al.: The Activated Joining of Tungsten, Welding Journal, 9/1963

[5] www.frisch-gmbh.de

[6] http://www.fujifilm.eu/eu/products/industrial-products/prescale/prescale-measurement-film/

[7] Chintapalli, P.: Diffusion bonding of large substrate MECS devices based on differential thermal expansion, Oregon State University thesis, 2006

[8] http://www.pvatepla.com/fileadmin/pdf/engPDF/Adhoc/2014/10-PM-Heisspresse-ENG.pdf

[9] www.metsomaterialstechnology.com

[10] www.kinzoku.co.jp

[11] www.bodycote.com

[12] www.ihi.co.jp/en/products/industrial_machinery/index.html

[13] http://www.alibaba.com/product free/113812847/Semi_Continuous_Diffusion_bonder_for _Titanium.html, 01.02.2013

[14] www.mut-jena.de

[15] www.fct-systeme.de

[16] www.formtech.de

[17] www.swan.ac.uk, (www.swan.ac.uk/engineering/materials/research/aerospacematerialsresearch/lowcostdiffusionbonding/), 31.05.2012

[18] www.gleeble.com

[19] Jahn, S.: Technologieentwicklung zur Herstellung variantenreicher innenstrukturierter Bauteile und Werkzeuge, Verlag Mainz, Germany, 2007

TLP Brazing of Titanium Alloys and Austenitic CrNi-Steel Using a Copper-Filler Deposited by Means of an Arc-PVD Process

W. Tillmann, C. Liu, M. Manka and L. Wojarski
Institute of Materials Engineering, Faculty of Mechanical Engineering, Dortmund University of Technology (TU Dortmund), Dortmund, Germany
wolfgang.tillmann@tu-dortmund.de, 0049-231-7552583

Abstract

The bonding of dissimilar materials in various fields of engineering increasingly gains significance In this regard, adequate manufacturing processes are required in order to provide hybrid concepts with tailored properties. In our previous work, commercial pure titanium (CP Ti) with CP Ti-AISI 1008 brazements were successfully produced by using the TLP-brazing process. In this study, the joining of the titanium alloy Ti6Al4V with the austenitic CrNi-steel AISI 304 was investigated. Copper layers, used as a filler material, were deposited onto the faying surfaces of the base material with varying thicknesses, using an arc-PVD coating process. The samples were brazed at 910°C for 60 and 30 min in a vacuum furnace. The microstructure and phase constitution of the brazed joints were analyzed metallographically as well as by means of scanning electron microscope (SEM) and energy dispersive X-ray (EDX). Furthermore, based on the experimental investigations and the mechanical properties of the brazements, which were determined in hardness and shear strength measurements, the diffusion coefficients were numerically calculated. Finally, the corrosion behavior of the joints was investigated using both immersion and electrochemical tests.

Introduction

Titanium alloys are characterized by a high strength-to-weight ratio, excellent fatigue resistance, and a high corrosion resistance. In consideration of these properties, titanium alloys are used in the chemical industry, aircraft industry, and in the field of nuclear power engineering [1-3]. With an increased use of hybrid bonds, successfully joining titanium with other technical materials, such as steel, in the above-mentioned applications is the key issue in order to use the full potential of the unique properties of titanium.

A review of the existing literature shows that diffusion welding has been successfully used to join titanium to itself and to steel alloys. For industrial applications, this method is not yet fully developed. The main requirement is that the bonding surface has to be carefully prepared. The impracticality of this method for mass production has therefore limited the usage of this process [4-7].

As a solution for this problem, brazing techniques can be used to join titanium and titanium to other metals. In this regard, the application of a filler metal leads to the formation of a liquid during the brazing process, which fills gaps and thus enables joining under low pressure or even without an external load. Furthermore, brazing enables to join dissimilar metals without severe distortions.

Different filler metals such as pure silver, silver based alloys, titanium-based alloys, aluminum-based alloys, and copper-based alloys were developed and utilized to braze titanium to different metals [8-12]. Moreover, the use of coating technologies, such as arc-PVD coating, to apply the filler alloy open up new avenues for difficult-to-join materials. Very thin filler alloy layers can be locally applied due to the precise dosage [13]. This is especially of interest for liquid phase bonding processes, where the amount of the filler alloy should suffice to fill the joint gap. However, at the same time, an excess amount of filler alloy significantly enlarges the isothermal solidification and can foster the formation of unfavorable intermetallic compounds.

In this regard, these investigations focus on the TLP bonding of the titanium alloy Ti6Al4V with the austenitic CrNi-steel AISI 304. Based on the results of our previous studies, pure copper layers were employed as a filler material due to their good compatibility to both substrates.

Experimental Procedures

Materials

The materials used for the experiments were the titanium alloy Ti6Al4V (with a thickness of 3 mm) and austenitic CrNi-steel AISI 304 (with a thickness of 2 mm). For subsequent microstructural analyses as well as for the shear strength tests, the sheets were cut into coupons with the dimension 30 mm x 25 mm. Prior to the PVD coating process, the surfaces of all samples were machined using different grinding papers of varying grades, before polishing them up to 1 μm. Finally, the samples were cleaned in an ultrasonic bath using ethanol, before they were placed into the arc-PVD device in order to apply copper thin films.

The surface of the samples was plasma-etched before the actual coating process was started. During the process, energized target ions are accelerated onto the surface of the substrate due to a high bias voltage and hence remove impurities on the surface. Besides the cleaning effect, an activation of the surface takes place, which increases the adherence between the substrate and the layer [13]. The coating parameters were adjusted to deposit a thin film of pure copper with an average thickness of less than 10 µm.

Brazing Condition

In order to position the pre-coated Ti6Al4V and CrNi-steel samples, and to guarantee an intimate contact between the joining couple during the entire brazing cycle, a stainless steel fixture was used. The samples were brazed in an overlap configuration (8 mm overlap) in a vacuum. All samples were brazed at 910°C with a dwell time of 30, respectively 60 min. In order to prevent non-uniform temperature distributions in the assembly, the samples were first heated up to 860°C and were held for 5 min. The joint pairings and used brazing parameters are given in Table 1.

Table 1: Joint pairings and brazing parameters

Joint	Deposition rate (Ah)	Brazing temperature (°C)	Dwell time (min)
Ti/St90-910-60	90	910	60
Ti/St90-910-30	90	910	30
Ti/St135-910-60	135	910	60
Ti/St135-910-30	135	910	30

Microstructural Observation and Mechanical Testing

The polished cross-sections of the joints were examined with a light optical microscope (SEM), equipped with an energy dispersive X-ray (EDX) for chemical analyses. Shear tests were employed to evaluate the strength of the joint of the brazed specimen as suggested by Harvey et al. [14]. The test was carried out at room temperature and the displacement rate was 0.5 mm/s. Five samples were used to calculate the average shear strength of the joint.

Electrochemical Measurements

The corrosion behavior of the joints was studied on cross-sections at room temperature. The working electrodes were cut and metallographically prepared. The experimental procedure included the open-circuit potential (OCP) and the potentiodynamic polarization experiments. The OCP was monitored for 3600 s and potentiodynamic test samples were polarized from -1500mV down to the OCP and to 2500mV above the OCP at a scan rate of 2 mV/s. The selected test solution was a 0.1 M sulfuric acid and the potentials were measured against an Ag/AgCl/Sat. KCl electrode. A platinum electrode, with an exposed area of 4 cm², was used as a counter electrode.

Immersion Tests

The immersion tests of brazed joints were carried out in a 6% $FeCl_3$ solution at 20°C. Like with the electrochemical measurements, the cross-sectional areas of the embedded samples were polished and had a dimension of 25 mm x 5mm. The samples were hung in the test solution for 24h. After immersion, the corroded samples were removed from the solution, cleaned with distilled water, and dried. The Morphologies of the attack in the joint zone were studied using a 3D microscope.

Results

Copper Coating

The cross-sections of the titanium and steel samples after PVD coating using a 90 Ah deposition rate are shown in Figure 1. The applied copper layers are dense and feature only small deviations of the thickness. The variation is caused by the process immanent formation of droplets during the deposition. The drops instantly solidify at the surface of the base materials. This leads to a sudden shrinkage of the target material, resulting in local peeling. The thickness of the layers of the titanium and steel substrates is almost equal and varies between 4 and 5 µm. By increasing the deposition rate to 135 Ah, the layer thickness increases to up to approximately 7 µm.

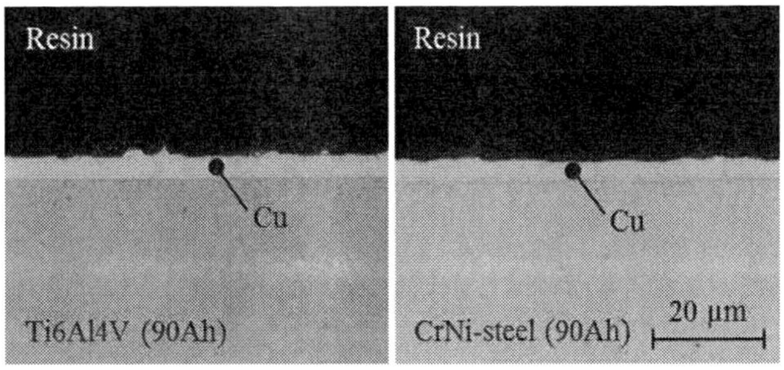

Figure 1: Cross-sections of pre-coated samples (deposition rate: 90Ah)

Microstructure and Phase Analysis

The obtained microstructures of the vacuum brazed Ti/St90-910-60 and Ti/St90-910-30 joints are shown in Figure 2. It is obvious that sound joints were achieved using both dwell times. Due to the solely small increase of the layer thickness by increasing the deposition rate to 135 Ah, a similar microstructure is apparent. A metallic bond was realized already after a dwell time of 30 min for all joints. The microstructure of the joint zone is clearly visible. Within the joint zone, mainly four characteristic microstructures could be. In the direction of titanium to steel, the first characteristic microstructure consists of a wide diffusion zone at the titanium side, which is followed by a grey phase and a light grey scattered phase. Because of the limited mutual solubility of copper, only a very thin intermetallic layer can be detected at the steel boundary.

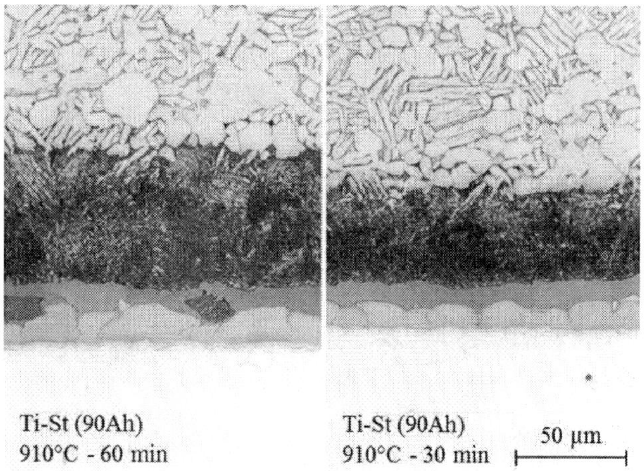

Figure 2: Cross-sections of joints brazed at 910°C with varying dwell times using a copper coating

Figure 3 shows the SEM microstructure of the brazed TiSt90-910-60 joint. The EDS analyses of the compositions of selected zones are summarized in Table 2.

The microstructures of the substrates were not influenced by the brazing process. Due to the pure solubility of copper in iron, the copper content is smaller than 1 at.-% in the steel substrate G and zone F. On the other hand, copper diffused deep into the titanium substrate. Reaction products with a high copper concentration indicate a good reactivity between copper and titanium. The titanium diffused to the steel side with a quite high concentration, as can be seen in zone F. The steel/copper interface showed the presence of Fe_2Ti. Even with the samples with the higher deposition rate of 135 Ah and a dwell time of 60 min, the thickness of this Fe_2Ti phase remains at about 3 µm. Zone D and zone E mainly differ in their iron concentration. According to the Cu-Ti-Fe ternary phase diagram and conforming to the EDS analysis, the presence of T2 ($Ti_{0.4}Fe_xCu_{0.6-x}$) is suggested in zone E and the TiCu phase was formed in the interior of the zone D. In zone C, B, and A, the concentration of copper decreases in the direction of the titanium substrate. The Ti-Cu binary phase diagram defines that β-Ti dissolves at 13.5 at. % copper at 1005°C, meanwhile α–Ti dissolves up to concentrations of 1.6 at. % copper at 790°C [15]. Therefore, the whole diffusion zone is an α-β Ti region. In zone C, thick dendritic phases with up to 12.4 at. % copper were formed. In comparison to zone B, much finer phases can been found due to a smaller copper concentration.

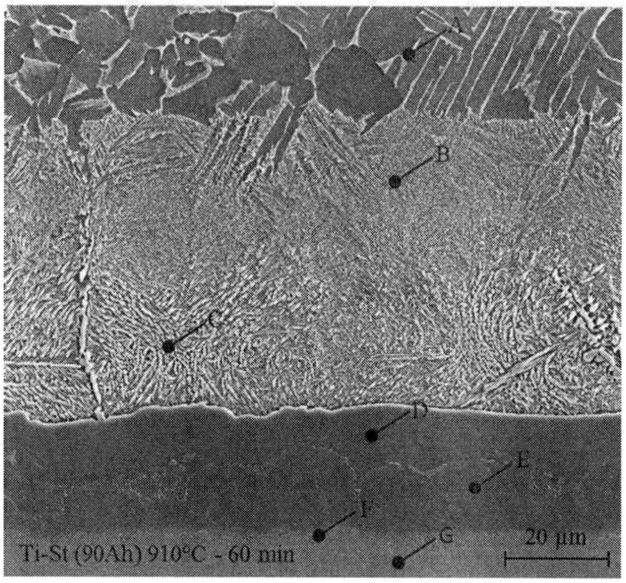

Figure 3: SEM microstructures of the joint with a 90 Ah deposition rate, brazed at 910°C with 60 min dwell time

Table 2: Chemical analyses at points shown in F

	Chemical analyse, at. %									
	Al	Si	S	Ti	V	Cr	Mn	Fe	Ni	Cu
A	10.4			76.3	7.68			0.97		4.65
B	11.1			77.9	3.15			0.49		7.34
C	9.02			74.5	3.17			0.95		12.4
D	2.35			57.8				0.72	0.71	36.2
E	6.24			40.3	1.92	0.73		9.73	1.64	39.4
F		4.18	1.72	25.3		12	1.02	51.3	3.92	0.58
G		1.02		1.65	0.44	20.6	1.49	67.1	7.05	0.61

Calculation of the Diffusion Coefficient

As the SEM analyses showed, the diffusion rate of copper into the titanium substrate is much faster at the beginning of the diffusion process, due to the formation of copper-titanium eutectic liquid phases. According to the distribution and formation of different phases, the diffusion coefficient for copper in titanium for dwell times of up to 60 minutes can be calculated quantitatively. In case of the brazed joint, using a 90 Ah deposition rate with a dwell time of 60 min, the concentration profile of copper in the brazed seam is shown in Figure 4.

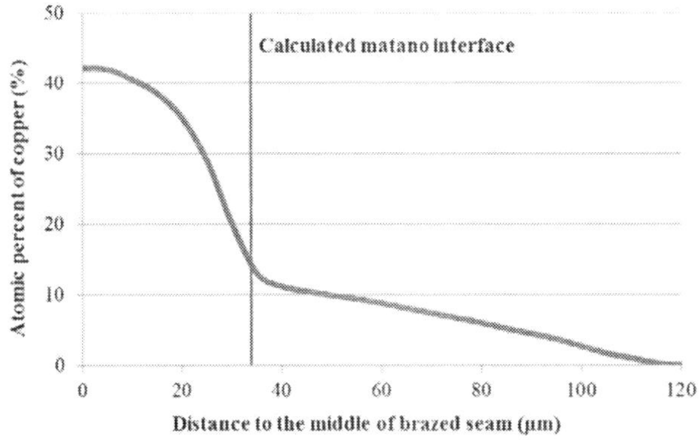

Figure 4: The concentration profile of copper in the brazing seam (in case of the brazed joint TiSi90-910-60)

The concentration-dependent diffusivity can be calculated using the Boltzmann–Matano method, which is widely used to analyze the concentration depth profiles measured after a diffusion time t. The corresponding mathematical equation (1) is given as follows.

$$D(C) = -\frac{1}{2t}\left(\frac{dx}{dc}\right)_{c=c'}\int_0^{c'} x\,dC \qquad (1)$$

On the other hand, the concentration profile of copper from Figure 9 can be described using a polynomial curve fitting and can be described with equation (2).

$$C = -8E - 12x^5 + 6E - 9x^4 - 2E - 6x^3 + 0.0003x^2 - 0.02x + 0.6574 \qquad (2)$$

Two points from the concentration profile of copper were selected as an example. The diffusion coefficients were calculated using equations (1) and (2), and listed in Table 3.

It is clearly visible that the diffusion coefficient depends on the copper concentration. At the position with a higher copper concentration, the diffusion coefficient for copper in titanium is higher as well. Due to the formation of a liquid phase and a grain boundary diffusion during the brazing process, the calculated diffusion coefficient was given as an approximate value. In comparison to diffusion coefficients from other binary systems in solid state diffusion process, the calculated diffusion coefficient in this case is much higher. Therefore, a very wide diffusion zone has formed after the 60 minutes brazing process.

Table 3: Calculation of concentration-dependent diffusion coefficient

	Distance to Matano interface	Concentration of copper (normalized)	t = 60 min T = 910°C
1	15µm	0.35	D = 6,4E-7 m²/s
2	50µm	0.08	D = 1,6E-7 m²/s

Hardness Measurements of the Joining Zones

Micro-hardness indentation measurements were performed to determine the effect of the hardness of the joint zone on the strength of the joints. The hardness distribution along selected indentations is shown in Figure 5. Because of the formation of Fe_2Ti, T2 and TiCu intermetallic compounds in the center of the brazed seam, especially close to the steel side, this area presented the highest hardness value. The high hardness at the interfaces, which can cause a brittle behavior of the joints and reduce the ductility, was determined. Hence, the tolerable plastic deformations could be limited.

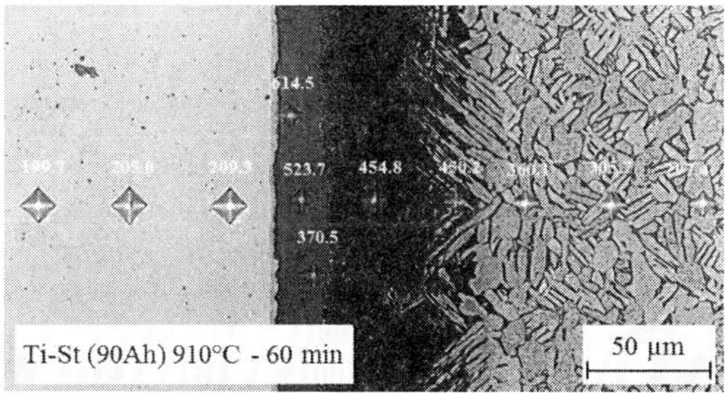

Figure 5: Hardness measurement (joints with 90 Ah deposition rate, brazed at 910 °C and with 60 min dwell time)

Mechanical Properties of Joints

The shear strength was calculated as the failure load divided by the overlap area. Figure 6 shows the average shear strength of all combinations. The average shear strength of the joint was higher when using a 60 min dwell time. The TiSt135-910-60 joint achieved a maximum shear strength of 53 MPa.

A higher copper content in the joint zone can decrease the resulting residual stress after the brazing process due to the significantly different thermal expansion coefficients of both base materials. Additionally, the high brazing temperature and long dwell time are beneficial for the diffusion process. However, the increase in thickness of intermetallic phases at the interfacial area, which leads to a high hardness, can limit the bonding strength.

The representative fracture morphologies at the steel side of the Ti/St joints with a 90 Ah deposition rate, brazed at 910°C for 60 min are presented in Figure 7. Four spots of different typical zones were selected and investigated.

The EDX analysis (Table 4) highlighted that an almost equal atomic percent of steel and titanium constituted the phase at the fracture surface. It can be verified that the fracture path follows the interior of the joint zone, mainly through zones E and F, as shown in Figure 3. This implies that despite its high hardness compared to other intermetallic compounds, the harmful Fe-Ti IMCs is the weakest phase in the joint.

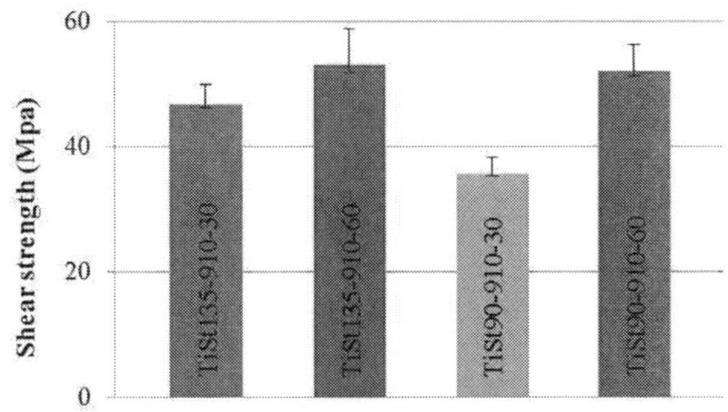

Figure 6: Shear strength of the brazed joints

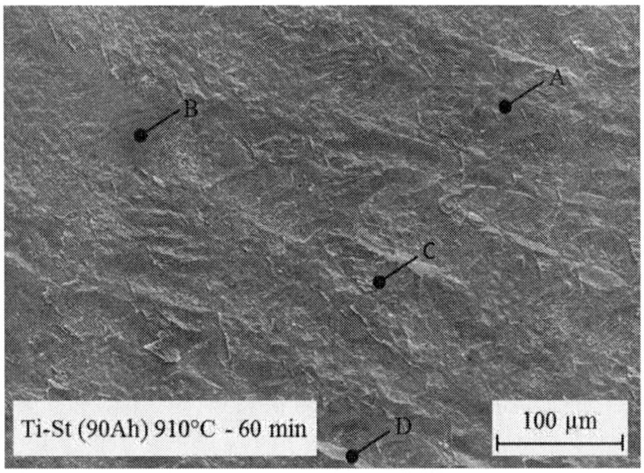

Figure 7: Fracture surface of a brazed joint at the steel side with a 90 Ah deposition rate, brazed at 910°C with a 60 min dwell time

Table 4: Chemical analyses at points shown in F

	Chemical analyses, at. %								
	Al	Si	Ti	V	Cr	Mn	Fe	Ni	Cu
A		3.31	31		8.96	0.86	52.4	3.48	
B	0.84	2.11	30.3		7.74	0.72	51.7	3.39	0.43
C	1.42		12.8	0.51	23.1	1.42	57.8	2.73	0.31
D	5.93		48.1	1.28	1.15		11.5	3.07	29

Corrosion Behavior

The corrosion potential as a function of time for the Ti/St joints at different brazing parameters is shown in Figure 8. It visualizes that after 3600 s, all the samples reach a stable OCP value between -0.27 and -0.25. Applying different thermal cycles during brazing leads to differing microstructures in the brazed area, and hence a varying content of the phases formed. Nevertheless, the OCPs measured in the joint are quite similar. This means that all joints showed a similar thermodynamic tendency in the corrosion medium. The samples with a 135 Ah deposition rate have a slightly higher OCP value because of the greater copper content in the joint zone than the samples with a 90 Ah deposition rate.

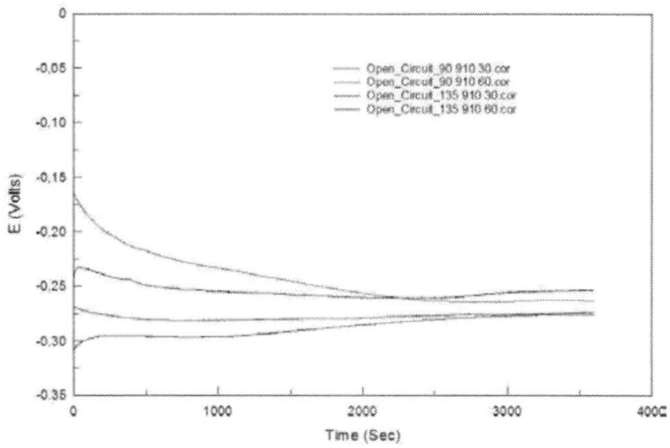

Figure 8: Open circuit potential measurements of Ti/St-joints

Quantitative information related to the kinetics of the corrosion reactions and the corrosion rate can be obtained from polarization curves. Figure 9 shows the cathodic and anodic polarization curves of the brazed joints. The joints brazed with different brazing parameters showed a more or less similar behavior in terms of the corrosion potential, with a small variation in the dissolution current density'. The joints brazed with a 30 min dwell time exhibited a relatively low current density compared to the joints brazed with a 60 min dwell time.

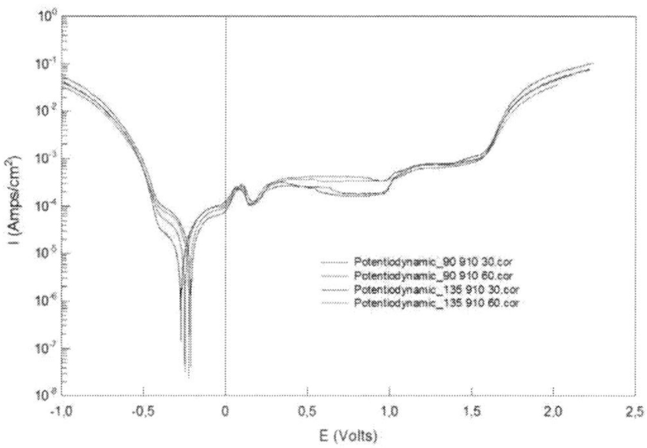

Figure 9: Potentiodynamic polarization curves of the brazed joints

Figure 10 shows the morphology of the TiSt90-910-60 joint after performing the immersion test. There is neither an obvious corrosion attack on the titanium, nor on the steel surface. The profile of the morphology of the joint, measured with a 3D microscope, shows a minimal difference in material loss between both substrates, which is caused by the sample preparation. In comparison to the hard titanium alloy, the relatively soft steel material was removed more rapidly while grinding and polishing the cross-section. Furthermore, no corrosion attack could be detected in the center of the joint zone. Only some corrosion products accumulated on the titanium surface near the joint zone. Hence, the brazements showed a high corrosion resistance against the $FeCl_3$ solution during the 24h test phase.

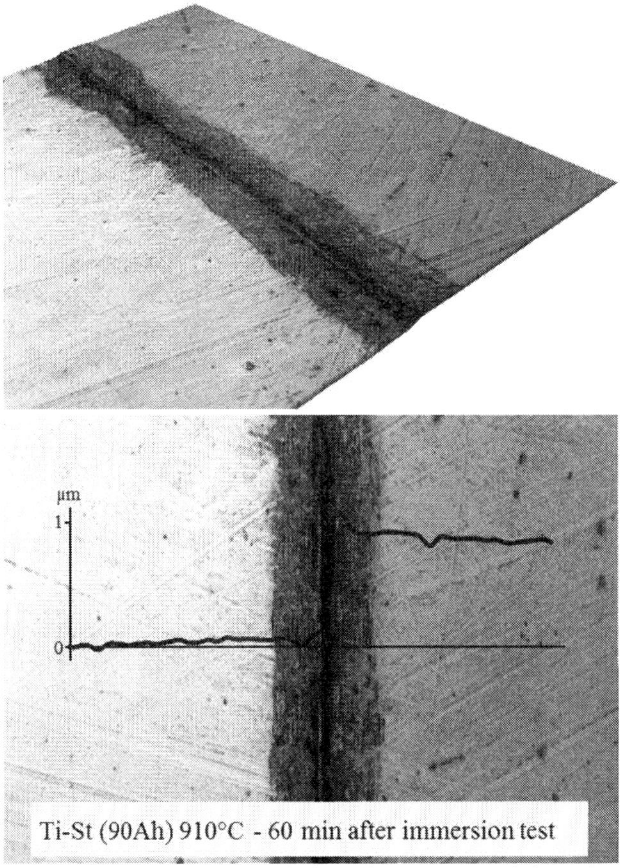

Figure 10: Morphology of the TiSt90-910-60 joint after performing an immersion test

Conclusions

The investigations showed that the titanium alloy Ti4Al4V and CrNi-steel AISI 304 could be successfully TLP brazed at 910°C, using varying dwell times. In order to obtain sound joints, a certain copper layer thickness is necessary. The interfacial microstructure evolution, the hardness distribution, the shear strength of the vacuum brazed joints, and the corrosion behaviors were evaluated. The most important conclusions are summarized below:

1. Sound Ti/St brazed joints were already obtained with copper coatings with a deposition rate of 90Ah and a dwell time of 30 min.
2. Various Cu-Ti and Fe-Ti intermetallic phases were detected at the interfacial area. In all cases, the diffusion of the copper coating into the titanium substrate is depended on the copper content and the dwell time. Meanwhile, intermetallic Fe-Ti phases were formed only as a very thin layer at the steel boundary.
3. The interfacial reaction layers between the base materials play a crucial role for the bonding strength of the joint. The joints brazed with a 60 min dwell time achieved a maximum strength of 53 MPa.
4. All joints show an excellent corrosion behavior in electrochemical measurements as well as in immersion tests. The widths of the diffusion zones and the compositions of the intermetallic phases have only a slight influence on the corrosion resistance.

Acknowledgments

The authors gratefully acknowledge the financial support from the German Research Foundation (DFG) within project No. TI343/31-2.

References

[1] Möller, F., Grden, M., Thomy, C. and Vollertsen, F., "Combined Laser Beam Welding and Brazing Process for Aluminium Titanium Hybrid Structures", Phys. Proc., 12, (2011), pp. 215–223.

[2] Ozaki, H., Ichioka, R., Matsuura, T. and Kutsuna, M., "Laser Roll Welding of Dissimilar Metal Joint of Titanium to Low Carbon Steel", Mater. Sci. Forum, 580–582, (2008), pp. 543–546.

[3] Jinkeun, O., Nack, J.K., Sunghak, L. and Eui, W.L., "Correlation of Fatigue Properties and Microstructure in Investment Cast Ti-6Al-4V Welds", Mater. Sci. Eng., A340, (2003), pp. 232–242.

[4] Lee, H., Yoon, J., Park, C.H., Ko, Y.G., Shin, D.H. and Lee, C.S., "A Study on Diffusion Bonding of Superplastic Ti–6Al–4V" ELI, Grade, J. Mater. Process. Technol., 187–188, (2007), pp. 526–529.

[5] Kurt, B., Orhan, N., Evin, E. and alik, A. C., "Diffusion Bonding Between Ti-6Al-4V Alloy and Ferritic Stainless Steel", Mater. Lett., 61, (2007), pp. 1747–1750.

[6] Sidyakin, V.A., Pechenkin, D.K., Arbuzov, V.M. and Khaustov, V.S., "Butt Welding of Steel-Titanium Pipe Transition Pieces", Weld. Int., 18(12), (2004), pp. 977–981.

[7] Qin, B., Sheng, G.M., Huang, J.W., Zhou, B., Qiu, S.Y. and Li, C., "Phase Transformation Diffusion Bonding of Titanium Alloy With Stainless Steel", Mater. Charact., 56(1), (2006), pp. 32–38.

[8] Kundu, S., Ghosh, M., Laik, A., Bhanu-murthy, K., G. B. Kale, and Chatterjee, S., ''Diffusion Bonding of Commercially Pure Titanium to 304 Stainless Steel Using Copper Interlayer'', Mater. Sci. Eng. A, Vol. 407, No. 1-2, (2005), pp. 154−160.

[9] Liu, C. C., Ou, C. L., and Shiue R. K., ''The Microstructural Observation and Wettability Study of Brazing Ti-6Al-4V and 304 Stainless Steel Using Three Braze Alloys'', J. Mater. Sci., Vol. 37, No. 12, (2002), pp. 2225−2235.

[10] Smorygo, O., Kim, J. S., Kim, M. D., and Eom, T. G., ''Evolution of Interlayer Microstructure and the Fracture Modes of the Zirconia/Cu-Ag-Ti Filler/Ti Active Brazing Joints'', Mater. Lett., Vol. 61, (2006), pp. 613-616.

[11] Knepper, P. and Lohwasser, D., ''High-Temperature Brazing of Titanium Structures'', DVS-Berichte, Vol. 208, (2001), pp. 83-88.

[12] Huang, X. and Richards, N. L., ''Activated Diffusion Brazing Technology for Manufacture of Titanium Honeycomb Structures'', Weld. J., Vol. 83, No. 3, (2004), pp. 73−81.
[13] Tillmann, W., Osmanda, A. M., and Wojarski, L., ''Diffusion Brazing of Al-alloys for Hybrid Structures'', Quart. J. Jpn. Weld. Soc., Vol. 27, No. 2, (2009), pp. 179−182.
[14] Harvey, J., Partridge, P.G., Snooke, C.L., "Diffusion bonding and testing of Al-alloy lap shear test pieces", J.. Mater. Sci., (1985), pp. 1009-1014
[15] Materials Science International Team MSIT®, Jialin, Y., De Keyzer, J., Kubaschewski, O., Schmid-Fetzer. R., Shcherban. O., Tomashik V., Tretyachenko, L., Ag-Cu-Ti (Silver-Copper-Titanium). G. Effenberg and S. Ilyenko, Eds., Springer Materials—The Landolt-Bo¨ rnstein Database http://www.springermaterials.com. doi:10.1007/10916070_15

Diffusion Bonding of a Single Crystal Superalloy to a Wrought Nickel-based Superalloy

Wei Fan*, Wei Mao, Yaoyong Cheng, Bo Chen, Xin Wu, Han Zhang
Welding and Plastic Forming Division, Beijing Institute of Aeronautical Materials, AVIC, Beijing 100095, China
The corresponding author: brooks21@126.com

Abstract

DD6 superalloy is a second generation single crystal superalloy developed in China, which has been used for fabricating turbine blades of advanced aeroengine. Nickel-based wrought superalloy GH4169 is a material candidate for turbine disk. Joining of DD6 superalloy to GH4169 superalloy is crucial for manufacturing turbine blisk. Induction heating rather than furnace heating was applied at the joined area to avoid the harmful effect on microstructure and mechanical properties of the joined base superalloy. High temperature processing would cause grain growth of the base superalloy. Sound joints were successfully achieved under the condition of 1120℃/30MPa/15min and 1140℃/30MPa/15min, respectively. It was found that a thin reaction layer had been formed at the joining interface. The reaction layer was mainly composed of γ solid solution out of strengthening phase, and the elements W, Ta and Co from the DD6 superalloy diffused more intensively at 1140℃ than that at 1120℃. The average tensile strength of joint bonded at 1120℃ and 1140℃ reached 522MPa and 678MPa at room temperature, respectively. In addition, the post-bonding heat treatment greatly improved the average room-temp joint strength to 955MPa.

Introduction

Due to the superior strength, plasticity and fracture toughness [1-3], GH4169 superalloy has been widely used in aircraft and aerospace industry. Typically, this superalloy can be used as a material candidate for turbine disk. As a second generation single crystal superalloy developed in China, DD6 superalloy, exhibits excellent high-temperature strength and overall performance. At present, DD6 superalloy serves for fabricating turbine blades of advanced aeroengines [4-6]. Therefore, the joining technology is crucial in manufacturing the turbine blisk. However, in the present the specific research on joining technology of the dissimilar superalloys is far insufficient. W.B.Han et.al carried on some researches on diffusion bonding GH4169 superalloy to GH4169 superalloy, but the joint strengths are not satisfactory: that is, average shear strength of diffusion bonded joints at room temperature was only 350MPa [7]. Other researchers got the brazed joints exhibiting average tensile strength of 269MPa at room temperature [8].

The solid solution temperature of GH4169 superalloy is 1065±10℃ [9]. It is scientifically considered that if the heating temperature of diffusion bonding excesses 1100℃, the joint properties would be undoubtedly improved, but the grains of the base superalloy would become coarse and this will deteriorate the base superalloy. To settle this contradiction, a method is to take induction heating rather than furnace heating at joined area to avoid the adverse effect on microstructure and mechanical properties of the base superalloy caused by high temperature processing. In this paper, the wrought superalloys GH4169 were diffusion bonded to the single crystal superalloys DD6 using induction heating at the joined area, and then part of them took a post-bonding heat treatment. The joint microstructures and the tensile fractured surfaces of the joints subjected to tensile test were analyzed.

Experiments

The base metals were GH4169 superalloy and DD6 single crystal superalloy with the compositions as shown in Table 1 and Table 2 [9].

Table1 Chemical compositions of GH4169 superalloy (wt, %)

C	Co	Mo	Cr	Nb	Al	Ti	Ni	Fe
≤0.08	≤1.0	2.8-3.3	17.0-21.0	4.75-5.50	0.3-0.7	0.75-1.15	50-55.0	Bal.

Table2 Chemical compositions of DD6 single crystal superalloy (wt, %)

Cr	Co	Mo	W	Nb	Al	Ta	Re	Hf	Ni
3.8-4.8	8.5-9.5	1.5-2.5	7.0-9.0	0-1.2	5.2-6.2	6.0-8.5	1.6-2.4	0.05-0.15	Bal.

DD6 superalloys and GH4169 superalloys were cut into cylinders with the dimension ofφ14×40mm as by spark cutting method. Then the specimens were mechanically polished using abrasive paper and ultrasonically cleaned in acetone. Subsequently, all the specimens were prepared in the form of butt joint, as shown in Figure1.

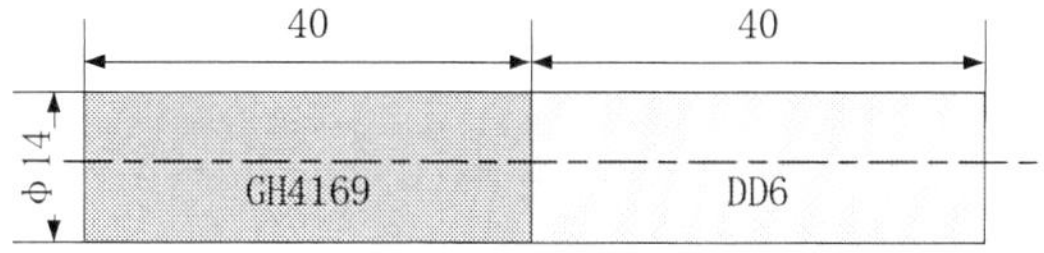

Figure1 Assembly diagram of the specimens

After assembling, the specimens were placed into a vacuum furnace and induction heated up to 1120℃/1140℃ at a vacuum pressure less than 5.0×10^{-3} Pa. The experiment was performed at 30MPa for 15min. Then the diffusion treatment 900℃/4h was carried for part of specimens, respectively. Subsequently, tensile strength of the joints at room temperature was tested, and joints microstructures were analyzed and determined by scanning electron microscope (SEM) equipped with energy-dispersive spectrometer (EDS).

Results and Discussion

Microstructure of the joint diffusion bonded

Figure2 shows the microstructure of the GH4169/DD6 joint diffusion bonded under the condition of 1120℃/30MPa /15min. The composition of phases of the joint was shown in Table 3. It is clear distinctly read from Figure2a that a sound bonded interface was obtained, without any pore, crack and unbonding, and there is no re-crystallization in DD6 single crystal superalloy. The diffusion bonding zone is mainly composed of light gray phase "B" with a thickness of about 10μm, while Table 3 shows the phase is rich in Ni, Cr, Co and Fe. So it can be confirmed as γ solid solution [10]. During the diffusion bonding process, the exclusive elements W and Co of DD6 single crystal superalloy and elements Fe and Nb from GH4169 superalloy diffused into the reaction zone where trace amount of element Si gathered. The white strip-shaped and small block-shaped phases near the centerline of the joint in GH4169 superalloy mainly containing C, Nb and fewer Ni, Ti, Cr could be inferred NbC and phase δ, which are major strengthen phases of GH4169 superalloy [7,11]. At 1020℃, phase δ and γ'' completely dissolved in phase γ, that's why diffusion reaction layer was out of strengthen phases. However, the elements' movement was so mild that the interface reaction was not intense enough, according to Table 3. DD6 single crystal superalloy elements Ta and Re even had not diffused into the reaction zone "B" (Figure2). It can be clearly seen from Figure3a and Figure2b that the diffusion distances of elements Co from DD6 single crystal superalloy and Fe from GH4169 superalloy diffusing are almost identical 30μm.

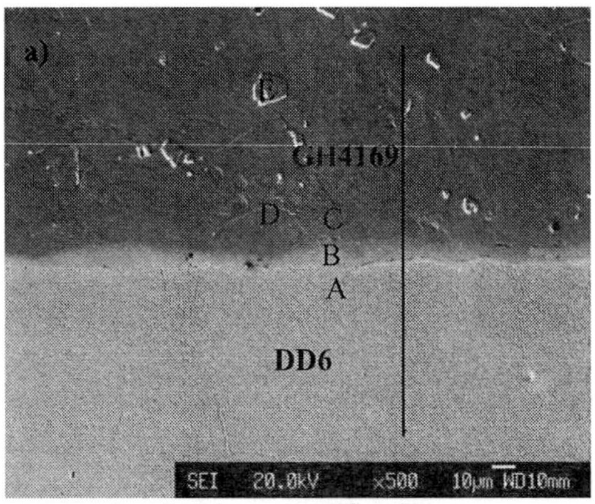

a) At a low magnification

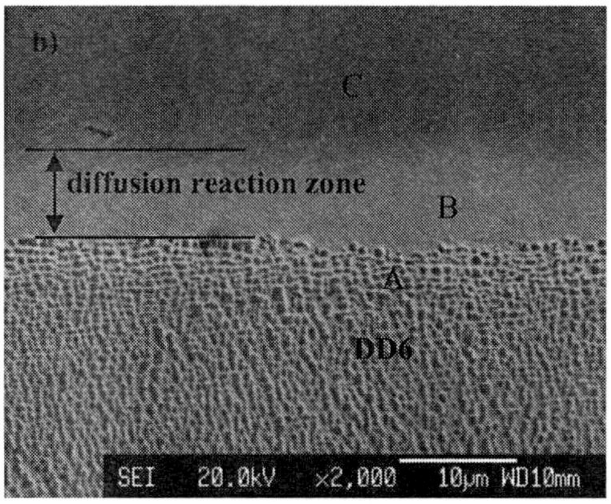

b) At a high magnification

Figure2 Microstructure of the diffusion bonded joint at 1120℃

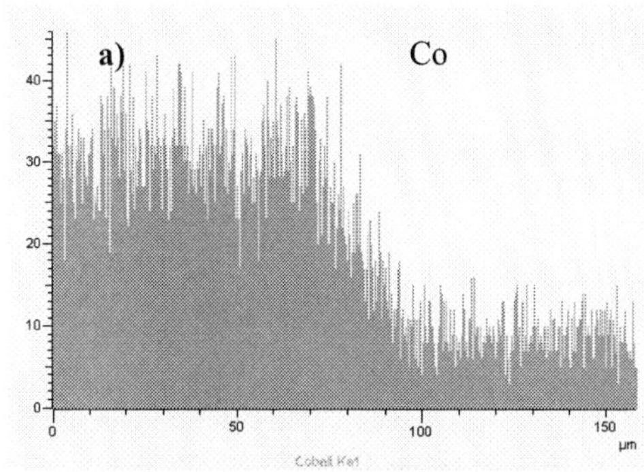

a) Distribution diagram of Co

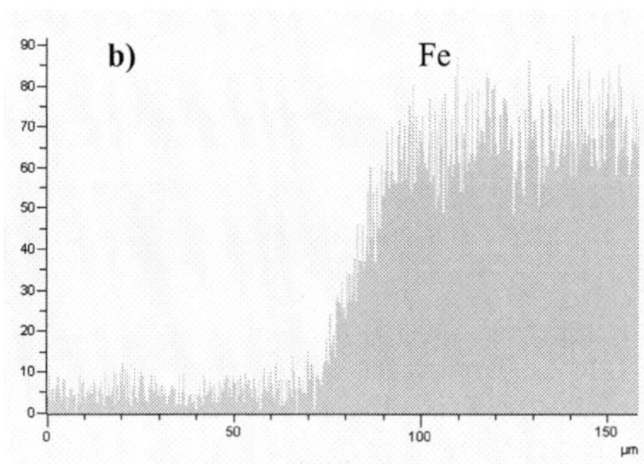

b) Distribution diagram of Fe

Figure3 Line distribution maps of elements Co and Fe (Figure2a)

To stimulate the element diffusion, the bonding temperature was elevated to 1140℃. Microstructure of the diffusion bonded joint under the condition of 1140℃/30MPa/15min is illustrated in Figure4 and the composition of phases can be seen from Table 4. A diffusion reaction zone is composed of distinctive two γ phases ("A" and "B") [10]. And it measures about 19μm, which is much wider than that in Figure2. According to the line distribution maps (Figure5), elements Co and Fe diffused nearly 40μm and 50μm, respectively. But there is an obvious phenomenon that elements Ta and Re were still absent in the diffusion reaction zone.

Table3 EDS analysis results for the mirozones marked in Figure2a (at/%)

zone	Cr	Fe	Co	W	Mo	Al	Ti	Si	Ni	Ta	Re	C	Nb	Deduced phase
A	5.16	-	9.67	2.82	1.16	11.45	-	-	66.33	2.08	1.05	0.28	-	DD6
B	12.14	8.30	5.92	1.41	1.60	6.79	0.71	1.96	58.80	-	-	-	2.37	γ phase
C	20.21	17.81	-	-	2.07	2.60	0.96	0.54	52.58	-	-		2.65	GH4169
D	2.73	1.80	-	-	-	-	8.41	-	4.43	-	-	43.61	38.0	NbC, δ
E	0.56	-	-	-	-	-	13.4	-	2.46	-	-	52.48	31.1	NbC, δ

Table4 EDS analysis results for the mirozones marked in Figure4a (at/%)

zone	Cr	Fe	Co	W	Mo	Al	Ti	Si	Ni	Ta	Re	C	Nb	Deduced phase
A	6.11	1.85	8.51	2.69	1.39	11.88	-	3.10	62.94	-	-	-	1.53	γ phase
B	12.70	8.40	6.50	1.85	2.07	6.65	0.75	-	58.60	-	-	-	2.48	γ phase
C	0.29	0.31	-	-	-	-	3.47	-	1.97			66.56	27.4	MC

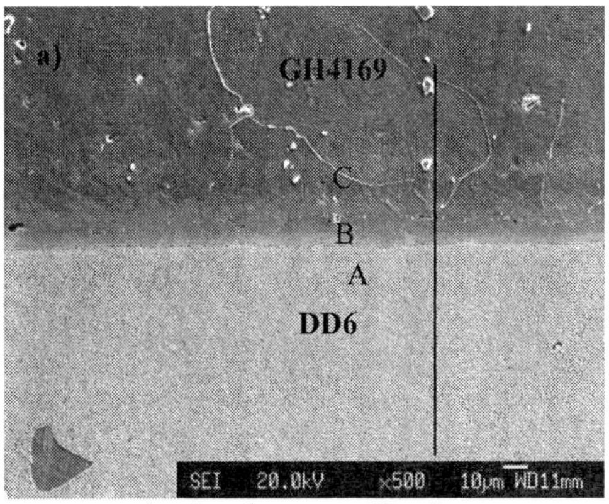

a) At a low magnification

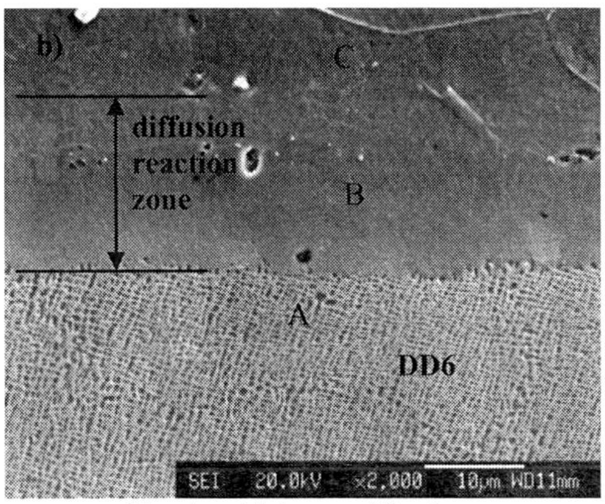

b) At a high magnification

Figure4 Microstructure of the diffusion bonded joint at 1140℃

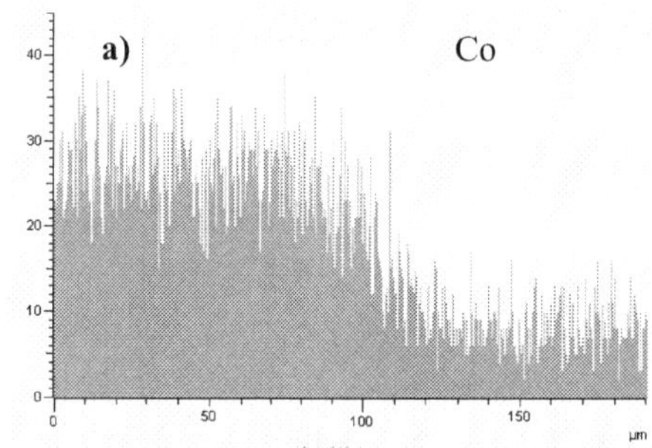

a) Distribution diagram of Co

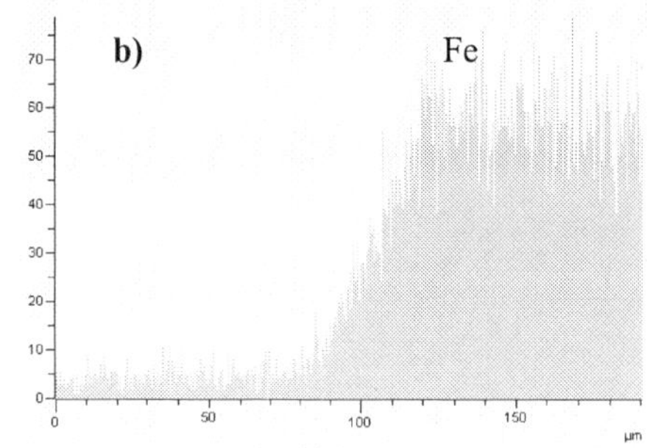

b) Distribution diagram of Fe

Figure5 Line distribution maps of elements Co and Fe (Figure4a)

Microstructure of the joint subjected to post-bonding heat treatment

Generally, post-bonding heat treatment is an effective way to stimulate element diffusion. The diffusion treatment of some specimens bonded at 1120℃ is 900℃ for 4h, followed by air cooling. Figure6 shows the heat treated joint microstructure and EDS analysis results for marked microzones are listed in

Table 5. It is clear that five distinctive layers constitute about 35μm width diffusion reaction zone. According to the second Fick's law, the elements Cr, Fe, Mo, Ti and Nb of GH4169 superalloy in higher concentration migrated to DD6 single crystal superalloy during the treatment process, while elements Co, W, Al, Ta, Re moved with an opposite direction. Due to the addition of elements Nb and Ti, a mass of phases γ' were promoted to be separated out in phase "B" which still kept sufficient amount of Al [10]. Then microstructure of phase "B" is characterized as a eutectic structure. The light gray phase "C" containing fewer elements Al and Ta can be inferred γ solid solution [12]. It can be concluded that zone "D" is the centerline of the joint where element Cr dismissed and Si gathered. The appearance of phase "E" and "F" varied for different contents of "Co". In addition, phase "F" can attribute distinctions from phase "G" to the absence of the strengthened phase caused by high temperature. Seen from the line distribution maps, elements Co and Fe both diffused more than 60μm.

Table5 EDS analysis results for the mirozones marked in Figure4a (at/%)

zone	Cr	Fe	Co	W	Mo	Al	Ti	Si	Ni	Ta	Re	C	Nb	Possible phase
A	3.70	-	8.73	3.46	0.70	9.83	-	-	71.51	1.43	0.62	-	-	DD6
B	4.27	1.37	6.82	3.11	0.93	9.50	0.27	-	71.28	1.44	-	-	1.00	γ + γ'
C	7.05	3.45	6.80	2.92	1.22	6.47	0.30	-	68.77	0.99	0.56	-	1.45	γ phase
D	4.69	6.67	4.56	1.25	1.18	5.14	0.42	0.46	68.23	-	-	-	1.64	γ phase
E	12.95	9.16	2.22	-	1.68	3.37	0.55	-	67.70	-	-	-	1.83	γ phase
F	15.68	12.07	-	-	1.62	1.57	0.68	-	66.23	-	-	-	2.14	γ phase
G	15.97	12.21	-	-	1.73	1.07	0.86	-	65.25	-	-	-	2.90	GH4169

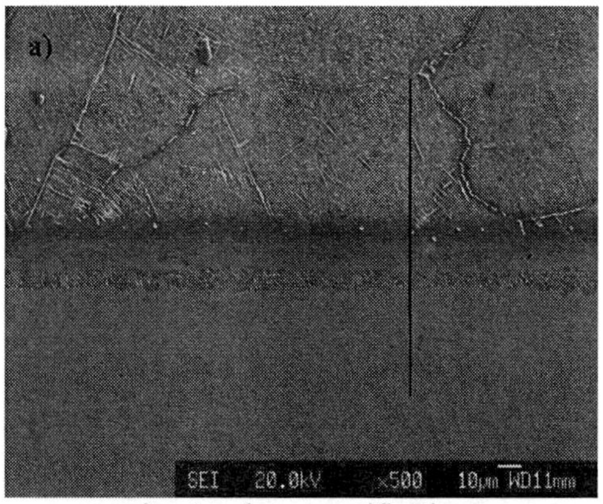

a) At a low magnification

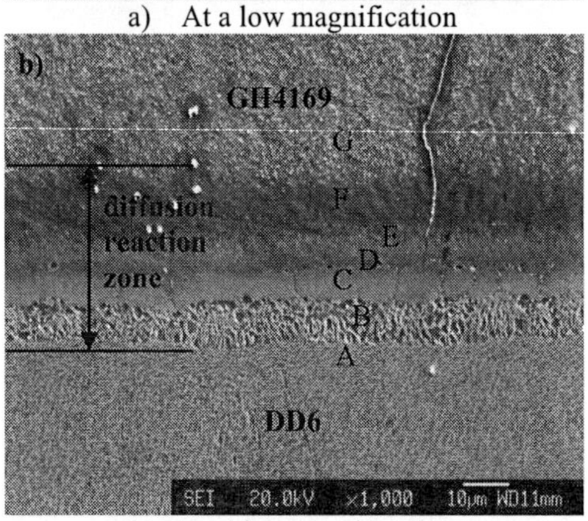

b) At a high magnification

Figure6 Microstructure of the joint subjected to post-bonding heat treatment

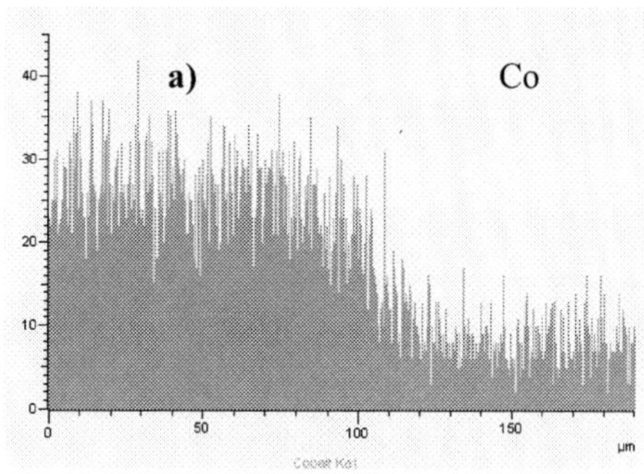

a) Distribution diagram of Co

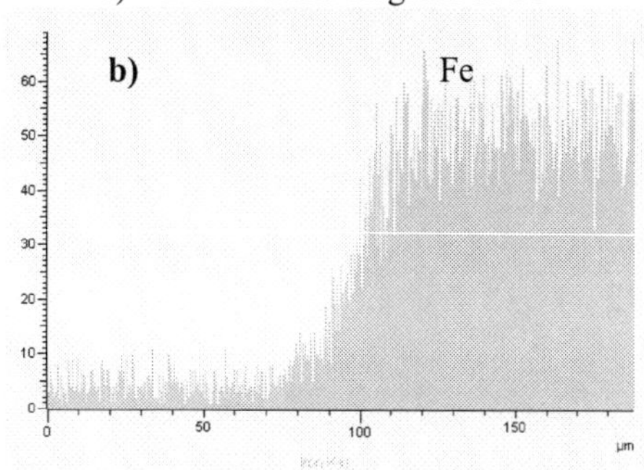

b) Distribution diagram of Fe

Figure7 Line distribution maps of elements Co and Fe (Figure6a)

Figure8 shows the comparison of diffusion reaction zone widths and element diffusion distances of the joints under different bonding conditions. These statistics reflect the extent of the diffusion reactions. Both the widest diffusion reaction zone and the longest elements diffusion distances belong to the heat treated joint. And the value of the joint bonded at 1140℃are all more than these of the joint bonded at1120℃. Conclusions can be reasonably drawn: Higher bonded temperature can stimulate the diffusion reaction for this dissimilar superalloys bonding. And the heat treatment had a positive effect on the diffusion reaction.

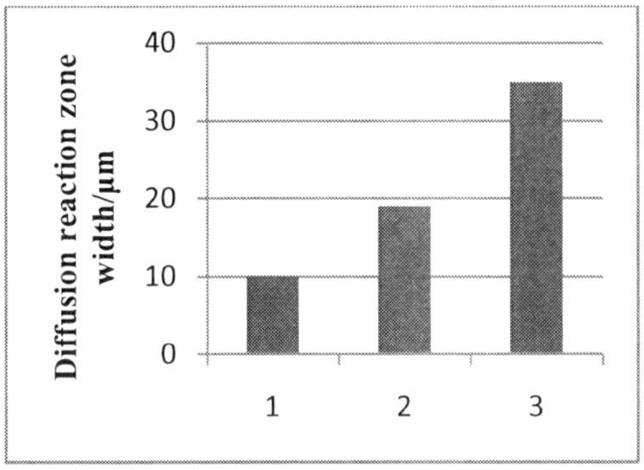

a) Diffusion reaction zone width(μm)

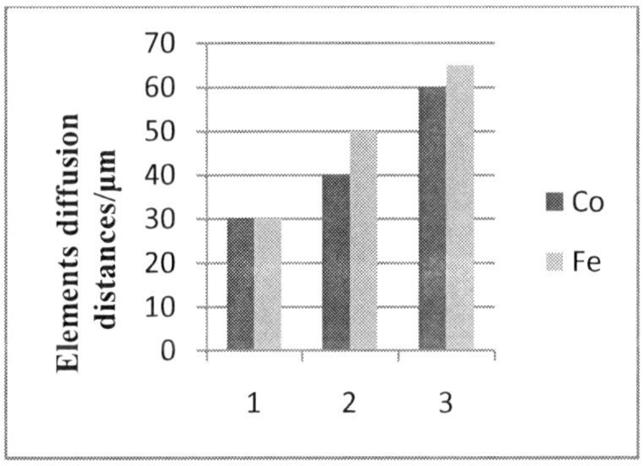

b) Element diffusion distances (μm)

Figure8 Comparison of diffusion reaction zone widths (a) and elements diffusion distances (b) of the joints under different bonding condition
(1Joint bonded at 1120℃, 2Joint bonded at 1140℃, 3 Joint subjected to the heat treatment)

Joint property

Figure9 presents the tensile strength and ductility of different joints at room temperature. The average tensile strength at room temperature of diffusion bonded joint under the condition of 1120℃/30MPa/15min and 1140℃/30MPa/15min reached 522MPa and 678MPa, respectively and the fractured position is the joint interface. Seen from Figure10, the fracture presents ductile characteristics and the average elongation rate of the joint bonded at 1140℃ reached 14.83%, much higher than that at 1120℃ 6.65%. After heat treatment, the average tensile strength at room temperature of the joint has greatly improved (955MPa). The fracture occurred in GH 4169 superalloy base metal, shown in Figure11. The joint elongation exhibits the elongation of GH4169 at this time. It can be speculated that the distinctions of dissimilar material (physical properties, chemical properties, mechanical properties, thermal properties and fractural properties) generate plenty of problems that worsened the strength of the joints [13]. But when the elements diffuse sufficiently, the composition of base metal near the interface gradually get close to the diffusion bonding seam. Then the differences of properties turned to decrease and correspondingly the joint strength certainly has been improved. Explain the initial steps in your preparation and analysis.

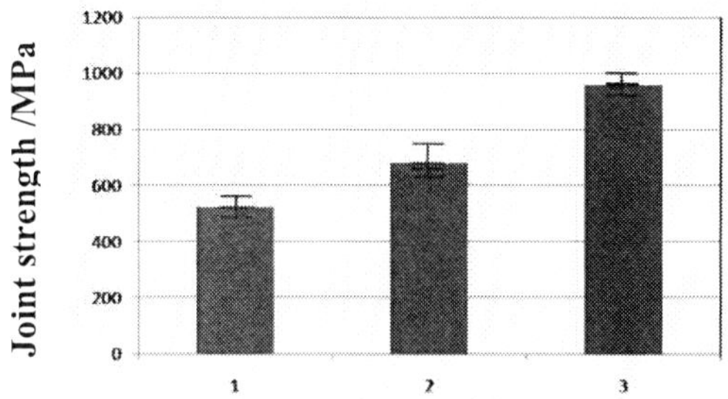

a) Tensile strength(MPa)

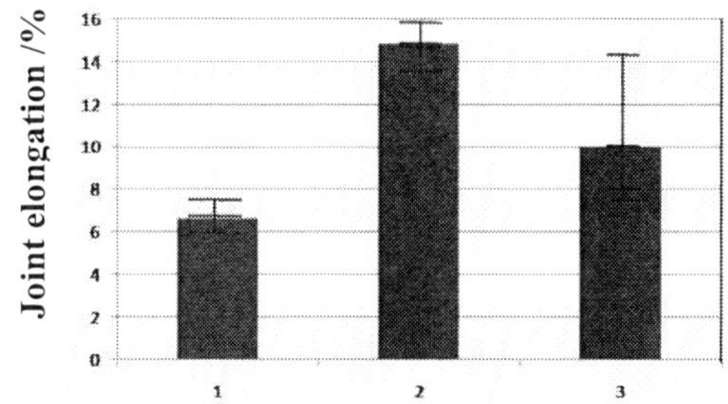

b) Elongation rate(%)

Figure 9 Comparison of tensile strength (a) and elongation rate (b) of the joints under different bonding condition
(1 Joint bonded at 1120℃ 2 Joint bonded at 1140℃ 3 Joint subjected to the heat treatment)

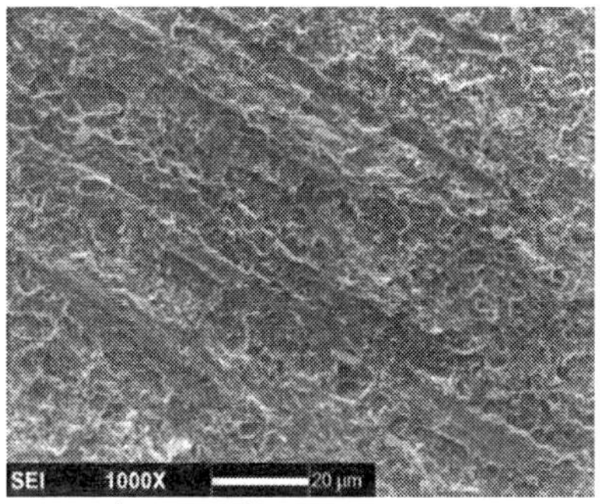

a) 1120℃/30MPa/30min

b) 1140℃/30MPa/30min

Figure10 Microscopic fracture morphology of joints with different diffusion bonding parameters

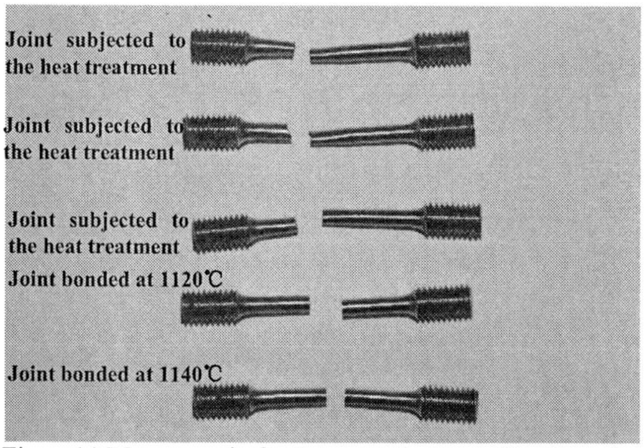

Figure11 Macroscopic fracture morphology of joints under different bonding condition

Conclusions

(1) Sound joints between GH4169 and DD6 were successfully achieved under the condition of 1120℃/30MPa/15min and 1140℃/30MPa/15min, respectively.

(2)Joint bonded at 1140℃ has a wider diffusion reaction zone(about 19 μ m) and longer elements （Co and Fe） diffusion distance than that bonded at 1120℃. And it was found that the reaction layer was mainly composed of γ solid solution, without strengthening phase.

(3)The average tensile strength at room temperature of diffusion bonded joint under the condition of 1120 ℃ /30MPa/15min and 1140℃/30MPa/15min reached 522MPa and 678MPa, respectively.

(4)The post-bonding heat treatment of 900℃/4h can greatly stimulate the diffusion reaction and improve mechanical properties. And the average tensile strength at room temperature reached 955MPa. The fracture occurred in GH 4169 superalloy base metal.

References

[1] R.L Kennedy, "Alloy-718 plus Superalloy for the Next Forty Years Sixth," International Symposium on Superalloy 718,626,706 and Derivatives, Pittsburgh, PA, October, 2005, pp.1-14.

[2] Dae-Ho J., Myung-Je C., Masahiro G. et al, "Effect of service exposure on fatigue crack propagation of Inconel 718 turbine disc material at elevated temperatures," Materials Characterization, Vol. 95, No. 0 (2014), pp. 232-244.

[3] Chang, S. H., Lee, S. C., Tang, T. P., "Evaluation of HIP pressure on Inconel 718 superalloy," International Journal Cast Metals Research, Vol. 19, No. 3 (2006), pp. 181-187.

[4] Minggao Y., Bocao L.et al, China Aeronautical Materials Handbook, China Standards Press, (Beijing, 2002), pp. 812- 818.

[5] Fengli S., Jiarong L., Shizhong L. et al, "Microstructure and Mechanical Property of Single Crystal Superalloy DD6 with Different States," Journal of Iron and Steel Research, Vol. 15, No. 1 (2003), pp. 23-25.

[6] Li J. R., Zhong Z. G., Tang D. Z., et al, "A low-cost second generation single crystal superalloy DD6," The Minerals, Metals & Materials Society. Vol. 1, No. 1, 2000, pp. 87-93.

[7] Wenbo H., Kaifeng Z., Guofeng W., "Research on microstructures and properties of diffusion bonding joints for Inconel 718 superalloy," Materials Science and Technology, Vol. 13. No. 3 (2005), pp. 308-311.

[8] A. Thomas, M. E. Wahabi, J. M. Cabrera, "High temperature deformation of Inconel 718," Journal of Materials Processing Technology, Vol. 1, No. 77 (2006), pp. 469-472.

[9] Minggao Y., Bocao L.et al, China Aeronautical Materials Handbook, China Standards Press, (Beijing, 2002), pp. 323-359.

[10]Yunrong Z., De T., Metallographic Investigation of Superalloys and Steels, National Defense Industry Press, (Beijing, 1999), pp. 5-10.

[11] Yan W, Lin L, Wenzhu S.et al, "Effect of solid-solution treatment on microstructure and performance of GH4169 superalloy," Transactions of Materials and Heat Treatment, Vol. 28, No. 7 (2007), pp. 8-10.

[12] Yuan Z, Wei M, Xiaohong L., "Microstructure and Mechanical Proper t ies of Single Crystal Superalloy DD6 Joint Brazed with BNi82CrSiB Filler Metal," Materials Engineering, Vol. 5, No. 16 (2007), pp. 46-49.

[13] Yanhua Z., Jiaqi L., Xiaoxue D., "Interface mechanics in the dissimilar material joints," Electric Welding Machine, Vol. 37, No. 7 (2007), pp. 33-38.

Measurement of Magnetic Flux on Modified Diamond Surface Using Vanadium

N.Tanaka
Tokyo Institute of technology, Graduate Student,
2-12-1 O-okayama Meguroku, Tokyo, Japan

T.Yamazaki, T.-T.Ikeshoji, A.Suzumura
Tokyo Institute of technology, Department of Mechanical and Aerospace Engineering
2-12-1 O-okayama Meguroku, Tokyo, Japan

Abstract

For joining diamond to a metal, a surface cleaning and a modification are necessary for diamond to adhere to metal by infrared ray heating with accreted molecular vibration. It is assumed that the magnetic properties of diamond change after the surface modified process using transition material elements. In this study, we measured a magnetic flux on a modified diamond surface when we use manganese or vanadium before and after the surface treatment. The silver paste mixture containing vanadium was printed on the diamond surface so that diamond does not cause a dewetting.

X-ray diffraction analysis revealed Mn_7C_3 and Mn_5C_2 formation on diamond (110) modified by manganese. From the result of the VSM measurement, B-H curve shows a hysteresis which means the modified surface by manganese has ferrimagnetism. We guess the formation of V_4C_3 on diamond (111) change magnetic property of the diamond (111), and would be measured magnetic flux by a Hall probe.

Introduction

Ornament diamond is used for jewel as the most valuable jewel for a long time, fixed it by claws. The physical joining by the method of the claws is not effective to align a lot of diamonds with various shapes. Thus, the chemical and metallurgical bonding would be replaced for its effectiveness in manufacturing of the diamond jewels, since the carbon has covalent bonds. Active metal brazing filler is used to make the joints. When we let this active metal react with carbon on a diamond, it formed reaction product with the diamond and the filler metal[1]. We chose Mn and V due to the results of strength test and X-ray diffraction analyses. According to the previous X-ray diffraction, Mn_xC_y (Mn_7C_3 and Mn_5C_2) had its highly oriented to (110), and V_xC_y (V_4C_3) had its highly orientated to (111) [2, 3]. At this time, manganese forms mono layer, and vanadium get into the diamond. Initial coordinates of V_4C_3 for numerical calculation are determined based on the V_8C_7 lattice space of NaCl type crystal. The initial atomic distance between V and C in V_4C_3 was set to 2.08 Angstroms (In Fig. 1, V^{3+} exist at the V_4C_3 surface). In addition, we conducted condensed infrared heating to enhance molecular vibration for the changing the binding of diamond surface. For this heating method, diamond was considered to be modified at the 770 K from previous study [4].

Then, in this study, we performed the susceptibility measurement for the magnetic field by the method of using simple magnetic flux measurement using the Hall probe. The magnetron number of V is approximately 0 in a calculation, but it is thought that Hall voltage is appeared in experiment. Hall probe is a magneto-metric sensor of semiconductor which can output magnetic field as voltage. The voltage between Hall probe occurs when electric current exists when magnetic field is applied to semiconductor. This voltage is called Hall voltage. However, it is necessary to amplify it with an amplifier because this output voltage is feeble. After that, we performed more detailed magnetic measurement using vibrating sample magnetometer and clarified a magnetic property.

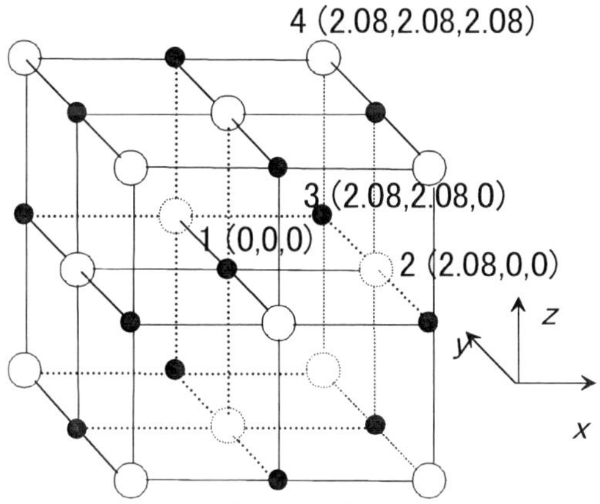

Figure 1 Coordinates of the numbering atoms in the NaCl type lattice model of reaction product V_4C_3 based on V_8C_7

Experimental

Specimen A in Figure 4 was produced using manganese powder (mesh size was less than 100 μm). This manganese powder was spread in a platinum cell 2mm in diameter. Then, artificial single crystal diamond (110) was mounted on the powder.

Specimen B in Figure 4 was produced using V and silver nano ink paste. Silver nano ink paste which was the mixture of micro silver particles and nano silver particles coated with organic film, dispersed into alcohol was used in this experiment. As for micro particles which diameter was about 1 μm and for nano particles was about 10 nm. Figure 2 shows the schematic illustration of specimen fixed the diamond to the alloy plate. At first, we produced silver-copper eutectic alloy plate into 15×70mm, and appropriate amount of the paste was applied on it. Vanadium hydride powder was screenprinted on it, and artificial single crystal diamond (111) was mounted on the paste. Then, the diamond was pressed lightly so that bonding surface contacted with the paste and the specimen was baked at 423 K in ambient air atmosphere.

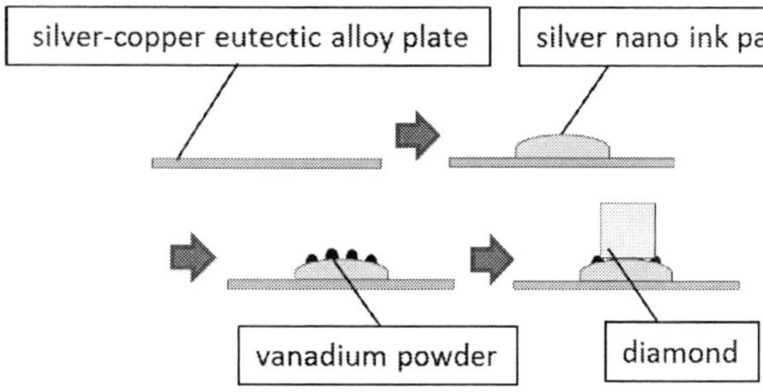

Figure 2 Schematic illustration of specimen jointed the diamond to the alloy plate

Then, these specimens were heated by infrared irradiation through diamond. The spot diameter was 10 mm, and it was focused on the silver ink. The schematic illustration of the infrared heating device is shown in Figure 3. The ultra-high vacuum chamber contains a viewport flange which the infrared radiation passed. The thermocouple that was fixed on the quartz stage, made contact with the back side of the eutectic plate, so as to measure its temperature without direct irradiation by infrared rays. And inside of the chamber was evacuated until it becomes to 10^{-5} Pa .The specimens were heated near 770 K, and then annealed.

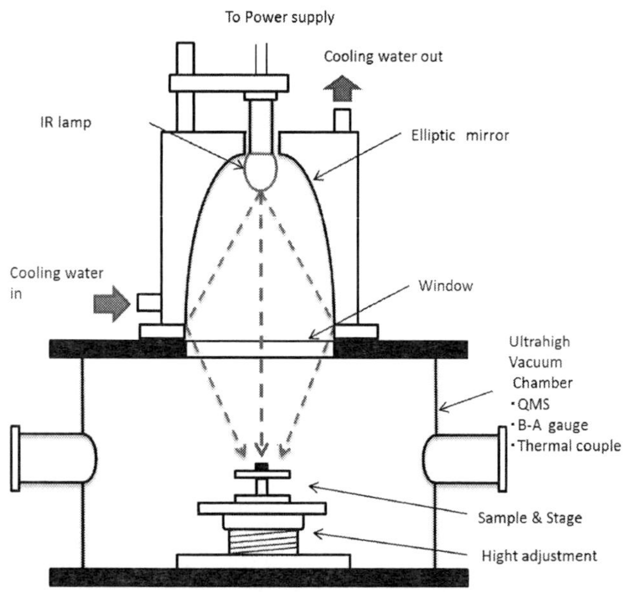

Figure 3 Infrared rays heating device

These specimens were fractured in the interface of the cell or the plate and they were measured potential differences by using simple magnetic flux measurement. This device has plural DC amplifiers and a differential amplifier which were connected. The Hall probe was fixed in a dewar flask. Then, we put diamond on at hall device and flow current to hall device. Hall voltage which was amplified was measured a by voltmeter. In addition, we measured at room temperature (about 289 K) and low temperature (about 270 K). When we measured at low temperature, the dewar flask was filled with liquid nitrogen.

And then, we measured the details of the magnetic property by VSM. In this time, we measured at room temperature.

Result

Figure 4 shows the Optical microscope image of specimens. The color of specimen A turned black, and both specimens had a metallic luster.

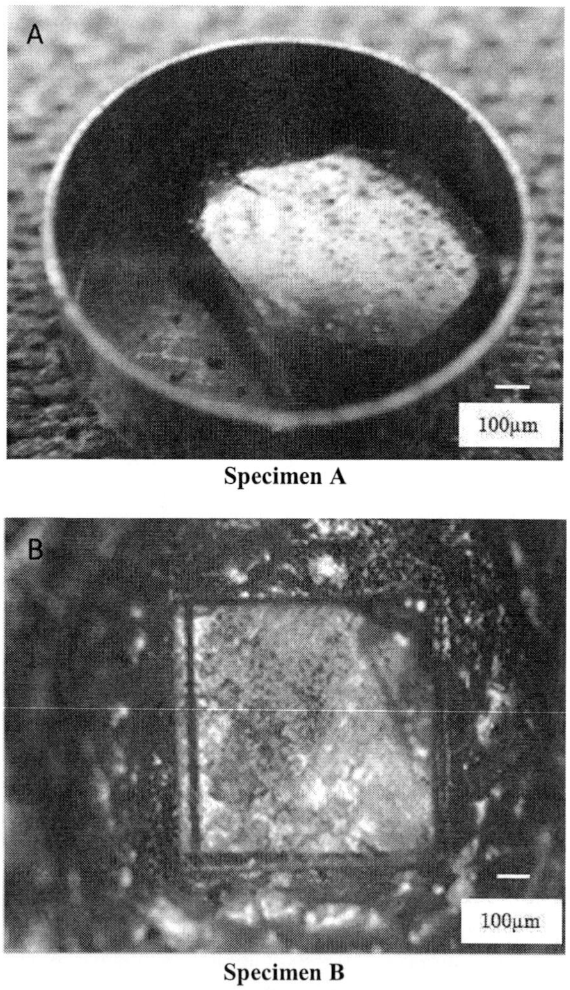

Figure 4 Optical microscope images of specimens A and B

Figure 5 shows the adherents on the diamonds interface after the fracture. The both specimens had metallic luster, seen especially around the deposited at specimen B. It is considered that nano particles of silver were adhered to the diamond by the improvement of free energy.

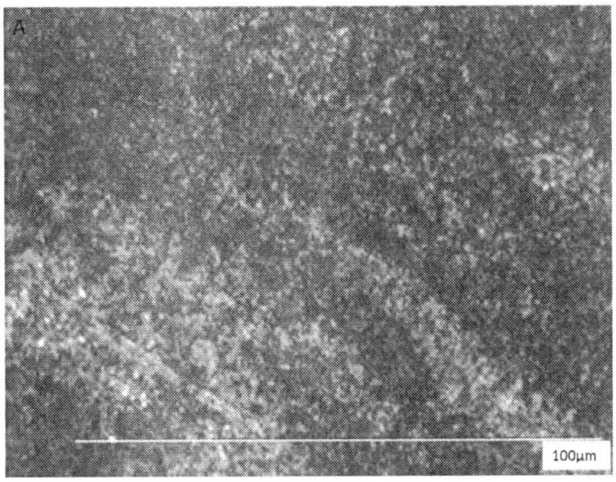

Specimen A

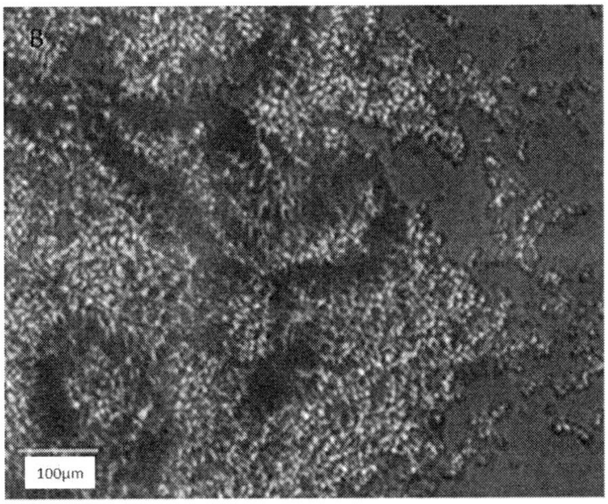

Specimen B

Figure 5 Optical microscope images of adherent on diamond interface

Table 1 and Table 2 show the results of simple magnetic flux measurement using the Hall probe. From these tables, the Hall voltage appeared for a clear difference at both temperatures at specimen B. In addition, it was feeble but the Hall voltage of specimen A was detected, too.

Table 1 The result of Hall probe voltage amplified in case of Specimen A

Sample	At room temperature (289 K)	At low temperature (270 K)
Nothing	174 mV	204 mV
Diamond (modifying surface)	175.5 mV	207.5mV

Table 2 The result of Hall probe voltage amplified in case of Specimen B

Sample	At room Temperature (289 K)	At low temperature (270 K)
Nothing	103 mV	99 mV
Diamond (non-modified surface)	107 mV	109 mV
Diamond (modified surface)	122 mV	122 mV

Figure 6 shows the result of the VSM. From this result of specimen A, though the diamond had originally structure of the diamagnetism, it had characteristic of spontaneous magnetization. This is because magnetic characteristic of the diamond surface was changed. The specimen B didn`t show the spontaneous magnetization from B-H curve. It was revealed that a magnetic property was not appeared by at least room temperature because magnetism is small.

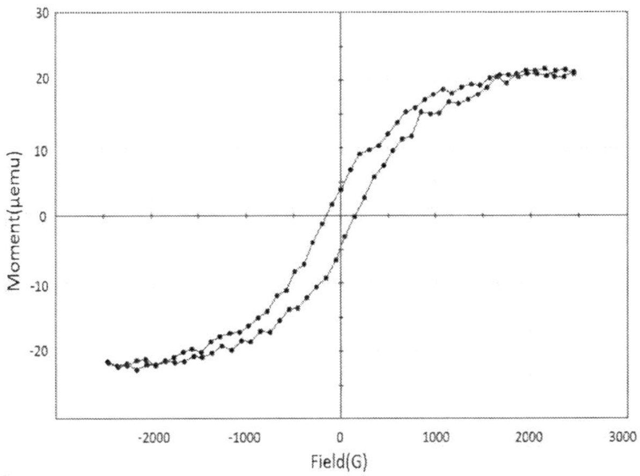

Specimen A

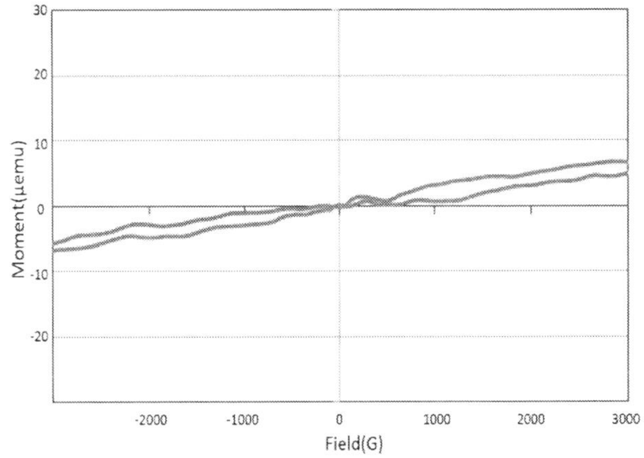

Specimen B

Figure 6 B-H curve of diamond modified by vanadium

From the results of simple magnetic measurement using Hall probe and VSM, the magnetic property existed by the modified surface using manganese and vanadium. When we used manganese, it is thought that magnetic characteristic of the diamond surface was changed. In addition, it is estimated that the magnetism is ferrimagnetism. However, the specimen used vanadium was not able to identify to the characteristic in VSM. It is thought that this is because the diamagnetism of the diamond is strongly detected when it was tested at room temperature. Being conscious of a Curie temperature, it will be necessary to analyze it in cryogenic temperature.

Discussion

1.Idea of silver coating on diamond surface by vanadium

To improve of the bonding and the refractive index of diamond interface, we add vanadium in the bonding interface as an electronic donor. Figure 7 shows the process. When we add vanadium, and heating, the vanadium ionizes. And then, vanadium which ionized moves to the diamond surface by improvement of intermolecular force and silver moves to these vanadium by surface energy. In this process, distance of the diamond and Ag particles getting small and diamond get high refractive index.

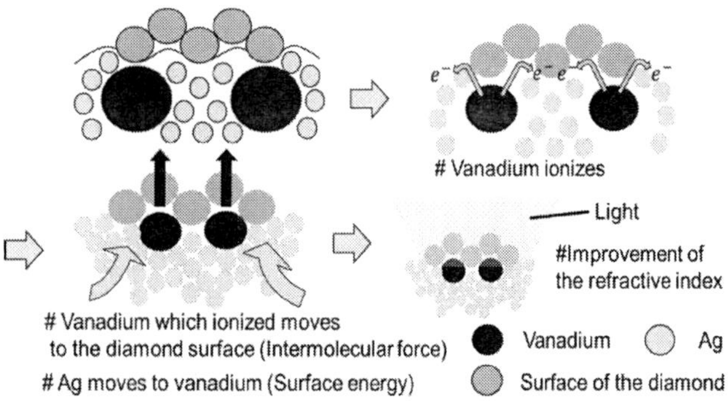

Figure 7 The process of silver coating on diamond surface by using vanadium

2. Effective magnetron number of 3d transition ion

The orbit angular momentum of the 3d transition ion which is related with magnetic moment is quenched in the value of effective magneton. Table 1 shows the value of effective magnetron number of 3d transition ion. P_j is the value of composition of orbit angular momentum and the spin angular momentum, P_s is the value of only for spin angular momentum and P_e is measured value. From this table, magnetic moment of 3d transition ion is depended on by only spin angular moment.

Table 3 The value of effective magneton number of 3d transition ion [5]

ion	orbital	P_j	P_s	P_e
Ti^{3+} , V^{4+}	$3d^1$	1.55	1.73	1.8
V^{3+}	$3d^2$	1.63	2.83	2.8
Cr^{3+} , V^{2+}	$3d^3$	0.77	3.87	3.8
Mn^{3+} , Cr^{2+}	$3d^4$	0	4.90	4.9
Fe^{3+} , Mn^{2+}	$3d^5$	5.92	5.92	5.9
Fe^{2+}	$3d^6$	6.70	4.90	5.4

Conclusion

For joining diamond to a metal, a surface cleaning and a modification are necessary for diamond to adhere to metal by infrared ray heating with accreted molecular vibration. It is assumed that the magnetic properties of diamond change after the surface modified process using transition material elements. In this study, we measured a magnetic flux on a modified diamond surface when we use manganese or vanadium before and after the surface treatment. The silver paste mixture containing vanadium was printed on the diamond surface so that diamond does not cause a dewetting. From previous X-ray diffraction, Mn_xC_y (Mn_7C_3 and Mn_5C_2) was found to closely correspond in its orientation to (110), and V_xC_y (V_4C_3) was understood that its orientation is high to (111).

We performed the susceptibility measurement for the magnetic field for manganese and vanadium by the method of using simple magnetic flux measurement using the Hall probe, The Hall voltage appeared for a clear difference at ordinary temperature and low temperature using liquid nitrogen. However, from the result of the VSM measurement, B-H curve didn`t show the spontaneous magnetization. It was revealed that a magnetic property was not appeared by at least ordinary temperature. It is thought that this is because the diamagnetism of the diamond is strongly detected when I tested it at normal temperature. Being conscious of a Curie temperature, it will be necessary to analyze it in cryogenic temperature in future.

References

[1]M.D.Drory, J.W.Hutchinson, "Diamond Coating of Titanium Alloys", Science, vol.263, (1994), pp.1753-1755

[2]H.Tsuda, T.Yamazaki, T.-T.Ileshoji, A.Suzumura, Change of magnetic properties on modified diamond surface with manganese powder, Interfacial Joining and Surface Technology, (2013), pp.107-108

[3]T.Yamazaki, A.Suzumura, T.-T.Ileshoji, T.Ishiguro, "Role of valence electron of vanadium carbide in metallization of diamond surface", Proceedings of the 4th International Brazing and Soldering Conference, Las vegas, NV, April. 2009 pp. 291-295

[4]T.Yamazaki et al, "Improvement of wettability and joint strength using Sn-In-Al-Si-V solder for diamond", 8th International Conference on Brazing, Aachen, Germany, June. 2007, pp.40-44

[5]H. Nose, T. Sato, The fundamentals of magnetism, Shokabo, (Tokyo, 1997), pp.34-35 (in Japanese)

Microstructural Characterization of Single Crystal Superalloy CMSX-4 Brazed Joints with BNi-2 and BNi-9 Filler Metals

Bryan Riggs, Boian Alexandrov, Avraham Benatar
Welding Engineering, Ohio State University, Columbus OH, U.S.A.

Ray Xu
Rolls-Royce, Indianapolis, IN, U.S.A.

Abstract

In this study, metallurgical characterization was performed on lap joint specimens of CMSX-4, vacuum furnace brazed with BNi-2 (AMS-4777) and BNi-9 (Amdry 775) filler metals. Both Ni-Cr-B based filler metals produced joints with two distinct microstructural regions, an isothermally solidified region and a multi-component center-line eutectic. Compositional gradients and elemental distribution across the joint were analyzed with Electron Probe Micro-Analysis (EPMA). Micro-hardness tests were performed across the joint area. Results showed very little diffusion of base metal elements into the joint and substantial boron diffusion into CMSX-4. The boron diffusion region had little effect on the gamma prime γ' strengthening precipitates, but showed a slight increase in micro-hardness. The data and methodology generated will provide a basis for thermodynamic simulations (discussed in a separate paper of this collection [1]) for prediction of joint microstructure in brazed turbine components.

Introduction

Brazed joints are commonly used for the manufacture and repair of high temperature gas turbine components made of nickel based superalloys [2]. Superalloys, like commercially available CMSX-4, are well known for their excellent high temperature properties due to the high volume fraction of gamma prime (γ') strengthening precipitates [3]. In brazing, filler metal with melting point depressant (MPDs) elements, such as boron or silicon, added to their composition melt at significantly lower temperatures than the γ/γ' base material. The diffusion of these elements into the base metal creates a strong metallurgical bond upon eutectic solidification of the filler metal. Typical microstructures of brazed joints consist of an isothermally solidified region that forms along the interface due to diffusion of MPDs across the interface and a centerline eutectic region that forms upon solidification containing brittle intermetallic phases [4]. When modeling and designing brazed joints it is important to account for the size and distribution of these microstructural regions, for their properties and for properties across the joint.

As a result, the main areas of investigation for this work includes: (i) composition and elemental distribution of intermetallic phases and regions, (ii) diffusion of MPDs in CMSX-4, (iii) micro-hardness properties across the joint and of individual phases.

Experimental Procedure

The same brazing procedure used for thermodynamic simulations in [1] was investigated. CMSX-4 single crystal nickel-based superalloy plate specimens were vacuum furnace brazed with Ni-Cr-B based filler metals, BNi-2 and BNi-9, in a lap shear configuration. Table 1, presents the chemical composition for these materials in wt%. These combinations were chosen to represent characteristic joints between single crystal superalloys brazed with a Ni-Cr-B based filler metals, which have a wide range of industrial applications. A laser tack weld was used to maintain a joint gap between 0.002-0.004" and initial brazing cycles were carried out in a high vacuum ($<10^{-5}$torr) furnace at Rolls Royce facilities with a 10 minute hold time at brazing temperatures of 1066°C and 1121°C for BNi-2 and BNi-9 respectively. Joint samples were cut, mounted and polished both normal to the joint and using a low angle micro-sectioning (LAMS) technique [5]. Optical and scanning electron microscopes were used for imaging, and compositional analysis was performed using Electron Probe Micro-Analysis (EPMA). Micro-hardness maps were also performed across the brazed regions.

Table 1: Chemical composition of materials investigated and liquidus/solidus temperatures for BNi-2 and BNi-9 filler metals.

Alloy	Composition (wt%)										
Superalloy Base Metal											
CMSX-4	Ni	Cr	Re	Co	Mo	W	Ta	Ti	Al	Si	Mn	Others
	balance	6.4	2.9	9.6	0.61	6.4	6.5	1.02	5.61	0.4	0.1	0.05 Zr, 0.1 Hf, 0.042 Fe, 0.0037 C
Brazing Filler Metal											
BNi-2 (AMS-4777)	Ni-7Cr-3.1B-4.5Si-3Fe											Liquidus/Solidus: 971°C/999°C
BNi-9 (Amdry 775)	Ni-15Cr-3.5B											Liquidus/Solidus: 1020°C/1055°C

Results and Discussion

Standard brazing cycles produced excellent joints for both filler metals with no visible porosity, other than a few micro-pores in the fillet regions. Figure 1 shows typical examples of the joint microstructure for both BNi-2 and BNi-9 filler metals. The isothermally solidified region was approximately 15-25 μm in width and appeared as a white phase along the joint interface with a multi-component centerline eutectic region. The BNi-2 eutectic region consisted of two distinct intermetallic phases within a multi-phase matrix while the BNi-9 joint showed only one bulky (~80μm in diameter) intermetallic phase in a finer eutectic matrix.

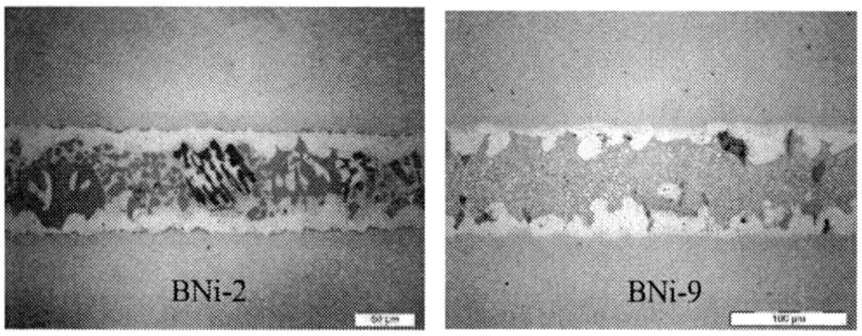

Figure 1: Final joint microstructure of CMSX-4 vacuum furnace brazed with BNi-2 and BNi-9 filler metals.

Low angle micro-sectioning (LAMS) applied to the narrow joints (~100 um) revealed nearly 20x more joint area than a normally sectioned sample shown in Figure 2. It should be noted that the volume fraction of the darker intermetallic phase was observed to increase when the LAMS technique was applied. This would indicate that the shape of this phase is oriented parallel to the joint interface and the stereology of these phases is an important consideration when characterizing brazed joint microstructures.

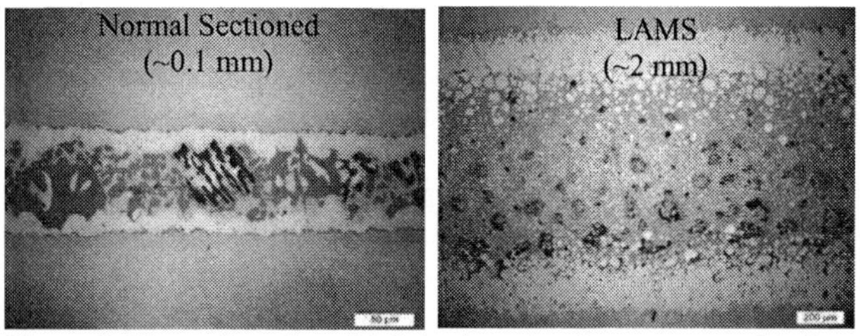

Figure 2: Normal sectioned joint microstructure compared to low angle micro-section (LAMS) for BNi-2 joint.

Figure 3 shows the results of the EPMA line scan across the BNi-2 brazed joint. The composition given in wt% is split into CMSX-4 elements (Al, Ta, W, Re, Ti, Co) on the top plot and BNi-2 elements (Ni, Cr, B, Si, and Fe) in the bottom plot. From these results the two intermetallic phases in the joint were identified as CrB (dark phase) and Ni_3B (gray) type borides. The eutectic matrix surrounding the Ni_3B and CrB borides was rich in BNi-2 elements and the isothermally solidified zone consisted of a Ni-rich solid solution (α-Ni) phase increased iron, chromium and silicon content. Average chemical compositions and approximate volume fraction of these four primary phases in the BNi-2 joints are listed in Table 2 and labeled in the BNi-2 microstructure in Figure 4. Phase compositions showed strong correlation with EPMA results reported by Vargas et al. on BNi-2/Ni couples [6]. CMSX-4 base metal elements were not detected in the joint

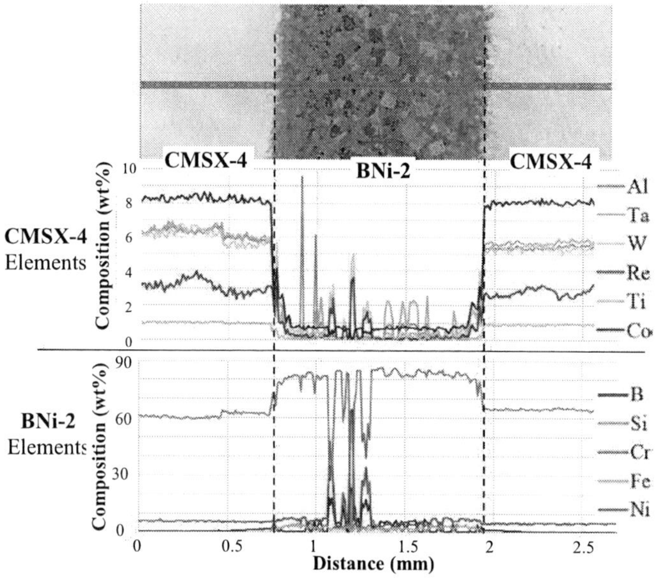

Figure 3: EPMA line scan across BNi-2 brazed joint for CMSX-4 elements (top) and BNi-2 elements (bottom).

region and their composition dropped immediately at the interface aside from a tungsten/rhenium enrichment in the CrB borides and tantalum enrichment in the Ni_3B type borides. The composition of titanium and aluminum, important elements for creating the γ' strengthening precipitates in superalloys, was significantly reduced in the joint region.

Figure 5 shows similar EPMA line scan results across the LAMS sample for the BNi-9 brazed joint. As expected, the Cr and B content increased in the joint area for this filler metal.

Table 2: Average composition and volume fraction of phases in BNi-2 brazed joint.

Phase	Amount (vol. fraction)	Chemical Composition (wt%)					
		Ni	B	Cr	Si	Fe	Others (W, Re, Al, Ti, Co, Mo)
CrB	0.13	1.9	24.3	62.2	0.01	0.4	11.1
Ni_3B	0.22	81.9	6.1	4.2	0.5	1.6	5.7
α-Ni	0.40	81.0	0.4	7.0	3.7	3.1	4.8
Eutectic	0.25	83.7	5.5	3.7	2.0	1.6	3.5

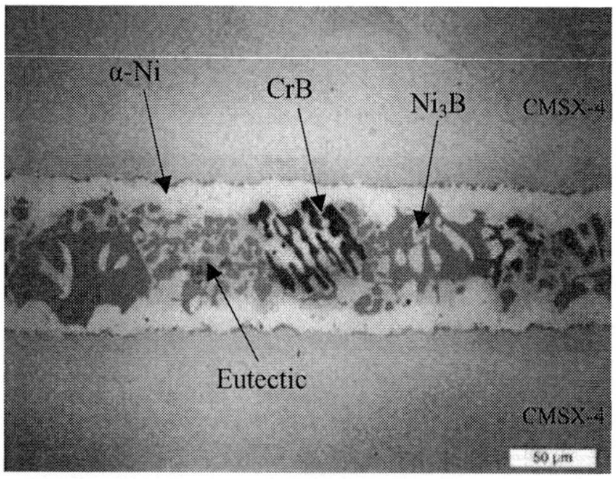

Figure 4: Identified phases in CMSX-4 brazed joint using BNi-2 filler metal.

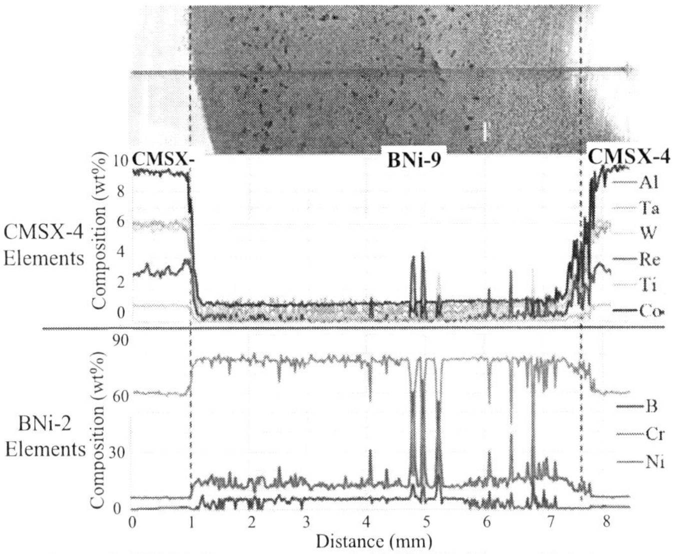

Figure 5: EPMA line scan across LAMS BNi-9 brazed joint.

The bulky intermetallic phase in these joints was identified as the same CrB type boride found in the BNi-2 joints. The rest of the joint consisted of a multi-component eutectic phase rich in Ni, Cr, and B and a similar α-Ni phase at the interface. These phases are labeled in the normal sectioned braze joint in Figure 6 and average composition and volume fraction of the BNi-9 phases are listed in Table 3. Again, CMSX-4 elements were not present in the joint region aside from rhenium and tungsten enrichment in the CrB borides.

The EPMA line scans across both joints show similar boron diffusion in the CMSX-4 base metal. Figure 7, shows the boron diffusion affected zone in CMSX-4 extends nearly 25 μm from the joint interface in both joints. Boride formation typically observed when brazing poly-crystalline Ni-base

Table 3: Average composition and volume fraction of phases in BNi-9 brazed joint.

Phase	Amount (vol. fraction)	Chemical Composition (wt%)					
		Ni	B	Cr	Si	Fe	Others (W, Re, Al, Ti, Co, Mo)
CrB	0.09	3.79	20.48	64.18	0.09	0.04	11.42
α-Ni	0.46	73.7	0.56	12.63	0.12	0.11	12.88
Eutectic	0.45	70.4	5.1	11.6	0.27	0.09	12.54

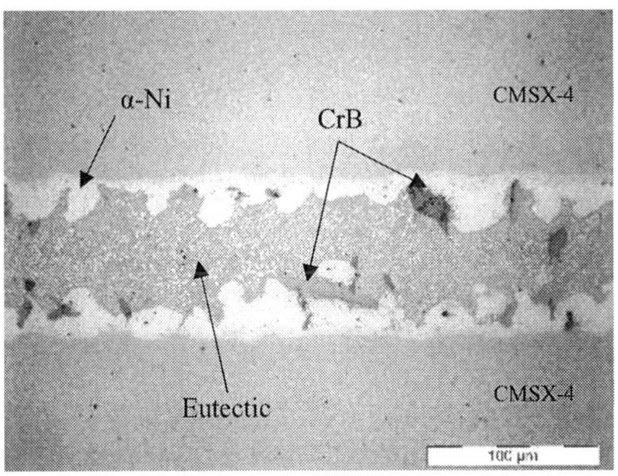

Figure 6: Identified phases in CMSX-4 brazed joint using BNi-9 filler metal.

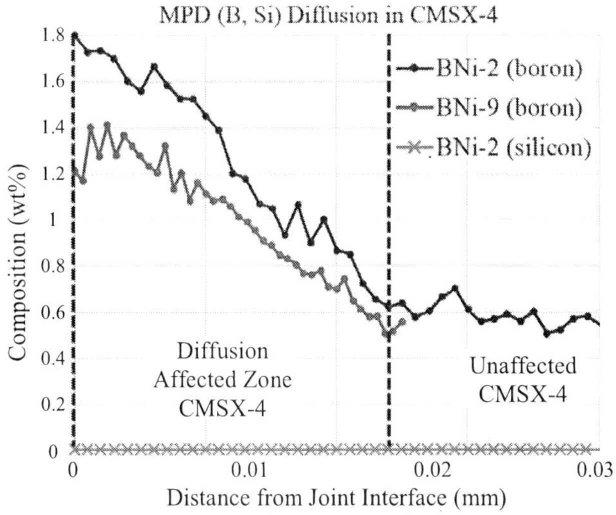

Figure 7: MPD composition in CMSX-4 near joint interface.

superalloys [7] was not observed in the CMSX-4 single crystal. Silicon which can act as an MPD in BNi-2 was not found to diffuse into CMSX-4 and remained enriched in the eutectic region. Gamma prime strengthening precipitates in CMSX-4 appear to be unaffected by the diffusion of boron. An SEM image of the etched BNi-2 brazed joint interface at high magnification in Figure 8 shows a consistent γ/γ' structure in the diffusion affected zone without precipitate coarsening or dissolution. A transition region of decreasing size and distribution of γ' was observed in the BNi-2 filler metal near the boundary line.

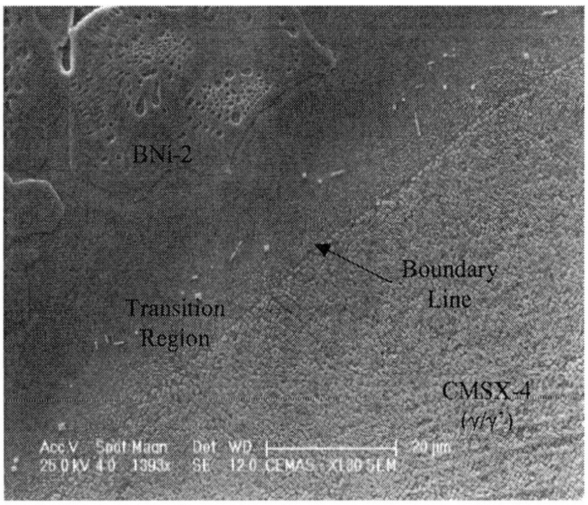

Figure 8: SEM image of CMSX-4/BNi-2 joint interface.

Figure 9 gives a contour plot of the micro-hardness measured in Rockwell C across the LAMS BNi-2 specimen with red areas indicating higher hardness and purple indicating low hardness. Similar profiles were made for BNi-9. In both joints a large mismatch between interlayer hardness properties was observed. The centerline eutectic constituent with CrB and Ni_3B borides was found to be extremely hard with RC of 63-68, while the α-Ni phase was much softer at 27 RC. The softening in the α-Ni region and the stress concentrations created by extremely hard intermetallics may need to be considered when designing these joints. Slightly higher micro-hardness values (50-55 RC) were found in the diffusion

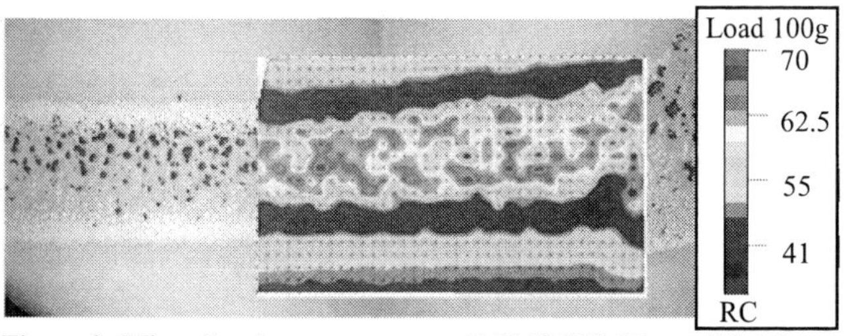

Figure 9: Micro-hardness map across LAMS BNi-2 brazed joint.

affected zone in CMSX-4 base metal regions affected by boron diffusion. Boron occupies interstitial lattice sites in the γ/γ' matrix which may account for the increased hardness in these regions[8]. This is an interesting result because a common concern when fusion joining nickel base superalloys is heat affected zone softening due to dissolution of γ' precipitates, grain growth or overaging [9]. With brazing instead of softening a harder diffusion affected zone is observed.

Conclusions

1. Vacuum furnace brazing CMSX-4 with BNi-2 and BNi-9 filler metals produced excellent joints consisting of isothermally solidified regions containing α-Ni solid solution phase and centerline eutectic region containing CrB, Ni_3B precipitates and a multi-component eutectic constituent.
2. Low angle micro-sectioning technique applied to brazed joints increases the observable joint area for improved characterization of the size and distribution of phases.
3. Boron diffusion was observed in the CMSX-4 base metal, but the γ/γ' structure remains unaffected aside from a local increase in hardness.
4. Micro-hardness testing showed a large mismatch in properties between the hard, brittle centerline eutectic and soft, ductile α-Ni phase.

Acknowledgments

This research has been supported by Rolls Royce and performed within the Center for Integrative Materials Joining Science for Energy Applications (CIMJSEA), a National Science Foundation, Industry-University Cooperative Research Center. EPMA work performed by Paul Carpenter at Washington University in St. Louis.

References

[1] Riggs, B. Alexandrov, B. Benatar, A. R, Xu. "Thermodynamic Simulations of Single Crystal Superalloy CMSX-4 Brazed Joints with BNi-2 and BNi-9 Filler Metals" IBSC 2015, The Ohio State University

[2] Rabinkin, A. "Overview: Brazing with (NiCoCr)-B-Si Amorphous Brazing Filler Metals: Alloys, Processing, Joint Structure, Properties and Applications." IBSC 2003, Honeywell Intl., Metglas Solutions.

[3] Donachie MJ, Donachi SJ. "Superalloys: A technical guide." S. 1: ASM International, Materials Park, 2002. 2nd edn.

[4] Brazing Handbook. Miami, FL: American Welding Society, 1991

[5] B.T. Alexandrov, J.C. Lippold, J.W. Sowards, A.T. Hope, and D.R. Saltzmann. 2013. Fusion Boundary Microstructure Evolution Associated with Embrittlement of Ni-base Alloys.

[6] J. Ruiz-Vargas, N. Siredey-Schwaller, N. Gey, P. Bocher. "Microstructure development during isothermal brazing of Ni/BNi-2 couples." 213, s.l. : Elsevier, 2013, Vol. Journal of Materials Processing Technology.

[7] V. Jalivand, H. Omidvar. "Microstructural evolution during transient liquid phase bonding of Inconel 738LC using AMS 4777 filler alloy." s.l. : ELSEVIER, 2013, Vol. Materials Characterization 75.

[8] Schnell, Alexander. "A study of the diffusion brazing process applied to the single crystal superalloy CMSX-4". Lausanne, EPFL : ÉCOLE POLYTECHNIQUE FÉDÉRALE DE LAUSANNE, 2004.

[9] DuPont, John N., John C. Lippold, and Samuel D. Kiser. "Welding Metallurgy and Weldability of Nickel-base Alloys." Hoboken, NJ: Wiley, 2009. Print.

Thermodynamic Simulations of Single Crystal Superalloy CMSX-4 Brazed Joints with BNi-2 and BNi-9 Filler Metals

Bryan Riggs, Boian Alexandrov, Avraham Benatar
Welding Engineering, Ohio State University, Columbus OH, U.S.A.

Ray Xu
Rolls-Royce, Indianapolis, IN, U.S.A.

Abstract

Brazed joints are commonly used in the manufacture and repair of high temperature gas turbine components made from nickel based superalloys. When designing and modeling brazed joints it is important to account for the micro-constituents that form as a result of the brazing process. This research aims at predicting these micro-constituents in vacuum furnace brazing CMSX-4 with two nickel braze filler metals BNi-2 and BNi-9. Metallurgical characterization was performed using light optical microscopy, Electron Probe Micro-Analysis (EPMA) and volume fraction point counting. Thermodynamic simulations were made using Thermo-Calc^{TM} software. The thermodynamic simulations predicted the same phases as were experimentally identified for both brazed joints.

Experimental Procedure

Cast CMSX-4 single crystal nickel-based superalloy plate specimens were vacuum furnace brazed with Ni-Cr-B based filler metals, BNi-2 and BNi-9, in a lap shear configuration. Table 1, presents the chemical composition for these materials in wt% and typical liquidus/solidus temperatures of the filler metals. These combinations were chosen to represent characteristic joints between single crystal superalloys brazed with Ni-Cr-B based filler metals, which have a wide range of industrial applications [1,2]. A laser tack weld was used to maintain a joint gap between 0.002-0.004" and brazing cycles were carried out in a high vacuum (<10^{-5}torr) furnace at Rolls Royce facilities with standard brazing cycles listed in Table 2. A 10 minute hold time at standard brazing temperatures (T_b) of 1066°C and 1121°C for BNi-2 and BNi-9 respectively was used.

Table 2: Vacuum furnace brazing cycles for BNi-2 and BNi-9.

Standard Brazing Cycles					
BNi-2 (AMS-4777)			**BNi-9 (Amdry 775)**		
Ramp C/min	T_b °C	Hold Time, min	Ramp C/min	T_b °C	Hold Time, min
11	538	20	11	538	20
15	899	20	15	927	20
15	1066	10	15	1121	10
Vacuum Cool	538	5	Vacuum Cool	538	
Gas fan cool			Gas fan cool	66	

Joint samples were cross-sectioned, mounted and polished using standard metallography techniques. Light optical microscopy was used for imaging and compositional analysis was performed with Electron Probe Micro-Analysis (EPMA) at Washington University in St. Louis. EPMA spot analyses

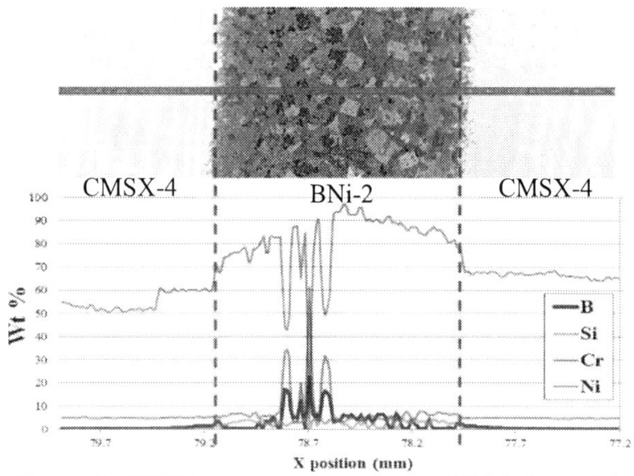

Figure 1: EPMA traverse scan across BNi-2 brazed joint used to determine experimental composition.

Table 1: Chemical composition of materials investigated and solidus/liquidus temperatures for brazing filler metals.

Alloy	Composition (wt%)											
Superalloy Base Metal												
CMSX-4	Ni	Cr	Re	Co	Mo	W	Ta	Ti	Al	Si	Mn	Others
	balance	6.4	2.9	9.6	0.61	6.4	6.5	1.02	5.61	0.4	0.1	0.05 Zr, 0.1 Hf, 0.042 Fe, 0.0037 C
Brazing Filler Metals												
BNi-2 (AMS-4777)	Ni-7Cr-3.1B-4.5Si-3Fe											Solidus/Liquidus: 971°C/999°C
BNi-9 (Amdry 775)	Ni-15Cr-3.5B											Solidus/Liquidus : 1020°C/1054°C

and line traverses were performed at 10 kV across both joints, as shown for the BNi-2 brazed joint in Figure 1. EPMA was used for more accurate detection of boron, the primary melting point depressant element in BNi-2 and BNi-9 filler metals.

Thermo-Calc is a commercially available thermo-dynamic software for prediction of equilibrium phases, composition and properties in multi-component alloys based on the CALPHAD technique [3]. Thermodynamic databases used for calculations have been developed for a wide range of alloys systems including steels, nickel-based superalloys, aluminum, magnesium and solder alloys. However, a thermodynamic database does not currently exist for braze alloys, instead all Thermo-Calc simulations were made using the nickel based superalloy thermodynamic database (TTNi7). Equilibrium solidification was assumed due to the slower heating and cooling rates associated with vacuum furnace brazing, however Thermo-Calc is also capable of calculating solute redistribution during non-equilibrium solidification through the Scheil equation. Equilibrium phases were calculated from room temperature (25°C) to 1300°C for both BNi-2 and BNi-9 compositions.

Results and Discussion

Vacuum furnace brazing with the BNi-9 filler metal produced a joint microstructure consisting of a single phase isothermally solidified zone (ISZ) and a multi-component centerline eutectic shown in Figure 2. The bulky intermetallic phase in the eutectic region was identified as CrB type borides through EPMA. The plot in Figure 2 shows the Thermo-Calc predicted equilibrium phases for the BNi-9 composition between 200-1200°C. Thermo-Calc predicts that the solidification path for BNi-9 liquid will begin as an FCC nickel phase (FCC_A1) at 1065°C followed by precipitation of a chromium boride labeled as MB_ORTH before reaching a eutectic temperature at 1010°C. At this temperature any remaining liquid will form a constituent composed of these two phases and a nickel boride (M3B) secondary phase. The predicted solidus and liquidus temperatures are within 10°C of values given in the AWS A5.8 Specification for Filler Metals for Brazing. Table 3, shows the chemical composition for these equilibrium phases predicted by Thermo-Calc at room temperature compared to the experimental compositions of the intermetallics within the actual brazed joint determined from EPMA scans. Based on this compositional analysis, the MB_ORTH phase corresponds to CrB type borides labeled for the BNi-9 microstructure in Figure 2. These phases were found to be extremely hard and non-uniformly distributed

Table 3: Comparison between Thermo-Calc predicted phase composition and EPMA experimental results for BNi-9 brazed joints.

Phase	Vol. Fraction	Chemical Composition (wt%)			
		Ni	B	Cr	Others
Experimental (EPMA)					
CrB	0.09	3.79	20.48	64.18	11.55
α-Ni	0.30	73.70	0.56	12.63	13.11
Eutectic	0.61	70.40	5.10	11.60	12.90
Calculated (Thermo-Calc) - 25°C					
CrB (MB_ORTH)	0.28	0.00	17.21	82.79	0.00
α-Ni (FCC_A1)	0.63	99.99	0.00	0.00	0.00
Ni_3B (M3B)	0.09	94.22	5.78	0.00	0.00

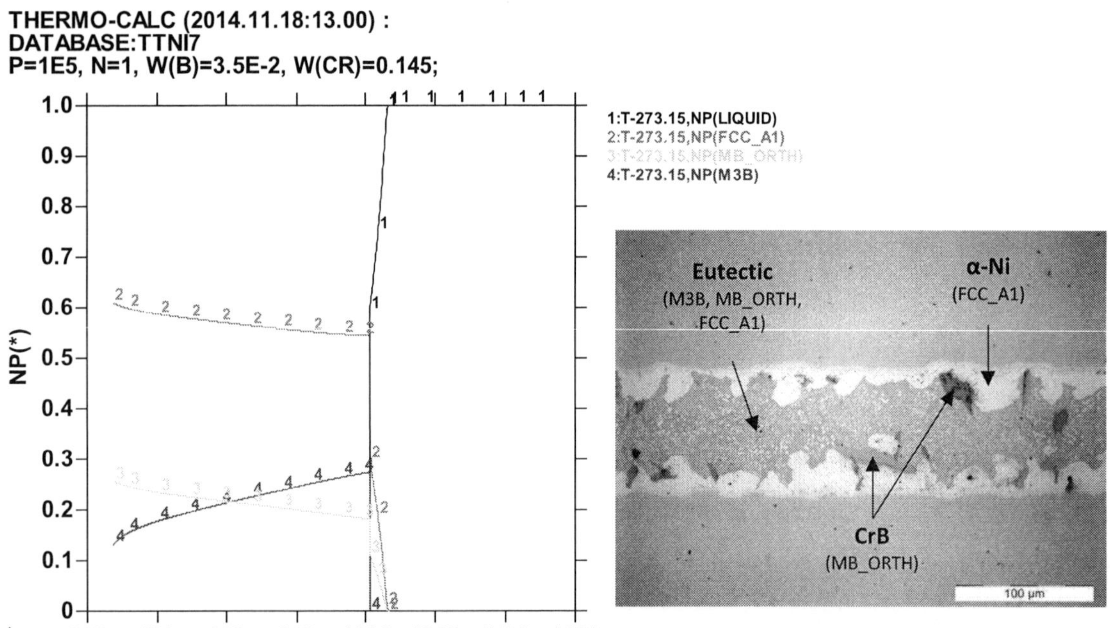

Figure 2: Equilibrium phases predicted with Thermo-Calc for BNi-9 filler metal and corresponding phases in the BNi-9 brazed joint; 1) Liquid 2) FCC_A1, α-Ni phase 3) MB_ORTH, CrB boride 4) M3B, Ni_3B boride.

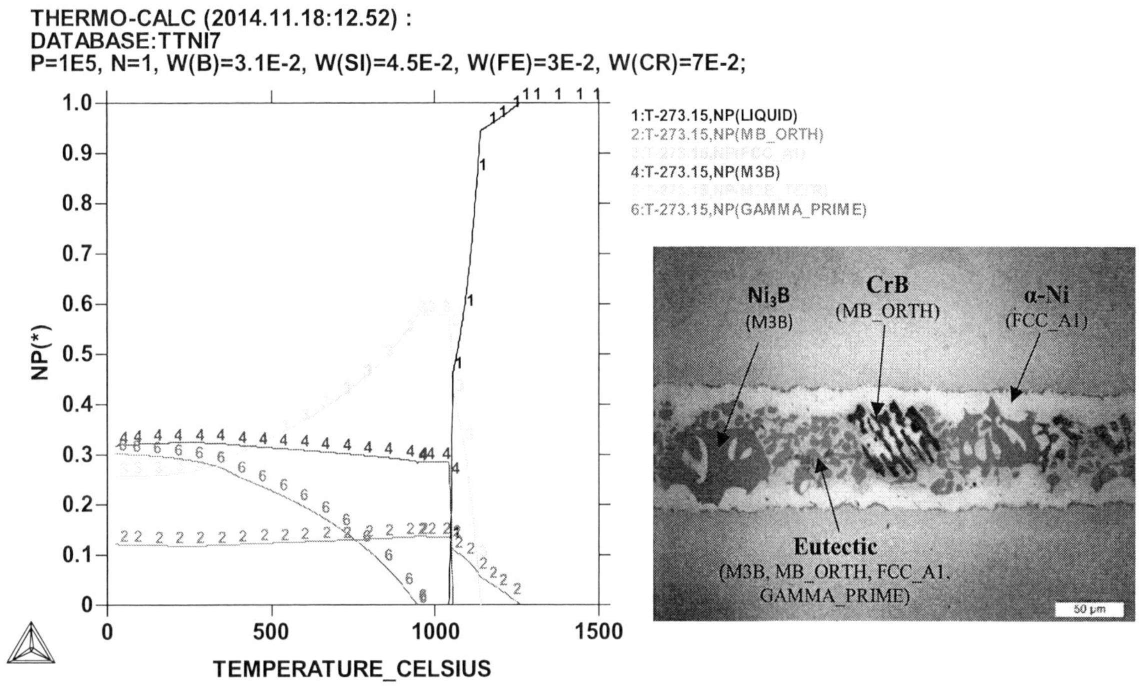

Figure 3: Equilibrium phases predicted with Thermo-Calc for BNi-2 filler metal and corresponding phases in the BNi-2 brazed joint; 1) Liquid 2) MB_ORTH, CrB boride 3) FCC_A1, α-Ni phase 4) M3B, Ni_3B boride 5) M2B_TETR phase 6) GAMMA_PRIMA, Ni_xSi_y

across the joint. The FCC_A1 phase corresponds to the white colored phase found primarily in the isothermally solidified region. Labeled in Figure 2, this soft, ductile FCC_A1 phase is rich in nickel and is often referred to as α-Ni. It forms along the joint interface at the brazing temperature as a result of base metal dissolution and diffusion of melting point depressant elements, boron in this case [5]. The formation of this isothermally solidified region is not captured in the equilibrium thermodynamic calculations, but may be accounted for through DICTRA diffusion based calculations [6]. The two-phase centerline eutectic constituent observed in the joint contained both the FCC_A1 phase and a phase rich in Ni, Cr and B. This constituent correlates well to the eutectic constituent predicted by Thermo-Calc which is a combination of the MB_ORTH, M3B and FCC_A1 phases. It should be noted that chemical composition and volume fraction of each phase were different between the thermodynamic prediction and experimental measurement using EPMA. This is to be expected since the brazing process is transient while the Thermo-Calc predicts equilibrium conditions.

The BNi-2 filler metal produced a similar joint microstructure again with both an isothermally solidified region and centerline eutectic phases. Shown in Figure 3, the eutectic microstructure was slightly different from the BNi-9 joint due to additional silicon and iron in the BNi-2 composition. EPMA identified the two primary intermetallic phases within the multi-component eutectic as Ni_3B and CrB type borides. The CrB borides found in this joint had similar shape and nearly identical composition to the CrB phase in the BNi-9

Table 4: Comparison between Thermo-Calc predicted phase composition and EPMA experimental results for BNi-9 brazed joints.

Method	Phase	Vol. Fraction	Chemical Composition (wt%)					
			Ni	B	Si	Fe	Cr	Others (W, Re, Al, Ti, Co, Mo
Experimental (EPMA)								
Experimental (EPMA)	CrB	0.13	1.94	24.28	0.01	0.46	62.18	11.13
	Ni_3B	0.22	81.88	6.02	0.46	1.64	4.25	5.75
	α-Ni	0.40	80.80	0.44	3.73	3.13	7.00	4.90
	Eutectic	0.25	83.70	5.44	2.00	1.64	3.74	3.48
Calculated (Thermo-Calc) - 25°C								
Calculated (Thermo-Calc)	CrB (MB_ORTH)	0.12	--	17.21	--	--	82.79	--
	Ni_3B (M3B)	0.33	94.22	5.78	--	--	--	--
	α-Ni (FCC_A1)	0.25	97.10	--	2.87	0.03	--	--
	GAMMA_PRIME	0.30	85.39	--	12.45	--	--	--

joint, indicating these are the same intermetallic phase. Thermo-Calc equilibrium calculations for the BNi-2 filler metal in Figure 3 confirm this and again the predicted phases are in agreement with the final joint microstructural phases that were determined with EPMA. The solidification of the BNi-2 liquid is slightly more complicated than the BNi-9 filler metal [5,7]. Thermo-Calc predicts solidification begins at 1260°C with precipitation of the chromium boride (MB_ORTH) followed by solidification of FCC_A1 nickel phase and precipitation of M3B nickel borides before a final eutectic temperature is reached at 1041°C. At this temperature the remaining BNi-2 filler metal transforms to a constituent consisting of FCC, MB, and M3B. These temperatures are significantly higher than the AWS A5.8 solidus (971°C) and liquidus (999°C) temperatures for this braze alloy. The solidification path is different from the BNi-9 filler metal which first solidifies as FCC_A1. As with BNi-9, the same phases are predicted for BNi-2 as found experimentally using EPMA, but the composition and volume fractions differ due to equilibrium modeling of the transient brazing process, see Table 4. The chemical composition of the Ni_3B intermetallic phase found in the BNi-2 joint corresponds to the M3B phase predicted, but not observed in the BNi-9 joint. Thermo-Calc predicts that this phase forms above the eutectic temperature in the BNi-2 alloy which may explain why Ni_3B was only observed in the BNi-2 brazed joint microstructure. A GAMMA_PRIME phase is predicted to precipitate in the solid state at 944°C. Thermo-Calc indicates this phase is Ni_xSi_y based compound as opposed to the $Ni_3(Al,Ti)$ strengthening precipitates typically found in nickel based superalloys. An increase in silicon content observed in the eutectic region suggests this phase forms within the multi-component eutectic constituent.

Conclusions

For both filler metals, equilibrium phases predicted with the nickel base superalloy database in Thermo-Calc were the same as found experimentally in the vacuum brazed joints using EPMA. Phases that are stable before the end of solidification (eutectic temperature) are more likely to form large precipitates as shown by the Ni_3B borides in the BNi-2 joints. The composition of phases determined through EPMA were similar to predicted phase compositions especially for the CrB and Ni_3B borides. Diffusion of melting point depressant elements during the brazing process is not captured in the equilibrium calculations, but can contribute to the solidification temperature range and formation of the isothermally solidified region. In general, equilibrium Thermo-Calc simulations allow for the prediction of micro-constituents in nickel based brazed joints and can be a useful tool when modeling and designing these joints or the brazing process. A thermodynamic database for braze alloys based on metallurgical characterization would improve accuracy in prediction of these eutectic systems and better account for diffusion of MPDs.

Acknowledgments

This research has been supported by Rolls Royce and performed within the Center for Integrative Materials Joining Science for Energy Applications (CIMJSEA), a National Science Foundation, Industry-University Cooperative Research Center. EPMA work performed by Paul Carpenter at Washington University in St. Louis and Paul Mason at Thermo-Calc assisted with the simulations.

References

[1] Rabinkin, A. "Overview: Brazing with (NiCoCr)-B-Si Amorphous Brazing Filler Metals: Alloys, Processing, Joint Structure, Properties and Applications." IBSC 2003, Honeywell Intl., Metglas Solutions.

[2] Brazing Handbook. Miami, FL: American Welding Society, 1991. Print.

[3] Andersson J.O., Helander T., Hoglund L., Shi P.F., and Sundman B. (2002) Thermo-Calc and DICTRA, Computational tools for materials science. Calphad, 26, 273-312

[4] Brazing and Soldering Products. Nickel Brazing Filler Metals. Lucas Milhaupt. Web. 15 Nov. 2014

[5] J. Ruiz-Vargas, N. Siredey-Schwaller, N. Gey, P. Bocher. "Microstructure development during isothermal brazing of Ni/BNi-2 couples." 213, s.l. : Elsevier, 2013, Vol. Journal of Materials Processing Technology.

[6] Schnell, Alexander. "A study of the diffusion brazing process applied to the single crystal superalloy CMSX-4". Lausanne, EPFL : ÉCOLE POLYTECHNIQUE FÉDÉRALE DE LAUSANNE, 2004.

[7] V. Jalivand, H. Omidvar. "Microstructural evolution during transient liquid phase bonding of Inconel 738LC using AMS 4777 filler alloy." s.l. : ELSEVIER, 2013, Vol. Materials Characterization 75.

Assessing and Reducing Thermal Residual Stresses in Metal-Ceramic Joints by Combined Experimental and Numerical investigations

T. Ivas, A. Lis, C. Leinenbach*
Empa, Swiss Federal Laboratories for Materials Science and Technology, Dübendorf, Switzerland
**Email: toni.ivas@empa.ch*

Abstract

Extreme operating environments require well evaluated and reliable joining techniques where in many cases only soldering or brazing can fulfil the ambitious demands. Both soldering and brazing are essential for joining dissimilar material combinations like metals to ceramics, applied e.g. in power generation, space technology and high power electronics. However, due to the large misfit of thermal expansion coefficients between dissimilar materials, residual stresses are introduced after processing.

This work investigates this problem by combining experimental and numerical methods to assess and minimize critical residual stresses mainly focusing on active brazing and Transient Liquid Phase Bonding (TLPB). Shear test experiments with joined assemblies provide a global shear strength value, and subsequent fractographic investigations identify the failure position and mode. Based on these experimental results, the shear testing conditions have been modelled using finite element methods FEM, which yields a critical local shear stress for a defined joining system. Coupled thermo-mechanical simulations are performed for metal-ceramic joints cooled down from process temperature and provided theoretical residual stresses within the joined assembly which can be compared to the critical stress state. Finally, a numerical parameter study is conducted that aims at reducing residual stresses in critical regions and consequently improving the reliability of joined components.

Introduction

Ceramics parts with complex shapes are expensive and difficult to produce. By joining ceramics to metals it is possible to overcome some of these drawbacks. Thus joining of the metal-ceramic parts is a crucial technology for future technological applications. For this, the ceramic part is metallized before joining to a metal to ensure good wetting of the ceramic surface by solder or braze filler alloy. Standard processes are moly-manganese or CVD/PVD coating methods. Active brazing is a process for joining ceramics to metals by direct chemical reaction between ceramic component and active braze alloy, thus eliminating costly metallization process. The inherent mismatch of thermal expansion between metal and ceramic material introduces significant residual stresses in the brazed joint which can lead to cracking of the interfacial region after cooling from the process temperature or to long term failure of the joint during service. In this study, we report on our investigations on two different bonding processes, Transient Liquid Phase Bonding (TLPB) and active brazing performed at comparably low and high process temperatures.

During TLPB, a low melting point interlayer is sandwiched between substrates with a higher melting point. After melting of the interlayer, the two materials react and new phases with higher melting point form. This phase transformation raises the overall remelting temperature of the joined assembly above the melting temperature of the initial interlayer arrangement and isothermal solidification proceeds until the liquid interlayer is entirely consumed [1]. As a result, the joined components can withstand higher service temperatures than the joining temperature. In the present study, the Ag-Sn interlayer system, which requires a process temperature above 232 °C (T_m,Sn) and has a remelting temperature of 480 °C (T_m,Ag_3Sn), was considered for TLPB [2-4]. This low temperature bonding process is of great interest for bonding dissimilar material combinations with large mismatch of thermal expansion coefficients. In the case of power modules for example, Si or SiC chips have to be joined to metallized ceramic insulators and the assembly has to withstand cyclic temperature loadings during operation [5-7]. Despite the low process temperature during TLPB (<300°C), thermally induced residual stresses have to be considered in the context of the corresponding strength of the joined layer [8,9], in particular since the resulting phase in the joining zone is intermetallic and relatively brittle. Therefore, this study presents a concept that combines experimental and theoretical investigations with the aim to optimize the power module architecture and to assess the influence of brazing defects regarding thermally induced residual stresses after processing.

In active brazing, the capillary flow of the filler material between faying surfaces of the joint is facilitated by reactive metals in the braze alloy, mostly Ti, Zr, Cr, and V, which form oxides at the interface between ceramic and braze layer. Such oxidation reaction forms high temperature and corrosion resistant oxides which are required for power generation, aerospace, and turbo-machine applications. The active brazing of Yttria Stabilized Zirconia (YSZ) to Ni-based superalloys has been already discussed in the literature [10,11]. As a part of this study we have investigated active brazing of YSZ and HastelloyX with commercially available Au-based active braze alloy Nioro-Ti® which is relevant e.g. for high temperature and corrosion resistant applications. In comparison with the intermetallic structure of TLP joints, active braze alloys are very ductile. However, active brazing is usually performed at comparably high temperatures (>700°C),

leading to significantly higher residual stresses than in the case of TLP bonded components.
In this work, the residual stresses in several metal-ceramic joints produced by TLPB and active brazing were calculated using FE methods with a special emphasize on the influence of additional metallic interlayers and defects.

Experimental Methods

Transient liquid phase bonding

TLP joining was performed using the Ag/Sn/Ag interlayer system. A Cu plate (110 x 60 x 4 mm) was electroplated with 14 μm of pure Ag for preliminary tests. After coating, smaller samples with dimensions 5 x 5 x 4 mm were produced by Electro Discharge Machining (EDM) for a basic microstructural investigation as well as for preliminary mechanical tests. The single components were assembled by positioning Sn foil with a thickness of 8 μm and applying a pressure of 0.2 MPa. The TLP process was conducted in Argon shielding gas atmosphere at an isothermal temperature of 280 °C for a holding time of 60 min. In addition, Ag coated Si chips were joined in the same way to Ag coated AlN substrates.
The mechanical strength of the TLP bonded Cu substrates was determined using a shear tester (Walter & Bai AG). Mean shear strength for the joined assemblies of 30 MPa ± 7 MPa was measured which corresponds to an external force of 750 N ± 175 N. Cu substrates were selected in order to minimize thermally induced residual stresses after TLP bonding due to the proper matching of CTE values between Cu and Ag. The results obtained from shear tests could thus be directly correlated to the actual mechanical strength of the TLP joint without considering the influence of residual stresses. A fracture surface analysis revealed that brittle fracture occurs within the Ag_3Sn IMC layer.

Active Brazing of Yttria Stabilized Zirconia on HastelloyX

A defined HastelloyX plate (HX) (20 x 10 x 3 mm) was processed by EDM machining. Subsequently, the plates were annealed at 800 °C for 45 min under vacuum at a pressure of 10^{-6} mbar. The ceramic parts were produced from powder (Tosoh Cooperation, Japan) with an averaged particle size of 53 nm consisting of ZrO_2 with 13 wt-% of Y_2O_3 commonly known as YSZ. The powder was pressed axisymmetrically into pellets of 10 mm diameter and thickness of 3 to 4 mm. The pellets were sintered at 1600 °C for 1h and the relative density of the YSZ was measured as more than 99.9% using the Archimedes method. The filler braze alloy is commercially available as Nioro-Ti® (Au 81.1 wt%, Ni 18.2 wt% Ti 0.7 wt%) and was purchased from Wesgo Metals (Hayward, CA, USA). The brazing of the YSZ to the HX plate was performed using a vacuum furnace at a pressure better than 10^{-5} mbar between 960 and 1000 °C. The optimized process parameters were identified at a brazing temperature of 980 °C and a dwell time of 5 min. In order to reduce the residual stresses caused by the thermal mismatch between YSZ ($10 \cdot 10^{-6}$ °C^{-1}) and HX ($15.6 \cdot 10^{-6}$ °C^{-1}), a slow cooling rate of 3°C/min was chosen.

The microstructural characterization of the brazed samples was done by Scanning Electron Microscopy (SEM) in combination with Energy Despersive X-ray Spectroscopy (EDS). An EDS line scan was performed across the interface between Nioro-Ti® and YSZ. As shown in Fig. 1 the line scan over the interface YSZ/Nioro-Ti reveals an increased amount of Ti of up to 40 at-%, which corresponds to the Ti_2O_3 phase.

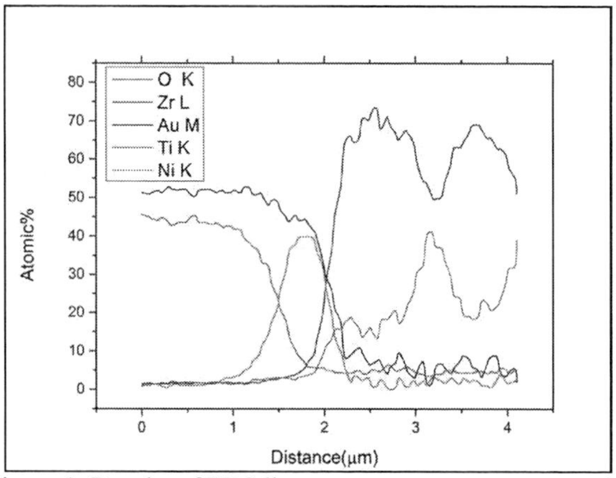

Figure 1: Results of EDS line scan measurements across the interface between YSZ and Nioro-Ti layers.

The titanium oxide phase is responsible for the interfacial bonding of YSZ to the braze alloy [12]. These layers continuously grow at higher temperatures due to reduction of YSZ and reach a thickness of several microns. The large residual stresses forming in the joint will strongly influence the durability of the brazed joint. A number of samples fractured directly after cooling. Subsequent fractographic investigations revealed crack propagation from the interface of Nioro-Ti®/YSZ at angle of 44° degrees.

Numerical Simulations

Thermally induced residual stresses in TLP Bonded power modules

In a first step, the critical shear stress leading to failure of the TLPB joints was determined. For this, the experimental conditions during shear testing were modelled by means of Finite Element Methods (FEM) with Abaqus 6.11 EF [1].

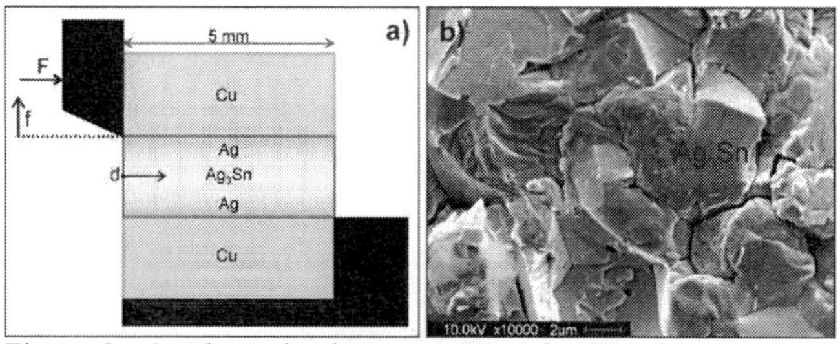

Figure 2: a) schematic shear test setup for TLP bonded Cu substrates, b) fracture surface of TLP bonded components.

[1] ABAQUS (2011) "ABAQUS Documentation", Dassault Systèmes, Providence, RI, USA.

A 2D FE model with plane strain constraints was set up according to the schematic illustration in Fig. 2 a) where the thickness of both the Ag_3Sn layer and the Ag layers was around 10 μm. Each of these layers has 8 triangular elements (CPE3) in thickness direction. Elasto-plastic material behavior was considered for Cu and Ag whereas the IMC layer was modelled with linear elastic properties. The implemented material parameters are summarized in Table 1.

Table 1: Summary of material parameters for FE modelling of residual stresses in TLP Bonded power modules.

	E	ν	σ_y	σ_{UTS}	ε_{fr}	CTE at RT	CTE at 300 °C
Ag	76	0.37	54	140	0.25	19.0	20.4
Cu	110	0.34	70	220	0.5	16.2	18.6
Ag_3Sn	82	0.35	-	-	-	18.0	21.1
Si	112	0.28	-	-	-	2.5	3.7
AlN	310	0.24	-	-	-	3.4	4.9

E-Elastic modulus in GPa, σ_y-Yield strength in MPa, σ_{UTS}-Ultimate tensile strength in MPa, CTE- Coefficient of thermal expansion in 10^{-6} °C^{-1}.

The experimentally determined external fracture force of 750 N was applied on one side of the Cu substrate and the influence of the parameter *f* representing the distance between the joining zone and the force axis was taken into account. Since it is known from experiments that distance *f* can vary between 0 and 50 μm, three different options were modelled with values for *f* of 0, 50 and 100 μm, respectively. According to the FE simulations a local shear stress peak value of 63 MPa occurs in the Ag_3Sn layer for $f = 50$ μm at an external load of 750 N. This value was therefore selected as local failure criterion $\tau_{12,crit}$ for further investigations on TLP bonded components with the Ag-Sn interlayer system.

The 2-dimensional modelling of the power module assembly is shown in Fig. 3 and represents the set up where an AlN insulator metallized with Cu is interconnected to a Si chip. The joint consists of two Ag metallization layers, one of them at the chip side (M1) and the other one on the substrate side (M2) which sandwich the Ag_3Sn IMC layer.

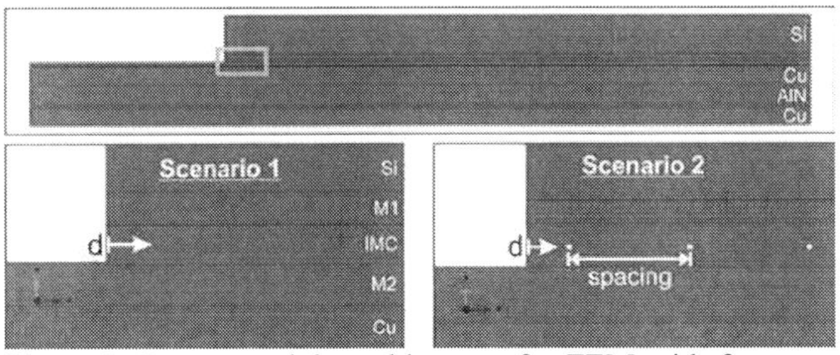

Figure 3: Power module architecture for FEM with focus on scenario 1 and 2.

In scenario 1, the influence of metallization layer M1 on the residual stress distribution along the IMC layer was analyzed for cooling down from 250 and 300 °C. Scenario 2 assessed the sensitivity of residual stresses to point defects (pores) with a size of 1 μm when metallization layers M1 and M2 are used and the assembly is cooled down from 250 and 300 °C. The first pore was positioned at 7 μm from the joint edge; further pores were placed 20 μm from each other. The model accounts for the fact that metallized Ag and Sn foil are initially only stacked on each other by choosing very low yield strength of 0.1 MPa for the Sn layer during heating up and thus allowing the material to deform plastically. Consequently, in the second step during cooling when the TLP process is finished and Sn is completely transformed into Ag_3Sn, the parameters from Table 1 were used. This modelling approach describes well the real conditions when the ceramic substrate and the Si chip can expand without any constraints during heating. During cooling down from the holding temperature, both components are constrained due to the TLP bond. The assembly was meshed with triangular, thermo-mechanically coupled elements (CPE3T) optimized for plane strain conditions. Failure within the IMC layer was defined when the local shear stress is greater than the critical local shear stress $\tau_{12,crit} = 63$ MPa, as derived from FE simulations. The calculated local shear stress τ_{12} within the brazed assembly was normalized to the critical shear value for better failure assessment:

$$\tau_{12,norm} = \frac{\tau_{12}}{\tau_{12,crit}} \quad (1)$$

Consequently, failure is defined for $\tau_{12,norm} > 1$.

Figure 4a) shows the calculated normalized shear stress for Scenario 1 up to an edge distance of 200 μm. The FE calculations suggest significantly different stress distribution with a changing metallization layer thickness (M1). At 250 °C, the highest residual shear stress peak with $\tau_{12,norm} = 0.6$ is calculated for an assembly without metallization M1. Interestingly, a thin buffer layer of only 2 μm reduces the peak value considerably to $\tau_{12,norm} = 0.2$ whereas using a metallization layer of 8 μm only decreases the peak value to around 0.39. None of the considered conditions suggest failure of the joint.

Additionally, the shear stress evolution along the path d was calculated for the case when the assembly is TLP bonded at a slightly elevated temperature of 300 °C. Without any metallization layer M1, the normalized shear stress increases to $\tau_{12,norm}$ of 0.76. However, placing an Ag layer of 8 μm between chip and IMC layer raises the normalized shear stress comparably less by only 0.04 to a value of 0.43. Stress relaxation can be observed for all cases when the residual shear stress approaches a value close to zero at a distance d of 150 μm from the edge.

In summary, the system benefits from the capability of the metallization layer M1 to deform plastically during cooling, which protects the IMC layer from the high CTE mismatch between the different layers.

Due to the transformation of Ag and Sn into Ag_3Sn IMC during TLPB, joining defects like small pores cannot be entirely avoided. Scenario 2 considers possible joining defects (pores) within the IMC layer and calculates the corresponding thermally induced residual stresses. The thickness of the metallization layer M1 was set to 8 μm. The calculated curves for this scenario are shown in Fig. 4 b). It is apparent that the shear stress sharply increases in the vicinity of each pore and reaches its maximum value of around 0.4 close to the joint edge (d < 100 μm). Towards the centre of the joint, the normalized shear stress oscillates between -0.2 and 0.2.

Changing the process temperature from 250 to 300 °C does not significantly influence the local stress state when metallization layer M1 is used.

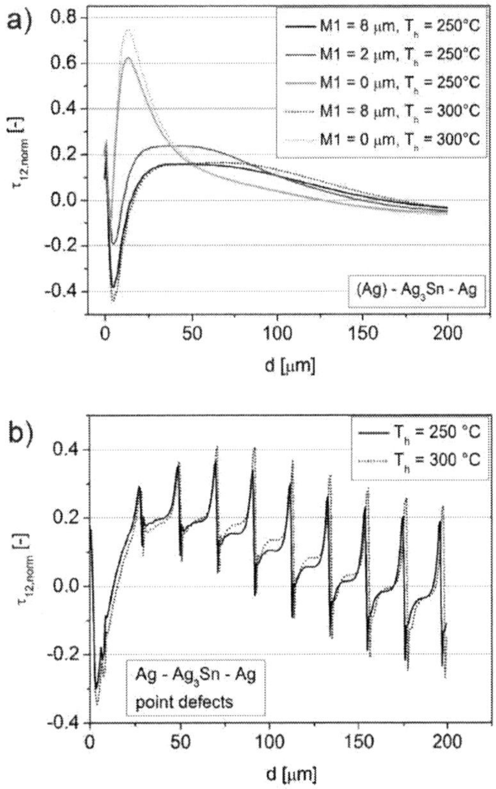

Figure 4: Calculated residual shear stress distribution along *d* for a) scenario 1 and b) scenario 2.

Since the IMC layer was clearly identified as the critical part of the assembly where failure is expected to occur, reduced residual stresses will facilitate longer lifetime of these modules. The process temperature was shown not to affect the residual stress state significantly when metallization M1 is left after the TLP reaction is completed. This is an important aspect in terms of process efficiency because kinetics for IMC formation is known to be faster at higher holding temperatures. Experimental verification of an enhanced performance during operation, i.e. cyclic thermal loading, has to be analyzed in a follow-up study.

Thermomechanical finite element analysis of active brazing process

The thermo-mechanical process of cooling after the brazing cycle was modelled using Abaqus FEM suite. The computational model is based on the experimental set up. The block-shaped ceramic part (4 x 4 x 3 mm) was placed on the HX plate (20 x 10 x 3 mm) with the brazing layer in between. Different dimensions of the ceramics and metal parts were investigated. Furthermore, several configurations with different metallic interlayers were calculated to gain insight in the distribution of the residual stresses as a function of the metallization system. It is known that reactive Ti components produce TiO_x phases at the interface with the oxide ceramics. However, these phases were not considered due to their small volume in comparison to the overall brazed joint. The braze alloy was assumed to keep its thermo-mechanical properties independent of the compositional changes during the brazing process. The thermo-mechanical material properties for the ceramic and metal parts were taken from the literature. Thermo-mechanical data for Hastelloy are given[2]. The filler alloy Nioro-Ti from Wesgo Metals was investigated by Stephens [14]. We considered that the filler metal has viscoplastic properties with isotropic hardening behaviour. The ceramic YSZ is modelled as an entirely elastic body. Heating and cooling in the vacuum furnace at 10^{-5} mbar is modelled by solving the heat equation:

$$\rho c \frac{\partial T}{\partial t} = -\nabla\left(\mathrm{k}\nabla \mathrm{T}\right) + \mathrm{S} \tag{2}$$

Where T is the temperature, t is time, ρ is the density, c is the specific thermal capacity, k is thermal conductivity and S is a heat source. The natural convection in the vacuum oven can be neglected due to very low pressure of the gas in the range of 10^{-5}-10^{-6} mbar. Most of the energy transport at higher temperatures is due to radiation between heating elements and brazed sample. The small size of the sample in comparison to the vacuum furnace ensures a uniform temperature profile within the sample. Therefore, the temperature of the brazed sample follows the measured temperature profile of the vacuum furnace in our simulations. Since brazing is performed at 980°C and cooling occurs with comparably low cooling rates, residual stresses can relax due to high temperature creep of the HX and in the braze filler alloy. This was considered by using Norton's steady state 5 power-law (PL) [15]:

$$\varepsilon_{ss} = A_o \exp\left(\frac{-Q_c}{kT}\right)\left(\frac{\sigma_{ss}}{E}\right)^n \tag{3}$$

Where A_o is a constant, k is Boltzmann constant, T is the temperature, E is Young's modulus and Q_c is the activation energy for creep. For many materials, the exponent n is close to 5 over a broad range of temperature and strain rates. For homologous temperatures lower than 0.5-0.6 T_m and high strains, the PL breaks down but it can be described with Garofalo's Sinh or the power law breakdown (PLB) relation:

$$\varepsilon_{ss} = A_1 \exp\left(\frac{-Q_c}{kT}\right)\left(\mathrm{Sinh}\left(B_1\frac{\sigma_{ss}}{E}\right)\right)^n \tag{4}$$

Where A_1 and B_1 are constants. Table 2 summarizes the material parameters used for modelling creep in HX and Nioro-Ti filler alloy. The model includes linear cooling from the solidus temperature of the Nioro-Ti filler alloy (1213 K) to room temperature (RT = 298 K). Additionally, metallic interlayers with good high-temperature ductility (Cu, Ni) or with CTE values in between the ones of the metallic and the ceramic substrate are considered for relieving the residual stresses during brazing.

[2] http://www.haynesintl.com

Table 2: Summary of the material parameters used for FEM modelling of residual stresses in active brazing process of YSZ and HX.

				Nioro-Ti				
Properties	E=110000	ν=0.42	ρ=1.74E-8	α=1.4E-5				
PLB	A_1= 494E6	B_1=0.00901	n=2.224	Q_c=210.282E6				
				Hastelloy X				
Properties			ρ = 7.8E-9	α=1.56E-5				
T	298	366	477	700	811	922	1033	1255
E	205000	203000	197000	184000	178000	170000	161000	141000
A						1.099E−21	1.995E−20	4.116E−16
n						6.61	7.12	6.02
				Ni				
Properties	E =199500	ν =0.312	ρ = 8.9E-9	α= 1.33E−5				
T	673	873	1073	1273				
A	2.24E-17	3.94E-14	2.68E-11	4.62E-9				
n	4.6	4.6	4.6	4.6				
				Cu				
Properties	E =110000	ν =0.34	ρ=8.96E-9	α= 1.65E−5				
T	673	723	773	823	903	973		
A	2.64E-12	1.62E-11	3.50E-11	7.25E-11	2.43E-9	4.09E-8		
n	3.57	3.36	3.39	3.38	2.66	2.26		
				YSZ				
Properties		ν =0.30	ρ=5.88E−9	α= 1.01E−5				
T	224	550.3	689	794	1074	1495.5	1576.6	1618.8
E	213000	221000	178600	178400	178600	175770	172320	170060

T temperature in K, α thermal expansion coefficient in K^{-1}, E elastic modulus in MPa, ρ density,
ν Poisson's ratio, A and n creep law, A_1, B_1 and Q_c PLB law.

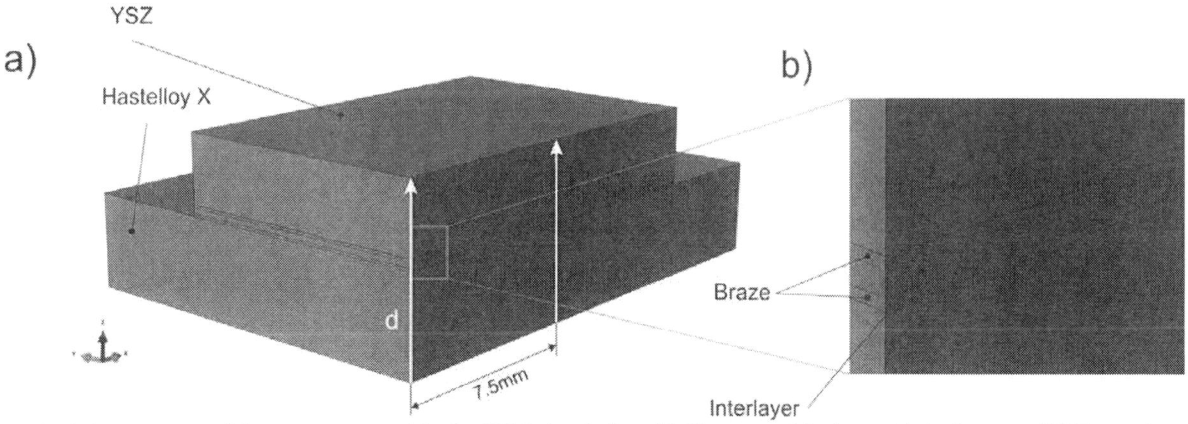

Figure 5: a) One-quarter of the geometry used in the FEM simulations b) Close up of the brazed joint between YSZ, two layers of braze filler alloy and interlayer. A white arrow indicates the path *d* along which are von Mises stresses is evaluated.

The 3-dimensional model of the brazed joint is given in Fig. 5. The joint consists of the YSZ ceramic on top and the HX metal part at the bottom. Several configurations with different interlayer materials and thicknesses were modelled and the influence on the residual stress distribution in the joining zone due to the CTE mismatch between YSZ and HX was considered.

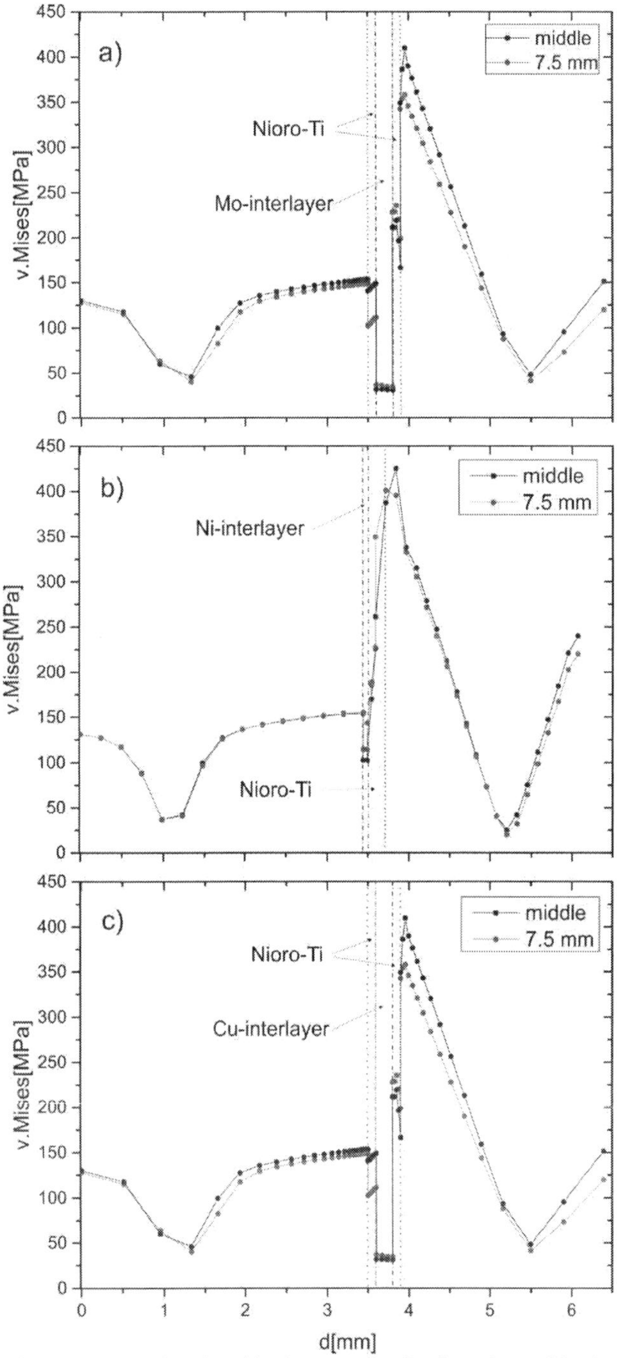

Figure 6: Calculated residual stress distribution along *d* in the middle and 7.5 mm from the middle of the joint for a) Mo-interlayer 200μm, b) Ni-interlayer 50μm and c) Cu-interlayer 200μm.

The von Mises stresses in the sample with Mo interlayer are shown in Fig. 6a). The highest residual von Mises stresses of up to 400 MPa occur in the centre of the brazed joint at the interface between the YSZ ceramic and the braze filler alloy.
The calculated residual stress distribution for a braze joint with Ni and Cu interlayer are shown in Fig. 6 b) and c). The brazed joint with Cu interlayer showed the overall smallest von Mises stress at the interface between ceramics and braze alloy. The results clearly show that both an interlayer with high ductility (Cu) and one with low thermal expansion coefficient (Mo) lead to a pronounced reduction of the residual stresses in the joint.

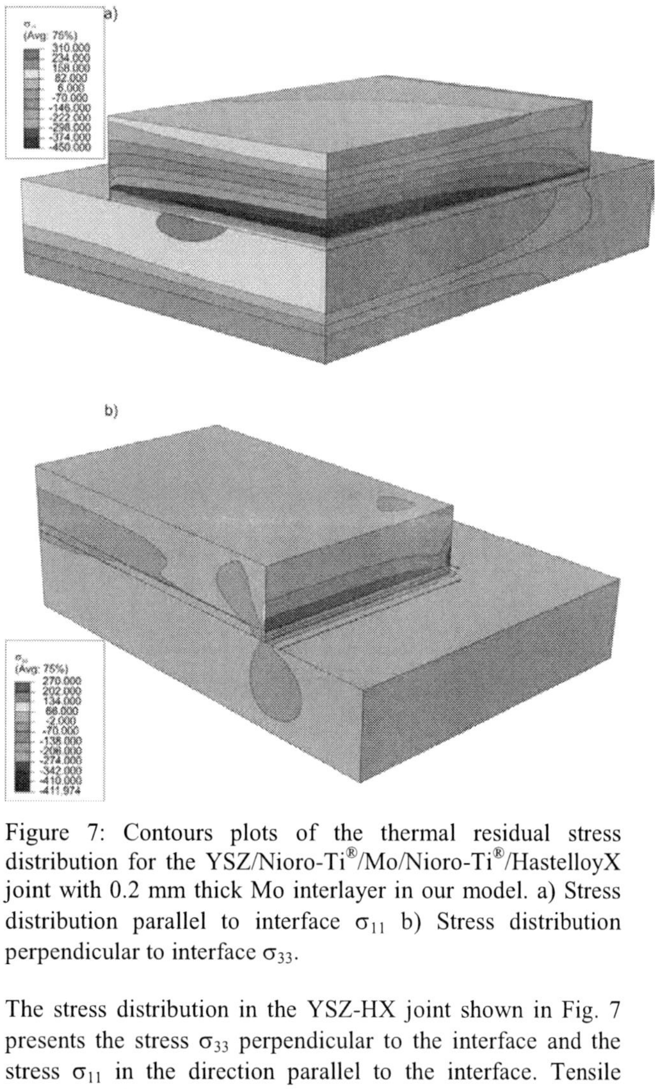

Figure 7: Contours plots of the thermal residual stress distribution for the YSZ/Nioro-Ti®/Mo/Nioro-Ti®/HastelloyX joint with 0.2 mm thick Mo interlayer in our model. a) Stress distribution parallel to interface σ_{11} b) Stress distribution perpendicular to interface σ_{33}.

The stress distribution in the YSZ-HX joint shown in Fig. 7 presents the stress σ_{33} perpendicular to the interface and the stress σ_{11} in the direction parallel to the interface. Tensile stresses at the ceramic side are highest at the free end of the joint with approximatively 250 MPa. The reduced fracture toughness of the ceramic coupled with significant tensile stress can induce cracks in the ceramic and will possibly lead to fracture. The direction of the highest tensile stress as calculated by FEM is in good agreement with the fractographic investigations mentioned in the previous chapter.
The flexural strength of the fully stabilized YSZ is ranging between 300 and 150 MPa depending on the sintering process,

impurity level and powder grain size. Thus in our study some of the samples fractured directly after cooling down from the brazing temperature (980 °C). However, partially stabilized YSZ which has up to five times flexural strength of the fully stabilized YSZ was successfully brazed using Cu and Ni-interlayer configurations. All ceramic parts brazed directly to HX using Nioro-Ti® braze showed significant crack propagation or complete failure after cooling. In the follow up study we will investigate the thermal cycling loading of the actively brazed YSZ and HX parts.

Conclusions

In the present work, experimental and numerical methods were combined and it could be shown that coupled thermo-mechanical simulations by FEM can be used for estimating the residual stresses in complex metal-ceramic joints. In particular, the residual stresses forming in TLP joints, which is performed at relatively low temperatures, and in active braze joints, which is performed at relatively high temperatures were calculated as a function of different material combinations and processing parameters as well as with regard to the presence of defects.

The following points were derived for TLP bonded components with Ag-Sn interlayers:

- The local stress state clearly benefits from a buffer metallization layer between chip and IMC layer
- Small pores within the joint are not critical
- Slight increase in the process temperature does not increase the residual stresses significantly

The active brazing of the YSZ with commercially gold-based braze alloy (Nioro-Ti®) investigated in this study showed the following:

- Mo, Ni and Cu metal foils with thicknesses of 50-200 μm between ceramics and metal significantly reduces thermal residual stresses.
- The best joint configuration in terms of residual stress distribution is reached with an interlayer of Cu foil and a thickness of 200 μm.
- Both strategies, using low expansion metal interlayers or ductile interlayers like Cu are beneficial to increase the lifetime of the joint.

The methods presented are useful tools for the optimization of metal-ceramic joints with regard to their application under extreme operating conditions.

References

[1] Cook, G.O., Sorensen, C.D., "Overview of transient liquid phase and partial transient liquid phase bonding," Journal of Materials Science, Vol.46, No. 16 (2011), pp.5305-5323.

[2] Gollas, B., Albering, J.H., Schmut, K., Pointner, V., Herber, R., Etzkorn, J., "Thin layer in situ XRD of electrodeposited Ag/Sn and Ag/In for low-temperature isothermal diffusion soldering," Intermetallics, Vol.16, No. 8 (2008), pp.962-968.

[3] Li, J.F., Agyakwa, P.A., Johnson, C.M., "Kinetics of Ag3Sn growth in Ag-Sn-Ag system during transient liquid phase soldering process," Acta Materialia, Vol.58, No. 9 (2010), pp.3429-3443.

[4] Lis, A.P., M. S.; Arroyave, R.; Leinenbach, C., "Early stage growth characteristics of Ag_3Sn intermetallic compounds during solid-solid and solid-liquid reactions in the Ag-Sn interlayer system: Experiments and simulations," Journal of Alloys and Compounds, Vol.617, No. (2014), pp.11.

[5] Eo, S.H., Kim, D.S., Jeong, H.J., Jang, J.H., "Fluxless eutectic bonding of GaAs-on-Si by using Ag/Sn solder," Electronic Materials Letters, Vol.9, No. 6 (2013), pp.787-791.

[6] Kim, J.S., Yokozuka, T., Lee, C.C., "Fluxless bonding of silicon to Ag-cladded copper using Sn-based alloys," Materials Science and Engineering a-Structural Materials Properties Microstructure and Processing, Vol.458, No. 1-2 (2007), pp.116-122.

[7] Takahashi, T., Komatsu, S., Kono, T., "Development of Ag3Sn Intermetallic Compound Joint for Power Semiconductor Devices," Electrochemical and Solid State Letters, Vol.12, No. 7 (2009), pp.H263-H265.

[8] Dudek, R., Auersperg, J., Michel, B., Reich, H., "Thermo-mechanical reliability aspects and finite element simulation in packaging," Proceedings of 5th Electronics Packaging Technology Conference, No. (2003), pp.440-449.

[9] Dudek, R., Bramer, B., Irsigler, R., Rzepka, S., Michel, B., "Thermo-Mechanical Reliability Assessment for 3D Through-Si Stacking," Eurosime 2009: Thermal, Mechanical and Multi-Physics Simulation and Experiments in Micro-Electronics and Micro-Systems, No. (2009), pp.474-480.

[10] Lin, K.-L., Singh, M., Asthana, R., "TEM characterization of Au-based alloys to join YSZ to steel for SOFC applications," Materials Characterization, Vol.63, No. (2012), pp.105-111.

[11] Lin, K.-L., Singh, M., Asthana, R., "Characterization of yttria-stabilized-zirconia/stainless steel joint interfaces with gold-based interlayers for solid oxide fuel cell applications," Journal of the European Ceramic Society, Vol.34, No. (2014), pp.355-372.

[12] Singh, M., Shpargel, T.P., Asthana, R., "Brazing of Stainless Steel to Yttria-Stabilized Zirconia Using Gold-Based Brazes for Solid Oxide Fuel Cell Applications," International Journal of Applied Ceramic Technology, Vol.4, No. (2007), pp.119–133.

[14] Stephens, J.J., "Elevated Temperature Creep Properties for Selected Active Metal Braze Alloys," Sandia National Labs., Albuquerque, NM (United States)No. SAND--96-1084C; CONF-970803--2 (1997).

[15] Norton, F.H., The Creep of Steel at High Temperatures, McGraw-Hill Book Company, (New York, 1929).

Capillary Enhanced Dissolution in Brazing

Wen Liu and Brian L. Barten*
Delphi Thermal Systems, Lockport Technical Center, Lockport, NY, USA
**Corresponding contact: wen.2.liu@delphi.com*

Abstract

During a brazing process, dissolution (also known as "erosion" or "guttering") of the solid base metal into the liquid filler metal is an inevitably present phenomenon which can be enhanced by surface tension driven molten metal flow. Severe dissolution associated with base metal thickness reduction can be detrimental for brazed products; therefore, it is important to understand and model the dissolution phenomenon. In this study, an analytical model is established which combines the mechanism of both dissolution and surface tension driven molten metal flow. The model explains the underlying physics of dissolution for a brazing process and associated impacts caused by the capillary action. Groups of aluminum heat exchanger samples were brazed by using industrial CAB furnaces and the dissolution was studied. The results show that the base metal dissolution can be largely impacted by controlling the potential of capillary molten clad flow, which supports the concept of the proposed model.

Introduction

Manufacturing of aluminum heat exchangers with high quality requires reliable metal joining technology to join together various components, such as tubes, fins and manifolds. Controlled atmosphere brazing (CAB) of aluminum is currently a leading technology for mass production of compact aluminum heat exchangers, such as radiators, condensers, evaporators and heaters. During a CAB process, dissolution is closely associated with the braze quality of the products. Severe dissolution may cause leakage, low burst pressure, braze joint degradation and induce more pronounced corrosion in the field, etc. [1, 2], and may also lead to unexpected failure of the heat exchanger [3]. Therefore, understanding, modeling and controlling the base metal dissolution into the filler metal during molten metal spreading over solid substrates at elevated temperatures are of importance.

Regardless of the fact that brazing is a highly developed process, the study on dissolution during brazing lacks theoretical analysis and understanding of the underlying physics. The dissolution of a base metal into the filler has been found to interact with many other factors such as capillary action, wetting and spreading of the molten fillers, as well as the microstructure and the joint strength [4-7]. Only a few studies focused on theoretical understanding and modelling of the dissolution in a brazing process [8-10], in which the dissolution in brazing is analytically determined to be controlled and/or impacted by major factors such as materials properties, brazing gap design and dwell time and peak temperatures.

During the furnace brazing process of an aluminum heat exchanger, all joints between various components are formed simultaneously by wetting and capillary action of the molten clad. Therefore a good understanding and modeling of the effects of the molten flow behavior on base metal dissolution appears to be critical. Few studies, however, on dissolution in brazing include the modeling of capillary action of the molten phase.

In this study, a dissolution phenomenon was registered in a mass production process for aluminum heat exchangers. Such dissolution is assumed to be enhanced by the capillary action of the molten clad. A theoretical model was established to combine both dissolution mechanism and the modeling of the capillary action of molten metal phase in order to better understand such a capillary impacted dissolution with capturing the underlying physics, thus providing possible and feasible solutions. The model suggests that a reduction of the dissolution level can be achieved by altering the capillary effects of the molten metal. A series of experimental studies in a mass production environment was conducted and the results support the concept that is proposed in the model.

Problem Formulation

A heat exchanger product made of aluminum alloy was brazed in a production CAB furnace, and consisted of the following components: folded tubes, air centers or fins, manifolds, end caps and a separator. The components with Al-Si cladding are folded tubes (with Al-7Si cladding), manifolds (with Al-10Si cladding) and end caps (with Al-10Si cladding). The temperature history of the brazed heat exchanger which was measured using Datapaq logger indicates that the peak temperature was 600 C +/- 2 °C and the time above Al-Si eutectic point was about 5 +/- 1 min, which is generally adequate and proper for Al-7Si and Al-10Si cladding to form a joint. However, an undesired level of dissolution between two folded tube legs was registered as shown in Figure 1. The dissolution resulted in an average of 43% (with a standard deviation of 14%) thickness reduction in tube legs for this specific heat exchanger configuration based on the metallographic examination of numerous production development samples. (The dissolution percentage used in this study is defined as the ratio of dissolution depth to original base metal depth.) Such a level of dissolution is highly undesirable for an aluminum heat exchanger product and needs to be reduced to a lower and acceptable level. In contrast, no or very slight dissolution was noticed for the brazing of other different heat exchanger configurations (i.e., with extruded tubes) which even featured higher peak temperatures (i.e., ~ 615 °C) and longer brazing dwell time.

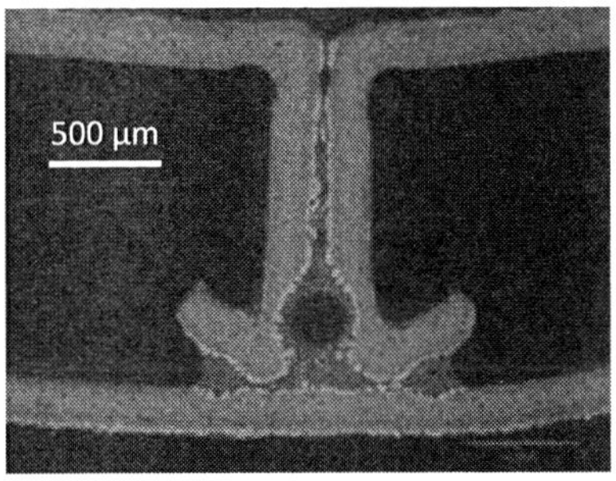

Figure 1. A representative cross section showing severe dissolution into folded tube legs in a folded tube in the heat exchanger product sample

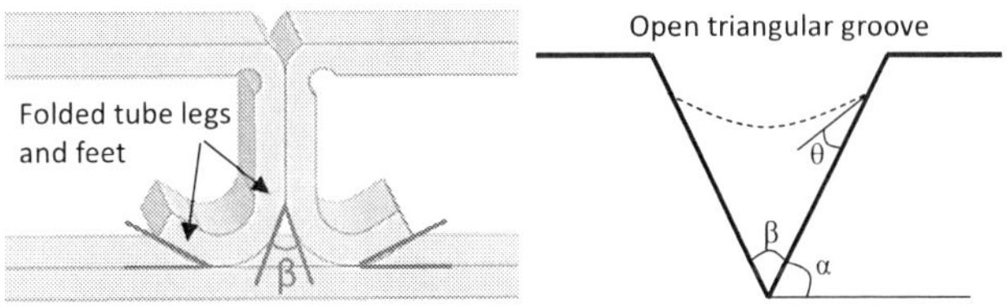

Figure 2. Open triangular groove embedded in the folded tube design (left); Angle definition in a standard open triangular groove (right).

It can be noticed that the geometrical features of a folded tubes, as shown in Figure 2, in fact provide multiple capillary channels for a liquid phase metal, i.e., molten Al-Si cladding in this case. Those indicated channels are considered as open triangular grooves for a capillary transport of the liquid phase [11] as they meet the criteria that the groove angle (β) must be smaller than ninety degrees minus two times of the liquid contact angel (θ): $\beta < 90° - 2\theta$ [12] (contact angle for molten clad θ is usually smaller than 20 degrees [13].) For such open triangular grooves and/or network of such grooves, the detailed analytical modelling of associated capillary actions of liquids including particularly molten metals, has been reported in depth by the author [14, 15, 16]. Due to a close proximity between the tube and the manifold domain, the free molten cladding with higher Si content in the manifold domain will be drawn by capillary into the open triangular channels in the folded tubes, Figure 2. With understanding the mechanism of dissolution reported by Li et al [10] which will be discussed in next section, the hypothesis is that the dissolution level in brazing such a heat exchanger product has been enhanced and/or impacted by the capillary action of the molten cladding that is induced by the geometrical features in this specific product as shown in Figure 2.

The object of this proposed study is to (i) explore and demonstrate theoretically the impact of general capillary action on base metal dissolution, and (ii) apply the concept of the analytical model to mitigate or solve the dissolution issue in a production CAB setting thus validating the concept of the model. The complete solution of the analytical model and the quantitative validation of the model against experimental data will be studied further in the future and addressed elsewhere.

Theory

Four major factors were found to be influential for dissolution in a brazing process at elevated temperature [9]: (i) brazing temperature (particularly the peak temperature), (ii) dwell time associated with brazing temperature, (iii) materials properties of the molten clad, and (iv) the volume of the molten clad "reservoir". The capillary action in brazing will directly and strongly influence the factor (iv). To be more specific, the volume of molten metal reservoir that interacts with base metal is increased by capillary driven movement of the molten phase.

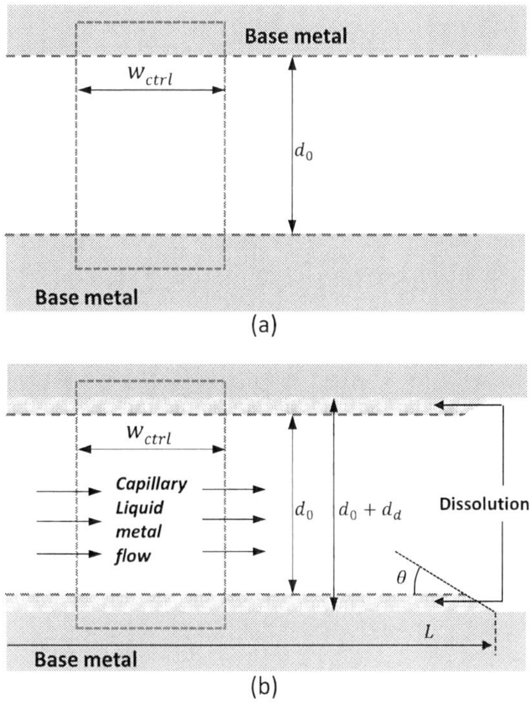

Figure 3. A general physical configuration of a dissolution process with a capillary action: (a) an initial braze gap and base metal without capillary filling of liquid metal; (b) a braze gap with liquid metal filling due to the capillary force and with dissolution taking place

Figure 3 offers a simplified sketch of the physical configuration of the dissolution process associated with capillary action between two parallel faying plates in a *two dimensional* fashion. The physical domain considered is the control volume confined with red dashed line. The width of the control volume w_{ctrl} is in the direction of the capillary flow, the original brazing joint gap is d_0, the dissolution depth is designated as d_d, the saturated concentration of the base material in the molten filler is n_s and the uniform concentration outside the dissolution boundary layer is n, the density of the base material is ρ_s. L represents the length of the capillary flowand θ is the associated contact angle. A similar physical configuration (but without capillary action) for the dissolution in a high temperature brazing process was reported by Li et al [9]. During the dissolution process, the original liquid/solid interface will move toward the base materials. The dimension of each physical entity for the model establishment is listed in Appendix A.

The molten metal, instead of being a static reservoir, is transported by capillary action, therefore the volume of molten metal that the base metal interacts with is increased. The mass conservation for the dissolution in the control volume can be expressed in a 2D fashion as

$$\rho_s d_d w_{ctrl} = n(d_0 + d_d)L \qquad (1)$$

An application of the Nernst-Brunner equation, which was adopted for soldering process [17], to the dissolution process leads to

$$\frac{dn}{dt} = K\frac{A}{V}(n_s - n) \qquad (2)$$

Where, $A = 2w_{ctrl}$ is the surface area of the solid in contact with the liquid phase that is associated with dissolution; V is the volume of the liquid that interacts with the solid phase.

The length of the capillary flow of the liquid between parallel plates, which is a function of material properties, capillary dimension and time [18], can be expressed as

$$L = \left[\frac{\sigma}{\mu}\frac{(d_0 + d_d)\cos\theta}{3}t\right]^{\frac{1}{2}} \qquad (3)$$

Where σ is the liquid-gas surface tension of the liquid phase and μ is the viscosity.

The total volume of liquid flow that interacts with solid in a 2D fashion can be expressed as in equation (4)

$$V = L(d_0 + d_d) = [\frac{\sigma}{\mu}\frac{(d_0 + d_d)^3\cos\theta}{3}t]^{1/2} \qquad (4)$$

Combining equation (1) and (2):

$$\frac{d}{dt}\left(\frac{\rho_s d_d w_{ctrl}}{(d_0 + d_d)L}\right) = K\frac{A}{V}\left(n_s - \frac{\rho_s d_d w_{ctrl}}{(d_0 + d_d)L}\right) \qquad (5)$$

Equations (3), (4) and (5) are the theoretical frame for a dissolution process that is impacted by a capillary action within a brazing gap. It should be pointed out that n_s and K is also a function of the brazing temperature. As the equation has no analytical solution, numerical analysis of the model will be further studied and discussed elsewhere. The solution of equations (3-5) also involves the dissolution distribution along the channel length which requires a series of further studies. The proposed theoretical frame can shed a light on studying theoretically the dissolution in brazing along with surface tension phenomena (i.e., capillary flow in this case.)

In order to evaluate the impact of capillary action on dissolution, a few simplifications and heuristic approaches have been made. For the length of the capillary flow L, assuming brazing dwell time is a fixed value for a specific brazing process t_{braze}, the final value of L therefore is actually a function of (σ/μ) which is determined by the materials properties. In the same vein, the approach is to utilize a constant equivalent capillary gap value (d_0+d_{braze}) for evaluating the length of the capillary flow. So a heuristic approach was made that, with a given brazing time t_{braze}, the length of the capillary flow in the model is considered as a final capillary flow distance L_{braze} which is a strong function of (σ/μ), and the associated liquid phase volume V will be a strong function of materials properties. Therefore the total volume of the filler metal that is transported by capillary action V_{braze} is assumed to be a function of surface tension to viscosity ratio, with a given brazing time t_{braze}.

$$V_{braze} \sim [\frac{\sigma}{\mu}\frac{(d_0 + d_{braze})^3\cos\theta}{3}t_{braze}]^{1/2} \qquad (6)$$

So the with initial condition that $d_d = 0$ with $t = 0$, and with the liquid phase volume V_{braze} to be a function of only materials properties, the integration of equation (1) and (2) from $t = 0$ to t will yield

$$d_d = \frac{n_s V_{braze}}{\rho_s w_{ctrl}}\left[1 - \exp\left(-\frac{2Kt}{V_{braze}}\right)\right] \qquad (7)$$

It should be noticed that d_d is a monotonically increasing function of V_{braze} with $V_{braze} \in (0,\infty)$, in another word, with decreased liquid phase volume V_{braze} the dissolution d_d will decrease accordingly. It should be pointed out that in equation (6) that with a fixed time and capillary gap, the volume of the liquid phase is controlled by surface tension to viscosity ratio thus $d_d \sim f((\sigma/\mu))$. Within a given brazing temperature range the viscosity change will be much more dominant than the surface tension change [19]. Therefore, the model suggests that an increase of viscosity of the liquid phase can reduce the volume of liquid phase associated with capillary action thus leading to a decreased dissolution depth. It is the authors' intention to apply the qualitative suggestion indicated by the model to a brazing process in an industrial setting for a suppressed dissolution.

To be more specific, for the purpose of reducing the dissolution depth in a controlled atmosphere brazing process as outlined in Problem Formulation section, a suppressed dissolution d_d requires a decreased V_{braze} thus a decreased (σ/μ). For Al-Si alloy cladding within a normal temperature range for CAB, a lower content of Si generally feature a smaller (σ/μ) comparing to higher Si content Al-Si alloy [19], thus switching cladding alloy from Al-10Si to Al-7Si is considered as a feasible solution for suppressing the dissolution.

Experiments, Results and Discussion

Brazing experiments that follow the theoretically deduced concept were conducted in order to (i) verify the concept in the established theory, and (ii) to manage to suppress the dissolution for production development samples.

In accordance with Al-Si phase diagram [20] and [19, 21], within the normal brazing temperature range (i.e., 577 °C to ~615 °C) for Al-Si clad, the surface tension to viscosity ratio of lower Si content cladding will be much smaller than that of high Si content cladding, i.e., Al-10Si. The goal of the experiments is to verify that the cladding with higher viscosity can result in a reduced dissolution level in the same heat exchanger configurations with the same brazing parameters.

Table 1. Cladding material used for heat exchangers in Group A and Group B

Group	Number of brazed heat exchangers	Manifolds clad alloy	Folded tube clad alloy	End cap clad alloy
A	6	Al-10Si	Al-7Si	Al-10Si
B	6	Al-7Si	Al-7Si	Al-7Si

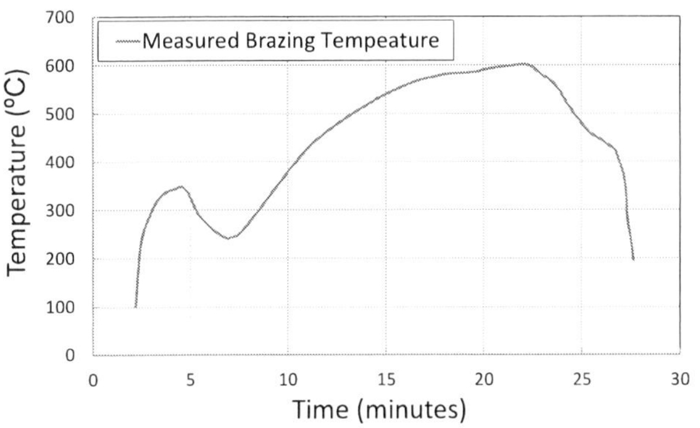

Figure 4. Representative brazing temperature history measured on heat exchanger products using Datapaq logger

Two groups of heat exchanger products (A and B) were manufactured (i.e., formed, folded, and stacked) in the same production line and brazed in the same production CAB furnace at the same time frame. As indicated in the table 1, 6 heat exchangers in Group A serve as the baseline that has a cladding composition of ~ Al-10Si on manifold and end cap as consistent with the original production setting. Another 6 heat exchangers in Group B feature a lower Si level cladding with the composition of ~ Al-7Si on manifold and end cap. They were brazed at the same time frame in the same CAB furnace with the same furnace settings. The brazing temperature profile was logged by Datapaq and is shown in Figure 4. The peak temperature was controlled to be 600 °C +/- 2 °C.

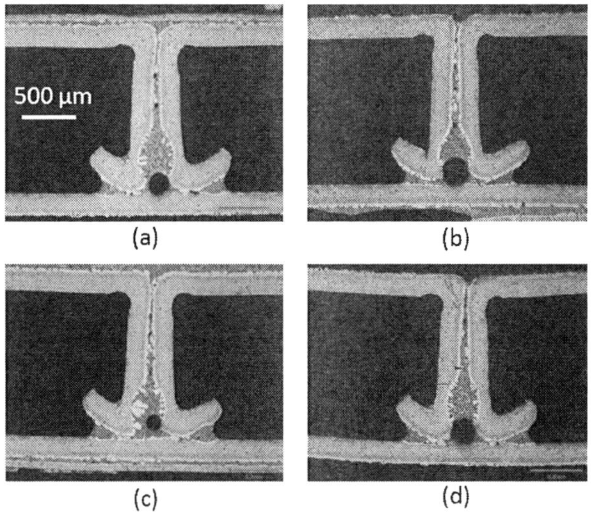

Figure 5. Representative metallographic results for tube base metal dissolution for baseline samples (Group A): (a) Sample A1-R2@15mm; (b) Sample A2-R1@5mm; (c) Sample A4-R2@10mm; (d) Sample A6-L2@5mm

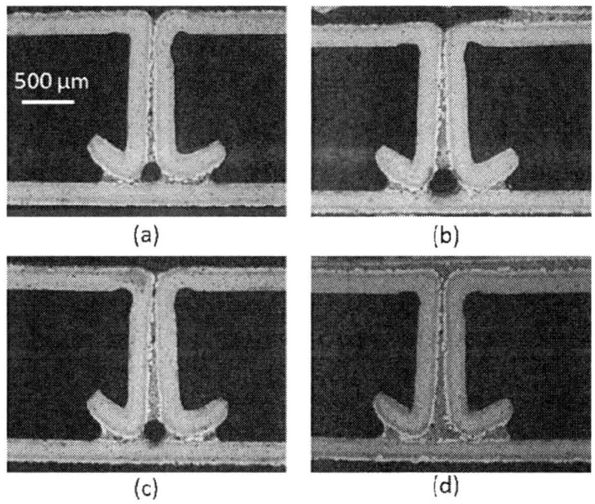

Figure 6. Representative metallographic results for tube base metal dissolution for Group B samples: (a) Sample B1-R1@10mm; (b) Sample B2-R2@15mm; (c) Sample B5-R1@5mm; (d) Sample B6-L2@5mm

Four folded tubes in the brazed heat exchangers were for dissolution study, i.e., 2 at each side of the heat exchanger, and those tubes are designated as R1 (the rightmost one), R2 (the one adjacent to R1), L1 (the leftmost one) and L2 (the one adjacent to L1). The cross sections of the folded tubes were taken at the locations at 5mm, 10 mm and 15mm away from the manifold domain for a metallographic study. The samples were successively ground and polished in order to reveal the microstructure for examining dissolution, which was shown in figure 5 and 6. Note that for simplification the metallographic samples are designated in the format such as "Sample A1-R2@15mm", which means the first sample in group A - the second tube on the right side - the cross section at 15mm away from manifold domain. The distribution of the base metal dissolution across the heat exchanger and along a folded tube will be addressed elsewhere.

Figure 5 and 6 show the representative metallographic images of tube leg dissolution for both Group A and Group B. It is noticed that Group A features a much more severe dissolution than that of Group B that has lower Si level in the clad alloy. For each group and each tube, the maximal dissolution levels (among locations of 5mm, 10mm and 15mm) on 6 cores were averaged in order to be compared. The results are shown in Figure 7, in which one can notice that the dissolution in Group B is suppressed to a very large extent and is much lower than that of Group A.

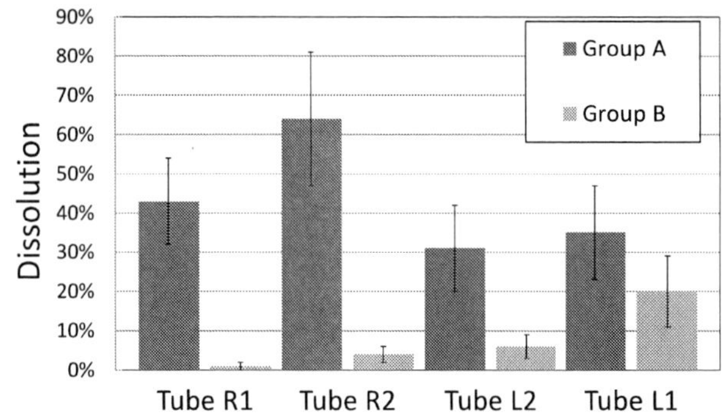

Figure 7. The average dissolution level on selected tubes for Group A and B

Table 2 shows the average dissolution values and standard deviation on those 4 selected tubes for each group. The results show that with clad having 10% Si on manifold and end caps, the dissolution level is found to be within a range from 35% to 64% with a standard deviation from 11% to 17%. In contrast with altering the manifold and end cap clad alloy to Al-7Si, the average dissolution dropped to a low and acceptable 1% to 20% level with a smaller deviation of 1% to 9%.

Table 2. Average dissolution level and standard deviation for selected 4 tubes in both Group A and B

Tube	Group	Avg. Dissolution	Standard Dev.
R1	A	43%	11%
	B	1%	1%
R2	A	64%	17%
	B	4%	2%
L1	A	31%	11%
	B	6%	3%
L2	A	35%	12%
	B	20%	9%

Therefore, it can be clearly concluded that utilization of lower Si content clad that features a lower surface tension to viscosity ratio suppressed significantly the base metal dissolution during the same brazing process, as suggested in the proposed model.

Conclusions

Severe dissolution in controlled atmosphere brazing (CAB) of aluminum may cause leaking, low burst pressure, braze joint degradation and induce more pronounced corrosion in the field. Therefore, understanding, modeling and control of base metal dissolution into the filler metal during molten metal spreading over solid substrates are of importance. Only a few previous studies focused on theoretical understanding of the dissolution in brazing, let alone the modelling that combines both dissolution and capillary actions.

In this study, a theoretical model has been established which combines the mechanism of both dissolution and surface tension driven molten metal flow. The model explains the underlying physics of dissolution for a brazing process and associated enhancements caused by the capillary action. The model suggests that with lowering the Si content in the Al-Si alloy cladding towards a reduced surface tension to viscosity ratio at elevated temperatures, the capillary action can be weakened thus achieving a suppressed dissolution.

To validate the concept in the model, two groups of aluminum heat exchanger samples were brazed by using an industrial CAB furnace and the dissolution was examined. (Note that those two groups of heat exchangers were manufactured and brazed using the same facilities in the same time frame.) The metallographic study of the dissolution show that the baseline group of the heat exchangers experienced an average dissolution up to 64% which is consistent with the results monitored for mass production development samples in a certain period of time. In contrast, the group of heat exchanger with altered Si content in the clad experienced a very low level of dissolution. Therefore the results show that the base metal dissolution can be largely impacted by controlling the potential of capillary molten clad flow, which supports the concept in the proposed model.

Acknowledgments

The authors acknowledge the support from Mark Zima and Dr. Prasad Kadle, Advanced Engineering Group, Delphi Thermal Systems, Lockport, NY, USA. The authors appreciate the support from Lorraine Anders and Daniel Wilson, General Motors, Lockport, NY, USA.

Appendix A: Nomenclature

The dimension for each symbol / physical entity in this paper is listed as follows:

A	solid area associated with dissolution	m^2
d_d	total dissolution distance	m
d_0	initial brazing gap	m
K	dissolution coefficient	m/s
n,	concentration	kg/m^3
n_s	saturated concentration	kg/m^3

L	length of the capillary flow	m
L_{braze}	L that is associated with t_{braze}	m
t	time	s
t_{braze}	a given specific brazing time	s
V	volume of the liquid phase	m^3
V_{braze}	V that is associated with t_{braze}	m^3
w_{ctrl}	width of the control volume	m
α	groove angle	degree
β	groove angle	degree
θ	contact angle	degree
μ	dynamic viscosity	Pa·s
ρ_s	density of the solid phase	kg/m^3
σ	surface tension	N/m

References

[1] Liu S, Olson DL, Martin GP. "Modeling of brazing processes that use coatings and interlayers." Weld J Vol.70 No. 8 (1991), pp. 207–15.

[2] Shiue RK, Wu SK,Chen SY. "Infrared brazing of TiAl intermetallic using BAg-8 braze alloy." Acta Materialia. Vol. 51, No. 3 (2003), pp. 1991-2004

[3] Solvay Fluro document, 2013, Nocolok encyclopedia, available online at: http://www.aluminiumbrazing.com/sponsor/nocolok/Files/PDFs/NOCOLOK-Encyclopedia-2013.pdf

[4] Wu Z. Y., Yeh T. Y., Shiue R. K. "The Effect of Substrate Dissolution in Brazing CP-Ti and Ti-15-3 Using Clad Ti-15Cu-15Ni Filler." ISIJ INTERNATIONAL. Vol. 50 No. 3 (2010), pp. 450-454

[5] Zhang Yang, Yan Jiu-chun, Chen Xiao-guang. "Ultrasonic dissolution of brazing of 55% SiCp/A356 composites." Transactions of Nonferrous Metals Society of China. Vol. 20 No.5 (2010), pp746-750

[6] Chada S; Fournelle RA; Laub W. "Copper substrate dissolution in eutectic Sn-Ag solder and its effect on microstructure. Journal of Electronic Materials." Vol. 29 No. 10 (2000), pp. 1214-1221

[7] Nylen M; Gustavsson U; Hutchinson B. "Mechanisms of Erosion during Brazing of Aluminium Alloys." Materials Science Forum. Vol. 396 No. 4 (2002), pp. 1585-1590

[8] Lashko NF, Lashko SV. Contact metallurgical processes in soldering metals. Moscow: Metallurgy Press; 1977

[9] Zhang X.P, Shi Y.W. "A dissolution model of base metal in liquid brazing filler metal during high temperature brazing." Scripta Materialia. Vol. 50 (2004), pp. 1003-1006

[10] Yulong Li, Wen Liu, Peng He and Dusan P. Sekulic, "Dissolution of TiAl alloy during a high temperature brazing", Journal of Materials Science, Vol. 48 No.15 (2013), pp 5247-5252

[11] Yost, F. G.; Rye, R. R.; Mann, J. A., Jr. "Solder wetting kinetics in narrow v-grooves." Acta Materilia, Vol. 45 (1997), pp. 5337–5345.

[12] Romero, L.A. and Yost, F.G. "Flow in an open channel capillary", J. Fluid Mech., Vol. 322 (1996), pp.109-129

[13] Dong, Fangxiao, "Kinetics of wedge-tee joint formation during brazing of an aluminum alloy under controlled atmosphere" Thesis and Dissertations – Mechanical Engineering. (2013), Paper 23. University of Kentucky.

[14] Wen Liu and Dusan P. Sekulic, "Capillary Driven Molten Metal Flow over Topographically Complex Surface", *Langmuir*, Vol. 27 No.11 (2011), pp. 6720-6730

[15] Wen Liu, Yulong Li, Yuguang Cai and Dusan P. Sekulic, "Capillary Rise of Liquids over a Microstructured Solid Surface", *Langmuir*, Vol. 27 No.23 (2011), pp. 14260-14266

[16] Liu, W., Lu, L., Li, Y., Cai, Y., Sekulic, D. P. "Preferential spreading of molten metal over an anisotropically microstructured surface." *EPL*. Vol 97 (2012), 46003.

[17] Mannan, S. and Clode, M.P. "Dissolution of solids in contact with liquid solder", Soldering and Surface Mount Technology, Vol. 16 (2004), pp. 31–33

[18] Mattew K. Schwiebert and William H. Leong, "Underfill Flow as Viscous Flow Between Parallel Plates Driven by Capillary Action", IEEE Transactions on Components, Packaging, and Manufacturing Technology – Part C, Vol. 19, No.2, April 1996

[19] Battezzati, L. and Green, A.L., The viscosity of liquid metals and alloys, Acta. Metal., Vol.37 No. 7 (1989), pp. 1791-1802

[20] Iida, T. and Guthrie, R., 1988. The Physical Properties of Liquid Metals, Clarendon Press.

[21] J. L. Murray, A. J. McAlister, The Al-Si (Aluminum-Silicon) system, Bulletin of Alloy Phase Diagrams ,February, Volume 5, Issue 1 (1984) pp 74-84

Strength of Brazed Joints and Metallurgical Testing of Cu-4Sn-6P and Cu-40Zn-1Sn-0.3Si Silver-free Filler Metals

Matthew J. Duffey, Jacob T. Marchal, Matthew R. Loney, Boian T. Alexandrov
The Ohio State University, Columbus, Ohio, USA
duffey.53@buckeyemail.osu.edu

Alexander E. Shapiro
Titanium Brazing, Inc., Columbus, Ohio, USA
ashapiro@titanium-brazing.com

Abstract

The results of shear strength testing and metallurgical characterization of two new Cu-based silver-free filler metals TiBraze®P14 (Cu-6P-4Sn wt.%) and TiBraze®LOK59-03 (Cu-40Zn-1Sn-0.3Si wt.%) designated for brazing copper, brass, low-carbon steel, stainless steel, and Ni-plated carbon steel are reported. Both studied filler metals are designated to replace some silver-based filler metals in brazing applications.

Shear strength and microstructure of brazed joints were analyzed and compared to standard silver-based BAg-1, BCuP-5, and a silver-free Blockade® (BCuP-9) filler metals that are currently used in industry.

TiBraze®LOK59-03 provides the best combination of strength, flow, and microstructure as compared to standard BAg-1 and BCuP-5 silver-containing filler metals. LOK59-03 is better suited to braze steels than the conventional silver-based BFM; it makes sound quality joints and can withstand high loads.

TiBraze®P14 provided sufficient strength of carbon steel brazed joints despite the iron phosphide layer at the interface. This low-temperature silver-free filler metal also exhibited good flow and strength of copper and brass brazed joints comparable with standard silver-based filler metals.

When used for joining copper or brass tubes with connectors, adapters, and faucets, both tested silver-free filler metals provide formation of quality, dense brazed joints with smooth fillets. Significant cost savings can be realized with silver-free brazing filler metals, warranted even in the case that sometimes joint overlap lengthening is required.

Introduction

Silver-based brazing filler metals (BFM) are widely used in the industry for many years, due to their excellent flow and strong joints of steels and copper alloys [Ref. 1]. However, high price of silver nowadays forced us to seek alternative, silver-free filler metals to decrease production costs. Silver-based filler metals costs have risen due to the cost of silver. BCuP-5, which contains 15% of silver, costs about $173/lb and BAg-1, which contains 45% of silver, costs about $490/lb. Alternative silver-free filler metals cost much less (around $40-60/lb).

Silver-free braze materials TiBraze®P14 (Cu-6P-4Sn wt.%) and TiBraze®LOK59-03 (Cu-40Zn-1Sn-0.3Si wt.%) are relatively new in the world of brazing and are becoming competitive in terms of mechanical properties and ability to produce acceptable fill geometries. Both filler metals contain copper and tin. The primary difference is that TiBraze®P14 utilizes phosphorus to suppress the melting temperature, while TiBraze®LOK59-03 utilizes zinc and silicon to suppress the melting temperature and for added strength [Ref. 2, 3].

Experimental Procedure

Two new silver-free filler metals: TiBraze®P14 in the form of rubber-binder paste and TiBraze®LOK59-03 rods Ø2.4 mm were evaluated to determine characteristics such as spreading area, joint strength, and microstructure. Table 1 contains chemical compositions of each filler metal.

Table 1 – Brazing filler metals, their compositions and melting ranges

Brazing Filler Metal	Chemical Composition, wt%	Melting temperature range,	
		°C	°F
TiBraze®LOK59-03	Cu-39.7Zn-(0.9-1.1)Sn-(0.2-0.4)Si	880-905	1616-1662
TiBraze®P14	Cu-(5.3-6)P-(3.5-4)Sn	635-680	1175-1256
BCuP-9	Cu-(6-7)Sn-(6-7)P-(0.01-0.4)Si	637-674	1178-1274
BCuP-5	Cu-(14.5-15.5)Ag-(4.8-5.2)P	643-802	1190-1475
BAg-1	Ag-(14-16)Cu-(14-18)Zn-24Cd	607-618	1125-1145

Table 2 - Average shear strengths of steel brazed joints, ksi (MPa)

Braze Filler Metal	1018 Carbon Steel	Ni-Plated 1018 Carbon Steel	Stainless Steel 304
TiBraze®LOK59-03	38.6 (266.3)	-	46.8 (322.9)
TiBraze®P14	19.9 (137.3)	7.6 (52.4)	-
BAg-1	35.2 (242.9)	22.2 (153.2)	33.0 (227.7)
BCuP-5	9.3 (64.2)	15.1 (104.2)	-
BCuP-9	12.3 (84.9)	13.3 (91.8)	-

Table 3 - Average shear strengths of copper and brass brazed joints, ksi (MPa)

Braze Filler Metal	Copper	Brass
TiBraze®LOK59-03	18.0 (124.2)	-
TiBraze®P14	22.2 (153.2)	29.1 (200.8)
BAg-1	25.0 (172.5)	34.5 (238.1)
BCuP-5	17.8 (122.8)	26.0 (179.4)
BCuP-9	21.2 (146.3)	22.2 (153.2)

Two standard silver-containing filler metals BAg-1, BCuP-5 and a silver-free BCuP-9 alloy were tested for comparison with new silver-free filler metals. All filler metals were tested on 1018 carbon steel, Ni-plated 1018 carbon steel, and stainless steel 304.

Brazing was performed in air by heating with a propane torch. The flux used during brazing was the boron-modified, black fluoroborate flux 601B/3411 (Superior Flux & Mfg. Co., Solon, OH).

Standard lap steel specimens were used for brazing and testing according to AWS C3.2M/C3.2:2008, Standard Methods for Evaluating the Strength of Brazed Joints. The base metal thickness was 0.125" (3.175 mm) and width was 0.5" (12.7 mm). Three to five specimens were fabricated for every combination of the base and braze metals that were subjected to tensile testing to determine shear strengths and evaluate their ductility.

Microstructure of brazed joints was evaluated by optical and electronic microscopy. Joints made with BAg-1 filler metal were etched using acetic acid (5 mL) and ethanol (100 mL) [Ref. 4]. Joints made with LOK59-03 were etched by the solution containing 4 g of $FeCl_3$, 25 ml of HCl, and 75 ml of water. For all other joints, a two step etching process took place. The first step was immersing the sample in a 4% solution of HNO_3 in water for 5-10 seconds. The second step was immersing the sample in a 5% solution of ammonium persulfate in water for 5-10 seconds. Following the two step procedure, the samples were washed with ethanol.

Also: thermodynamic simulation software ThermoCalc™ and ternary phase diagrams [Ref. 5,6] were used to determine phase composition of the joint metal.

Results and Discussion

Average shear strengths of steel brazed joints made with all tested filler metals are presented in Table 2, and shear strength of copper and brass brazed joints - in Table 3.

Brazing with P14 resulted in sufficient shear strength with 1018 carbon steel (19.9 ksi), although it was lower than the average shear strength of LOK59-03 and BAg-1. P14 produced lower shear strengths than competitive silver-phosphorus and silver-free filler metals BCuP-5 and BCuP-9 on brazing of Ni-plated carbon steel.

LOK59-03 exhibited excellent shear strength on stainless steel 304 and 1018 carbon steel. The average shear strength of stainless steel joints was 46.8 ksi and carbon steel joints 38.6 ksi. This shear strength exceeded the averaged shear strength of silver-based filler metal BAg-1, which contained an averaged shear strength of 33 ksi. This implies that LOK59-03 is more effective than competitive brazing filler metals on brazing of steels.

Brazed joints of copper and, especially, brass--exhibited relatively high strength, the second highest value after the standard silver-based alloy BAg-1. That makes the silver-free P14 alloy a very prospective substitute for silver. At least, 15%-silver containing BCuP-5 can be completely substituted by P14 in such applications as the manufacture of copper tubing and brass faucet brazed joints.

Figures 1, 2 and 3 are images of a copper and brass joints made with the P14 filler metal. Quality of this joint is characterized by dense joint metal and smooth, defect-free fillets.

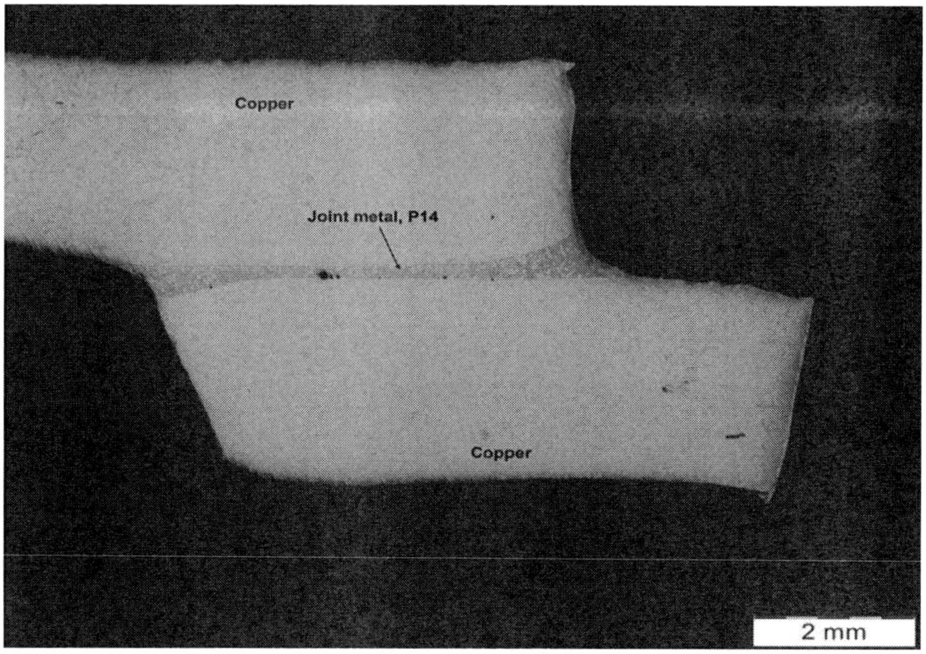

Figure 1 Macrostructure of copper + P14 brazed joint, x12.5

Figure 2 Macrostructure of brass + P14 brazed joint, x100

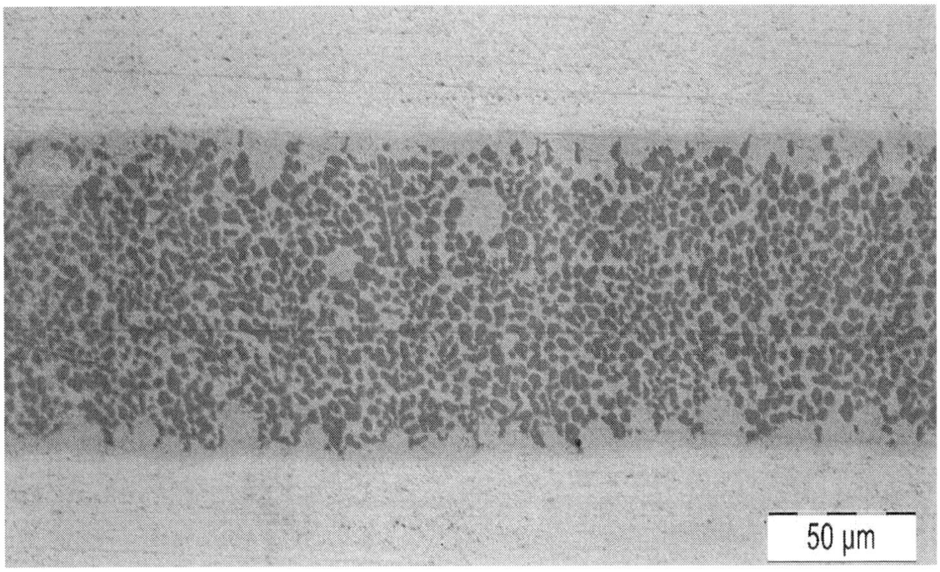

Figure 3 Microstructure of brass + P14 brazed joint, x500

The phase composition of the joint metal consists of a FCC phase of a ternary solid solution Cu-(8.5-11)Zn-2Sn wt.% and an eutectic containing Cu_3P and Cu_3Sn components. The microstructure displays that the FCC phase grows along the joint interface with the base metal, similar to the copper base metal joint. The eutectic phase dominates the middle of the joint. Fig. 2 and 3 portray a smooth fillet and a uniform microstructure of joint metal. Considering the relatively high strength of the brazed joints, as well as joint and fillet formation, this base metal and filler metal combination is good for brazing brass parts.

Figure 4 is an image of a joint made with 1018 carbon steel and TiBraze®P14 silver-free filler metal and Figure 5 is an image of a 1018 carbon steel joint made with TiBraze®LOK59-03 silver-free filler metal. The joint metal consists of the FCC phase, which is the Cu-40Zn brass alloy, and BCC phase that is Cu-40Zn-8.6Sn bronze alloy.

There is more FCC phase in the joint than BCC phase but both phases are distributed uniformly and that results in the high strength of brazed joint. The interface of the 1018 carbon steel + LOK59-03 joint is characterized by formation of a discontinuous FeSi intermetallic layer, due to a reaction of silicon with steel.

Silver-free brazing filler metals P14 and LOK59-03 are suitable as substitutes for silver-based alloys in joining hydraulic, pneumatic, and lubrication pipelines; and they can be widely used in the manufacture of automobiles, refrigerators, house water pipes, machinery equipment, electronics, etc. Typical examples of brazed joints of copper, steel, and brass tubes made with these silver-free filler metals are presented in Figures 6, 7.

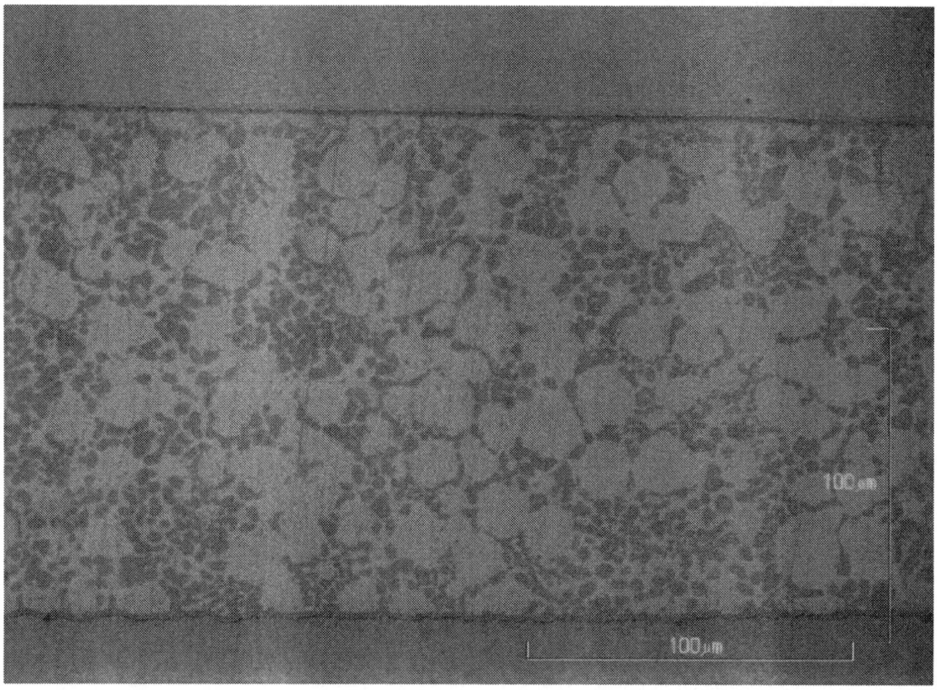

Figure 4 Microstructure of carbon steel +P14 brazed joint, x100

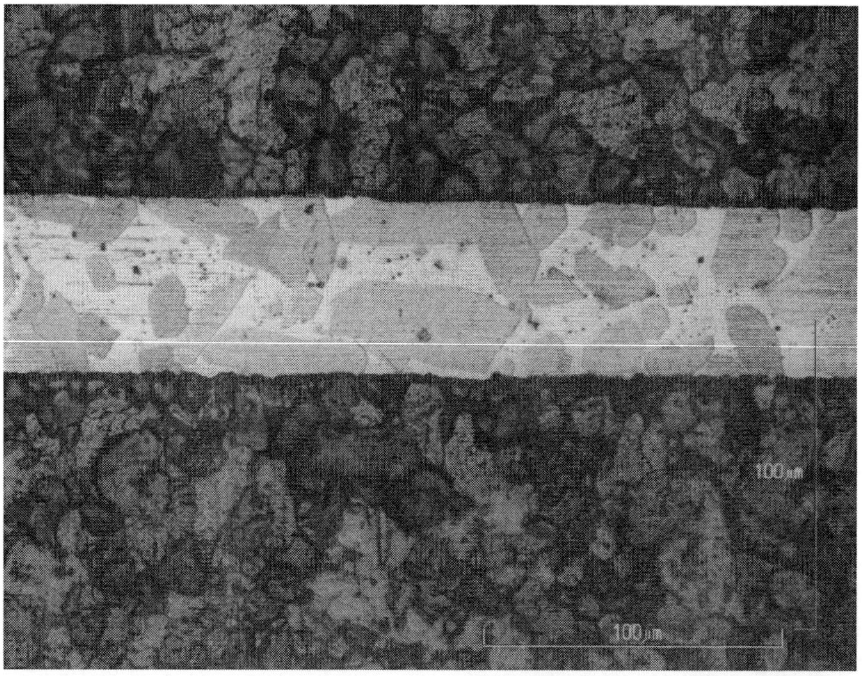

Figure 5 Microstructure of carbon steel +LOK59-03 brazed joint, x500

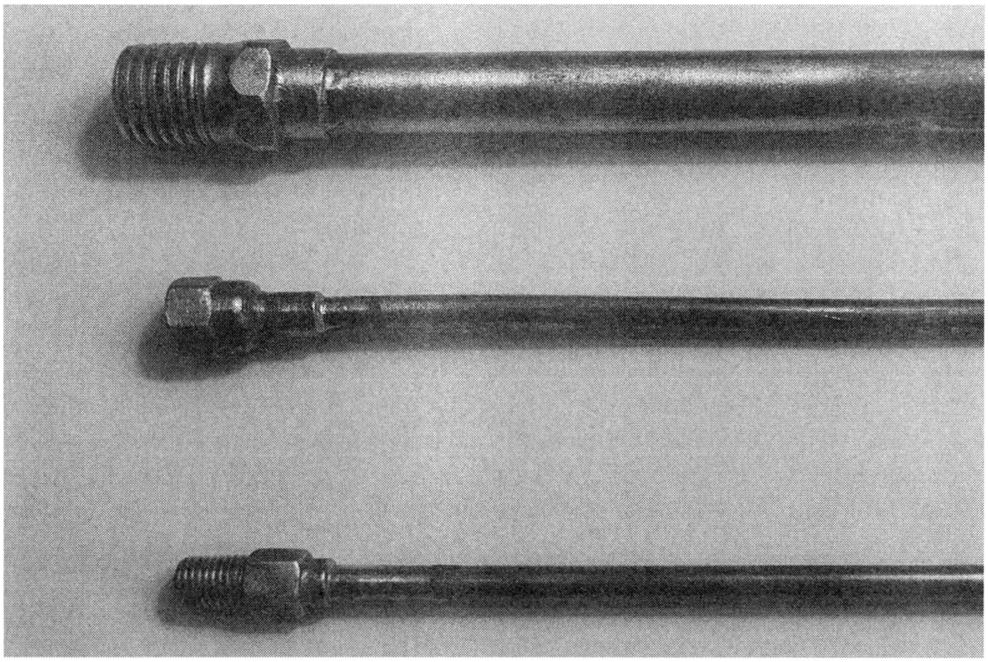

Figure 6 Examples of copper and steel pipe joints brazed with silver-free P14 and LOK59-03 filler metals

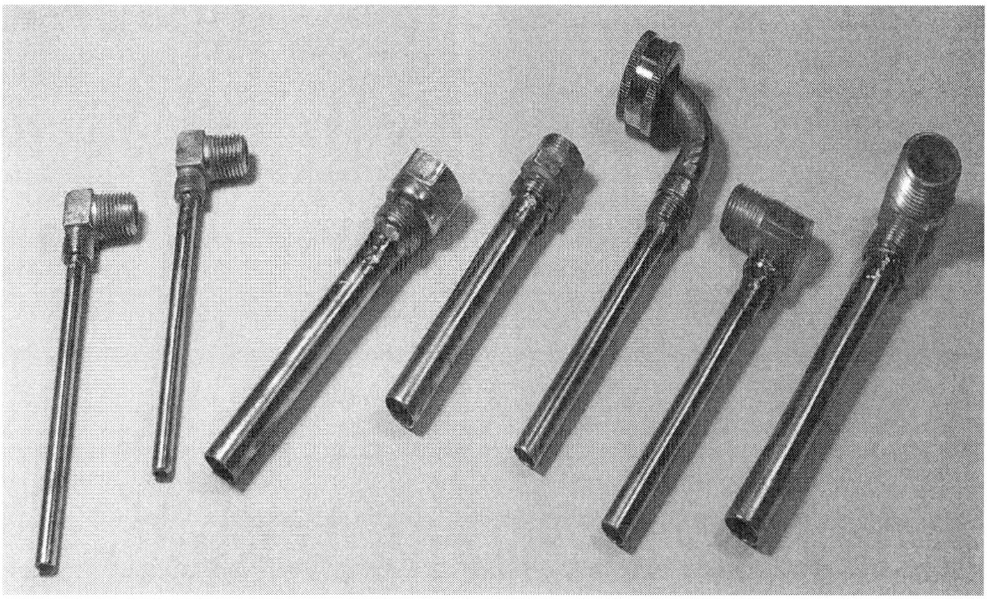

Figure 7 Examples of brass pipe and faucet joints brazed with silver-free P14 filler metal

Conclusions

1. Silver-free brazing filler metals are capable of meeting strength requirements in copper and brass brazed joints. Brazing with the low-temperature silver-free filler metal TiBraze®P14 (Cu-4Sn-6P) resulted in sufficient shear strengths with copper and, especially, with brass-- the second highest value after the standard silver-based alloy BAg-1. That makes the silver-free P14 alloy a very prospective substitute for silver. At least, 15%-silver containing BCuP-5 can be completely substituted for by P14 in such applications as the manufacture of copper tubing and brass faucets brazed joints.

2. TiBraze®LOK59-03 (Cu-40Zn-1Sn-0.3Si) filler metal provides the best combination of strength and flow as compared to standard BAg-1 and BCuP-5 silver-containing filler metals. LOK59-03 is better suited to braze steels than the conventional silver-based BFM; it makes sound quality joints and can withstand high loads. LOK59-03 is recommended for industrial testing and approval of brazing stainless steel and carbon steel.

3. Unlike other copper-phosphorus filler metals, low-temperature silver-free filler metal TiBraze®P14 (Cu-4Sn-6P) provided sufficient strength of carbon steel brazed joints despite the iron phosphide layer at the interface.

4. When used for joining copper or brass tubes with connectors, adapters, and faucets, both tested silver-free filler metals provide formation of quality, dense brazed joints with smooth fillets. Significant cost savings can be realized with silver-free brazing filler metals, warranted even in the case that sometimes joint overlap lengthening is required.

References:

1. Peaslee R. L., Fortuna D., and Fink D. A. 2007. Brazing Filler Metals, Brazing Handbook, 5th edition, Chapter 3,Miami, FL: American Welding Society, 67-99

2. Pashkov I. N., Ilina I. I., and Shapiro A. E. 2006. Properties and applications of Cu-based silver-free brazing filler metals made by rapid solidification technique, Proc. of 3rd Int. Brazing and Soldering Conference IBSC-2006, San-Antonio, TX, 157-166

3. Ilina I.I., Pashkov I.N., and Shapiro A.E. 2006. Copper-based silver-free brazing filler metals, US Patent Application 60/789461

4. Shapiro, A. E. 2014. Etching procedure and reagents for revealing microstructure of brazed or soldered joints, "Brazing Q & A", Welding Journal, vol. 93, No. 2, 26-27

5. Muromachi S., Watanabe H., Tomimoto S. 2006. Cu-P-Sn Phase Diagram, ASM Alloy Phase Diagrams Center, P. Villars, editor-in-chief; H. Okamoto and K. Cenzual, section editors, ASM International, Materials Park, OH

6. Chou C., Chen S. 2006. Cu-Sn-Zn Phase Diagram, ASM Alloy Phase Diagrams Center, P. Villars, editor-in-chief; H. Okamoto and K. Cenzual, section editors, ASM International, Materials Park, OH

Single and Double Reaction Layer Formation Using Reduced Active Element Containing Brazing Filler Metals

Charles A. Walker, Greg Bishop, Dennis J. DeSmet, Michael Brumbach
Sandia National Laboratories, Albuquerque, New Mexico, USA
cawalke@sandia.gov, 505.844.7103

Abstract

Active braze filler metals allow nonmetals to be joined to metals and nonmetals without the use of metallization coatings. This is accomplished by including sufficient active element to form a continuous reaction layer on each nonmetal surface resulting in strong, hermetic brazements. The reaction layer is made possible by the inclusion of small quantities, usually 1-3 weight percent, of a Group IVA metal into the braze filler metal composition. Because active filler metal compositions are limited and designed to work with various materials systems, the amount of active element is often more than required resulting in thick, brittle reaction layers or reactions with base materials that lead to excessive braze filler metal flow. Reduced quantities of the active elements titanium and zirconium have been added to standard braze filler metals to determine the quantities required to form continuous reaction layers without compromising tensile strength or promoting excess filler metal wetting/flow.

Sandia National Laboratories is a multi-program laboratory managed and operated by Sandia Corporation, a wholly owned subsidiary of Lockheed Martin Corporation, for the U.S. Department of Energy's National Nuclear Security Administration under Contract DE-AC04-94AL85000.

Introduction

Active brazing is different than conventional brazing. Conventional brazing is performed on metallic substrates using a filler metal that melts at a temperature lower than that of the substrates. To join metals to nonmetals it is necessary to provide a continuous metallic surface on the nonmetal in order to promote wetting, thus enabling the braze filler metal to form a chemical/structural bond. Active brazing uses the "active element" within it to form a metallic surface on the nonmetal while in the molten state which is subsequently joined to the metallic substrate by the remainder of the filler metal. Figure 1 shows a simplified sketch that illustrates these concepts for a metal brazed to a nonmetal using conventional and active braze filler metals. A closer inspection of both assemblies post-brazing would reveal that a metal layer had reacted with the nonmetal forming a metallic interface with which the liquid brazing filler metal had wetted and formed a metallurgical bond. However, as shown in Figure 2, the manners in which the two different brazing methods accomplish this vary greatly in both the number and complexity of the required steps. Perhaps most importantly, for the conventional brazing process the metallic surface is applied to the nonmetal substrate prior to brazing while the active brazing method forms the required reaction layer and metallic surface in-situ.

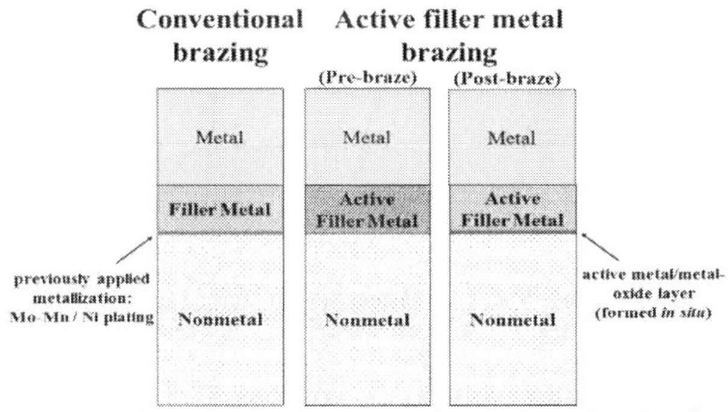

Figure 1. Schematic showing conventionally brazed and active filler metal brazed metal/nonmetal sample.

While Figure 2 does clearly indicate that the active brazing process is much simpler overall to perform than the conventional brazing method for metal-nonmetal or nonmetal-nonmetal assemblies, in practice it is not possible or practical to braze everything using the active brazing method. As has been previously reported [1-3] not all metal-nonmetal brazement designs are feasible for active metal brazing, such as those that require the liquid filler metal to be drawn between the faying surfaces by capillary action or those that do not allow the brazing filler metal to be preplaced between the brazement faying surfaces.

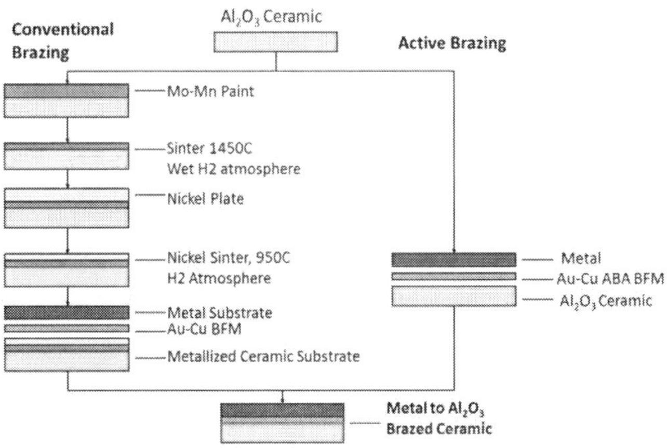

Figure 2. Conventional and active brazing comparison.

Active brazing is a viable alternative to conventional brazing whenever traditional metallization techniques are difficult to perform, i.e. small features, irregular shapes, large ceramic piece parts, etc. Also, active brazing is suitable when there are low numbers of parts being built or small-lot production, where automated processes are not cost-effective, or practical to set-up. Third, the active brazing alternative can be used when the joint geometry permits pre-placement of the braze filler metal (butt, lap, shear joints, etc.). And finally, the active filler metal approach is simplest to implement when filler metals are available in active alloy compositions as that matches the conventional composition counterparts.

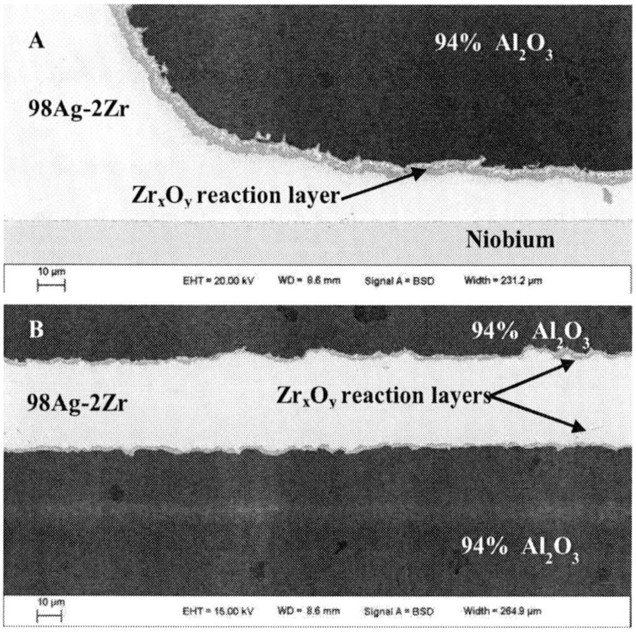

Figure 3. SEM images showing (A) one thick, single, continuous reaction layer in a 94% Al_2O_3 -Nb sample; and (B) two reaction layers formed on 94% Al_2O_3 interfaces using 98Ag-2Zr active brazing filler metal.

After performing a survey of available filler metal compositions it becomes quite clear that the last point is one of the primary limiting factors as to why more users do not consider replacing conventional brazing plus metallization with active brazing processes. The quantity of commercially available active filler metal compositions is limited to only a few compositions versus far more conventional alloy types to cover the 600°C - 1200°C temperature range of brazing applications [4, 5]

Certainly, it is a case of supply-and-demand. It is not cost effective for a supplier to develop a product that has a limited market. On the other hand, the users cannot develop actively brazed products when suitable filler metals are not available. Because of this situation, the user is forced to apply an active filler metal that is not optimized for the brazement, thereby posing a potential reliability concern. For example, brazing metals to nonmetals with a filler metal produces a joint having one reaction layer as seen in Figure 3A. However, the active braze alloy was optimized for two reaction layers in joints between nonmetals. The reaction layers are considerably thinner in the latter case as shown in Figure 3B. Excess active element leads to the formation of too thick of a reaction layer (Figure 3A) that jeopardizes joint strength and hermeticity. Excessive reactions can lead to excessive braze filler metal flow ("run-out").

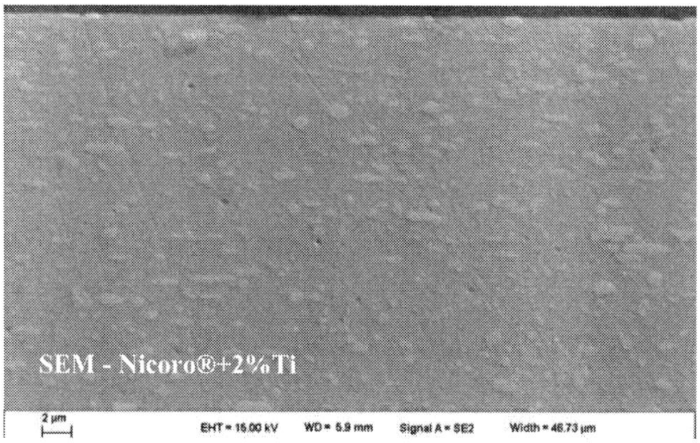

Figure 4. SEM image of as-fabricated 62Cu-35Au-2Ti-1Ni active braze filler metal.

The fabrication process for active brazing filler metals is much the same as those used for conventional filler metal alloys, the result being the active element is distributed throughout the bulk of the brazing filler metal as shown in Figures 4 and 5. Figure 4 is a image captured by a scanning electron microscope (SEM) of a commercially available 62Cu-35Au-2Ti-1Ni (weight percent) active brazing filler metal.

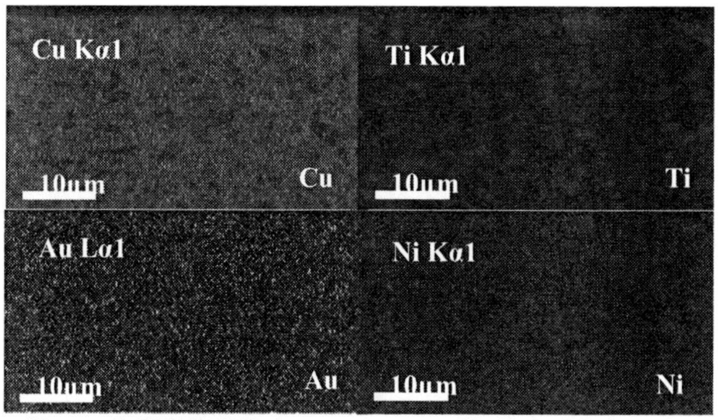

Figure 5. SEM/EDS elemental maps of bulk as-fabricated 62Cu-35Au-2Ti-1Ni active braze filler metal.

Figure 5 displays energy dispersive spectroscopy (EDS) images captured on an SEM. The EDS images give a qualitative assessment of the elements that are present in a sample and how those elements are dispersed in the microstructure. During the brazing process an active element would react with the nonmetal or oxide surface(s). Figure 6 is an SEM image of two 94% alumina samples that were brazed using the same 62Cu-35Au-2Ti-1Ni active braze filler metal and subsequently cross-sectioned and polished by metallographic techniques. The lighter phases seen at the braze filler metal/alumina ceramic interfaces and within the bulk filler metal are titanium rich. Figure 7 makes this more readily apparent as the EDS maps clearly show the unreacted titanium within the bulk filler metal as well as titanium that successfully migrated through the liquid filler metal to the oxide ceramic surface where it reacted to form the Ti_xO_y reaction layer that gives the brazement the both strength and hermeticity. Excess titanium active element not used to form the reaction layer with the nonmetal surface(s) may remain within the bulk filler metal, forming Ti_xNi and $AuTi_x$ compounds, or reacting with the metal substrate.

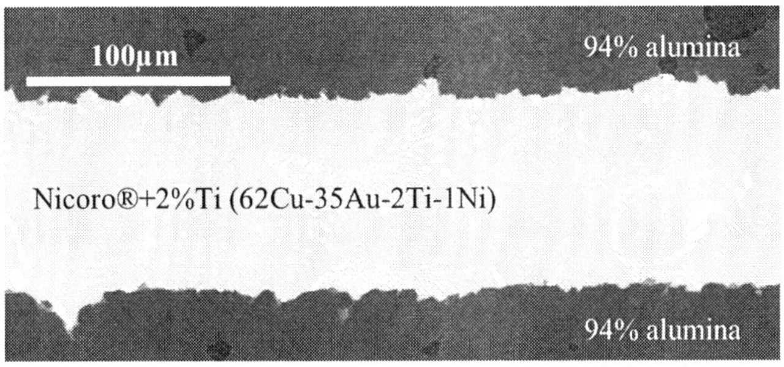

Figure 6. SEM Image, Vacuum brazed 94% Al_2O_3 using 62Cu-35Au-2Ti-1Ni active braze filler metal

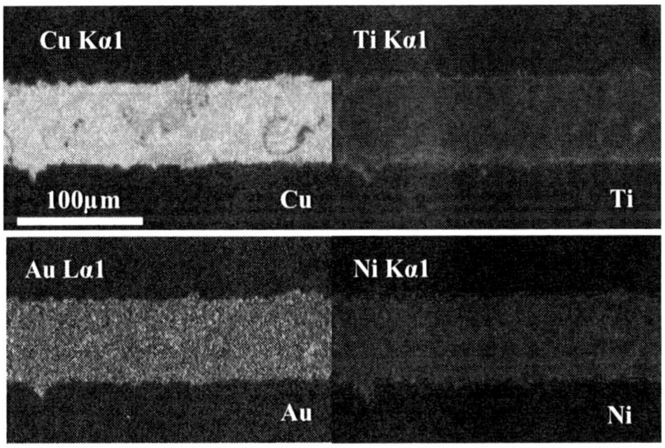

Figure 7. SEM/EDS elemental maps, Nicoro®+2%Ti active braze filler metal.

Closer inspection of the EDS maps in Figure 7 show this phenomena where the remaining or excess titanium within the bulk filler metal has reacted with the nickel also present in the filler metal, which though not deleterious in these samples has been shown to produce undesired effects for other material combinations. Specific active element/metal substrate combinations have been shown to react excessively with the base metals, effectively reducing the active element available to generate a desirable reaction layer at the nonmetal [6, 7].

Braze filler metal selection and modification

In order to determine the optimum quantity of active element required for ceramic to ceramic brazements as well as metal to ceramic brazed assemblies, varying amounts of the active elements titanium and zirconium were deposited onto conventional brazing filler metal using the physical vapor deposition method, direct current (DC) sputtering. A noble protective layer of either silver or gold was deposited over the active element layer to prevent unwanted oxidation when exposed to air. For the first trials, sputtered layer thicknesses of 0.5μm were selected for both the active element and the noble metal. These initial thicknesses were chosen because previous efforts to join metals to nonmetals using a thin-film metallization system demonstrated [8] (shown in Figures 8 and 9) that hermetic, high-strength metal-nonmetal brazements can be made using sputter deposited active element layers only 0.25μm - 0.5μm thick deposited directly onto the faying surfaces of bare alumina ceramic substrates. It was hypothesized that a similar thickness of active element deposited onto the brazing filler metal would be sufficient to form a sufficient reaction layer at the alumina ceramic interface.

Table 1 – Active element sputtered thickness equivalent

Brazing Filler Metal	Thickness inches (μm)	Weight percentage of active element	Sputtered layer equivalent, (μm)
BVAg-0 +0.626Zr	0.002 (50.8)	0.6	0.50
BVAg-0+0.4Zr	0.003 (76.2)	0.4	0.50
BAu-3 +0.35Ti	0.002 (50.8)	0.35	0.50
BAu-3 +0.235Ti	0.003 (76.2)	0.235	0.50
98Ag-2Zr	0.002 (50.8)	2.0	1.68
98Ag-2Zr	0.003 (76.2)	2.0	2.52
Nicoro®+2%Ti	0.002 (50.8)	2.0	2.90
Nicoro®+2%Ti	0.003 (76.2)	2.0	4.35

The two commercially fabricated active brazing filler metals evaluated in this study are 98Ag-2Zr (weight percent) and Nicoro®+2%Ti (62Cu-35Au-2Ti-1Ni, weight percent), provided by the Wesgo Metals Division of Morgan Advanced Ceramics[1]. A series of calculations were performed to correlate active element sputtered layer thickness to weight percentage, the results of which are displayed in Table 1. It has been reported that approximately fully dense metallic zirconium films have been formed using DC magnetron sputtering methods [9]. The calculations used for this study assume that the sputtered layers for zirconium and titanium will be fully (100%) dense. With this knowledge, 0.5μm thick

[1] Nicoro® is a registered trademark of MTC-Wesgo Metals.

titanium and zirconium active element layers were DC sputtered onto one side only of 0.002" (50.8μm) or 0.003" (76.2μm) thick silver (99.99%) or BAu-3 (35Au-62Cu-3Ni) conventional brazing filler metal materials.

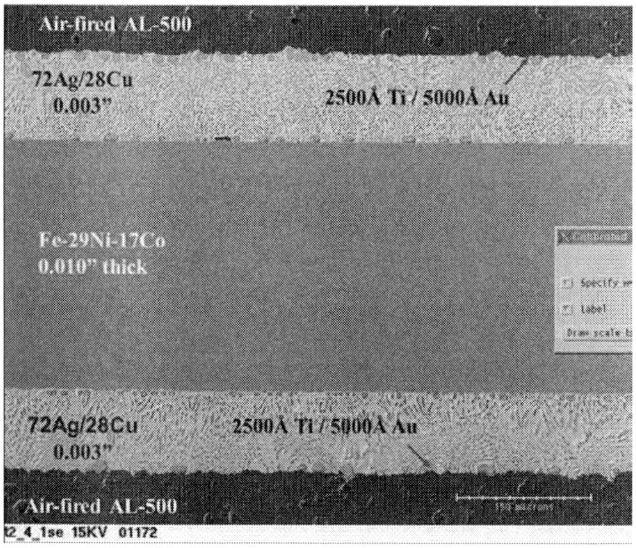

Figure 8. Metal/Al_2O_3 brazement using thin-film metallization deposited directly onto the alumina ceramic.

The different weight percentages displayed in Table 1 for similar titanium or zirconium thickness additions are due to the differences in their respective densities as well as whether the brazing filler metals are gold-based or silver-based, which greatly affects the bulk filler metal density.

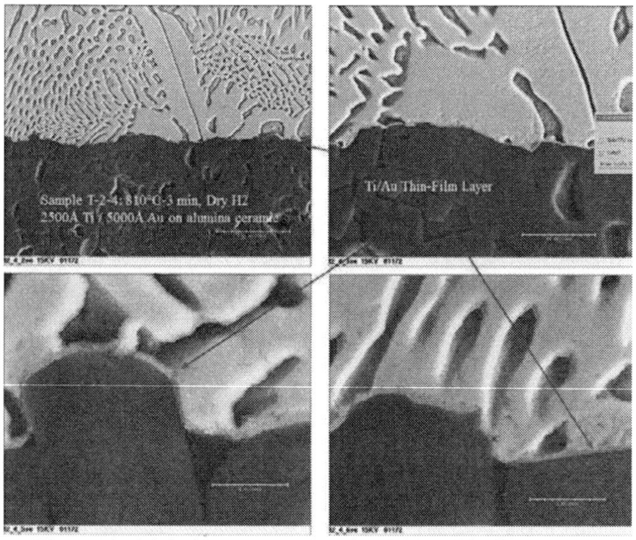

Figure 9. Al_2O_3 brazement showing Ti/Au thin-film layer metallization deposited directly onto the alumina ceramic.

Double reaction-layer test samples

Brazing preforms correctly sized for ASTM-F19 tensile buttons were punched from the materials listed in Table 1. The tensile specimens were fabricated without a metal interlayer in order to evaluate the effectiveness of forming two simultaneous reaction layers at the ceramic interfaces.

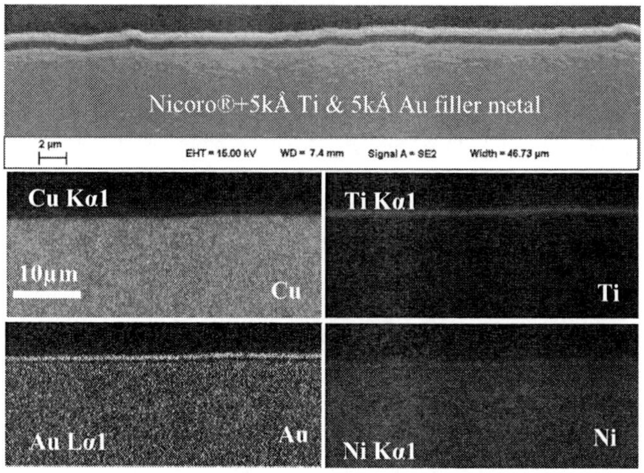

Figure 10. SEM image (upper) and EDS maps (lower) showing sputtered thin-film layers on BAu-3 braze filler metal.

Filler metal preforms having outer diameters of 0.625" (1.59cm) and 0.400" (1.01cm) inner diameters were fabricated with the sputtered titanium or zirconium (plus noble metal coating). The corresponding SEM images and EDS maps of the modified filler metals are shown in Figures 10 & 11.

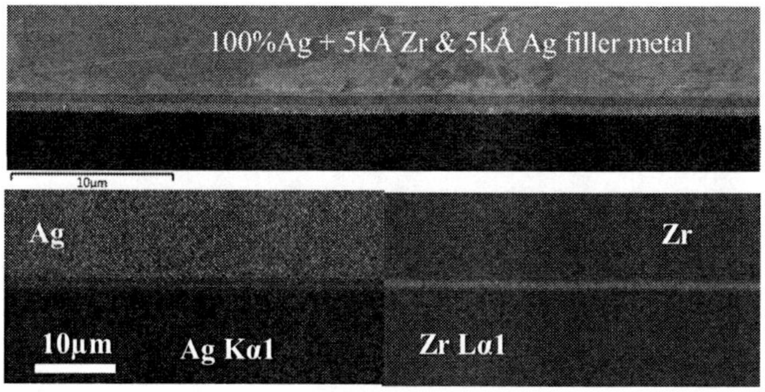

Figure 11. (Upper) SEM image; (Lower) EDS maps showing sputtered thin-film layers on pure silver braze filler metal.

Brazing test sample preparation and geometries. Three types of sample geometries were chosen to test the modified brazing filler metals: ASTM F-19 ceramic test specimens made from 94% alumina; 94% alumina discs; and 94% alumina cylinders brazed to metal discs, either commercially pure niobium or the Fe-29Ni-17Co alloy, Kovar® [11]. The proprietary 94ND10 alumina ceramic samples had a roughness value, R_A, less than 0.8μm (32μin). ASTM-F19 tensile buttons and the alumina cylinder-metal disc assemblies were helium mass-spectrometer leak tested to assess hermeticity. A positive leak check result, i.e. no detectable helium leak, is an indication that a continuous reaction layer formed on the alumina surface(s). Additionally, the ASTM-F19 samples brazed with the modified braze filler metals were tensile tested and compared to the samples made using commercially available active filler metals (four samples per filler metal composition). The 94% alumina cylinders brazed to either commercially pure niobium or Kovar were also analyzed to determine how effective the modified filler metals were at forming a single continuous reaction layer.

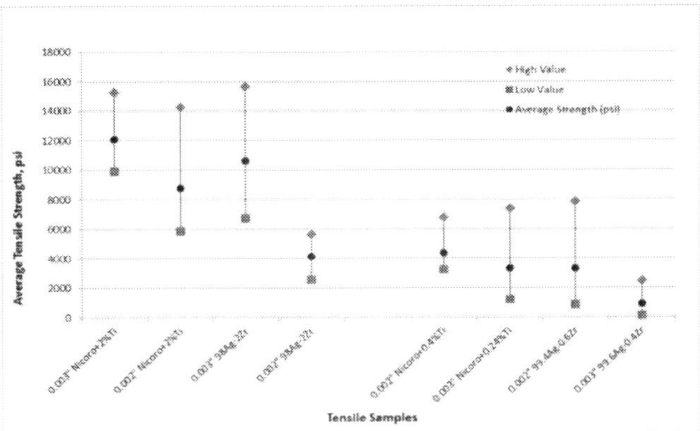

Figure 12. Tensile test comparison of 0.5μm thick Ti and Zr thin-films and standard active brazing alloys.

Brazing equipment, brazing process parameters and test equipment. The vacuum brazing process was performed in an R.D. Brew top-loading high-vacuum furnace, using a 10" cryogenic pump for the primary pump. A NIST-traceable calibrated ionization gauge tube is used to monitor absolute pressure during the brazing thermal-cycle. Furnace rate of rise is less than 1.5 E-08 Torr/s (4.8E-04 Pa-l/s).

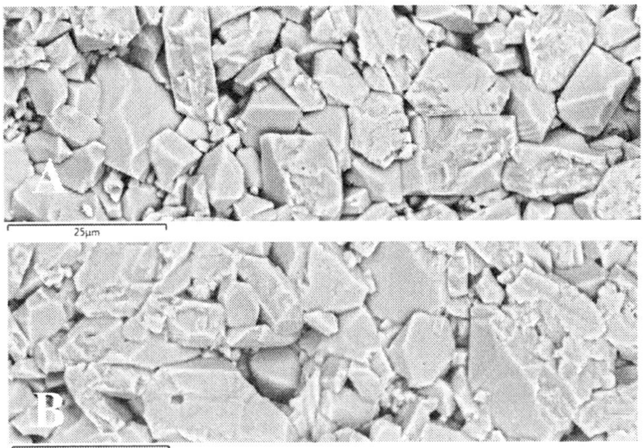

Figure 13. (A) 94% Al_2O_3 in the as-received condition; (B) 94% Al_2O_3 air-fired at 1200°C for 1 hour.

The temperature cycle is not started until the vacuum furnace base pressure is less than 5E-06 Torr. The thermal cycle used was as follows: 10°C/min from room temperature to T_X, soak for 10 minutes; ramp at 5°C/min to T_{Peak}, soak 5 minutes; Cool at 25°C/min to T_Y, soak 0 minutes; cool to room temperature at a rate ≤10°C/min. T_X is 975°C for the gold-based filler metal and 925°C for the silver based material; T_{Peak} is 1035°C for the gold-based filler metal and 985°C for the silver based material while T_Y is 950°C for the gold-based filler metal and 925°C for the silver based material. Following the cooling to room temperature and subsequent unloading and post-brazing visual inspection, the tensile button and metal-ceramic cylinder assemblies were leak checked using a Pfeiffer QualyTest™ Select HLT 265[2] helium mass-spectrometer leak detector (MSLD) capable of detecting leaks as low as 5E-12 atm-cc/sec (~5E-10 Pa-l/s). Any detectable helium leak was determined to have failed the leak test. Following visual inspection and leak testing, the ASTM-F19 tensile buttons were pull tested using a 22 kip MTS servohydraulic test frame at a crosshead speed of 8.38E-06m/s (3.3E-04 in/sec) in accordance with AWS standards [10]. The tensile strength was determined using the cross-sectional area of the tensile button faying surfaces, 1.11E-04m^2 (1.80E-01in^2).

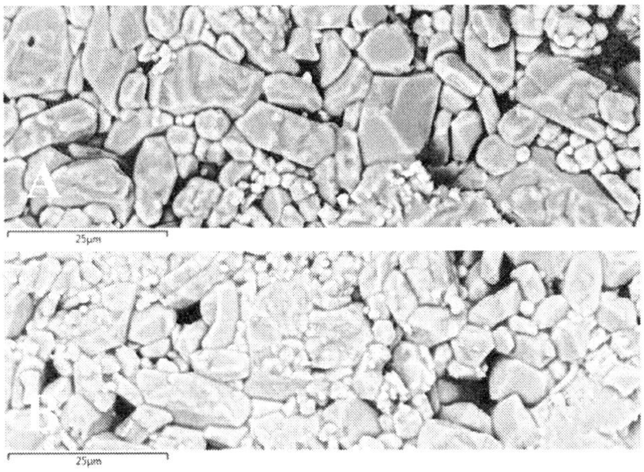

Figure 14. (A) 94% Al_2O_3 resintered at 1575°C for 2 hours (air); (B) 94% Al_2O_3 resintered at 1500°C for 1 hour (wet H_2).

Post-brazing leak testing and tensile testing revealed anomalies not typically observed when using the commercially fabricated active brazing filer metals. Eleven of the sixteen samples brazed with the commercially manufactured active braze filler metals failed the helium leak test.

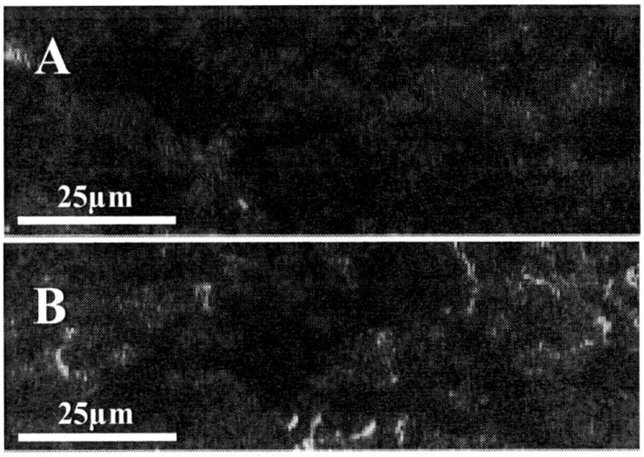

Figure 15. (A) EDS map showing the silica present on the surface in the as-received condition; (B) EDS map showing the silica present in the sample air-fired at 1200°C for 1 hour.

Additionally, as displayed in Figure 12, there are wide disparities in strengths of the tensile buttons fabricated using commercially manufactured active braze filler metals. Figure 12 also shows that with the exception of the 50.8μm thick

---

[2] QualyTest™ is a trademark of INFICON AG Liechtenstein

98Ag-2Zr brazing filler metal, the tensile strengths of the ASTM F-19 tensile buttons brazed with the modified braze filler metals were considerably lower than the samples brazed with the commercially produced filler metals. Further investigations into possible causes of the variation of tensile strength revealed that improper sintering parameters were used; this step is typically performed in air on the ceramics prior to brazing. The peak temperature of 1200°C is lower than the recommended 1575°C. Also, a 1 hour period was used instead of the standard 2 hours.

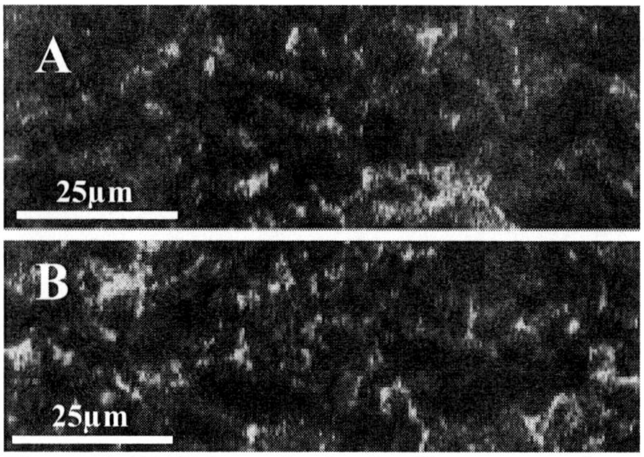

Figure 16. (A) EDS map showing the silica present on the surface after air-firing at 1575°C for 2 hours; (B) EDS map showing silica present wet H_2 firing at 1500C for 1 hour (B).

Previous studies [1] have shown that resintering 94% alumina ceramic at the recommended temperatures at 1575°C in air (2 hours) or 1500°C in wet hydrogen of 28°C dew point (1 hour) result in hermetic and strong brazements [14,15]. These conditions heal surface damage introduced on the faying surfaces during final grinding or polishing operations. An explanation for the poor brazement performance of the tensile button samples resintered at the 1200°C temperatures is that the resintering temperature was too low to melt and reflow the glassy phase between the alumina grains. As seen in Figure 13 the there is little difference between upper image A, which shows the as-received alumina, and lower image B, which was obtained from alumina fired at 1200°C,. Figure 14 shows high magnification SEM images of the surfaces of alumina pieces that had been resintered at 1575°C (A) and 1500°C (B). The effects of the greater firing temperatures are obvious as both the upper and lower images in Figure 14 show the alumina grains having smoothed, rounded edges as compared to the angular edges in Figure 13. Similarly, when comparing silicon EDS maps of the same four samples, the differences between the as-received sample (Figure 15A) and the sample air-fired at 1200°C (Figure 15B) are minimal while Figure 16A and B indicate that much more silicon has migrated to the sample surfaces.

Metallographic analysis of the two reaction layer samples brazed with the modified braze filler metals confirmed that the primary cause of the poor helium leak test performance and tensile test results were insufficient, noncontiguous reaction layer formation at the alumina ceramic/braze filler metal interfaces. As seen in Figure 17, the initial titanium trial layer thickness, 0.5μm, deposited on the gold-based brazing filler metal was insufficient to form a continuous reaction layer on either of the ceramic-filler metal interfaces.

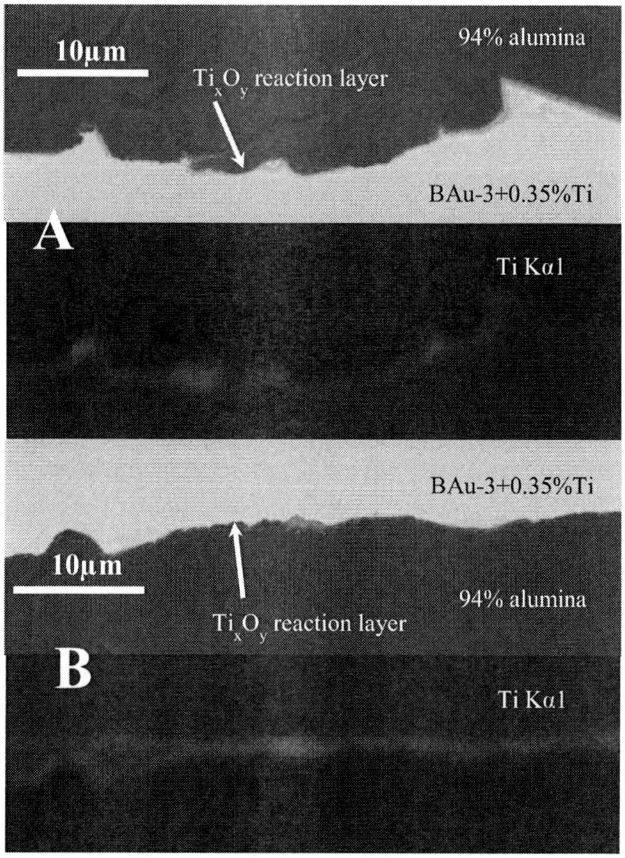

Figure 17. (A) Interfaces of a 94% Al_2O_3/94% Al_2O_3 assembly brazed with BAu-3+0.35%Ti brazing filler metal. (B) Titanium (Kα1) EDS map of active-braze reaction.

The 50.8μm thick brazing filler metal, while having the same titanium deposition thickness as the 76.2μm thick brazing foil has 50 weight percent more active element (0.35% vs. 0.235%). While only the 50.8μm thick sample is shown (Figure 17) in both cross-sectioned thicknesses, the ceramic faying surface adjacent to the sputtered titanium layer appeared to have a more continuous layer than was formed on the faying surface away from the thin-film deposition. This was expected as the limited supply of titanium should react readily with the nearest oxide surface. Visual examination of the ASTM-F19 tensile specimens post-testing revealed that in all cases, the alumina button away from the sputtered film was the side where failure initiated. This failure mode was confirmed at a later date using an SEM analysis. The examination of samples brazed using the modified brazing filler having 0.5μm of zirconium and 0.5μm of silver applied to the 50.8μm and 76.2μm pure silver filler metal shows unique differences from the BAu-3 filler metal modified with titanium. Figure 18 shows that the ceramic faying surface in contact with the side of the filler having the zirconium coating formed a nearly continuous Zr_xO_y reaction layer with the alumina ceramic surface. On the other hand, the opposite interface has little or no Zr_xO_y reaction layer.

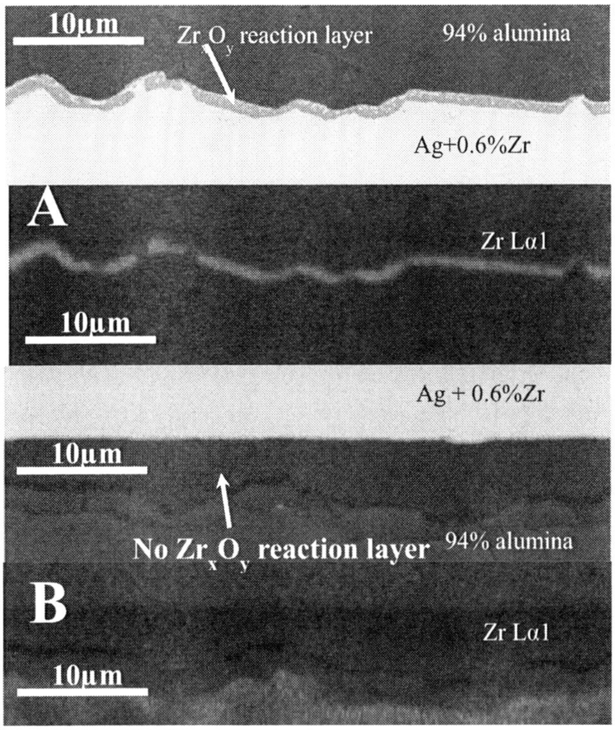

Figure 18. (A) Interfaces of a 94% Al_2O_3/94% Al_2O_3 assembly brazed with Ag+0.6% Zr brazing filler metal. (B) Zr (Lα1) EDS Map of active-braze reaction.

This is in contrast with the modified BAu-3 material where titanium and a Ti_xO_y reaction layer is evident on both faying surfaces of the brazements It is believed that the differences in the atomic sizes of tanium and zirconium and the quantities of the respective element atoms required to make a finite layer thickness explain this observation.

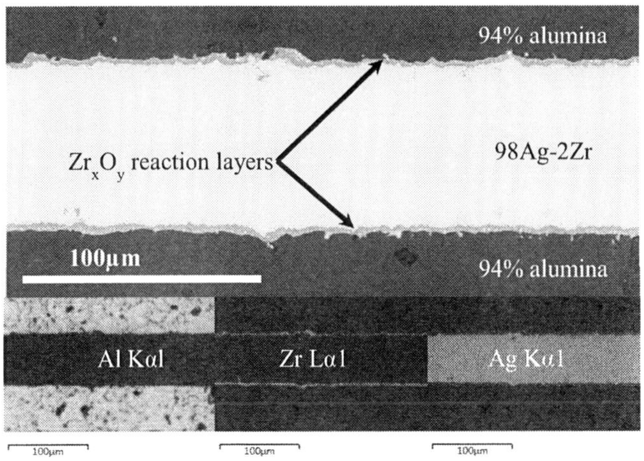

Figure 19. Top: SEM image showing both interfaces of a 94% Al_2O_3/94% Al_2O_3 assembly brazed 98Ag-2Zr active brazing filler metal. Bottom: EDS maps showing Al, Zr, and Ag.

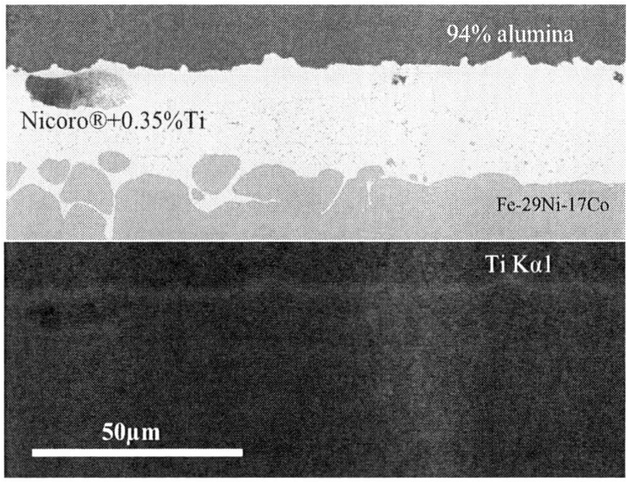

Figure 20. Top: SEM image showing interfaces of a 94% Al_2O_3/Fe-29Ni-17Co assembly brazed using BAu-3+0.35%Ti active brazing filler metal. Bottom: EDS map showing titanium (Kα1) in the completed brazement.

Recall from Table 1 that a 0.5μm layer of zirconium amounts to a 0.6 weight percentage for 50.8μm thick silver foil and 0.4 weight percentage for the 76.2μm pure silver filler metal compared to 0.35 and 0.235 weight percentages of titanium for the modified 50.8μm and 76.2μm thick BAu-3 braze filler metal. These differences are due to the gold-based filler metal having a higher density than the silver based material and because titanium, used to modify the gold-based material, is less dense than zirconium.

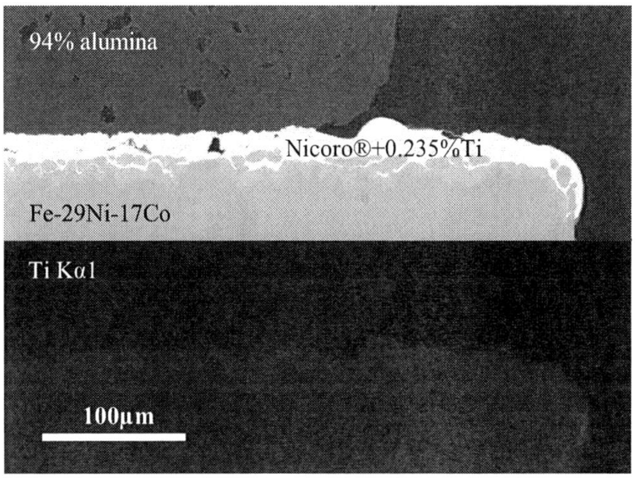

Figure 21. Top: SEM image showing the interfaces of a 94% Al_2O_3/Fe-29Ni-17Co alloy assembly brazed with BAu-3+0.235%Ti active brazing filler metal. Bottom: EDS map showing titanium in the completed brazement.

This impact is more fully appreciated when comparing the thicknesses of the two active element materials required to make a material equivilent to the commercially manufactured active brazing filler metal materials, which is generally 2 weight percent. To make a 98Ag-2Zr material, a zirconium coating 1.6μm thick is required, slightly more than three time times the material used in this study. When using titanium as the active element, a thin-film that is 2.9μm thick is required to make the equivilent active braze filler metal counterpart, BAu-3+2%Ti. That thickness is a factor of almost six times greater than what was used initially. Because of the differences in atomic densities of the two materials, 14.1 g/mol Zr versus 10.64 g/mol Ti, for any given deposition thickness, there will be approximately 33 percent more titanium atoms than zirconium atoms. Additional, or excess, titanium active filler metal, is more readily observed than zirconium when comparing ceramic-ceramic samples brazed with the two commercially available active filler metals, 98Ag-2Zr, shown in Figure 19, and Nicoro®+2%Ti, as displayed previously in Figures 6 and 7. Note that Figure 19 shows that even with 2 weight percent of zirconium in the brazing filler metal, all is utilized by the two reaction layers, unlike the titanium in Figures 6 and 7.

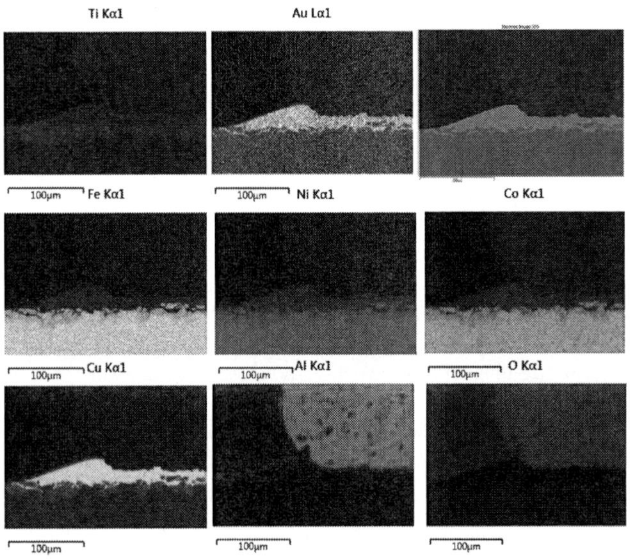

Figure 22. SEM image (upper right) showing interfaces of a 94% Al_2O_3/Fe-29Ni-17Co alloy assembly brazed with Nicoro®+2%Ti filler metal. Other images are EDS maps showing titanium distribution and base metal dissolution.

Single reaction-layer test samples

Single reaction layer samples were brazed with the modified braze filler metals and commercial active braze alloys using the same brazing thermal profiles. For each of the filler metal compositions, samples were fabricated using 94% alumina ceramic cylinders and either niobium or Fe-29Ni-17Co alloy washers. The brazing filler metal was preplaced between the metal and ceramic faying surfaces with the sputtered thin-film layers, if applicable, adjacent to the metal surface. Previous experiments have confirmed that the titanium or zirconium active element readily diffused through the molten filler metal and reacted with the oxide surface. Metallographic analysis (Figures 20 -22) showed that the Fe-29Ni-17Co alloy base metal substrates were prone to dissolution by all of the titanium-containing filler metal compositions.

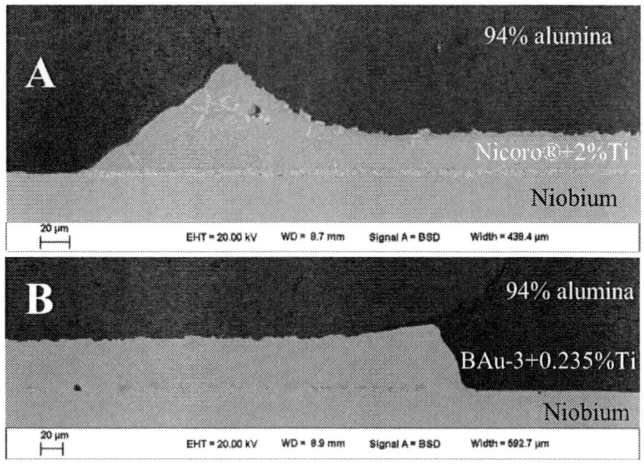

Figure 23. SEM back-scattered images showing interfaces of a 94% Al_2O_3/Nb assembly brazed using (A) Nicoro®+2%Ti and (B) BAu-3+0.235%Ti active brazing filler metal.

As both BAu-3 filler metal thicknesses used in these samples had the same quantity of titanium added, the amount of Fe-29Ni-17Co alloy erosion appeared similar in the samples. For the two weight percent titanium filler metal, however, the erosion appears to be more severe. As represented in Figure 23, the titanium bearing active brazing filler metals did not appear to have any deleterious effects on the niobium base metal samples evaluated for this study.

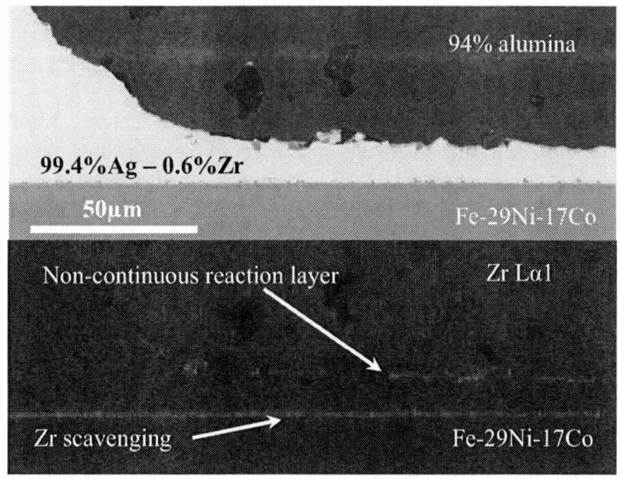

Figure 24. Top: SEM image showing a 94% Al_2O_3/Fe-29Ni-17Co alloy interface brazed using Ag-0.6%Zr filler metal. Bottom: EDS map showing the incomplete Zr_xO_y reaction layer formation and scavenging of the Zr.

SEM images and EDS maps of the cross-sectioned samples made with the modified BAu-3 brazing filler metals exhibit more nickel segregation and interaction with the niobium surfaces than do the samples made with the Nicoro®+2%Ti active brazing alloy. This behavior is expected since the commercial filler metal contains only one weight percent nickel, versus three weight percent for the modified brazing filler metals.

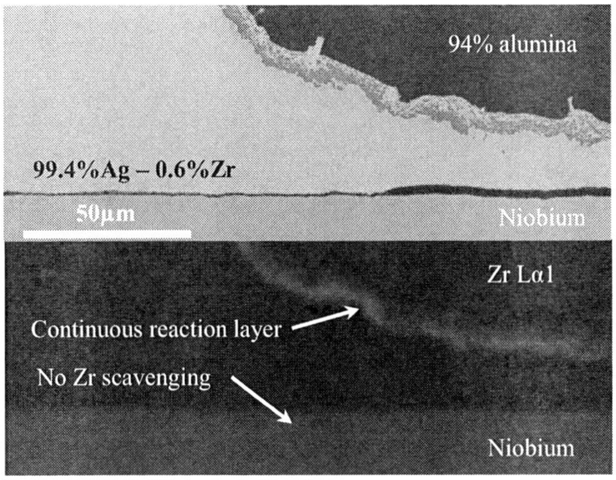

Figure 25. Top: SEM image showing a 94% Al_2O_3/Nb washer interface brazed with Ag-0.6%Zr filler metal. Bottom: EDS map showing a continuous Zr_xO_y reaction layer formation.

The final samples to be discussed are the alumina ceramic brazed to either Fe-29Ni-17Co or niobium discs using pure silver braze filler metal modified with either 0.4 or 0.6 weight percent zirconium to form the active brazing filler metal, 98Ag-2Zr. The zirconium containing filler metals, both those made by modifying pure silver and the commercial material, displayed minimal erosion of the Fe-29Ni-17Co substrates and none on the niobium base materials as shown in Figures 24-27. The interface reaction layers formed with zirconium tended to be thicker and more continuous than those formed with titanium. Measurements taken using SEM software indicate Zr_xO_y measurements of 4-6µm compared to 2-3µm for the Ti_xO_y reaction layers. Also it was observed that, on the zirconium bearing samples but not the titanium active element brazements, there is a continuous silica layer along the Zr_xO_y reaction layer. This layer is visible in many of the zirconium containing samples, but is most easily seen in the commercial 98Ag-2Zr material as shown in Figures 26 and 27, where SEM/EDS analysis revealed that the alumina ceramic glassy-phase material, silica, had formed a completely continuous reaction with the zirconium, independent of which metal substrate was utilized in the braze joints. The silicon layer is more pronounced on the niobium base metal samples. This is believed to be due to the tendency of a portion of the zirconium to be scavenged by the Fe-29Ni-17Co, shown in Figures 24 and 26, leaving less material to form a reaction with the alumina or silica material.

Conclusions

Active brazing filler metals were fabricated by depositing thin-films of titanium or zirconium on to conventional brazing filler metals. The sputter-deposited active element layers were prevented from oxidation by depositing a thin layer of noble metal (silver or gold) over them prior to exposure to the atmosphere. The modified BAu-3 active brazing filler metal contained either 0.235 or 0.35 weight percent titanium, while the modified silver active brazing filler metal contained 0.4 or 0.6 weight percent zirconium. Brazed assemblies were fabricated to have two reaction layers by using either 94% alumina ceramic ASTM F-19 tensile buttons or 94% alumina discs. Although Ti_xO_y or Zr_xO_y reaction layers formed on both interfaces of each of the test samples, the samples were not hermetic of either pedigree.

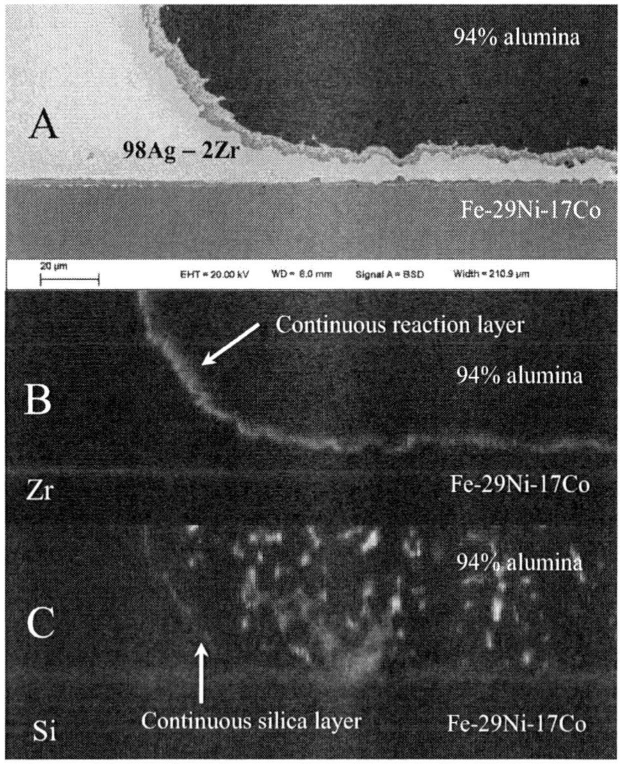

Figure 26. (A) SEM image of a 94% Al_2O_3/Fe-29Ni-17Co alloy interface brazed using 98Ag-2Zr filler metal. (B & C) EDS maps showing Zr_xO_y and Si_xO_y reaction layers.

SEM/EDS analysis confirmed that the reaction layers formed were not continuous on either of the surfaces, as compared to samples fabricated with commercially manufactured active brazing filler metals containing two weight percent of titanium or zirconium.

The average tensile strengths of the samples brazed using the two weight percent active element were superior to the strengths of those fabricated using the modified filler metals having reduced active element percentages. Future test samples should include greater concentrations of active element and also amended filler metal preparation methods as the analysis of the metallographic cross-sections also showed the tendency of the sputter deposited active element to preferentially form a thicker, more continuous reaction layer on the faying surface it was positioned adjacent to during the prebrazing fixturing process.

Of the samples fabricated so as to form a reaction layer at only a single interface, those utilizing Fe-29Ni-17Co metal substrates were shown to scavenge both of the active elements, titanium and zirconium, more readily than the commercially pure niobium metal. While scavenging and base metal erosion was observed on all the titanium containing filler metals, positioning the titanium or zirconium away from the metal substrate during brazement fixturing process should reduce observed scavenging and base metal erosion. Regarding the two active elements, titanium was found to be more likely to be scavenged, or react unfavorably with the base metals than did zirconium.

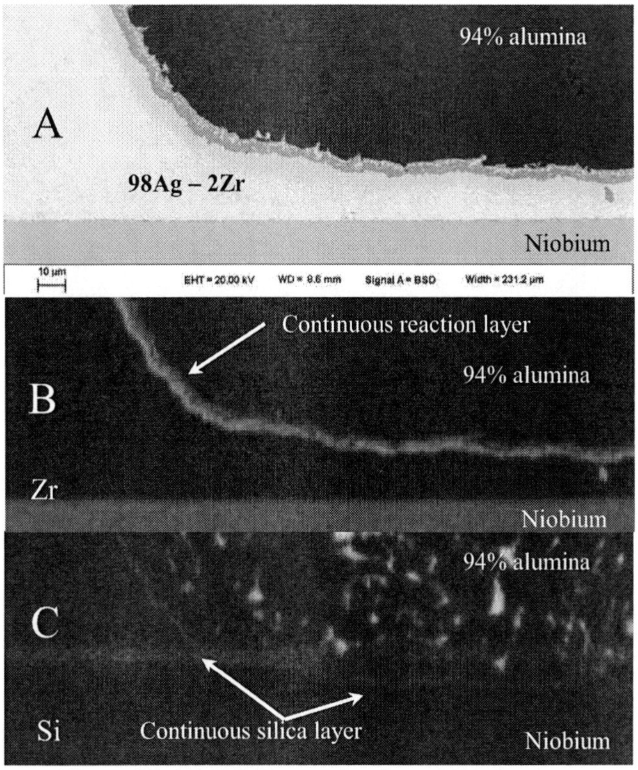

Figure 27. (A) SEM image showing a 94% alumina ceramic/Nb washer assembly brazed using 98Ag-2Zr filler metal. (B & C) EDS maps showing complete reaction layer formation of Zr_xO_y and Si_xO_y on the alumina surface.

While not quantified as part of this study, the materials combination observed to have the greatest amount of base metal erosion and active element scavenging was 94% alumina ceramic and Fe-29Ni-17Co brazed with titanium containing filler metal. Conversely, the samples showing the least base metal erosion and scavenging were 94% alumina ceramic and niobium brazed with zirconium containing filler metal. This latter combination also appeared to form a more continuous reaction layer with the silica material within the alumina ceramic glassy phase. Mixed helium leak detection results and overall poor tensile button strengths of the tensile buttons confirmed the importance of properly resintering alumina ceramic substrates following the grinding and/or polishing process prior to brazing. Firing temperatures of 1500°C for one hour (wet hydrogen) and 1575°C (air) for two hours, were adequate to reflow the glassy phase of the alumina ceramic samples used in this study.

Acknowledgments

The authors wish to recognize Amy Allen, Tom Crenshaw, Alice Kilgo, Bonnie McKenzie, and Shelly Williams who provided the metallographic sample preparation, tensile testing and the SEM/EDS imaging and analysis for this study.

References

[1] Walker, C A, Hodges, V C "Comparison of metal-ceramic brazing methods". Welding Journal, 87, 10, (2008) 43-50
[2] Brazing handbook, 5th Edition, American Welding Society, (Miami, 2007), pp. 464-466.
[3] Sekulic, D.P, Editor, Advances in Brazing: Science Technology and Applications, Woodhead Publishing Limited (Oxford, 2013) pp. 503-508
[4] LucasMilhaupt Global Brazing Solutions. 2012. Brazing Filler Metals. http://www.lucasmilhaupt.com/en-US/products/fillermetals/1/. [Accessed 08 December 14].
[5] Wesgo Metals. 2009. Precious Metal Filler Metals. http://www.wesgometals.com/products-materials/. [Accessed 08 December 14].
[6] Vianco, P. T. et al, "Titanium Scavenging in Ag-Cu-Ti Active Braze Joints," Welding Journal, (2003) 82 (10), pp 268-S to 277-S.
[7] Brazing handbook, 5th Edition, American Welding Society, (Miami, 2007), pp. 467-468.
[8] Walker, C.A. et al, "High-temperature Joining of Low-temperature Co-fired Ceramic," Proc of the 3rd IBSC, San Antonio, TX, April, 2006, pp 54-59
[9] Venkataraj, S. et al, "Structural and optical properties of thin zirconium oxide films prepared by reactive direct current magnetron sputtering", Journal of Applied Physics, 92, 7, (2002), pp 3599-3607
[10] AWS C3.2M/3.2:2008: Standard method for evaluating the strength of brazed joints, American Welding Society, (Miami, 2008).
[11] Mizuhara H, Huebel E and Oyama T "High-reliability joining of ceramic to metal", Ceramic Bulletin, 68, 9, (1989) 1591–1599.
[12] Mizuhara H, Mally, K, "Ceramic-to-Metal Joining with Active Brazing Filler Metal", Welding Journal, Oct 1985 pp. 27-32

Properties of new Ag-Cu-Zn-Sn Filler Metals with a 10 Weight-% Reduction of Silver and same Liquidus Temperature

Daniel Schnee, Gunther Wiehl, Sebastian Starck, Max Schimpfermann
Umicore AG & Co. KG, Business Unit Technical Materials, Business Line BrazeTec, Hanau, Germany
Daniel.schnee@eu.umicore.com

Abstract

BrazeTec BlueBraze is a new Ag-Cu-Zn-Sn filler metal system with a 10 weight-% reduction of silver and same liquidus temperature as corresponding standard alloy. In this paper latest results of shear strength tests according AWS and EN standards, high-cycle fatigue and corrosion resistance test in comparison to standard Ag-Cu-Zn-Sn brazing filler metals will be presented.

Introduction

Low melting silver brazing filler metals are used in heating, ventilation, air-conditioning and refrigeration industry (HVACR) to join different parent materials like copper, brass, steel and stainless steel. Due to the strong price increase and fluctuation of precious metals since 2009 silver content is an important factor in raw material perspective and price calculation for produced parts. Up to now, with reduction of silver content, the manufacturers have to face higher brazing temperatures leading to an increase of process times, energy costs and change on parent material properties [1], [2], [3]. Additionally the brazing quality and safety is reduced with increasing temperatures.

Within a three years internal R&D program Umicore developed a new cadmium free alloy system, which has a low liquidus temperature at a given silver content, good deformability at ambient temperature and sufficient ductility in brazed joints. The basis of the product series are the alloy components Ag-Cu-Zn-Sn in new ratios, with manganese as new alloying element and additions of silicon and indium. With this, BlueBraze named, innovating brazing filler metal system, alternative filler metals with 10 weight-% less silver but same liquidus and brazing temperatures are available for manufacturers in HVACR industry; please refer to table 1 and table 2.

The performance of these new alloys has been evaluated in several tests. The evaluation included wetting investigations, metallographic examinations, joint strength and dynamic load tests and corrosion resistance [4]. These tests outlined a similar reliability of BrazeTec BlueBraze products compared with standard alloys and a slightly nobler behavior in chloride-ion containing environments due to manganese content and scale formation.

Current standard silver brazing filler metals are listed in table 1; corresponding BrazeTec BlueBraze alternatives are shown in table 2.

Table 1: Standard silver brazing filler metals with their codes according ISO 17672 [5], AWS 5.8M [6] and ISO 3677 [7] are listed. The temperature ranges are according [4].

Code acc. ISO 17672	Code acc. AWS 5.8	Code acc. ISO 3677	Silver content [wt.-%]
Ag 145	BAg-36	B-Ag45CuZnSn-645/**695**	45
Ag 140	BAg-28	B-Ag40CuZnSn-665/**725**	40
Ag 134	-	B-Cu36AgZnSn-655/**745**	34
Ag 130	-	B-Cu36ZnAgSn-675/**760**	30

Table 2: BrazeTec BlueBraze products with their codes according ISO 3677 [7] and silver content. The temperature ranges are according [4].

BrazeTec Name	Code acc. ISO 3677	Silver content [wt.-%]
BlueBraze 3510	B-Ag35CuZnMnSn(Si)-680/**700**	35
BlueBraze 3010	B-Cu38AgZnMnSn(Si)-690/**730**	30
BlueBraze 2410	B-Cu44AgZnMnSn(Si)-690/**750**	24
BlueBraze 2010	B-Cu43ZnAgMnIn(Si)-710/**765**	20

In this paper latest results of shear strength tests according AWS C3.2M/C3.2 [8] and EN DIN EN 12797 [9] standards, high-cycle fatigue and corrosion resistance test in comparison to standard Ag-Cu-Zn-Sn brazing filler metals will be presented. Additionally experiences from several customer applications around the world will be discussed to demonstrate, that these new filler metals are already used in industrial mass production.

Experimental Procedures

Shear Strength according AWS C3.2M/C3.2

Blanks of carbon steel S235JR [10] (1015(SAE)[11]) with dimensions of 130 x 30 mm and thickness of 3.0 mm were spot welded to ensure a brazing gap of 0.2 mm. Overlap distances were varied according table 2 of the mentioned standard.

The specimen were manually brazed by torch using rods of Ag 134, BlueBraze 2410 and flux BrazeTec h paste (FH 10 acc. DIN EN 1045 [12]; FB3-A acc. AWS A5.31 [13]).

Afterwards the specimen was machined to the configuration and tolerances given in figure 5 of AWS C3.2M/C3.2.

Finished specimen and non-machined ones were tested with Zwick Roell universal testing machine "Zwick 1474" with a speed of 20 mm/min.

Shear Strength according DIN EN 12797

The shear strength was tested by joining a 9.7 mm diameter cylinder of carbon steel S235JR with a ring of the same material with interior diameter of 9.9 mm. Caused by the bolted connections of cylinder and ring, the brazing gap was fixed and centered. All specimen were manual brazed with an acetylene oxygen torch and BrazeTec h paste. The tested overlap of the joint was 3 mm. A universal testing machine "Zwick 1474" with a speed of 1 mm/min. was used for testing the specimen.

Figure 1 shows on left the specimen before brazing, on the right the specimen before testing. The screw thread is drilled off and therefore the brazing gap remains.

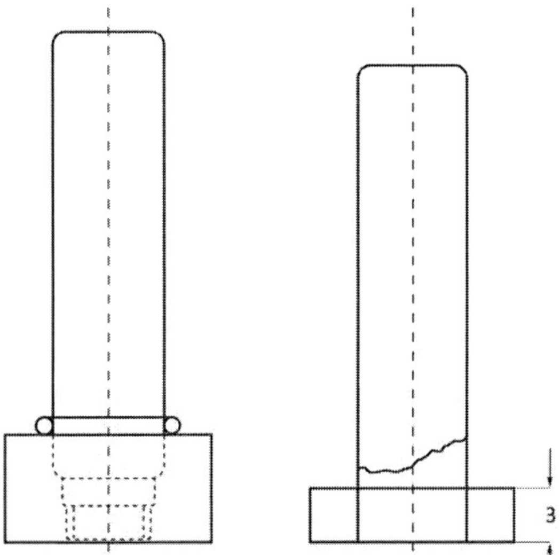

Figure 1: Specimen acc. DIN EN 12797 (shear test).

High-Cycle Fatigue Tests

The used specimen is shown in Figure 2. The thickness of the S235JR steel plates was 3 mm and therefore an overlap of 12 mm (4T) was designed according the rules for lap joints [14][15].

The specimen were spot welded to ensure a homogeneous brazing gap of 0.2 mm. The joint was brazed by acetylene-oxygen torch using universal flux BrazeTec h paste. Afterwards the parts were machined and put on a "Schenk PWY" bending test device.

Pretests, including tensile strength and measurements with strain gauges, were performed to evaluate machine setting ensuring desired stress level in pure steel specimen. The brazed overlapping steel plates were tested with evaluated settings. Due to the geometric design local stresses might be elevated.

Brazed parts were tested with bending stress loads of 100 to 300 MPa until rupture or cycle numbers of minimum 1,000,000.

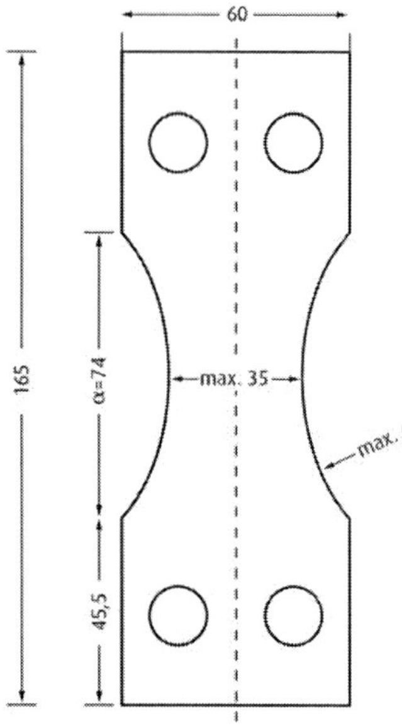

Figure 2: Specimen of high-cycle fatigue test.

Corrosion Resistance Tests

To evaluate the corrosion resistance of brazing filler metals in a practical way, tests were conducted in a climate chamber. The objective was to simulate a high stress condition, like in tropical areas with a periodically direct contact of water, like e.g. humidity and rain. During a period of 4 weeks, the climate conditions changed periodically. During the whole time the test specimen were stored at 60 °C and a humidity of 95 %. Twice a week, the specimen were wetted with distilled water until they were dripping wet.

The specimen itself were carried out in two different geometries, as T-Joint (length 80 mm) and as Pipe/bush joint (diameter 18 mm); please refer to figure 3. As base materials different materials were used a carbon steel S235JR, stainless steels 1.4301 (X5CrNi18-10; 304(SAE)) and 1.4571 (X6CrNiMoTi17-2-2; 316 Ti (AISI)) and pure copper.

The specimen were brazed by flame with an acetylene-oxygen torch and flux BrazeTec h paste. The brazing filler metal (rods, diameter 1.5 mm) was added after the flux was molten and active. The whole brazing process took app. 30 seconds. After the part cooled down to room temperature, the flux residues where washed off with water. After the testing time of 4 weeks, the specimen were examined and cross sections were prepared. At the cross section, the type of corrosion and the depth of the corrosion front in braze metal layer were evaluated.

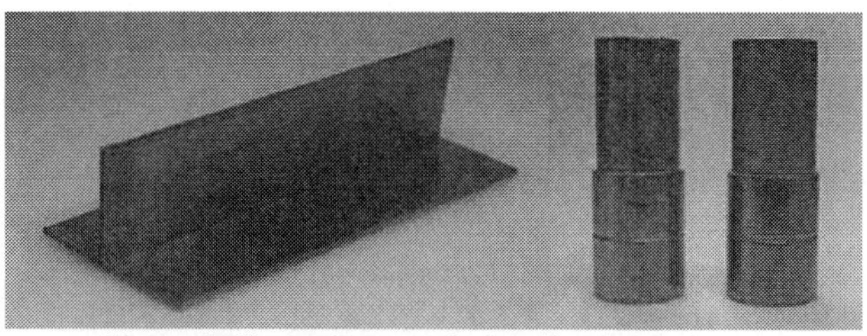

Figure 3: T-specimen and pipe / bush specimen.

Results

Shear Strength according AWS C3.2M/C3.2

Brazing filler metals Ag 134 and BlueBraze 2410 were tested in single lap shear test. The specimen were machined after brazing to width of 12.5 mm in the overlap area according figure 5 of AWS C3.2M standard. Additionally parts were tested in origin width of used steel sheets of 30.0 mm. Overlap was changed from 6 mm (2T) to 15 mm (5T) during the test. Tables 3 and 4 showing the reached shear strength of specimen and position of failure: parent material (P) or braze metal (B).

Table 3: Lap shear test of Ag 134 according AWS C3.2M on steel S235JR and different specimen widths.

Overlap	Strength at 12.5 mm [MPa]	Strength at 30.0 mm [MPa]
2T	154 (B)	144 (B/P)
3T	120 (B)	100 (P)
4T	91 (B/P)	88 (P)
5T	67 (P)	60 (P)

Table 4: Lap shear test of BlueBraze 2410 according AWS C3.2M on steel S235JR and different specimen widths.

Overlap	Strength at 12.5 mm [MPa]	Strength at 30.0 mm [MPa]
2T	134 (B)	130 (B/P)
3T	104 (B)	100 (P)
4T	82 (B/P)	75 (P)
5T	55 (P)	60 (P)

The amount of specimen per setting varies from two to five. Therefore typical brazing imperfections, e.g. flux inclusion, can have strong impact onto the results. Considering these influences, one can see no significant differences between the tested filler metal systems; especially on 30.0 mm samples the values are quite close and according to [4] BlueBraze 2410 showed a strength of 172 MPa vs. Ag 134 with 150 MPa at an overlap ratio of 1T.

Shear strength values of 12.5 mm and 30.0 mm appear to be similar, which is in line with Flom [16].
Significant difference can be seen in the position of failure between the tested widths. At 12.5 mm the crack occurred mostly in braze metal layer, starting at the end of braze metal fillet on steel plate surface and running along the diffusion zone braze metal/steel. After testing the steel plates deformed in 45 angle (figure 4), like expected for acting opposite forces. Reaching overlap ratios of 4T the failure can be found either in braze metal or parent material far away from brazing joint.
At 30.0 mm sample width, failure mode is mainly in parent material. Only at 2T both positions could be found.

Figure 4: Lap shear specimen of Ag 134 (4T) failed in parent material.

Strength test plots show difference curve progression. Parent material of wide specimen started to flow at approx. 200 – 230 MPa and cracked constantly at 308 MPa when converting measured maximum forces with cross-sectional area of origin steel sheet. Braze metal layer is forced to loads of 50 – 80 MPa at this start of deformation. In contrast steel parent material is forced only to 180 MPa for parts failed in braze metal layer, converting maximum force into stress of cross-sectional area of origin steel sheet.
12.5 mm specimen started to deform at approx. 280 MPa load on steel sheet and > 100 MPa in braze metal layer. When this layer failed at 2T to 4T parts, the parent material load was already in range of 310 to 400 MPa, calculated by maximum test force and cross-sectional area of origin steel sheet.

Width of specimen does not indicate influence onto total shear strength value, but position of crack. The yield strength (σ_y) of S235JR is at 235 MPa and therefore massive deformation appears on 12.5 mm samples which will effect stress load in braze metal layer and especially in diffusion interlayer. Flom [16] found by test and finite element analysis an ultimate Mises strength of his tested silver brazing filler metal at 434 MPa. The diffusion zone layer of parent material in this simulation showed already high Mises stresses, while the cross-sectional area was stressed less than in the used 12.5 mm specimen of this investigation.

Figure 4 summarizes the test results according AWS lap shear test of both filler metals and widths of specimen.

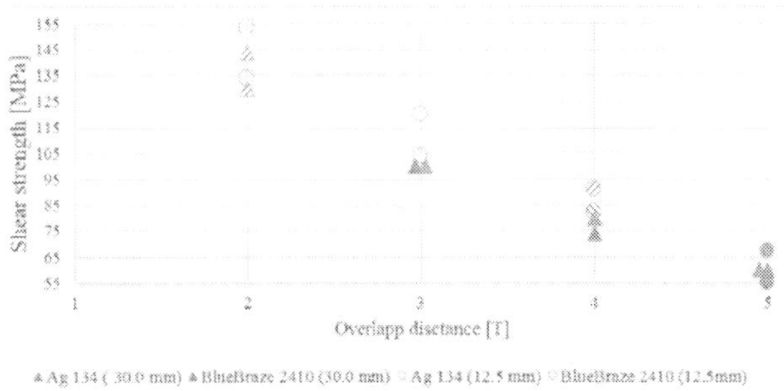

Figure 4: Average stress as function of overlap distance on width. Open symbols represent failure in the braze metal; filled symbols represent failure in the parent material; patterned represent failures occurred in both.

Shear Strength according DIN EN 12797

Due to the design of specimen according DIN EN 12797 a uniform brazing gap between centered cylinder and brazed disc is ensured. Machining after brazing enables defined test overlap. Due to fixture of specimen in strength test equipment the stresses acting directly in braze metal layer; please refer to figure 5.

Figure 5: Shear strength specimen of Ag 134 after testing.

The results of the shear strength tests according EN 12797:2000 on steel S235JR are summarized in table 5. The values are based on minimum 5 specimen.

The calculated average shear strength of the standard Ag-Cu-Zn-Sn-alloys is in a range of 179 – 190 MPa with standard deviations of 18 – 30 MPa. The new filler metal system has an average shear strength of 172 – 185 MPa with 7 – 25 MPa deviation. Ag 134 had the lowest measurement (139 MPa) in the standard BFM system and Ag 130 the highest (240 MPa). In the new system the extreme values, 130 to 212 MPa, were found in one alloy - BrazeTec BlueBraze 3510.

Especially this last figures demonstrate clearly the strong influence of the process, e.g. heat distribution and flux inclusions during manual flame brazing, onto the final shear strength of the filler metal.

Table 5: Shear strength results according DIN EN 12797 on steel S235JR (3mm).

Filler metals	Average [MPa]	Standard deviation [MPa]
Ag 145	179	23
BlueBraze 3510	172	25
Ag 140	186	18
BlueBraze 3010	179	7
Ag 134	179	18
BlueBraze 2410	185	11
Ag 130	190	30
BlueBraze 2010	181	15

Although the number of specimen per filler metal was limited, the low differences in the average strength and the large shear strength range in each filler metal, it can be stated, that both filler metal systems show same shear strength in the tests.

High-Cycle Fatigue Tests

Based on the measurements Wöhler curves were plotted for Ag 134 (figure 5) and BlueBraze 2410 (figure 6). Both curves are similar to plain steel and therefore the concentration factor caused by brazing joint should be 1.

Specimen of Ag 134 and BlueBraze 2410 were tested with loads of 100 to 300 MPa. Only one sample of Ag 134, stressed with 300 MPa, failed after 10,000 cycles in the braze metal (marked as X). All other cracks occurred in the steel plates (marked as X), starting at the heat effected zone or at the end of braze metal fillet on steel plate surface.

At 300 MPa the number of cycles had been in range of 9,200 - 63,000 cycles, increasing to 70,000 - 190,000 at a load 250 MPa. Decreasing to 200 MPa the cycle numbers increased to 197,500 - 496,000 brazed with Ag 134 and 506,000 - 932,000 brazed with BlueBraze 2410. This was the only test setting showing difference between the filler metals. But due to the limited numbers of specimen (3 parts) this has limited statistic relevance.

At 100 MPa no specimen failed and marked as "o" in the diagram. The test was stopped at cycle number between 1,000,000 and 1,350,000. To prove the assumption of endurance limit, Ag 134 was tested up to 15,900,000 and BlueBraze 2410 up to 21 million cycles without failure.

These test demonstrated the strength of both test brazing filler metals. Amazingly the sample passing 21 million cycles show a big imperfection with missing braze metal of approx. 30 % in the center, caused by higher capillary attraction in the fillet during brazing [17].

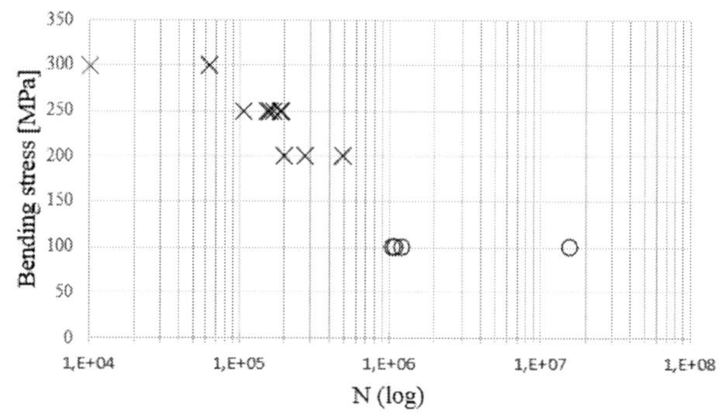

Figure 5: Wöhler curve of Ag 134.

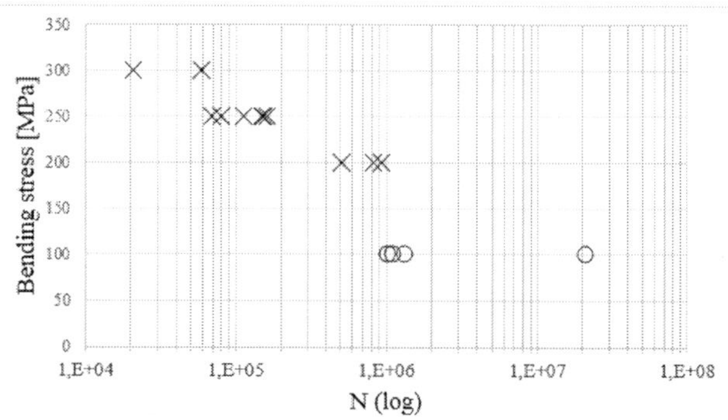

Figure 6: Wöhler curve of BrazeTec BlueBraze 2410.

Corrosion Resistance Tests

After four weeks test period all specimen showed a more or less corroded or patinated surface on the base material as well on the braze metal. The appearance of the braze metals were the same, independent from the used alloy system. As one can see in the figures 7 & 8, the test conditions were so strong, that even the base materials corroded (S235JR) or patinated black colored (1.4301).

Figure 7: Specimen after four weeks test run (left S235JR; right 1.4301) brazed with Ag 134.

Figure 8: Specimen after four weeks test run (left S235JR; right 1.4301) brazed with BrazeTec BlueBraze 3510.

The measured depth was within 2 and 20 µm. The primarily ruling corrosion type in the braze metal was the surface corrosion. In the cross sections, one could detect, that the copper rich phase of the braze metal is more prone to corrosion than the silver phase.

All braze metals showed a dark greyish seam consisting out of corrosion products of the copper (figure 9). A separation of the braze metal and the base material could not be observed with both filler metal systems.

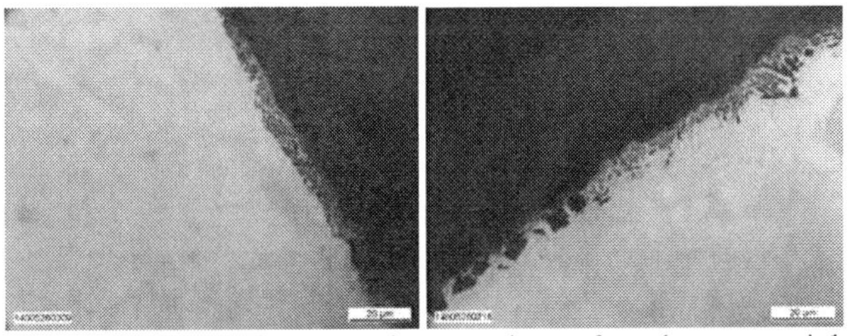

Figure 9: Corroded copper phase on the surface; base material 1.4571 (left: BrazeTec BlueBraze 3510; right: Ag 134).

Obviously the test showed, that the lowest rate of corrosion was detected in combination with the base material S235JR; please refer to table 6. In this case the standard alloy system seems to be a little bit more stable. But one should consider, that observed rates are still on a very low level compared to the other tested base materials (fig. 10). The reason for the comparatively low corrosion rate in the braze metal can be explained with relatively ignoble behavior of the steel, which is indicated by the strong formation of oxides on the surface. The other base materials are nobler, which ends in a better stability against corrosion. In a case of a corrosional attack the braze metal will be oxidized more easily. The evaluation of the cross sections showed, that the depth of the corrosion front had its minimum in the brazed carbon steel specimen.

Table 6: Maximum corrosion depth of the tested filler metals.

Code acc. ISO 17672 or BrazeTec name	Maximum depth of corrosion [µm]				
	S235JR	1.4301	1.4571	Copper / Copper	Copper / Steel
	T-Joint	T-Joint	T-Joint	pipe / bush	pipe / bush
Ag 145	2	15	14	16	5
Ag 140	2	5	7	18	20
Ag 134	3	18	15	10	10
Ag 130	4	8	7	14	5
Average Ag-Cu-Zn-Sn	**3**	**12**	**11**	**15**	**10**
BlueBraze 3510	1.5	11	9	22	10
BlueBraze 3010	4	12	6	15	10
BlueBraze 2410	8	9	20	10	5
BlueBraze 2010	7	11	10	14	5
Average BrazeTec BlueBraze	**6**	**11**	**11**	**15**	**8**

As one can see in table 6, no trend regarding the corrosion depth could be observed within one brazing filler metal group. Also the silver content did not influence the result verifiable. Because of this finding, it was decided to calculate the average of the corrosion depth within one group to get comparable data for the two tested filler metal groups. The comparison of the averages ended in a behavior, which is absolutely comparable for the two groups.

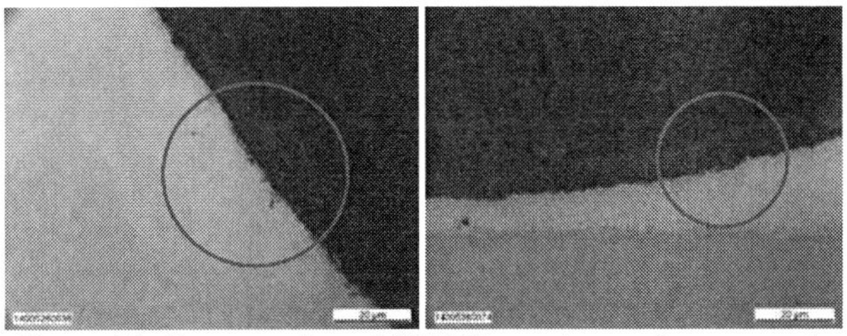

Figure 10: Cross section of S235JR specimen after test cycle: left: Ag 134; right: BrazeTec BlueBraze 3510.

Results in the Field

Ag-Cu-Zn-Sn filler metals are used in various applications in HVACR industry, often in combination with Cu-P or Ag-Cu-P brazing alloys. The first tests in field was therefore concentrated on this industry and following applications:

- Compressor
- Refrigerator
- Heat pumps
- Valves

The brazing filler metals are used in different shapes as wires, rods, rings and pre-forms together with general purpose (white) fluxes in form of pastes, powders or coated onto rods

or rings. Occasionally gaseous fluxes are used exclusive or in combination with inorganic flux mixtures. The main heat source is flame, either in manual or automated processes, followed by induction heating.
Especially in automated processes, BrazeTec BlueBraze could prove the claimed property of similar liquidus / brazing temperature. The new material could be used on several brazing machines around the world without the change of any settings, providing a perfect filled joint.
In manual brazing application, the brazer commented on different melting characteristics and different colour. The smaller melting range of new BrazeTec BlueBraze products and a later initial melting when applying the rod onto the joint during heating causes a different melting impression.
The coloration after brazing is caused by the manganese content of the filler metals, and not known in the HVACR industry before. The appearance is influenced by brazing temperature, maximum part temperature and part cooling speed, as reaction starts after removing reducing flame.

Compressor

In compressor production, the main consumption of brazing filler metals is for joining copper tubes and steel shell. In machine flame processes, all tested BrazeTec BlueBraze products could replace the before or parallel running standard brazing filler metal by same settings. All customer commented that this was never achieved before with brazing filler metals containing 5 – 10 wt.-% less silver. Strength test and first leakage tests results indicate same performance of BlueBraze products. After standard cleaning procedures, the brownish layer disappeared giving bright and shiny fillet, which enables following painting process. Initial production tests are scheduled for final evaluation.

Additional application area is inside pipe connections, brazed manually by torch. BrazeTec BlueBraze 2410 is used already in mass production since middle of 2014 with same production process stability.

Refrigerator

Several pipe connections between compressor and heat exchangers are brazed manually by flame using copper-phosphorous brazing filler metals for copper-to-copper joints and silver brazing alloys for various steel and copper connections. Fluxes can be already coated onto the used rods, applied by hot rodding into powders or applied as pastes onto the joints.
The new filler metals show same gap filling performance at all tested joints around the world. The degree of filling depends on joint design, experience of brazer and process conditions. Oversized lap lengths can not be filled properly by any brazing filler metal and will consume more material as needed.
Performed strength test at brazed steel and copper tubes showed similar figures. The needed force to achieve crack in copper or steel parent material is listed in table 7. All tested joints, brazed with standard filler metal or BrazeTec BlueBraze, did not fail.

Table 7: Tensile test results of brazed refrigerator tubes.

Filler metals	Max. force steel to steel [N]	Max. force steel to copper [N]
Ag 134	2570	1810
BlueBraze 2410	2650	1800
Ag 130	2460	1820
BlueBraze 2010	2570	1800

Long time production tests showed the same results. After short adaption period, due to different melting characteristic and color of joint after brazing, the workers achieved the same performance level.
E.g. in one of the first production tests of four hours, zero to maximum two leakages occur at the different stations leading to a total rate of below 0.2 %. These leakages were produced in the first hour of the test, in which the brazers had to get convertible with the new material. Next initial production tests with more brazes and over 10,000 brazing joints resulting in a leakage rate of approx. 0.1 %.
Most important for successful brazing is homogeneous heating of the brazing joint and to avoid overheating of parent materials, flux and filler metal.

Heat pumps

Stainless steel plate heat exchanger to copper tubes and other tube joints are brazed manually by flame (figure 11). Commonly flux coated brazing rods are used in Europe for this application. Other options are use of flux paste prior brazing or hot rodding.

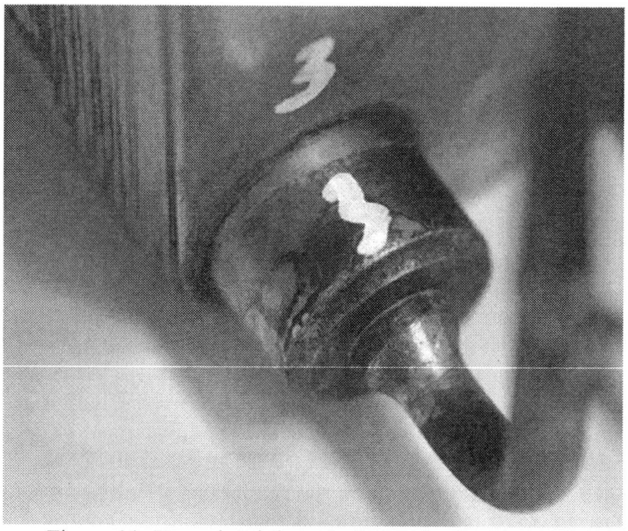

Figure 11: Brazed stainless steel plate heat exchanger.

Flux coated product passed already all production tests and further investigation, e.g. cross sections (figure 12), showing no difference in the performance, at first customer. Mass production already started successfully.

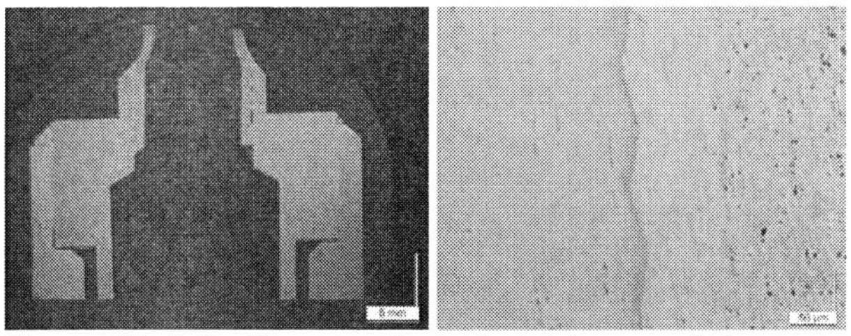

Figure 12: left: Cross section of stainless steel, brass and copper parent materials braze with BlueBraze 3510. Right: cross section of stainless and brass.

Valves

Refrigeration valves are made of steel, brass and copper components. All steel combinations are brazed silver brazing filler metals using flux pastes. The parts are brazed either by flame or induction heating.

Tests with BlueBraze at different companies lead to same gap filling results without any change of process parameters.

First customer starts to use BlueBraze 2410 pre-forms in mass production in 2014. So far more than 100.000 pieces are brazed in the induction heating process without any leakage in the pressure test during quality control.

Conclusions

Performed shear strength test showed differences between the DIN EN 12797 and AWS C3.2M standards.

AWS C3.2M specimen parent material influences the test results; investigated brazing filler metal systems appeared similar. Total shear strength values of origin width (30.0 mm) and machined (12.5 mm) are similar, but position of cracks varies. Wide samples failed above 2T overlap always in parent material, while 12.5 mm samples cracked below 4T always in braze metal layer. Deformation and stress load of parent material seems to influence braze metal layer and initiate cracks starting at the end of braze metal fillet on steel plate surface.

Design and fixture of specimen in strength test equipment of DIN EN 12797 ensures acting of applied forces directly in braze metal layer. According to [14] a shear strength value of 100 MPa could be used for design and calculation for overlapped brazing joints. Shear strength tests according DIN EN 12797 shows similar results for both tested filler metal system. The measured average shear strength were in range of 170 – 190 MPa. The tests showed definitely the strong influence of the process, e.g. heat distribution and flux inclusions during manual flame brazing, onto shear strength results. This influence is stronger than the differences between the filler metal systems.

Calculation formula of brazed joints uses yield strength of parent materials. As produced parts will be designed to operate on the upper yield point in maximum, the stresses inside the braze metal is much lower. In case of steel S235JR with ultimate strength (σ_{ts}) of 400 MPa and yield strength (σ_y) of 235 MPa the maximum stress with and overlap of 4T will be 59 MPa. This is three times below the measured shear strength in the braze metal.

The high-cycle fatigue specimen of steel S235JR with overlap of 4T showed only one specimen of Ag 134 failing in braze metal at 300 MPa bending stress. The failures on all other samples tested between 200 and 300 MPa occurred in the steel plates close to braze joint. Bending stresses of 100 MPa did not lead to any cracks. Ag 134 was stopped at more than 15 million cycles and BlueBraze 2410 at 21 million cycles without failure.

"21 million cycle specimen" showed an imperfection of approx. 40 % in the center. According to Flom [16] strength of 5T lap joint containing a 1T wide center flaw is the same as the strength of 4T joint containing no flaws. In case of high-cycle fatigue sample the 4T lap joint can be seen as 3T joint. This will lead to a maximum stress in the braze metal of approx. 79 MPa, which is still below applied bending stress and maximum shear strength of investigated filler metals. This rough estimation explains passing of this sample in high-cycle fatigue tests.

All the performed strength tests showed similar results between the standard Ag-Cu-Zn-Sn and the new developed Ag-Cu-Zn- Mn-Sn(In) system. Correct joint design and homogeneous heat distribution are essential for successful brazing in common.

Especially the high cycle fatigue test demonstrates the absolute usability of the new filler metal system in HVACR application. The joints withstand increased stress levels and failed in steel plates. In the application very often steel is joined with copper or copper alloys. These additional parent material is softer and therefore the impact onto braze metal is lower. As the toughest combination (steel to steel) was successfully tested, other combination could be used unconsidered.

The rate of corrosion was on a low level and both systems were absolutely comparable. Even within one alloy group, one was not able to detect a trend of the rate regarding the composition or the silver content. By assessing the results, one should consider that the results show a punctual view about the appearance after 4 weeks. One can assume that e.g. patina or stable oxide layers are able to reduce the rate of corrosion in a long run test.

All performed tests in laboratories and the experience in the field on various customers' sites in production applications demonstrated same results and performance of new silver brazing filler metals. These products enables the customer to ensure savings on silver costs with less impact of metal prices fluctuation onto final products without any change of process or production capacity.

References

[1] American Welding Society C3 Committee on Brazing and soldering, Brazing Handbook, American Welding Society (Miami, 2007), pp. 76; 92-95.

[2] Roberts, P., Industrial Brazing Practice, CRC Press (Boca Raton, 2013), pp. 56-62.

[3] Dorn, L., Hartlöten und Hochtemperaturlöten, expert Verlag (Renningen, 2007), pp. 58.

[4] Schnee, D., "Development of Ag-Cu-Zn-Sn Brazing Filler Metals with a 10 Weight-% Reduction of Silver and Same Liquidus Temperature", *Proc. 2014 International Conference on Brazing, Soldering and Special Joining Technologies*, Beijing, China, June 2014, pp. 53-58.

[5] DIN EN ISO 17672:2010-11, Brazing – Filler metals; German version EN ISO 17672:2010, DIN (Berlin, 2010).

[6] AWS 5.8M/A5.8:2011, Specification for Filler Metals for Brazing and Braze Welding, American Welding Society (Miami, 2011).

[7] ISO 3677:1995-04, Filler Metal for Soft Soldering, Brazing and Braze Welding – Designation; German version EN ISO 3677:1995, DIN (Berlin, 1995).

[8] AWS C3.2M/C3.2:2008, Standard Method for Evaluating the Strength of Brazed Joints, American Welding Society (Miami, 2008).

[9] DIN EN 12797:2000-12, Brazing – Destructive Tests of Brazed Joints; German version EN 12797:2000, DIN (Berlin, 2000).

[10] DIN EN 10025-2:2005-04, Hot rolled Products of Structural Steels - Part 2: Technical Delivery Conditions for Non-alloy Structural Steels; German version EN 10025-02:2004, DIN (Berlin, 2004).

[11] SAE J403:2014, Chemical Compositions of SAE Carbon Steels, SAE International (Warrendale, 2014).

[12] DIN EN 1045:1997-08, Brazing – Fluxes for Brazing – Classification and Technical Delivery Conditions; German version EN 1045:1997, DIN (Berlin, 1997).

[13] ANSI/AWS A5.31-92R:1992, Specification for Fluxes for Brazing and Braze Welding, American Welding Society (Miami, 1992).

[14] Wittel, H., Roloff/Matek Maschinenelemente, Vieweg+Teubner (Wiesbaden, 2011), pp 109-111.

[15] Schnee, D., "Hartlote und Flussmittel", *Proc. 2014 Hartlöten und Hochtemperaturlöten,* Esslingen, Germany, February 2014, pp. 72-74.

[16] Flom, Y., "Flaw tolerance in lap shear brazed joints", *Proc. 2004 International Conference on Brazing, High Temperature Brazing and Diffusion Welding,* Aachen, Germany, June 2004, pp. 89-93.

[17] Schnee, D., Principles of Brazing and Soldering, Umicore (Hanau, 2013), pp 13-14.

New Silver based Brazing Filler Metals for brazing Cemented Carbides with a highly decreased Level of Silver and low Brazing Temperatures

Max Schimpfermann, Gunther Wiehl, Daniel Schnee, Philipp Rädecker
Umicore AG & Co. KG, Business Unit Technical Materials, Rodenbacher Chaussee 4, 63457 Hanau-Wolfgang, Germany
Email: max.schimpfermann@eu.umicore.com

Abstract

The steadily rising silver price of the last 10 years has also increased the price of finished tools. Reducing the silver content of brazing filler metals provides direct cost saving. Especially the tooling industry has a high potential of saving silver by using optimized brazing filler metals.

Currently special silver based brazing filler metals with a composition of Ag-Cu-Zn-(Mn)-Ni with a silver content of app. 50-60 wt.-% and a processing temperature of app. 710 – 720°C are commonly used.

Aim of this research was to elaborate and analyze different brazing filler metals based on silver and copper with a reduction of more than 10 wt.-% of silver compared to standard filler metals without significantly influencing any brazing process parameters and shear strength of the joint.
During development the melting range of brazing filler metals were determined via DSC analysis. Additionally brazing tests were conducted with selected geometries and parent materials to determine the shear strength of the joint. Finally structural examinations were conducted via WDS-analysis to elaborate structure-property-relationships between shear strength and brazing joints microstructure.

Introduction

In the industrial application field of tool manufacturing (e.g. saw blades) often brazing is used as joining technology. The advantage of brazing compared to welding is that only low temperatures are required to join the cutting material (e.g. cemented carbides such as Co-WC) to the parent material (e.g. steel blades). The lower process temperature facilitate the conservation of the previous annealed microstructure and therefore hardness of e.g. steel blades. So a low-temperature brazing filler metal, which allows to braze at or below steel's ferrite- to-austenite transition temperature of 723°C (in the following denominated as "AC1 temperature"), is required to minimize microstructural changes in the steel part involved. In previous years such a low brazing temperature was achieved by using the standardized brazing filler metal Ag 449 acc. to DIN EN ISO 17672 [1] (B-Ag 22 acc. to AWS 5.8 [10]). The filler metal was developed based on Ag-Cu-Zn-Mn alloy family, wherein the silver content is 49 wt.-% and nickel of 4.5 wt.-% was added [2].
In DIN EN ISO 17672 a melting range of 680 °C to 705 °C is indicated for the standardized Ag 449 [1]. Its shear strength for joints between cemented carbides and steels varies between is 250-300 MPa [2].
Another standard brazing filler metal for joining tool materials is Ag 450 acc. to DIN EN ISO 17672 (B-Ag 24 acc. to AWS 5.8). This manganese free filler metal on the basis Ag-Cu-Zn has a silver content of 50 wt.-% with an addition of 2 % nickel and a melting range of 660 °C to 715 °C.
Table 1 summarizes the description of the two standardized brazing filler metals according to different international standards.

Table 1: Standard silver brazing filler metals with their codes according ISO 17672 [1], AWS 5.8M [10] and ISO 3677 [11].

Code acc. ISO 17672	Code acc. AWS 5.8	Code acc. ISO 3677	Silver content [wt.-%]
Ag 450	BAg-24	B-Ag50CuZnSn-660/**715***	50
Ag 449	BAg-22	B-Ag49CuZnMnNi-680/**705***	49

*Melting range acc. to DIN EN ISO 17672

In this publication it is investigated if Ag-Cu-Zn-Mn filler metals with a reduced silver content of 28 wt.-% and adjusted solidus and liquidus temperatures can generate similar joint strengths during brazing at 720 °C, like Ag 449 and Ag 450.

The state of the art is, that there are different options to lower the melting range of Ag-Cu alloys or vice versa lower the silver content with a constant melting range. Often alloys of the system Ag-Cu-Zn are used. While increasing zinc content the price of the brazing filler metal and melting range can be decreased [3][4]. On the other hand, ductility is reduced and susceptibility to oxidation increased, which leads to a maximum concentration of zinc technically feasible in the alloys [3][5]. The reduction of silver content leads to increased liquidus temperatures. For maintaining the brazing temperature the activity of the other elements have to be increased

relatively to silver. The effects of the variation of the activity of tin, gallium and indium in Ag-Cu-Zn is already investigated [7]. The effects of the variation of manganese content on the wetting properties are elaborated in [3][4][5]. High levels of manganese affect, similar to zinc, the ductility of the material and lead to embrittlement [3]. So balancing the zinc, manganese and indium concentrations in order to get low melting ranges and required material ductility is crucial.
Besides the above mentioned elements, nickel is attributed a special role: It can affect positively the wetting behavior of the brazing filler metal [3][7][5] and influence other properties such as liquidus temperature, spreading behavior and especially the shear strength of brazing joints. Nickel has a very high affinity for cobalt [7]. This was shown in comparison tests with and without nickel content in system Ag-Cu-Zn by Yaoita [7]. The spreading behavior was improved by addition of nickel [7] while liquidus temperature increased. In the literature it is not clearly elaborated how nickel affects the shear strength of the brazing joints. The majority of the literature [3][4][5][6] impute nickel to improve shear strength and ductility of the joints. Nickel is insoluble at temperatures around 700 °C in silver crystals [8]. It accumulates very well in copper [6], which leads to a solid solution hardening of the copper crystals. The good solubility of nickel in copper stabilizes the fcc-phase relative to bcc-phase [7]. However Yaoita et al. also found at high brazing temperatures, that nickel is lowering the shear strength of the brazing joint nonetheless higher volume fractions of ductile fcc-copper phase is formed [7]. The underlying mechanism, Yaoita et al. are assuming, is based on the diffusion of cobalt from carbide matrix into brazing filler metal according to the relatively long brazing times (several hundred seconds) compared to standard induction brazing processes. So the melted brazing filler metal generates in the cemented carbide a predetermined breaking point due to cobalt-binder depletion [6]. Therefore additionally added cobalt is attributed an increasing of shear strength in system Ag-Cu-Zn [7]. It was justified by the fact, that the cobalt containing brazing filler metal decreases the concentration gradient between the brazing filler metal and cemented carbide. Thereby the diffusion of cobalt from cemented carbide matrix is minimized. [7]. To obtain the highest possible shear strength is one of the most important goals in the development of the new brazing filler metal for tooling industry.

During the R&D work, different systems and brazing filler metals were tested under consideration of the described state of knowledge. By creating more and more data, it was necessary to describe especially also the effect of nickel onto the silver reduced system Cu-Ag28-Zn-Mn. So the influence of nickel onto phase formation during brazing and resulting shear strength of the joint and furthermore the affect of nickel onto the general melting behavior, solidus and liquidus temperature, was analyzed.

The current article will also consider the relationship between shear strength and melting behavior of brazing filler metals in common.
Table 2 shows the chemical composition of commonly used silver based brazing filler metals, which are used in the tooling industry. The alternative new product with a silver content of 28 % (in the following denominated as BFM 28) is described in addition.

Table 2: Composition of standard silver brazing filler metals and BFM 28

Code acc. ISO 17672	Code acc. AWS 5.8	Composition [Weight %]					
		Ag	Cu	Zn	Mn	Ni	In
Ag 450	BAg-24	50	20	28		2	
Ag 449	BAg-22	49	16	23	7.5	4.5	
BFM 28		28	39	20	10	1	2

Experimental Procedures

Density
The density of a brazing filler metal was measured according to the Archimedes principle with a sample size of 100 g.

Quantitative Microstructure Analysis
For a qualitative description of the chemical character of phases involved in the brazing filler metals Wavelength - dispersive spectroscopy (WDS) was used.
The quantitative analysis of liquid phase fractions was performed with the program "Imagic - IMS Client" on back-scattered electron images with metallographic cross sections of rapidly quenched microstructures. The method is based on differences in the Z-contrast (material contrast) of the back-scattered electrons between quenched solid and liquid phase fraction in the cross sections.

DSC Analysis
The DSC diagrams of the tested brazing filler metals were measured with Mettler Toledo's "DSC /TGA 1" in aluminum oxide crucible under argon protective atmosphere. The heating rate was set to 10 K/min and a sample amount of 150-200 mg was used. The melting range was elaborated as described by Schnee et al. [9].

Specimen and Brazing Trials
For a comparison the brazing joint's shear strength, a test specimen was defined as follows: A non-coated fine grain cemented carbide segment of the type K10 according to DIN ISO 513 [13] (HC 10; Source CERATIZIT) with dimensions 8 x 8 x 4 mm was brazed onto a 1.2210 (115CrV3) steel pin with dimensions of 8 x 8 x 30 mm. The compositions of the base materials are given in Table 3. The brazed specimen for the shear tests is shown in Figure 1.

All brazing trials were conducted in standard air atmosphere by using flux to reduce the oxide surface layers of the parent materials. A water based FH12 flux paste (BrazeTec spezial h Paste), according to DIN EN 1045 [12], was used throughout all test arrangements. As flux residues are hygroscopic and corrosive, these were immediately removed with tap water after the brazing process.

Table 3: Parent materials

	Cemented carbide K 10	Steel 1.2210 (115CrV3)
Fe		Rest
C		1.10 – 1.25 %
Si		0.15 – 0.30 %
Mn		0.20 – 0.40 %
P		0 00 – 0.03 %
S		0.00 – 0.03 %
Cr		0.50 – 0.48 %
V		0.07 - 0.12 %
WC	94.4 %	
Co	5.6 %	
Grain size	0.8 – 1.3 μm	

For all tests 8 x 8 mm foil pieces (thickness 0.2 mm) were punched from brazing filler metal sheet. Before the assembly, flux was applied by brushing on steel side and onto cemented carbide side. For the brazing itself, the brazing filler metal was placed between the two parent materials, so the flux was wetting both parent materials and the brazing filler metal piece before brazing.

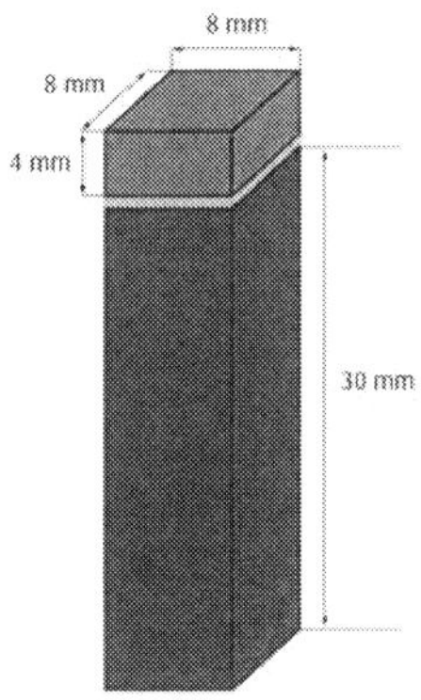

Figure 1: Design and dimensions of the specimen.

The correct positioning of the cemented carbide (even during the brazing process) in the x- and z-axis direction on the top of the steel, was ensured by using a support with pins on two sides of the cemented carbide, Figure 2.

For brazing a heating system a high frequency (3.5 kW) induction brazing machine was used with an induction frequency of 1.2 MHz. The inductor coil was square shaped with an edge length of 13 mm. So the distance between the specimen surface and the inductor coil was adjusted to 2.5 mm on each side. The y-axis positioning of the inductor coil was also defined. The coil was in line with the upper end of the steel part, Figure 2. Therefore the main energy was inducted into the steel. The cemented carbide was heated mainly by heat transport from steel.

Figure 2: High frequency brazing equipment with fixed specimen.

The brazing cycle was divided into two steps. First a pre-heating step up to 600 °C and a holding time of 1 second was used followed by a second brazing step with 3 seconds at the defined brazing temperature of 720 °C. After brazing, the specimen cooled down freely to room temperature. To monitor and to control the temperature during the brazing cycle, a pyrometer was used. The measuring point was focused on the surface of the cemented carbide (red dot in Figure 2). For the pyrometer measurement the upper surface of the cemented carbide was coated with special temperature-resistant emission-calibrated lacquer (emission factor 0.93). The lacquer had the effect, that the emission rates of the base materials are more stable and well-known. Therefore temperature measurement is more accurate and repetitious. In addition the device was protected against external scattered light to avoid misinterpretation by the pyrometer.

From time to time, the temperature was also measured with a thermocouple (1 mm diameter) inserted close to the brazing gap to verify the pyrometer measurements. The couple was placed in a blind hole in the steel and or the cemented carbide as near as possible to the joint surface with a distance of 4 mm from the edges. The comparison showed a homogeneous repetitious temperature distribution in the joint area due to all, previously described, standardization measures.

Shear Strength of Brazing Joint

One main parameter, besides the melting range of the brazing filler metal, is the shear strength of the joint. For this purpose a manual shear testing machine was used with a maximum load of 40 kN (Figure 3). The steel body side of the specimen was placed in a tailor-made fixation. The steel protruded 0.4 mm out of the support into the direction of the punch to prevent the support against damages and finally to minimize incorrect measurements, due to direct contact of cemented carbide and support. The design of the used punch correlated with the dimension of the cemented carbide (8 x 4 mm) for a planar and homogeneous pressure application, Figure 4. In the end the fixed setup allowed to maximize the ratio of the force into the y-direction and reduced bending moment of the x-direction at once. So the measured value corresponded with the maximum shear force and can be used for calculation of the maximum shear strength. At least 10 specimen were evaluated per filler metal variation to provide appropriate statistics to the evaluation.

Figure 3: Shear testing machine

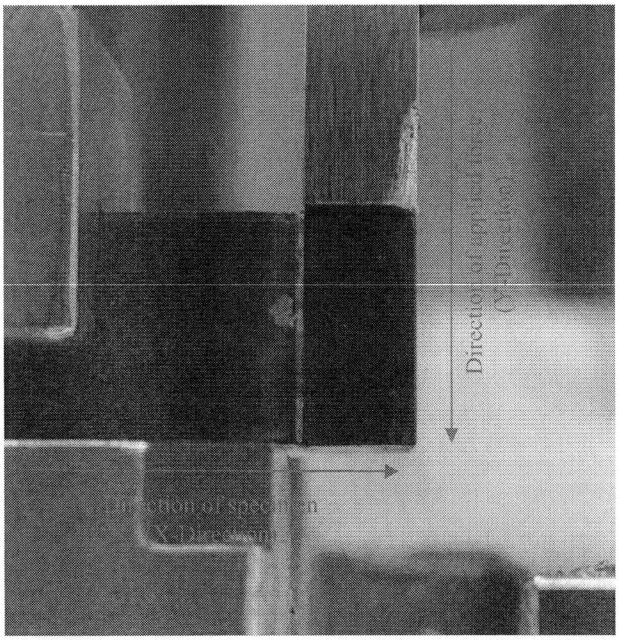

Figure 4: Setting of shear testing machine with fixture for specimen and design of punch.

Results

Density

The density showed a trend of being reduced by decreasing silver content (density of silver 10.5 g/cm^3). An additional effect is based on the addition of manganese with a density of 7.4 g/cm^3, which reduces the final filler metal density as well very effective.

Summarized BFM 28 has a 5 % respectively 10 % lower density compared to the standard brazing filler metals Ag 449 / Ag 450 (Figure 5).

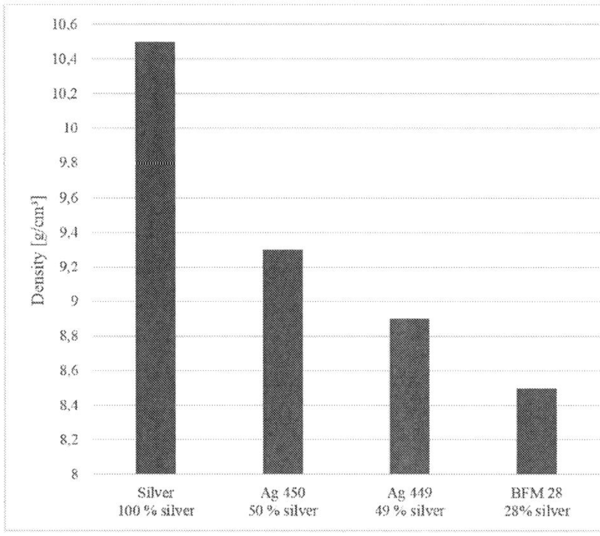

Figure 5: Density of silver, standard brazing filler metals compared to BFM 28.

Phase Description of Brazed Joints

With Wavelength - dispersive spectroscopy (WDS) three braze metals were analyzed: Ag 449 (Figure 6), Ag 450 (Figure 7) and BFM 28 (Figure 8). A difference in microstructure of the brazing joints between the standardized brazing filler metals and BFM 28 was found. As it could be expected from the chemical concentration the fraction of primary-grown crystals (often copper rich crystals) is distinctive higher for BFM 28 (Figure 8, red marker) than for Ag 449 and Ag 450.

In all investigated brazing filler metals the primary phase is surrounded by a silver rich "phase". This "phase" in fact is not a single phase. It seems to consist out of two or more phases. Therefore it is called in this paper "silver rich phase mixture" and the concentration measured by WDS of this silver rich phase mixture can be considered as an average composition of several very fine distributed phases.

The morphology of the primary-grown crystals is different between the standardized brazing filler metals and BFM 28: In Ag 449 and Ag 450 few Mn-Ni-rich (Figure 6, red marker) respectively Cu-Ni-rich (Figure 7; red marker) dendrites have formed while in BFM 28 a high abundance of spherical Cu-Ni-rich crystals (Figure 8, red marker) were observed.

The difference is reasoned by the different nucleation behavior during solidification of the different filler metals and the relative position of the brazing temperature relative to the individual liquidus temperature of the brazing filler metals.

For BFM 28 the spherical Cu-Ni-rich crystals represent the remaining, not-melted solid fraction at brazing temperature. Therefore their concentration is rich in copper and nickel and depleted by the other alloying elements that are located in the re-solidified liquid – namely the silver rich phase mixture.
The later chapters contain also more information on the melting range and liquid fraction of the Ag28 base system at brazing temperature.

On the contrary Ag 449 and Ag 450 were fully liquid respectively close to 100 % liquid at brazing temperature. Therefore in the microstructure the primary-grown crystals are dendrites that re-solidified at first from the liquid. Their nucleation can take place at the interfaces to the parent materials or within the brazing filler metal liquid (red markers in Figure 6 & Figure 7).
In the case of Ag 449 these dendrites contain much manganese and nickel as a result of the low copper activity in the filler metal. For the manganese free Ag 450 brazing filler metal the copper and nickel concentrations are dominating the dendrite composition.

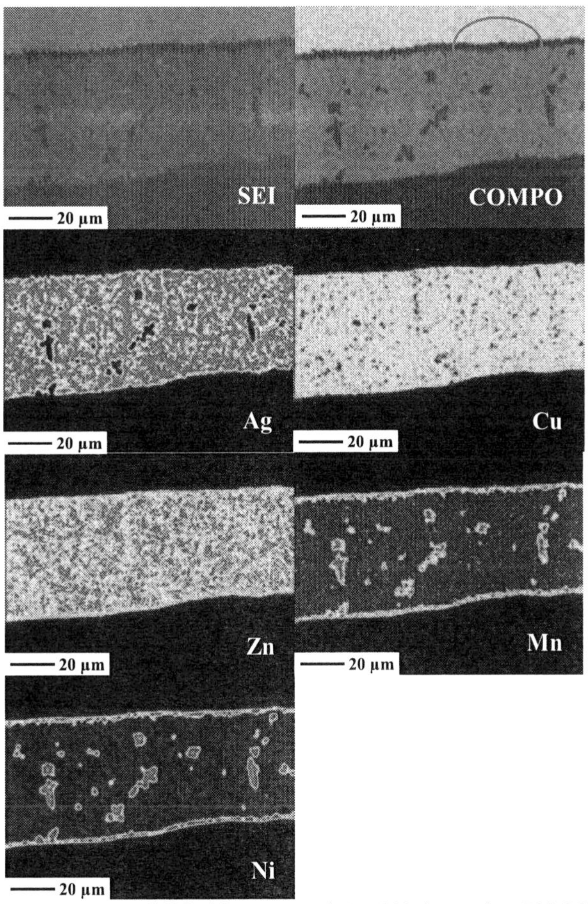

Figure 6: Re-solidified braze metal Ag 449; brazed at 720 °C.

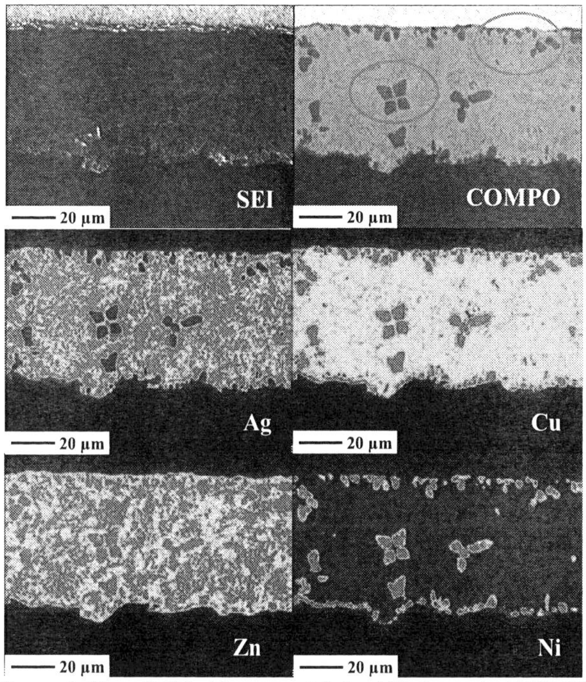

Figure 7: Re-solidified braze metal Ag 450; brazed at 720 °C.

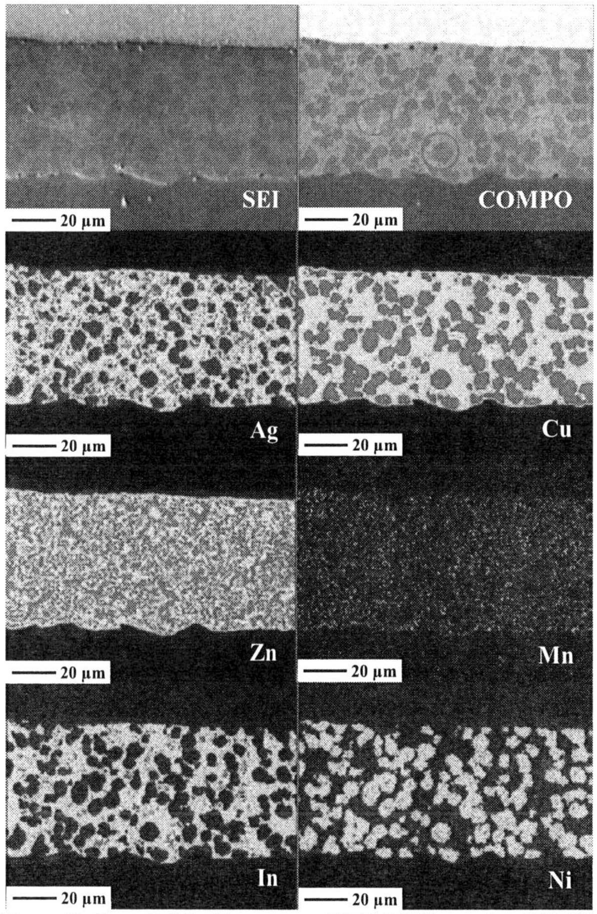

Figure 8: Re-solidified braze metal BFM 28 brazed at 720 °C.

Melting Range and Brazing Temperature

As mentioned already in the introduction, it was necessary to understand and to describe the influence of a nickel addition onto the CuAg28Zn20Mn10 base system (in the following denominated as "Ag28 base system") relating the melting range (and later on also onto the shear strength). Therefore an experimental series was started on the described Ag28 base alloy. Within this series, different nickel concentrations of 0% up to 9 %, were determined.

As measured with DSC analysis, the addition of nickel surprisingly lowered the solidus temperatures as well as raised as expected the liquidus temperatures in the system.

Therefore the melting range increases by increasing the nickel content in the alloy (Figure 9).

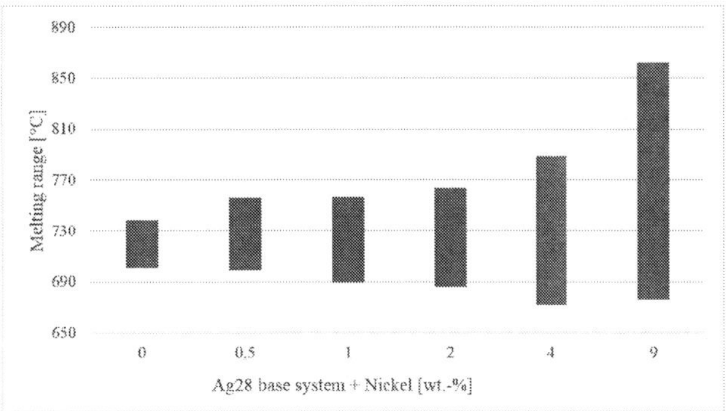

Figure 9: Melting range of Ag28 base system with increasing nickel content.

Because of the above described behavior, a technical limit of a nickel addition is given, by the increasing liquidus temperature and by the limited brazing temperature of 720°C. During the development the nickel content of the silver reduced brazing filler metal was fixed to 1 weight %.

The melting range of the developed brazing filler metal BFM 28 is 680 – 760 °C. These values were measured via DSC analysis. To provide an informative basis for the direct comparison to the described standard brazing filler metals, they have to be measured as well with the same DSC setup because melting range values differ from standard to standard for these identical brazing filler metals. In summary, the measured values showed slight variations to the values of the standards (e.g. DIN EN 17672; AWS 5.8) and cannot be dismissed (Table 4).

So compared to the standard brazing filler metals Ag 449, the solidus temperature of the brazing filler metal BFM 28 is the same. On the other hand the liquidus temperature increased slightly up to 760 °C compared to 705 °C (Ag 449) and 730 °C in the case of Ag 450.

Due to the completely new composition of the silver reduced brazing filler metal, the brazing temperature had to be defined and determined as well.

According to the standard AWS 5.8, Table A1, the minimum brazing temperature correlates with the liquidus temperature of the silver based brazing filler metal. This indicates, that the brazing filler metal should be completely liquid for the brazing task. As we elaborated by combining the melting behavior on cemented carbide surfaces and the resulting shear strength of the described specimen as a function of the brazing temperature, a brazing task can also be processed with BFM 28, Ag 449 and Ag 450 within the melting range without a significant loss of shear strength in the joint. Also a minimum working temperature was described by the revised standard DIN 8505, where a brazing filler metal should wet. According to this standard a temperature of 690 °C (for Ag 449), per definition within the melting range, was published in the past. So this situation of brazing within the melting range is not completely new and is additionally also known from the copper phosphorous brazing filler metals for copper / copper brazing tasks in addition.

Table 4: Melting ranges of brazing filler metals compared to BFM 28 according to different standards and measured by DSC and their and brazing temperatures.

	Standard / measurement	$T_{Melting}$ app. [°C]	$T_{Brazing}$ app. [°C]
Ag 450 (B Ag-24)	DIN EN ISO 17672	660 - 715	-
	AWS 5.8	660 - 707	min. 707
	DSC measurement	670 - 730	700
Ag 449 (B Ag-22)	DIN EN ISO 17672	680 – 705	-
	AWS 5.8	680 – 699	min. 699
	DSC measurement	680 – 705	min. 690
BFM 28	DSC measurement	680 – 760	710

Liquid Fraction at Defined Brazing Temperature

To understand, why a brazing process can also be successful within the melting range, an examination regarding the ratio of remaining solid phase to liquid phase was carried out at a brazing temperature of 720 °C. Figure 10a) shows the microstructure of the Ag28 base system as an example with an addition of 2 % of nickel. The brazing filler metal was quenched from 720 °C to room temperature in order to freeze the microstructure. Rounded dark crystals (Cu-Ni-rich phase from WDS analysis), which are enclosed in a light gray phase (silver rich phase mixture from WDS analysis) can be observed in the quenched microstructure as a result of the coexistence of liquid and remaining solid phase at 720°C. The rounded dark crystals are interpreted as the solid phase because of their round shape and their high copper and nickel concentration (compare WDS results, Figure 8). In image analysis, these crystals were colored orange (Figure 10 b). The light gray phase is interpreted as the former liquid phase at 720 °C.

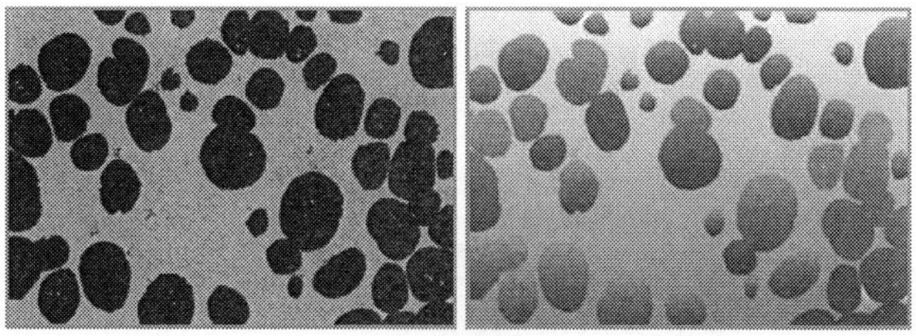

(a) (b)

Figure 10: (a) microstructure of Ag28 base system with 2 %Ni; (b) orange colored phase fraction of remaining solid phase in microstructure.

Due to a constant solidus temperature and an only slightly liquidus temperature increase (corresponds with an only little additional remaining solid fraction near liquidus temperature according to DSC analysis) by adding nickel up to 0.5 %, the fraction of the re-solidified silver rich phase mixture was, as expected, on a nearly constant level compared to the nickel free alloy (Figure 11).
Rising nickel content up to 1 %, the re-solidified, formerly liquid, fraction had a maximum because of an ideal relation of a reduced solidus temperature combined with an only slightly increased liquidus temperature.
The fraction of the re-solidified phase and therefore of the liquid fraction at 720 °C decreases by increasing the nickel contents from 1 % to 9 % in the described system.
The trigger for this behavior is founded in the strongly increasing liquidus temperature which goes hand in hand with a rising nickel content, as one can see in Figure 9.

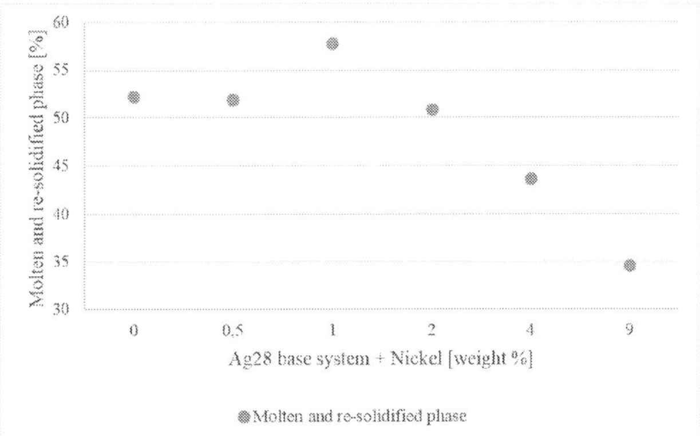

Figure 11: Ratio of molten phase at 720 °C for the Ag28 base system with increasing nickel content.

Shear Strength at Brazing Temperature of 720 °C
The shear strength observed for increasing nickel content has a maximum approximately at Ag28 base system with 1 % nickel for the investigated alloys. For higher nickel contents the shear strength drops (Figure 12). It should be additionally mentioned, that the increase in shear strength of the 9 % nickel sample maybe random due to its high standard deviation. Therefore the samples with 4 % and 9 % of nickel are approximately at one level.

The highest shear strength was determined at the Ag28 base system with an addition of 1 % of nickel. As specified in Figure 12, this composition had the highest amount of molten phase simultaneously.
So it can be concluded, that the shear strength directly depends on the liquid phase fraction at brazing temperature as a first approximation.

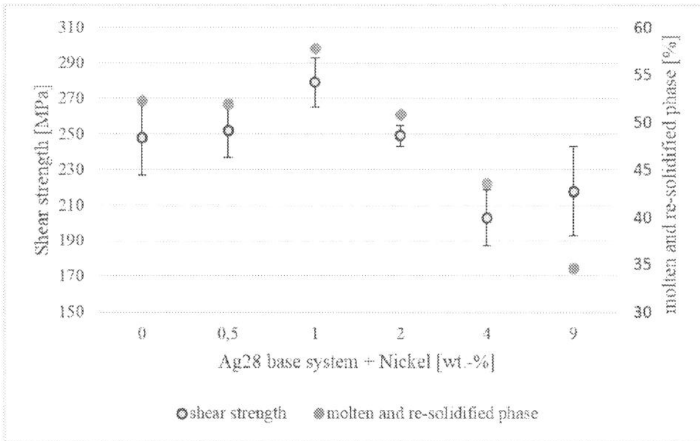

Figure 12: Shear strength and molten phase amount depending on the nickel content.

The shear strength of the final product BFM 28 was evaluated in direct comparison to the mentioned standard brazing filler metals at a brazing temperature of 720 °C.
As one can see in Figure 13, the introduced low silver containing silver based brazing filler metal showed with an average of 280 MPa a very high shear value associated with a very low standard deviation under the defined test conditions.
Especially in contrast to Ag 449 the standard deviation is 30 % less with BFM 28. Compared to the described performance, the brazing filler metal Ag 449 showed a shear strength on an equal level whereas Ag 450 leads to shear strength values 10 % below the new brazing filler metal. At this point the shear strength analysis was expanded by one filler metal in addition to get the described values directly related to another low silver containing system. This alternative and standardized brazing filler metal (for tool applications) is Ag 427[1] with a silver content of 27 %. This system however leads to 25 % lower shear strength with an average of 210 MPas. Furthermore the test run for Ag 427 showed also a minimum value of 180 MPa.

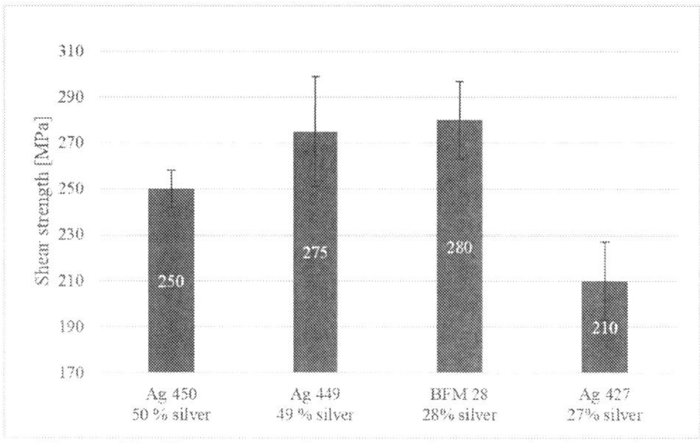

Figure 13: Shear strength, measured at samples brazed at 720°C.

Conclusions

It could be shown that shear strength is influenced strongly by the alloying of nickel into the described brazing filler metal system with 28 % of silver. As factors that are mainly responsible for the shear strength can be considered: Solid solution hardening of copper crystals and liquid fraction of the brazing filler metal at brazing temperature.

It is known, that with increasing manganese content the solid solution hardening of copper increases. [14]. Because nickel has a similar solubility inside the copper crystals as manganese and the substitution of copper atoms by nickel leads to a distortion of the crystal lattice, dislocation movement should require also more energy which leads to enhanced cold work strengthening of copper for increasing nickel contents [15]. However, no attempts for measuring this property were carried out in this study. It is assumed that this effect takes place also in the Ag28 base system

In the study it was found that the measured liquid fraction at brazing temperature, determined by the values of solidus and liquidus temperature, and the brazing filler metal concentration have a significant influence on the shear strength. Its maximum of liquid phase correlates with the maximum in shear strength. This indicates that the liquid fraction is the dominating factor over the solid solution hardening of copper. The moderate drop of the shear strength for higher nickel contents (4-9 %) could be interpreted as compensation effect by the solid solution hardening of nickel in copper against the falling liquid fraction at brazing temperature. However it has to be considered also that shear strength of the Ag28 base system with 9 % nickel has the highest standard deviation of the mean compared to the other measurements in the nickel series.

Therefore it is concluded that the maximum in shear strength at the given brazing temperature reflects the contrary influence of raising liquidus temperature (and therefore liquid fraction) and increased solid solution hardening on the shear strength of the joints by adding nickel to the brazing filler material.

For alloys containing more nickel, higher shear strengths would be possible if higher temperatures would be permitted by the application. However, this is of no interest for the brazing of cemented carbides on saw blades, because the ferrite- to -austenite transition temperature (AC1 temperature; 723 °C) of steel should not be exceeded in the process.

The brazing filler metal BFM 28 with the nominal composition Cu38Ag28Zn20Mn10Ni1In2 has in addition to its significant lower precious metals content the technical advantages over Ag 449 and Ag 450 that its density is much lower and its mean shear strength is slightly higher with a lower standard deviation. This means that a user can expect with BFM 28 a comparable or higher and maybe more important, a more reliable strength of their brazing joints in the tested field of cemented carbides brazing.

The lower density can be considered in combination with the lower silver content as a two-fold cost saving of the precious metal silver.

Brazing filler metal Ag 427, with a comparable silver content like the presented BFM 28, performed on a significant lower level regarding the shear strength at the evaluated brazing temperature of 720 °C.

In addition to the R&D laboratory work, first brazing trials with circular saw blades using BFM 28 were carried out under industrial conditions. For this performed trials at users sites, the brazing conditions were identical compared to brazing operations with Ag 449 brazing filler metal. After verifying that these conditions are ideal for brazing with Ag 449, the braze filler metal was replaced directly with BFM 28. The visual inspection of the brazed joints were subjective good. Following shear tests verified a comparable strength level of the joints with even a lower standard deviation of the shear test, mean values of BFM 28 comparable with Ag 449 and Ag 450.
Detailed experimental results is beyond the scope of this paper and will be presented in future publication.

Summarized it can be noted that the newly developed low silver containing brazing filler metal BFM 28 is an excellent alternative to existing brazing filler metals used in tool industry at much lower costs. It can be used with the existing process parameters and provides same or slightly better shear strength results as Ag 449 and Ag 450.

Acknowledgments

The authors would like to acknowledge Dr. Michael Magin and Sven Rassbach from Ceratizit Luxembourg S.à r.l for their support.

References

[1] DIN EN ISO 17672:2010-11, Brazing – Filler metals, German version, DIN (Berlin, 2010).

[2] BrazeTec Delivery Program, Umicore AG & Co. KG, (Hanau, 2013).

[3] Lashko, N.F., Lashko-Avakayn, S.V., Brazing and Soldering of Metals, S. Monson (Jerusalem, 1961), pp. 90-101.

[4] Frings, H.-P., „Untersuchungen im System Silber-Kupfer-Zink-Silicium als Grundlage für die Entwicklung von Loten für Hartmetall-Stahl-Verbindungen", *Technische Hochschule Aachen* (Aachen, 1976).

[5] Beck, et al., Edelmetall-Taschenbuch; 2nd Edition, Hüthig-Verlag (Heidelberg 1995), pp. 492-493.

[6] Zhongmin, L. et al., "Study on Microstructure and Property of Brazed Joint of AgCuZn-X(Ga, Sn, In, Ni) Brazing Alloy," *Rare Metal Materials and Engineering*, Vol. 39, (2010) pp. 397-400.

[7] Yaoita, S., Watanabe, T, Sasaki, T., "Effects of Ni and Co elements in filler metals in Ag brazing of cemented carbide," *Materials Research Innovations,* Vol. 17 (2013), pp. 142-147.

[8] Singleton, M., Nash, P., Binary Alloy Phase Diagrams, Vol. 1, ASM Intl (Ohio, 1990), pp. 64-66.

[9] Schnee, D., Wiehl, G., Starck, S., "Development of Ag-Cu-Zn-Sn brazing filler metals with a 10 weight-% reduction of silver and same liquidus temperatur," *2014 International conference on Brazing, Soldering and special Joining Technologies*, Beijing, China, June 2014, pp. 53-58.

[10] AWS 5.8M/A5.8:2011, Specification for Filler Metals for Brazing and Braze Welding, American Welding Society (Miami, 2011).

[11] ISO 3677:1995-04, Filler metal for soft soldering, brazing and braze welding, German version, DIN (Berlin, 2000).

[12] DIN EN 1045:1997-08, Brazing – Fluxes for Brazing – Classification and technical delivery conditions, German version, DIN (Berlin, 1997).

[13] DIN ISO 513: 2014, Classification and application of hard cutting materials for metal removal with defined cutting edges, DIN (Berlin, 2005)

[14] http://www.tf.uni-kiel.de/matwis/amat/mw1_ge/kap_8/backbone/r8_3_2.html, (04.11.2014)

[15] Askeland, D.R., Materialwissenschaften, Spektrum Akademischer Verlag GmbH (Heidelberg, 1996)

High-Temperature Resistant Co-Based Brazing Fillers

T. Uhlig, S. Weis, B. Wielage and G. Wagner
Technische Universitaet Chemnitz, Institute of Materials Science and Engineering, Chemnitz, Germany
thomas.uhlig@mb.tu-chemnitz.de

S. Schuberth
CeWOTec gGmbH – Chemnitzer Werkstoff- und OberflächenTechnik, Chemnitz, Germany
stefan.schuberth@cewotec.de

Abstract

Co-based materials are widely used in high-temperature applications like aircraft industry, industrial furnaces or gas turbine manufacturing. The main requirements on the Co-super alloys are high-temperature and corrosion resistance. To produce complex structures, it is necessary to join these alloys. Especially in case of corrosive service conditions the composition of the brazing filler should be similar to the base material. In this work Co-based brazing fillers for high-temperature service conditions are developed. The reduction of the melting temperature is achieved via addition of the hard phase forming refractory metals Ta, Ti, Zr and C. Brazing tests are carried out on Co-based alloy Mar M 509 using induction heating. The mechanical properties are determined by tensile tests.

Introduction

Co-based materials are widely used in applications like the aircraft industry, industrial furnace applications or gas turbine manufacturing. Since the 1960's these alloys play an important economic role in the fabrication of high-temperature resistant components [1]. The common requirements for the so called Co superalloys are good high-temperature strength as well as resistance against corrosion and creep. In order to save material and costs it is preferable to fabricate complex structures consisting of multiple parts. For joining the superalloys it is necessary to employ corrosion- and creep-resistant brazing fillers. Until now commercially nickel-, gold-, or palladium-based fillers are used to join Co-based superalloys [2]. In order to lower the melting temperature, the Ni-based materials often contain high amounts of boron, silicon or phosphorous. Especially in case of brazing gaps exceeding a thickness of 50 - 80 µm, these elements form brittle intermetallics in the center of the joint. Because of their higher tolerance on gap geometry, expensive noble metal based fillers are preferred in some industrial processes [3-4]. Nevertheless especially gold-based fillers behave disadvantageous due to erosion effects at the grain boundaries of the base materials and the local difference in electrochemical potential [5]. To improve the corrosion resistance, the chemical composition of the base material and the brazing filler should be similar. Furthermore the use of brazing fillers of the same base material improves the homogenization of mechanical properties within the joints. The aim of this study is the development of alternative Co-based brazing fillers for use under high-temperature corrosive service conditions. The alloying concept is based on results of Shurin et.al. from the 1990s [6-7]. The published data indicates that there is a lowest melting composition within the quasi-ternary system Co-$ZrC_{0.81}$-$TaC_{0.82}$ at 1250 °C that can be used as starting point for the filler development. As a first step of investigation, the melting behavior of alloys from this system is determined by DSC, the results are published in [8]. The near eutectic composition with a carbide content of 9 wt% (Zr,Ta)C melts at 1350 °C, Fig.1. The results of DSC are in good agreement with the microstructure of the alloys. For use as brazing filler a lower melting point is aspired. Because of the discrepancy to literature data, further alloying is necessary. This paper will deal with alloying concept, brazing procedure and mechanical properties of the produced joints.

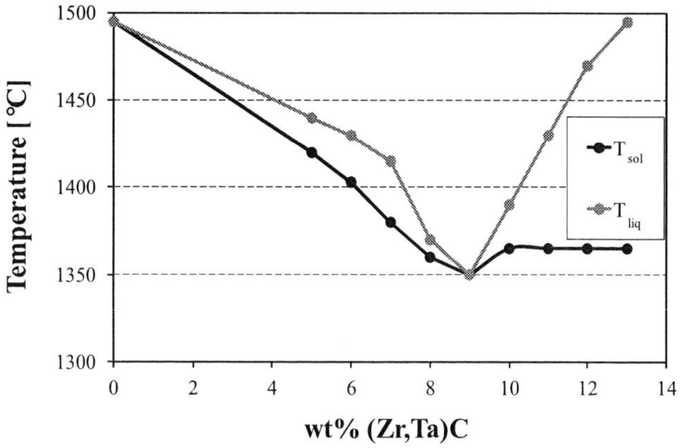

Figure 1: Melting behavior of CoZrTaC-alloys as a function of carbide content, determined by DSC [8]

Experimental Procedure

The alloys are prepared by melting in an electric arc furnace on a water cooled copper crucible using a non-consumable tungsten electrode, Fig. 2. The furnace is evacuated to 10^{-2} Pa and refilled with high purity argon (purity 99.9999 vol%) in order to create a slight overpressure of about 1.1x 10^5 Pa. Metallic elements with a purity of not less than 99.8 % and spectrally pure graphite are used as starting materials. The intended weight of the specimens is about 10 g. On the basis of the inert melting procedure, the melting losses can be determined by comparing the charge weight and the resulting alloy weight. Mass losses are measured after each step of alloying using a laboratory scale with an accuracy of 1 mg. After homogenizing the samples by continuous flipping over and re-melting, suction-casting in cylindrical molds is carried out.

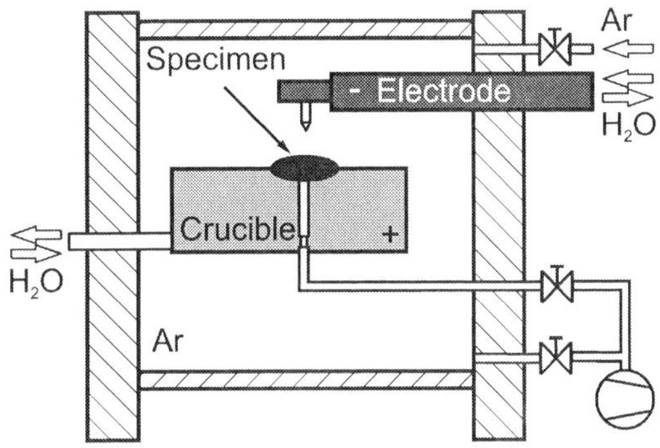

Figure 2: Schematic drawing of the employed electric arc furnace

The specimens are cut into slices by a metallographic cut-off machine. In order to ensure a good homogeneity within the samples, X-ray fluorescence analysis is carried out at each cut surface. The acceleration voltage is adjusted at 30 keV and a collimator of 2 mm is used in order to cover a relatively big surface area of the specimen. The measuring procedure can be used to detect elements with atomic numbers above 13 (Al), consequently no light elements like carbon are detectable. Due to the strong chemical affinity of carbon to the refractory metals, homogeneity can be ensured by measuring the metal contents. Parts of the specimens are hot mounted, ground and polished using grinding consumables and polishing cloths. Vickers hardness is measured on the polished surfaces with a load of 2.94 N. Differential scanning calorimetry is carried out using a NETZSCH STA 449 F1 simultaneous thermal analyzer. The maximum attainable temperature is 1650 °C, high purity argon (99.9999 vol%) is used as protective gas. The weighed portion of specimens for DSC is about 50 mg. Because of the high thermodynamic stability of intermetallics, high heating and cooling rates need to be employed to be able to distinguish the liquidus reaction from base line. The rates are set to 40 K/min. The onset- and peak temperatures of the heating- and cooling curves are used to determine the melting behavior of the alloys. As first step of alloying, elements which constitute eutectic systems with Cobalt are chosen and alloyed to the near-eutectic composition CoZrTaC9 [8]. The utilized contents are summarized in Tab. 1.

Table 1: Contents of alloying elements

Alloying element	Content [wt%]		
Al	2.5	5	10
B	0.5	1	1.5
Sn	5	10	34
Ti	2.5	5	10

Results

After cutting the suction-cast specimens and checking the homogeneity, micro-sections are made and the Vickers hardness is determined on the ground and polished surfaces. As can be seen in Fig 3, the texture significantly changes in case of Al- and Sn-containing alloys. The addition of B and Ti doesn't cause major changes.

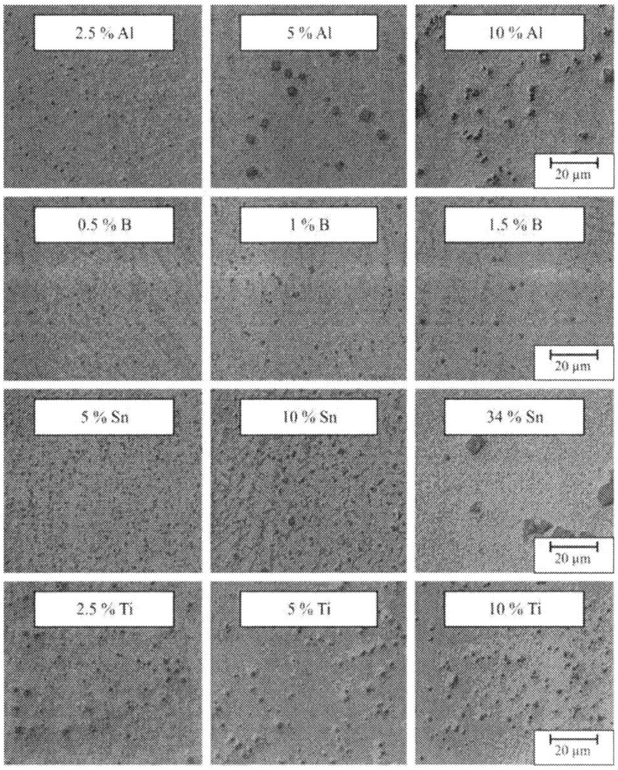

Figure 3: Micro-sections of Al-, B-, Sn- and Ti-alloyed near-eutectic composition CoZrTaC9

In correlation with the texture, the hardness of B- and Ti-containing near-eutectic alloy CoZrTaC9 doesn't change significantly with increasing B- and Ti-content, Fig. 4. Al- and Sn-alloyed compositions show a sharp increase of hardness for the highest investigated Al- and Sn-contents, Fig. 5. At lower contents there is no significant influence. With regard to the intended mechanical properties of brazing fillers, a high hardness is not desired. So the maximum Sn- and Al-contents are limited.

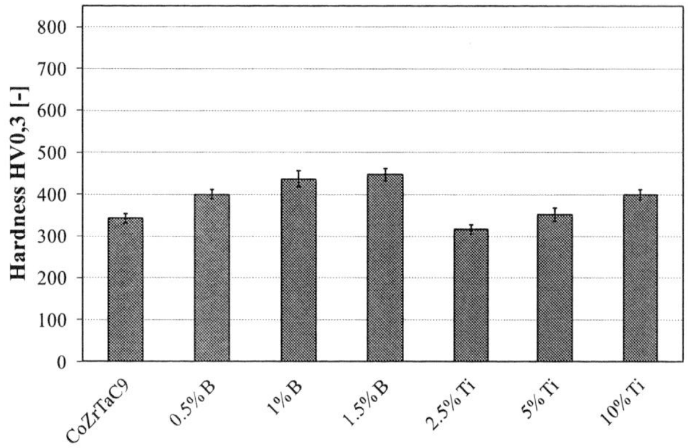

Figure 4: Vickers hardness of B- and Ti-alloyed CoZrTaC9

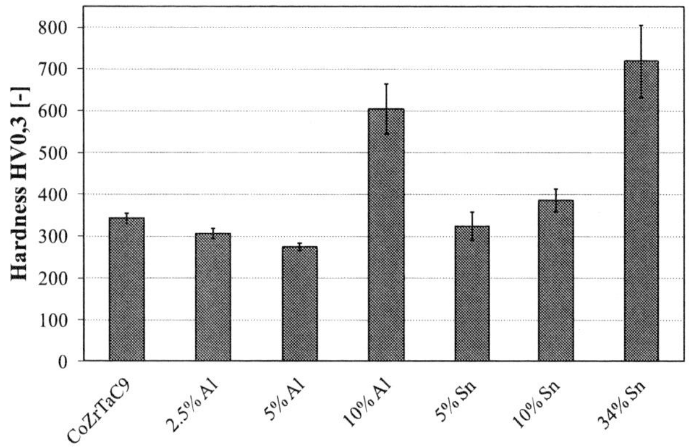

Figure 5: Vickers hardness of Al- and Sn-alloyed CoZrTaC9

The melting behavior is determined by DSC using the onset- and peak temperatures on heating and cooling. Alloying with Al doesn't influence the melting temperatures significantly, a second phase transition occurs at about 1200 °C. The measured temperatures for B-, Sn- and Ti-containing alloys are presented in the following figures. Due to the high thermodynamic stability of the intermetallics the slope of the melting peak is flat, so the onset on heating can't be determined accurately. That's why it is plotted blurred in the diagrams. As can be seen in Fig. 6, the addition of B to the near eutectic alloy CoZrTaC9 decreases the liquidus and the solidus temperature slightly. Furthermore a phase transition occurs at approximately 1100 °C. This temperature correlates with literature data dealing with the binary eutectic (L↔Co+Co_3B) [9].

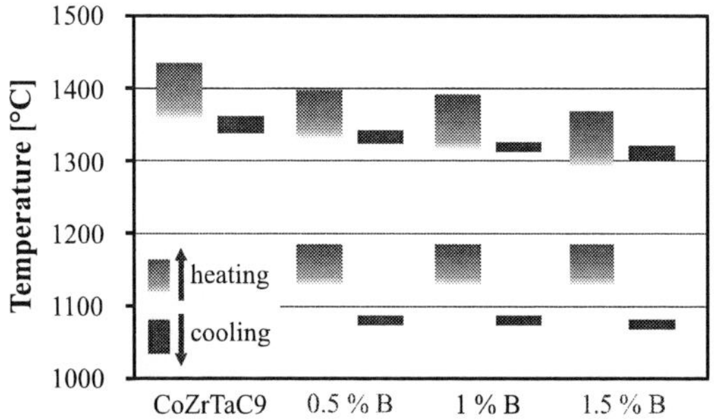

Figure 6: Influence of B on the melting of CoZrTaC9

The influence of Sn on the melting behavior of the alloy CoZrTaC9 is shown in Fig. 7. The liquidus temperature decreases with increasing Sn-content, a second phase transition takes place at about 1100 °C. At 34 wt% Sn liquidus- and solidus temperature coincide. This value is in good accordance to literature data on the binary system Co-Sn, Fig. 8 [9].

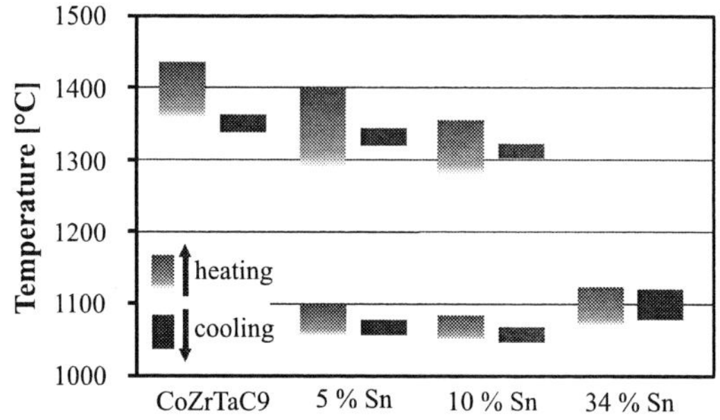

Figure 7: Influence of Sn on the melting of CoZrTaC9

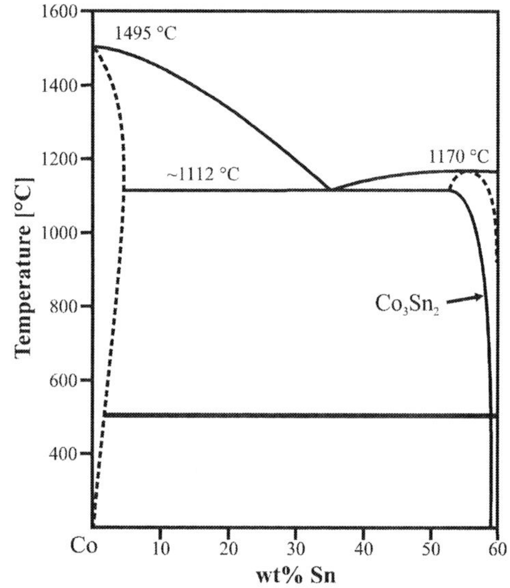

Figure 8: Binary phase diagram Co-Sn [9]

The addition of Ti leads to an almost linear decrease of liquidus- and solidus temperature with increasing Ti-content, Fig. 9. All alloys can be characterized by a melting range of about 200 K. Alloys containing 10 wt% begin to melt at 1150 °C. The wide melting range indicates that a liquidus temperature below 1200 °C can be achieved by adapting the composition of CoZrTaC-Ti.

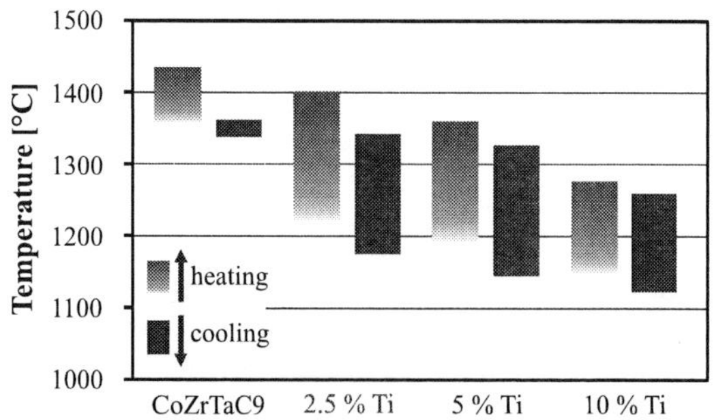

Figure 9: Influence of Ti on the melting of CoZrTaC9

Thermodynamic simulations of alloy CoZrTaC9+10 wt% Ti, which are carried out by a project partner, indicate a solidus temperature of about 1150 °C, which is in good agreement with DSC results. Furthermore the formation of primary TiC, which stays solid up to high temperatures, is assumed. The carbide is also observable metallographically. After subtracting out the TiC from CoZrTaC9+10 wt% Ti, an alloy with 8 wt% Ti and a lower carbon content is identified. The melting range is reduced by adapting the Zr/Ta ratio afterwards. The liquidus temperatures of CoTi8C0,16 alloys are plotted as function of Zr/Ta ratio in Fig. 10. The composition CoTi8Zr8Ta4C0,16 melts near-eutectic at about 1160 °C.

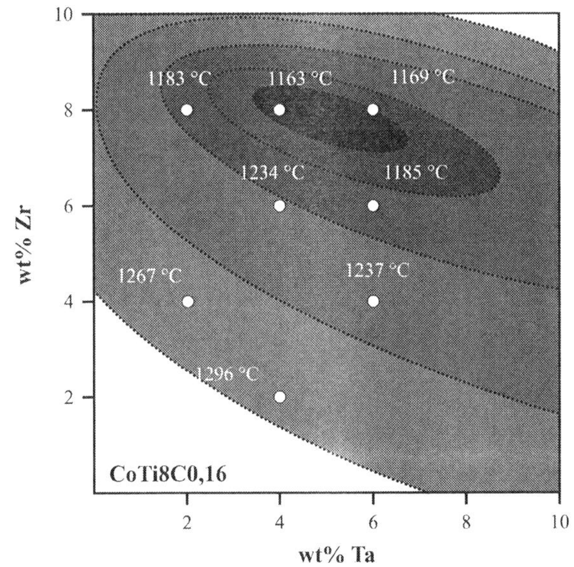

Figure 10: Liquidus temperatures of CoTi8C0.16 alloys as function of Zr- and Ta-content (white), estimated liquidus surface (dashed)

For further investigation, including brazing experiments and the mechanical tests, 3 alloys are chosen, Tab. 2. CoTiZrTaC-Sn has a slightly lower melting temperature, CoTiZrTaC-Ni is used to generate a partial austenitic lattice structure that should lead to a better ductility of the joint.

Table 2: Alloys used for brazing experiments

Alloy	Composition [wt%]
CoTiZrTaC	CoTi8Zr8Ta4C0.16
CoTiZrTaC-Sn	CoTi8Zr8Ta4C0.16+10Sn
CoTiZrTaC-Ni	CoTi8Zr8Ta4C0.16+20Ni

All joints are produced by induction brazing in vacuum at a pressure of < 10^{-2} Pa, a temperature of 1225 °C and a holding time of 120 s. Mar M 509 is used as base material. The experimental setup is shown in Fig. 11a. The samples with dimensions of 4 x 6 x 20 mm³ are brazed face to face in order to produce tensile samples with a length of 40 mm. The area near the brazed joint is shaped tapered to a cross-section of 2 x 4 mm² by milling. Two holes are drilled into the samples to apply the force in tensile tests, Fig 11b.

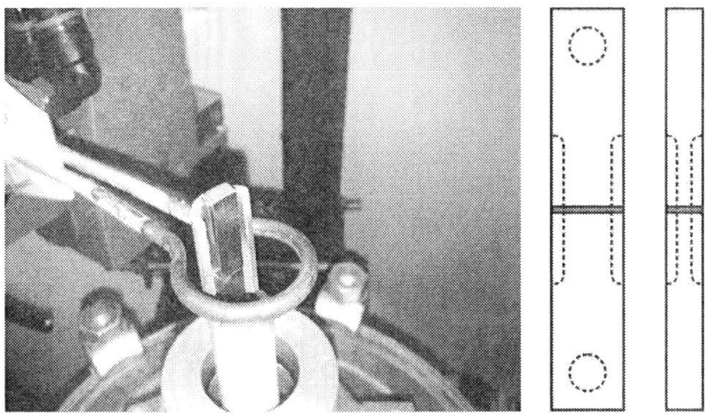

(a) (b)

Figure 11: a) Experimental setup for induction brazing
b) Brazed specimen prior to milling

After mechanical processing, monotonic tensile tests are carried out with a constant strain rate of 10^{-3}. The results are summarized in Fig. 12. The measured stresses using alloy CoTiZrTaC are relatively low, approx. 120 MPa, and can be characterized by a high standard deviation. CoTiZrTaC-Sn reaches a higher stress level, approx. 280 MPa, with very small standard deviation. The nickel alloyed brazing filler achieves the highest strength level, approx. 500 MPa, which is close to the yield strength of the base material [1].

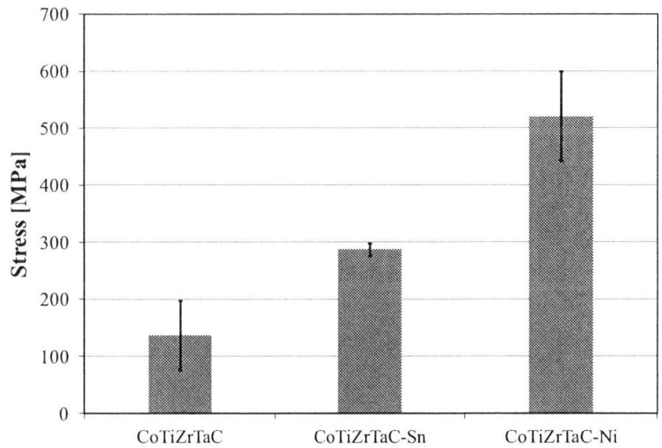

Figure 12: Rupture stresses of filler alloys determined by monotonic tensile tests

The fractured surfaces of the samples brazed with the different alloys are shown in Fig. 13. While joints produced with filler CoTiZrTaC show delamination between braze metal and base material, joints brazed with alloys CoTiZrTaC-Sn as well as CoTiZrTaC-Ni fail within the braze metal.

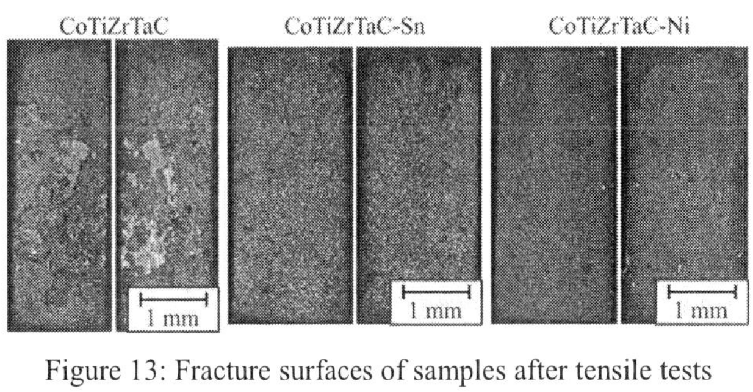

Figure 13: Fracture surfaces of samples after tensile tests

Conclusions

The influence of the alloying elements Al, B, Sn and Ti on the melting temperature of Co-based brazing filler, based on the quasi-ternary system CoZrTaC, is investigated. The addition of Al and Sn to the near eutectic alloy CoZrTaC9 leads to a sharp increase in hardness at high contents, that is not desirable for brazing fillers. B is known to form brittle phases and doesn't influence the melting behavior significantly. Alloying with Ti leads to an almost linear decrease of the liquidus temperature with increasing Ti-content without influencing the texture or hardness negatively. The wide melting range of these alloys indicates that a liquidus temperature below 1200 °C can be reached by adapting the composition within this system. This is achieved by adjusting the Zr-, Ta- and C-contents. The adapted composition CoTi8Zr8Ta4C0.16 has a melting temperature of 1163 °C. Brazed joints using the Ni-alloyed brazing filler CoTiZrTaC-Ni and the superalloy MAR M 509 at 1225 °C achieve mechanical properties comparable to the base material.

Acknowledgments

Part of the research was carried out within the scope of the project # 17622 BR funded by the Bundesministerium für Wirtschaft und Energie (BMWi) via the Arbeitsgemeinschaft industrieller Forschungsvereinigungen "Otto von Guericke" e.V. The authors thank the BMWi for financing this project and the project partners for carrying out the thermodynamic simulations.

References

[1] Davies, J. R., Nickel, cobalt and their alloys, ASM International, (2000)
[2] Heikinheimo. L., Miglietti. W. et al, "Brazing of Co- and Ni-based superalloys using new amorphous brazing filler metals" DVS-Berichte 212 (2001). pp. 30-34
[3] Greaves, W., "Cost effective repairs for IGT vanes" Proceedings of the 3rd IBSC, 2006, pp 392-394
[4] Falkenberg. S.; Watermeyer. K., „Die Untersuchung des Lötverhaltens von auf Ti-Al-Phasen aufgebauten Hochtemperaturwerkstoffen", Sonderbände der Praktischen Metallographie, (1995), pp. 401-404
[5] Donachie, M. J.; Donachie, S. J.: Superalloys: a technical guide, ASM International, (2002)
[6] Shurin. A. K.; Dmitrieva. G. P. et al, "Phase equilibria in Co-Me'C-Me"C alloys. 1: Systems with three-phase eutectic equilibrium" Powder Metallurgy and Metal Ceramics, (1996), pp. 615–620
[7] Shurin. A. K.; Dmitrieva. G. P et al, "Phase Equilibria in Co-Me'C-Me"C alloys. II: Systems with four-phase eutectic equilibrium". Powder Metallurgy and Metal Ceramics, (1997), pp. 193–196
[8] Uhlig, T.; Weis, S. et al, "Co-based brazing fillers", Welding Technology Review, (2013), pp. 64-67
[9] Massalski, T.B.; Okamoto, H., Binary Alloy Phase Diagrams, ASM International, (1990)

Wettability and the joint strength of Ag-Cu-Sn brazing filler metals on mild steel base metal

D.K. Basri, L.Sisamouth, Y.Farazila
Department of Mechanical Engineering, Faculty of Engineering, University of Malaya, Kuala Lumpur, Malaysia
basri.dk@gmail.com

M A. Hassan
Department of Mechanical Engineering, Faculty of Engineering, Assiut University, Assiut, Egypt

T.Ariga
Department of Materials Science, Faculty of Engineering, Tokai University, Hiratsuka, Japan

Abstract

The effect of tin concentration in silver base Ag-Cu-Sn filler metals on wettability, brazed-joint microstructures and shear strength were investigated. The wettability test on mild steel was based on a classical sessile drop method. Shear strength of the brazed joints was evaluated from a lap joint. The increment of tin content in the filler metals was found to cause poor wetting when they were used in brazing mild steel to mild steel. The interface between the base metal and the filler metal was of sharp boundary and can be considered as non-reactive wetting. The empirical equation to determine the contact angle with less than 10% percentage error was proposed. As tin content increased, the shear strength of mild steel/Ag-Cu-Sn/mild steel brazed joints became lower due to low interfacial adhesion strength and the increased amount of the $Cu_{41}Sn_{11}$ phase in the braze layer.

Introduction

Silver brazing alloys especially those based on Ag-Cu-Zn-Cd system are those of general purpose applications and is widely used for brazing steel, copper, brass, gun metal, tin bronze, aluminum bronze, manganese bronze, silver based contacts, nickel based alloys and nickel silver of all types [1, 2].

The addition of cadmium to silver-copper-zinc system reduces the solidus temperature and the melting range; reduces silver content; and improves the fluidity of the alloys [2]. However, the main problem associated with cadmium-containing silver brazing filler metals is the highly toxic cadmium oxide fume produced in dangerous concentrations during brazing processes [3-6]. Therefore, a series of cadmium-free silver-based brazing filler metals have been developed by replacing the cadmium in the Ag-Cu-Zn-Cd system with tin, indium, gallium and other elements [7]. However, the zinc that is contained in those filler metals is a volatile material, thus making it unsuitable for vacuum brazing [8]. Therefore, it is of interest to develop cadmium- and zinc-free brazing filler metals. In our previous work [9], the ternary Ag-Cu-In system was investigated. Ag-Cu-Sn filler alloys are also attractive because tin in this system has low melting point and low vapor pressure. Literature review showed that there have been only few investigations into these alloys, for example, the measurement of surface tension and density of the alloys [10], and the investigation of the alloys in the copper-rich corner of the ternary Ag-Cu-Sn [11] . However, the examination of the alloys in the silver-rich corner of Ag-Cu-Sn system remains open.

In this study, the performance of ternary Ag-Cu-Sn alloys with different compositions was examined. The effects of tin on wettability of the Ag-Cu-Sn brazing filler metals were investigated. In addition, the brazed-joint microstructures and shear strength were examined.

Experimental procedure

The brazing filler metals used in the experiments were Ag-Cu-Sn alloys. The nominal composition of Ag-Cu-Sn brazing filler metals and their solidus and liquidus temperatures obtained from differential thermal analysis (DTA) are tabulated in Table 1. Silver content was fixed at 60 wt.% and tin content was varied from 5 to 25 wt.%. The selected base metal was SS400 mild steel which have the melting point of 1430(°C).

The wettability test on mild steel was based on a classical sessile drop method. Three Ag-Cu-Sn filler metals (Sn = 5, 10, 15) along with the Ag-Cu eutectic alloy (table 1) were selected in this test. The surface of mild steel base metal plates of 30 mm x 30 mm x 2.5 mm in dimension were polished progressively down to 800 grit with silica paper in order to remove any oxide layer, contaminants, and to provide smooth mating surface. Then the mild steel plates and brazing filler metals were cleaned with acetone in an ultrasonic bath. In the wetting test, the filler alloy (0.2 g) was put on the plate and the AWS FB3A flux was applied on the plate and filler alloy. The wetting test was carried out in a tube furnace at a temperature of 50°C above the liquidus temperatures of the brazing filler metals. The inert gas was also introduced into the furnace at a rate of 2 l/min. The specimens were dwelled at the brazing temperature for 5 minutes and were subsequently cooled to room temperature. The wetting contact angle which was used as a quantitative measure of wettability was recorded. There were three replicates in each test condition.

Shear strength of the brazed joints was evaluated from a lap joint as shown in Fig. 1. The nominal joint gap of 50 µm was carefully set using a special jigging system (Fig. 2). The specimens for shear strength test were brazed in the same condition with the specimen for wettability test. The shear strength test of brazed specimen was conducted using an INSTRON Universal Test Machine. In this test, joints were subjected to shear tension load.

After brazing, the specimens for metallographic examination were cross-sectioned by means of abrasive cutting. After mounted, the specimens were ground progressively with silica papers from 400 to 2000 grit. The specimens were subsequently polished by alumina suspension (0.05 µm) in order to obtain a mirror surface. Cross-sectional microstructure inspection and analysis of brazed joints were performed using an optical microscope, a SHIMADZU electron probe micro-analyzer (Shimadzu EPMA-1600), and a scanning electron microscope (FEI Quanta 200F FE-SEM) equipped with energy dispersive X-ray (INCA Energy 400).

Table 1. Nominal composition (wt. %) and property of Ag-Cu-Sn filler metals

No.	Ag	Cu	Sn	Solidus (°C)	Liquidus (°C)
1	60	35	5	680	767
2(BAg-18)	60	30	10	602	718
3	60	25	15	580	680
4	60	20	20	521	640
5	60	15	25	514	609
6(BAg-8)	72	28	-	780	780

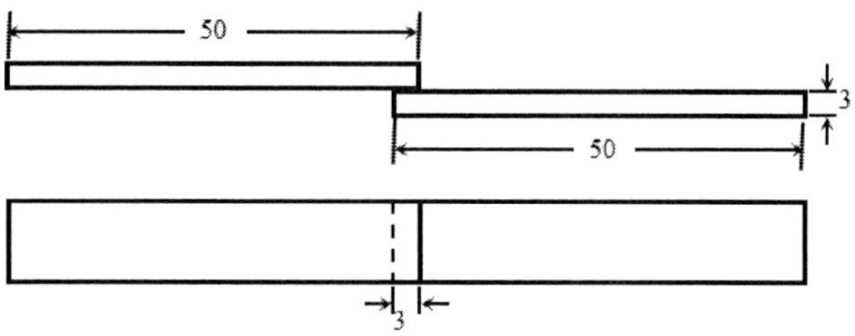

Fig. 1. Shear test specimen (mm)

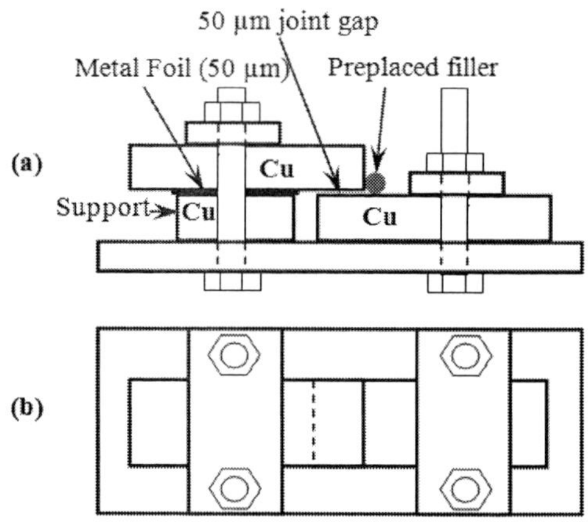

Fig. 2. Schematic illustration of a jig system: (a) front view (b) top view.

Results and discussion

Fig. 3 shows the wetting contact angles of eutectic Ag-Cu and Ag-Cu-Sn filler metals on mild steel. The contact angle of 60Ag-35Cu-5Sn filler metals on mild steel was 18.5°, which was higher than that of the BAg-8, 12°. The increase of tin content was detrimental to wetting property of the filler metals. The average contact angles of 60Ag-30Cu-10Sn and 60Ag-25Cu-15Sn increased by 58% and 82% compared to that of the 60Ag-35Cu-5Sn filler metal.

Fig. 4 shows an SEM image and EDX analysis of mild steel/60Ag-35Cu-5Sn/mild steel brazed joint. The interface between mild steel and the filler metal was almost sharp. The brazed layer comprised copper-rich solid solution, silver solid solution and $Cu_{41}Sn_{11}$ phase. Iron was present in the copper-rich solid solution phase but not in the silver solid solution phase. The presence of iron in the brazed layer indicates that there was minor dissolution of the base metal during brazing.

The SEM and EDX analysis of mild steel/60Ag-25Cu-15Sn/mild steel brazed joint is shown in Fig. 5. It was revealed that when the tin content in the filler metal was raised, the amount of the $Cu_{41}Sn_{11}$ intermetallic phase significantly increased. However, the interface between the braze layer and the base metal remained sharp. Protsenko *et al.* [12] demonstrated the role of intermetallic in wetting in the metallic system by comparing the wetting behavior of lead and tin on steel base metal. They found that the contact angle of lead was high (close to 90°) and did not change with time. From microstructure observation, no reactivity was observed at the interface between lead and iron. On the contrary, the wetting angle of molten filler metal on steel significantly decreased with respect to time and good wetting was found to be attributed to the intermetallic compound formation at the interface between tin and iron ($FeSn_2$). In the present investigation, although the amount of tin in the filler metal was increased, tin did not enter into reaction with iron, but formed a compound with copper in the brazed layer. This is illustrated in Fig. 5 where the compound at the braze layer near the interface was found to be $Cu_{41}Sn_{11}$, although a very small amount of iron appeared in this phase. This was probably because tin is more chemically favorable to form a compound with copper than with iron. Since the interfaces between mild steel base metal and the filler metals were sharp, the dissolution of base metal was negligible, and there was no compound formation at the interfaces, wetting of mild steel by Ag-Cu-Sn filler metals could be regarded as a non-reactive wetting case. Based on the classical model of wetting, Young's equation:

$$\cos\theta = \frac{\gamma_{SV} - \gamma_{SL}}{\gamma_{LV}} \qquad \text{(Eq.1)}$$

where, θ is the contact angle; γ_{SL} is surface tension between the liquid and the solid; γ_{LV} is surface tension between the liquid droplet and atmosphere; and γ_{SV} is surface tension between the substrate and the atmosphere. Obviously, the contact angle can be minimized by increasing γ_{SV} and decreasing γ_{SL} and γ_{LV}. The term γ_{SL}, in particular, which is a constant at a fixed temperature for a particular solid-liquid combination can be reduced by changing the composition of the materials system. It is therefore indicated that the increase of tin content in the Ag-Cu-Sn filler metals led to the increase of surface tension between solid and liquid (γ_{SL}), which consequently increased the contact angle as illustrated in Fig. 3.

It has been shown that for binary metal systems where non-reactive wetting occurred (solubility is less than 0.1%), Young's equation (eq.1) can be expressed empirically in terms of the melting point of the solid metal (base metal) and the melting point of the liquid metal (filler metal) as [13]:

$$\cos\theta = 1 + k\left[\frac{T_{lm}}{T_{sm}} - 1\right] \qquad \text{(Eq.2)}$$

where k is an experimental correction function, T_{sm} is the melting point of the solid metal (base metal) and T_{lm} is the melting point of the solid metal. For the contact angle of eutectic Ag-Cu metal (BAg-8) on mild steel base metal, the value of k were obtained through data fitting in this experiment is equal to 0.05. The result obtained in this empirical equation (Eq.2) is shown in Fig. 3. For ternary system such as Ag-Cu-Sn filler metals used in this study, the wetting equation cannot be truncated to Eq.1. The value of the solidus and the liquidus temperature need to be considered in replacement of the value of melting point of the liquid metal. Therefore, in order to consider the effect of the liquidus temperature and the solidus temperature, Eq.2 is modified as:

$$\cos\theta = 1 + kx\left[\frac{\frac{T^S + T^L}{2}}{T_{sm}} - 1\right] \qquad \text{(Eq.3)}$$

where T^S and T^L is the solidus and liquidus temperature of the filler metals respectively and x is the amount of Sn content (wt.%). By using Eq.3 and through experimental data fitting, the value of k was obtained to be 2.3 for each Ag-Cu-Sn filler materials. The calculated value of contact angle obtained from empirical Eq.3 is shown in Fig.3. It is found that the error for the proposed empirical equation compared to the experimental value is approximately less than 10%. It is therefore concluded that the equations are effective for predicting the contact angle.

Fig. 6 presents the relationship between shear strength of mild steel brazed joint and tin contents in Ag-Cu-Sn filler metals. Additionally, the shear strength of a mild steel joint brazed with the Ag-Cu eutectic filler metal is presented. The average shear strength of mild steel/60Ag-35Cu-5Sn/mild steel brazed joint was 248 MPa, slightly inferior to that of mild steel/BAg-8/mild steel brazed joint, 259 MPa. The average shear strength dropped by 17.7 % and 31.7 % when tin content was increased to 10%, and 15%, respectively. It obviously implies that the increment of tin content was detrimental to the shear strength of brazed joints. According to Miller and Schwaneke [14], the average shear strength of mild steel to mild steel brazed joints made with cadmium-containing filler alloys BAg-1a and BAg-3 were 203 and 240 MPa, respectively. This shows that 60Ag-35Cu-5Sn and BAg-3 filler metals gave equivalent joint shear strength and the 60Ag-30Cu-10Sn filler metal yielded similar strength as BAg-1a. The low brazed joint strength resulted from two factors: the increased amount of hard and brittle intermetallic phase in the brazed layer and very sharp interface between mild steel and the brazed layers. According to Miller and Schwaneke [14], when this sharp boundary is formed, the joint is usually weak because it is more of a mechanical than a metallurgical bond. Based on Young-Dupre's equation, the work of adhesion is expressed by:

$$W_a = \gamma_{LV}(1 + \cos\theta) \qquad \text{(Eq.4)}$$

This equation clearly shows that the work of adhesion is low when wetting contact angle is high. In the wetting test, the values of contact angle were high when tin content in the filler metals increased, thus decreasing adhesion strength.

Fig. 7 depicts typical fracture locations during shear test. For brazed joints made with 60Ag-35Cu-5Sn and 60Ag-30Cu-10Sn filler metals, the fracture occurred at the brazed layer, indicating that the adhesion strength was stronger than at the brazed layer and the low shear strength was due to the increase of intermetallic compound. In the case of mild steel/60Ag-25Cu-15Sn/mild steel joint, the failure was at the interface although the brazed layer contained relatively large amounts of intermetallic phase. This was due to the adhesion strength being even weaker than the brazed layer.

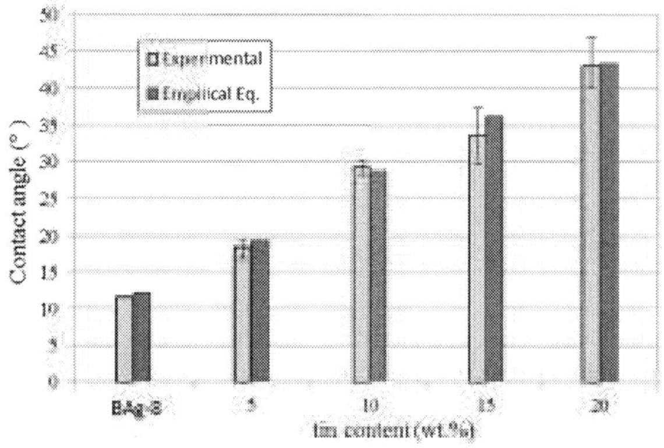

Fig. 3. Effect of tin content on contact angle.

Zone	Element (Atomic percent)			
	Ag	Cu	Sn	Fe
1	4.60	93.45	0.51	1.44
2	81.76	13.52	4.93	-
3	2.13	79.35	18.14	0.37

Fig. 4. SEM image and EDX analysis of mild steel/60Ag-35Cu-5Sn/mild steel joint

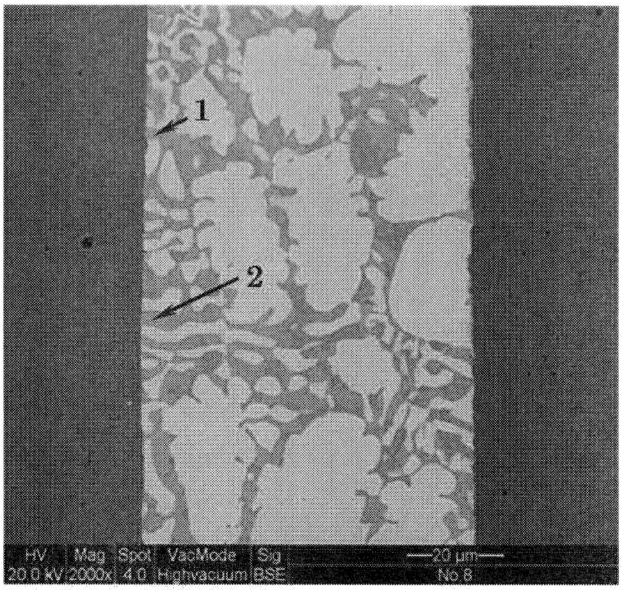

Zone	Element (Atomic percent)			
	Ag	Cu	Sn	Fe
1	2.76	73.18	20.82	3.25
2	3.45	74.40	19.61	2.54

Fig. 5. BSE image and EDX analysis of mild steel/60Ag-25Cu-15Sn/mild steel joint

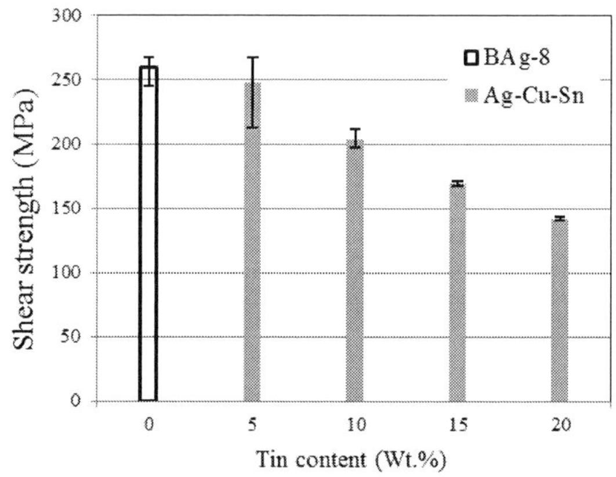

Fig. 6. Effect of tin content on shear strength of mild steel/mild steel brazed joints.

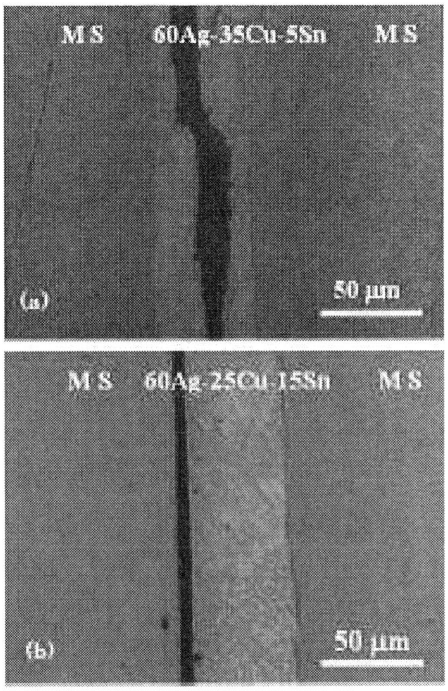

Fig. 7 Failure locations in shear strength test: (a) at brazed layer (b) at interface

Conclusion

The examination of the alloys in the silver-rich corner of Ag-Cu-Sn system was examined in this study. The effect of tin on wettability, brazed-joint microstructures and mechanical properties were investigated. The increment of tin content in the filler metals caused poor wetting on mild steel base metal and the interface between the base metal and the filler metal was of sharp boundary. The empirical equation to determine the contact angle with less than 10% percentage error was proposed. As tin content increased, the shear strength of mild steel/Ag-Cu-Sn/mild steel brazed joints became weaker due to low interfacial adhesion strength and the increased amount of the $Cu_{41}Sn_{11}$ phase in the braze layer.

Acknowledgement

This project was funded under Fundamental Research Grant Scheme (FRGS), grant no: FP024-2012A.

References

[1] S. C. Dev and C. S. Sivaramakrishnan, "An indigenous technology for a silver brazing alloy," *Materials & Design,* vol. 17, (1996), pp. 75-78.

[2] M. M. Schwartz, "Brazing," ASM International (Ohio, 2003), pp. 200-201.

[3] B. D. Thorne and P. J. Hewitt, "Generation of cadmium fume from alloy surfaces," *Annals of Occupational Hygiene,* vol. 29, (1985), pp. 181-189.

[4] B. D. Thorne and P. J. Hewitt, "Chemical composition and particle morphology of brazing fume from cadmium brazing alloy and fluoroborate flux," *Ann Occup Hyg,* 1988), pp. 489-497.

[5] T. Nishida, M. Inagaki, *et al.*, "Behavior of cadmium on brazing - Problems associated with fumes.," *Japan Welding Society,* vol. 4, (1986), pp. 592-596.

[6] T. Nishida, K. Kimura, *et al.*, "Behavior of Cadmium on Brazing - Problems Associated with Fumes, Flux and Brazed Parts.," *Japan Welding Society,* vol. 12, (1994), pp. 485-494.

[7] P. F. Timmins, "Development of nontoxic Ag-based brazing alloys," *Welding Journal* vol. 73, (1994), pp. 31-33.

[8] A. W. Society, "Brazing Handbook," American Welding Society (Miami, 1991), pp. 581-584.

[9] L. Sisamouth, M. Hamdi, *et al.*, "Investigation of gap filling ability of Ag–Cu–In brazing filler metals," *Journal of Alloys and Compounds,* vol. 504, (2010), pp. 325-329.

[10] I. Kaban, S. Gruner, *et al.*, "Surface tension and density in liquid Ag–Cu–Sn alloys," *Journal of Non-Crystalline Solids,* vol. 353, (2007), pp. 3717-3721.

[11] Z. He and L. Ding, "Investigation on Ag-Cu-Sn brazing filler metals," *Materials Chemistry and Physics,* vol. 49, (1997), pp. 1-6.

[12] P. Protsenko, A. Terlain, *et al.*, "The role of intermetallics in wetting in metallic systems," *Scripta Materialia,* vol. 45, (2001), pp. 1439-1445.

[13] D. M. Jacobson and G. Humpston, "Principles of brazing," ASM International (OH) (Ohio, 2005), pp. 16-19.

[14] V. Miller and A. Schwaneke, "Interfacial compositions of silver filler metals on copper, brass and steel," *Welding Journal,* vol. 57, (1978), pp. 303-310.

Interface Microstructure and Mechanical Properties of the brazed C_f/LAS Composites Joint

Duo Liu, Xiaoguo Song and Jicai Feng
Harbin Institute of Technology at Weihai, Shandong Provincial Key Laboratory of Special Welding Technology, Weihai, China
liuduo0376@163.com

Abstract

In the present work, a commercially available Ag-Cu-Ti braze alloy was used to join continuous carbon fiber reinforced lithium aluminosilicate glass–ceramics matrix composites (C_f/LAS composites). According to the experimental results and observations after brazing, the interface structure can be expressed as C_f/LAS composites/$TiSi_2$/Cu_2Ti_4O/TiCu/ Ag(s,s)+Cu(s,s)/TiCu/Cu_2Ti_4O/$TiSi_2$/C_f/LAS composites. With the increased brazing temperature or holding time, the thickness of interfacial reaction layer near C_f/LAS composites increases. And the thickness of the reaction layer should be kept at an optimal value to avoid the influence of residual stress. According to the testing results of mechanical properties, the maximum shear strength is obtained for the joint brazed at 900℃ for 10 min.

Introduction

Continuous carbon fiber reinforced lithium aluminosilicate (C_f/LAS) glass-ceramic matrix composites[1] have been studied intensively in recent years because of their low or even negative coefficient of thermal expansion (CTE) and good electrical performance[2]. C_f/LAS composites have great potentials for application in the fields of microelectronics[3], optical precision instruments[4] and so on. However, the brittleness of C_f/LAS composites makes it difficult to manufacture large or complex structure, which hinders their application seriously. Joining single or simple structures together is considered as a good solution for this problem. Among the joining methods, brazing is suitable for C_f/LAS composites bonding due to the low deformation and little negative weight of brazed joint[5-7]. So far, the research of brazing of C_f/LAS composites is quite scarce. Therefore, active filler alloy is utilized to braze C_f/LAS composites in the present work, and the microstructure and mechanical properties of brazed joints are investigated as well.

Experimental procedure

C_f/LAS composites were brazed with a commercially obtained Ag-21Cu-4.5Ti (wt.%) braze alloy in the experiment. The microstructure of the composites is shown in Figure 1. C_f/LAS composites have been fabricated by sol-gel process and hot pressing technique. According to Figure 1, vertical direction of carbon fiber is selected to take this micrograph. The black spots are cross sections of carbon fibers. The surfaces to be brazed were polished and then ultrasonically cleaned by acetone before vacuum brazing. The size of C_f/LAS composites for mechanical testing specimens is 8mm×5mm×5mm and 30mm×10mm×3.5mm, respectively.

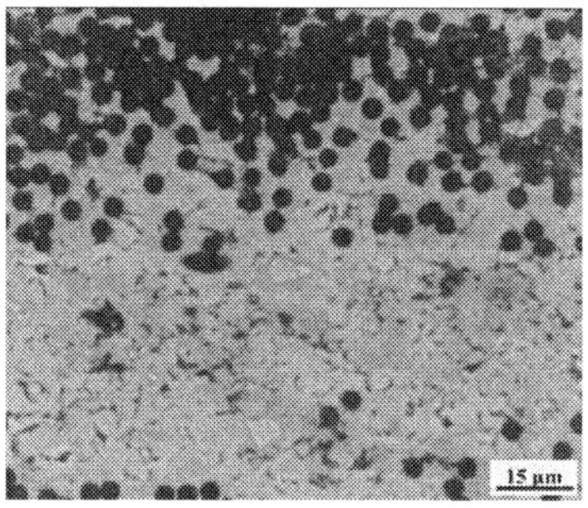

Figure1 Microstructure of C_f/ LAS composites

The brazing was performed in a vacuum furnace under the vacuum of 3×10^{-4}Pa. Universal testing machine was used to test the shear strength of the brazed joints. The testing specimens were pressed by the machine with a constant speed of 0.5mm/min. Interfacial microstructure and fracture surface were examined with scanning electron microscope (SEM) equipped with an energy dispersive spectrometer (EDS) and an electron probe X-ray microanalysis (EPMA). The interfacial structure of brazed joint was identified by XRD.

Results and discussion

Figure 2 shows the typical interfacial microstructure of the C_f/LAS composites joint brazed at 900℃ for 10 min. Good metallurgical combinations were formed between the filler metal and both sides of the base material, and a continuous interfacial layer was observed. Defect-free brazed joints were obtained, and no cracks and pores were found in the joints.

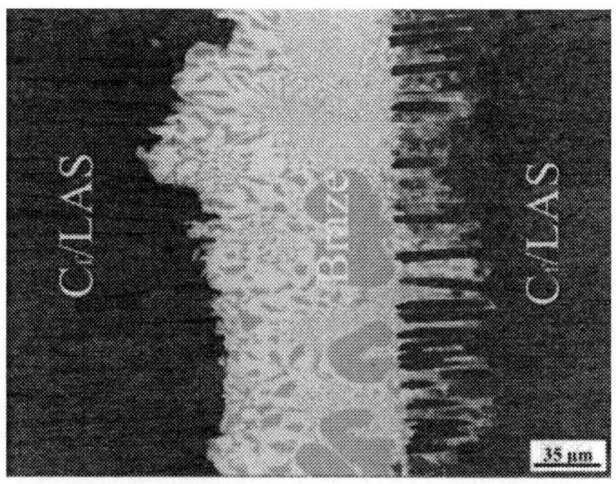

Figure 2 Microstructure of C_f/ LAS/AgCuTi/C_f/ LAS joint

Interface morphology of one side is enlarged in Figure 3. Due to the reaction between element Ti and C_f/LAS composite, a small amount of carbon fibers are inserted into brazing seam. Moreover, a staggered microstructure was formed by the penetration of liquid filler into the micro-holes and micro-cracks of LAS composite which could increase the reaction area of composite. The joint consisted of three reaction zones as shown in Figure 3. Zone I (brazing seam), Zone II (staggered area of carbon fiber and braze) and Zone III (reaction zone adjacent to LAS composite). Major elements at each spot in Figure 3 detected by EDS are listed in Table 1. According to the results of EDS and XRD, Zone I was consisted of Ag-based solid solution (light grey phase) and Cu-based solid solution (grey phase). Point A and D in Zone III can be determined to be Cu_2Ti_4O and $TiSi_2$, respectively.

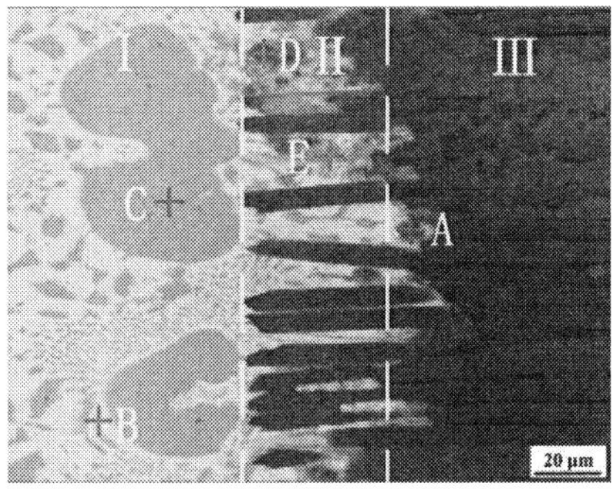

Figure 3 One side of C_f/ LAS/AgCuTi/C_f/ LAS joint

Table1 EDS analysis results of brazed joint(at.%)

Site	Ag	Cu	Ti	C	O	Si	Al	Possible phase
A	11.43	13.51	44.36	8.75	14.02	3.21	4.72	Cu_2Ti_4O
B	72.83	7.55	6.56	2.72	4.22	3.78	2.44	Ag(s,s)
C	3.68	78.43	5.91	7.89	2.53	1.09	0.47	Cu(s,s)
D	2.52	37.66	40.23	3.24	5.73	8.32	2.31	TiCu
E	9.09	5.05	23.38	2.97	9.03	45.85	4.63	$TiSi_2$

The joints brazed at higher temperature and longer holding time are shown in Figure 4.

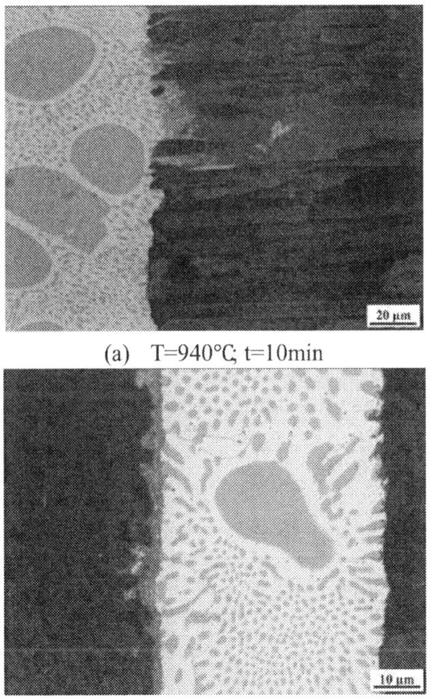

(a) T=940°C; t=10min

(b) T=900°C; t=15min

Figure 4 Microstructure of C_f/ LAS/AgCuTi/C_f/ LAS joint

As the brazing temperature raised and holding time increased, a large amount of Ti and Ti-Cu-O compounds were produced by the excessive reaction between composite substrate and filler. Two kinds of compounds gathering in the composite boundary would aggravate the brittleness of reaction layer which decreases the strength of brazed joints.

The fracture after shear test is shown in Figure 5 and Figure 6. Failure of the C_f/ LAS joints using Ag-Cu-Ti filler alloy usually occurs in the ceramic substrate near the braze interface. For the joint of C_f/LAS composites brazed at 900℃ for 10 min and at 860℃ for 10 min, fiber pull-out is observed on the fracture surface (Figure 5 and Figure 6). It is indicated that the reinforcing effect of C_f is significant and the continuous fibers play an important role in bearing the load. Fiber pull-out is found on the fracture surface of the joint brazed at 900℃ for 10 min (Figure 6), however, the pull-out length is far shorter than that of the joint brazed at 860℃ for 10 min (Figure 5).

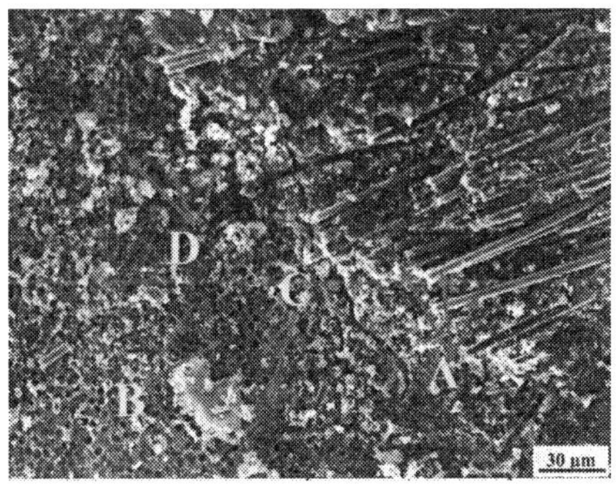

Figure 5 C_f/LAS /AgCuTi/C_f/LAS fracture morphology, T=860℃; t=10min

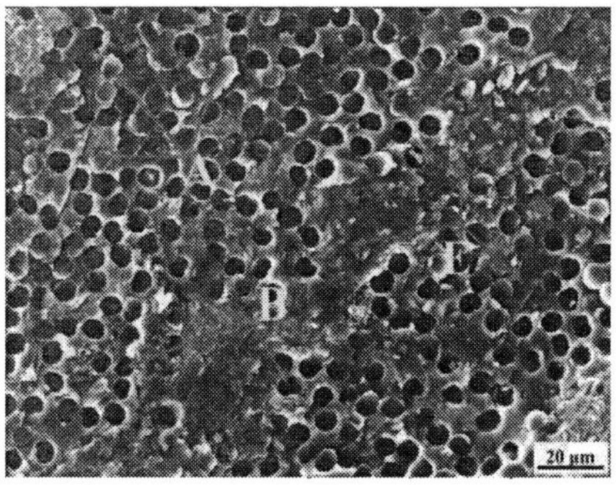

Figure 6 C_f/LAS /AgCuTi/C_f/LAS fracture morphology, T=900℃; t=10min

For the joint brazed at 860℃ for 10 min, the fracture surface is uneven and strewn with fiber pull-out hole. For the joint brazed at 900℃ for 10 min, the fracture surface is smooth and intensive fiber pull-out hole could be observed (Figure 6).

To further analyze the fracture surfaces, the EDS result of corresponding points is shown in Table 2. According to the EDS results of spot A and B in Table 2 and the previous studies, the phase in this reaction layer can be identified as C_f. Figure 5 shows that there are lots of TiCu(D). The point C and E in the fractured surfaces (as shown in Figure 5 and Figure 6) are Ag(s,s) and Cu(s,s) respectively. It is indicated that the joint brazed at 900℃ for 10 min is better than that brazed at 860℃ for 10 min. That's because when the brazing temperature is below 900℃, the atomic diffusion amount is less, and insufficient reaction occurs between composites and filler metal, active Ti can not react with composites completely and form a strong interface bonding.

Table 2 EDS analysis results of fractured surfaces (at.%)

Site	Ag	Cu	Ti	C	O	Si	Al	Possible phase
A	5.76	8.51	13.03	62.35	4.02	3.29	3.04	C_f
B	7.47	7.79	13.91	61.08	4.22	2.79	2.74	C_f
C	72.83	7.55	6.56	2.72	4.22	3.78	2.44	Ag(s,s)
D	12.52	25.75	24.37	16.11	5.91	12.32	3.02	TiCu
E	6.76	66.54	5.10	13.81	2.66	2.37	2.76	Cu(s,s)

Conclusions

C_f/LAS composites were successfully joined with Ag-Cu-Ti braze alloy. The following conclusions may be drawn:

1. The interface structure of the C_f/LAS /AgCuTi /C_f/LAS joint brazed at 900 ℃ for 10 min is C_f/LAS composites/$TiSi_2$/Cu_2Ti_4O/TiCu/Ag(s,s)+Cu(s,s)/TiCu/Cu_2Ti_4O/$TiSi_2$/C_f/LAS composites.
2. The maximum shear strength of the joints is 8.4 MPa acquired at 900℃ for 10 min.

Acknowledgments

This work was supported by Natural Scientific Research Innovation Foundation in Harbin Institute of Technology (HIT.NSRIF.201119) and China Postdoctoral Science Foundation (2013M541367).

References

[1] Luo F, Zhou W C, Jiao H., *et al*, "Preparation and microwave dielectrical properties of Li_2O-Al_2O_3-SiO_2 glass ceramic used as an impedance transfer layer of micro-wave absorbing materials," *Acta Aeronauticaet Astro-nautica Sinica*, Vol. 24, No. 3 (2003), pp. 273-277.

[2] Beal G H, Pinchney L R., "Nanophase glass ceramics," *Journal of the American Ceramic Society*, Vol. 82, No. 1 (1999), pp. 15-16.

[3] Hou Z X., "Crystallization behavior study on the Li_2O-Al_2O_3-SiO_2 system glass-ceramics," *Ordnance Material Science and Engineering*, Vol. 30, No. 6 (2008), pp. 42-46.

[4] Liu D, Zhang L X, Feng J C., *et al*, "SiO_2 glass ceramic and the activity of TC4 titanium alloy brazing," *Transactions of the China Welding Institution*, Vol. 30, No. 2 (2009), pp. 117-120.

[5] Zhao L, Zhang L X, Tia X Y., *et al*, "Electroless plating on the surface of modified silica fiber composites with invar alloy vacuum brazing," *Transactions of the China Welding Institution*, Vol. 31, No. 9 (2010), pp. 9-12.

[6] Feng G J, Li Z R, Xu K., *et al*, "SiC ceramic vacuum braze welding head interface structure and mechanism analysis," *Transactions of the China Welding Institution*, Vol. 35, No. 1 (2014), pp. 13-16.

[7] Liu J K, Qi J L, Cao J, Wang Z J., *et al*, "DD3 high temperature alloy with Ti_3AlC_2 ceramic vacuum brazing," *Transactions of the China Welding Institution*, Vol. 35, No. 3 (2014), pp. 41-44.

Brazing of Carbon-Carbon Composites and GH4169 Ni-based Super-alloy using BNi-2 as the Interlayer

Shuai Huang, Wei Guo, Ying Zhu, Cheng Ma, Ping Qu, Hui Kang
Beijing University of Aeronautics and Astronautics, Beijing, China
Email:hshuai987@163.com

Abstract

Carbon-carbon (C/C) composites and nickel-based super-alloy GH4169 were brazed at different brazing temperature, different holding time using the BNi-2 filler as interlayer. The microstructure of the joint between C/C composites and GH4169 was investigated by scanning electron microscopy, the shear fracture and the microhardness of the joint was tested by thermal simulated testing machine and rigiscan. The maximum shear strength of the joints was 28.1 MPa with the brazing temperature of 1170℃ for 60 minutes, joint fracture occurred in the C/C composite material near the solder side. The hardness distribution of the joint presented a trend with high value at the middle and low value at the both sides, this is due to that the weld joint have hard brittle inter-metallic compounds, which was caused by brazing process.

Introduction

C/C composites have successfully demonstrated their potential in aerospace and automotive applications because of their low density, excellent oxidation resistance, and high-temperature strength [1, 2]. In most cases, the applications of C-C composites require joining of C-C composites to metals. As a metal-based material, nickel-based super-alloy GH4169 is regarded as a high mechanical strength and high-temperature resistance material compared with other materials which are used in aerospace applications [3]. Hence, the reliable joining of C/C composites and nickel-based super-alloy GH4169 will be helpful for extending their applications [4].

Brazing has been developed as a main method for joining composites to metals. However, there are usually significant differences between C/C composite and GH4169 in physical and chemical properties. In particular, large difference in thermal expansion coefficients lead to difficulties in the joining process. Silver filler metals have been previously reported to be used for joining C/C composite andhigh temperature resistant alloys. Guo et al. [5, 6] used silver filler brazing the C/C composite and TC4. Qin et al. [7, 8] studied the brazing technology of joining C/C composite and TC4 with AgCuTi fillers or SiC particle-reinforced AgCuTi filler. Huang et al. [9, 10] demonstrated that carbon fiber reinforced SiC can be successfully brazed to TC4 using AgCuTi alloy powder or Ag-Cu-Ti-(Ti+C) mixed powder as filler metals. The brazed joint between C/C composite and high temperature resistant alloys can achieve good integrity. However, such brazed joint cannot work under high temperature due to the relatively low melting temperature of filler metal compared with the base materials.

In this work, vacuum brazing of the C/C composites and superalloy GH4169 with BNi-2 filler was performed. The mechanical properties and microstructure analysis of the joints were discussed.

Experimental procedures

The C/C composite used in the experiment was semi-3D C/C composite which was fabricated with needle felt and carbon fiber cloth. The nickel-base super-alloy used in the experiment was GH4169 block, the The chemical composition of GH4169 is listed in Table 1.

Table 1. Ingredients of GH4169 (wt%).

Co	Cr	Al	Ti	Ni	Mo	C	Mn	Si
≤1.0	17-21	0.2-0.8	0.65-1.55	50-55	2.8-3.3	≤0.08	≤0.35	≤0.35

The BNi-2 powder was used in the brazing experiment (the nominal compositions of BNi-2 is Ni-7Cr-3.2B-4.5Si-3Fe-0.06). The C/C composites and GH4169 were cut into 10mm ×5mm×5mm rectangular blocks by wire-cutting for brazing. BNi-2 filler metal powders were made into paste form using organic resin. The specimen assembly was shown in Fig. 1.

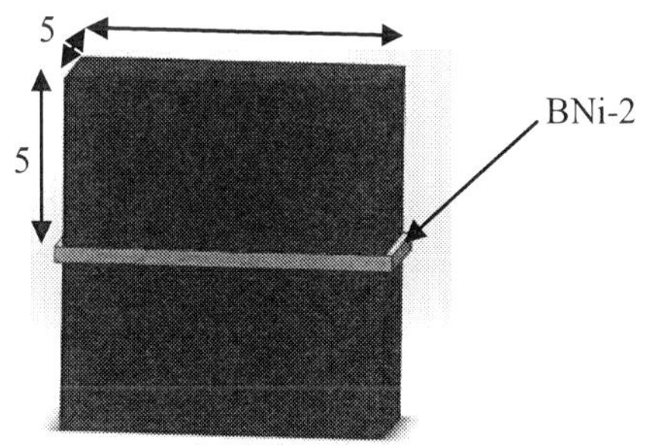

Figure 1. The assemble sketch map of brazing C/C composites and GH4169

All the brazed surfaces of the base material were polished using SiC paper up to 2000 grit and ultrasonically cleaned by ethanol and acetone prior to brazing. The base material and composite interlayer were prepared as sandwiched assembly. Brazing temperatures of 1130- 1210℃, holding time of 20-80 minutes and a vacuum of less than 2.0×10^{-3}Pa were used. The heating rate of the brazing processwas set at 30℃/min. The cooling rate of the brazing process was set at 5℃/min. The specimens were mounted and polished for microstructure evaluation by scanning electron microscopy (SEM). The brazed specimens were put in a special hard alloy test clamp for shear test shown in Fig. 2. Shear tests were performed at room temperature on GLEEBLE-1500 thermomechanical simulator machine to evaluate the shear strength of brazed joints with a constant speed of 1 mm/min. Three specimens were used to obtain the average shear strength of the brazed joints at different brazed parameters.

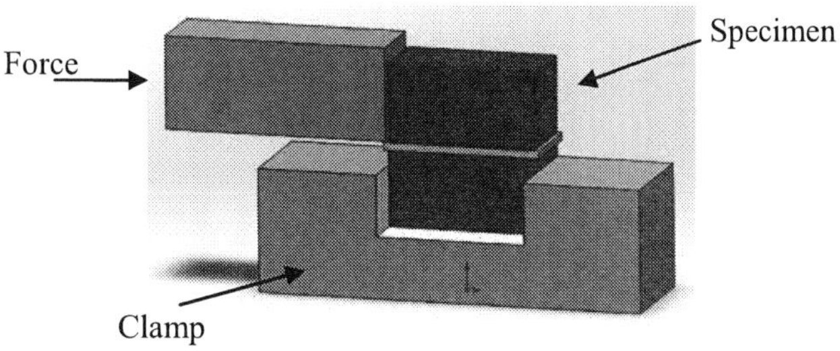

Figure 2. The assemble sketch map of shear test

Results and discussion

2.1 Microstructure of the brazed joints

Fig. 3 shows the microstructures of C/C composites/GH4169 joints brazed at 1170℃ for 60 minutes. A number of different areas can be seen in this figure. To facilitate the analysis, the dark grey continuous layer, shallow gray area in the region, and white block areas respectively called A, B, C in Fig. 3(a). Fig. 3(b) was the SEM figure of the solder layer, the small white lumps were marked D, small dark grey lumps were marked E. Fig. 3(c) was the SEM figure near the side of the GH4169, and find some small needle-like phase in figure marked as F.

From Fig. 3, it can be seen that BNi-2 solder can connect C/C composites with nickel-based super-alloy GH4169 obviously. On this side of the C/C composites is found no holes or cracks and other defects; solder in the middle layer composition is evener, almost no defects; at the side of the high temperature alloy, solder and composition of alloy diffusion is more sufficient, the connection effect is also very well. In order to further analyze the above A, B, C, D, E, F marked points, we still need to EDS spectrum analysis. As shown in Table 2 for joint interface of EDS spectrum analysis results.

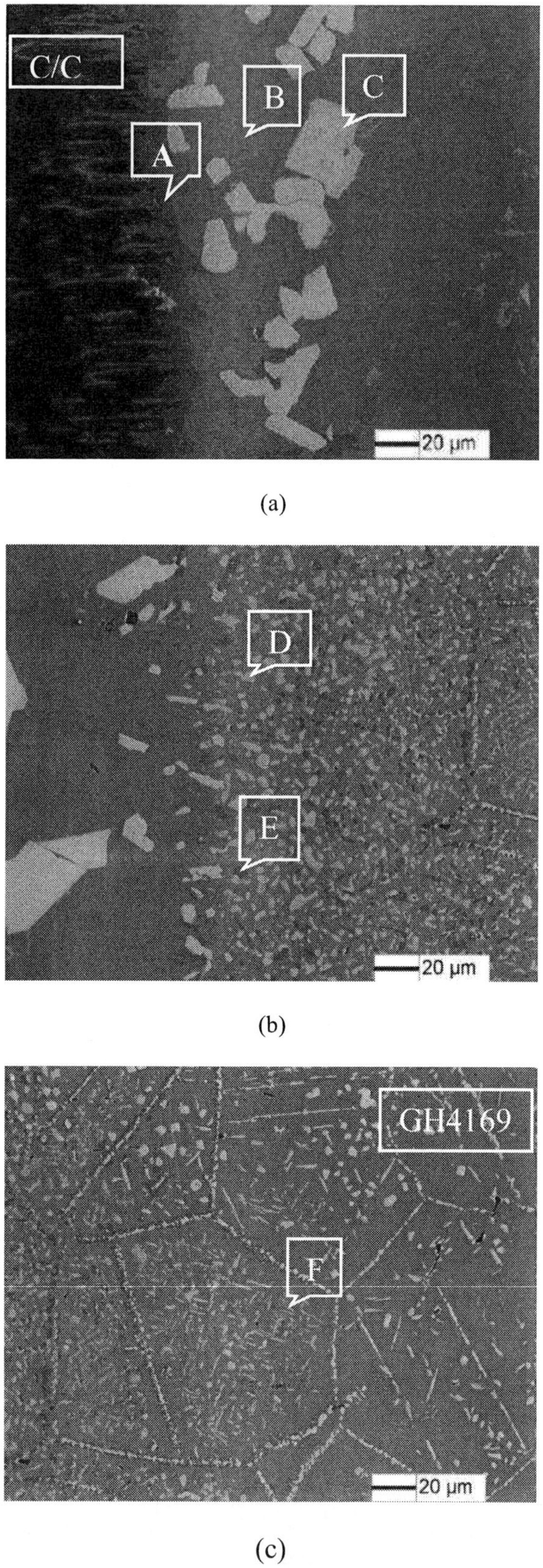

Figure 3. SEM of the joint brazed 1170℃ for 60 minutes: micrograph of (a) near the side of the C/C composites;(b) solder layer;(c) near the side of the GH4169.

Through the analysis of EDS spectrum diagram, it can be learn that in phase A was C, Ni and Cr elements were given priority, in phase B was given priority to with Ni; In the phase C was given priority to with Si, Cr, Ni, and Mo. Combining joint fracture of XRD graph, we can conclude the main ingredients in area of A is Cr7C3 and Ni solid solution; main ingredients in B is Ni solid solution; C in the major composition is complicated, because of the XRD analysis failed to find a suitable compound contains the ingredients, but contain more Si, Cr, Ni and Mo element, the phase near the C/C composites interface generation, can form between the elements of complex compounds, so can be written as $M_xNi_ySi_z$ (M was unknow phase, may contain Mo and Cr). Phase D was speculated to be $Cr_6Ni_{16}Si_7$ and Ni solid solution according to the EDS results. Phase E was speculated to be Cr_3C_2 and Ni solid solution according to the EDS results. Phase E was complicated, so can be written as M_xB_y (M was unknow phase, may contain Mo and Cr).

In addition, GH4169 was used in this experiment and solder was BNi-2, so it can be investigated in EDS spectrum high content of Si, Cr and Fe and other elements.

Table 2. The EDS spectrum analysis results of joint interface

Phase	A	B	C	D	E	F
C	12.20	3.10	5.93	8.14	8.73	5.92
Si		2.73	13.56	12.94	1.56	2.48
Cr	79.32	6.70	17.98	18.17	53.41	19.8
Fe	0.45	7.11	0.74	1.05	2.60	0.50
Ni	1.07	78.98	37.24	36.79	27.28	61.45
Nb	0.47	0.17	1.06	1.40	0.19	0.16
Mo	6.48	1.21	23.49	21.52	6.23	9.68
Possible phase	Cr_7C_3	Ni(s,s)	$M_xNi_ySi_z$ (M=Mo Cr)	$M_xNi_ySi_z$ (M=Mo Cr)	Cr_3C_2	M_xB_y (M=Mo Cr)

From Fig. 3, it can be seen that from diffusion effect area to the isothermal solidification area, Ni and Si content gradually increased, Cr gradually reduce, this is due to the solder and the base metal of Si, B, Fe and Cr in isothermal solidification process for mutual diffusion. Due to the formation of γ-Ni solid solution, at the same time, from the isothermal solidification area to the joint center for melting the middle tier of the surplus element such as discharge of B, Cr, Si, generated the eutectic in the center of the joint organization and intermetallic compound. From Fig. 3(a) is visible in the isothermal solidification section organization, there are two forms of compounds were irregular black and light grey block continuous organization, combining shown in Table 2, compound feature point analysis of the Cr - B are intermetallic phase and eutectic γ-Ni solid solution phase, and the γ-Ni interstitial solid solution phase detected between rich Ni silicide phase; Isothermal solidification area element distribution mainly for Ni, Fe, Si and Cr, combined with Table 2 is mainly generated in the region of rich silicon Ni based solid solution phase.

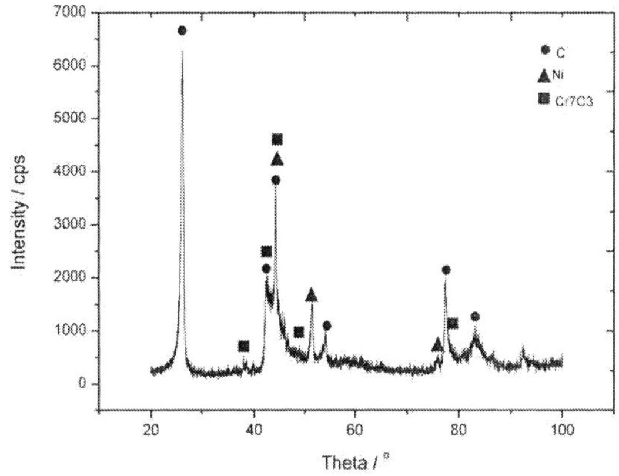

(a)

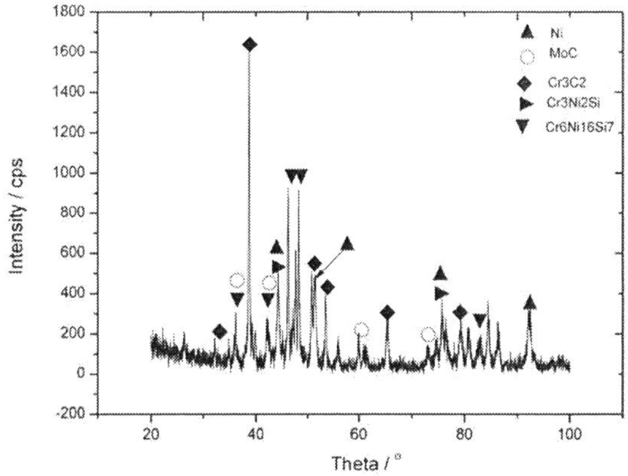

(b)

Figure 4. XRD pattern of the joint at (a) C/C composites side; (b)GH4169 side brazed at 1170℃ for 60 minutes

2.2 Fracture morphology of the brazed joints

Fig. 5 and Fig. 6 were the SEM photos of joint fracture at the side of the C/C composites and on one side of the nickel-based superalloy GH4169.

From the above figures, the fracture surface of joint was very uneven, the screen can be seen a lot of fibre was pulled up and pull out pits left by pull out after. In addition, some parallel carbon fiber fracture can be observed, its bond with some other substances, generally for the spread of the solder.

For specific research fracture of the phase composition and fracture mechanism, the side of C/C composites and the side of the fracture nickel-based super-alloy were taken by EDS spectrum analysis, in the SEM photo tagging A, B, C, D, E and F six points, six points in the energy spectrum analysis of the results as shown in Table 3.

It can be seen from Table 3, on the side of C/C composite fracture are mainly C, Si, Cr, Mo and so on elements, and the main fracture at the side of the nickel base superalloy are C, Ni and Fe and other elements.

Figure 5. Joint fracture SEM Figure close to C/C composites side

Combined with the SEM photos, it can be seen that near the side of the nickel base superalloy Cr_7C_3 exist. For the diffusion zone spectrum analysis results show that both contain more Cr, Mo, W and Ni four elements.

Because B atomic radius is small, brazing will spread to the matrix of GH4169 was investigated, and B atomic at high temperature is easy to be with three elements Cr, Mo, W forming boride, remember to MB (M = Cr, Mo, W). The near side of the main considerations of C/C composites with C and Ni, Ni is the solid solution phase and more carbide.

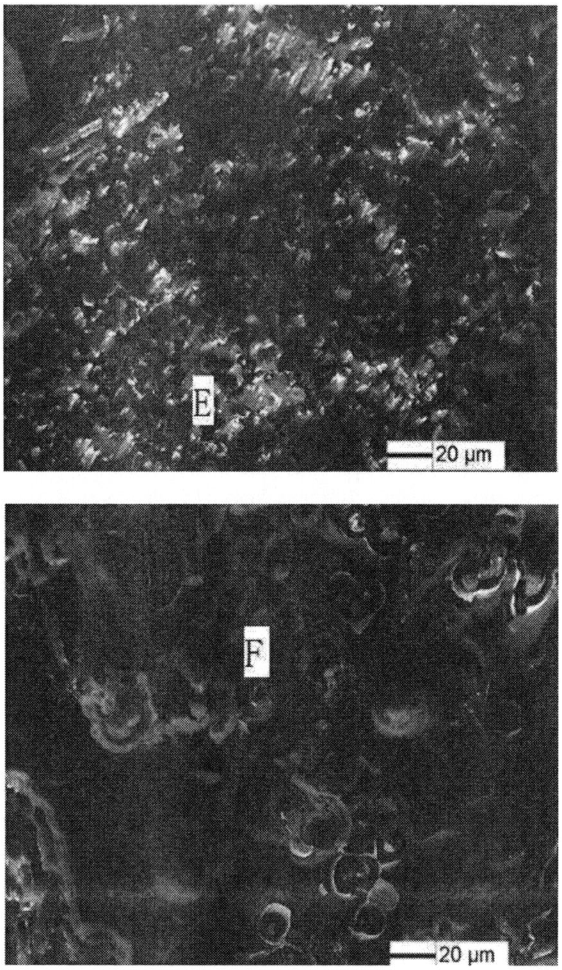

Figure 6. Joint fracture SEM Figure close to C/C composites side

Table 3. The fracture morphology EDS spectrum analysis results

element	A	B	C	D	E	F
C	62.84	83.47	98.52	5.93	66.21	80.55
Si	0.81	0.35	---	13.56	0.45	0.63
Cr	1.30	1.55	---	17.98	3.44	0.91
Fe	6.21	2.40	0.50	0.74	5.52	4.12
Ni	28.08	11.33	0.98	37.24	23.30	12.30
Nb	0.13	---	---	1.06	0.11	0.20
Mo	0.61	0.90	---	23.49	0.97	1.29

2.3 Shear strength of the brazed joints

The variation of shear strength of the C/C composites /GH4169 brazed joints with different brazing parameters are shown in Fig. 7 (a) and (b). It was shown that the shear strength of joints increased when the brazing temperature was from 1130 ℃ to 1170 ℃ and then decreased at the temperature range from 1170 ℃ to 1210 ℃.

When the brazing temperature was 1170 ℃, the joint shear strength increased with the holding time from 20 to 60 minutes and then dramatically decreased from 60 to 80 minutes, as shown in Fig. 7(b).

The maximum shear strength is 28.1 MPa brazed at 1170℃ for 60 minutes. The joint shear strength was well dependent on the joint microstructure. The situation of fracture surface is related to the orientation of carbon fiber. When carbon fiber is parallel to the joining surface, the joints are fractured in the composites. When carbon fiber is vertical to the joining surface, the joints are mainly fractured in the C/C.

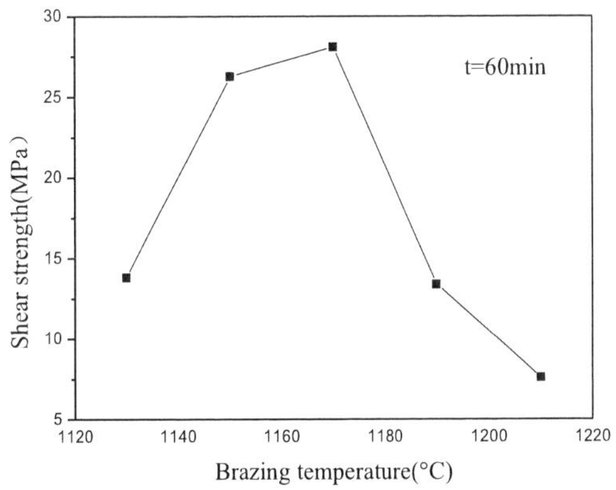

(a) Effects of the brazing temperature (t=60minutes)

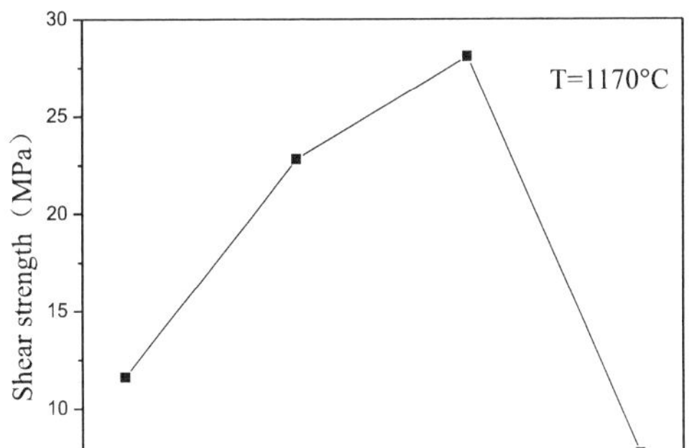

(b) Effects of the holding time (T=1170℃)

Figure 7. Effects of brazing parameters on shear strength of the joints.

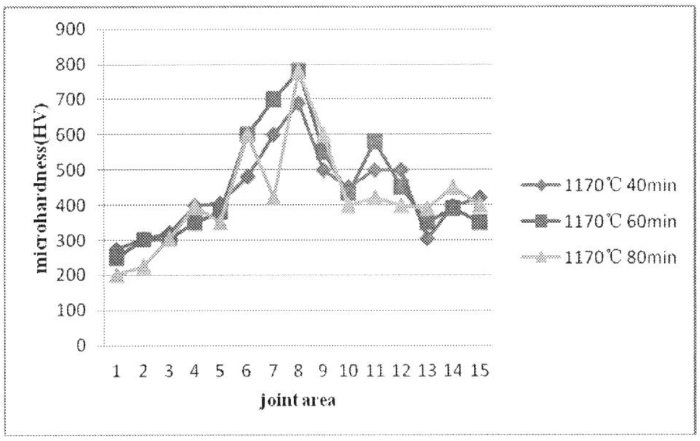

(a)

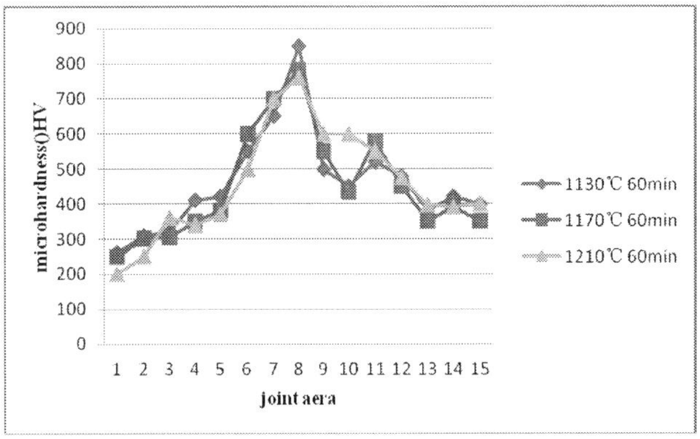

(b)

Figure 8. The hardness distribution of joint under different brazing parameters

The microhardness measurement is along the middle line from the C/C composite material to nickel base super-alloy GH4169, in order to get the hardness distribution. As shown in Fig. 8 for different brazing joint parameters obtained under the hardness distribution of the joints, which can be seen, in the middle tier of the solder hardness is bigger than in other parts obviously, estimates that formed the hardness of the larger material here.

For the hardness of joint, the area that near C/C composites is low, but its hardness is nearly three times of C/C composite material, it can be seen in the process of brazing, filler metals of elements in the diffusion into the C fibers formed in the various compounds, make the hardness in the region has greatly improved. In the near area of nickel-based superalloy GH4169 was investigated that the hardness was close to the base superalloy GH4169. In the middle of the joint has the highest hardness, roughly twice than the nickel-based superalloy. From component analysis, in this area formed rigid intermetallic compounds and eutectic organization in the process of brazing, so this area overall high hardness.

We can see from these components exist many elements from the middle layer spread on both sides. So combining with the joint fracture and fracture morphology, the fracture mode of joint was brittle fracture.

Conclusions

From observation of the microstructures and mechanical properties of C/C composites/GH4169 brazed joints, the results are as follow:

(1) Using BNi-2 solder can braze C/C composites and nickel base super-alloy GH4169, found that can achieve better connections. Braze welding joint can be determined through the analysis of the shear strength in the brazing temperature 1170 ℃, holding time for 60 min reached the highest under the brazing process parameters of 28.1 MPa.

(2) Brazing joint has the highest microhardness shows in the middle of the joint, and the closer the base metal the lower hardness. this is due to that the weld joint have hard brittle inter-metallic compounds, which was caused by brazing process.

Acknowledgments

The research was supported by International Science and Technology Cooperation Program of China (no.2013DFR50590), National Natural Science Foundation of China (no.11372019).

References

[1] Tiesong Lin, Minxuan Yang, Peng He. et al, "Effect of in situ synthesized TiB whisker on microstructure and mechanical properties of carbon-carbon composite and TiB/Ti-6Al-4V composite joint," Materials and Design, Vol. 32, (2011), pp. 4553-4558.

[2] M. Singh, R. Asthana, T.P. Shpargel, "Brazing of Carbon-Carbon composites to Cu-clad molybdenum for thermal management applications," Materials Science and Engineering A, 452-453(2007), pp. 699-704.

[3] Youqiong Qin, Jicai Feng, "Active brazing carbon/carbon composite to TC4 with Cu and Mo composite interlayers," Materials Science and Engineering A, Vol.525, (2009), pp. 181-185.

[4] Y.H. Kong, R.Y. Liu, L.X. Xie. et al, "Effects of Different Heat Treatments on Microstructures and Greep Properties of GH4169 Superalloy," JMEPEG, Vol.22, (2013), pp. 1371-1377.

[5] Wei Guo, Ying Zhu, Lin Wang. et al, "Microstructure evolution and mechanical properties of vacum-brazed C/C composite with AgCuTi foil," Materials Science and Engineering A,Vol.564, (2013), pp.192-198.

[6] Wei Guo, Lin ang, Ying Zhu. et al, "Microstructure and mechanical properties of C/C composite/TC4 joint with in active AgCu filler metal," Ceramics International, DOI: http://dx.doi.org/10.1016/j.ceramint.2015.02.006

[7] Youqiong Qin, Jicai Feng. "Microstrure and mechanical properties of C/C composites/TC4 joint using AgCuTi filler metal," Materials Science and Engineering A,Vol. 454, (2007), pp. 322-327.

[8] Youqiong Qin, Zhishui Yu. "Joining of C/C composite to TC4 using SiC particle-reinforced brazing alloy," Materials Characterization, Vol. 61, (2010), pp. 635-639.

[9] Jinhui Xiong, Jihua Huang. et al. "Brazing of carbon fiber reinforced SiC composite and TC4 using Ag-Cu-Ti active brazing alloy," Materials Science and Engineering A,Vol. 527, (2010), pp. 1096-1101.

[10] Bing Cui, Jihua Huang. et al. "Reaction-composite brazing of carbon fiber reinforced SiC composite and TC4 using Ag-Cu-Ti-(Ti+C) mixed powder," Materials Science and Engineering A,Vol. 562, (2013), pp. 203-210

Joining of Kovar to Alumina and to Low-Temperature Co-Fired Ceramics (LTCC)

Andreas Pönicke, Jochen Schilm, Adrian Goldberg, Uwe Partsch, and Alexander Michaelis
Fraunhofer IKTS, Fraunhofer Institute for Ceramic Technologies and Systems, Winterbergstraße 28, D-01277 Dresden, Germany
Jochen.Schilm@ikts.fraunhofer.de, +49-351-2553-7824

Abstract

For integration of ceramic sensor elements based on low-temperature co-fired ceramics (LTCC) into standardized steel housings it is necessary to compensate the mismatching thermal expansion behavior. Therefore balancing elements made of Kovar® (Fe-29Ni-17Co) and alumina ceramic (Al_2O_3) can be used. However, these elements have to be joined hermetically to each other and to the LTCC sensors. In this study, brazing experiments were performed for combinations of Kovar/Al_2O_3 and Kovar/LTCC with Ag-Cu-Ti and Ag-Cu-In-Ti based commercial braze filler metals, Cusil-ABA® and Incusil®-ABA, respectively. For both active braze filler metals optimized processing parameters were investigated to realize hermetic Kovar/Al_2O_3 and Kovar/LTCC joints.

Introduction

LTCC Sensors
Low temperature co-fired ceramics (LTCC) is a well-known technology for highly integrated, reliable, and high temperature stable microelectronic packages in mobile communication or for automotive, space, or medical applications [1]. Due to its linear stress/strain behavior and its ability for integration of three-dimensional shapes like diaphragms, channels, and cavities, LTCC complies with all requirements for the integration of mechanical structures e.g. for pressure sensors [2-4]. At Fraunhofer IKTS a new piezo-resistive pressure sensor concept was developed [5, 6]. The new design principle allows the sensor cell to be fully mechanical decoupled and stress free inside the sensor frame. Only thin LTCC cantilevers containing micro channels are used for the pressure connection of the sensor cell. Thick film resistors, screen printed on the surface of the sensor cell, are connected to a Wheatstone bridge and measure the strain caused by the deflection of the LTCC diaphragm due to pressure differences in- and outside of the cell. The LTCC technology easily enables variable sensor geometries e.g. different diaphragm thicknesses for different pressure ranges by using different tape thicknesses and thus all types of pressure sensors (relative, absolute, differential) constructed. In comparison with classic steel or ceramic based pressure sensors such LTCC based pressure sensors have many advantages like a very linear pressure dependent signal behavior together with an excellent high temperature and long-term stability [4]. Furthermore, LTCC is a high volume technology which helps to produce sensor elements in a cost effective manner as all components of the sensor system (sensor frame, electronic components) can be integrated in one LTCC based multilayer substrate. However, for testing, calibration, and subsequent application the LTCC sensor elements have to be connected gas-tight to the measuring equipment.

Scientific Approach

Integration of LTCC Based Sensor Elements
In most cases standardized steel housings are used to ensure a gas-tight connection of the LTCC sensors towards the system. However, the integration of LTCC based sensor elements for high temperature applications require suitable interconnect technologies. For example, if the LTCC sensors and the steel housings are glued directly together, then the maximum operation temperature of the sensor is limited to the glass transition temperature of the epoxy resin which is in most cases below 190°C. Also soldering offers no reliable option as the joint strength decreases rapidly at higher temperatures due to interdiffusion processes what in turn results in a loss of gas-tightness. For an increased thermal stability other joining technologies like glass sealing or brazing can be used. However, in this case the mismatching coefficients of thermal expansion of LTCC and steel will limit the lifetime of the integrated sensor elements as thermal cycles will initiate cracks along the sealing or inside the sensor. To overcome these problems, a stepwise integration of the LTCC based sensor elements into steel housings was developed. Our approach offers the opportunity to outbalance the different thermal expansion coefficients of LTCC and steel, and to increase the operation range of the sensor assembly to high temperatures up to 300°C. The integration concept is schematically shown in Figure 1.

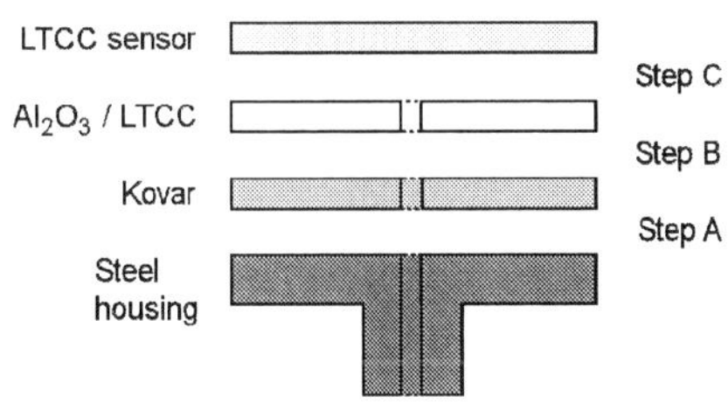

Figure 1 Stepwise integration of a LTCC based sensor element into steel housing.

The integration approach of LTCC based sensors can be divided into three steps. In *step A* a balancing element made of a nickel-cobalt ferrous alloy i.e. Kovar® (Fe-29wt%Ni-17wt%Co) with a coefficient of thermal expansion that closely matches that of the ceramic materials is joined to the steel housing. Appropriate technologies to realize hermetic and non-warping joints are welding or brazing. The following *step B* is necessary in order to create a joinable surface for *step C* and will be described later in detail. In *step B* the bonding of a ceramic layer made from alumina (Al_2O_3) or LTCC offers the possibility to integrate the LTCC based sensor elements in *step C* by sealing with a glass based solder or other techniques.

Joining of Ceramics to Metals

To join ceramics and metals together several direct bonding techniques are available to produce hermetic joints [7]. Besides techniques without any liquid phase like diffusion bonding or friction welding direct bonding methods utilizing a liquid phase based on adhesives, braze filler metals or glass solders can be used. For the desired application, where in joining *step B* a planar joint between Kovar and Al_2O_3 or LTCC has to be realized which is suitable for operation temperatures up to 300°C, brazing is the appropriate method. A common way to braze ceramics and metals is the metallization of the ceramic prior to the brazing process since then the metallized ceramics can be brazed to metals without any active braze filler metal [8]. However, the metallization processes imply several individual process steps with makes them complicated and expensive. In contrast with the active metal brazing technique ceramics can be directly brazed to metals or themselves without any additional metallization steps [9]. Active brazing alloys are based on filler metals like Ag, Cu, or Au and contain low fractions of so called active species (i.e. Ti, Zr, Hf). These elements enhance wetting of the ceramic surface during brazing in an oxygen-free environment using protective atmospheres or vacuum ($p < 10^{-4}$ mbar).

Brazing of Kovar to Alumina

In literature active metal brazing of Kovar to Al_2O_3 is much more investigated than the brazing of Kovar to LTCC. A good overview on this problem can be found in [10]. When Al_2O_3 ceramics are brazed to themselves with Ag-Cu-Ti active braze filler metals the joint microstructure shows a bi-layered reaction zone at the interface between the Al_2O_3 and the brazing alloy. This reaction layer consists of a Ti-rich oxide layer with a thickness below 1 µm completely covering the alumina interface and a second, a few microns thick mixed oxide layer containing Ti, Cu, and Al [11, 12]. Al_2O_3/Al_2O_3 joints prepared in this manner are hermetic and reach joint strengths of > 95 % of the strength of the base material. If now one of the Al_2O_3 pieces is substituted by Kovar the results changes. The Kovar/Al_2O_3 joints showed no high strength values and reached only 40 % of the initial Al_2O_3/Al_2O_3 joint strength [13]. Additionally, Kovar/Al_2O_3 joints made by active metal brazing showed poor hermeticity. Microstructural analysis revealed that in non-hermetic joints no continuous Ti_xO_y reaction layers were formed at the interface between Al_2O_3 and the brazing alloy [14]. This phenomenon of titanium scavenging can be explained by the dissolution of nickel and iron from Kovar in the molten braze and the strong affinity of nickel to titanium [15]. Thus during the brazing process Ni and Ti react to form intermetallic compounds while the activity of Ti in the melt is decreased and the formation of the necessary reaction zone at the Al_2O_3 interface is suppressed [16].

To improve the joint strength of Kovar/Al_2O_3 and to prevent the formation of intermetallic compounds different barrier layer concepts were developed. Mo-, Ni-, and Mo-Ni-coatings on Kovar were tested leading to increased bending strength of the joints with more than 80 % of the base material [13]. Vianco *et al* investigated the influence of Mo thickness and found that brazing of Kovar/Al_2O_3 with a 500 µm thick Mo-barrier layer yielded to the best hermeticity performance and strength [17]. Other methods to improve the joint strength and gas-tightness of Kovar/Al_2O_3 joints include magnetron sputtering of Ti-layers on or mechanical metallization of the alumina surface [18, 19]. Besides introducing barrier layers Wielage *et al* showed that an improvement of the joints is possible, if induction brazing with much shorter brazing times compared to conventional furnace brazing is applied [20]. Microstructural analysis showed a remarkable reduction of intermetallic compounds and a decrease of the reaction layer thickness between Kovar and the brazing alloy.

Brazing of Kovar to LTCC

For the desired application and workflow in *step B* it seems to be interesting to use LTCC as joining partner for Kovar instead of Al_2O_3. However, in comparison to alumina only less information on brazing of LTCC is available. One approach to braze LTCC is to use a metallized LTCC in combination with a non-active braze filler metal. For this purpose Keusseyan and Dilday investigated the brazeability of Cu-, Ag-, and Au-based thick film metallization layers and concluded that for LTCC/metal joints the brazing temperature should be limited to 500°C in order to minimize the mechanical stresses caused by the mismatching thermal expansion coefficients [21]. Another approach, followed by Walker *et al*, was the investigation of PVD thin-film coatings like Ti-Au, Ti-Pt, and others [22]. While brazing with a Ag-Cu-In braze filler metal the Ti-Pt thin-films yielded the best hermeticity performance and highest strength. However, on active metal brazing of LTCC without further metallization layers only one study was found in literature. But hermetic joints of LTCC and Kovar were only be possible if the LTCC was ground and re-fired prior to the brazing process [22].

As shown, brazing of LTCC is only less investigated and no systematic results are given in literature which let on judge about the suitability of the active metal brazing technique for LTCC. In the case of Al_2O_3 besides Ag-Cu-Ti active braze filler metals no brazing alloys with lower brazing temperatures were tested. Thus in the present study the commercial available active braze filler metal Incusil®-ABA is investigated to provide brazing parameters as means to obtain hermetically brazed Kovar/Al_2O_3 and Kovar/LTCC joints. For comparison with literature joining of Al_2O_3 and LTCC to Kovar with Cusil-ABA® was investigated as well.

Experimental Procedure

Materials

The LTCC sensors based on Du Pont's Green Tape 951 system, as this material system is a fair compromise in comparison to other ceramic co-firing materials regarding Young's Modulus and fracture strength. For joining experiments LTCC samples were made by laminating three tape layers followed by a firing process with a peak temperature of 850°C similar to the one described in [4]. After firing the sintered LTCC had a thickness of 630 µm and was cut into single samples (7 x 7 mm^2) with a dicing saw. The surface roughness of the as-fired LTCC was $R_a < 0.36$ µm. The alumina ceramic was obtained in thick film standard quality with an Al_2O_3 content of 96 % (Rubalit® 708 S, CeramTec, Marktredwitz, Germany), a thickness of 250 µm, and a surface roughness of $R_a < 0.36$ µm. Samples were prepared by laser scribing and subsequent breaking along the scribed lines. As metallic joining partners samples with the size 15 x 15 mm^2 were laser cut from a 1 mm thick Kovar sheet metal with a surface roughness of $R_a < 0.4$ µm. The surfaces of all samples were degreased prior to assembly.

Brazing Technology

For joining of the ceramics to Kovar two types of braze filler metals provided by Wesgo Metals, Hayward, CA, USA were used, i.e. Ag-35.25wt%Cu-1.75wt%Ti (Cusil-ABA®) and Ag-27.25wt%Cu-12.5wt%In-1.25wt%Ti (Incusil®-ABA). These alloys were applied in the form of laser cut foils with a thickness of 50 µm. Brazing was carried out in a full-metal vacuum furnace with molybdenum heating elements at a pressure $< 1 \cdot 10^{-5}$ mbar. The following process cycle was used for brazing: heat from room temperature to 550°C (for Incusil-ABA) or 700°C (Cusil-ABA) at a rate of 10 K/min, hold for 20 min to obtain a temperature homogenization inside the furnace, further heating-up to the desired brazing temperature, hold of brazing temperature for 10 min, and then cool down to room temperature at a rate of 5 K/min to 400°C with subsequent furnace cooling. The brazing temperature was 755°C for Incusil-ABA and was varied between 810°C and 850°C for Cusil-ABA in order to investigate the influence of brazing temperature on gas-tightness and microstructure of the joined assemblies.

Characterization

Upon brazing, the gas-tightness of each brazed assembly was measured using a helium leak detector (Phoenix XL30, Oerlikon Leybold Vacuum, Cologne, Germany). A joint with a helium leak rate of $1 \cdot 10^{-8}$ mbar·l/s or better was considered as hermetic. Microstructural analysis of the joints was performed on polished cross-sectioned samples using a scanning electron microscope (abbr. SEM, NVision 40, Carl Zeiss SMT, Oberkochen, Germany). The micrographs shown in this paper were recorded in the element specific back scattered electron mode. Additionally, the scanning electron microscope is equipped with an energy dispersive X-ray analysis system (abbr. EDX, Inca x-sight, Oxford Instruments, Abingdon, England), which allows for a quantitative detection of elements.

Results and Discussion

Brazing with Incusil-ABA

Active metal brazing of LTCC and Al_2O_3 to Kovar with Incusil-ABA at 755°C for 10 min yield in both cases to hermetically sealed assemblies. Surprisingly, we were able to realize hermetic joints of as-fired LTCC and Kovar which is in sharp contrast to the results of Walker *et al* [22]. SEM images of the microstructures of Kovar/Al_2O_3 and Kovar/LTCC joints brazed with Incusil-ABA are shown in Fig. 2 and 3, respectively. These micrographs show the typical structure of the Ag-Cu-In eutectic with a Ag-rich phase (white regions, with dissolved In and Cu) and a Cu-rich phase (grey regions) together with enclosed intermetallic phases in the brazing alloy (dark grey regions) and reaction layers on both interfaces.

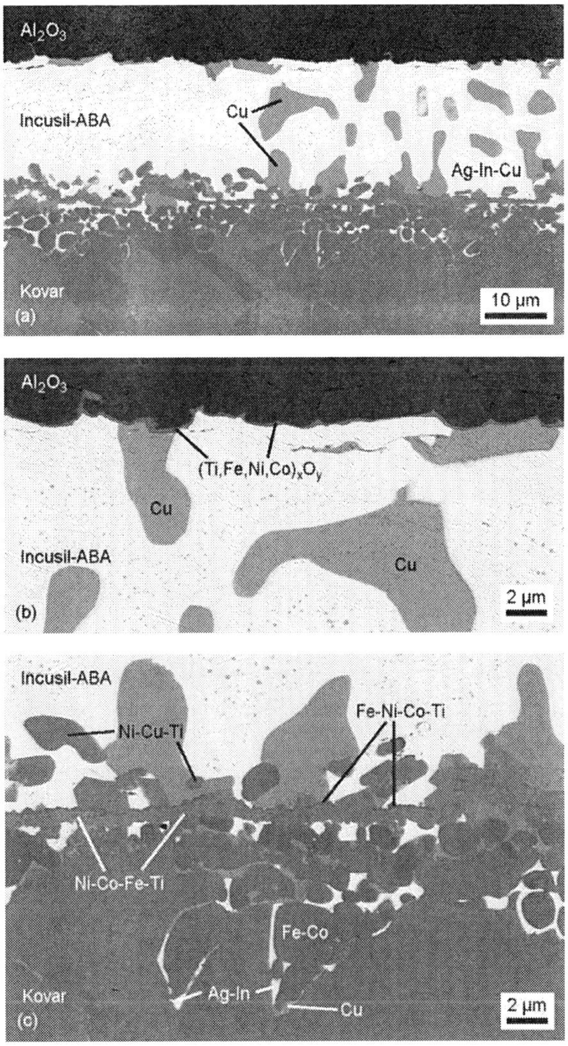

Figure. 2 SEM image of Kovar/Al_2O_3 joints brazed with Incusil-ABA at 755°C for 10 min.

Figures 2b and 2c are the enlarged images from Fig. 2a displaying the reaction layers at the interface between Al_2O_3 and the brazing alloy, and between Kovar and the brazing alloy, respectively. At the interface Al_2O_3/brazing alloy a very thin reaction layer with submicron thickness was formed. This reaction layer completely covers the alumina interface yielding to a mean helium lake rate of $6 \cdot 10^{-10}$ mbar·l/s. The main constituents of the reaction layer are Ti and O, but also elements of the Kovar, i.e., Ni, Fe, and Co, are detected. This suggests that the constituents of the Fe-Ni-Co alloy show a strong affinity to Ti even at lower temperatures than usually investigated in literature [14, 15]. The strong reactivity of Fe, Ni, and Co with Ti shaped the interface between Kovar and the brazing alloy as several intermetallic compounds like $(Fe,Ni,Co)_xTi_y$ with a high amount of Ni (abbr. Ni-Co-Fe-Ti) or Fe (abbr. Fe-Ni-Co-Ti) are observed. These intermetallic phases form a small band which meanders parallel to the Kovar surface. Further away from the interface in the brazing alloy Ni-Cu-Ti compounds are visible. In addition, down to a depth of 25 µm Ag, In, and Cu from the brazing alloy are found at the grain boundaries of the Kovar and along Fe-Co grains which are depleted of Ni.

In comparison with Kovar/Al_2O_3 joints, the microstructure of Kovar/LTCC joints with Incusil-ABA looks similar. Figures 3b and 3c are the enlarged images from Fig. 3a showing the interfaces between LTCC and the brazing alloy, and between Kovar and the brazing alloy, respectively. Again, through diffusion of Fe, Ni, and Co and their reaction with the active element Ti intermetallic compounds were formed in the brazing alloy and along the interface towards Kovar. However, in the micrographs two differences in comparison with Kovar/Al_2O_3 joints are found. At first, the very thin $(Ti,Fe,Ni,Co)_xO_y$ reaction layer at the interface LTCC/brazing alloy contains traces of Si and Pb, the main constituents of the glass phase of the LTCC. Secondly, the reaction layer is non-continuous with some pores where the brazing alloy was not able wet the LTCC completely. However, all brazed assemblies were hermetic with a mean helium lake rate of $4 \cdot 10^{-9}$ mbar·l/s.

Brazing with Cusil-ABA

Active metal brazing of Al_2O_3 and LTCC to Kovar with Cusil-ABA was performed at three different brazing temperatures for a minimum of six samples for each brazing condition. Table 1 summarizes the obtained hermeticity data. While brazing of Al_2O_3 yield in all cases to hermetically sealed assemblies, brazing of LTCC gave rather different results.

Table 1 Hermeticity after brazing with Cusil-ABA as a function of brazing parameters.

Joint setup	Peak process temperature	Hermetic joints (fraction)
Al_2O_3/Kovar	810°C	5/6
Al_2O_3/Kovar	830°C	12/12
Al_2O_3/Kovar	850°C	6/6
LTCC/Kovar	810°C	2/6
LTCC/Kovar	830°C	6/6
LTCC/Kovar	850°C	Not tested

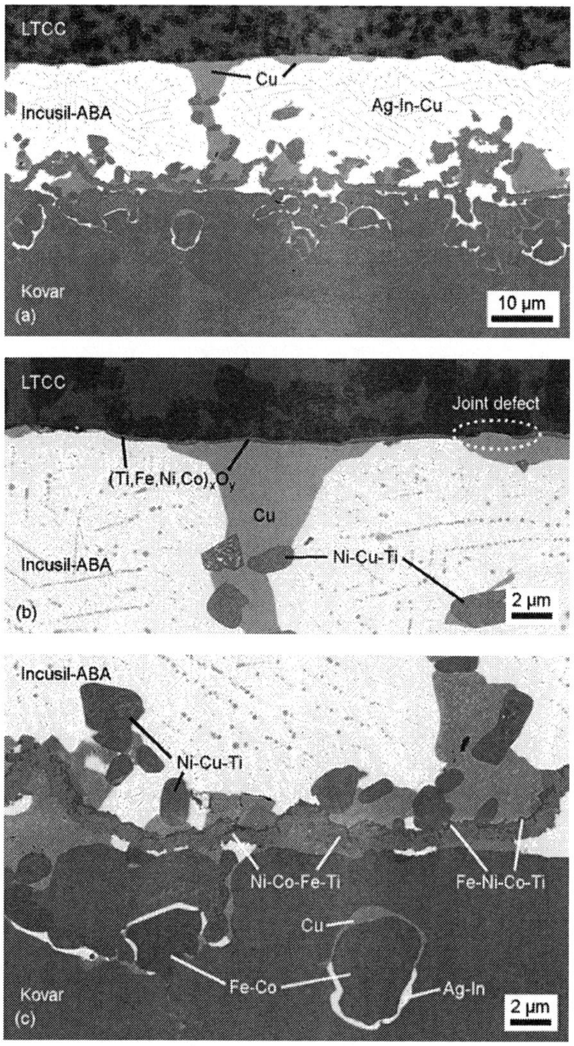

Figure 3 Scanning electron micrographs of Kovar/LTCC joints brazed with Incusil-ABA at 755°C for 10 min.

After brazing of LTCC/Kovar joints at a temperature of 810°C only a few of the assemblies were hermetic. Microstructural investigation showed, that the interface between LTCC and the brazing alloy is weakly bonded because only a non-continuous and very thin reaction layer was formed. Thus at this brazing temperature no reliable joining is possible. If the brazing temperature is increased to 830°C all brazed LTCC/Kovar joints showed gas-tightness due to the formation of a continuous reaction layer at the interface between LTCC and Kovar which is shown later in detail. In contrast to Al_2O_3 brazing of LTCC at 850°C was not tried as the LTCC is sintered at this temperature and the stability of the ceramic material is limited.

Scanning electron micrographs of the microstructures of Kovar/Al_2O_3 and Kovar/LTCC joints brazed with Cusil-ABA at 830°C are shown in Fig. 4 and 5, respectively. In these micrographs the brazing alloy displays the typically structure of the Ag-Cu eutectic with a Ag-rich phase (white region) and a Cu-rich phase (grey regions). Furthermore, the formation of reaction layers on both interfaces is visible. These reaction layers are shown in more detail in Fig. 4b and 4c.

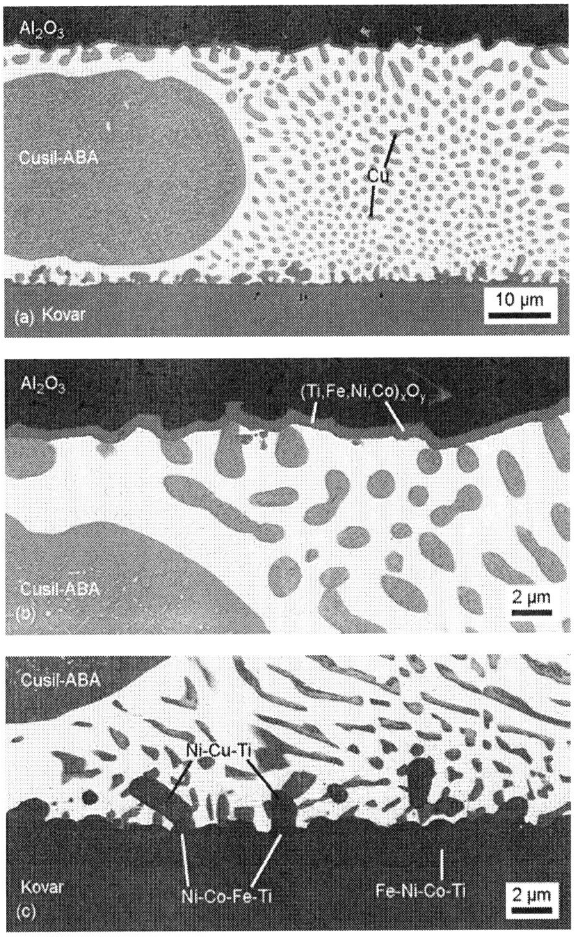

Figure 4 Scanning electron micrographs of Kovar/Al_2O_3 joints brazed with Cusil-ABA at 830°C for 10 min.

The active element Ti formed a continuous reaction layer with a thickness of 0.7 to 1 µm bordering the interface between Al_2O_3 and the brazing alloy. The reaction layer consists of Ti and O with minor amounts of Ni, Fe, and Co. The elemental composition is the same as observed for Kovar/Al_2O_3 joints brazed with Incusil-ABA. At the interface between Kovar and the brazing alloy an up to 3 µm thick reaction layer with multiple phases was formed. The main phase comprises of a Fe-rich intermetallic compound (Fe-Ni-Co-Ti) that covers the interface of the Kovar completely. Adjacent to the Fe-Ni-Co-Ti phase a second Ni-rich phase (Ni-Co-Fe-Ti) was found. A third intermetallic phase composed of Ni, Cu and Ti is observed in the brazing alloy. The microstructural analysis of Kovar/Al_2O_3 joints brazed with Cusil-ABA and Incusil-ABA showed, that during the active metal brazing processing similar phases were formed in the brazing seam. However, in the case of Cusil-ABA the intermetallic compounds are attached to the Kovar (Fig. 4b) whereas in the case of Incusil-ABA a lacework phase was formed (Fig. 2b). Additionally, while brazing with Cusil-ABA no penetration of the Kovar along the grain boundaries by the brazing alloy was observed. This leads to the conclusion that the reactivity of Fe, Ni, and Co with the active element Ti in the braze filler metals is enhanced because of the presence of indium or of the lower melting temperature of the Ag-Cu-In eutectic.

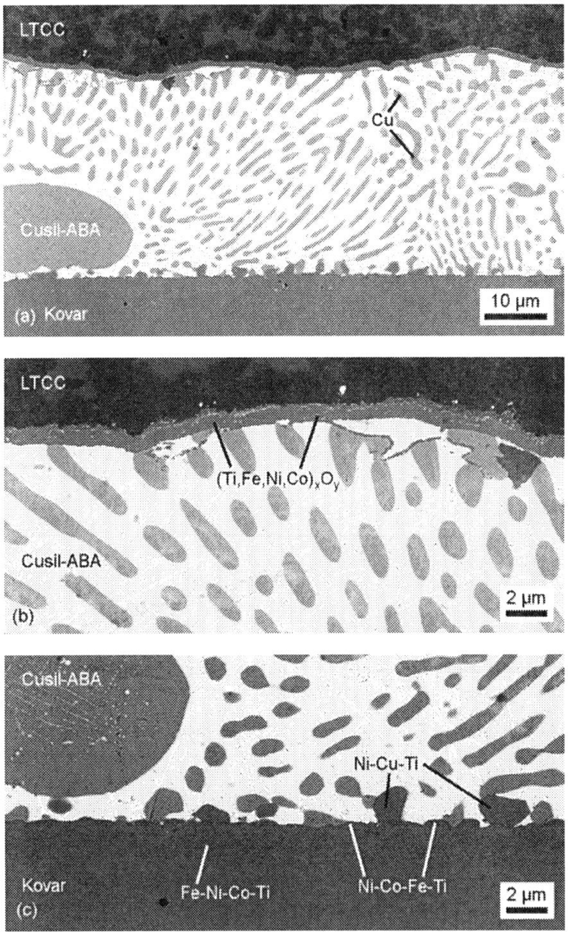

Figure 5 Scanning electron micrographs of Kovar/LTCC joints brazed with Cusil-ABA at 830°C for 10 min.

The microstructure of Kovar/LTCC joints brazed with Cusil-ABA is similar to the microstructure of Kovar/Al_2O_3 joints. Figures 5b and 5c are the enlarged images from Fig. 5a displaying the reaction layers at the interface between LTCC and the brazing alloy, and between Kovar and the brazing alloy, respectively. At the interface between LTCC and the brazing alloy a nearly 1 µm thick reaction layer was formed. Besides the main constituents Ti, and O, the EDX analysis revealed the presence of Fe, Ni, and Co from Kovar and of minor traces of Si and Pb from the glass phase of the LTCC.

Noteworthy, the composition of the interfacial reaction layer bordering the LTCC interface is quite similar to the one found after brazing of LTCC and Kovar with Incusil-ABA. However, the higher brazing temperature for Cusil-ABA yielded in a much thicker reaction layer than for samples brazed with Incusil-ABA. The comparison of the interface between Kovar and the brazing alloy after brazing to Al_2O_3 (Fig. 4c) and LTCC (Fig. 5c) showed no difference in microstructural appearance like thickness, phases formed, or elemental composition.

Conclusions

The present work was focused on the joining of Kovar with alumina and LTCC as part of on approach to integrate LTCC based sensors into steel housings. While using a commercially available active braze filler metals (Cusil-ABA, Incusil-ABA) under certain conditions both ceramic types were hermetically sealed towards Kovar. Hermetic joining of Al_2O_3 to Kovar was possible with Incusil-ABA and with Cusil-ABA for all investigated temperatures. Additionally, brazing of LTCC to Kovar was possible and shown for the first time. At 755°C with Incusil-ABA hermetic LTCC/Kovar joints were realized. With Cusil-ABA joints were hermetically sealed at brazing temperatures > 810°C. However, in all cases microstructural analysis revealed the development of intermetallic compounds, that might be brittle, but their influence on the joint strength is unclear and will be investigated in the future.

Acknowledgments

The authors thank Felix Köhler, Birgit Manhica, Maria Striegler, and Sabine Fischer for sample preparation, helium leak rate measurement, and scanning electron microscopy.

References

[1] Peterson, K. A. *et al*, "LTCC in microelectronics, microsystems, and sensors," *Proc. 15th International Conference on Mixed Design of Integrated Circuits and Systems,* Poznan, Poland, June 19-21, 2008, pp. 23-37.

[2] Zarnik, M. S. *et al*, "An LTCC-based capacitive pressure sensor with a digital output," *Informacije MIDEM*, Vol. 40, No. 1 (2010), pp. 74-81.

[3] Partsch, U. *et al*, "LTCC-based sensors for mechanical quantities," *Informacije MIDEM*, Vol. 42, No. 4 (2012), pp. 260-271.

[4] Fournier, Y. *et al*, "Integrated LTCC pressure/ flow/ temperature multisensor for compressed air diagnostics," *Sensors*, Vol. 10, No. 12 (2010), pp. 11156-11173.

[5] U. Partsch, U.S. Patent 8,079,268

[6] Partsch, U. *et al*, "LTCC based sensors for mechanical quantities," *Proc. 16th European Microelectronics and Packaging Conference & Exhibition,* Oulu, Finland, June 17-20, 2007, pp. 381-388.

[7] Fernie, J. A. *et al*, "Joining of engineering ceramics," *Int Mater Rev*, Vol. 54, No. 5 (2009), pp. 283-331.

[8] Nascimento, R. M. *et al*, "Review Article: Recent advances in metal-ceramic brazing," *Cerâmica*, Vol. 49, No. 312 (2003), pp. 178-198.

[9] Walker, C. A. and Hodges, V. C., "Comparing metal-ceramic brazing methods," *Weld J*, Vol. 87, No. 10 (2008), pp. 43-50.

[10] Walker, C. A., "Metal-nonmetal brazing for electrical, packaging and structural applications," in Sekulić, D. P. (Edt), Advances in Brazing: Science, Technology and Applications, Woodhead Publishing (Oxford, 2013), pp. 498-524.

[11] Stephens, J. J. *et al*, "Reaction layers and mechanisms for a Ti-activated braze on sapphire," *Metall Mater Trans A*, Vol. 34, No. 12 (2003), pp. 2963-2972.

[12] Lin, K.-L. *et al*, "Interfacial characterization of alumina-to-alumina joints fabricated using silver-copper-titanium interlayers," *Mater Char*, Vol. 90 (2014), pp. 40-51.

[13] Hahn, S. *et al*, "A study of the reliability of brazed Al_2O_3 joint systems," *IEEE Trans Compon Packag Manuf Tech C*, Vol. 21, No. 3 (1998), pp. 211-216.

[14] Stephens, J. J. *et al*, "Microstructure and performance of Kovar/alumina joints made with silver-copper base active metal braze alloys," in Vianco, P. T. (Edt) and Singh, M. (Edt), Advanced brazing and soldering technologies, ASM International (Materials Park, 2000), pp. 240-251.

[15] Vianco, P. T. *et al*, "Titanium scavenging in Ag-Cu-Ti active braze joints," *Weld J*, Vol. 82, No. 9 (2003), pp. 268s-277s.

[16] Arróyave, R. and Eagar, T. W., "Metal substrate effects on the thermochemistry of active brazing interfaces," *Acta Mater*, Vol. 51, No. 16 (2003), pp. 4871-4880.

[17] Vianco, P. T. *et al*, "A barrier layer approach to limit Ti scavenging in FeNiCo/Ag-Cu-Ti/Al_2O_3 active braze joints," *Weld J*, Vol. 82, No. 9 (2003), pp. 252s-262s.

[18] Zhu, W. *et al*, "Effects of Ti thickness on microstructure and mechanical properties of alumina-Kovar joints brazed with Ag-Pd/Ti filler," *Ceram Int*, Vol. 40, No. 4 (2014), pp. 5699-5705.

[19] Nascimento, R. M. *et al*, "Interface microstructure of alumina mechanically metallized with Ti brazed to Fe-Ni-Co using different fillers," *Mater Sci Eng A*, Vol. 466, No. 1-2 (2007), pp. 195-200.

[20] Wielage, B. *et al*, "Induction brazing of alumina and zirconia with various metals," *Proc. 5th International Brazing and Soldering Conference,* Las Vegas, NV, USA, April 22-25, 2012, pp. 101-108.

[21] Keusseyan, R. L. and Dilday, J. L., "Development of brazing interconnection to low thermal expansion glass-ceramics for high performance multichip packaging," *Proc. 43rd Conference on Electronic Components and Technology*, Orlando, FL, USA, June 1-4, 1993, pp. 896-903.

[22] Walker, C. A. *et al*, "High-temperature joining of low-temperature co-fired ceramics," *Proc. 3th International Brazing and Soldering Conference,* San Antonio, TX, USA, April 24-26, 2006, pp. 54-59.

Wettability and Brazing of Iono–Ionocovalent Ceramic Materials by Metal Alloys Containing Electronegative Elements

Naidich Yu. V., Krasovskyy V. P., Durov O. V., Sydorenko T. V.
Frantsevitch Institute for Problems of Materials Science of NAS of Ukraine, Kyiv, Ukraine
e-mail: naidich@ipms.kiev.ua

Abstract

High wettability of materials to be joined by brazing alloys determines both very process of joint formation and its high adhesion strength. But ceramics are as a rule hardly wetted by standard brazing alloys. To increase wetting activity traditionally metal additions Ti, Zr, Nb and others with significant chemical affinity to atoms of solid phase must be introduced in brazing alloy. But as to iono-ionocovalent ceramics there is alternative type of active additions to govern wettability: namely electronegative elements of VI–VII groups of periodic system (O, S, Se, F, Cl, Br) with high affinity to electron. Such additions being dissolved in suitable metal strongly decrease surface and interface energy (liquid/gas, liquid/solid) in contrast e.g. Ti, which changes only liquid/solid interface energy.

In this work physicochemical interface processes and brazing technology were considered and developed with using AgCuO-alloys for oxides (Al_2O_3), fluorides (CaF_2), perovskite substrates; and with using In(Sn,Pb)–S(Se)-alloys for sulphide and selenide (ZnS, ZnSe). On the base of our investigations the soldered windows of optical and infrared-light transparent for special cryostats, discharge devices and models of capacitors were manufactured.

Introduction

The development of dissimilar materials (including ceramics) joining by brazing or soldering is an important problem in advanced materials science and technology.

Wettability processes are among the main factors in brazing and soldering procedure. High wettability of materials to be joined by brazing filler alloys determines both the process of joint formation and its high adhesion strength.

Theoretically, the liquid metal brazing (soldering) alloy will fill the brazing gap if the wetting contact angle value will be $\theta<90$. However, for practice, this condition must be formulated more precisely: contact angle must be in the range 15-20°. But ceramics [1–4] are as a rule hardly wetted by standard brazing alloys. In order to increase wetting activity, traditionally metal additions Ti, Zr, Nb and others with significant chemical affinity for atoms of solid phase must be introduced in the base brazing alloy.

But as to ionic or ionocovalent ceramics there is and alternative type of active additions to govern wettability: namely the use of the nonmetal electronegative elements of VI–VII groups of the periodic system (O, S, Se, F, Cl, Br) with high affinity for electrons. Such additions being dissolved in suitable metal strongly decrease surface and interface energy (liquid/gas, liquid/solid). These phenomena were investigated in our earlier works beginning in the 1960s and generalized in [1, 2] particularly for AgCuO alloys.

In the present work, physicochemical interface processes in alloys containing electronegative elements have been investigated in detail, and short review of early studies in this area was done. Brazing technologies were elaborated and developed using Ag-Cu-O-alloys for oxides (Al_2O_3 and others), $BaTiO_3$–substrates and other perovskite materials joining; and using In(Sn, Pb)–S(Se)-alloys for sulfide and selenide (ZnS, ZnSe) soldering. The "cross" brazing process: oxygen containing filler for fluoride substrates joining was studied as well.

General theoretical comments

The electronegative elements O, S, Se and F, Cl, Br – elements of VI "a" and VII "a" groups of periodic system with high affinity for electrons -- are presented in Table 1.

Positive affinity value means energy liberation when an electron is joined to an atom. Negative affinity value means energetic expenditure when formation of O^{2-}, or S^{2-}, or Se^{2-} ions takes place. But we have to take into account that when ionic compound is forming energetic expenditure is compensated by electrostatic interaction (attraction) of oppositely charged ions, that is in both cases ionic compounds metal-electronegative element are stable substrates.

Wetting of ionic or ionocovalent solids will be increased considerably by addition to the base inert metal of elements with high affinity for electrons, for example oxygen.

The elements of VI group (first of all oxygen) decrease both surface (liquid/gas interface) and liquid/solid interface energy. Likely the same behavior shall be observed for elements of VII group as well.

Table 1. The affinity for electron of elements VI "a" and VII "a" groups [5]

VI "a"			VII "a"		
atom	ion	affinity for electron, eV	atom	ion	affinity for electron, eV
O	O^- O^{2-}	2.93 – 6.76	F	F^-	3,62
S	S^- S^{2-}	1.04 – 3.47	Cl	Cl^-	3,82
Se	Se^- Se^{2-}	1.7 – 4.28	Br	Br^-	3,54

The oxygen effect on the capillary and adhesive properties of ceramic-metal contact systems was investigated and discussed earlier for many systems (Cu-O-Al_2O_3, Cu-O-MgO, Ni-O-

Al_2O_3, Ag-O-Al_2O_3, Ag-Cu-O-Al_2O_3) [1–3]. The Ag-Cu-O system is especially interesting as it allows carrying out experiments or technological processes in the air medium.

Some authors [6–9] have confirmed a strong influence of oxygen on surface tension of metals and alloys and on wetting of oxides by them. W. D. Kingery [6], A. Passerone [7] and K. Ogino [8–9] have established high surface and interface activity of oxygen in iron-contacted aluminum oxide. The perovskite materials in contact with Ag–Cu–O system were considered in [10]. In works [11–13], the contact interaction and wetting in Ag–Cu–O/oxides systems were investigated and also it is underlined, that the similar effect can be observed in such systems as Ag–V_2O_5, Pt–Nb_2O_5.

In order to understand the nature and reasons for such action of electronegative elements, let us consider the most studied systems: oxygen containing metal melts contacted with solid oxides. The mechanism of oxygen surface and wetting activity is as follows [2]. In the liquid metal, oxygen exists in the form of near to O^{2-} ions and can form complex metal-oxygen (Me^{2+}–O^{2-}). Positive metallic ion of the complex is adsorbed on a negatively charged solid surface of ionic substrate. Localization of external electrons of metal atom at oxygen must weaken the metallic bond of metal ion (Me^{2+}) with other neutral metal atoms due to excluding of bonding electron from metal–metal bond. So the bond between the metallic ion Me^{2+} of metal-oxygen complex and other metal atom must be weaker than metallic bond of atoms with each other: bond energy Complex-Me < bond energy Me-Me. It is the thermodynamic condition for metal-oxygen complex surface activity. The particles in fluid (atoms, molecules), which faintly bonded to other particles, are ousted from the bulk to surface of fluid. Adsorption of positive metal-ions on the negative charged oxide surface results in higher adhesion. For these reasons, oxygen containing melts wet the surface of solid oxides and other ionic compounds well. So in the mechanism of wettability stimulation, the role of electronegative elements lies in their accepting of valence electrons of metal atom and metal ion formation (Me^{+} or Me^{2+}). In such treatment oxygen role of not specific, it can be other nonmetal elements with high affinity for electrons: sulfur, selenium, halogens (see Table 1) or complex anion (SO_4^{2-}, PO_4^{3-} etc.). Such generalization has been done in [14] and developed in [2].

The investigations, which have been carried out by a number of scientists [15–21], have confirmed the given assertion mentioned above about universal action of electronegative elements. In particular, it was marked that the sulfur and selenium dissolved in metals of iron group and copper lower a surface tension of these metals. High adsorption of sulfur and selenium ions in gas/liquid interface was observed by K. Ogino [15], investigating behavior of elements VIA group in iron, and they have established their high surface activity in metal, which grows from oxygen to sulfur, selenium and tellurium. We can suppose that in these cases the process and mechanism of capillary activity will be the same like for oxygen. At dissolution in metal of oxygen, sulfur, selenium the ions ($X = O^{2-}$, S^{2-}, Se^{2-}) and complexes Me^{2+}–X^{2-}, will arise.

Observing the action of different electronegative additions, which stimulate wettability in different contact systems, cross influence ("cross effect") is manifested: oxygen increases wetting of fluoride solids, chlorine acts by analogy in oxide system see (Figure 1). The ability of additives to accept an electron is the main factor. It results in positive metal ions arising that is bonded with negative charged surface of ionic compound in all cases.

The analysis and systematization of experimental data about adhesion, wetting, and interface activity of electronegative elements dissolved in liquid metals allowed to develop theoretical, physicochemical bases of these processes and methods to govern by them.

There are difficulties in the explanation of surface and adhesion activity of the different electronegative elements.

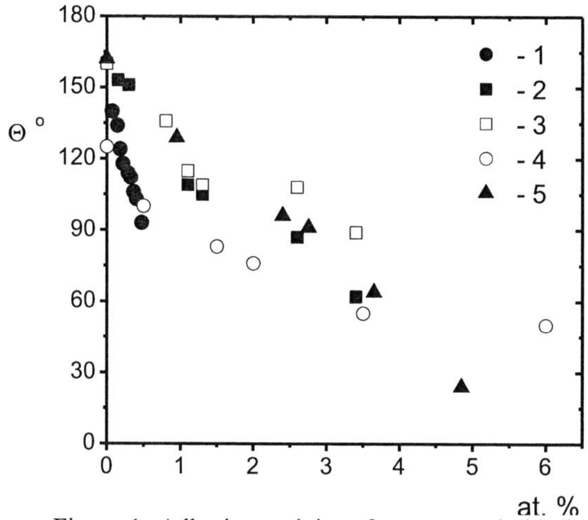

Figure 1. Adhesion activity of oxygen and chlorine: 1 – Ni-O_2-Al_2O_3 (1770 K, vacuum); 2 – Cu-O_2-Al_2O_3 (1420 K, vacuum); 3 – Cu-O_2-MgO (1420 K, vacuum); 4 – (Ag-28Cu)-O_2-CaF_2 (1420 K, air); 5 – Bi-Cl-UO_2 (1070 K, [22]).

On the basis of literature data (attempts to explain the wetting activity of electronegative elements) and our own theoretical notions and experimental results, basic parameters determining the adhesion and surface properties of such substances (electronegative elements) can be formulated.

1. The base digital criterion is the element affinity for electrons. Data are represented in Table 1.
2. One should take into account that the objects considered are solid materials with ionic (ionic-covalent) type of chemical bond.
3. At interaction of electronegative elements to metal melt (dissolution), the chemical forces of ionic character are determinative.
4. The cation-anion interaction is complicated by polarization of anion by cation and vice versa.
5. The displacement of an electron orbital at polarization usually more significant for anion of large dimension, it increases the local density of the electron cloud and covalent component of chemical bond appears in cation-anion system when electrons closing in. The bond in $Me^{+} - X^{-}$ complexes also have some part of covalent bond.
6. Theoretical calculations of interatomic interaction [23] showed that the elements of VI and VII groups of periodic system are forming compounds with metals (for example Ag, Zn, In, Sn give such compounds as AgF, AgCl, AgBr, ZnO, ZnS, ZnSe) where non-ionic share in general bond force decrease in order: $F^{-} < Cl^{-} < Br^{-}$ and $O^{2-} < S < Se^{2-}$.

7. The consideration of metal-oxygen systems shows that growth of bond ionicity in the $Me^{2+} - O^{2-}$ complex means large localization of electrons near oxygen ion and increasing of bond strength, it increases the surface activity of complex. But when the Me – O bond force is very high the bond energy of complex-metal can be so low that formation of new phase - solid metal oxide which insoluble in liquid metal - results. So for high enough surface- and adhesion activity of electronegative element, the optimal ratio of bonds strength in $Me^{2+} - O^{2-}$ complex and bond in complex-metal is necessary.

Experimental procedures and brazing technology development

The aim of these works is to perform new experimental investigations of wetting processes with use of electronegative elements and to enlarge a range of the new (not studied earlier) materials to be joined. The Ag-Cu-O melts as brazing alloys and solder melts containing S, Se will be under study.

Wettability studies were carried out by a sessile drop method in vacuum (~ 10^{-4} Pa), in air media, under pure oxygen and helium atmosphere at temperature 870 – 1370 K. The wetting contact angles and surface energy at the liquid–gas interface were measured. The following solid substrates were chosen for wettability measuring and brazing processes: Al_2O_3-ceramics and monocrystals, ZrO_2 (stabilized by 3.5 % (mol.) Y_2O_3) and HfO_2-ceramics, perovskite materials ($BaTiO_3$, $SrTiO_3$, $Ba_{0.7}Sr_{0.3}TiO_3$), $AlPO_4$, CaF_2, ZnSe. The substrates were mechanically polished by <0.1 μm diamond powder. The average surface roughness value (R_a) was less than 0.01 μm for ZnSe and less than 0.02 μm for $BaTiO_3$, ZrO_2, Al_2O_3. All chosen solid materials are important in practice and interesting from the theoretical view point. Especially it concerns perovskites, ZrO_2 ceramics and ZnSe. For wetting experiments the liquid metals Cu, Ag (purity 99.999%), In, Sn, Pb (purity ≥ 99.995%) and binary alloys of the Cu–Ag system were used. As additives, sulfur and selenium were added to metals (In, Sn, Pb) during experiment. The silver-copper alloys were obtained by vacuum melting in a graphite crucible. The measurement accuracy of contact angles was ± 2°; for surface tension these values were ± 10 mJ/m^2.

Interface microstructures were studied by optical microscopy and SEM, phase composition was obtained by X-ray method.

Results and discussion

The physicochemical interface processes and brazing technology for salt-like substrates ($BaTiO_3$, other perovskites) and some oxides.

In this work, the wetting and contact interaction of various ceramic materials including oxygen-containing compounds ($BaTiO_3$, ZrO_2, Al_2O_3) with Ag–Cu melts were investigated in air at high temperatures. It should be noted that at first oxygen ceramic materials brazing technology by Ag–Cu alloys in air media has been elaborated in our earlier work in 1961 [24]. Now there are versions of this technology by using pure oxygen media.

The results of our experiments show that the oxygen in air, which was dissolved in the Ag–Cu melts and oxidize the copper, improves strongly the wettability of different ionic and ionic-covalent compounds, including the stoichiometric barium titanate (Figure 2) and other perovskite materials, ZrO_2, HfO_2, etc.

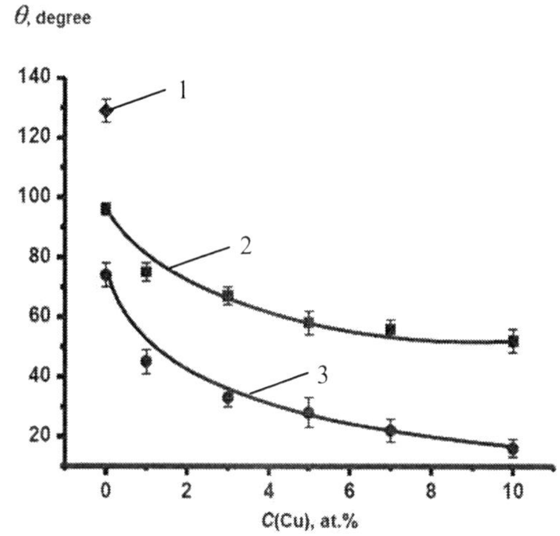

Figure 2. Dependence of contact angle for the ferroelectric $BaTiO_3$ ceramic by Ag–Cu–O melt in air and pure oxygen at concentration of copper at 1250 K [26]: 1 – pure Ag (vacuum), 2 – Ag–Cu (air), 3 – Ag-Cu (oxygen).

This is especially important as it allows safe ferroelectric properties of barium titanate (stoichiometric compounds) at high temperature (air media condition). During heating of $BaTiO_3$ in vacuum, its stoichiometric properties will be lost [25]. On the surface of barium titanate ceramics, the contact angle with the silver in vacuum at 1250 K was equal to 129°, the same value in air was 96°, and in pure oxygen – 75°. Copper addition to silver melt (~10 at.%) leads to considerable contact angle decrease to 45–47° (in air) and to almost complete spreading of the alloy (θ ≈ 5–10°) in the pure oxygen atmosphere. The temperature increase intensifies the wetting process; contact angle decreases to 25–30° in air (with 10 at.% of Cu in liquid Ag). For pure oxygen, full spreading can be reached with Cu content at about 6–7 at.%. The data of wetting in air of different ceramic materials with Ag—Cu alloys are presented in Table 2 and Figures 2–3.

Table 2. Data of wetting in air of different ceramic materials by Ag–Cu alloys, T = 1270 K

Material	Contact angle, degree	
	Ag–5Cu	Ag–39Cu
Al_2O_3	80	40
CaF_2	81	80
HfO_2	44	28
ZrO_2	54	46
$BaTiO_3$	56	43
$Ba_{0.7}Sr_{0.3}TiO_3$	46	–
$SrTiO_3$	44	–
$AlPO_4$	15	3

Adsorption of metal-ions on the negative charged oxide surface results in high adhesion. For these reasons, silver-copper-oxygen melts well wet the surface of solid oxides in air – so they can be successfully used for brazing and metallization of different refractory materials.

Analysis of the ceramics/Ag-Cu-O alloy interface shows the presence of a transitive zone ~3–10 μm thick, which is an obvious result of copper diffusion into the ceramic substrate in air.

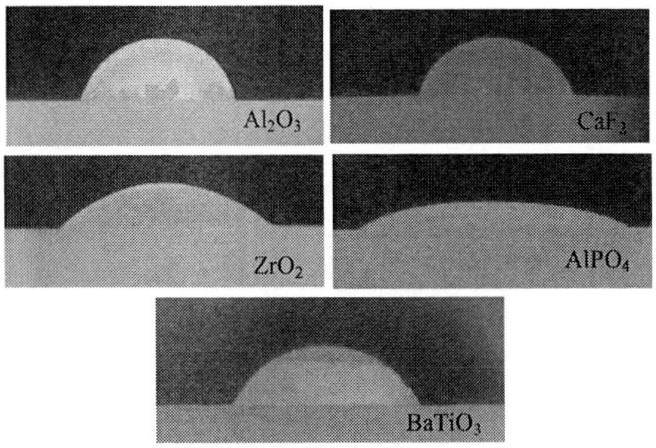

Figure 3. Wetting of ceramic materials by Ag–5Cu melt (T = 1270 K).

The microstructure of the transition layers at the contact interface ceramic-to-metal was investigated. It was shown that the contact metal-ceramic zone contains a high concentration of CuO, which provides good wetting in the system (Figure 4). It also remarkable that for Al_2O_3 copper reach interlayer is significantly discontinuous, for ZrO_2 and $BaTiO_3$ it is more. These differences in interface structure explain the variant wetting in studied systems: "island" interface structure provides higher contact angles than uniform interlayers.

To verify the CuO presence in systems after annealing, the mixes of powders of silver, copper and studied materials (Al_2O_3, ZrO_2 and $BaTiO_3$) were prepared, tablets were pressed from it, and annealed in air at 1270 K.

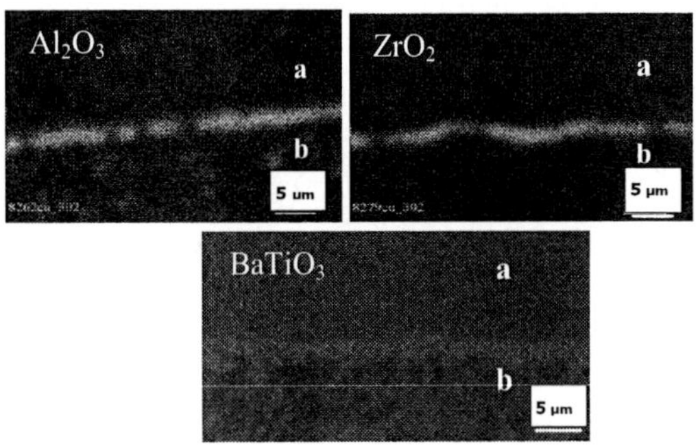

Figure 4. Microstructure of metal-ceramics contact area (a – ceramics, b – alloy) at ×1000: characteristic emission of elements at ceramics/molten metal interface with copper.

The samples obtained were studied by the X-ray method and presence of CuO was confirmed (Figure 5). The presence of the new phase Al_2CuO_4 for Al_2O_3-Ag-Cu-O system also should be noted: probably just the reaction of Al_2O_3 to CuO leads to discontinuous interlayer formation and so to worse wetting of Al_2O_3 by Ag-Cu-O melt than that for other materials. In other systems, for example $BaTiO_3$-Ag-Cu-O (Figure 5 b), formation of such new phases was not observed, so intensive interaction of the melt to the substrate is not present and CuO interlayer is continuous.

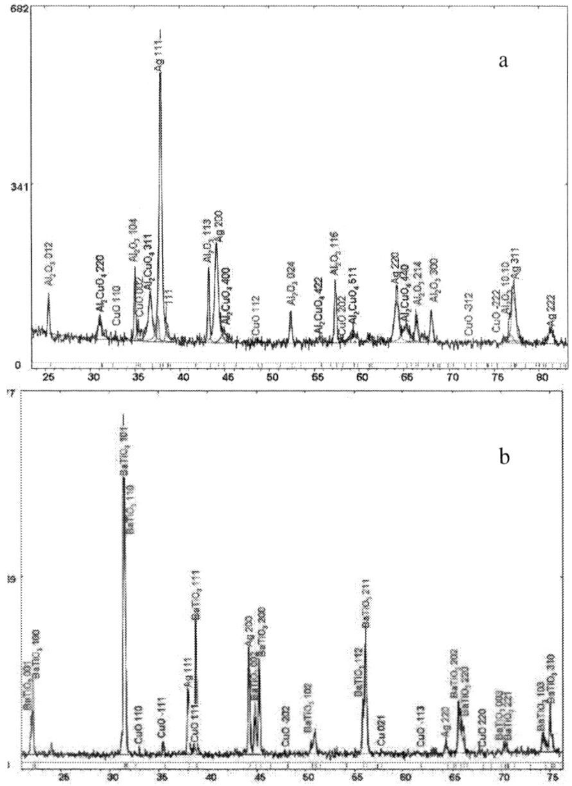

Figure 5. Diffractograms of pressed mix of silver, copper and ceramic powder compositions (1000 °C, 15 min): a – Al_2O_3; b – $BaTiO_3$.

The process of joining (brazing) ferroelectric $BaTiO_3$ ceramics can be performed in air. The brazing alloy Ag-10Cu was used for joining $BaTiO_3$ ceramics in air atmosphere.

Samples of the perovskite ceramics brazed with Ag-Cu-O alloy were obtained (Figures 6, 7).

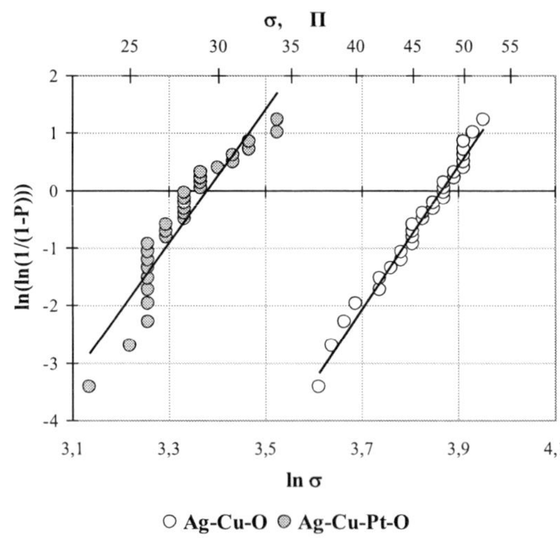

Figure 6. Weibull's graph of shear strength for ferroelectric $BaTiO_3$ ceramics samples brazed in air [26].

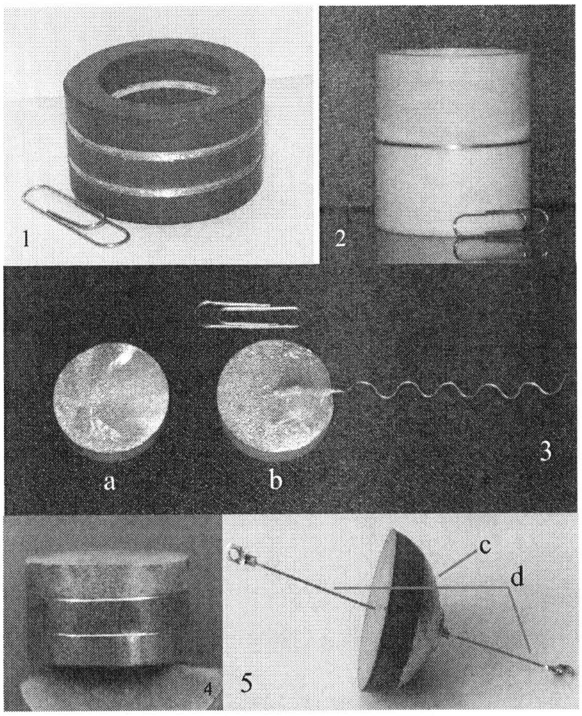

Figure 7. Samples of ceramics which were brazed or metalized by metal-oxygen technology: 1 – component of electron-beam device consist of Al_2O_3-ceramic rings brazed by Ag-Cu-O filler metal; 2 – Al_2O_3-Pt-Al_2O_3 assembly brazed by Ag-Cu-O filler metal; 3 – $BaTiO_3$-ceramic metalized by Ag–Cu–O (a) and metalized by Ag–Cu–O with simultaneous brazing of platinum current lead (b); 4 – $BaTiO_3$-ceramic brazed using Ag–Cu–O filler metal; 5 – brazed components for special devises based on perovskite material (c – $BaTiO_3$-ceramics metalized by Ag-Cu-O filler metal, d – Pt electrodes).

Wetting and soldering of zinc selenide. Zinc selenide (ZnSe) is a widely used material: as luminofor, semiconductor, in laser technique, or special optical material, transparent in a wide spectrum – from a visible to an infrared light. In this later case, the ZnSe windows (monocrystals or ceramic) for cryogenic and gas discharge devices are especially important. Such windows must be joined to base metal or nonmetal (silica glass) parts; the resulting joints must meet severe technical requirements:

- stability and reliability of all soldered joints at low temperatures (the cryostats are working at temperatures of liquid nitrogen);
- high vacuum tightness of soldered joints (leakage in the device must not exceed 10^{-9} l·μm/sec);
- integrity of soldered joints during thermocyclic loadings from temperature of liquid nitrogen up to temperature of boiling of water, i.e. 77–370 K).

The problem of soldering ceramics to metals becomes complicated, first of all, due to the low mechanical strength of the ZnSe–ceramics. Diffusion welding of ZnS–ceramics with copper in argon at temperature 1123 K, under pressure 8–10 MPa was offered [27]. Preliminary sulfurizing of a metal surface before welding to increase strength of welded joining is recommended.

Moreover, high residual stresses in the weld joint owing to a difference of the temperature expansion coefficients of zinc sulfide (α=7.85·10^{-6} 1/K) and copper (α=16.6·10^{-6} 1/K), reduces the stresses of joining. Diffusion welding can be carried out at lower temperatures (up to 870 K), using an aluminum gasket that allows to preserving the required optical transparency of these ceramics. However, high-pressure is applied in this case too, and this is a negative factor, as mentioned above.

The adhesion-active soldering by using a highly plastic metal is the most reliable method. Use of electronegative elements (S, Se in this case) as wetting promoters that are applied to such plastic metals as In, Sn, Pb is suitable for ZnSe soldering, and this is the only method for ZnSe/metal joining.

Wetting of ZnSe–substrates by different metals and alloys was studied to develop soldering technology and to learn what is an optimal solder alloy composition. The fusible metals: In, Sn, Pb and S, Se additions were used. Wetting and soldering researches were carried out in helium gas at the pressure 0.5·10^5 Pa and at temperature 970 K. The wetting data and the profile of an In–Se droplet on the ZnSe substrate are presented in Figures 8 –10.

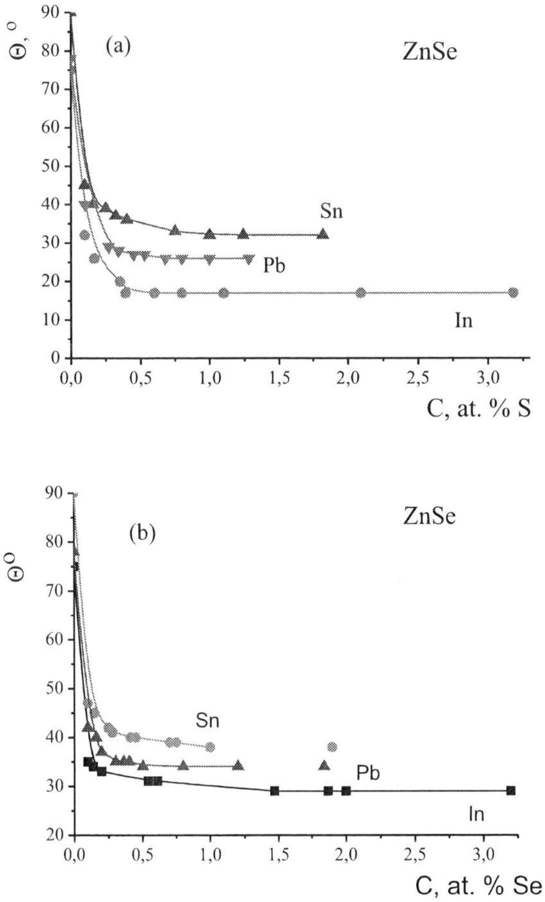

Figure 8. Wetting contact angle in Me (In, Sn, Pb)–S, Se/ZnSe systems (helium medium at pressure 0.5·10^5 Pa, temperature 970 K): a – Me-S/ZnSe, b – Me-Se/ZnSe.

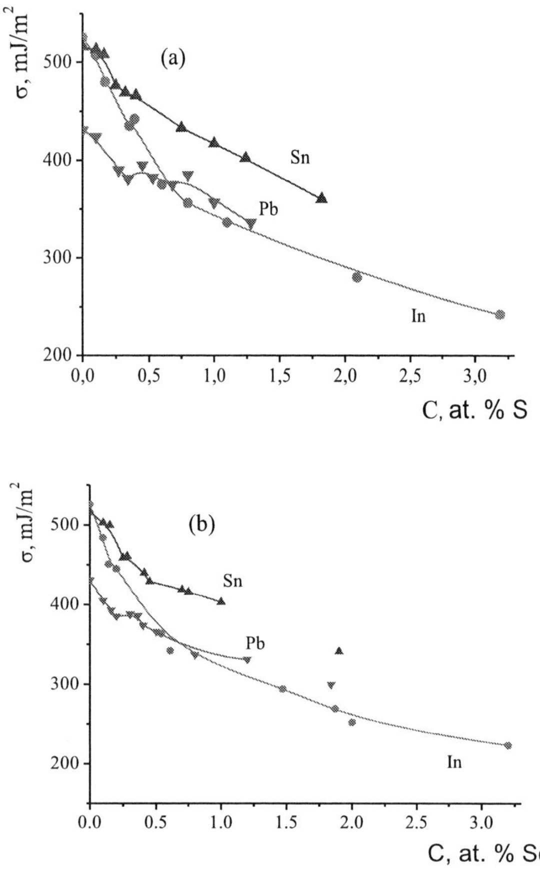

Figure 9. Surface tension of (In, Sn, Pb)–S, Se alloys (helium medium at pressure $0.5 \cdot 10^5$ Pa, temperature 970 K): a – Me-S/ZnSe, b – Me-Se/ZnSe.

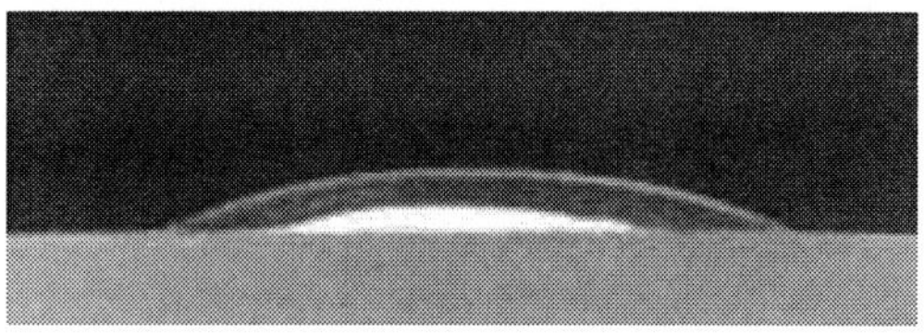

Figure 10. The profile of an In–Se droplet on ZnSe substrate (helium gas at pressure $0.5 \cdot 10^5$ Pa, temperature 970 K).

Contact angle sharply decreases (wetting improves already at addition 0.1 at.% of S, Se). The surface tension of all alloys is reduced as well [19–21]. Sulfur is a stronger adhesive- and surface active element in comparison with selenium. Their behavior can be explained similarly to behavior of oxygen in a copper-silver alloy.

On this basis, the soldering technological processes for joining optical ZnSe ceramics with ferronickel alloys (Fe–42 wt.% Ni; Kovar: Fe–29Ni–17Co wt.%) was created. These materials have sufficiently close temperature expansion coefficients at soldering temperature (these values for ZnSe and for Kovar are $\alpha=7.57 \cdot 10^{-6}$ 1/K and $\alpha=6.5 \cdot 10^{-6}$ 1/K, respectively). Structure of soldered windows (Figure 11), the method of brazing, joining of highly plastic adhesion-active solders with S and Se (which are protected by USSR patents [28, 29]), is developed.

In–Se solders, which were preliminarily prepared by melting in a graphite crucible at 970 K in helium at pressure $0.5 \cdot 10^5$ Pa, are used.

Titanium is also an adhesion-active metal for ZnSe, however, soldering of ZnSe by titanium-containing filler metals leads up to deterioration of optical properties of the ceramics in this case. Thus the titanium-containing solders are not useful for soldering of ZnSe-ceramics, and consequently, the technology that uses solders containing sulfur and selenium has been applied.

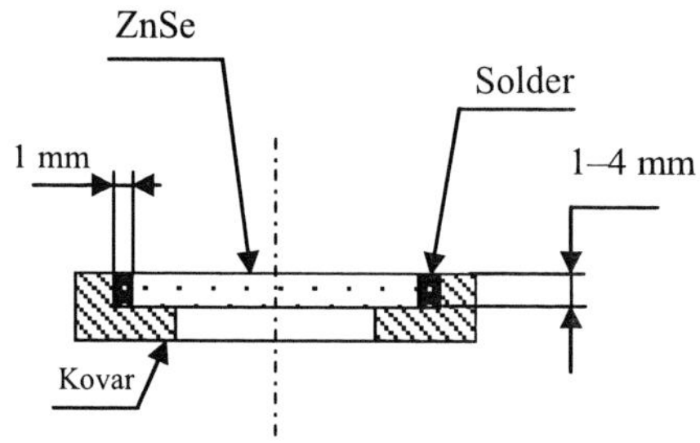

Figure 11. Structure of joining by soldering.

The contact angle dependences for the titanium- and sulfur-containing systems are compared in Figure 12. Values of angles at 970 K are approximately close. The mechanism of S and Se action to ZnSe is similar to action of oxygen to oxide ceramics. Soldered windows of optical and infrared light transparent parts for special cryogenic and gas discharge devices in diameter up to 12–50 mm (Figure 13) were manufactured using a In-S solder. Vacuum tightness of soldered joints is 10^{-9} l·μm/sec. Thermal cyclic loading of soldered assembly on a mode 300 – 370 – 77 – 370 – 300 K has been carried out. 10 cycles of tests were made; vacuum tightness of soldered joint was tested after each cycle step. Deterioration of vacuum tightness did not occur. A breaking test of samples has been made as well. Cohesion destruction of products took place through the body of ceramics.

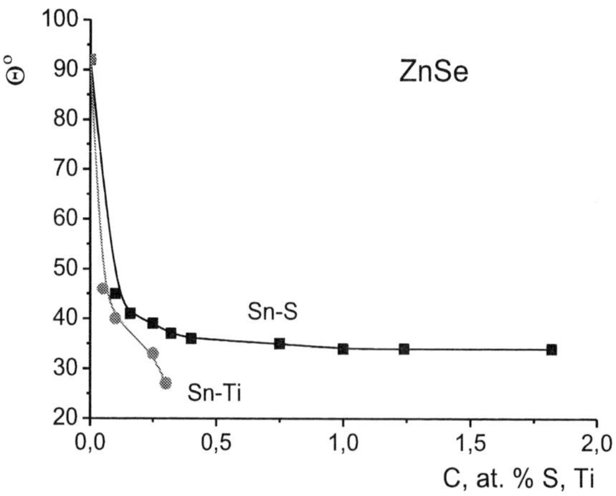

Figure 12. Wetting of ZnSe by Sn–S and Sn–Ti melts in helium gas at pressure $0.5 \cdot 10^5$ Pa and temperature 970 K.

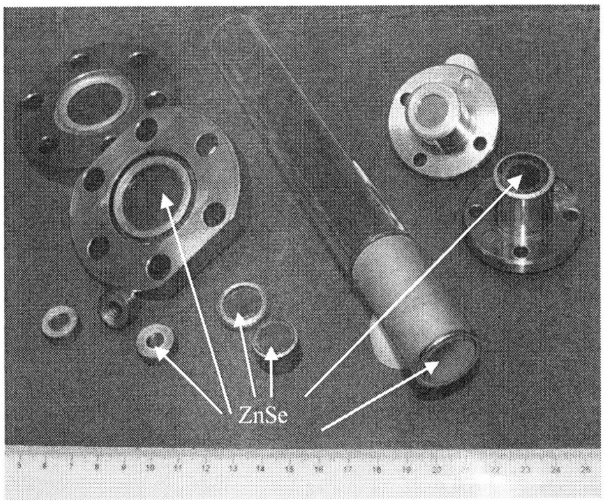

Figure 13. Soldered windows, that are optical and infrared light transparent, for special cryogenic and gas discharge devices.

Conclusions

The method of brazing (soldering) alloys to increase its wetting-adhesion ability, by use of electronegative elements with high affinity for electrons, is developed. Elements of the VI – VII groups of the periodic system (O, S, Se; F, Cl, Br) were introduced into filler metals.

This method proposed for joining ceramic materials with iono-ionocovalent nature is an alternative to the traditional method where chemically-active additions (Ti, Zr, V and other) are used.

Physicochemical interface processes in alloys containing electronegative elements have been investigated. Wetting is different for different materials; contact angles varies from 15° for $AlPO_4$ to 80° for Al_2O_3 in experiments with Ag–5Cu initial alloy. It is explained by the difference in interaction of oxidized copper and substrate.

Brazing technologies were elaborated with using AgCuO-alloys for oxides (Al_2O_3, HfO_2, ZrO_2), $BaTiO_3$, $Ba_{0,7}Sr_{0,3}TiO_3$, $SrTiO_3$, $AlPO_4$–substrates; and with using In (Sn, Pb)–S (Se)-alloys for zinc sulfide and selenide (ZnS, ZnSe) joining. The "cross" brazing process: oxygen-containing (Ag-Cu-O) filler for CaF_2 substrate joining was studied as well. Strength of joints obtained by this technology is nearly 90% of strength of joints made from monolithic material.

The metal-oxygen technology of brazing oxide materials and perovskites significantly simplifies the joining technological process because all operations are performed in air. This technology allows preservation of the stoichiometric composition of perovskite and unique electrical properties of ferroelectrics.

The developed soldering technology of joining ZnSe-ceramics by In-S (Se) fillers has allowed getting a vacuum tightness of joined assemblies that work in the temperature range of 77-300K. No deterioration of optical properties of the ceramics was observed.

The soldered windows of optical and infrared light transparent materials were manufactured for optical cryostats.

Acknowledgments

The authors wish to thank Dr. Karpets M. V. and PhD. Koval A. Y. for microstructure interface investigations.

Reference

[1] Naidich Yu. V. Contact Phenomena in Metal Melts, Naukova Dumka (Kyiv, 1972), 196 p.
[2] Naidich, Yu. V., Progress in Surface Membrane Science, 14, Academic Press, Inc. (New York, 1981), pp. 353-484.
[3] O'Brian T. E. et al, "Effect of Oxygen on the Reaction between Copper and Sapphire," J. Amer. Ceram. Soc. Vol. 57, No. 8 (1974), pp. 323-333.
[4] Durov O. V. et al, "Wetting of some nonstoichiometric oxide ceramic materials by liquid metals," Proc E-MRS Fall Meeting, 2012, pp. 184-189.
[5] Goronovskyy I. T. et al, Brief Handbook in Chemistry, Naukova Dumka (Kyiv, 1974), 833 p.
[6] Halden F. A. et al, "Surface Tension at Elevated Temperatures. II. Effect of C, N, O and S on Liquid Iron Surface Tension and Interfacial Energy with Al_2O_3" J. Phys. Chem., Vol. 59, No. 6 (1955), pp. 557-559.
[7] Beruto D. et al, "Refrectory Oxides, High Temperature Solid–Gas and Solid–Liquid Behavior," Oxides and Oxide films: NY & Basel, Vol. 6 (1981), pp. 1-83.
[8] Ogino K. et al, "Effect of Oxygen on a Surface Tension of Liquid Co and Co – Fe Alloys, and also on Wettability of Al_2O_3," J. Jap. Inst. Metals, Vol. 46, No. 10 (1982), pp. 957- 962.
[9] Taimatsu H. et al, "Effect of Oxygen on Wettability of Magnesium Oxide by Liquid Nickel," J. Jap. Inst. Metals, Vol. 50, No. 2 (1986), pp. 176- 180.
[10] Erskine K. M. et al, "Brazing Perovskite Ceramics with Silver/Copper Oxide Braze Alloys," J. Mater. Sci., Vol. 37 (2002), pp. 1705-1709.
[11] Shao Z. B. et al, "Equilibrium Phase-Diagrams in the Systems PbO-Ag and CuO-Ag," J. Am. Ceram. Soc., Vol. 76, No. 10 (1993), pp. 2663-2664.
[12] Roth R. S. et al, eds. Phase Diagrams for Ceramists, Volume VI. The American Ceramic Society, Westerville (Ohio 1987).
[13] Hardy J. S. et al, "Development of High-Temperature Air Braze Filler Metals for Use in Two-step Joining and Sealing Processes," Weld Journal, Vol. 87 (2008), pp. 195-201.
[14] Naidich Yu. V. Surface Phenomena in Metals, Kabardin-Balkarian University (Nalchik, 1965). pp. 30-39.
[15] Ogino K. et al, "Effect of Selenium and Tellurium on the Surface Tension of Molten Iron and the Wettability of Alumina by Molten Iron," Transactions Iron and Steel Inst. J., Vol. 23 (1983), pp. 234-239.
[16] Gupt K. M. et al, "Study of Effects of the Concentration and Temperature on Surface Tension of Fe–O–S System," Transactions of Indian Inst. Metals, Vol. 29, No. 4 (1976), pp. 286-291.
[17] Kingery W. D., "Surface Tension at Elevated Temperatures. IV. Surface Tension of Fe–Se and Fe–Te Alloys," J. Phys. Chem., Vol. 62, No. 7 (1958), pp. 878-879.

[18] Sveshkov Yu. V. et al, "Surface Activity of Selenium and Tellurium in Iron–Chromium–Nickel Melts," New institutes. Ferrous metallurgy, No. 5. (1974) pp. 5-7.
[19] Naidich Yu. V. et al, "Influence of Sulfur and Selenium on Wettability by Metal Melts of Zinc Selenide and Zinc Sulfide," Powder Metallurgy and Soviet Metal Ceramics, No. 11 (1986), pp. 69-72.
[20] Naidich Yu. V. et al, "Surface Tension and Density of Fusible Metals Melts with Sulfur and Selenium," Surface. Physics, chemistry, mechanics, No. 5 (1990), pp. 155-156.
[21] Naidich Yu. V. et al, "Effect Chalcogenes on Molten Metals Capillary Properties," Proc 1st Internat. Conf , Bratislava, Slovakia, 1995, pp.133-138.
[22] Bradhurst D. H. et al, "The Surface Properties of Liquid Lead in Contact with Uranium Dioxide," J. Phys. Chem., Vol. 63, No. 9 (1959), pp. 1486-1488.
[23] Barnard A. K., Theoretical Basis of Inorganic Chemistry, McGaw-Hill Publishing Lim. (N.Y. – London – Toronto, 1965).
[24] Naidich Yu. V. and Eremenko V. N., Patent USSR № 149020, 1962.
[25] Xu, Y., Ferroelectric Materials and Their Applications, North-Holland (Amsterdam, 1991), 391 p.
[26] Sydorenko T. V. et al, "Wettability by Liquid Metals, Metallization, and Brazing of Barium Titanate Ceramics," Weld. Journal, Vol. 93 (2014), pp. 6-14.
[27] Kazakov N. F., Diffusion welding of metals, Mashunostroenie (Moscow, 1976), 312 p.
[28] Naidich Yu. V., Chuvashov Yu. N., and Krasovskyy V. P., Patent USSR 1290636, 15.10.1986.
[29] Naidich Yu. V., Kolesnichenko G. A., Chuvashov Yu. N, Krasovskyy V. P., and Kostjuk B. D., Patent USSR 1327383, 01.04.1987.

Soldering of Aluminum with Zn-Al Alloys: Metallography of Joints Formation and Technological Features

Igor N. Pashkov, Alexey A. Anohin
PBSU, Moscow, Russia
paikaspp@gmail.com

Viacheslav E. Bazhenov, Mikhail V. Pikunov
NUST MISiS, Moscow, Russia
v.e.bagenov@gmail.com

Abstract

Zn-Al alloys are positioned as a pretty good alternative for aluminum alloys soldering due to good corrosion resistance and mechanical properties but especially comparatively low temperature of joint formation. A lot of researches of microstructure were carried out to explain properties upon different technical parameters impact as soldering temperature, exposure time and solder composition. The fact that properties of soldered joints depended of interaction of base metal with liquid solder was determined as more important. Researches of joints made with different form of soldering alloys as wire and foil were carried out to study join formation in case of filler metal use and external application. Increasing of joint porosity because of shrinkage was determined corresponding to increase of Al content in soldering alloy. Some technological ways are needed to avoid negatives mentioned above.

Introduction

Aluminum soldering is a complex technological process. Firstly, it is necessary to remove an oxide film on a base metal surface for the solder wetting and spreading provision. Secondly, due to the low temperature soldering of aluminum and its alloys, it is difficult to determine visually the aluminum surface temperature. It requires the presence of certain skills and experience of the worker, while hand-operated gas-torch soldering, for the burn-through of base metal prevention. Therefore, the aluminum soldering is implemented in furnaces mostly, and sometimes in salt-baths [1].

A variety of products of aluminum and its alloys, an increasingly widening application of aluminum in the heat-exchangers, refrigerators, mechanical engineering parts, aviation technique production, dictates the necessity of application not only the furnace soldering of aluminum, but also the hand-operated gas-flame and induction soldering, as well.

The necessity of joining by soldering of aluminum based alloys, alloyed with magnesium and copper, results in the aspiration to carry out the joining process at low temperatures in comparison with the soldering by means of conventional solders based on eutectic alloy Al12Si. Such an aspiration is observed in numerous operations on repair, mounting, elimination of defects after the furnace soldering of aluminum parts, especially of the thin-walled constructions.

The soldering of aluminum with the low-temperature Sn-based solders is possible also; however, the corrosion resistance of joint limits the application range of such soldering. Therefore, the solders on the basis of Zn arise a great interest, in particular, Zn-Al alloys, which allow soldering of aluminum alloys at temperatures of 420-570°C [2-3]. Zeng, et al. [4] specify the low cost, easy to manufacture, no IMC formed under equilibrium as Zn-Al solders advantages, and as their deficiencies – high corrosiveness, poor solder acceptance. However, zinc solders have not been spread widely.

In our opinion, the reasons of the limited applicability consisted in the absence of the effective soldering flux at temperatures of 420-500°C and the unsatisfactory quality of joints (high shrinkage porosity, base metal erosion). Therefore, the soldering by means of such kinds of solders was often limited to abrasive soldering or to the ultrasonic soldering [5, 6]. In many aspects, the first reason has been solved with the appearance of non-corrosion flux on the basis of fluoroaluminate cesium. This flux has provided the hand-operated gas-flame and induction soldering at temperatures of 420-470°C implementation [7]. The elimination of the second reason is possible by means of understanding of processes, proceeding while contact of the liquid solder with the base metal, their interaction at soldering temperatures and finally, the soldered seam solidification. The structure formation of joint determines its mechanical and exploitation properties. Abou El-khair, et al. [8] studied the strength and wearing resistance of ZA8, ZA12 and ZA27 alloys, containing 8, 12 and 27% Al respectively. It has been shown, that the ultimate strength and wearing resistance raise along with the aluminum content increasing, at the same time in work [7] the maximum strength of the seam has been achieved in case of Zn-2%Al solder usage. The understanding of paths of achievement of the soldered seam optimal structure allows the correct determination of soldering process parameters and techniques while soldering with Zn-based solder alloys. The present work is dedicated to the analysis of structure formation of the joint by means of solders with the aluminum content of 4, 15 and 22 wt. %.

Materials and Methods

Materials

Zn-4%Al, Zn-15%Al, Zn-22%Al (wt. pct.) solders in the form of 3 mm diameter bars have been prepared in the resistance furnace. Aluminum (99.7%Al), zinc (99.95%Zn) have been used as a charge material. As a base material for the soldered seams microstructure analysis and the technological properties determination, the deformed billet of aluminum (99.7%Al) have been used, for the strength properties determination Al4.5Cu1.5Mg0.6Mn alloy in the form of 4 mm thickness sheet with σB=450 MPa have been used. While soldering, the flux on the basis of cesium fluoroaluminate ($CsAlF_4$) in a paste form has been used.

Soldering procedure of samples for studying the interaction between the solder melt and the base metal

The soldering of aluminum with Zn-4%Al and Zn-15%Al (wt. %.) solders has been investigated at the experimental device, presented at Fig. 1. Solder was placed on the aluminum sample edge. Flux was placed nearby. 0.2 mm gap value was fixed with steel spacers, which were installed between lower and upper samples. The load of 2 kg was applied to the upper sample. The whole assemblage was placed within the field of double-coil inductor with the graphite ring screen for the uniform heating provision. The assemblage was heated up to the necessary temperature and the isothermal exposure was implemented. While soldering with Zn-4%Al solder, the isothermal exposure was carried out at temperatures of 420 and 450°C, and while soldering with Zn-15%Al solder – at 460 and 500°C. The temperature control was implemented by means of the thermocouple, placed in the hole at the bottom side of the aluminum sample. The exposure time was 30 and 180 s. Afterwards, the soldered samples were sliced in a cross-section direction and the polished section of joint were prepared, which then were analyzed by EDS.

Investigation of solders flowing ability into a gap

Aluminum tubes of two sizes - ∅16x1.2 mm and ∅10x1 mm were chosen as specimens. The specimens of 25 mm in length were cut across and connected to each other along an axis by means of the screw on top of the specimen (Fig. 2). The specimen with a thermocouple, attached to it by means of wire, was fastened to a rod with a vertical movement rack gear. The assemblage was installed at the muffle furnace.

The depth of sinking of the sample into melt was checked by means of a ruler on the moving device. Depth of specimen sink into melt was equal to 2 mm. The specimen exposure time in melt was 10 seconds. The alloy and specimen temperatures during the experiment are presented at Table 2.

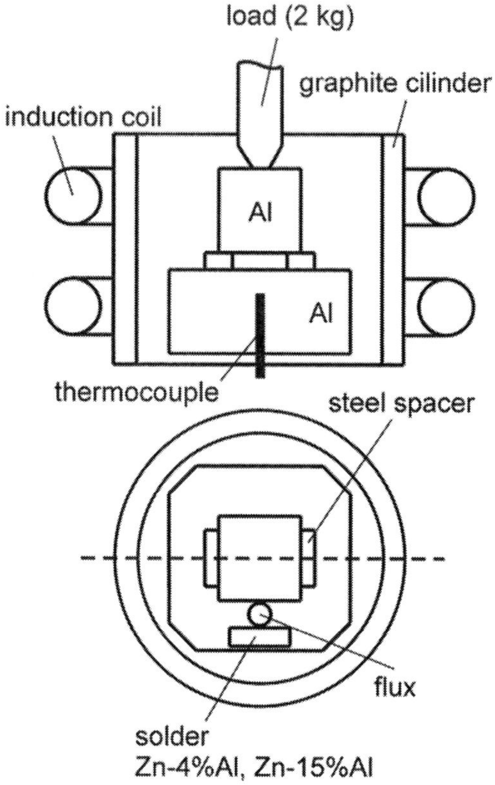

Figure 1. - Scheme of the experimental device for investigation of the interaction of Zn-Al solders with aluminum

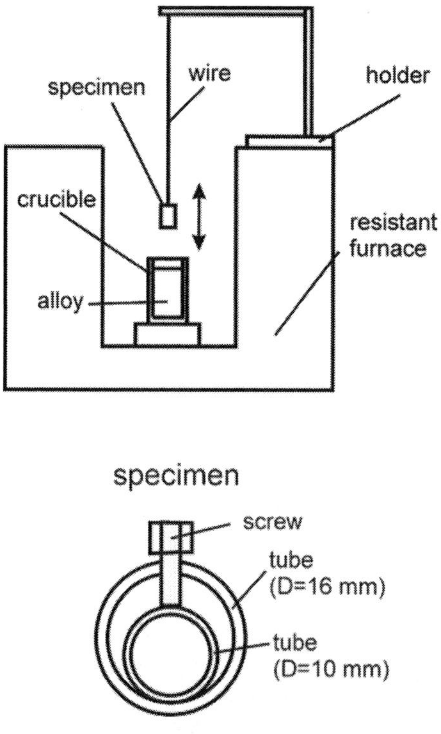

Figure 2. - Schema of the experimental device for the investigation of solders flowing into a gap

Table 2 - Alloy and the specimen temperatures during experiment

Sample #	Solder alloy	T_{alloy}, ° C	T_{spec}, ° C
1	Zn–4 %Al	430	420
2	Zn–15 %Al	480	440
3	Zn–22 %Al	550	520

The soldered specimens were sliced in a cross-section direction with a step of 3 mm for the purpose of the solder rising height value definition for various gaps. Polished sections were produced of specimens for the analysis by means of SEM.

Preparations of soldered samples for the mechanical tests

For the influence definition of the Zn-based solder composition on the shearing strength of joint, the samples have been soldered, and mechanical tests have been carried out in accordance with Russian state standard GOST 28830-90. Base metal plates have been cleaned from the lubrication residues and dirt prior to the soldering. Due to the low solidus temperature of Al4.5Cu1.5Mg0.6Mn alloy, it seemed impossible to implement the soldering with Al-22%Zn (wt. %.) solder alloy.

SEM and EDS analysis

Microstructure investigations and also the Energy-dispersive X-ray spectroscopy analysis of the obtained samples have been carried out by means of TESCAN VEGA LMH scanning electron microscope with LaB_6 cathode and the system of energy dispersive X-ray analysis of Oxford Instruments Advanced AZtecEnergy. The etching of samples had not been implemented. The soldered seam composition was determined within the square of 200x200 μm by EDS analysis. 3 measurements have been implemented per each seam.

Results and discussion

Investigation of the interaction between Zn-Al solder melt and the base metal

The soldered seams microstructures, obtained with the use of Zn-4%Al solder at the soldering temperature of 450°C and exposure time of 30 seconds, Zn-15%Al at the temperature of 460°C and exposure time of 30 and 180 seconds are presented at Fig. 3. It is possible to see, that in the seam obtained with the usage of Zn-4%Al solder the seam width is much more, than at soldering with Zn-15% Al solder that testifies of the considerable dissolution of the base material. It was revealed also, that seams, which were soldered with Zn-4%Al solder at high temperature, had considerable shrinkage porosity. The aluminum content in seams obtained at different soldering parameters is presented at Fig. 4. The aluminum content in seam increases along with the soldering temperature increasing that is in accordance with the Zn-Al phase diagram. The phase diagram was obtained by means of Thermo-Calc software. TCBIN thermodynamic database was used. The aluminum contents in the seam increases along with the exposure time, at that, in the seam, formed at the temperature of 500°C, the difference in aluminum content exceeds 10 %. The special attention should be paid to the fact, that the aluminum content in seams considerably exceeds the aluminum content in liquid at soldering temperature (Fig. 5).

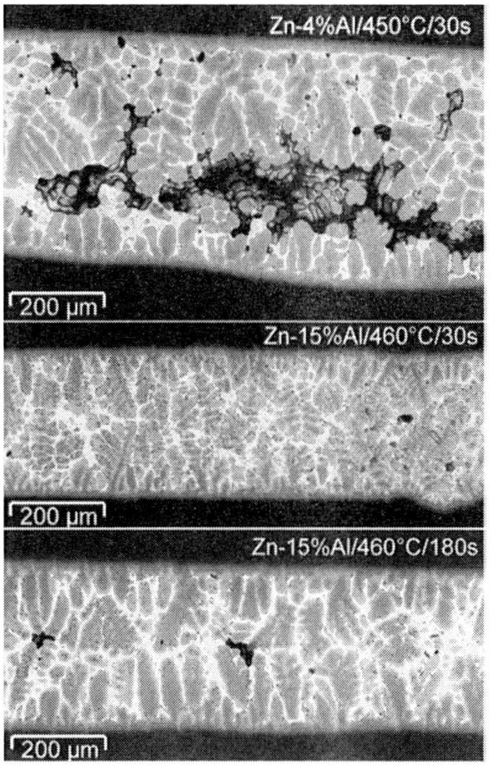

Figure 3. – Seam microstructures, obtained at aluminum soldering with Zn-4%Al solder at temperature of 450°C and exposure time of 30 seconds (top); Zn-15%Al solder at temperature of 460°C and exposure time of 30 seconds (medium) and 180 seconds (bottom)

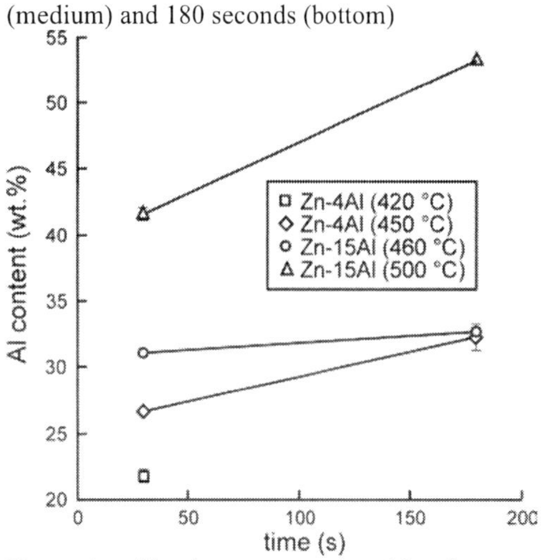

Figure 4. – Aluminum content in soldered seams

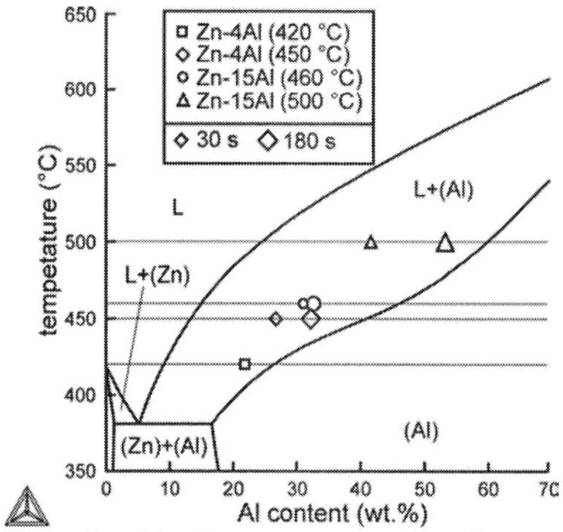

Figure 5. – Aluminum content in soldered seams placed at Zn-Al phase diagram.

Thus, the strong erosion of the base material by the solder melt is observed. Along with the increasing of temperature and time of soldering, the base metal dissolution increased, that resulted in the solidification temperature interval increasing and the shrinkage porosity formation in seams (Figure 3), and consequently to their tightness failure. Therefore, while designing the soldering technological process the attention should be paid to the soldering parameters choice and their observance. The optimal results were obtained at small values of the solder overheating above liquidus temperatures and at short times of the isothermal exposure while soldering.

Flowing of solder into a gap

The relation of maximum filled gap to solder raising height is presented at Fig. 6. It is clear from the diagram, that the aluminum content decreasing in solder results in the decreasing of the gap maximum admissible value. Therefore, in order to fill the large gapes, for example, at soldering while repairing, plugging of defects is possible by means of solders with the aluminum content more than 15 %. The solder with the aluminum content of 4 % may be used at the tight assemblage of connections and minimal gaps only.

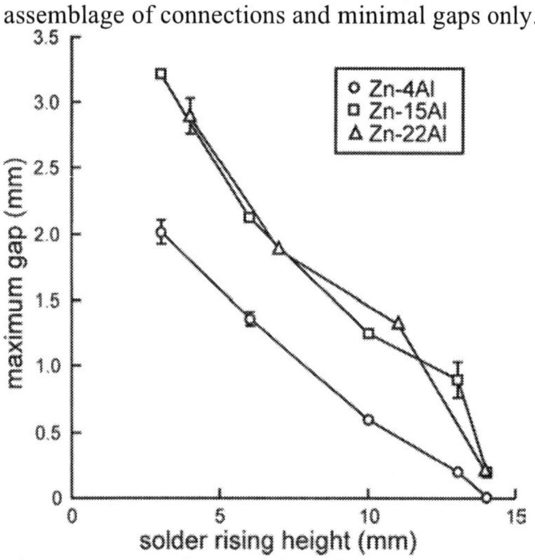

Figure 6. Dependence of the filled gap maximum value on the solder rising height

It was typical, that the shrinkage porosity became noticeable in small gaps along with the aluminum content increasing. Its quantity grew along with the aluminum content growth in the solder. The aluminum content in soldered seam was24, 27 and 37 % for the seams soldered with Zn-4%Al, Zn-15%Al and Zn-22 %Al accordingly, that is associated with the soldering temperature increasing. The minimal value of porosity in small gaps occurred while soldering with Zn-4%Al solder. The difference in structure inside the porosity formation zone should be noted (Fig. 7). Solders with the high aluminum content after dissolution of the base material solidified with columnar dendrite arms, growing from base metal surface, directed towards each other. Obviously, it resulted in difficulty of the liquid feeding while seams solidification and, hence, towards the shrinkage porosity formation.

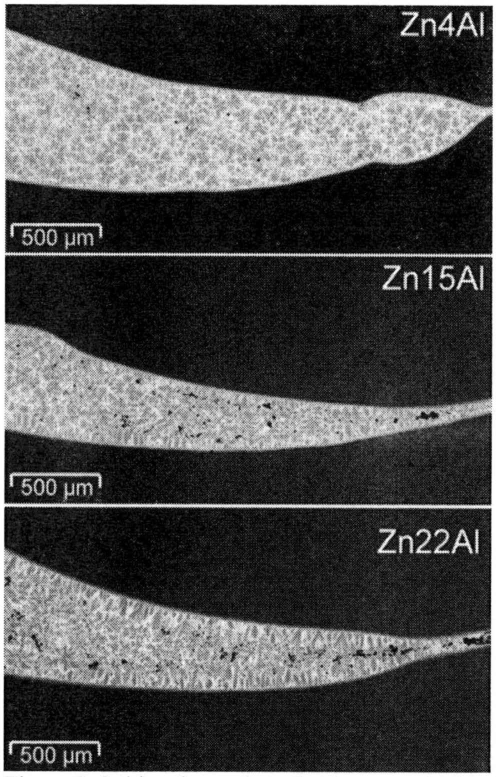

Figure 7. Soldered samples microstructure inside the small gap area

Soldered joints strength

The soldered joints strength is well correlated with the other authors' data [5]. The shearing strength for connections, soldered with Zn-4%Al and Zn-15%Al solders, made 125 and 150 MPa accordingly.

Conclusions

1. While soldering of aluminum with the Zn-Al solders, the strong dissolution of the base metal and the growth of aluminum content in soldered seams is observed. The maximal dissolution is observed while using Zn-4%Al solder. Along with aluminum content increasing in solder, its content grows in soldered seam that is associated with the soldering temperature increasing. The aluminum content in seams may

reach 55 wt. %. This process is intensified at isothermal exposure time prolongation.

2. The intense dissolution of base metal by the solder melt leads to the solder composition changing within a seam, and in accordance with the phase diagram, to the non-equilibrium solidification temperature interval increasing, that results in the shrinkage porosity formation in seams.

3. The ability to fill the increased gaps grows along with the aluminum content increasing in solder. The Zn-base solder with 4 % of aluminum may be used for the tight assemblage of soldered parts with the small gaps only.

4. Solidification of solders in small gaps is accompanied by the formation of porosity, which increases along with the growth of aluminum content in solder alloy. Solidification of Zn-base solders with 15 and 22 % Al is accompanied by the formation of large columnar dendrite arms, directed towards each other, that reduces the possibility of liquid feeding while solidification.

5. While soldering with Zn-Al solders, the sufficient overheating above liquidus temperature and the long time of exposure should be avoided. The increasing of the joint cooling rate after soldering or the gradients presence in the cooling rate of joining parts may promote to the shrinkage porosity decrease.

References

[1] ASM Handbook Volume 6: Welding, Brazing, and Soldering. ASM International, 1993.

[2] J. Pstrus, P. Fima, T. Gancarz. "Wetting of Cu and Al by Sn-Zn and Zn-Al Eutectic Alloys". *Journal of Materials Engineering and Performance, 2012, vol. 21, pp. 606–613.*

[3] M. Movahedi, A.H. Kokabi, H.R. Madaah Hosseini. "An investigation on the soldering of Al 3003/Zn sheets". *Materials characterization, 2009, vol. 60, pp. 441–446.*

[4] G. Zeng, S. McDonald, K. Nogita. "Development of high-temperature solders: Review". *Microelectronics Reliability, 2012, vol. 52, pp. 1306–1322.*

[5] Y. Xiao, H. Ji, M. Li, J. Kim, H. Kim. "Microstructure and joint properties of ultrasonically brazed Al alloy joints using a Zn–Al hypereutectic filler metal". *Materials and Design, 2013, vol. 47, pp. 717–724.*

[6] T. Nagaoka, Y. Morisada, M. Fukusumi, T. Takemoto. "Selection of soldering temperature for ultrasonic-assisted soldering of 5056 aluminum alloy using Zn–Al system solders". *Journal of Materials Processing Technology, 2011, vol. 211, pp. 1534–1539.*

[7] W. Dai, S. Xue, J. Lou, Y. Lou, S. Wang. "Torch brazing 3003 aluminum alloy with Zn-Al filler metal". *Trans. Nonferrous Met. Soc. China, 2012, vol. 22, pp. 30–35.*

[8] M.T. Abou El-khair, A. Daoud, A. Ismail. "Effect of different Al contents on the microstructure, tensile and wear properties of Zn-based alloy". *Materials Letters, 2004, vol. 58, pp. 1754–1760.*

Influence of Production Steps on Aluminium Surfaces and Brazing Results

Hofmann, E., Zschetzsche, J, Füssel, U.
Technische Universität Dresden, Dresden, Germany

Introduction

During processing of parts a material surface can be affected by each step in the production chain. The surface itself consists of various layers, such as contamination, adsorption and reaction layer. Whereas contaminations and adsorbed layers can be removed easily, the reaction layer is harder to treat. One form of reaction layer can be an oxide layer. Aluminium forms natural oxide on its surface as soon as it is in contact with oxygen. The oxide layer impairs successful wetting on the surface.

During processing of brazing parts, the oxide layer can be modified. In order to examinate the influence of the process steps on the oxide layer each step was considered individually. Whereas the oxide layer is characterized by thickness with XPS measurement, the brazeability is characterized with brazing tests in a laboratory shielding and brazing gas furnace.

Aluminium oxide layer and brazing

Brazing of aluminium is carried out in shielding or vacuum furnaces, but also by using flame or induction. The filler metal can be applied with different technologies; in furnace brazing filler metal cladded sheets are state of the art. Main influences on wetting and filling of the brazing fit for aluminium brazing are the oxide layer, grease, oil and contaminations on the surface.

In contact with oxygen containing atmosphere aluminium builds a firm, nm thick oxide layer. When humidity is present in the atmosphere, the oxide layer is bisected: a firm layer consisting of aluminium oxide on the base material, covered with a porous layer of aluminium hydroxide [1].

Raised humidities or condensation lead to growth of aluminium hydroxide. The oxide layer as a whole affects the wetting.

In order to activate the surface for brazing, different methods are used: in shielding gas furnaces flux is applied, whereas in vacuum brazing fracturing of the oxide layer occurs due to different thermal expansion coefficients and then enables wetting and can therefore be carried out without flux.

Influence of transport and storage

It is known that the oxide thickness is critical for the filler metal flow in a shielding gas furnace with flux. For production a correlation between oxide layer thickness and flux dose is assumed: if the oxide layer thickness increases from 4 nm to 22 nm, the flux dose has to increase from 2 g/m^2 to at least 5 g/m^2 to achieve a high-quality joint.

This correlation is also applicable for vacuum brazing: if the oxide layer is thicker than 20 nm, the brazeability is affected and incorrect joints may occur.

Transport and processing can affect the oxide layer regarding chemical composition of the oxide, contamination and surface deposits.

Main aspects of the investigation are influences due to transport and storage on the oxide layer and the corresponding brazing results.

Terms of transport

Filler metal plated aluminium sheets are mostly transported as a coil. This can be done by ship, truck or train. During transport high humidity and condensation can occur. This has an effect on the natural aluminium oxide layer. In order to avoid growth of the layer, the surfaces can be oiled or inhibitor can be added to the packaging.

A typical course of humidity and temperature during transport within Germany is given in Figure 1.

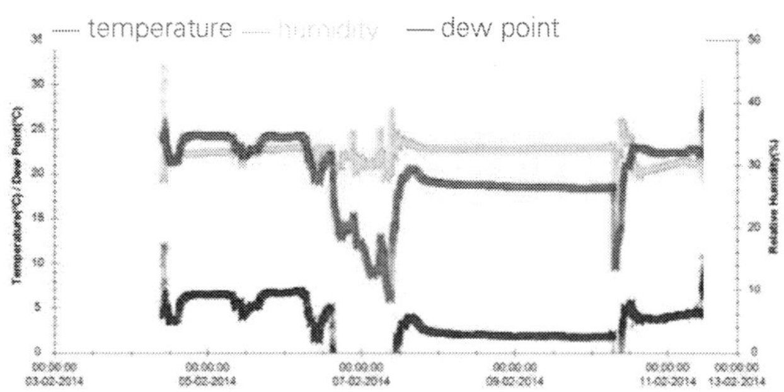

Figure 1: Temperature, humidity and dew point during transport

The mean values of three transports show that maximum was 53,8 %, minimum 22,3 % whereas the maximum temperature was 27,6 °C and the minimum was 7 °C. To obtain information whether condensation occurred, the temperature difference between the atmosphere and the metal is also important. Presuming the relative humidity is 50 %, the temperature difference between atmosphere and metal has to be 10-13 K [2] so that condensation occurs.

Terms of storage

Condensation can also occur during storage of the material. Incorrect storage of coils is especially problematic. If

condensation appears on the surface it can creep between the sheets due to capillary effect.
If the relative humidity is below 100 %, corrosion can be detected as white stripes on the surface of the metal. In contrast, relative humidity of 100 % leads to dark spots and areas on the surface [3]. In these areas, the oxide thickness is raised and wetting is difficult.

Brazing tests

In order to show the influence on the surface of aluminium, furnace brazing has been carried out. Samples were brazed directly after the transport as well after storage under a condensate climate for a definite amount of time.
Brazing was carried out in a shielding gas furnace (Figure 2) and in a vacuum furnace.

Brazing in a shielding gas furnace

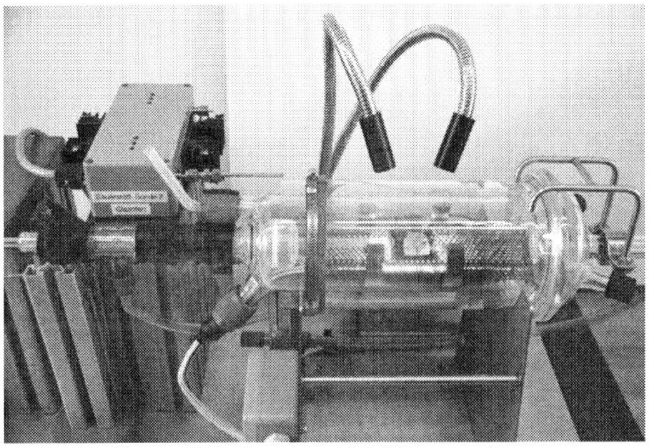

Figure 2: Laboratory shielding gas furnace

As a shielding gas technical nitrogen (class 5.0) was used. The material (base material: EN AW 3003, filler metal: EN AW 4045) was coated with flux (5 g/m^2). The material was shipped directly after annealing; small oil residues might be existent on the surface.

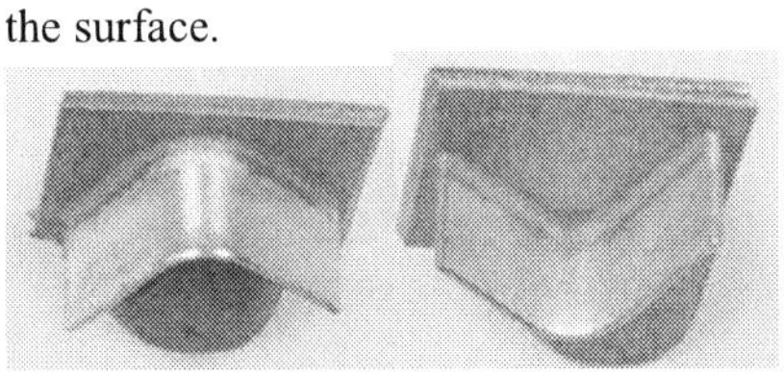

Figure 3: Examle sample (brazed directly after arrival)

The results of the brazing tests carried out directly after the arrival of the material show very good brazeability of the material (fillet filling: 100 %, Figure 3).
After storing the material under condensate climate for 5 and 10 days, brazing tests were carried as well.

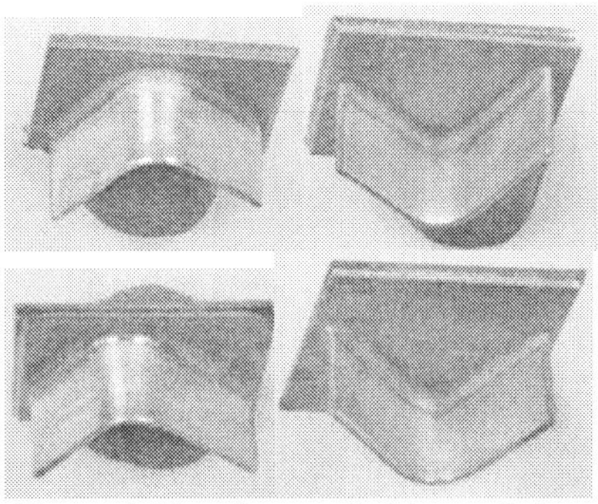

Figure 4: Examle sample (above: 5 day storage, below 10 day storage in condensate climate)

The results (examples see Figure 4) show very good fillet filling (100 %).

Vacuum brazing

Brazing in vacuum furnace was carried out with samples made of EN AW 3005 as base material and EN AW 4104 as filler metal.

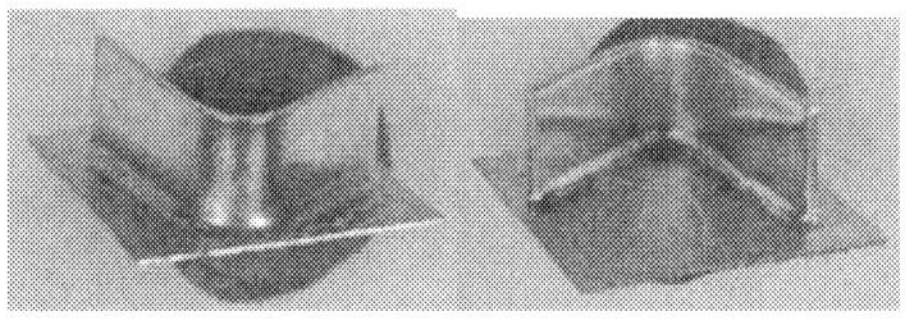

Figure 5: Example sample (brazed directly after arrival)

As shown in Figure 5, the brazing result is very good (fillet filling 100 %). First tests were carried out after storage in condensate climate (Figure 6). They also showed good brazing results.

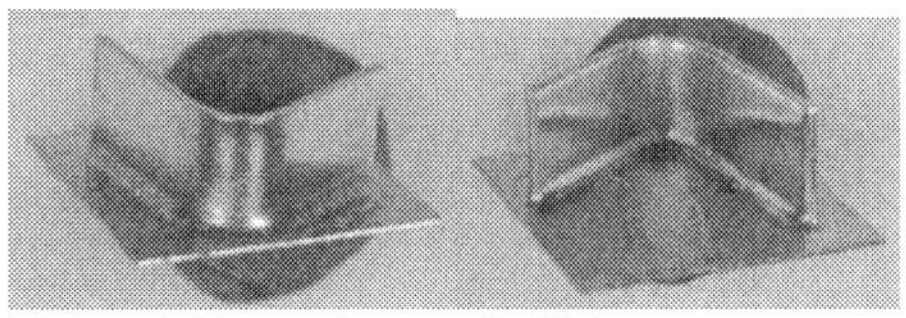

Figure 6: Example sample (5 day storage storage in condensate climate)

Longer storing leads to higher oxide thickness. Current tests show that especially vacuum brazing is sensitive to the thickness. After storing the material for 20 days the brazeability declined.

Discussion and outlook

Both transport and storage have an influence on the surface of aluminium. Especially the presence of water leads to an increase in oxide layer thickness, which complicates wetting during brazing. First brazing tests have been carried out after transport and storage in condensate climate. Both, brazing in a shielding gas furnace with flux and in a vacuum furnace, lead to very good brazing results. In a shielding gas furnace flux

activates the surface of aluminium, so that wetting is enabled. Brazing in a vacuum furnace is more sensitive to the thickness of the aluminium oxide layer.
In further experiments the influence of process oils has been tested both on the actual brazeability as well as its property to growth of the oxide during storing.
Longer storage (thicker oxide layer) may pose more difficulties for brazing with flux as well as vacuum brazing.
Additional investigations using XPS to determine the oxide layer will establish a correlation between processing conditions, oxide layer thickness and brazeability.

Acknowledgments

This project has been funded by the German Federal Ministry for Economic Affairs and Energy (IGF-Vorhaben 17748 BR).

References

[1] ZÄHR, J., Einfluss des Oberflächenzustandes auf das thermische Fügen von Aluminium, (Dresden, 2010)
[2] Aluminium-Verband Schweiz, Fachgruppe Halbzeug, Transport und Lagerung von Aluminium-Halbfabrikaten
[3] GODARD, H. P., "Oxide Film Growth over Five Years on some Aluminium Sheet Alloys in Air of Varying Humidity at Room Temperature", Journal of the Electrochemical Society. 114 (1967) pp. 354-356

Arc Brazing of Aluminum Matrix Composites and Stainless Steel

S. Weis, M. Elßner, B. Wielage and G. Wagner
Technische Universitaet Chemnitz, Institute of Materials Science and Engineering, Chemnitz, Germany
sebastian.weis@mb.tu-chemnitz.de

Abstract

Suitable joining technologies are necessary for the application of aluminum matrix composites (AMC). Especially dissimilar joints like between AMC and stainless steel are of interest. In this work the arc brazing of this material combination is investigated. The results of wetting tests are described and discussed. Wetting angles and microstructure of the interfaces are regarded.

Introduction

Aluminum matrix composites are the subject of current research work in the field of light-weight materials. Due to the combination of the advantages of aluminum alloys and ceramics a high specific strength and increased wear resistance are achieved [1, 2]. This results in numerous applications in aerospace and automotive industry as well as plant manufacturing. For example thermal expansion matched heat transfer components can be mentioned. To realize this application an appropriate joining technology for similar (AMC - AMC) and dissimilar joints (AMC - stainless steel) is necessary.

In the past, the suitability of different joining technologies was investigated. By using the TIG (Tungsten Inert Gas) welding process aluminum steel joints are realizable. However, brittle intermetallics like $Al_{13}Fe_4$, Al_5Fe_2, Al_2Fe and Al_5Fe_4 occur in the joint. To reduce the formation of the intermetallic phases, the steel is coated with zinc [3]. The TIG welding of AMC/ AMC-joints leads to an inhomogeneous welding area. An unequal distribution of the reinforcing particles and the intermetallics occurred. The properties of these joints are insufficient [4]. Thus, the aim is to reduce the thermal induced damage of the AMC developing during the joining process. Otherwise ineligible porosities and inhomogeneities, which reduce the strength, occur [5]. Furthermore, there is a risk of particle degradation at high temperature followed again by the formation of undesirable brittle intermetallics. Acicular aluminum carbides (Al_4C_3) were detected within the joint of SiC particle reinforced AMC [6]. This results in an up to 50% decreased joining strength [5]. Friction stir welding is one possible technology to avoid these negative effects [7]. But the tool is affected by the reinforcement particles where a considerable wear of the tool occurs during the welding process. Hence, the process conditions change continuously, which does not lead to reproducible results [5]. A promising alternative is arc brazing. Due to the lower melting temperature of the filler in comparison to TIG welding it is suitable for joining of AMC and stainless steel. With an appropriate process control particle degradation can be widely prevented. Moreover the appearing temperature gradient due to the selective heating by arc leads to rupture of the oxide layers on the base material and allows wetting. The tendency to form brittle intermetallics can be reduced by an adapted alloying concept of the filler.

Al-Si [8, 9, 10] and Zn-Al [11, 12] based fillers are commonly used for brazing of aluminum alloys. Both filler types have specific disadvantages. The high operating temperature of Al-Si-fillers limits the number of brazable aluminum alloys. That can be explained by a decreasing solidus temperature with increasing content of alloying elements in the aluminum base material. Zinc-based fillers cannot be utilized for vacuum brazing due to the high vapor pressure of Zn. Also the corrosion behavior is very poor.

In this paper, a new filler based on the ternary system Al-Ag-Cu for arc brazing is investigated. The eutectic composition of 40 wt% Al, 40 wt% Ag and 20 wt% Cu (named AlAgCu) has a melting temperature of 506 °C, measured by Differential Scanning Calorimetry (DSC) [13].

Experimental Procedure

The used base material EN AW-2017 + 10 vol% $(SiC)_P$ is a powder metallurgically produced AMC. The feedstock is mechanically alloyed in a high energy ball mill. Hot isostatic pressing and extruding is carried out afterwards. The generated rods with a cross-section of 15 x 15 mm² are divided into 15 mm long sections. The cut surfaces are ground and polished ($R_z < 1$ µm). The used stainless steel sheets (AISI 304L resp. X2CrNi18-10) have a thickness of 1.5 mm and the sample dimensions are 50 x 50 mm². The AlAgCu filler is smelted in a continuous caster using a B-Ag72Cu28 (wt%) master alloy with addition Cu and Al. The casting is done in an Ar atmosphere. The produced rod has a diameter of 14 mm. To proof the proposed composition the rod is cut into sections. The homogeneity of each section is qualitatively determined by X-ray fluorescence analysis (XRF) comparing the content of Cu and Ag. For wetting tests a constant amount of filler (0.1 g) is used. The wetting test consists of two steps. In the first step the arc is ignited next to the applied filler and subsequently moved about it. All tests are done in an inert atmosphere (Ar). A TIG welding source of max 250 A is used. The distance between the electrode and the sample surface is approx. 10 mm. The arc is interrupted after wetting of the filler on the substrate or after 4 s. That should avoid the partly melting of the base material. Two different current intensities

(40 A and 45 A) are tested. The characterization of the wetting samples is performed by light microscopy (LM) and scanning electron microscopy (SEM). The composition of the phases and the diffusion zones are examined by energy dispersive X-ray spectroscopy (EDXS) and X-ray diffraction analysis (XRD). The wetting angles are measured at the cross-sections using a digital imaging tool.

Results and Discussion

In Literature wetting is categorized into various fields: spreading, very good wetting, good wetting, poor wetting and no wetting. The limits are discussed controversially by the experts. A wetting angle of 90° is an important limit. Some authors make a distinction between good wetting (< 90°) and poor wetting (> 90°) [14, 15]. Another definition distinguishes between wetting (< 90°) and no wetting (> 90°) [16, 17]. Almost all experts confirm about the optimal angle for brazing < 30° [14, 17, 18]. In this work the first mentioned definition is preferred. Diffusion between filler and substrate at higher wetting angles can be observed, too. The measured wetting angles are shown in Fig. 1. AMC and stainless steel could be wetted successfully by the filler AlAgCu. The angles on AMC are quite different from those on the stainless steel. While the AMC possesses a very good behavior (< 30°), the stainless steel wets poor (approx. 90 - 100°). The influence of the current intensities seems less significant.

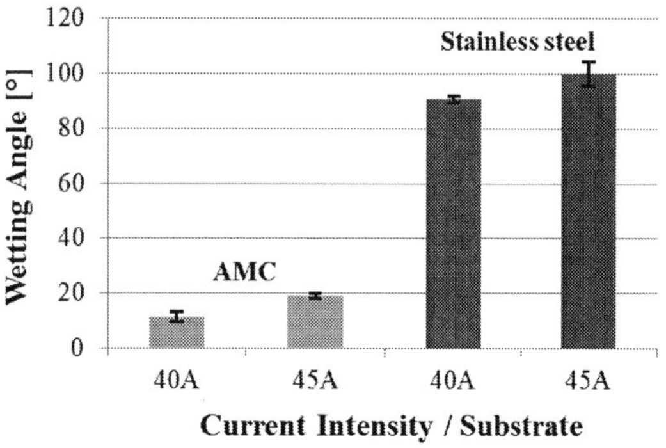

Figure 1: Wetting angles on AMC and stainless steel under variation of the intensity of current

SEM investigations of wetting samples of AlAgCu on AMC show a large diffusion zone, Fig. 2 and Fig. 3. The results of the EDXS measurements are listed in Tab. 1. Similar compositions occur at the measuring points 1 and 4. In point 1 a eutectic texture can be found. XRD analysis in Fig 4 suggests a ternary eutectic consisting of $CuAl_2$, Ag_2Al and Al solid solution. These results confirm with literature data [19, 20].

The measurement results of point 2 show a primary solidified phase with a comparable composition to point 1 and 4. In addition, twinning can be observed in this type of phase. A significant difference in composition is present at point 3. XRD (Fig. 4) shows a stoichiometry of Al_7Ag_3. In the binary phase diagram of Al and Ag no fitting intermetallic exists. Probably it is an Al solid solution with a high Ag content. This assumption is based on the fact that the pattern of the aluminum is almost completely covered by Al_7Ag_3. The diffusion zone (point 4) includes Si. This can be attributed to the AMC. XRD also detects SiC reinforcing particles from the base material.

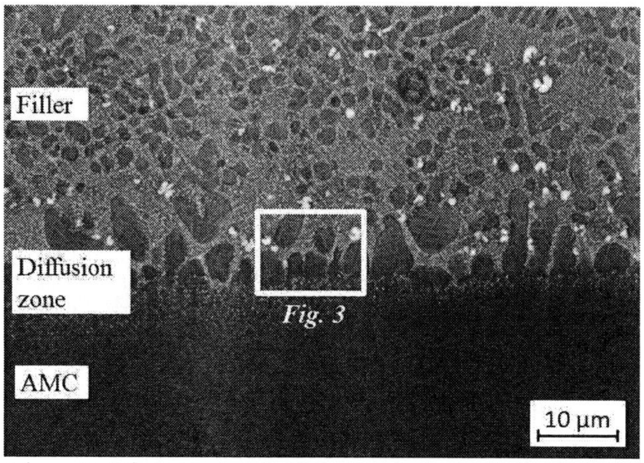

Figure 2: Microstructure (SEM) of AlAgCu wetted on AMC, current intensity 40 A

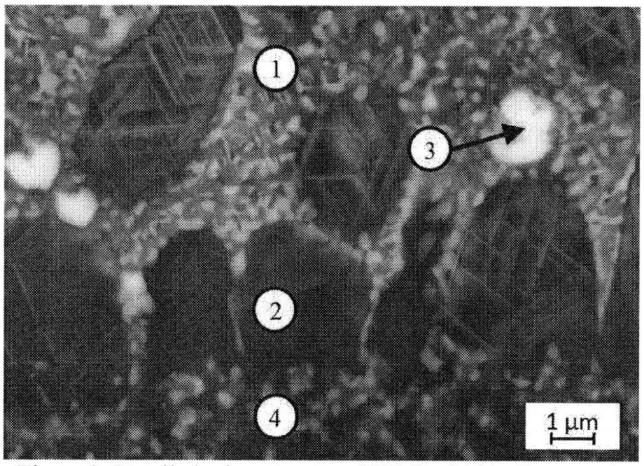

Figure 3: Detailed microstructure (SEM) of AlAgCu wetted on AMC, current intensity 40 A, EDXS Tab. 1

Table 1: EDXS results of measurement points in Fig. 3

	1	2	3	4
Element	at.-%	at.-%	at.-%	at.-%
Al	84	80	62	81
Ag	10	14	30	9
Cu	6	6	8	6
Si				4

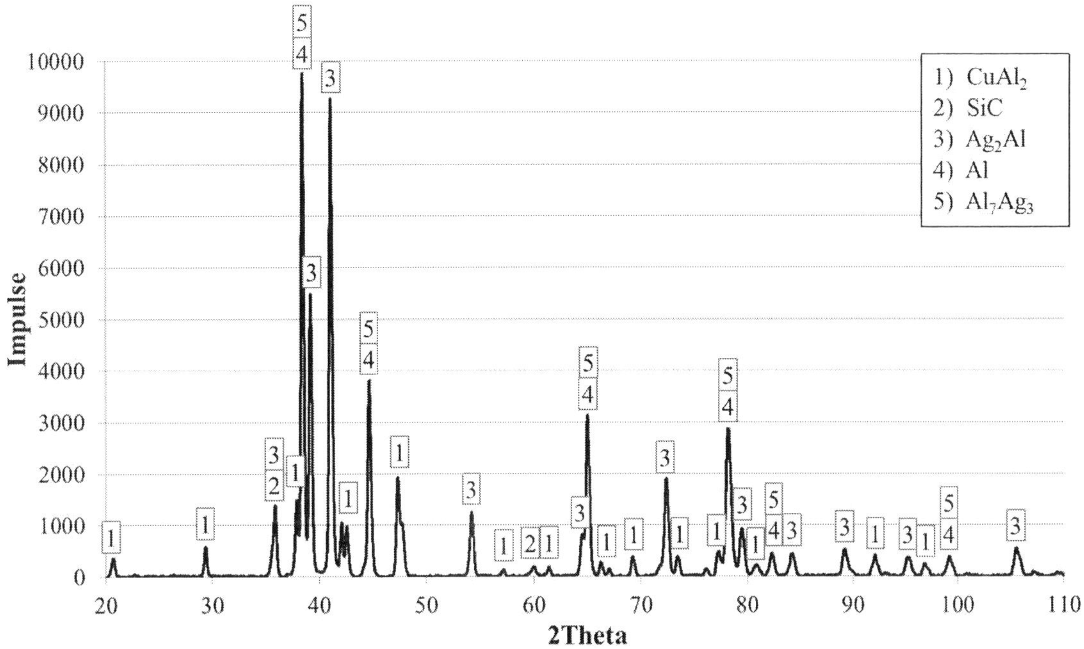

Figure 4: XRD pattern of AlAgCu on AMC

The microstructure of the AlAgCu wetted stainless steel samples is quite different, Fig. 5 and Fig. 6, in comparison to the AMC. A compact intermetallic phase seam next to the steel interface can be observed. A columnar solidified structure arises between the compact seam and the eutectic. Isolated cracks occur in the at least 10 µm thick seam. Investigations of Schmid and Wetter show that a compact phase seam with a thickness of approx. 10 µm and more affects the joining strength negatively [21]. Moreover, it is assumed that a further increasing thickness leads to residual stresses inducing failure on cooling. This can be explained by a difference in the thermal expansion (CTE, filler: $23.4 \cdot 10^{-6} K^{-1}$, stainless steel: $18 \cdot 10^{-6} K^{-1}$, 100 - 400 °C) and the brittleness of the intermetallics [22].

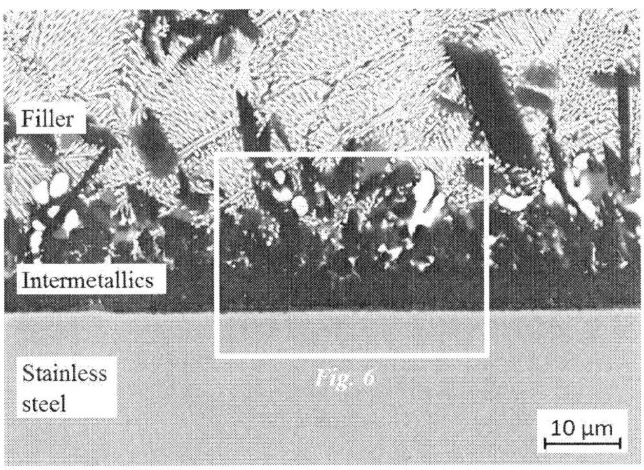

Figure 5: Microstructure (SEM) of AlAgCu wetted on stainless steel, current intensity 40 A

EDXS and XRD analyses are carried out similar to the AMC wetting samples, Tab. 2 and Fig. 7. In point 1 an Al- and Ag-rich phase is present. At point 2, the Al and Fe contents measured by EDXS indicate the intermetallic $Al_{13}Fe_4$ referenced in literature. The compact seam (point 3) probably consists of the intermetallic Al_2Fe containing Cr and Ni from the base material [23].

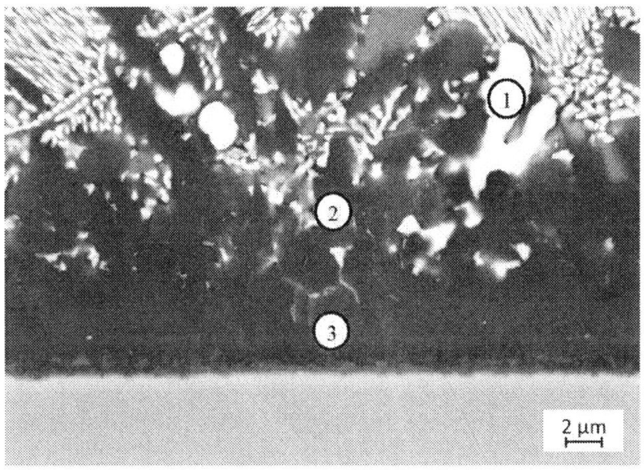

Figure 6:Detailed microstructure (SEM) of AlAgCu wetted on stainless steel, current intensity 40 A, EDXS Tab. 2

The XRD pattern shows the intermetallics $CuAl_2$ and Ag_2Al as well as Al solid solution, Fig. 7, which is also detected on AMC. In addition a solid solution of Fe, Ni and Cu with austenitic lattice is determined. The detected $AlFe_3$ is the solid solution of Al and Fe in a specific chemical composition. In the XRD pattern the quantity of this phase is low.

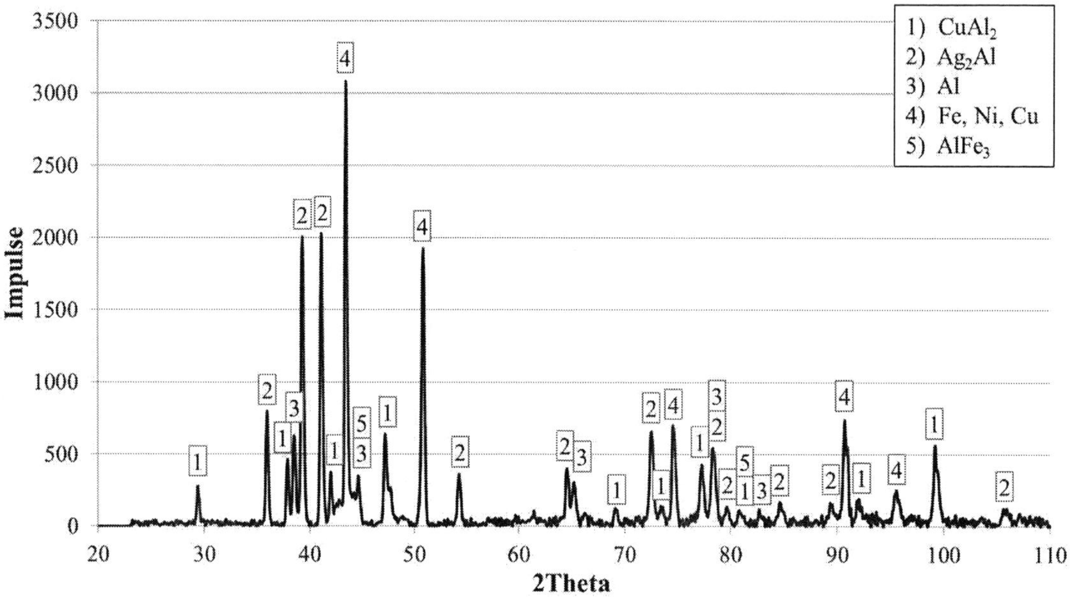

Figure 7: XRD pattern of AlAgCu on stainless steel

Table 2: EDXS results of measurement points in Fig. 6

	1	2	3
Element	at.-%	at.-%	at.-%
Al	52	75	52
Ag	40	1	
Cu	5	3	2
Fe	2	17	33
Cr	1	3	9
Ni		1	4

Conclusions

The results of the wetting experiments in this work show, that arc brazing using a eutectic AlAgCu filler is suitable for joining AMC to AMC or AMC to stainless steel. The wetting tests demonstrate an excellent wettability of the AMC. For stainless steel an optimization is necessary to improve the behavior. This can be achieved by a pretreatment of the stainless steel surface. Different current intensities do not influence the wetting angles significantly. SEM investigations in combination with EDXS and subsequent XRD analyses show an intermetallic seam at the interface of filler to AMC as well as stainless steel. Depending on the thickness of this seam, cracks can occur. According to literature data a thickness up to 10 µm is tolerable regarding to the joining strength. The filler solidifies eutectic, consisting of $CuAl_2$, Ag_2Al and an Al solid solution. Based on the results of the wetting tests AMC – stainless steel joints can be realized in further experiments.

Acknowledgements

This work is supported by the DFG, German Research Foundation as part of the collaborative research center SFB 692 „Hochfeste aluminiumbasierte Leichtbauwerkstoffe für Sicherheitsbauteile (HALS)“.

References

[1] Xiu, Z., Yang, W., Chen, G., Jiang, L., Mac, K., Wu, G., “Microstructure and tensile properties of Si_3N_4p/2024Al composite fabricated by pressure infiltration method,” Mat. a. Designs, 33 (2012), pp. 350-358.

[2] Sajjadi, S. A., Ezatpour, H. R., Parizi, M. T., “Comparison of microstructure and mechanical properties of A356 aluminium alloy/Al2O3 composites fabricated by stir and compo-casting processes,” Mat. a. Designs, 34 (2012), pp. 106-111.

[3] Steiners, M., Höcker, F., “Einfluss der Beschichtungen beim stoffschlüssigen Lichtbogenfügen von Stahl mit Aluminium,” Mat. u. Werkst., 38 (2007), pp. 559-564.

[4] Ureña, A., Escalera, M. D., Gil, L., “Influence of interface reactions on fracture mechanisms in TIG arc-welded aluminium matrix composites,” Comp. Scien. and Tech., 60 (2000), pp. 613-622.

[5] Prater, T., “Solid-state joining of metal matrix composites: a survey of challenges and potential solutions,” Mat. a. Man. Proc., 26 (2011), pp. 636-648.

[6] Lean, P. P., Gil, L., Ureña, A., “Dissimilar welds between unreinforced AA6082 and AA6092/SiC/25p composite by pulsed-MIG arc welding using reinforced filler alloys (Al-5Mg and Al-Si),” J. of Mat. Proc. Tech., 143-144 (2003), pp. 846-850.

[7] Wan, L., Huang, Y., Lv, Z., Lv, S., Feng, J., "Effect of self-support friction stir welding on microstructure and microhardness of 6082-T6 aluminum alloy joint," Mat. a. Design, 55 (2014), pp. 197-203.

[8] Qin, G., Lei, Z., Su, Y., Fu, B., Meng, X., Lin, S., "Large spot laser assisted GMA brazing-fusion welding of aluminium alloy to galvanized steel," J. of Mat. Proc. Tech., 214 (2014), pp. 2684-2692.

[9] Song, J. L., Lin, S. B., Yang, C. L., Ma, G. C., Liu, H., "Spreading behavior and microstructure characteristics of dissimilar metals TIG welding-brazing of aluminum alloy to stainless steel," Mat. Scien. a. Eng.: A, 509 (2009), pp. 31-40.

[10] Wielage, B., Klose, H., "Das Aluminiumlöten von Wärmetauschern," DVS-Berichte, 166 (1995), pp. 88-90.

[11] Wielage, B., Martinez, L., "Aluminiumlöten bei 550°C – Eigenschaften von ZnAl-Verbindungen," DVS-Berichte, 212 (2001), pp. 214-217.

[12] Wielage, B., Trommer, F., "Löten von Aluminium mit Zinkbasisloten," Schweiß. u. Schneid., 5 (2003), pp. 273-275.

[13] Kumar, K. C. H., Arkens, O., "Silver – Aluminium – Copper," Springer Materials, – The Landolt-Börnstein Database (2014).

[14] Dorn, L. et al, Hartlöten und Hochtemperaturlöten, expert Verl. (Renningen, 2007), pp. 14.

[15] Yuan, Y., Lee, T.R., Contact Angle and Wetting Properties, Springer-Verlag (Berlin Heidelberg, 2013), pp. 3.

[16] Zaremba, P., Hart- und Hochtemperaturlöten, DVS-Verlag (Düsseldorf, 1988), pp. 12.

[17] N.N., Brazing Handbook, American Welding Society (Miami, 1991), pp. 3.

[18] Müller, W., Müller, J.-U., Löttechnik – Leitfaden für die Praxis, Dt. Verlag für Schweißtechnik (Düsseldorf, 1995), pp. 13.

[19] Liu, S., Zhao S., Zhang Q., "Phase Diagram of the Aluminium-Copper-SilverAlloy System," Acta Metall. Sin., 19 (1983), pp. 70-73.

[20] Witusiewicz, V.T., Hecht, U., Fries, S.G., Rex S., "The Ag–Al–Cu system: II. A thermodynamic evaluation of the ternary system," J. o. Alloys a. Comp., 387 (2005), pp. 217-227.

[21] Schmid, G., Wetter, H., "Stanzniet- und Hybridfügeverfahren im Karosseriebau," Join-Tec, (2005), pp. 42–57.

[22] Data sheet 1.4301, URL:http://www.ll-adelsdorf.com/pdfdoku/Edelstahl/1.4301.pdf; 13.06.2014.

[23] Springer, H., Kostka, A., Dos Santos, J.F., Raabe, D., "Influence of intermetallic phases and Kirkendall-porosity on the mechanical properties of joints between steel and aluminium alloys," Mat. Scien. a. Eng.: A, 528 (2011), pp. 4630-4642.

Low Temperature Controlled Atmosphere Brazing of Aluminum Matrix Composite

Hui Zhao[1], Stefan Elbel[1, 2], Pega Hrnjak[1,2]
[1]Creative Thermal Solutions, Inc., Urbana, IL 61802, USA
[2] University of Illinois at Urbana-Champaign, Urbana, IL 61801, USA
Email: hui.zhao@creativethermalsolutions.com

Abstract

Aluminum matrix composite (AMC) uses aluminum alloy as metal matrix and ceramic particles or fibers as reinforcement. AMC materials feature many attractive properties such as excellent mechanical properties, high thermal and electrical conductivities and low coefficient of thermal expansion (CTE). However, joining of AMC materials using traditional aluminum brazing techniques is usually difficult: (1) filler metal generally shows poor wettability on AMC materials due to the exposure of ceramic phase on substrate surface; (2) high brazing temperature leads to degradation of the composite materials properties. Therefore, development of reliable joining technologies for aluminum matrix composite is essential for the practical engineering application of the materials. In this paper, the feasibility of using low melting temperature filler metal such as Zn-Al alloys to braze AMC materials under inert gas atmosphere are explored. Wetting behavior of liquid filler metal on AMC substrates is experimentally studied using brazing visualization systems. Particle (Al_2O_3) reinforced aluminum matrix composites are tested as AMC substrate materials. The brazed joint microstructure and integrity are metallurgically examined. This study shows promising results of controlled atmosphere brazing of AMC materials at a low temperature level.

Introduction

Metal matrix composites (MMCs) are advanced materials that utilize ceramic fibers or particles mixed in metals to enhance mechanical properties and maintain desired metallic characteristics such as high conductivity and good machinability [1,2]. Aluminum matrix composite (AMC) is one of the most widely used MMC materials in commercial applications. Alumina (Al_2O_3) and silicon carbide (SiC) are often used as reinforcement phases for AMCs [3]. The reinforcement ceramics can be in the form of short fiber, whisker or particle. Particle reinforced aluminum matrix composites are less expensive, therefore suitable for commercial applications. The material properties can be tailored to meet specific application requirement. For example, excellent thermal and electrical properties of SiC particle reinforced AMCs make them very promising candidates for electronic packaging and thermal management applications [4].

Among various joining methods, brazing is one of the proper choices for AMCs bonding. However, the addition of ceramic reinforcement in aluminum makes the joining of AMCs materials difficult because of the general poor wettability of liquid filler metal on ceramic phases. In addition, high temperature heating in a brazing process can lead to rapid deterioration of material properties due to undesired reactions between the composite phases [5]. Figure 1 illustrates one example of aluminum matrix composite – A359 alloy composite that contains 30vol% SiC particle reinforcement (manufactured by MC-21, Inc., NV).

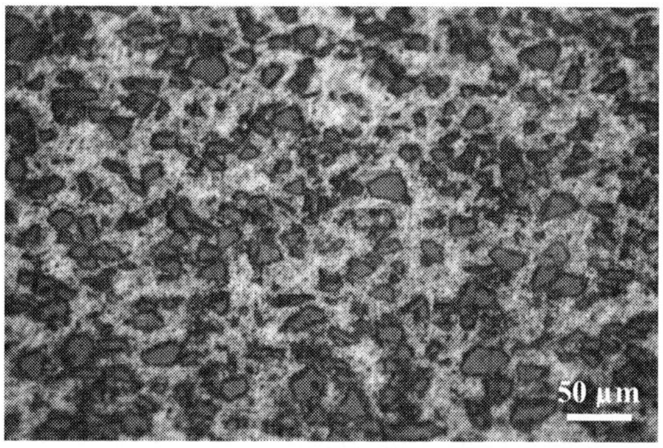

Figure 1 Aluminum matrix composite A359/30%SiC_p

In theory, conventional aluminum brazing processes should be applicable to join aluminum matrix materials. However, the commonly used Al-Si series filler metals for aluminum brazing have relatively high melting temperatures. The required brazing temperature is in the range of 580-615 °C, which is very close to the melting temperatures of the aluminum matrix alloys. Brazing AMCs at such a high temperature level can lead to significant reduction of material strength. Therefore, it is essential to look for reliable brazing processes that take place at relatively low temperature level. Brazing at lower temperature can also provide other advantages such as energy saving on brazing equipment and higher efficiency for mass production.

Recent studies have shown that Zn-Al alloy is a good candidate of low temperature brazing of aluminum and aluminum matrix composite. Studies using Zn-Al filler metal to join aluminum matrix materials with assistance of ultrasound technology have been reported, where the joining temperature is in the range of 400-530 °C [6-8]. It has been demonstrated that low volume percentage (<13%) SiC particle reinforced AA2014 matrix composite can be brazed successfully using Zn-3Al filler metal at a temperature level around 450 °C [6]. An extensive spreading of the Al-Zn filler metal on a SiC particle reinforced AA6061 matrix composite has also been reported [7]. Brazing temperatures of these applications are much lower than the traditional aluminum brazing temperatures when Al-Si alloys are used as filler metals. However, it has been reported that Zn rich filler metals have the tendency to penetrate deep into the Al matrix grain boundary, and Al matrix dissolution by molten filler metal can occur due to the large solubility of Al in Zn [9].

In this study, wettability of Zn-15Al alloy on an Al_2O_3 particle reinforced aluminum composite under controlled atmosphere brazing (CAB) conditions is studied. An application of using such low temperature process in joining AMC plate and aluminum fin is also presented. The wetting and brazing processes are visualized using transparent heating chambers. The brazed joint integrity is evaluated by metallographic analysis.

AMC and filler metal materials

Coupon plates of the particle reinforced aluminum matrix composites are used as substrate for wetting and brazing tests. The materials compositions are AA6092 matrix alloy reinforced with 15vol% of spheroidized Al_2O_3 particles (manufactured by Gamma Technology LLC, CA). Coupons with a thickness around 0.8 mm were cut from an extruded AMC rod. Filler metals used in this study is Zn-15wt%Al alloy which has a relatively low melting temperature range (381 - 451 °C).

The 6xxx matrix aluminum alloys contains Mg as major alloying element, which provide the desired mechanical strength of the material. However, the addition of Mg also adds difficulties in brazing under atmosphere conditions due to the formation of complicated oxidation compound on material surface during heating. A proper selection of flux is essential to the success of brazing in atmosphere conditions. Recent development in non-corrosive cesium (Cs) bearing flux has shown promising results in assisting brazing of Mg-bearing aluminum alloys [10]. In this study, a cesium fluoride flux (manufactured by Solvay Fluor GmbH, Germany) with a low melting temperature range is used for CAB brazing tests of AMCs.

Experiment set up and testing facilities

Sessile drop wetting tests were performed using both a transparent furnace and a high temperature heating stage system. The transparent chambers allow direct observations of the furnace brazing processes. All brazing tests were performed under tightly controlled nitrogen atmosphere.

In addition to the wetting tests, brazing experiments were performed to monitor the joint formation process between an aluminum fin and the AMC substrate. A schematics of the brazing sample set up is illustrated in Figure 2. A video camera is used to capture a series of phenomena including filler metal melting, meniscus formation, and filler metal solidification. A thin piece of Zn-15Al filler foil was sandwiched between the aluminum fin and the AMC coupon plate. Non-corrosive Cs bearing flux is used for the purpose of removing oxide film on metal surfaces during brazing.

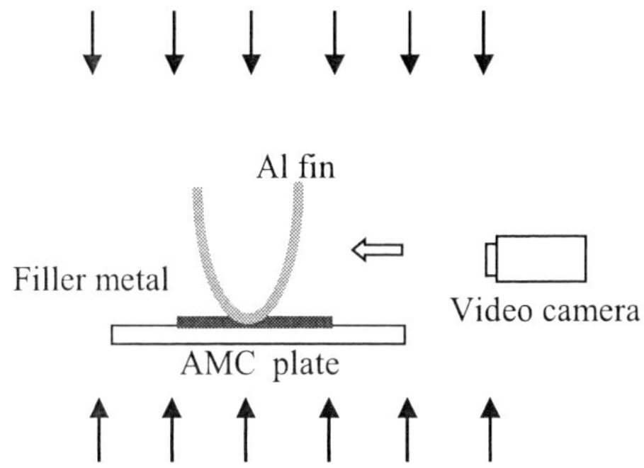

Figure 2 Plate/fin brazing set up

Wettability of Zn-15Al filler metal on AA6092/15%Al_2O_{3p} composite

A previous study [11] has shown good wettability of Al-12Si filler metal on the AA6092/15%Al_2O_{3p} coupon plate. However, the relative high brazing temperature may limit its application in AMC joining process. In this study, wettability of the low melting temperature filler metal Zn-15Al on the AMC material is first performed using a heating stage microscopy system. A small piece of the Zn-15Al alloy disc (diameter ~ 1mm, thickness ~ 0.35 mm) was place on an AMC coupon and heated by a mini-heater inside the heating stage chamber. The control temperature was measured by a thermocouple attached to the sample holder. Continuous flow of pure N_2 gas was supplied to mimic the CAB brazing condition. A video camera attached on a Mitutoyo FM microscope was used to capture the molten filler metal wetting and spreading behaviors.

Figure 3 presents a few extracted images of the heating stage wettability test at different stages of the filler metal melting process. In Figure 3(a), both the solid filler metal disc and the AMC substrate are covered with flux powders. As the sample is heated above filler metal melting temperature, the filler metal start to melt and dewet slightly on the substrate surface, as illustrated in Figure 3(b). The incomplete melting of the flux on substrate surface may have caused such an initial dewetting phenomenon. As the heating continues, flux on the substrate melts completely and liquid filler metal spreads extensively on the AMC surface. The peak temperature (measured on sample holder, which is generally higher than the actual sample temperature due to thermal gradient) in this test is 470 °C. Figure 3(c) illustrates the filler metal upon solidification. The final wetting area on the substrate is approximately 2 times of the original filler disc area.

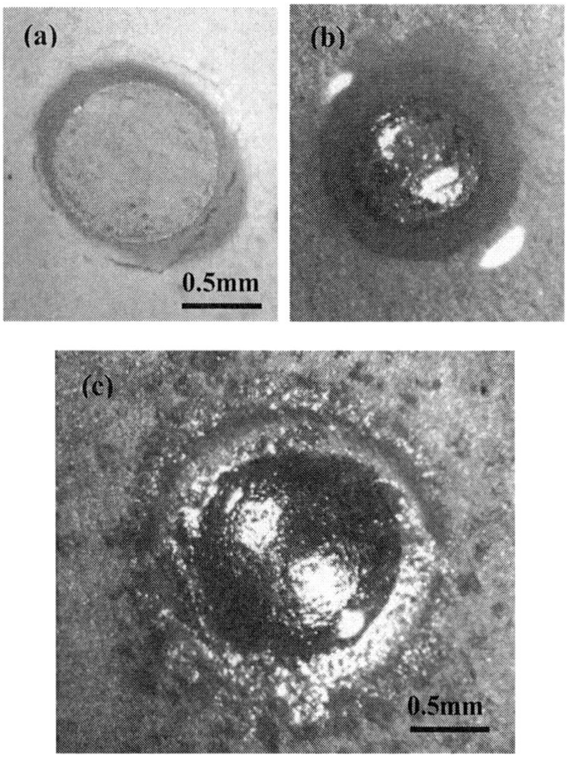

Figure 3 Wettability test using heating stage microscope: (a) before filler metal melting; (b) filler metal melting; (c) spreading of filler metal.

The heating stage microscopy system provides an opportunity for detailed observation of the transport phenomena involved in the AMC brazing process. The videos captured during the wetting test can also be used to accurately measure wetting dynamics of the liquid filler metal, i.e., wetting area vs. time relation. Such information are important for understanding flowability of the filler metal during brazing/solder processes. An in-depth study on wetting kinetics of Zn-Al filler metals on AMC surfaces will be presented in a future publication.

To investigate the feasibility of using Zn-Al alloy in brazing AMC materials by the CAB process, a series of brazing experiments were performed using a transparent CAB furnace. A piece of Zn-15%Al rod (ϕ 2mm, height around 3mm) was placed on a 6092/15%Al_2O_3 coupon plate. The sample was heated inside the glass furnace by radiation with a ramp rate around 25 °C/min. The control temperature was measured by a thermocouple attached to the AMC coupon plate. Nitrogen gas was purged throughout the process to maintain the dew point below -40 °C. Cs bearing flux was applied on both filler and base materials. The peak control temperature is approximately 450 °C. Figure 4 illustrates a few extracted images from the captured video. It is clear that the molten filler metal shows good wettability on the given alumina particle reinforced composite. The flux effectively promotes wetting of the liquid filler metal on the AMC substrate. This result confirms the observation from the heating stage wetting test which is performed in a much smaller sample scale.

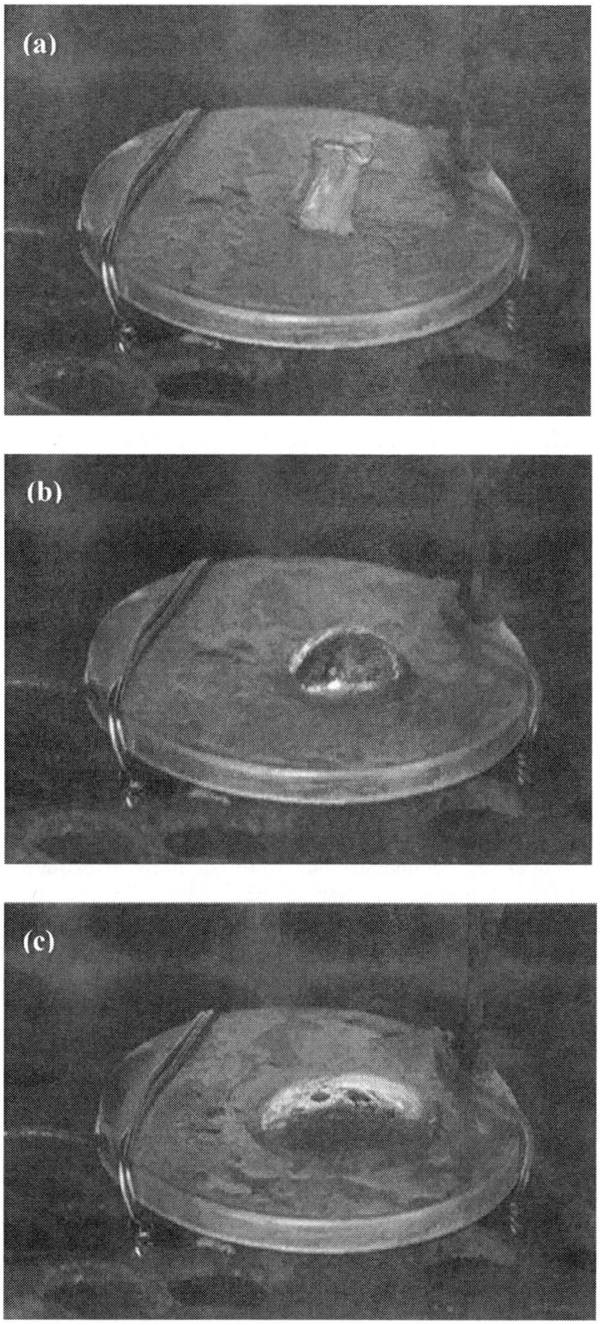

Figure 4 Wettablity test in a transparent CAB furnace: (a) before filler metal melting; (b) onset of filler melting; (c) extensive spreading of filler metal on the AMC surface.

From the wetting tests, it is found that the spreading of Zn-15Al filler on the 6092/15%Al_2O_{3p} surface is much more extensive than on the A359/30%SiCp substrate which has been presented in a previous study [10]. In general, both matrix alloys used in these studies, i.e., the AA6092 and A359 alloys are difficult to braze in CAB conditions because they contain Mg as major alloying element. It is known that Mg has a "poisoning" effect on the flux which reduce the flux capability in removing oxide film. However, the low melting temperature cesium fluoride flux assists good wetting of Zn-Al filler metal on the AMC substrates in both cases. A better wetting of Zn-Al filler on the AA6092/15%Al_2O_3 composite than on the A359/30%SiC composite may be caused by following factors: (1) smaller percentage of the ceramic reinforcement in the AA6092/Al_2O_3 composite; and (2) matrix alloy compositions and characteristics are different.

Plate/fin joint formation

The wetting tests of Zn-15Al alloy on AA6092/15%Al_2O_3 composite surface show promising results in low temperature brazing of AMC materials. Therefore, additional joint formation experiments were preformed to mimic practical applications. One of the commercial applications of particle reinforce AMC materials is for thermal management devices. Therefore, a plate/fin joint assembly is prepared for brazing tests. The experimental set up has been illustrated in Figure 2. The brazing experiments of joining an AMC plate and AA3003 fins are performed in the transparent CAB furnace. A control thermocouple is inserted between the vertical fins. One of the samples is illustrated in Figure. 5. In Figure 5(a), the sample is heated and filler metal is still in the solid state. In Figure 5 (b) filler metal starts to melt and wets both the fin and the AMC surface, eventually forms a meniscus shaped joint at peak temperature approximately 460 °C.

To exam the brazed joint integrity, one of the brazed fin/plate samples was cut and metallurgically polished to reveal the joint microstructure. Microscope images of the resolidified joint are presented in Figure 6.

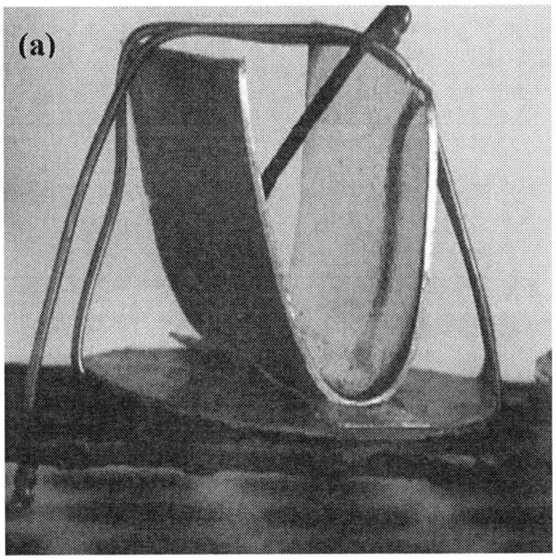

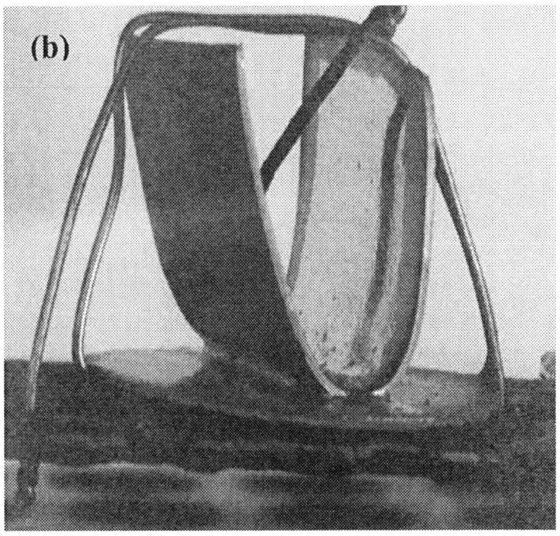

Figure 5 Plate/fin joint formation test in a transparent CAB furnace: (a) before filler metal melting; (b) liquid filler metal forms joint between the fin and the plate.

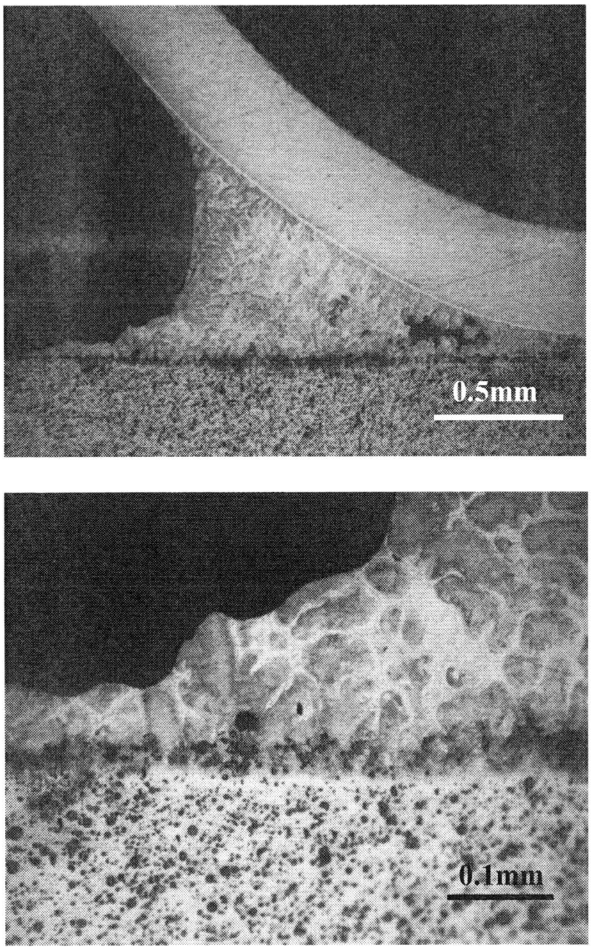

Figure 6 Microstructure of brazed joint: (a) left side of the fin/plate joint; (b) interface between the filler metal and the AMC plate.

From Figure 6(a), overall good joint integrity between the AA3003 fin and the AMC plate is observed. However, some porosity inside the joint can be identified at random locations. The existence of voids in brazed joint may be related to Zn loss and flux residue as consequences of the brazing operation. It is known the Zn has high vapor pressure and tends to evaporate at brazing temperature. The loss of Zn at high temperature during brazing can cause large joint porosity and unsatisfactory joint integrity. A good control of brazing temperature is important to ensure the success of brazing with Zn-Al filler metals. The optimization of flux load in CAB furnace brazing needs to further investigated not only due to the fact that Cs-bearing flux is relatively expensive, but also because excessive flux residue can be trapped in brazed joints and reduce the joint quality.

In Figure 6(b), closed view of the interface between Zn-Al filler metal and the AMC surface shows good metallurigcial bond between the two materials. Small portion of the ceramic particles seem to transport into the filler metal during liquid/solid interaction. However, no significant substrate dissolution was observed at the interface. Serious substrate dissolution/erosion can cause problems in brazing application, such as reduced mechanical strength and device leakage.

Conclusions

The development of aluminum matrix composites for commercial applications requires reliable joining method at relative low temperature to protect the AMC materials from degradation. In this study, a series of wetting and joint formation tests of Zn-15Al filler metal on $AA6092/Al_2O_3$ particle reinforced composite were performed under controlled atmosphere brazing conditions. Good wettability of the filler metal on the AMC surface at below 500 ^oC brazing temperature is confirmed. An $AA6092/Al_2O_{3p}$ coupon plate and AA3003 fin were successfully brazed using the transparent CAB furnace. However, small percentage of voids is identified in the brazed joints. Future study will focus on optimization of the brazing conditions including brazing temperature, flux load and etc. This preliminary study shows promising results on using Zn-15Al filler metal in furnace brazing of the $AA6092/Al_2O_3$ composite materials. Such an application provide at least two advantages in joining AMC materials: (1) the low brazing temperature not only reduces energy consumption, but also significantly reduces risk of composite material degradation; (2) CAB brazing provides an economic option for mass production of the AMC products.

References

[1] Clyne, T.W. and Withers P. J., An Introduction to Metal Matrix Composites, Cambridge University Press (Cambridge, 1993)

[2] Chawla, N. and Chawla, K.K., Metal Matrix Composites, Springer (2005).

[3] Rajan, T.P.D, Pillai, R.M., Pai, B.C., "Review, Reinforcement Coatings and Interfaces in Aluminum Metal Matrix Composites," Journal of Material Science, Vol. 33 (1998), pp 3491-3503.

[4] M.A.Occhionero, R.W. Adams, K.P. Fennessy, "A New Substrate for Electronic Packaging: Aluminum Silicon Carbide (AlSiC) Composites", Proceedings of the Forth Annual Portable by Design Conference, Electronics Design, March 24-27 (1997), pp 398-403.

[5] Robertson A.R., Miller M.F., Maikish C.R., "Soldering and Brazing of Advanced Metal-matrix Structures," Welding Journal, Vol.54(10) (1973), pp 446s-453s.

[6] Urena, A, Gil, L., Escriche, E., Gomez de Salazar, J.M., Escalera, M.D., "High Temperature Soldering of SiC Particulate Aluminum Matrix Composite (series 2000) using Zn-Al filler alloys". Science and Technology of Welding and Joining, Vol. 6 (2001), pp 1-11.

[7] Z.W. Xu, J.C. Yan, C. Wang, and S.Q. Yang, "Substrate Oxide Undermining by a Zn-Al Alloy during Wetting of Alumina Reinforced 6061 Al Matrix Composite," Materials Chemistry and Physics, Vol.112 (2008), pp 831-837.

[8] Zhang Y., Yan, J.C., Wu, Q., "Ultrasonic Brazing of High Fraction Volume of SiC Particulate Reinforced Aluminium Matrix Composites," Materials Science and Technology, Vol. 25(3) (2009), pp. 379-382.

[9] Zhang Q.Y. and Zhuang H.S. editors, Brazing and Soldering Manual, China Machine Press, (Beijing, China, 2006), in Chinese.

[10] Garcia José, C. Massoulier and P.Faille, "Brazeability of Aluminum Alloys containing Magnesium by CAB Process using Cesium Flux", SAE paper #2001-01-1763 (2001).

[11] Zhao H., Elbel S. and Hrnjak P., "Wetting Behavior of Al Based Filler Metals on Aluminum Matrix Composites," 2014 Interantional Conference on Brazing, Soldering and Special joining technologies, June 9-13, (Beijing, China, 2014).

Influence of Substrate Properties on Fatigue Performance of Brazed Steel Joints

Christian Leinenbach, Michael Koster, Christoph Kenel, Wookjin Lee
Empa-Swiss Federal Laboratories for Materials Science and technology, Duebendorf, Switzerland
christian.leinenbach@empa.ch

Abstract

In this work, the fatigue behavior of brazed steel joints with different elastic-plastic properties and geometries was investigated by experiments and numerical analysis. The base material was the steel AISI 316 C-NM with AuNi18 as filler metal. Two types of brazed samples, round shaped specimens and T joints, were prepared and subsequently differently heat treated two to study the effect of different elastic-plastic properties of the base material. Moreover, defined defect geometries were introduced to some samples in the braze layer by electrical discharge machining to investigate the influence of defects.

The fatigue behavior of the samples was studied at R = 0.1. The cyclic deformation behavior and the fatigue lifetimes were investigated by Digital Image Correlations. Detailed microstructural analyses and Finite Element calculations were also carried out. The results of this work underline the significant influence of the elastic-plastic properties of base material on the fatigue behavior and can be used for the appropriate design of brazed components.

Introduction

Brazing has gained increasing importance as a favorable joining technology for many seminal applications in industry in the last years. By definition, brazing is performed by heating an assembly over the melting point of the filler metal placed between two substrates without reaching the melting point of the substrate material. The liquid filler metal wets the surfaces of the substrate material and fills the joint gap. Subsequent adhesion and diffusion processes during the cooling of the assembly are essential for the final joint strength. The general differentiation between soldering and brazing is made according to the process temperatures used for the joining process. The above described process at T < 450 °C is referred to as soldering, whereas it is named brazing at T > 450 °C. In several applications, brazing plays an essential role because the thermal stress of the joining partners and the processing times can be reduced, compared to e.g. welding. Furthermore, brazing allows joining of dissimilar materials as e.g. metals and ceramics at narrow tolerances. With the use of advanced brazing technologies, brazed joints can also be used for thermally and mechanically heavily loaded components in chemical engineering, power generation and for the production of power electronic components [1-3].

From a microstructural point of view, brazed joints form heterogeneous systems, consisting of base material, filler metal and diffusion zone. Under mechanical loading, the properties of brazed joints vary significantly from those of the individual joining partners. Therefore, the complex deformation behavior of the brazed joint is characterized by geometrical and microstructural interactions as e.g. by different elastic-plastic properties of substrate material and thin braze layer. Uniaxial loading and the constrained deformation of the thin filler alloy layer can lead to a triaxial stress state which strongly influences the joint performance [4-6]. In the scope of a previous work [7], a procedure which allows estimating the influence of defects on the strength and failure of brazed joints was developed. In addition, a rather unusual fatigue crack growth behavior of brazed steel joints, i.e. extremely steep da/dN-ΔK curves in comparison with the ones of bulk materials, was observed [4]. It could be shown that under defined boundary conditions, the influence of defects on the fatigue lifetime can be estimated based on the stress intensity factor caused by the defect [8].

The main objective of this work is to study how the brazing defect influences the fatigue performance of the joint, in terms of elastic-plastic behavior of the base material. For this, fatigue experiments were performed with specimens of AISI 316 C-NM martensitic steel brazed with AuNi18 as a filler metal. The effects of specimen geometries, heat treatment and defect on the fatigue behavior were investigated.

Generally, fatigue of metallic materials is closely related to the formation of (local) irreversible strains in the bulk. In homogenous materials, the strains are relatively equally distributed over a macroscopic length scale and can therefore be averaged using integrating measurement techniques, e.g. with extensometers. However, plastic strains in brazed joints are usually localized in a small volume in or next to the brazing zone. The thickness of braze layers is usually in the range of 50-100 μm, so conventional strain measuring techniques cannot be applied reasonably. To measure local strains during cyclic loading with a sufficiently high resolution, Digital Image Correlation (DIC) measurements were performed. The technique was applied to monitor the fatigue induced damage evolution in brazed steel joints during cyclic loading, in particular under the influence of brazing defects. Furthermore, the local stress fields and plastic deformations during the fatigue tests were analyzed by numerical FE models to obtain information on the underlying failure mechanisms.

Testing materials and experimental setup

The fatigue tests in this work were performed with brazed joints of the stainless steel AISI CA 6-NM (X3CrNiMo 13-4). As shown in Table 1, the main alloying elements are Chromium, Nickel and Molybdenum. This material provides a high strength at a relatively large fracture strain and its mechanical properties can be varied significantly with different heat treatments. As a filler metal, foils of AuNi18 with a thickness of 100 µm were used. The filler metal exhibits a good corrosion resistance, wettability and a melting point of 955 °C.

Table 1: Chemical composition of the base material

Element	Cr	Ni	Mo	Mn	Cu	Co	V	Fe
wt. %	12.4	3.5	0.5	0.7	0.2	0.2	0.1	82.3

Brazing was performed at T ≈ 1000 °C with H_2 as a reducing shielding gas. To investigate the influence of the specimen geometry, round specimens and T-joint specimens as shown in Fig. 1a were used for the cyclic loading experiments. The round dog-bone shaped specimens have a minimum diameter in the of 8 mm, whereas the T-joint specimens are characterized by a T-shaped braze area with a cross section of 8 × 16 mm^2. In some T-joint specimens, artificial defects were introduced to investigate the interaction of defects on the fatigue behavior. Two different shapes of the artificial defects, straight defects with sizes of a = 2 mm and semi-elliptical defects with a/b = 2/3 and a size of a = 3 mm, as shown in Figure 1b, were investigated. The defects were introduced into the braze layer by electric discharge machining (EDM).

To investigate the effect of different elastic-plastic properties of the base material, two different heat treatments were performed after the brazing process (designated HTM 1 and HTM 2). In HTM 1, the specimens were annealed by two temperature steps at relatively high temperatures, while in the HTM 2 only one annealing step is performed at lower temperature than in HTM 1. Therefore, in total 8 types of specimens with 4 different specimen geometries (round, defect free T joint, T-joint with straight and semi-elliptical defects) along with two different heat treatments (HTM 1 and 2), were used for the fatigue tests.

Load controlled cyclic loading experiments were performed with a sinusoidal load amplitude on a servo hydraulic testing facility. The fatigue experiments were performed at R = 0.1 with a testing frequency of 2 Hz ≤ f_{test} ≤ 5 Hz until specimen failure. For a lifetime oriented analysis of the fatigue behavior, the maximum loading amplitude of the assumed defect-free cross section is plotted versus the number of cycles to fracture (S,N-curves). The cyclic deformation behavior of the brazed specimens was characterized on an event-oriented approach by measuring stress-strain hysteresis curves during the experiments. For round specimen, strain acquisition was performed by a clip-on extensometer with a gauge length of 10 mm. In the case of the T-joint specimens the DIC was performed by applying a random speckle pattern on the surface of a specimen and comparing the patterns in different fatigue states. To allow monitoring the accumulated fatigue damage by DIC during the experiment, a high speed camera model MotionXtra HG-100K by Redlake with a 200 mm macro lens was used for image acquisition. The images were recorded at a frequency of 500 Hz with an exposure time of 100 µs. The random speckle pattern was applied by airbrush and provides an average speckle size of 45 µm. Image correlation was performed with the software Moiree Analysis V0.950©. To investigate the accumulation of fatigue induced damage, the strains in loading direction were measured by comparing the specimens in the unloaded state before the fatigue experiment with several defined fatigue states during the fatigue tests.

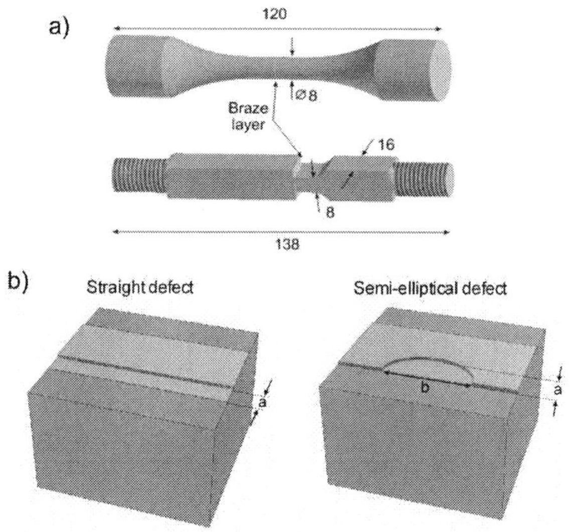

Figure 1: Geometries of round and T-joint specimens (a) and artificial defects (b)

Results and discussions

Quasi-static tensile tests of the base materials and the brazed specimens were carried out to investigate the basic mechanical properties of the brazed samples. Table 2 lists the mechanical properties of the base material and the brazed samples. For the T-joint samples, only ultimate tensile strengths (σ_{UTS}) were evaluated from the tests. The results show that HTM 1 is characterized by a high ductility, while HTM 2 provides better tensile strength and a reduced elongation to fracture than HTM 1. When comparing the properties of the brazed round samples with the ones of the base material, the presence of a braze layer leads to a decrease of the elongation to fracture but the yield strength and the ultimate strength remain nearly constant after both HTM 1 and HTM 2.

The influence of brazing defect on the fracture strength of the T-joints mechanical properties was more pronounced in HTM 2 than in the HTM 1. As shown in Table 2, for the defect containing samples the HTM 1 provided higher strength than the HTM 2, in spite of their less strength of the base material and the defect free joints. In the present case, a semi-elliptical defect leads to σ_{UTS} = 700 for HTM 1 and to σ_{UTS} = 648 MPa for HTM 2. The decrease of σ_{UTS} is even more pronounced for straight defects: a straight defect leads to σ^{UTS} = 614 MPa for HTM 1 and to σ^{UTS} = 453 MPa for HTM 2 (Table 2).

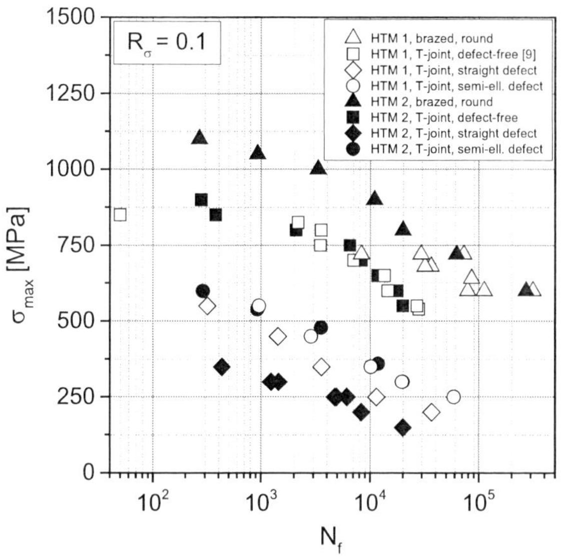

Figure 2: S,N-curve of defect-free and defect containing specimens

Figure 2 shows the maximum applied nominal load amplitude (σ_{max}) (i.e. for an assumed defect-free cross section) as a function of the numbers of cycles to failure (N_f). The results clearly indicate that the round specimens provide the best performance, both after HTM 1 and 2. At high number of load cycles ($N_f > 5\cdot10^4$), the brazed specimens after HTM 1 and HTM 2 provide comparable fatigue strengths. At high load amplitudes (>700°MPa), HTM 2 shows a better performance, which is related to the superior yield and tensile strength. The results for the brazed round specimens generally show that the joints provide comparable fatigue lifetimes if σ_{max} does not exceed the yield strength (σ_y) of the base material. The change from round to T-joint specimen geometry leads to similar results: Defect-free T-joint specimens provide comparable fatigue lifetimes for HTM 1 and HTM 2, when $\sigma_{max} < \sigma_y$. Compared to round specimens T-joint specimens generally provide lower fatigue lifetimes, which can be explained by the notch effect.

As could be expected, defects lead to a decrease of the fatigue performance. Interestingly, this effect was more pronounced for HTM 2 than for HTM 1. As shown in Fig. 2, , the specimens treated by HTM 1 (open rhombuses and circles in Fig. 2) show higher σ_{max} than those treated by HTM 1 (filled rhombuses and circles in Fig.2) at comparable N_f when a defect is present.

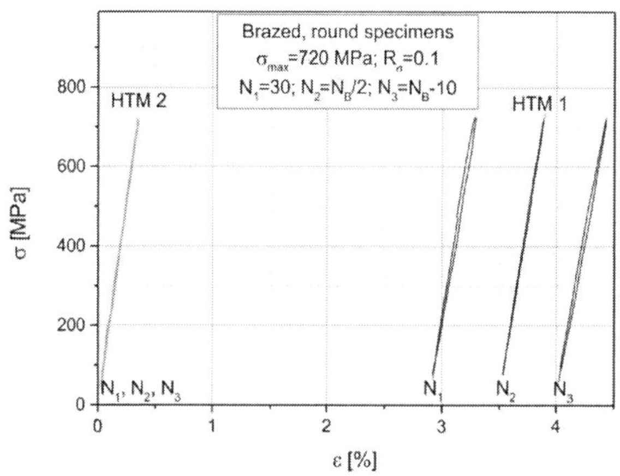

Figure 3: Hysteresis loops for brazed round specimens for selected cycles with R = 0.1 and σ_{max} = 720 MPa

To obtain a better understanding of thy cyclic deformation behavior, stress-strain hysteresis measurements were performed with brazed round specimens. At σ_{max} = 720 MPa failure occurred after N_f = 72574 cycles for HTM 1 and after N_f = 62293 cycles for HTM 2. The results show that the plastic mean strain strongly depends on the heat treatment. The hysteresis loops shown in Fig. 3 indicate that the HTM 1 specimen is characterized by pronounced cyclic creep (ratcheting), whereas the hysteresis loops remain at the same level for the HTM 2specimen in. In addition, the cyclic loading leads to fast strain accumulation for HTM 1 especially during the first 30 loading cycles. The fatigue induced strains

Table 2: Quasi-static tensile mechanical properties of the base materials and the brazed samples

Specimen geometry	Heat-Treatment	Young Modulus [MPa]	σ_y [MPa]	σ_{UTS} [MPa]
Round, base material	HTM 1	203	726	844
Round, braze joint		200	725	841
T-joint, defect-free		-	-	847
T-joint with straight defect		-	-	614
T-joint with semi-ellipsoidal defect		-	-	700
Round, base material	HTM 2	204	1017	1224
Round, braze joint		203	1020	1220
T-joint, defect-free		-	-	1120
T-joint with straight defect		-	-	453
T-joint with semi-ellipsoidal defect		-	-	648

increase continuously until fracture. The mean strain reaches 74 % after only 30 loading cycles and 87 % after $N_B/2$ of the mean strain at failure (where N_B is the number of the last cycle before the fracture, $N_B = N_f - 1$). For HTM 2, the hysteresis loop remains at a $\varepsilon_{mean} \approx 0.2$ % during the whole experiment. This result shows that for the brazed specimens in HTM 2, cyclic loads at $\sigma_{max} = 720$ MPa do not lead to significant fatigue induced strain accumulation. Generally, it could be expected that HTM 1 leads to a higher plastic mean strain at failure due to increased ductility, but it is remarkable that it provides comparable fatigue lifetimes to HTM 2.

As discussed above with the results presented in Fig. 2, HTM 1 provides even better fatigue strength than HTM 2, when defects exist. At $\sigma_{max} = 250$ MPa, failure occurred after 11419 cycles for the HTM 1 specimen and after 4909 cycles for the HTM 2 specimen. In Fig. 4 the results of the DIC on the T-joint specimens with straight defects after HTM 1 and 2 are compared. The DIC results were taken from two defined fatigue states at: 80% of the fatigue lifetime ($0.8 \cdot N_b$, Fig. 4a and c) and the cycle before the fracture occurs (N_b, Fig. 4b and d), for the specimens treated by HTM 1 and 2, respectively. At $N = N_b$, a long crack has formed and pronounced strain fields have developed ahead of the crack tip for HTM 1, reaching far into the base material, while plastic strain in the HTM 2 specimen is limited to a small volume in the braze layer near the defect. Specimen failure occurred shortly after crack initiation without any pronounced strain accumulation. These results indicate that the specimens treated by HTM 1 show a better tolerance towards structural hot spots such as defects or cracks compared to HTM 2.

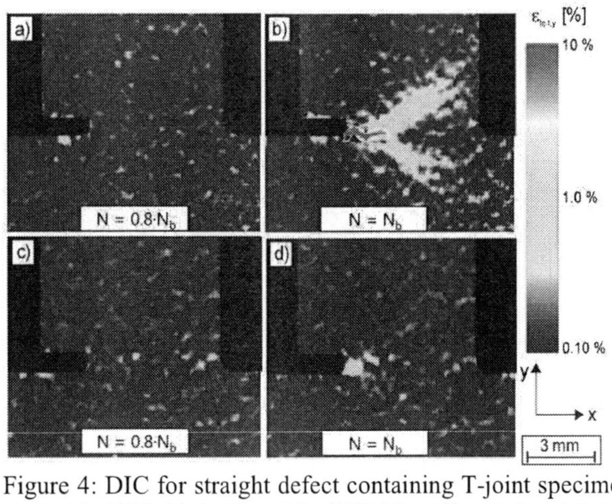

Figure 4: DIC for straight defect containing T-joint specimens treated by HTM 1 (a, b) and by HTM 2 c, d) at defined fatigue states

Brazed joints are generally characterized by a complex deformation behavior. Due to the different elastic moduli and plastic properties of the base material and the filler alloy, constraining effects leading to hydrostatic compressive stresses occur in the braze layer. The rigid steel substrates are expected to restrict the yielding of the filler metal despite the filler has lower yield strength than the steel. This constraining effect may cause complex local triaxial stress states in the vicinity of cracks or defects.

To investigate this complex stress state at the state before failure by cyclic loading, 3D FE simulations were carried out. The FE model considered the geometry of the braze zone of the straight defect containing T-joint specimen, by approximately 350,000 quadratic tetrahedral or brick elements with a size of 0.05-0.4 mm. In the model, elastic-plastic material behavior was assumed for both the base material and filer metal.

The FE simulations were performed with the commercial FE software Abaqus V6.13 by fixing the bottom of the model in vertical direction and applying a tensile stress of 250 MPa on the top surface. Between the braze layer and the base material, ideal bonding was assumed. Furthermore, the model takes account of the fatigue cracks at its state before the specimen failure. Based on the fractographic investigation of a real fatigue specimen, several interfacial elements on the upper interface between the braze layer and the base material were disconnected to simulate a realistic fatigue crack. Therefore, the top part of the braze layer (where the fatigue crack proceeds) is only partly connected with the overlaying base material.

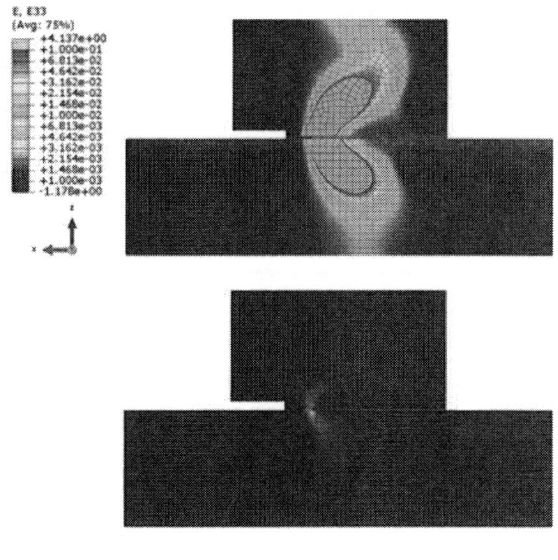
Figure 5: Side view of normal elastic strain fields in the loading direction near crack tips at 250 MPa, from 3D FE simulations of straight defect containing T-joint specimens at the state before failure for HTM 1(a) and HTM 2 (b). The loading direction is vertical in the figure.

The calculated strains in loading direction at a peak stress of 250 MPa are shown in Fig. 5. The results of the FE measurements are comparable to those from the DIC measurements. For HTM 1, the strain distribution provides the wing-like shape that was already determined by DIC. In contrast, strain accumulation occurs only locally at the crack tip in the zone around the braze layer for HTM 2. The results of the FE simulations indicate that the deformation of the braze layer and the surrounding base material is more restricted by HTM 2 comparing to HTM 1. A higher yield strength of the base material in HTM 2 leads to a more pronounced constraining effect which limits the lateral deformation of the braze layer in direction perpendicular to the loading, and consequently, leads to localized stress near the crack tip.

For both heat treatments, crack initiation processes are expected to be similar because of the same geometry and the limited plastic deformation in the base materials, as shown in the DIC of the middle stage of the cyclic tests in Fig. 4. However, the crack propagation process was completely different for the two different heat treatments. For HTM 2, a rather sudden fracture after crack initiation is suggested by the DIC, while for HTM 1 the crack seems to propagate over a larger number of cycles. This difference can be explained by the overall stresses that are not only composed of the stresses in loading direction, but also by the stresses that result from the constraining effect. Those stresses are located in the interface of the braze layer and are more pronounced for the less ductile material, as can be seen in Fig. 5. Therefore, crack growth is accelerated by the high crack initiation threshold and the crack length prior to failure is less pronounced for HTM 2, compared to HTM 1 - at the same loading amplitude. Furthermore, the increased ductility leads to a wider distribution of the resulting strains in the steel substrate – as was shown by DIC – and also to a more homogenous distribution of fatigue induced damage.

Summary

In this work, the fatigue behavior of brazed steel joints in different elastic-plastic properties and geometries was investigated by experiments and numerical analysis. The results showed that the brazed alloy with lower strength but higher ductility provides better fatigue resistance and fatigue lifetime compared with the alloy which has much higher strength by the heat treatment, especially when the defect is introduced. DIC measurements and FE simulations have shown that the improved ductility of the base material allows the distribution of the fatigue induced strains in a larger volume of the specimen. The ductile base material reduces the localization of the stress on the crack tip and consequently leads superior fatigue performance than the brittle material, especially when the brazing defect exists.

Acknowledgments

The authors gratefully acknowledge MAN Diesel & Turbo AG Zürich for the financial support of the project

References

[1] Messler, R.W., "The challenges for joining to keep pace with advancing materials and designs" *Mater Design* Vol. 16, No. 5 (1995) pp. 261-269.

[2] Boretius, M. "The potential of vacuum brazing - State of the art and developments from the perspective of a service contractor" *Adv Eng Mater* Vol. 8, No. 3 (2006) pp. 158-161.

[3] Sanchez, L. *et al*, "Development of high precision joints in particle accelerator components performed by vacuum brazing" *J Mater Process Tech* Vol. 211, No. 8 (2011) pp. 1379-1385.

[4] Nowacki, J. and Swider, P., "Producibility of brazed high-dimension centrifugal compressor impellers" *J Mater Process Tech* Vol. 133, No. 1-2 (2003), pp. 174-180.

[5] Schindler, H.J. and Leinenbach, C. "Mechanics of fatigue crack growth in a bonding interface" *Eng Fract Mech* Vol. 89 (2012) pp. 52-64.

[6] Leinenbach, C. *et al*, "Quasistatic fracture behaviour and defect assessment of brazed soft martensitic stainless steel joints" *Eng Fail Anal* Vol. 17 No. 3 (2010) pp. 672-682.

[7] Lis, A. *et al*, "Numerical and Experimental Investigations on the Defect Tolerance of Brazed Steel Joints" *MP-Mater Test* Vol. 54, No. 9 (2012) pp. 605-611.

[8] Koster, M. *et al*, "Fatigue and cyclic deformation behavior of brazed steel joints" *Mater. Sci. Eng A* Vol. 581 (2013) pp. 90-97.

Microstructure and FractureMorphology of Mo-Cu/Cr18-Ni8 Brazing Joint with Ni-Cr-P Filler Metal

Li Yajiang, Wang Juan, Liu Kun, Zheng Deshuang

Key Laboratory for Liquid-Solid Structural Evolution and Processing of Materials, Ministry of Education, Shandong University

Jinan 250061, Shandong Province, China

Liu Peng

School of Materials Science and Engineering, Shandong Jianzhu University, Jinan 250101, China

Abstract

Mo-Cu composite and Cr18-Ni8 stainless steel was brazed with Ni-Cr-P filler metal in a vacuum of 10^{-4} Pa and a Mo-Cu/Cr18-Ni8 joint was obtained. Microstructure in Mo-Cu/Cr18-Ni8 joint was investigated by field-emission scanning electron microscope (FE-SEM) with energy dispersive spectrometer (EDS). Shear strength of Mo-Cu/Cr18-Ni8 lap joint was measured by electromechanical universal testing machine. An excellent Mo-Cu/Cr18-Ni8 joint with a shear strength of 155 MPa was achieved at 980℃ for 20 min. Brazed joint was mainly comprised of eutectic structure in the center of brazing seam, matrix structure and lump structure. Ni-Cu(Mo) and Ni-Fe solid solution were at the interface beside Mo-Cu composite and Cr18-Ni8 stainless steel, respectively. Shear fracture exhibited mixed ductile-brittle fracture feature with trans-granular fracture, ductile dimples and tearing edges. Fracture originated from the interface between brazing seam and Mo-Cu composite.

Introduction

High density Mo-Cu composite has unique properties such as high electric conductivity, high thermal conductivity, low thermal expansion coefficient, excellent mechanical property at elevated temperature. Consequently, Mo-Cu composite has been widely used in the fields of aerospace, nuclear and electronics industry [1-3]. The key factor to affect joint quality is microstructure coarsening caused by weld thermal cycle and the performance failure due to gas impurities when joining Mo-Cu composite. And even interstitial impurities would form segregation at grain boundary under the condition of fast cooling by fusion welding process [4-6].

Recently, many efforts have been made to join molybdenum or its alloys with electron beam welding (EBW) [7], GTAW [8], friction welding [9] and vacuum brazing [10-12]. The major drawback of fusion welding is the large susceptibility of the joint to crack and porosity due to an impure welding atmosphere and poor ductility at room temperature due to impurities and grain growth in weld metal. For the welding of molybdenum with different materials, brazing process was conducted in many works. The filler metals used were Pd-Ag [10], Ag-Cu [11] and Ag-Cu-Ti [12]. These brazed joint were suitable to the application field at temperature less than 800 ℃.

Ni-based filler metals provide joints with excellent corrosion resistance and strength at elevated temperature. Phosphorus in Ni-Cr-P filler metal may reduce the melting point and improves the wettability while chromium provides high oxidation resistance [13, 14]. In order to expand the applications of Mo-Cu composite at high temperature more than 800 ℃, vacuum brazing with Ni-Cr-P filler metal was used for brazing Mo-Cu composite and Cr18-Ni8 stainless steel in this study. Field-emission scanning electron microscope (FE-SEM) and energy dispersive spectrometer (EDS) were used to investigate the microstructure characteristics of Mo-Cu/Cr18-Ni8 stainless steel brazed joint. The shear strength was measured by shearing measurement. The fracture pattern and characteristics were analyzed so that it can provide a theoretical foundation and experimental basis for brazing of Mo-Cu composite in the application at high temperature.

Experiments

Mo-Cu composite and Cr18-Ni8 stainless steel were used as test materials, the plate thickness was 2.5 mm. Pasty Ni-Cr-P filler metal was prepared to join Mo-Cu composite and Cr18-Ni8 stainless steel. The compositions of Ni-Cr-P filler metal were Cr 14%, Ni 76%, P 9.9% and Si 0.1%. The liquidus temperature was about 890 ℃. The chemical compositions of Mo-Cu composite were about Mo 70% and Cu 30% and the compositions of Cr18-Ni8 stainless steel were C 0.08%, Cr 18.1%, Ni 8.5%, Mn 1.8%, Si 0.8% and Fe balance.

Before vacuum brazing, the oxides and greasy dirt on the faying surface of Mo-Cu composite and Cr18-Ni8 stainless steel were removed by 360# and 800# SiC paper, cleaning in alcohol. The pasty Ni-Cr-P filler metal was overlaid on the surface prepared for brazing. The gap was 30 μm between two base metals. The brazing was conducted at 980 ℃ for 20 min in a vacuum of 10^{-4} Pa. The cooling process was performed in the vacuum chamber by circulating water until the temperature reached 100 ℃ and then it was taken out from vacuum furnace.

The Mo-Cu/Cr18-Ni8 stainless steel joint specimens were obtained by wire cutting method. The metallographic samples were prepared by being grounded, polished and etched in a solution of aqua regia for 5~10 s. Microstructure characteristics in the brazed joints were analyzed using HITACHI SU-70 field-emission scanning electron microscope (FE-SEM). Chemical composition was detected by the INCAX-SIGHT energy-dispersive spectroscope (EDS). The shear strength was measured using CMT 5105 testing machine and the fracture morphology was observed by SEM.

Results and Discussion

Microstructure in braze seam

Fig.1 shows the microstructure characteristics of Mo-Cu/Cr18-Ni8 stainless steel joint. Ni-Cr-P filler metal had a good wetting with Mo-Cu composite and Cr18-Ni8 stainless steel. There were good bonding interfaces without crack or micro-hole between Mo-Cu composite and Cr18-Ni8 stainless steel with Ni-Cr-P filler metal. The microstructure of Mo-Cu/Cr18-Ni8 stainless steel joint was mainly comprised of three parts: eutectic structure (labelled by A) in the center of braze seam, matrix structure (labelled by B) and lump structure (labelled by C), as shown in Fig.1a. The magnification of macrostructure in braze seam was shown in Fig.1b and different phases were marked by points from 1 to 4. The corresponding chemical compositions of special phases were analyzed by EDS and the results are given in Table1.

Structure (A) was consisted of matrix (point 2) and precipitated phase (point 1). According to the Ni-P phase diagram, there exists a eutectic reaction of the γ-Ni solid solution and the Ni_3P at 880℃ with a composition of P 11% [15]. Therefore, structure (A) was identified as binary eutectic of γ-Ni solid solution with Ni_3P inferred from the chemical compositions of point 1 and point 2. There were finer massive precipitations (labeled by point 3) on the matrix structure (B) and the content of P (20.3%) in point 3 was much more than that of Ni-P eutectic point. So, matrix structure (B) was Ni-Cr-P solid solution with dispersive distribution of Ni_5P_2, Cr3P precipitation. Due to the mutual diffusion of different elements between Mo-Cu composite and Ni-Cr-P filler metal, the chemical compositions of lump structure (C) (point 4) mainly consisted of Ni 61.7%, Cu 12.7%, Cr 10.3%, Mo 9.8% with barely P.

The inter-diffusion of elements Ni, Cr, Mo and Cu was observed between filler metal and base metal forming the brazing seam. Ni was distributed uniformly through the whole brazed joint. The diffusion of P was less due to its large diameter. So P was mainly distributed in Ni-P eutectic structure and matrix structure.

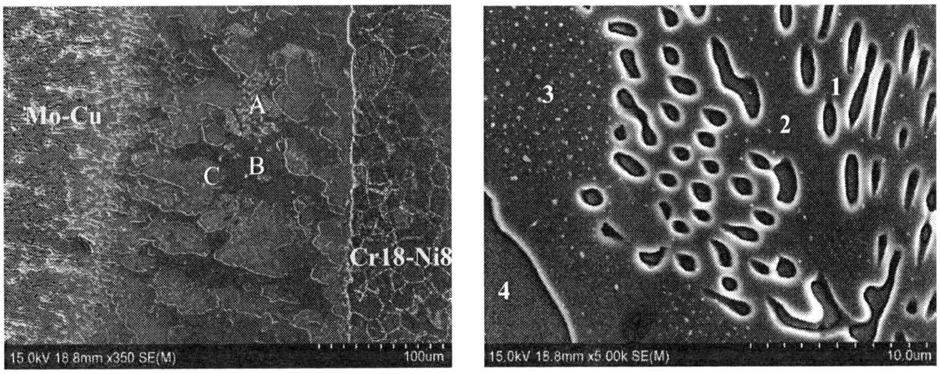

(a) Bonding joint (b) Braze seam

Figure 1 Microstructure characteristics of Mo-Cu/Cr18-Ni8 stainless steel joint

Table 1 Chemical compositions of the points (wt.%)

Point	P	Cr	Ni	Cu	Fe	Mo
1	5.3	6.0	67.0	15.0	2.8	3.9
2	17.2	5.8	74.1	—	1.0	1.9
3	20.3	18.8	35.2	—	0.9	13.8
4	0.5	10.3	61.7	12.7	5.0	9.8

Interface characteristics

Fig.2 shows the bonding interface characteristics. There existed Ni-base solid solution layer at the interface between braze seam and Cr18-Ni8 stainless steel (see Fig.2a).

As the inter-diffusion of Ni, P in filler metal and Cr, Fe in base metal, Ni-Fe solid solution first heterogeneous nucleated and grew up along the interface. As a result, chemical composition range in the filler metal deviated from that of eutectic composition. One phase with the same composition as pro-eutectic Ni-Fe solid solution grew up cling to it and formed the Ni-Fe solid solution at the interface between brazing seam and Cr18-Ni8 stainless steel. The other phase Ni(Cr, Fe)-P solid solution was formed. At the Cr18-Ni8 stainless steel side, some white precipitations were distributed at the grain boundary (shown in Fig.2b). This is resulted from the formation of $Cr_{23}C_6$ during the sensitization temperature range (600~850 ℃). Thus, white precipitations at grain boundaries are suggested to be $Cr_{23}C_6$.

Melted filler metal can permeate into Mo-Cu composite to form a jagged bonding interface (Fig.2c) during brazing. This is beneficial to enlarge contact area and obtain high strength brazed joint. There existed inter-diffusion of Cu in Mo-Cu composite and Ni in filler metal due to the concentration difference. Thus, Cu and Ni can form complete Ni-Cu solid solution at the interface between brazing seam and Mo-Cu composite. Mo particle in Mo-Cu composite was surrounded by Ni-Cu solid solution.

Fig.2d shows a good metallurgical combination between Mo particles and Ni-Cu solid solution. Due to the certain solid solubility between Ni and Mo, Ni-Cr-P filler metal had a good wetting on Mo-Cu composite. As the melting point of molybdenum was up to 2610 ℃, molten filler metal first heterogeneously nucleated and then grew up along original boundary surface in cooling process

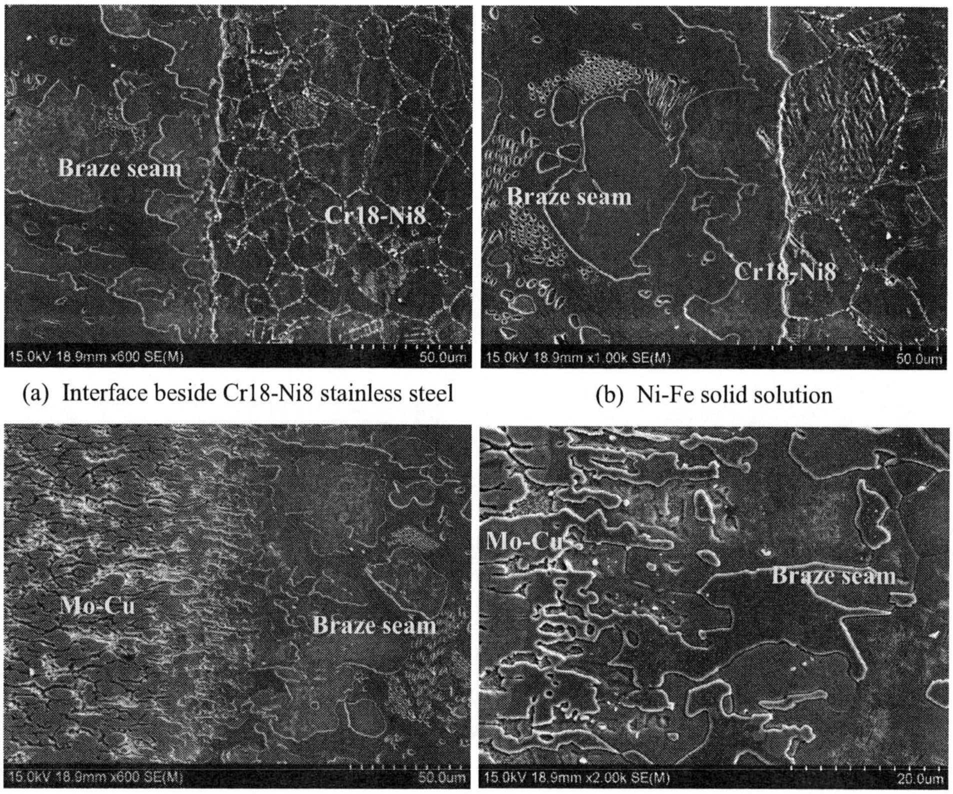

(a) Interface beside Cr18-Ni8 stainless steel (b) Ni-Fe solid solution

(c) Interface beside Mo-Cu composite (d) Ni-Cu solid solution

Figure 2 Interfacial characteristics of Mo-Cu/Cr18-Ni8 stainless steel brazed joint

Shear fracture

The average shear strength of Mo-Cu/Cr18-Ni8 joint was measured to be 155 MPa and the shear fracture morphology was shown in Fig.3. The shear fractography exhibits mixed ductile-brittle fracture feature with trans-granular fracture, ductile dimples and tearing edges presenting certain plasticity, shown in Fig.3a,b.

It was determined that shear fracture originated from trans-granular zone (see Fig.3c) and the composition analysis was shown in Fig.3d. Trans-granular zone locates at the brazing seam near Mo-Cu composite inferred from composition analyses. P aggregated at trans-granular zone and easily formed brittle compounds (Ni_5P_2, Cr_3P) with Ni or Cr. Those brittle compounds caused stress concentration and become the fracture sources. Under external shear force, the interface between brazing seam and Mo-Cu composite began to slip and then crack propagated to the braze seam until Mo-Cu/Cr18-Ni8 stainless steel joint ruptured completely. The Ni_5P_2, Cr_3P brittle precipitations and eutectic structure in the center of braze seam provided a path for the crack propagation.

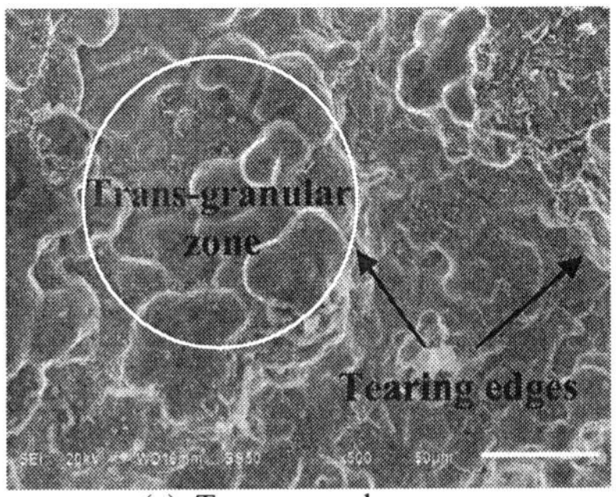

(a) Trans-granular zone

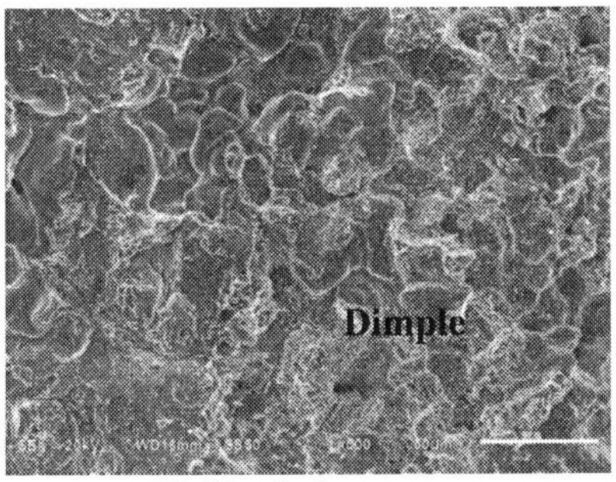

(b) Dimple feature

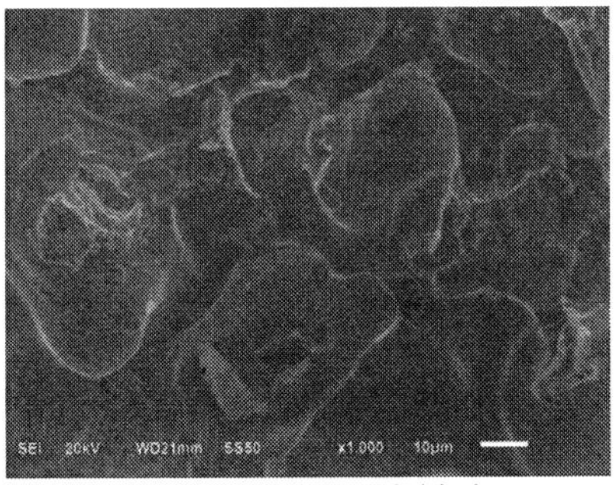

(c) Location of crack initiation

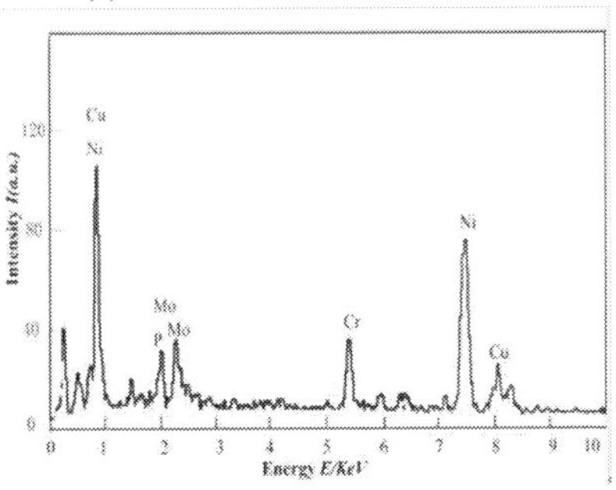

(d) Composition of shear fracture

Figure 3 Shear fracture morphology of Mo-Cu/Cr18-Ni8 stainless steel brazed joint

Conclusions

Vacuum brazing of Mo-Cu composite and Cr18-Ni8 stainless steel was performed with Ni-Cr-P filler metal at at 980 ℃ for 20 min in a vacuum of 10^{-4} Pa. Shear strength of brazed joint was about 155 MPa. Microstructure in the braze seam was comprised of Ni-Cr-P solid solution and binary eutectic including γ-Ni with Ni_3P. There are Ni-Cu(Mo) and Ni-Fe solid solution at the interface beside Mo-Cu composite and Cr18-Ni8 stainless steel, respectively. Shear fracture of Mo-Cu/Cr18-Ni8 stainless steel brazed joint exhibited mixed ductile-brittle fracture feature with trans-granular fracture, ductile dimples and tearing edges. Fracture originated from the interface between brazing seam and Mo-Cu composite and it propagated to the braze seam due to the formation of brittle Ni_5P_2 and Cr_3P precipitation.

References

[1] Tikhii G A, Kachalin N L, Belova V P. "Study of a Mo-Cu pseudo-alloy obtained from mechanically activated charge," Metal Science and Heat Treatment, Vol. 49, No. 3 (2007), pp. 448-452.

[2] Zhou W P. "The applications of molybdenum and molybdenum alloys," China Metal Bulletin, Vol. 42, (2008), pp. 34-37.

[3] Aguilar C, Ordonez S, Marin J, et al. "Study and methods of analysis of mechanically alloyed Cu-Mo powders," Materials Science and Engineering A, Vol. 464, (2007), pp. 288-294.

[4] Cockeram B V, Ohriner E K, Byun T S, et al. "Weldable ductile molybdenum alloy development," Journal of Nuclear Materials, Vol. 382, No. 2 (2008), pp. 229-241.

[5] Li X, Zhang P Y. "Analysis of influence factors in welding of molybdenum alloys," Annual Report for China Institute of Atomic Energy, Vol. 21, No. 1 (1999), pp. 57-59.

[6] Zhang D Y. "Effect on molybdenum and molybdenum alloy of oxygen, nitrogen and carbon," China Molybdenum Industry, Vol. 27, No. 2 (2003), pp. 20-25.

[7] Morito F. "Effect of heat treatment on mechanical behavior of electron beam welded sintered molybdenum," Journal of Nuclear Materials, Vol. 165, No. 2 (1989), pp. 142-148.

[8] Jiang Q L, Li Y J, Puchkov U A, et al. "Microstructure characteristics in TIG welded joint of Mo-Cu composite and 18-8 stainless steel," International Journal of Refractory Metal and Hard Materials, Vol. 28, No. 3 (2010), pp. 429-433.

[9] Ambroziak A. "Friction welding of molybdenum to molybdenum and to other metals," International Journal of Refractory Metal and Hard Materials, Vol. 29, No.4 (2011), pp.462-469.

[10] Yi M. "Brazing of singal-crystalline Mo by using Pd-20Ag alloy," China Molybdenum Industry, Vol. 19, No. 2, (1995), pp. 22-25.

[11] Chan H Y, Liaw D W, Shiue R W. "The microstructural observation of brazing Ti-6Al-4V and TZM using the BAg-8 braze alloy," International Journal of Refractory Metal and Hard Materials, Vol. 22, No. 1, (2004), pp. 27-33.

[12] Wang J, Li Y J, Zheng D S. "Microstructure and shear strength of Mo-Cu/Cr18-Ni8 brazing joint in vacuum," Vacuum, Vol. 86, No. 12 (2012), pp. 2054-2058.

[13] Stratford G, Battenbough A, Lee L et al. "Comparing High-Temperature Nickel Brazing Filler Metals," Welding Journal, Vol. 3, (2011), pp.54-59.

[14] Maciel T M, Chang I T, Strangwood M, et al. "Brazing of Al_2O_3 with rapidly solidified melt spun Ni-Cr-P alloy ribbons," Materials Science Forum, Vol. 530, (2006), pp. 473-477.

[15] Yeh M S, Chuang T H. "Effects of applied pressure on the brazing of superplastic INCONEL 718 superalloy,". Metallurgical and Materials Transactions A, Vol. 28, No. 6 (1997), pp. 1367-1376.

Aging of Reactive Air Brazed Ceramic/Metal Joints in Dependence of the Filler Metals

K. Bobzin, M. Öte, S. Wiesner
Surface Engineering Institute, RWTH Aachen University, Aachen, Germany
wiesner@iot.rwth-aachen.de

Abstract

Since the development of reactive air brazing the applicability of this technology in demanding devices for energy generation has been examined. Here, ceramic/metal joints have to fulfill challenging demands; i.e. chemical stability in oxidizing and reducing atmospheres, suitable mechanical and thermal properties at elevated temperatures and low porosity.

The aging behavior of Ag8Cu, the standard filler metal for reactive air brazing, has been examined before. However, the aging behavior of recently developed fillers has not yet been analyzed thoroughly. In order to improve the understanding about the aging processes and mechanisms the changing of the different areas of brazed 3YSZ/Crofer 22 H joints, e.g. Ag matrix, oxides and interfaces, was analyzed. The growth of the reaction zones during aging was mostly related to the content of oxide forming metals in the filler metal. At least a slight growth could be observed in all samples. The steel was affected by the use of a Ti-containing filler metal. This behavior was further examined by means of differential scanning calorimetry and thermogravimetric analysis. The porosity of several joints increased notably and most samples showed a higher hardness in comparison to that in the as-brazed state. The Cu-free Ag2Fe2Si2Al seems to have the highest aging stability of the analyzed filler metals.

Introduction

The high potential of the reactive air brazing (RAB) as a possible joining process for innovative ceramic/metal components for the energy generation such as solid oxide fuel cells and gas separation membranes has been proven in the last years [1,2,3]. These applications are challenging as the operation temperatures are around 800 °C for the fuel cells and around 850 °C for the membranes. Furthermore, the joints in the fuel cells are working in dual atmosphere conditions with an oxygen-rich and a hydrogen-rich side and are subject to mechanical loads [4].

RAB filler metals are based on a noble metal, mostly Ag, and contain oxide forming metals, typically Cu. During the heating phase of the brazing process the Cu oxidizes and can form, depending on the Cu content, an oxide layer on the ceramic. This enhances the wetting behavior of the Ag melt on the ceramic joining partner. If the Cu content is low, the wetting of the ceramic depends mostly on the oxygen content in the Ag melt [5,6]. In the case of ceramic/metal joining partners partially strong reactions occur between elements of the braze and the base metals leading to inhomogenous and distinct reaction layers at the steel interface [7]. The analysis of these reactions by means of differential scanning calorimetry (DSC) supported the development of new Cu-containing filler metals inducing less intense reactions, for example Ag4Cu4Ni. Moreover, the use of the Cu-free brazes Ag2Fe2Si2Al and Ag0.5Al0.5Ti led to very thin reaction layers on the steel [8].

The aging behaviour of Ag4Cu and Ag8Cu, which are the filler metals typically used in RAB applications, has been analyzed in several studies. In ceramic/ceramic joints neither microstructure nor strength of the joints were changed notably by a heat treatment at 800 °C in air. In zirconia/Crofer 22 APU joints, used for solid oxide fuel cells, the growth of mixed oxides consisting of steel and braze elements could be observed. The failure of the joints mostly occurred in the oxidic zone at the interface of the base materials indicating that this zone is the critical area of the joint [9,10]. It was shown that the thickness and the growth of the interfacial layers during aging, particular at the steel side, determine the strength of the joints [11]. Contrary to these joints the critical area of $Ba_{0.5}Sr_{0.5}Co_{0.8}Fe_{0.2}O_{3-\delta}$/AISI 314 joints, used for gas separation membrane technology, aged at 850 °C is the interface between ceramic and Ag3Cu braze [12].

Based on these results it can be stated that the compositions of the base and filler materials are important influencing factors for the aging behavior of the joints as they determine the structure and growth of the interfacial zone of mixed oxides. Therefore, it can be expected that the new filler metals Ag4Cu4Ni, Ag2Fe2Si2Al and Ag0.5Al0.5Ti lead to different aging behaviors of the joints. Brazed joints with the three filler metals are examined before and after a heat treatment to determine the change of microstructure and compositon caused by the occuring aging processes at elevated temperatures. Furthermore, it can be assumed that the growth of the oxide layer during aging also depends on the availability of oxygen at the interface between braze and base material. In order to gain further information about the influence of the availablity of oxygen on the aging processes and especially on the growth of the interfacial oxide layer, different types of samples were analyzed where the oxygen availability at the interface is partly restricted by the addition of filler metal and a ceramic joining partner.

Experimental

In order to examine the aging behavior of the different areas of the joints and especially the influence of the oxygen availability on the growth of the interfacial oxide layer during the heat treatment, three different types of samples were prepared as it is listed below:

- type 1: thin layer of oxide forming filler components (e.g. Cu/Ni, Fe/Si/Al, Al/Ti) on base metal
- type 2: filler metal on base material
- type 3: complete 3YSZ/Crofer 22 H joint

The oxygen availability at the interface is reduced from a high availability in sample type 1 to sample type 3 where oxygen can only diffuse from the sides of the samples to the oxidic interface layer. Samples of all types were produced for the three filler metals Ag4Cu4Ni, Ag2Fe2Si2Al and Ag0.5Al0.5Ti. As base materials the ferritic stainless steel Crofer 22 H (X1CrWNbTiLa22) and the ceramic 3YSZ (ZrO_2 with 3 at.% Y_2O_3) were used.

The filler metals Ag4Cu4Ni and Ag2Fe2Si2Al (values in at.%) were used as pastes. For the preparation of the paste pure elemental powders were mixed with 10 wt.-% of an ethanol based organic binder (Innobraze B2). The layer of oxide forming filler components for sample type 1 was also prepared by using a paste made of the elements in powder form. The ratio of the atomic amount of the elements was 1:1 for Ag/Ni and Al/Ti and 1:1:1 for Ag2Fe2Si2Al. The filler metal Ag0.5Al0.5Ti was applied as a foil with a thickness of around 100 µm. The foil was prepared by arc melting and rolling. Before brazing the steel and ceramic parts were cleaned for 5 min in an ethanol ultrasonic bath.

The brazing process is conducted in the air furnace with a heating rate of 10 K/min and a holding time of 18 min at brazing temperature. The brazing temperature was chosen according to the filler metal: 970 °C for Ag4Cu4Ni and Ag0.5Al0.5Ti and 1050 °C for Ag2Fe2Si2Al. The type 1 samples on which only the thin layer of oxide forming filler components was applied were also heat treated at the corresponding temperatures. The cooling in the furnace was uncontrolled. In order to simulate the operating conditions in a solid oxide fuel cell, one series of samples was aged after brazing at 800 °C for 500 h in air.

After brazing and after aging, cross sectional cuts of the samples were prepared metallographically and analyzed by using a light microscope and a scanning electron microscope (Zeiss Leo 1530) with a device for energy dispersive X-ray spectroscopy (EDX) (Bruker Quantax 200; resolution: 133 eV). Additionally microhardness mesurements (Micromet 1, Bühler) were done with a load of 10 g (HV 0.01) for 10 s according to DIN EN ISO 6507-1. The values stated in this work are an arithmetic average of at least 10 measurements.

Furthermore, combined DSC and thermogravimetric (TG) measurements were done in the system SETSYS Evolution (SETARAM Instruments). Miniaturized samples of the type 1 (total weight of Crofer 22 H and paste: 45 - 75 mg) were heated up in air while heat flow and weight change were recorded.

Results and Discussion

Ag4Cu4Ni

Figure 1 shows the cross sections of the three types of Ag4Cu4Ni/Crofer 22 H samples in the as-brazed state (A - C) and after aging (D - F). In sample type 1 (Fig. 1A+D) the oxide forming elements of the filler metal (Cu and Ni) were applied on the steel and heat treated at 970 °C in compliance with the brazing process. Afterwards, the Cu/Ni oxide layer is very porous. A two-layered reaction zone is formed in the Crofer 22 H which has an irregular thickness between 10 and 30 µm. A mixed Cr/Fe oxide layer can be found as a thin inner layer. This is the typical protection layer of the steel. The porous outer layer consists of mixed oxides made of Cu, Ni, Cr, Fe and Mn. After aging (Fig. 1D) the majority of the Cu/Ni oxide layer is detached from the substrate. The separation does not occur at the interface of the steel substrate but in the Cu/Ni oxide layer. The thickness of the reaction zone in the steel is on the same level as in the as-brazed state. Sintering processes during aging lead to a less porous structure of the reaction zone. It still consists of a Cr/Fe oxide inner layer and a complex outer layer made of steel elements as well as Cu and Ni.

After brazing type 2 samples (Fig. 1B) show an inhomogeneous, layered reaction zone spreading from the steel interface into the Ag matrix. The formation of this reaction layer has been investigated more closely before by differential scanning calorimetry. It has been shown that an exothermic reactions above 965 °C leads to the formation of the oxide zone on the steel. Due to the integration of Ni in this zone the reaction is less intense than in pure AgCu fillers [7,8]. The thickness of the layer close to the steel side is around 15 µm and the thickness of the layer deeper in the Ag matrix is around 20 µm. Both layers consist of complex mixed Cr/Fe/Cu/Ni/Mn oxides. Due to the limited availability of oxide forming elements in the filler metal and expected lower oxygen content, the steel is not strongly affected as in the type 1 samples showing only a slightly irregular interface. The innermost layer of the reaction zone is the steel's protective layer consisting of Cr and Fe oxides with a small amount of Mn. In addition to the oxides at the steel side round Cu/Ni oxides with small amounts of Cr can be found in the braze. The majority of these oxides formed close to the rim of the braze, an area with a high amount of soluted oxygen. Furthermore, small precipitations of CuO appear in the braze. After the aging process (Fig. 1E) the thickness and composition of the reaction zone is nearly the same compared to the as-brazed state. As before, a densification of the reaction zone can be noted probably due to sintering processes during aging. Voluminous pores are scattered in the braze of the aged samples. It can be assumed that the pores develop by diffusion or evaporation processes regarding their steplike structure. As

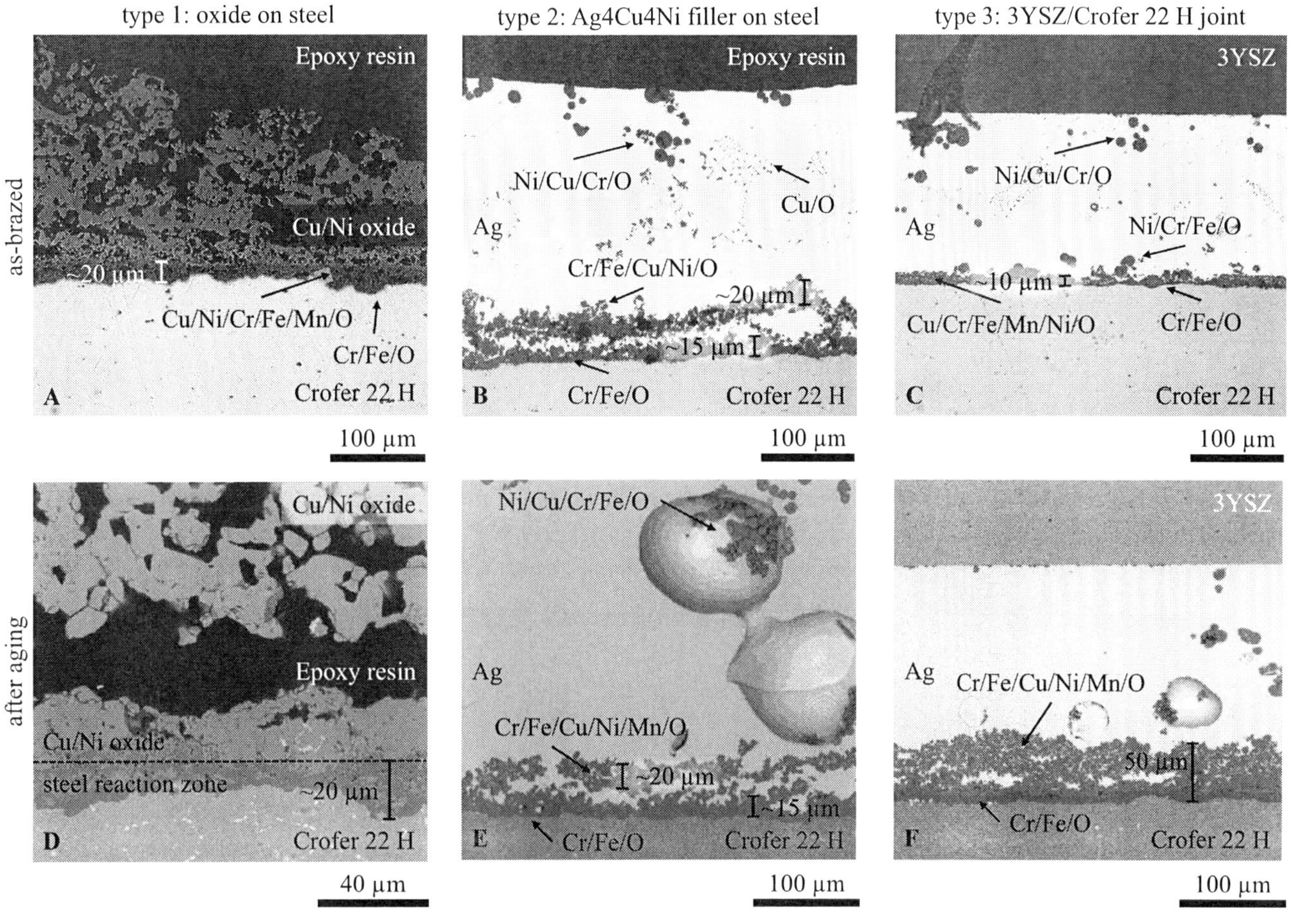

Figure 1: Cross sections of Ag4Cu4Ni/Crofer 22 H samples: A + D: sample type 1: Cu/Ni oxide layer on steel; B + E: sample type 2: Ag4Cu4Ni filler metal on steel; -C + F: sample type 3: complete 3YSZ/Crofer 22 H joint; A - C: as-brazed state, D - F: after aging

evaporation of Ag from RAB joints is a known phenomenon [18], the samples will be further analyzed to investigate the mechanisms leading to the porosity. At the rim of the pores accumulations of Ni/Cu/Cr/Fe oxides can be found. It seems that the Ag matrix around them evaporates or diffuses while the oxides remain unaffected. The CuO precipitations found in the samples in the as-brazed state are dissolved during the aging process.

The hardness of the Ag matrix increases from (50 ± 7) HV 0.01 to (85 ± 7) HV 0.01 during aging. It is improbable that this increase is due to the formation of a solid solution for example with oxygen as proposed in [6] or due to elements of the steel diffusing into the Ag matrix as several EDX measurements confirmed that the braze matrix consists only of Ag. A possible explanation is the change from a single crystalline to a polycrystalline microstructure in the joint during aging as shown by Skiera et al. for samples after rapid thermal cycling. [13]. Although the number of grains can be expected to be relatively low after isothermal aging, a decrease of the grain size generally leads to a higher hardness of the material as has been also shown for Ag and Ag alloys [17].

The reaction zone in the type 3 samples after brazing (Fig. 1C) is with a thickness of around 13 μm notably thinner than that of the sample type 2. In accordance with the results of a previous study about the Ag4Cu4Ni filler metal [14], it can be assumed that the thin reaction zone is caused by the low amount of oxygen available. Furthermore, several round Ni/Cr/Fe oxides can be found near the interface of the steel. After aging (Fig. 1F) the reaction zone is much thicker than in the as-brazed state. The inner Cr/Fe oxide layer has a thickness of around 10 μm and the outer layer a thickness of 40 to 50 μm. It consists of small particles of mixed Fe/Cr/Cu/Mn oxides. These are more dispersed and resemble the reaction zone in the type 2 samples. This growth of the reaction zone during aging can be due to the diffusion of oxygen through the Ag matrix. This supports the assumption that the reaction zone growth depends on the oxygen availability. The pore formation and the dissolution of the CuO precipitations observed in the type 2 samples can be observed here as well. Both phenomena do not seem to be significantly dependent on the amount of oxide forming filler elements or that of oxygen.

In these samples an increase of the hardness from (54 ± 4) HV 0.01 in the as-brazed state to (80 ± 7) HV 0.01

after aging is detected. As mentioned above, an influence of oxygen availability on the hardness seems to be improbable. The possible change of microstructure leading to the higher hardness will be analyzed further.

Ag0.5Al0.5Ti

Figure 2 shows the cross sections of a type 1 sample with Al/Ti oxide on Crofer 22 H in the as-brazed state (Fig. 2A) and a type 2 sample after aging (Fig. 2B). An enlarged detail of the sample, the reaction zone at the steel/braze interface, is shown in Fig. 2C. Contrary to the Cu/Ni oxide sample shown in Fig. 1A the interaction between the Al/Ti oxides and the steel does not lead to a continuous reaction zone at the steel interface. Instead, grains of the steel are detached from the substrate and can be found in the very porous oxide layer. The average depth of this dissolving process is around 110 µm. During the aging process the steel was not further affected.

In the as-brazed state the type 2 and 3 samples are characterized by a very low porosity, thin ribbons of Al/Cr/Ti oxides along the grain boundaries of the Ag matrix and a thin oxide layer on the steel. Only small amounts of Al can be found in this natural oxide layer of the steel. This is in accordance with differential scanning calorimetry measurements done in a previous study [15]. It could be seen that the filler metal on Crofer 22 H had a melting point of around 945 °C which is the melting point of pure Ag in air. On that account, it could be assumed that Ti and Al oxidize before the Ag melts, thus reducing their influence on the melting behavior. As no exothermic reaction occurs only the oxide layer of the steel can be found in the cross sections.

During aging the thickness of the reaction zone rises from around 8 µm up to 20 µm (Fig. 2B). Furthermore, its structure is more inhomogeneous. It consists of three different phases as can be seen in detail in Fig. 2C. The dark phase consists mainly of the elements of the steel's own oxide layer, but also Si from the steel and Ag. The lighter grey phase contains Cr and Fe from the substrate as well as Ag and Ti from the filler metal. The third phase bordering on the Ag matrix is made of Ag, Cr, Mn, Ti and Fe. It also appears in an acicular form known from other studies about Ag, Ag4Cu and Ag0.5Al filler metals [16]. There these similar structures consisted of mixed Ag/Cr/Mn oxides and grew notably during aging process up to 3000 hours. The addition of Ti to the filler metal used here seems to influence the formation of the phase formation and composition and thus the aging behavior of the joint. Samples aged for longer times will be analyzed to understand the influence of the Ti better. The strong dissolution of the steel substrate shown in Fig. 2A does not occur in the sample types 2 and 3 possibly due to the very small amount of Al and Ti in the filler metal. However, the steel substrate is affected as voids form at the interface of the steel during aging, especially at the beginning of the grain boundaries. Furthermore, the thin Cr/Fe(/Mn) oxide layer found on the Crofer 22 H when using the Ag4Cu4Ni filler cannot be found in the Ag0.5Al0.5Ti samples. Normally this thin oxide layer also acts a protection layer against diffusion of steel elements. As it cannot be found in the Ag0.5AlTi samples, Si can diffuse from the substrate during aging and can be integrated in the phases of the reaction zone and is additionally found in the oxide ribbons in the Ag matrix.

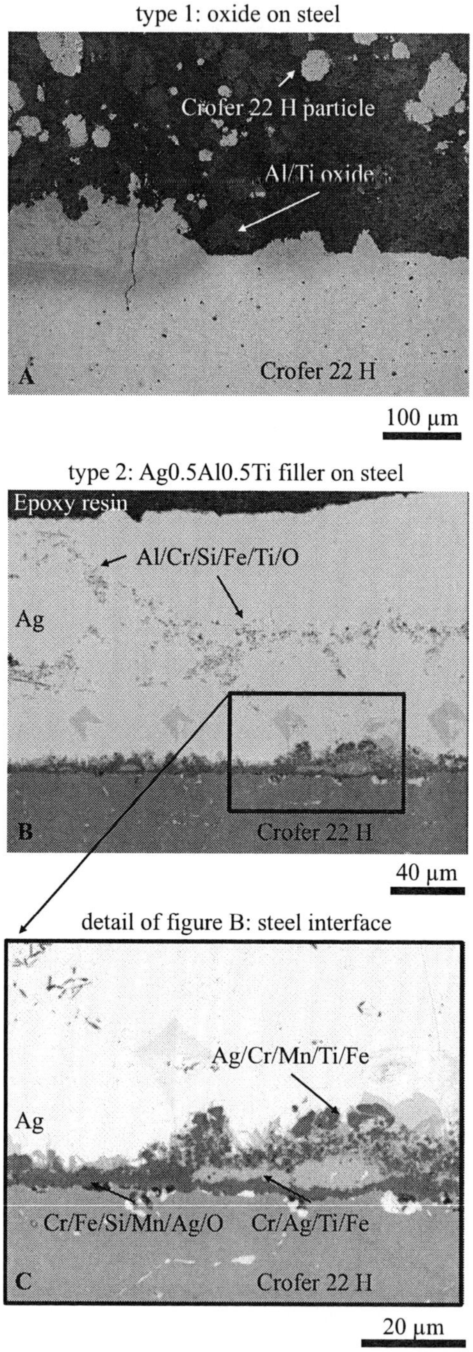

Figure 2: Cross sections of Ag0.5Al0.5Ti/Crofer 22 H samples: A: sample type 1: Al/Ti oxide on steel, as brazed; B: sample type 2: Ag0.5Al0.5Ti filler metal on steel after aging; C: detail of figure 2B: steel interlayer after brazing

All these observations lead to the assumption that Ti seems to be responsible for the affection of the steel and the associated changed structure and composition of the reaction zone and its aging behavior. Previously done DSC measurements support this assumption as the heat flow curve of Ag0.5Al0.5Ti differs from the curve of the Ti-free Ag0.5Al filler metal [8,15]. To further examine the reaction of the steel, combined DSC and thermogravimetric analysis of miniaturized type 1 samples were carried out for this work. The results of the measurements, heat flow and weight change in dependence on the temperature, are shown in Fig. 3 for the heat treatment of Al/Ti and Ti paste on Crofer 22 H. It can be seen that an exothermal reaction begins at around 550 °C and that the reaction is more intense in the sample on which only Ti was applied. As the weight of the samples increases simultaneously it can be assumed that this is an oxidation reaction. The small endothermic peak at 650 °C in the curve of the Al/Ti oxide has to be examined further.

The hardness of the Ag matrix increased from (49 ± 4) HV0,01 in the as-brazed state to (62 ± 5) HV0,01 after aging showing the same trend as the Ag4Cu4Ni samples. As before, the expected increase of the number of grains in the brazed joint can be a possible explanation. In addition to other measurements, analyses by electron backscatter diffraction can be used to prove this theory.

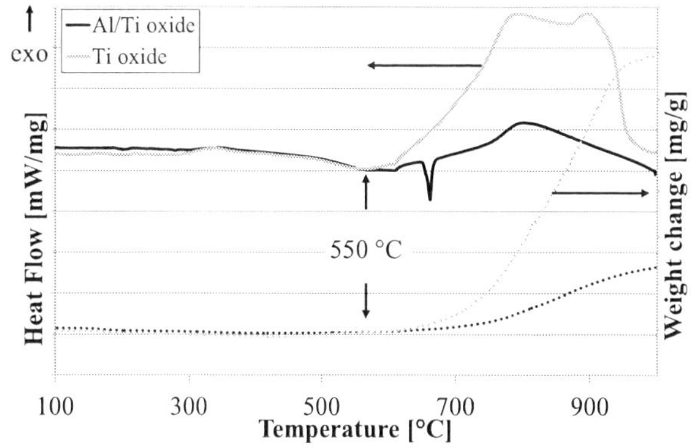

Figure 3: DSC/TG results: heat flow and weight change of Al/Ti and Ti oxide layer on Crofer 22 H during heating between 100 °C and 1000 °C

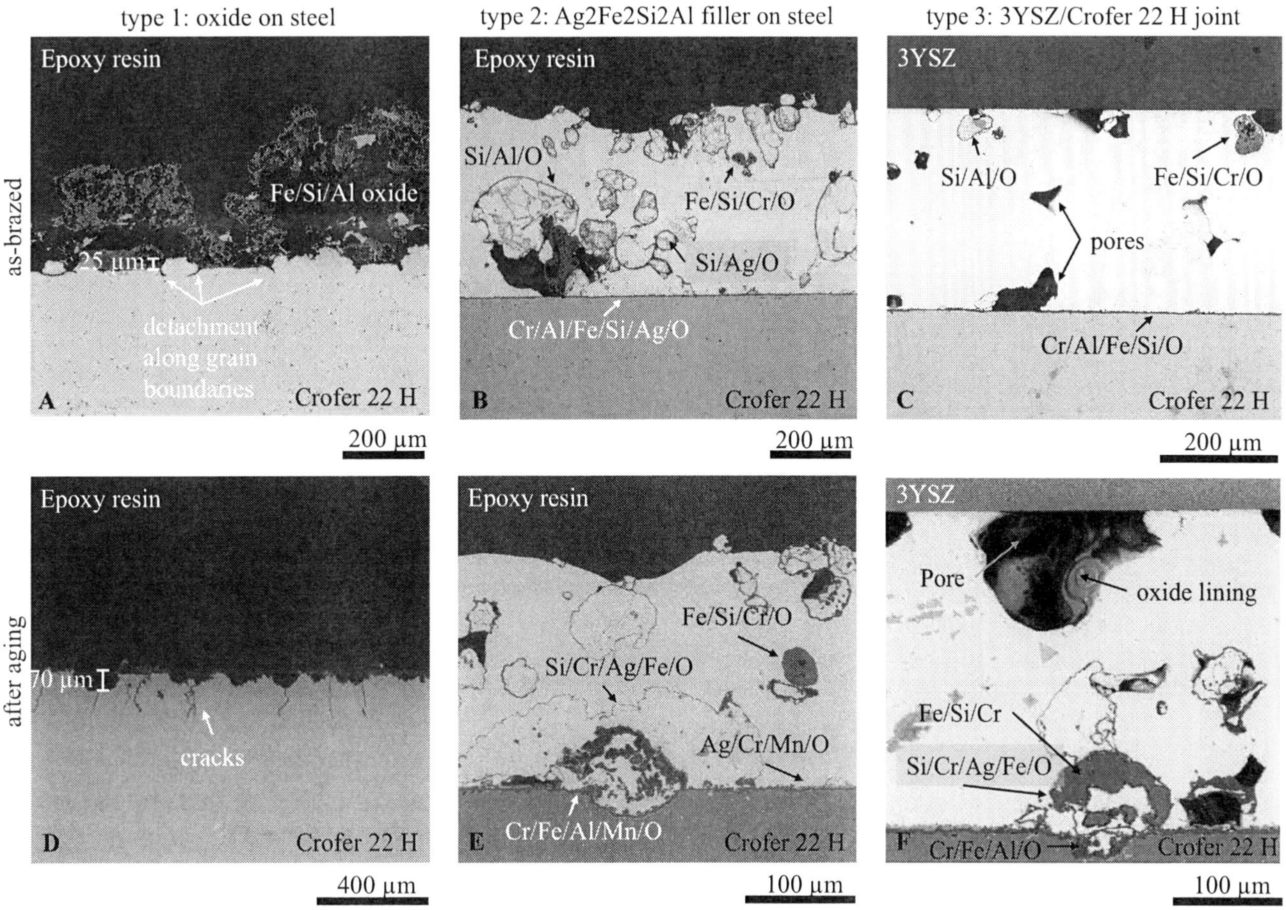

Figure 4: Cross sections of Ag2Fe2Si2Al/Crofer 22 H samples: A + D: sample type 1: Fe/Si/Al oxide layer on steel; B + E: sample type 2: Ag2Fe2Si2Al filler metal on steel; C + F: sample type 3: complete 3YSZ/Crofer 22 H joint; A - C: as-brazed state, D - F: after aging

Ag2Fe2Si2Al

Figure 4 shows the cross sections of the three types of Ag2Fe2Si2Al/Crofer 22 H samples in the as-brazed state (A - C) and after aging (D - F). The layer of oxide forming elements of the filler metal (Fe, Si and Al) in sample type 1 (Fig. 4A) is highly porous. Similar to the Ag0.5Al0.5Ti samples no continuous reaction zone is formed on the steel. Instead grains of the substrate are detached from the surface. It is obvious that this process proceeds along the grain boundaries of the substrate. The depth of the affected zone has a thickness of around 25 μm. After aging the effect is more pronounced with a depth of penetration of around 70 μm (Fig. 4D). Furthermore, cracks with lengths of over 100 μm can be found starting at places where grains of the steel have been detached. They could be induced by elements of the applied Fe/Si/Al paste diffusing along the grain boundaries. When the elements oxidize during aging, stresses and therefore cracks could arise in the material due the increase of volume of the particles. As the oxide layer is brittle, porous and not bonded strongly to the base material it is probable that this layer is detached from the steel when the sample is removed from the furnace.

Samples of the types 2 and 3 (Fig. 4B, C, E, F) are characterized by an Ag matrix with different precipitations. These are mostly areas separated by a Si/Al oxide shell along the grain boundaries from the matrix. To some extent elements of the steel, mostly Cr, can be found in this shell. Inside of the precipitations, surrounded by the oxide shell, Si oxides or mixtures of Si and Ag can be found. As shown in previous studies these areas can influence the solidification behavior of the filler metal [15]. Furthermore, pure Si or particles made of Fe, Si and Cr can be found in the braze. These can be caused by agglomerated powder in the paste or agglomeration processes during the heating stage. Another feature of these sample types is the relative high porosity whose percentage does not change notably during aging. This can be the result of the high contour accuracy of the filler metal caused by the oxide shells and ribbons in the material. These oxides constrict the movement of the molten Ag so that not all cavities caused for example by the evaporating binder of the paste or the inherent porosity in the powder based paste can be filled by the melt. Around these pores a majority of the oxide shells can be found. As those shells are also more numerous along the outer sides of the type 2 samples, areas of higher oxygen availability, it can be assumed that the formation of the oxide shells depends strongly on the oxygen amount. As the oxygen availability is lower in the type 3 samples less oxide shells can be observed in the joints.

The reaction zone of the type 2 and 3 samples is around 5 μm in the as-brazed state and consists of Cr and Fe, the typical elements of the steel oxide layer, but contains also Al and Ag from the filler metal (Fig. 4B + C). After brazing this oxide zone has a thickness of around 15 μm and shows two different layers: the typical inner oxide layer in which Al is embedded and the aforementioned acicular phase made of Ag and Cr with small amounts of Mn (Fig. 4E + F). A further growth of this phase can be expected when the sample is held for longer periods at elevated temperatures [16]. The composition of the oxide shells changed as more elements from the steel, mostly Cr and Fe, are integrated. The pore shown in Fig. 4F has a steplike form remotely similar to that in Fig. 1E. Here the steps are not formed by Ag but by oxides lining the outer walls of the pore.

Conclusions

The analysis of the Ag4Cu4Ni samples shows that the formation of the reaction layer on the steel and of the Cu/Ni in the matrix depends on the oxygen availability. When the amount of oxygen is low as in the type 3 sample during brazing, the reaction zone is small. During aging, oxygen can diffuse through the Ag matrix leading to a growth of the reaction layer. Apart from that the reaction zone does not change notably during aging. Based on this a relatively high stability of the joints with Ag4Cu4Ni during aging in air can be expected. The influence of the formation of the big pores and the increase of hardness on the mechanical properties has to be investigated.

The substrate of the Ag0.5Al0.5Ti samples seems to be affected by the Ti. This affection leads to a stronger diffusion of steel elements from the substrate into the braze and to a complex structure of the reaction zone. The effect of Ti during further aging processes has to be examined more thoroughly as it could lead to a stronger formation of voids in the substrate and an increased growth of the reaction zone. Both effects can decrease the possible life expectancy of the joint and affect negatively its mechanical properties.

The samples with the Ag2Fe2Si2Al filler metal have the smallest reaction zones even after aging. The oxide layer of the steel acts as a protective barrier so that only small amounts of steel elements are integrated in the oxide shells of the braze. These shells as well as the reaction zone on the steel are stable during aging so that a good aging stability of the joints can be expected. The growth of the Ag/Cr-based phase will be further analyzed for longer aging times. As the high porosity of the joints can affect the mechanical properties further optimization of the brazing process is needed.

Acknowledgments

This research is supported by the German Research Foundation (DFG) under contract No. PAK 524. Special thanks are addressed to our project partners IWM and ACCESS from RWTH Aachen University and IEK-2 from Research Centre Jülich.

References

[1] Weil, K.S. et al., "Reactive Air Brazing: A Novel Method of Sealing SOFCs and Other Solid-State Electrochemical Devices", Electrochem. Solid-State Lett., Vol. 8, No. 2 (2005), A133-A136.

[2] Weil, K.S. et al., "Brazing as a means of sealing ceramic membranes for use in advanced coal gasification processes", Fuel, Vol. 85 (2006), pp. 156-162.

[3] Kaletsch, A. et al., "Joining Oxygen Transport Membranes by Reactive Air Brazing", Proc. 5th International Brazing and Soldering Conf, Las Vegas, NV, April 2012, pp.437-443.

[4] Brandenberg, J. et al., "Dual-Gas Stability of RAB-Joints in SOFC Applications", Adv. Eng. Mat., DOI: 10.1002/adem.201400098 (2014)

[5] Friant, J.R. et al., "Transition in Wetting Behavior Between Liquid Ag-CuO Alloys and Al_2O_3 Substrates", J. Am. Ceram. Soc., Vol. 95, No. 5 (2012), pp. 1549-1555.

[6] Chatain, D. et al., "New Experimental Setup for Wettability Characterization under Monitored Oxygen Activity: II, Wettability of Sapphire by Silver-Oxygen melts", J. Am. Ceram. Soc., Vol. 77, No. 1 (1994), pp. 197-201.

[7] Bobzin, K. et al., "Thermochemistry of brazing ceramics and metals in air", Int. J. Mat. Res., Vol. 102, No. 8 (2011), pp. 972-976.

[8] Bobzin, K. et al., "Development of New Brazing Fillers and Process Variants for Reactive Air Brazing (RAB) of Electrochemical Devices", Proc. 5th International Brazing and Soldering Conf, Las Vegas, NV, April 2012, pp. 433-436.

[9] Kim, J.Y. et al., "High-temperature tolerance of the silver-copper oxide braze in reducing and oxidizing atmospheres", J. Mater. Res., Vol. 21, No. 6 (2006), pp. 1434-1441.

[10] Kuhn, B. et al., "Effect of isothermal aging on the mechanical performance of brazed ceramic/metal joints for planar SOFC-stacks", Int. J. Hydrogen Energy., Vol. 35 (2010), pp. 9158-9165.

[11] Poenicke, A. et al., "Mechanical Properties of Reactive Air Brazed Ceramic-Metal Joints for SOFC", Proc. 5th International Brazing and Soldering Conf, Las Vegas, NV, April 2012, pp. 450-457.

[12] Kaletsch, A. et al., "Effect of Aging on Microstructure and Mechanical Strength of Reactive Air Brazed BSCF/AISI 314-Joints", Adv. Eng. Mat., DOI: 10.1002/adem.201400102 (2014).

[13] Skiera, E. et al., "Mechanical Properties of Reactive Air Brazed (RAB) Metal/Ceramic Joints: Tailored Microstructure for Thermal Cycling Resistance", Adv. Eng. Mat., DOI: 10.1002/adem.201400073 (2014).

[14] Bobzin, K. et al., "Entwicklung neuer Lote und Prozesse zum Fügen von Keramik-Metall-Verbunden mittels Reactive Air Brazing", Proc. 14th Werkstofftechnisches Kolloquium, Chemnitz, Germany, September 2011, pp. 237-244.

[15] Bobzin, K. et al., "Untersuchung des Aufschmelz-, Reaktions- und Benetzungsverhalten von neuen Loten für das Reactive Air Brazing", Proc. 16th Werkstofftechnisches Kolloquium, Chemnitz, Germany, September 2013, pp. 161-169

[16] Li, C., "Mechanical behavior and microstructure of reactive-air-brazed metal/ceramic joints", Dissertation, 2013.

[17] Frye, J. H. and Hume-Rothery, W., "The hardness of primary solid solutions with special reference to alloys of silver", Proc. of the Royal Society of London. Series A. Mathematical and Physical Sciences, Vol. 181, No. 984 (1942), pp. 1-14.

[18] Poenicke, A., "Löten von Keramik-Metall-Verbunden mit Reaktivloten an Luft und deren Alterungsverhalten bei hohen Temperaturen", Dissertation, 2013.

AUTHOR INDEX

AUTHOR INDEX

American Welding Society
8669 NW 36 Street, #130
Miami, Florida 33166-6672
USA

CURRAN ASSOCIATES INC.
proceedings
.com

ISBN 978-1-5108-8729-9